Lecture Notes in Computer Science 1940
Edited by G. Goos, J. Hartmanis and J. van Leeuwen

Springer

Berlin
Heidelberg
New York
Barcelona
Hong Kong
London
Milan
Paris
Singapore
Tokyo

Mateo Valero Kazuki Joe
Masaru Kitsuregawa Hidehiko Tanaka (Eds.)

High Performance Computing

Third International Symposium, ISHPC 2000
Tokyo, Japan, October 16-18, 2000
Proceedings

Series Editors

Gerhard Goos, Karlsruhe University, Germany
Juris Hartmanis, Cornell University, NY, USA
Jan van Leeuwen, Utrecht University, The Netherlands

Volume Editors

Mateo Valero
Universidad Politecnica de Catalunya
Departamento de Arquitectura de Computadores, Spain
E-mail: mateo@ac.upc.es

Kazuki Joe
Nara Women's University
Department of Information and Computer Sciences, Japan
E-mail: joe@ics.nara-wu.ac.jp

Masaru Kitsuregawa
University of Tokyo
Institute of Industrial Science
Center for Conceptual Information Processing Research, Japan
E-mail: kitsure@tkl.iis.u-tokyo.ac.jp

Hidehiko Tanaka
University of Tokyo
Graduate School of Engineering
Electrical Engineering Department, Japan
E-mail: tanaka@mtl.t.u-tokyo.ac.jp

Cataloging-in-Publication Data applied for

Die Deutsche Bibliothek - CIP-Einheitsaufnahme

High performance computing : third international symposium ;
proceedings / ISHPC 2000, Tokyo, Japan, October 16 - 18, 2000. Mateo
Valero ... (ed.). - Berlin ; Heidelberg ; New York ; Barcelona ; Hong
Kong ; London ; Milan ; Paris ; Singapore ; Tokyo : Springer, 2000
(Lecture notes in computer science ; Vol. 1940)
ISBN 3-540-41128-3

CR Subject Classification (1998): D.1-2, F.2, E.4, G.1-4, J.1-2, J.3, J.6, I.6

ISSN 0302-9743
ISBN 3-540-41128-3 Springer-Verlag Berlin Heidelberg New York

Springer-Verlag Berlin Heidelberg New York
a member of BertelsmannSpringer Science+Business Media GmbH

Printed in Germany

Typesetting: Camera-ready by author, data conversion by DA-TeX Gerd Blumenstein
Printed on acid-free paper SPIN: 10781323 06/3142 5 4 3 2 1 0

Preface

I wish to welcome all of you to the International Symposium on High Performance Computing 2000 (ISHPC 2000) in the megalopolis of Tokyo. After having two great successes with ISHPC'97 (Fukuoka, November 1997) and ISHPC'99 (Kyoto, May 1999), many people have requested that the symposium would be held in the capital of Japan and we have agreed.

I am very pleased to serve as Conference Chair at a time when high performance computing (HPC) has a significant influence on computer science and technology. In particular, HPC has had and will continue to have a significant impact on the advanced technologies of the "IT" revolution. The many conferences and symposiums that are held on the subject around the world are an indication of the importance of this area and the interest of the research community.

One of the goals of this symposium is to provide a forum for the discussion of all aspects of HPC (from system architecture to real applications) in a more informal and personal fashion. Today we are delighted to have this symposium, which includes excellent invited talks, tutorials and workshops, as well as high quality technical papers.

In recent years, the goals, purpose and methodology of HPC have changed drastically. HPC with high-cost, high-power consumption and difficult-to-use interfaces will no longer attract users. We should instead use what the IT revolution of the present and near future gives us: highly integrated processors and extremely fast internet. Mobile and wearable computing is already commonplace and the combination with multimedia and various database applications is promising. Therefore we would like to treat HPC technologies as systems and applications for low-end users as well as conventional high-end users, where we can find a bigger market. In this symposium, we will discuss the direction of such HPC technologies with hardware, software and applications specialists.

This symposium would not have been possible without the significant help of many people who devoted resources and time. I thank all of those who have worked diligently to make the ISHPC 2000 a great success. In particular I would like to thank the Organizing Chair, Masaru Kitsuregawa of the University of Tokyo, and all members of the organizing committee, who contributed very significantly to the planning and organization of ISHPC 2000. I must also thank the Program Chair, Mateo Valero of the Technical University of Catalunya, and the program committee members who assembled an excellent program comprising a very interesting collection of contributed papers from many countries.

A last note of thanks goes to the Kao Foundation for Arts and Science, the Inoue Foundation for Science, the Telecommunications Advancement Foundation and Sumisho Electronics Co. Ltd for sponsoring the symposium.

October 2000 Hidehiko Tanaka

Foreword

The 3rd International Symposium on High Performance Computing (ISHPC 2000 held in Tokyo, Japan, 16–18 October 2000) was thoughtfully planned, organized, and supported by the ISHPC Organizing Committee and collaborative organizations.

The ISHPC 2000 Program consists of two keynote speeches, several invited talks, two workshops on OpenMP and Simulation-Visualization, a tutorial on OpenMP, and several technical sessions covering theoretical and applied research topics on high performance computing which are representative of the current research activities in industry and academia. Participants and contributors to this symposium represent a cross section of the research community and major laboratories in this area, including the European Center for Parallelism of Barcelona of the Polytechnical University of Catalunya (UPC), the Center for Supercomputing Research and Development of the University of Illinois at Urbana-Champaign (UIUC), the Maui High Performance Computing Center, the Kansai Research Establishment of Japan Atomic Energy Research Institute, Japan Society for Simulation Technology, SIGARCH and SIGHPC of Information Processing Society Japan, and the Society for Massively Parallel Processing.

All of us on the program committee wish to thank the authors who submitted papers to ISHPC 2000. We received 53 technical contributions from 17 countries. Each paper received at least three peer reviews and, based on the evaluation process, the program committee selected fifteen regular (12-page) papers. Since several additional papers received favorable reviews, the program committee recommended a poster session comprised of short papers. Sixteen contributions were selected as short (8-page) papers for presentation in the poster session and inclusion in the proceedings.

The program committee also recommended two awards for regular papers: a distinguished paper award and a best student paper award. The distinguished paper award was given to "Processor Mechanisms for Software Shared Memory" by Nicholas Carter, and the best student paper award was given to "Limits of Task-Based Parallelism in Irregular Applications" by Barbara Kreaseck.

ISHPC 2000 has collaborated closely with two workshops: the International Workshop on OpenMP: Experiences and Implementations (WOMPEI) organized by Eduard Ayguade of the Technical University of Catalunya, and the International Workshop on Simulation and Visualization (IWSV) organized by Katsunobu Nishihara of Osaka University. Invitation-based submission was adopted by both workshops. The ISHPC 2000 program committee decided to include all papers of WOMPEI and IWSV in the proceedings of ISHPC 2000.

We hope that the final program will be of significant interest to the participants and will serve as a launching pad for interaction and debate on technical issues among the attendees.

October 2000 Mateo Valero

Foreword from WOMPEI

First of all, we would like to thank the ISHPC Organizing Committee for giving us the opportunity to organize WOMPEI as part of the symposium. The workshop consists of one invited talk and eight contributed papers (four from Japan, two from the United States and two from Europe). They report some of the current research and development activities related to tools and compilers for OpenMP, as well as experiences in the use of the language. The workshop includes a panel discussion (shared with ISHPC) on Programming Models for New Architectures. We would also like to thank the Program Committee and the OpenMP ARB for their support in this initiative. Finally, thanks go to the Real World Computing Partnership for the financial support to WOMPEI. We hope that the program will be of interest to the OpenMP community and will serve as a forum for discussion on technical and practical issues related to the current specification.

E. Ayguade (Technical University of Catalunya),
H. Kasahara (Waseda University) and
M. Sato (Real World Computing Partnership)

Foreword from IWSV

Recent rapid and incredible improvement of HPC technologies has encouraged numerical computation users to use larger and therefore more practical simulations. The problem such high-end users face is how to analyze or even understand the results calculated with huge computation times. The promising solution to this problem is the use of visualization.

IWSV is organized as part of ISHPC 2000 and consists of 11 contributed papers and abstracts. We would like to thank the ISHPC 2000 Organizing Committee for providing us with this opportunity. We would also like to thank the ISHPC 2000 Program Committee for having IWSV papers and abstracts included in the proceedings, which we did not expect.

We hope that IWSV will be of fruitful interest to ISHPC 2000 participants and will indicate a future direction of collaboration between numerical computation and visualization researchers.

K. Nishihara (Osaka University),
K. Koyamada (Iwate Prefectural University) and
Y. Ueshima (Japan Atomic Energy Research Institute)

Organization

ISHPC 2000 is organized by the ISHPC Organizing Committee in cooperation with the European Center for Parallelism of Barcelona of the Polytechnical University of Catalunya (UPC), the Center for Supercomputing Research and Development of the University of Illinois at Urbana-Champaign (UIUC), the Maui High Performance Computing Center, the Kansai Research Establishment of Japan Atomic Energy Research Institute, Japan Society for Simulation Technology, SIGARCH and SIGHPC of Information Processing Society Japan, and the Society for Massively Parallel Processing.

Executive Committee

General Chair:	Hidehiko Tanaka (U. Tokyo, Japan)
Program Chair:	Mateo Valero (UPC, Spain)
Program Co-chair:	Jim Smith (U. Wisconsin, US)
	Constantine Polychronopoulos (UIUC, US)
	Hironori Kasahara (Waseda U., Japan)
Organizing Chair:	Masaru Kitsuregawa (U. Tokyo, Japan)
Publication & Treasuary Chair:	Kazuki Joe (NWU, Japan)
Treasuary Co-chair:	Toshinori Sato (KIT, Japan)
Local Arrangement Chair:	Hironori Nakajo (TUAT, Japan)
Poster Session Chair:	Hironori Nakajo (TUAT, Japan)
Workshop Chair:	Eduard Ayguade (UPC, Spain)
	Katsunobu Nishihara (Osaka U., Japan)

Organizing Committee

Eduard Ayguade (UPC)
Yasuhiro Inagami (Hitatch)
Yasunori Kimura (Fujitsu)
Steve Lumetta (UIUC)
Mitaro Namiki (TUAT)
Yoshiki Seo (NEC)
Ou Yamamoto (TEU)
Hiroki Honda (UEC)
Kazuki Joe (NWU)
Tomohiro Kudoh (RWCP)
Hironori Nakajo (TUAT)
Toshinori Sato (KIT)
Chau-Wen Tseng (UMD)

Program Committee

Yutaka Akiyama (RWCP)
Hideharu Amano (Keio U.)
Hamid Arabnia (Geogea U.)
Utpal Banerjee (Intel)
Taisuke Boku (U. Tsukuba)
George Cybenko (Dartmouth)
Michel Dubois (USC)
Rudolf Eigenmann (Purdue U)
Joel Emer (Compaq)
Skevos Evripidou (U. Cyprus)
Ophir Frieder (IIT)
Mario Furnari (CNR)
Stratis Gallopoulos (U. Patras)
Dennis Gannon (U. Indianna)
Guang Gao (U. Delaware)
Antonio Gonzalez (UPC)
Thomas Gross (ETHZ/CMU)
Mohammad Haghighat (Intel)
Hiroki Honda (UEC)
Elias Houstis (Purdue U.)
Yasuhiro Inagami (Hitatchi)
Kazuki Joe (NWU)
Yasunori Kimura (Fujitsu)
Yoshitoshi Kunieda (Wakayama U.)
Jesus Labarta (UPC, Spain)
Monica Lam (Stanford)
Hans Luethi (ETHZ)
Allen Malony (U. Oregon)
Hideo Matsuda (Osaka U.)
Mitsunori Miki (Doshisha U.)
Prasant Mohapatra (MSU)
Jose Moreira (IBM)
Shin-ichiro Mori (Kyoto U.)
Hironori Nakajo (TUAT)
Takashi Nakamura (NAL)
Hiroshi Nakasima (TUT)
Alex Nicolau (UCI)
Michael L. Norman (UIUC)
Theodore Papatheodorou (U. Patras)
John Rice (Purdue U.)
Eric Rotenberg (NCSU)
Youcef Saad (UMN)
Mitsuhisa Sato (RWCP)
Yoshiki Seo (NEC)
Guri Sohi (U. Wisconsin)
Peter R. Taylor (SDSC)
Chau-Wen Tseng (UMD)
Dean Tullsen (UCSD)
Sriram Vajapeyam (IIS)
Alex Veidenbaum (UCI)
Harvey J. Wassermann (LosAlamos)
Harry Wijshoff (Leiden U.)
Tao Yang (UCSB)
Mitsuo Yokokawa (JAERI)
Hans Zima (U. Vienna)

Referees

T. Araki
L. D. Cerio
A. Chowdhury
L. Chu
J. Duato
T. Erlebach
A. Funahashi
P. Grun
T. Hanawa
T. Kamachi
M. Kawaba
E. Laure
J. Lu
A.D. Malony
P. Marcuello
W. Martins
E. Mehofer
O.G. Monakhov
E.A. Monakhova
S. Mukherjee
E. Nunohiro
P. Ranganathan
M. Satoh
K. Shen
S. Tambat
W. Tang
H. Tang
T. Tarui
J. Torres
C.-W. Tseng
T. Uehara

Table of Contents

I. Invited Papers

II. Compilers, Architectures and Evaluation

III. Algorithms, Models and Applications

IV. Short Papers

V. International Workshop on OpenMP: Experiences and Implementations (WOMPEI)

VI. International Workshop on Simulation and Visualization (IWSV)

Instruction Level Distributed Processing: Adapting to Future Technology

J. E. Smith

Dept. of Elect. and Comp. Engr.
1415 Johnson Drive
Univ. of Wisconsin
Madison, WI 53706

1. Introduction

For the past two decades, the emphasis in processor microarchitecture has been on instruction level parallelism (ILP) -- or in increasing performance by increasing the number of "instructions per cycle". In striving for higher ILP, there has been an ongoing evolution from pipelining to superscalar, with researchers pushing toward increasingly wide superscalar. Emphasis has been placed on wider instruction fetch, higher instruction issue rates, larger instruction windows, and increasing use of prediction and speculation. This trend has led led to very complex, hardware-intensive processors.

This trend is based on "exploiting" technology improvements. The ever-increasing transistor budgets have left researchers with the view that the big challenge is to consume transistors in some fashion, i.e. "how are we going to use a billion transistors?" [1]. Starting from this viewpoint, it is not surprising that the result is hardware-intensive and complex. Furthermore, the complexity is not just critical path lengths and transistor counts; there is also high intellectual complexity -- from attempts to squeeze performance out of second and third order effects.

Historically, computer architecture innovation has done more than exploit technology; it has also been used to accommodate technology shifts. A good example is cache memories. In the late 1970s, RAM cycle times were as fast microprocessor cycle times. If main memory can be accessed in a single cycle, there is no need for a cache. However, over the years, RAM speeds have not kept up with processor speeds and increasingly complex cache hierarchies have filled the gap. Architecture innovation was used to avoid tremendous slow downs - by adapting to the shift in processor/RAM technologies.

There are a number of significant technology shifts taking place right now. Wire delays are coming to dominate transistor delays [2], static power will soon catch up and pass dynamic power in importance [3]. Fast transistors will no longer be "free", at least when in terms of power consumption. There are also shifts in applications -- toward commercial memory-based applications, object oriented dynamic linking, and multiple threads.

2. Implications: Instruction Level Distributed Processing

Given the above, the next trend in microarchitecture will likely be Instruction Level Distributed Processing (ILDP). The processor will consist of a number of distributed functional units, each fairly simple with a very high frequency clock cycle. There

M. Valero et al. (Eds.): ISHPC 2000, LNCS 1940, pp. 1-6, 2000.

will likely be multiple clock domains. Global interconnections will be point-to-point with delays of a clock cycle or greater. Partitioning the system to accommodate these delays will be a significant part of the microarchitecture design effort. There may be relatively little low-level speculation (to keep the transistor counts low and the clock frequency high) -- determinism is inherently simpler than prediction and recovery.

2.1. Dependence-based Microarchitecture

One type of ILDP processors consists of clustered "dependence-based" microarchitectures [4]. The 21264 [5] is a commercial example, but a very early and little known example was an uncompleted Cray-2 design [6]. In these microarchitectures, processing units are organized into clusters and dependent instructions are steered to the same cluster for processing.

The 21264 microarchitecture there are two clusters, with some instructions routed to each at issue time. Results produced in one cluster require an additional clock cycle to be routed to the other. In the 21264, data dependences tend to steer communication instructions to the same cluster. Although there is additional inter-cluster delay, the faster clock cycle compensates for the delay and leads to higher overall performance.

In general a dependence based design may be divided into several clusters, cache processing can be separated from instruction processing, integer processing can be separated from floating point, etc. (See Fig. 1) In a dependence-based designs, dependent instructions are collected together, so instruction control logic within a cluster is likely to be simplified, because there is no need to look for independence if it is known not to exist.

2.2. Heterogeneous ILDP

Another model for ILDP is heterogeneous processors where a simple core pipeline is surrounded by outlying "helper engines" (Fig. 2). These helper engines are not in the

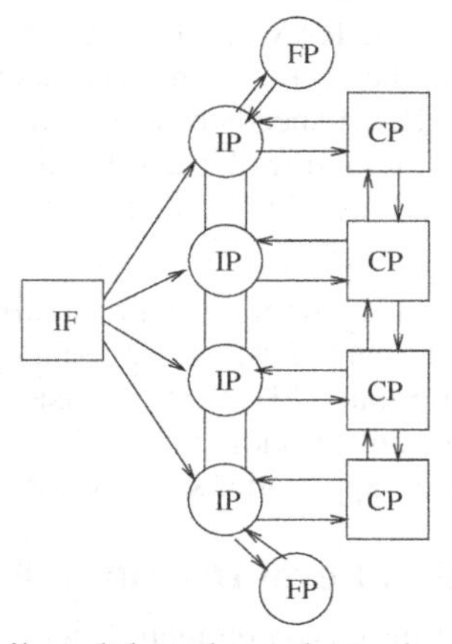

Fig. 1. A distributed data-dependent microarchitecture.

critical processing path, so they have non-critical communication delays with respect to the main pipeline, and may even use slower transistors.

Examples of helper engines include the pre-load engine of Roth and Sohi [7] where pointer chasing can be performed by a special processing unit. Another is the branch engine of Reinman et al. [8]. An even more advanced helper engine is the instruction co-processor described by Chou and Shen [9]. Helper engines have also been proposed for garbage collection [10] and correctness checking [11].

3. Co-Designed Virtual Machines

Providing important support for the ILDP paradigm is the trend toward dynamic optimizing software and virtual machine technologies. A co-designed virtual machine is a combination of hardware and software that implements an instruction set, the virtual ISA [12,13,14]. Part of this implementation is hardware -- which supports an implementation specific instruction set (Implementation Instruction Set Architecture, I-ISA). The other part of the implementation is in software -- which translates the virtual instruction set architecture (V-ISA) to the I-ISA and which provides the capability of dynamically re-optimizing a program. A co-designed VM is a way of giving hardware implementors a layer of software. This software layer liberates the hardware designer from supporting a legacy V-ISA purely in hardware. It also provides greater flexibility in managing the resources that make up a ILDP microarchitecture.

With ILDP, resources must be managed with a high level view. The distributed processing elements must be coordinated, and Instructions and data must be routed in such a way that resource usage is balanced and communication delays (among dependent instructions) are minimized -- as with any distributed system. This could demand high complexity hardware, if hardware alone were given responsibility. For

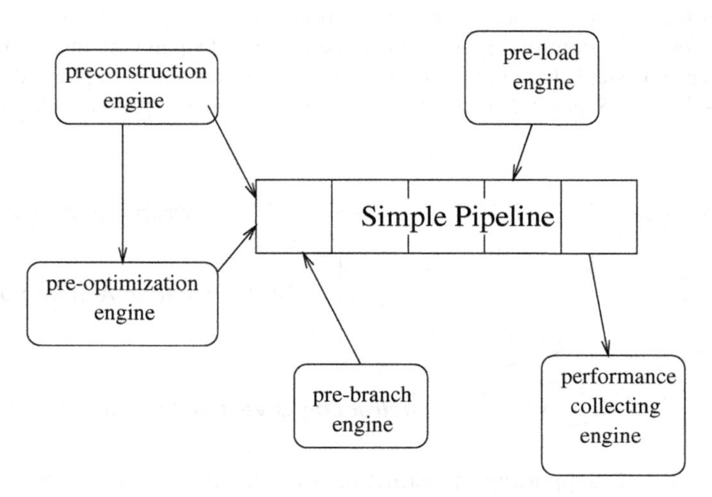

Fig. 2. A heterogenous ILDP chip architecture.

example, the hardware must be aware of some elements of program structure, such as data and control flow. However, traditional window-based methods give hardware only a restricted view of the program, and the hardware would have to re-construct program structure information by viewing the instruction stream as it flows by.

Hence, in the co-designed VM paradigm, software is responsible for determining program structure, dynamically re-optimizing code, and making complex decisions regarding management of ILDP. Hardware implements the lower level performance features that are managed by software. Hardware also collects dynamic performance information and may trigger software when "unexpected" conditions occur. Besides performance, the VM can be used for managing resources to reduce power requirements [13] and to implement fault tolerance.

4. New Instruction Sets

Thus far, the discussion has been about microarchitectures, but there are also instruction set implications. Instruction sets should be optimized for ILDP. Using VM technology enables new instruction sets at the implementation level. Features of new instruction sets should include focus on communication and dependence and emphasis on small, fast implementation structures, including caches and registers. For example, variable length instructions lead to smaller instruction footprints and smaller caches.

Most recent instruction sets, including RISC instructions sets, and especially VLIW instruction sets, have emphasized computation and independence. The view was that higher parallelism could be achieved by focusing on computation aspects of instruction sets and on placing independent instructions in proximity either at compile time or during execution time. For ILDP, however, instruction sets should be targeted at communication and dependence. That is, communications should be easily expressed and dependent instructions should be placed in proximity, to reduce communication delays. For example, a stack-based instruction set is an example of and ISA that places the focus on communication and dependence. Dependent instructions communicate via the stack top; hence, communication is naturally expressed. Furthermore stack-based ISAs tend to have a small instruction footprints. Although stack

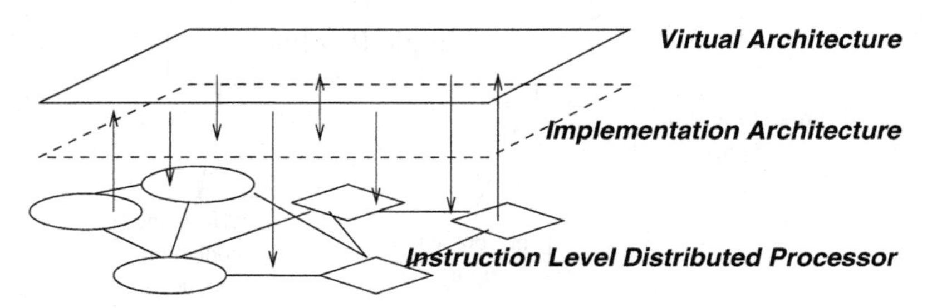

Fig. 3. Supporting an instruction level distributed processor with a co-designed Virtual Machine.

instructions sets may have disadvantages, e.g. they may lead to more memory operations, they do illustrate the point that dependence can be expressed in a clear way via the ISA.

5. Summary and Conclusions

In summary, technology shifts are forcing shifts in microarchitectures. Instruction level distributed processing will organized around small and fast core processors. These microarchitectures will contain distributed resources. Helper processors may be distributed around the main processing elements. These processors can execute highly parallel tasks and can be built from slower transistors to restrict static power consumption. Hence, the emphasis may shift from processor architecture to chip architecture where distribution and interconnection of resources will be key.

Virtual machines fit very nicely in this environment. In effect, hardware designers can be given a layer of software that can be used to coordinate the distributed hardware resources and perform dynamic optimization from a higher level perspective than is available in hardware alone. Finally, it is once again time that we reconsider instruction sets with the focus on communication and dependence. New instructions sets are needed to mesh with ILDP implementations and they are enabled by the VM paradigm which makes legacy compatibility unimportant at the I-ISA level.

Acknowledgements

This work was supported by National Science Foundation grant CCR-9900610, by IBM Corporation, Sun Microsystems, and Intel Corporation. This support is gratefully acknowledged.

References

1 Doug Burger and James R. Goodman, eds., "Billion Transistor Architectures", special issue, *IEEE Computer*, Sept. 1997.

2 V. Agarwal, M. S. Hrishikesh, S. W. Keckler, D. Burger, "Clock Rate versus IPC: The End of the Road for Conventional Microarchitectures," *27th Int. Symp. on Computer Architecture,*, pp. 248-259, June 2000.

3 S. Thompson, P. Packan, and M. Bohr, "MOS Scaling: Transistor Challenges for the 21st Century," *Intel Technology Journal*, Q3, 1998.

4 S. Palacharla, N. Jouppi, J. E. Smith, "Complexity-Effective Superscalar Processors," *24th Int. Symp. on Computer Architecture,*, pp. 206-218, June 1997.

5 Anonymous, *Cray-2 Central Processor*, unpublished document, 1979.

6 L. Gwennap, "Digital 21264 Sets New Standard," *Microprocessor Report,* pp. 11-16, Oct. 1996.

7 A. Roth and G. Sohi, "Effective Jump-Pointer Prefetching for Linked Data Structures," *26th Int. Symp. on Computer Architecture,*, pp. 111-121, May 1999.

8 8. G. Reinman, T. Austin, B. Calder, "A Scalable Front-End Architecture for Fast Instruction Delivery," *26th Int. Symposium on Computer Architecture*, pp. 234-245, May 1999.

9 Yuan Chou and J. P. Shen, "Instruction Path Coprocessors," *27th Int. Symposium on Computer Architecture*, pp. 270-281, June 2000.

10 Timothy Heil and J. E. Smith, "Concurrent Garbage Collection Using Hardware-Assisted Profiling," *International Symposium on Memory Management (ISMM)*, October 2000.

11 T. Austin, "DIVA: A Reliable Substrate for Deep Submicron Microarchitecture Design," *32nd Int. Symposium on Microarchitecture*, pp. 196-297, Nov. 1999.

12 K. Ebcioglu and E. R. Altman, "DAISY: Dynamic Compilation for 100% Architecture Compatibility," *24th Int. Symp. on Computer Architecture,*, June 1997.

13 A. Klaiber, "The Technology Behind Crusoe Processors," *Transmeta Technical Brief*, 2000.

14 J. E. Smith, T. Heil, S. Sastry, T. Bezenek, "Achieving High Performance via Co-Designed Virtual Machines," *Intl. Workshop on Innovative Architecture for Future Generation High-Performance Processors and Systems,*, pp. 77-84, Oct. 1998.

15 D. H. Albonesi, "The Inherent Energy Efficiency of Complexity-Adaptive Processors," *1998 Power-Driven Microarchitecture Workshop*, pp. 107-112, June 1998.

Macroservers: An Object-Based Programming and Execution Model for Processor-in-Memory Arrays

Hans P. Zima[1] and Thomas L. Sterling[2]

[1] Institute for Software Science, University of Vienna, Austria
[2] Center for Advanced Computing Research (CACR), California Institute of Technology, Pasadena, California, U.S.A.
{zima,tron}@cacr.caltech.edu

Abstract. The emergence of semiconductor fabrication technology allowing a tight coupling between high-density DRAM and CMOS logic on the same chip has led to the important new class of Processor-in-Memory (PIM) architectures. Recent developments provide powerful parallel processing capabilities on the chip, exploiting the facility to load wide words in single memory accesses and supporting complex address manipulations in the memory. Furthermore, large arrays of PIMs can be arranged into massively parallel architectures. In this paper, we outline the salient features of PIM architectures and describe the design of an object-based programming and execution model centered on the notion of *macroservers*. While generally adhering to the conventional framework of object-based computation, macroservers provide special support for the efficient control of program execution in a PIM array. This includes features for specifying the distribution and alignment of data in virtual object space, the binding of threads to data, and a future-based synchronization mechanism. We provide a number of motivating examples and give a short overview of implementation considerations.

1 Introduction

"Processor in Memory" or PIM technology and architecture has emerged as one of the most important domains of parallel computer architecture research and development. It is being pursued as a means of accelerating conventional systems for array processing [22] and for manipulating irregular data structures [10]. It is being considered as a basis for scalable spaceborne computing [23], as smart memory to manage systems resources in a hybrid technology multithreaded architecture for ultra-scale computing [24], and most recently as the means for achieving Petaflops performance [14]. PIM exploits recent advances in semiconductor fabrication processes that enables the integration of DRAM cell blocks and CMOS logic on the same chip. The benefit of PIM structures is that processing logic can have direct access to the memory block row buffers at an internal memory bandwidth on the order of 100 Gbps yielding the potential performance of 10 Gips (32-bit operands) on a

H. Yasuda (Ed.): IWAN 2000, LNCS 1942, pp. 7–25, 2000.

memory chip with a 16 Mbyte capacity. Because of the efficiencies derived from staying on-chip, power consumption can be an order of magnitude lower than comparable performance with conventional microprocessor based systems. But the dramatic advances in performance will be derived from arrays of tightly coupled PIM chips in the hundreds or thousands, either alone, or in conjunction with external microprocessors. Such systems could deliver low Teraflops scale peak performance within the next couple of years at a cost of only a few million dollars (or less than $1M if mass produced) and possibly a Petaflops, at least for some applications, in five years.

The challenge to realizing the extraordinary potential of arrays of PIM is not simply the interesting problem of the basic on-chip structure and processor architecture but also the methodology for coordinating the synthesis of as much as a million PIM processors to engage in concert in the solution of a single parallel application. A large PIM array is not simply another MPP, it is a new balance of processing and memory in a new organization. Its local operation and global emergent behavior will be a direct reflection of a shared highly parallel system-wide model of computation that governs the execution and interactions of the PIM processors and chips. Such a computing paradigm must treat the semantic requirements of the whole system even as it derives its processing capabilities from the local mechanisms of the individual parts. A synergy of cooperating elements is to be accomplished through this shared execution model.

PIM differs significantly from more common MPP structures in several key ways. The ratio of computation performance to associated memory capacity is much higher. Access bandwidth (to on-chip memory) is a hundred times greater. And latency is lower by a factor of two to four while logic clock speeds are approximately half that of the highest speed microprocessors. Like clusters, PIM favors data oriented computing where operations are scheduled and performed at the site of the data, and tasks are often moved from one PIM to another depending on where the argument data is rather than moving the data. PIM processor utilization is less important than memory bandwidth. A natural organization of computation on a PIM array is a binding of tasks and data segments logically to coincide with physical data allocation while making remote service requests where data is non-local. This is very similar to evolving practices for accomplishing tasks on the Web including the use of Java and encourages an object-oriented approach to managing the logical tasks and physical resources of the PIM array.

This paper presents a strategy for relating the physical resources of next generation PIM arrays to the logical requirements of user defined applications. The strategy is embodied in an intermediate form of an execution model that provides the generalized abstractions of both local and global computation in a unified framework. The principal abstract entity of the proposed model is the *macroserver*, a distributed agent of state and action. It complements the concept of the *microserver*, a purely local agent [3]. This early work explores one possible model that is object based in a manner highly suitable to PIM structures but of a sufficiently high level with task virtualization that aggregations of PIM nodes can be cooperatively applied to a segment of parallel computation without phase changes in representations (as would be found with Open MP combined with MPI).

The next section describes PIM architectures including the likely direction of their evolution over the next one to three years. Then, in Sections 3 and 4, a description of macroservers, a PIM-oriented object-based distributed execution model is presented. Section 5 discusses the implications of this model for its implementation on the PIM including those that may drive architecture advances. The paper concludes with a summary of the model's features, guided by a set of requirements, in Section 6, and an outlook to future work required to achieve the promise of this approach, in Section 7.

2 Processor in Memory

For more than a decade, research experiments have been conducted with semiconductor devices that merged both logic and static RAM cell blocks on the same chips. Even earlier, simple processors and small blocks of SRAM could be found on simple control processors for embedded applications and of course modern microprocessors include high speed SRAM on chip for level 1 caches. But it was not until recently that industrial semiconductor fabrication processes made possible tightly coupled combinations of logic with DRAM cell blocks bringing relatively large memory capacities to PIM design. A host of research projects has been undertaken to explore the design and application space of PIM (many under DARPA sponsorship) culminating in the recent IBM announcement to build a Petaflops scale PIM array for the application of protein folding.

The opportunity of PIM is primarily one of bandwidth. Typical memory parts access a row of memory from a memory block and then select a subsegment of the row of bits to be sent to a requesting processor through the external interface. While newer generations of memory chips are improving effective bandwidth, PIMs make possible immediate access to all the bits of a memory row acquired through the sense amps. Processing logic, placed at the row buffer, can operate on all the data read (typically 64 32-bit words) in a single memory access under favorable conditions. While a number of PIM proposals plan to use previously developed processor cores to be "dropped into" the die, PIM offers important opportunities for new processor architecture design that simplifies operation, lowers development cost and time, and greatly improves efficiency and performance over classical processor architecture. Many of the mechanisms incorporated in today's processors are largely unnecessary in a PIM processor. At the same time, effective manipulation of the very wide words available on the PIM imply the need for augmented instruction sets.

PIM chips include several major subsystems, some of them replicated as space is available:

- memory blocks
- processor control
- wide ALU and data path/register set
- shared functional units
- external interfaces

Typically PIMs are organized into sets of memory block/processor pairs while sharing some larger functional units and the external interfaces among them [17]. Detailed design studies suggest that PIM processors comprise less than 20% of the available chip real estate while the memory capacity has access to more than half of the total space. Approximately a third of the die area is used for external I/O interface and control as well as shared functional units. This is an excellent ratio and emphasizes the value of optimizing for bandwidth utilization rather than processor throughput. An important advantage of the PIM approach is the ability to operate on all bits in a given row simultaneously. A new generation of very wide ALU and corresponding instruction sets exploit this high memory bandwidth to accomplish the equivalent of many conventional operations (e.g. 32-bit integer) in a single cycle. An example of such a wide ALU is the ASAP ISA developed at the University of Notre Dame and used in such experimental PIM designs as Shamrock and MIND. Other fundamental advances over previous generation PIMs are also in development to provide unprecedented capability and applicability. Among the most important of these are on-PIM virtual to physical address translation, message driven computation, and multithreading.

Virtual-to-Physical Address Translation Early PIM designs have been very simple assuming a physically addressed memory and often a SIMD control structure [9]. But such basic designs are limited in their applicability to a narrow range of problems. One requirement not satisfied by such designs is the ability to manipulate irregular data structures. This requires the handling of user virtual addresses embedded in the structure meta-data. PIM virtual to physical address translation is key to extending PIM into this more generalized domain. Translation Lookaside Buffers can be of some assistance but they are limited in scalability and may not be the best solution. Virtual address translation is also important for protection in the context of multitasking systems. Address translation mechanisms are being provided for both the USC DIVA chip and the HTMT MIND chip. As discussed later in Section 5, alternative approaches to PIM address translation have been developed that are both efficient and scalable including set associative and in-situ techniques.

Message-Driven Computation A second important advance for PIM architecture is message driven computation. Like simple memories, PIMs acquire external requests to access and manipulate the contents of memory cells. Unlike simple memories, PIMs may have to perform complex sequences of operations on the contents of memory defined by user application or supervisor service routines. Mechanisms are necessary that provide efficient response to complex requests while maintaining generality. Message driven computation assumes a sophisticated protocol and on-chip fast interpretation mechanisms that quickly identify both the operation sequence to be performed and the data rows upon which to be operated. A general message driven low-level infrastructure goes beyond interactions between system processors and the incorporated PIMs, it permits direct PIM to PIM interactions and control without system processor intervention. This reduces the impact of the system processors as a bottleneck and allows the PIMs to exploit data level parallelism at the fine grain level intrinsic to pointer linked sparse and irregular data structures. Both the USC DIVA chip and the HTMT MIND chip will incorporate "parcel" message driven computation while the IBM Blue Gene chip will permit direct PIM to PIM communications as well.

Multithreading A third important advance is the incorporation of multithreading into the PIM processor architecture. Although counter intuitive, multithreading actually greatly simplifies processor design rather than further complicating it because it provides a uniform hardware methodology for dynamically managing physical processor resources and virtual application tasks. Multithreading is also important because it permits rapid response to incoming service requests with low overhead context switching and also enables overlapping of computation, communication, and memory access activities, thus achieving much higher utilization and efficiency of these important resources. Multithreading also provides some latency hiding to local shared functional units, on-chip memory (for other processor/memory nodes on the chip), and remote service requests to external chips. The IBM Blue Gene chip and the HTMT MIND chip both will incorporate multithreading.

Advanced PIM structures like MIND, DIVA, and Blue Gene require a sophisticated execution model that binds the actions of the independent processor/memory pairs distributed throughout the PIM array into a single coherent parallel/distributed computation. Some major requirements for this execution model are the following:

1. Features for structuring and managing the global name space.
2. Control of object and data allocation, distribution, and alignment, with special support for sparse and irregular structures, as well as dynamic load balancing.
3. A general thread model, as a basis for expressing a range of parallel and distributed execution strategies, with a facility for binding threads to data.
4. Support for an efficient mapping of the model's features to the underlying microserver/parcel mechanism and the operating system nucleus.

Additional requirements, which will not be further discussed in this paper, include the logical interface to I/O, protection issues, recovery from failed tasks and exceptional conditions, and the development of an API for code specification.

Added to this should be the desirable features of hierarchy of abstraction, encapsulation, and modularity as well as simplicity and uniformity. The distributed execution model must employ as its basis underlying mechanisms that can be implemented efficiently while retaining substantial generality and extensibility. The following model in its current inchoate state addresses most of these requirements.

3 Macroservers: A Brief Overview

We begin our concrete discussion of the macroserver model by providing a brief overview of the key concepts – macroserver classes and objects, state variables, methods and threads. Along the way we touch the relationship with the *microserver* model introduced by work at the University of Notre Dame and in the DIVA project [3,20].

A *macroserver* is an object that comes into existence by being *created* as an instantiation of a parameterized template called a *macroserver class.* Such a class contains declarations of variables and a set of methods defining its "behavior". While the hardware architecture provides a shared address space, the discipline imposed by the object-based framework requires all accesses to external data to be performed via method calls, optionally controlled through a set of access privileges. At the time a macroserver is created, a region in the virtual PIM array memory is allocated to the new object. This allocation can be explicitly controlled in the model, by either directly specifying the region or *aligning* the object with an already existing one. A reference to the created object can be assigned to a new type of variable, called *macroserver variable,* which can act as a handle to the object.

At any point in time, a macroserver (object) is associated with a state space in which a set of asynchronous threads is operating, each of which being the result of the spawning of a method. The data structures of a macroserver can be distributed across the memory region allocated to it. We provide explicit functionality for specifying the initial distribution of data and their incremental redistribution depending on dynamically arising conditions. While the basic ideas of this feature originate from data parallel languages [5,13], we have generalized this concept to include arbitrary distribution functions and to apply to general data structures such as those provided by LISP lists. Furthermore, the model offers functionality for controlling the location in memory where a thread is to be executed. Such bindings can be established dynamically and are particularly important for linking threads to data on which they operate as well as for dealing with irregular computations.

Threads are lightweight; they execute asynchronously as long as not subject to synchronization. Mutual exclusion can be controlled via atomic methods. A macroserver whose methods are all atomic is a *monitor* and can be used as a flexible instrument for scheduling access to resources. A "small" monitor can be associated with each element of a large data structure (such as a reservation system), co-allocating the set of variables required by the monitor with the associated element. This provides the basis for performing the scheduling in a highly efficient way in the ASAP.

Threads can be synchronized using condition variables or futures. *Condition variables* [12] provide a simple and efficient low-level mechanism allowing threads to wait for synchronization conditions or signal their validity. *Future variables*, which are related to the futures in Multilisp [11], can be bound to threads and used for implicit or explicit synchronization based upon the status of the thread and a potential value yielded by it.

Figure 1 uses a code fragment for a producer/consumer problem with bounded buffer to illustrate some of the general features of the model. We describe a macroserver class, *buffer_template*, which is parameterized with an integer *size*. The class contains declarations for a data array *fifo* – the buffer data structure –, a number of related auxiliary variables, and two condition variables. Three methods – *put*, *get*, and *get_count* – are introduced, the first two of them being declared atomic. We show how a macroserver object can be created from that class, and how to obtain access to the associated methods.

Our notation is based on Fortran 95 [8] and ad-hoc syntax extensions mainly motivated by HPF [13] and Opus [7]. Note that we use this notation only as a means for communication in this paper, without any intent of specifying a concrete language syntax, but rather focusing on semantic constructs which can be embedded into a range of programming languages.

4 Key Features of the Model

In this section, we outline those features of the model that were specifically designed to support the efficient execution of parallel programs on massively parallel PIM arrays. We focus on the following topics, which are discussed in the subsections below.

- control of object allocation
- distribution and alignment of data
- thread model
- synchronization

4.1 Object Allocation

Some next-generation PIM architectures such as the MIND chip under development incorporate multithreading mechanisms for very low overhead task context

```
MACROSERVER CLASS buffer_template(size)   ! declaration of the
                                          ! macroserver class buffer_template
  INTEGER :: size          ! declaration of the class parameter
 ! Declarations of the class variables:
  REAL          :: fifo(0:size-1)
  INTEGER       :: count = 0
  INTEGER       :: px=0, cx=0
  CONDITION :: c_empty, c_full
    ...
 CONTAINS  ! Here follow the method declarations

  ATOMIC METHOD put(x)        ! put an element into the buffer
      REAL :: x
    ...
  END

  ATOMIC REAL METHOD get()    ! get an element from the buffer
    ...
  END

  INTEGER METHOD get_count() ! get the value of count
    ...
  END

    ...
END MACROSERVER CLASS buffer_template

 ! Main program:
INTEGER   buffersize
MACROSERVER (buffer_template) my_buffer  ! declaration of the macroserver
                                            ! variable my_buffer
READ (buffersize)
my_buffer= CREATE (buffer_template,buffersize) IN  M(region)  ! Creation of a
     ! macroserver object as an instance of class buffer_template, in a region of
     ! virtual memory. A reference to that object is assigned to my_buffer.
  ...
 ! A producer thread putting an item in the buffer:
CALL my_buffer%put(...)   ! Synchronous call of the method put in the macroserver
                                        ! object associated with my_buffer
  ...
```

Fig. 1. Skeleton for a producer/consumer problem

switching. A multithreaded architecture such as the TERA MTA [4] employs many concurrent threads to hide memory access latency. In contrast, multithreading in PIMs is employed primarily as a resource management mechanism supporting dynamic parallel resource control with the objective of overlapping local computation and communication to maximize effective memory bandwidth and external I/O throughput. While the effects of latency are mitigated to some degree by the use of a small number of threads, latency in PIMs is primarily addressed by the tight coupling of wide ALU to the memory row buffers, by alignment of data placement, and by directed computation at the site of the argument data.

Moreover, due to the relatively small size of the memory on a single PIM chip, large data structures must be distributed. As a consequence, our model provides features that allow the explicit control of the allocation of objects in virtual memory, either directly or relative to other objects, and the distribution of data in object space.

Consider the creation of a new macroserver object. The size and structure of the data belonging to this object and the properties of the threads to be operating on these data may imply constraints regarding the size of the memory region to be allocated and an alignment with other, already allocated objects. Such constraints can be specified in the create statement. An example illustrating an explicit assignment of a region in virtual PIM memory has been shown in Figure 1. A variant of this construct would allocate the new object in the same region as an already existing object:

my_buffer = **CREATE** (buffer_template, ...) **ALIGN** (my_other_buffer)

One PIM node in the memory region allocated to a macroserver is distinguished as its *home*. This denotes a central location where the "metadata" needed for the manipulation of the object are stored, exploiting the ability of the ASAP to deal with compact data structures that fit into a wide word in a particularly efficient way [1].

4.2 Distribution and Alignment of Data

A variable that is used in a computation must be allocated in the region allocated to its macroserver. We call the function describing the mapping from the basic constituents of the variable to locations in the PIM region the *distribution* of the variable. Macroservers allow explicit control of distribution and alignment.

Variable distributions have been defined in a number of programming language extensions, mostly in the context of SIMD architectures and data parallel languages targeted towards distributed-memory multiprocessors (DMMPs). Examples include Vienna Fortran, HPF, and HPF+ [5,13,6]. We generalize these

[1] For complex data structures the metadata will be usually organized in a hierarchical manner rather than with the simple structure suggested here.

approaches as discussed below. Some of these ideas have been also proposed for *actor* languages [21,16].

- Our approach applies not only to the traditional "flat" Fortran data structures such as multidimensional arrays but also covers arbitrary list structures such as those in LISP. This extension is essential since the hardware-supported microserver mechanism for PIMs allows highly efficient processing of variant-structure irregular data, as outlined in the "list crawler" example discussed in [3].
- We generalize the *distribution functions* proposed in Vienna Fortran to "distribution objects", which allow arbitrary mappings of data components to a PIM memory region, and in addition provide special support for sparse matrix representations. Such mappings can be performed dynamically, and they may be established in a partitioning tool which is linked to the macroserver. This generalization is crucial for the processing of irregular problems.
- More formally, we associate each variable v with an **index domain**, $\mathbf{I}$, which serves to provide a unique "name" for each component of v. For example, if v is a *simple* variable such as logical, integer, or real, then we choose for the index domain the singleton set $\mathbf{I} = \{1\}$. If $v = A(1:n, 1:m)$ is a two-dimensional array, then we can define $\mathbf{I} = [1:n] \times [1:m]$. Similarly, we can use strings of the form $i_1.i_2 \ldots .i_k$, where all i_j are between 1 and n, to access a leaf of an n-ary tree of height k.
  For fixed-structure variables (such as those in Pascal, Fortran, or C which do not involve pointers), the index domain can be determined at the time of allocation, and is invariant thereafter. Otherwise, such as for LISP data structures, the index domain changes according to the incremental modifications of the data structure.
  Based on index domains, we can define the distribution of a variable v with index domain $\mathbf{I}$ as a total function $\delta^v : \mathbf{I} \to \mathcal{R}$, where $\mathcal{R}$ denotes the region in PIM memory allocated to the macroserver to which the variable belongs. Here, we disregard replication for reasons of simplicity. Given v, δ^v and a particular memory region, $R \subseteq \mathcal{R}$, the set of elements of v mapped to R is called the **distribution segment** of R: $\{\mathbf{i} \in \mathbf{I} \mid \delta^v(\mathbf{i}) \in R\}$. R is called the **home** of all elements of v in the distribution segment.
- The distribution mechanism is complemented by a simple facility for the alignment of data structures in PIM memory. Distribution and alignment are designed as fully dynamic mechanisms, allowing redistribution and realignment at execution time, depending on decisions made at runtime.

4.3 Threads

At any time, zero or more threads may exist in a given macroserver; different threads – within one or different macroservers – may execute asynchronously in parallel unless subject to mutual exclusion or synchronization constraints. All threads belonging to a macroserver are considered peers, having the same rights and sharing all resources allocated to the macroserver.

At the time of thread creation, a future variable [11] may be bound to the thread. Such a variable can be used to make inquiries about the status of the thread, retrieve its attributes, synchronize the thread with other threads, and access its value after termination.

In contrast to UNIX processes, threads are lightweight, living in user space. Depending on the actual method a thread is executing, it may be ultra lightweight, carrying its context entirely in registers. On the other hand, a thread may be a significant computation, such as a sparse matrix vector multiply, with many levels of parallel subthreads. The model provides a range of thread attributes that allows a classification of threads according their weight and other characteristics. Examples include properties such as "non-preemptive" and "detached" [19].

Threads can be either spawned individually, or as members of a group. The generality and simplicity of the basic threading model allows in principle the dynamic construction of arbitrarily linked structures; it can be also used to establish special disciplines under which a group of threads may operate. An important special case is a set of threads that cooperate according to the *Single-Program-Multiple-Data (SPMD)* execution model that was originally developed for DMMPs. The SPMD model is a paradigm supporting "loosely synchronous" parallel computation which we perceive as one of the important disciplines for programming massively parallel PIM arrays. For example, linear algebra operations on large distributed dense or sparse data structures are particularly suitable for this paradigm and can be implemented in a similar way as for DMMPs. However, the efficient data management facilities in the ASAP and the ability to resolve indirect accesses directly in the memory allow a more efficient implementation for PIM arrays.

At the time a thread is created it can be bound to a specific location in the memory region associated with its macroserver. Usually that location is not specified explicitly but rather indirectly, referring to the home of a data structure on which the thread is to operate. This facility of binding computations to data is orthogonal to the variable distribution functionality, but can be easily used in conjunction with it.

We use a variant of the Fortran 95 forall statement to indicate the parallel creation of a set of threads all executing the same method. Consider the following simple example:

FORALL THREADS (I=1:100, J=1:100, **ON HOME** (A(I,J)))
    F(I,J)= **SPAWN** (intra_block_transpose,I,J)

This statement has the following effect:

- 10000 threads, say $\mathtt{t}(I, J), 1 \leq I, J \leq 100$ are created in parallel.

- Each thread $\mathbf{t}(I, J)$ activates method *intra_block_transpose* with arguments I and J.
- For each I and J, thread $\mathbf{t}(I, J)$ is executed in the ASAP which is the home of array element $A(I, J)$.
- For each I and J, the future variable $F(I, J)$ is assigned a reference to $\mathbf{t}(I, J)$.

The above example is a simplified code fragment for a parallel algorithm that performs the transposition of a block-distributed two-dimensional matrix in two steps, first transposing whole blocks, and then performing an intra-block transposition [18]. In the code fragement of Fig. 2, matrix A is dynamically allocated in the PIM memory, depending on a runtime-determined size specification. The blocksize is chosen in such a way that each block fits completely into one wide word of the ASAP (8 by 8 elements). We make the simplifying assumption that 8 divides the number of elements in one dimension of the matrix.

We illustrate the second step of this algorithm. Parallelism is controlled by explicit creation and synchronization of threads. Each thread executing a call to *intra_block_transpose* can be mapped to a permutation operation directly supported in the ASAP hardware [3].

4.4 Synchronization

Our model supports mutual exclusion and condition synchronization [2], in both cases following a conventional low-level approach. *Mutual exclusion* guarantees atomic access of threads to a given resource, leaving the order in which blocked threads are released implementation specific. It can be expressed using *atomic methods* similar to those in CC++ [15] or Java. *Condition synchronization* is modeled after Hoare's monitors [12]. *Condition variables* can be associated with programmed synchronization conditions; threads whose synchronization condition at a given point is not satisfied can suspend themselves with respect to a condition variable, waiting for other threads to change the state in such a way that the condition becomes true.

The example code in Fig. 3 sketches the declaration of a monitor *scheduler_template* that can be used for controlling the access of a set of reader and writer threads to a data resource. Here, **WAIT** c blocks a thread with respect to condition variable c, **EMPTY** c is a predicate that yields **true** iff no thread is blocked with respect to c, and **SIGNAL** c releases a thread from the queue associated with c, if it is not empty. Each reader access must be enclosed by the calls *begin_read* and *end_read*, and analogously for writers. The rest of the program illustrates how in a large data structure (in this case, a flight reservation system) each element can be associated with its own monitor; these monitors could be parameterized. Furthermore, the alignment clause attached to the create statement guarantees that the variables of each monitor are co-located with the associated element of the flight data base. This allows highly efficient processing of the synchronization in the ASAP.

```
PROGRAM MATRIX_TRANSPOSE

MACROSERVER CLASS t_class
   INTEGER :: N
   FUTURE, ALLOCATABLE :: F(,)    ! future variable array
   REAL,ALLOCATABLE,DISTRIBUTE( BLOCK(8) , BLOCK(8) )::A(,)
 CONTAINS

  METHOD initialize()
   ...
  METHOD intra_block_transpose(II,JJ)
    INTEGER :: II,JJ
    INTEGER :: I, J
    FORALL (I=II:II+7,J=JJ:JJ+7) A(I,J) = A(J,I)
  END intra_block_transpose

   ...
END MACROSERVER CLASS t_class

! Main Program
INTEGER :: II, JJ
MACROSERVER (t_class) my_transpose = CREATE (t_class)
CALL my_transpose%initialize()
     ...          ! exchange whole blocks
! for each block, activate intra_block_transpose as a separate thread. The thread
! associated with (II, JJ) executes in the PIM storing the matrix block A(II, JJ).
FORALL THREADS (II=1:N-7:8, JJ=1:N-7:8, ON HOME (A(II,JJ)))
    F(II,JJ)= SPAWN (intra_block_transpose,II,JJ)
WAIT ( ALL (F))                    ! barrier
...
END PROGRAM MATRIX_TRANSPOSE
```

Fig. 2. Matrix Transpose

In addition to the synchronization primitives discussed above, our model proposes special support for synchronization based on future variables similar to those introduced in Multilisp [11]. This takes two forms, explicit and implicit.

Explicit synchronization can be formulated via a version of the wait statement that can be applied to a logical expression depending on futures. Wait can also be used in the context of a forall threads statement, as shown in the example of Figure 2.

Implicit synchronization is automatically provided if a typed future variable occurs in an expression context that requires a value of that type.

```
MACROSERVER CLASS MONITOR scheduler_template
   INTEGER wr=0, ww=0
   CONDITION c_R, c_W
  CONTAINS
   METHOD begin_read()
    DO WHILE ((ww GT 0) AND (NOT EMPTY(c_W))) WAIT c_R
    wr = wr + 1
   END begin_read

   METHOD end_read()
    wr = wr-1
    IF wr == 0 SIGNAL(c_W)
   END end_read

   METHOD begin_write()
      ...
   METHOD end_write()
      ...

END MACROSERVER CLASS scheduler_template

 ! Main program
INTEGER :: N, I, K
TYPE flight_record          ! sketch of data structure for a flight record
  INTEGER :: date, time, flight_number, create_status
  MACROSERVER(scheduler_template) my_scheduler
   ...
END TYPE flight_record
   ...
TYPE(flight_record), ALLOCATABLE,... ONTO M(region)::flights(:)
   ...
ALLOCATE(flights(N))
   ...
DO I=1,N
 flights(I)%my_scheduler= CREATE(scheduler_template,create_status(I))
                                            ALIGN(HOME(flights(I)))
    ...
END DO
...
 ! write access to element flight(K) in a thread:
K=...
CALL flights(K)%my_scheduler%begin_write()
    ! write
CALL flights(K)%my_scheduler%end_write()
...
```

Fig. 3. Fine-grain scheduling

5 Implications for Implementation

The semantic constructs of the execution model have been carefully selected both to provide a powerful framework for specifying and managing parallel computations and to permit efficient implementation of the primitive mechanisms comprising it. In this section, we discuss some of the most important implementation issues underlying the basic support for the execution model. Effective performance of the basic primitives ensures overall computational efficiency.

Parcels are messages that convey both data identifiers and task specifiers that direct work at remote sites and determine a continuation path of execution upon completion. Similar to active messages, parcels direct all inter-PIM transactions. An arriving parcel may cause something as simple as a memory read operation to be performed or something as complex as a search through all PIM memory for a particular key field value. Parcel assimilation therefore becomes a critical factor in the effective performance obtainable from a given PIM. Parcels must be variable length such that commands for primitive operations can be acquired quickly with only the longer executing tasks requiring longer parcel packets. A single access can take as little as 40 nanoseconds. Assuming 128-bit parcels for operation, address, and data (for writes in and read outs) and 1 Gbps unidirectional pins, read and/or write operations can be streamed using byte serial input and output parcel communications, sustaining peak memory bandwidth. More complex operations will demand somewhat longer packets but involve multiple memory accesses, thus overlapping communications with computation. Parcels can be stored in row-width registers and processed directly by the wide ALU of the PIM processor.

Address translation from virtual to physical may be made efficient through a combination of techniques. One uses set associative techniques similar to some cache schemes. A virtual page address is restricted to one or a small set of PIMs which contain the rest of the address translation through a local page table. Thus, any virtual page address reference can be directed to the correct PIM (or PIMs) and the detailed location within the PIM set is determined by a simple table look up. Associative scanning is performed efficiently because a row wide simple ALU allows comparisons of all bits or fields in a row simultaneously at memory cycle rates. These wide registers can be used as pseudo TLBs although they are in the register name space of the PIM processor ISA and therefore can be the target of general processing. Such registers can be used to temporarily hold the physical addresses of heavily used methods. Because the opcode and page offsets on a chip are relatively few, a single row register might hold 128 translation entries which could be checked in a single logic cycle of 10 nanoseconds, quite possibly covering all procedures stored on the PIM. A second technique to be employed is in-situ address translation. A complex data structure comprising a tree of pointer links includes both the virtual address of the pointer and the physical address. As a logical node in the data structure is retrieved from memory, the wide row permits both physical and virtual addresses to be fetched simultaneously. As page migration occurs relatively slowly, updating the contents of physical pointer fields can be done with acceptable overhead.

The multithreading scheduling mechanism allows the processing of one parcel request on register data to be conducted while the memory access for another parcel is being undertaken and quite possibly one or more floating point operations are propagating through the shared pipelined units. The multithreading mechanism's general fine grain resource management adapts to the random incidence of parcels and method execution requirements to provide efficient system operation. Each active thread is represented in a thread management register. A wide register is dedicated to each thread, holding the equivalent of 64 4-byte words.

Synchronization is very general as an array of compound atomic operations can be programmed as simple methods. A special lock is set for such operations such that thread context switches do not interfere with atomicity, ensuring correctness. The future-based synchronization operations require more than simple bit toggling. The execution record of a thread bound to a future variable stores pointers to sources of access requests when the solicited value has yet to be produced. Upon completion of the necessary computation or value retrieval, the buffered pointer redirects the result value to the required locations. Thus a future involves a complex data structure and its creation, manipulation, and elimination upon completion. This can be a very expensive operation using conventional processors and memories. But the PIM architecture with direct access to a 256-byte record (single row) allows single cycle (memory cycle) create, update, and reclaim operations making this form of synchronization very efficient.

6 Revisiting the Requirements

We started our discussion of the macroserver model by indicating four major requirements at the end of Section 2. The material in Sections 3 and 4 addressed these issues on a point-to-point basis in the context of the description of macroserver semantics. Here we summarize this discussion, guided by the set of requirements.

Structuring and Managing the Name Space

PIM arrays support a global name space. The macroserver model structures this space in a number of ways. First macroserver classes encapsulate data and methods, providing a "local" scope of naming. Second, each macroserver object establishes a dynamic naming scope for its instance of data and methods. Finally (this is a point not further mentioned in the paper), each macroserver has a set of "acquaintances" [1] and related access privileges, which allow a system designer to establish domains of protection.

Object and Data Allocation

Controlling the allocation, distribution, and alignment of objects and data is a key requirement for the model. Macroservers provide the following support for these features:

- At the time of object creation, constraints for the allocation of the object representation may be specified. Such constraints include size of representation, explicit specification of an area in the PIM array, or alignment with an already existing object.
- A general mechanism for the distribution of data in the memory region allocated to an object has been developed. This includes arbitrary mappings from elements of a data structure to PIMs and provides special support for sparse arrays.
- Distributions may be determined at runtime, supporting dynamic allocation and re-allocation of data and providing support for dynamic load balancing strategies.

Thread Model

Asynchronous threads are generated by the spawning of a method. At any given time, a set of lightweight threads, which can be activations of the same or of different methods, may operate in a macroserver.

The "weight" of threads may cover a broad range, from ultra-lightweight – with the context being kept completely in registers – to fairly heavy computations such as a conjugent gradient algorithm.

Threads can be spawned as single entities or as elements of a group. The simplicity of the basic concept and its generality allow for a range of parallel execution strategies to be formulated on top of the basic model.

An important aspect supporting locality of operation is a feature that allows the binding of threads to data: at the time of thread generation, a constraint for the locus of execution of this thread in PIM memory can be specified.

Efficiency of Implementation

A key requirement for an execution model is performance: the ability to map the elements of the model efficiently to the underlying PIM array. The macroserver model addresses this issue by providing a sophisticated set of control mechanisms (many of which are too low-level to be reflected in a high-level language specification). A more detailed discussion of implementation support issues was conducted in the previous section.

7 Conclusion

This paper discussed macroservers, an object-based programming and execution model for Processor-in-Memory arrays. Based upon the major requirements as dictated by the salient properties of PIM-based systems such as HTMT MIND, the DIVA chip, and IBM's Blue Gene, we worked out the key properties of the model addressing these requirements. A more detailed description of the macroserver model can be found in [26]; its application to irregular problems is dealt with in [27].

Future work will focus on the following tasks:

- completing a full specification of the model
- applying the model to a range of application problems, in particular sparse matrix operations and unstructured grid algorithms,
- developing an implementation study, and
- re-addressing a number of research issues such as
  - higher-level synchronization mechanisms,
  - the management of thread groups and related collective operations,
  - support for recovery mechanisms in view of failing nodes or exceptional software conditions, and
  - the I/O interface.

In addition, we will study the interface of the model to high-level languages and related compiler and runtime support issues.

Acknowledgements

The authors wish to recognize the important contributions and ideas by the University of Notre Dame including Peter Kogge, Jay Brockman, and Vince Freeh as well as those of the USC ISI group led by Mary Hall and John Granacki.

References

1. G. A.Agha. ACTORS: A Model of Concurrent Computation in Distributed Systems. *MIT Press*, 1986. 22
2. G. R.Andrews. Concurrent Programming. Principles and Practice. *Benjamin/Cummings*, 1991. 18
3. J. B.Brockman,P. M.Kogge,V. W.Freeh,S. K.Kuntz, and T. L.Sterling. Microservers: A New Memory Semantics for Massively Parallel Computing. *Proceedings ACM International Conference on Supercomputing (ICS'99)*, June 1999. 9, 12, 16, 18
4. G.Alverson, P.Briggs, S.Coatney, S.Kahan, and R.Korry. Tera Hardware-Software Cooperation. *Proc.Supercomputing 1997*, San Jose, CA, November 1997. 15
5. B. Chapman, P. Mehrotra, and H. Zima. Programming in Vienna Fortran. *Scientific Programming*, 1(1):31–50, Fall 1992. 12, 15
6. B. Chapman, P. Mehrotra, and H. Zima. Extending HPF for Advanced Data Parallel Applications. *IEEE Parallel and Distributed Technology*, Fall 1994, pp. 59-70. 15
7. B. Chapman, M. Haines, P. Mehrotra, J. Van Rosendale, and H. Zima. Opus: A Coordination Language for Multidisciplinary Applications. *Scientific Programming*, 1998. 13

8. J. C.Adams, W. S.Brainerd, J. T.Martin, B. T.Smith, and J. L.Wagener. Fortran 95 Handbook. Complete ISO/ANSI Reference. *The MIT Press*, 1997. 13
9. M.Gokhale,W.Holmes,and K.Iobst. Processing in Memory: The Terasys Massively Parallel PIM Array. *IEEE Computer 28(4)*,pp.23-31,1995. 10
10. M.Hall,J.Koller,P.Diniz,J.Chame,J.Draper, J.LaCoss, J.Granacki, J.Brockman, A.Srivastava, W.Athas, V.Freeh, J.Shin, and J.Park. Mapping Irregular Applications to DIVA, a PIM-Based Data Intensive Architecture. *Proceedings SC'99*, November 1999. 7
11. R. H.Halstead,Jr. Multilisp: A Language for Concurrent Symbolic Computation. *ACM Transactions on Programming Languages and Systems (TOPLAS), 7(4),501-538*, October 1985. 13, 17, 19
12. C. A. R.Hoare. Monitors: An Operating Systems Structuring Concept. *Comm.ACM* 17(10),pp.549-557,1974. In:*(C. A. R.Hoare and R. H.Perrott,Eds.) Operating System Techniques*, Academic Press,pp.61-71, 1972. 13, 18
13. High Performance Fortran Forum. *High Performance Fortran Language Specification, Version 2.0*, January 1997. 12, 13, 15
14. *http://www.ibm.com/news/1999/12/06.phtml* 7
15. C.Kesselman. CC++. In: G. V.Wilson and P.Lu (Eds.): Parallel Programming Using CC++, Chapter 3, pp.91–130, *The MIT Press*, 1996. 18
16. W.Kim. Thal: An Actor System for Efficient and Scalable Concurrent Computing. *Ph.D.Thesis, University of Illinois at Urbana-Champaign*, 1997. 16
17. P. M.Kogge. The EXECUBE Approach to Massively Parallel Processing. In: *Proc.1994 Conference on Parallel Processing*, Chicago, August 1994. 10
18. V.Kumar,A.Grama,A.Gupta,and G.Karypis. Introduction to Parallel Computing. Design and Analysis of Algorithms. *The Benjamin/Cummings Publishing Company*, 1994. 18
19. B.Nichols,D.Buttlar,and J.Proulx Farrell. Pthreads Programming. *O'Reilly*,1998. 17
20. Notre Dame University. Final Report: PIM Architecture Design and Supporting Trade Studies for the HTMT Project. *PIM Development Group, HTMT Project*, University of Notre Dame, September 1999. 12
21. R.Panwar and G.Agha. A Methodology for Programming Scalable Architectures. *Journal of Parallel and Distributed Computing*, 22(3),pp.479-487, September 1994. 16
22. D.Patterson et al. A Case for Intelligent DRAM: IRAM. *IEEE Micro*, April 1997. 7
23. T.Sterling and P.Kogge. An Advanced PIM Architecture for Spaceborne Computing. *Proc.IEEE Aerospace Conference*, March 1998. 7
24. T.Sterling and L.Bergman. A Design Analysis of a Hybrid Technology Multithreaded Architecture for Petaflops Scale Computation. *Proceedings ACM International Conference on Supercomputing (ICS'99)*, June 1999. 7
25. H. Zima, P. Brezany, B. Chapman, P. Mehrotra, and A. Schwald. Vienna Fortran – a language specification. Internal Report 21, *ICASE, NASA Langley Research Center*, Hampton, VA, March 1992.
26. H.Zima and T.Sterling. Macroservers: A Programming and Execution Model for DRAM PIM Arrays. Technical Report CACR-182, *Center for Advanced Computing Research (CACR), California Institute of Technology*, Pasadena, CA, and TR 2000-1, *Institute for Software Science, University of Vienna.* 24
27. H.Zima and T.Sterling. Support for Irregular Computations in Massively Parallel PIM Arrays, Using an Object-Based Execution Model. *Proc.Irregular'2000*, Cancun, Mexico, May 2000. 24

The New DRAM Interfaces: SDRAM, RDRAM and Variants

Brian Davis[1], Bruce Jacob[2], Trevor Mudge[1]

[1] Electrical Engineering & Computer Science, University of Michigan,
Ann Arbor, MI 48109
{tnm, btdavis}@eecs.umich.edu

[2] Electrical & Computer Engineering, University of Maryland at College Park,
College Park, MD 20742
blj@eng.umd.edu

Abstract. For the past two decades, developments in DRAM technology, the primary technology for the main memory of computers, have been directed towards increasing density. As a result 256 M-bit memory chips are now commonplace, and we can expect to see systems shipping in volume with 1 G-bit memory chips within the next two years. Although densities of DRAMs have quadrupled every 3 years, access speed has improved much less dramatically. This is in contrast to developments in processor technology where speeds have doubled nearly every two years. The resulting "memory gap" has been widely commented on. The solution to this gap until recently has been to use caches. In the past several years, DRAM manufacturers have explored new DRAM structures that could help reduce this gap, and reduce the reliance on complex multilevel caches. The new structures have not changed the basic storage array that forms the core of a DRAM; the key changes are in the interfaces. This paper presents an overview of these new DRAM structures.

1 Introduction

For the past two decades developments in DRAM technology, the primary technology for the main memory of computers, have been directed towards increasing density. As a result 256 M-bit memory chips are now commonplace, and we can expect to see systems shipping in volume with 1 G-bit memory chips within the next two years. Many of the volume applications for DRAM, particularly low cost PCs, will thus require only one or two chips for their primary memory. Ironically, then, the technical success of the commodity DRAM manufacturers is likely to reduce their future profits, because the fall in units shipped is unlikely to be made up by unit price increases.

Although densities of DRAMs have quadrupled every 3 years, access speed has improved much less dramatically. This is in contrast to developments in processor technology where speeds have doubled nearly every two years. The resulting "memory gap" has been widely commented on. The solution to this gap until recently has been to use

M. Valero et al. (Eds.): ISHPC 2000, LNCS 1940, pp. 26-31, 2000.

caches. In the past several years, DRAM manufacturers have explored new DRAM structures that could help reduce this gap, and reduce the reliance on complex multilevel caches. These structures are primarily new interfaces that allow for much higher bandwidths to and from chips. In some cases the new structures provide lower latencies too. These changes may also solve the problem of diminishing profits be allowing DRAM companies to charge a premium for the new parts. This paper will overview some of the more common new DRAM interfaces.

2 DRAM Architectures — Background

DRAMs are currently the primary memory device of choice because they provide the lowest cost per bit and greatest density among solid-state memory technologies. When the 8086 was introduced, DRAM were roughly matched in cycle time with microprocessors. However, as we noted, since this time processor speeds have improved at a rate of 80% annually, DRAM speeds have improved at a rate of 7% annually [1]. In order to reduce the performance impact of this gap, multiple levels of caches have been added, and processors have been designed to prefetch and tolerate latency.

The traditional asynchronous DRAM interface underwent some limited changes in response to this growing gap. Examples were fast-page-mode (FPM), extended-data-out (EDO), and burst-EDO (BEDO), each provided faster cycle times if accesses were from the same row and thus more bandwidth than the predecessor.

In recent years more dramatic changes to the interface have been made. These built on the idea, first exploited in fast-page-mode devices, that each read to a DRAM actually reads a complete row of bits or word line from the DRAM core into an array of sense amps. The traditional asynchronous DRAM interface would then select through a multiplexer a small number of these bits (x1, x4, x8 being typical). This is clearly wasteful and by clocking the interface it is possible to serially read out the entire row or parts of the row. This describes the now popular synchronous DRAM (SDRAM). the basic two-dimensional array of bit cells that forms the core of every DRAM is unchanged, although its density can continue to undergo improvement. The clock runs much faster than the access time and thus the bandwidth of memory accesses are greatly increased provided they are to the same word line.

There are now a number of improvements and variants on the basic single-data-rate (SDR) SDRAM. These include double-data-rate (DDR) SDRAM, direct Rambus DRAM (DRDRAM), and DDR2 which is under development by a number of manufacturers. DDR signalling is simply a clocking enhancement where data is driven and received on both the rising and the falling edge of the clock. DRDRAM includes high speed DDR signalling over a relatively narrow bus to reduce pin count and a high level of banking on each chip. The core array itself can be subdivided at the expense of some area. This multibanking allow for several outstanding requests to be in flight at the same time, providing an opportunity to increase bandwidth through pipelining of the unique bank requests. Additional core enhancements, which may be applied to many of these new interface specifications include Virtual Channel (VC) caching, Enhanced Memory System (EMS) caching and Fast-Cycle (FC) core pipelining. These additional improvements provide some form of caching of recent sense amp data. Even with these significant redesigns, the cycle time — as measured by end-to-end access latency — has continued to improve at a rate significantly lower than microprocessor performance. These redesigns have been successful at improving bandwidth, but latency continues to be a constrained by the area impact and cost pressures on DRAM core architectures [2].

In the next section we will give an introduction to some of the most popular new DRAMs.

3 The New DRAM Interfaces

The following sections will briefly discuss a variety of DRAM architectures. The first four are shown in Table 1. The min/max access latencies given in this table assume

Table 1: Synchronous DRAM Interface Characteristics

	PC100	DDR266 (PC2100)	DDR2	DRDRAM
Potential Bandwidth	0.8 GB/s	2.133 GB/s	3.2 GB/s	1.6 GB/s
Interface Signals	64(72) data 168 pins	64(72) data 168 pins	64(72) data 184 pins	16(18) data 184 pins
Interface Frequency	100 MHz	133 MHz	200 MHz	400MHz
Latency Range	30-90 nS	18.8-64 nS	17.5-42.6 nS	35-80 nS

that the access being scheduled has no bus utilization conflicts with other accesses.

The last two sections (ESDRAM and FCDRAM) discuss core enhancements which can be applied to a core which can be mated to almost any interface, and the last section covering graphics DRAM reflects the special considerations that this application requests from DRAM.

3.1 SDR SDRAM (PC100)

The majority of desktop PC's shipped in 1Q2000 use SDR PC100 SDRAM. SDR DRAM devices are currently available at 133 Mhz (PC133). The frequency endpoint for this line of SDRAMs is in question, though PC150 and PC166 are almost certain to be developed. As we noted, SDRAM is a synchronous adaptation of the prior asynchronous FPM and EDO DRAM architectures that streams or burst s data out under the synchronously with a clock provided the data is all from the same row of the core. The length of the burst is programmable up to the maximum size of the row. The clock (133 MHz in this case) typically runs nearly an order of magnitude faster than the access time of the core. As such, SDRAM is the first DRAM architecture with support for access concurrency on a single shared bus. Earlier non-synchronous DRAM had to support access concurrency via externally controlled interleaving.

3.2 DDR SDRAM (DDR266)

Earlier we noted that DDR SDRAM differs from SDR in that unique data is driven and sampled at both the rising and falling edges of the clock signal. This effectively doubles the data bandwidth of the bus compared to an SDR SDRAM running at the same clock frequency. DDR266 devices are very similar to SDR SDRAM in all other characteristics. They use the same signalling technology, the same interface specification, and the same pinouts on the DIMM carriers. The JEDEC specification for DDR266 devices provides for a number of "CAS-latency" speed grades. Chipsets are currently under development for DDR266 SDRAM and are expected to reach the market in 4Q2000.

3.3 DDR2

The DDR2 specification under development by the JEDEC 42.3 Future DRAM Task Group is intended to be the follow-on device specification to DDR SDRAM. While DDR2 will have a new pin-interface, and signalling method (SSTL), it will leverage much of the existing engineering behind current SDRAM. The initial speed for DDR2 parts will be 200 MHz in a bussed environment, and 400Mhz in a point-to-point application, with data transitioning on both edges of the clock [3]. Beyond strictly the advancement of clock speed, DDR2 has a number of interface changes intended to enable faster clock speeds or higher bus utilization. The lower interface voltage (1.8 V), differential clocking and micro-BGA packaging are all intended to support a higher clock rate on the bus. Specifying a write latency equal to the read latency minus one (WL = RL-1) provides a time profile for both read and write transactions that enables easier pipelining of the two transaction types, and thus higher bus utilization. Similarly, the addition of a programmable additive latency (AL) postpones the transmission of a CAS from the interface to the core. This is typically referred to as a "posted-CAS" transaction. This enables the RAS and CAS of a transaction to be transmitted by the controller on adjacent bus cycles. Non-zero usage of the AL parameter is best paired with a closed-page-autoprecharge controller policy, because otherwise open-page-hits incur an unnecessary latency penalty. The burst length on DDR2 has been fixed at 4 data bus cycles. This is seen as a method to simplify the driver/receiver logic, at the expense of heavier loading on the address signals of the DRAM bus [4]. The DDR2 specification is not finalized, but the information contained here is based upon the most recent drafts for DDR2 devices and conversations with JEDEC members.

3.4 Direct Rambus (DRDRAM)

Direct Rambus DRAM (DRDRAM) devices use a 400 Mhz 3-byte-wide channel (2 for data, 1 for addresses/commands). DRDRAM devices use DDR signalling, implying a maximum bandwidth of 1.6 G-bytes/s, and these devices have many banks in relation to SDRAM devices of the same size. Each sense-amp, and thus row buffer, is shared between adjacent banks. This implies that adjacent banks cannot simultaneously maintain an open-page, or maintain an open-page while a neighboring bank performs an access. The increased number of banks for a fixed address space has the result of increasing ability to pipeline accesses due to the reduced probability of sequential accesses mapping into the same bank. The sharing of sense-amps increases the row-buffer miss rate as compared to having one open row per bank, but it reduces the cost by reducing the die area occupied by the row buffer [5].

3.5 ESDRAM

A number of proposals have been made for adding a small amount of SRAM cache onto the DRAM device. Perhaps the most straightforward approach is advocated by Enhanced Memory Systems (EMS). The caching structure proposed by EMS is a single direct-mapped SRAM cache line, the same size as the DRAM row, associated with each bank. This allows the device to service accesses to the most recently accessed row, regardless of whether refresh has occurred and enables the precharge of the DRAM array to be done in the background without affecting the contents of this row-cache. This architecture also supports a no-write-transfer mode within a series of interspersed read and write accesses. The no-write-transfer mode allows writes to occur through the sense-amps, without affecting the data currently being held in the cache-line associated with that bank [6]. This approach may be applied to any DRAM interface, PC100 interface parts are currently available and DDR2 parts have been proposed.

3.6 FCDRAM

Fast Cycle DRAM (FCRAM) developed by Fujitsu is an enhancement to SDRAM which allows for faster repetitive access to a single bank. This is accomplished by dividing the array not only into multiple banks but also small blocks within a bank. This decreases each block's access time due to reduced capacitance, and enables pipelining of requests to the same bank. Multistage pipelining of the core array hides precharge, allowing it to occur simultaneously with input-signal latching and data transfer to the output latch. FCDRAM is currently sampling in 64M-bit quantities, utilizing the JEDEC standard DDR SDRAM interface, but is hampered by a significant price premium based upon the die area overhead of this technique [7]. Fujitsu is currently sampling FCDRAM devices which utilize both SDR and DDR SDRAM interfaces, additionally low-power devices targeted at the notebook design space are available.

4 Conclusions

DRAM are widely used because they provide a highly cost effective storage solution. While there are a number of proposals for technology to replace DRAM, such as SRAM, magnetic RAM (MRAM) [8] or optical storage [9], the new DRAM technologies remain the volatile memory of choice for the foreseeable future. Increasing the performance of DRAM by employing onboard cache, and interfaces with higher utilization or smaller banks may impact the cost of the devices, but it could also significantly increase the performance which system designers are able to extract from the primary memory system.

The DRAM industry has been very conservative about changing the structure of DRAMs in even the most minor fashion. There has been a dramatic change in this attitude in the past few years and we are now seeing a wide variety of new organizations being offered. Whether one will prevail and create a new standard commodity part remains to be seen.

References

[1] W. Wulf and S. McKee. 1995 "Hitting the Memory Wall: Implications of the Obvious." ACM Computer Architecture News. Vol 23, No. 1. March 1995.

[2] V. Cuppu, B. Jacob, B. Davis, and T. Mudge. 1999. "A performance comparison of contemporary DRAM architectures." In Proc. 26th Annual International Symposium on Computer Architecture (ISCA'99), pages 222–233, Atlanta GA.

[3] JEDEC 2000. Joint Electronic Device Engineering Council; 42.3 Future DRAM Task Group. http://www.jedec.org/

[4] B. Davis, T. Mudge, B. Jacob, V. Cuppu. 2000. "DDR2 and low-latency Variants." In Proc. Solving the Memory Wall Workshop, held in conjunction with the 27th International Symposium on Computer Architecture (ISCA-2000). Vancouver BC, Canada, June 2000.

[5] IBM. 1999. "128Mb Direct RDRAM". International Business Machines, http://www.chips.ibm.com/products/memory/19L3262/19L3262.pdf

[6] EMS. 2000. "64Mbit - Enhanced SDRAM". Enhanced Memory Systems, http://www.edram.com/Library/datasheets/SM2603,2604pb_r1.8.pdf.

[7] Fujitsu. 2000. "8 x 256K x 32 BIT Double Data Rate (DDR) FCRAM." Fujitsu Semiconductor, http://www.fujitsumicro.com/memory/fcram.htm.

[8] "Motorola believes MRAM could replace nearly all memory technologies." Semiconductor Business News, May 10, 2000. http://www.semibiznews.com/story/OEG20000510S0031

[9] C-3D. 2000. "C3D Data Storage Technology." Constellation 3D, http://www.c-3d.net/tech.htm

Blue Gene

Henry S. Warren, Jr.

IBM Thomas J. Watson Research Center,
Yorktown Heights, New York 10598, USA
hankw@us.ibm.com

Abstract. In December 1999, IBM Research announced a five-year, $100M research project, code named Blue Gene, to build a petaflop computer which will be used primarily for research in computational biology. This computer will be 100 times faster than the current fastest supercomputer, ASCI White. The Blue Gene project has the potential to revolutionize research in high-performance computing and in computational biology.
To reach a petaflop, Blue Gene interconnects approximately one million identical and simple processors, each capable of executing at a rate of one gigaflop. Each of the 25 processors on a single chip contains a half megabyte of embedded DRAM, which is shared among those processors. They communicate through a system of high speed orthogonal opposing rings. The approximately 40,000 chips communicate by message passing. The configuration is suitable for highly parallel problems that do not require huge amounts of memory. Such problems can be found in computational biology, high-end visualization, computational fluid dynamics, and other areas.
This talk will be primarily about the Blue Gene hardware and system software. We will also briefly discuss the protein folding application.

M. Valero et al. (Eds.): ISHPC 2000, LNCS 1940, p. 32, 2000.

Earth Simulator Project in Japan

Seeking a Guide Line for the Symbiosis between the Earth and Human Beings
- Visualizing an Aspect of the Future of the Earth by a Supercomputer -

Keiji Tani

Earth Simulator Research and Development Center,
Japan Atomic Energy Research Institute
Sumitomo-Hamamatsu-cho Bldg. 10F
1-18-16 Hamamatsu-cho, Minato-ku, Tokyo, 105-0013 Japan

Abstract. The Science and Technology Agency of Japan has proposed a project to promote studies for global change prediction by an integrated three-in-one research and development approach: earth observation, basic research, and computer simulation. As part of the project, we are developing an ultra-fast computer, the "Earth Simulator", with a sustained speed of more than 5 TFLOPS for an atmospheric circulation code. The "Earth Simulator" is a MIMD type distributed memory parallel system in which 640 processor nodes are connected via fast single-stage crossbar network. Earth node consists of 8 vector-type arithmetic processors which are tightly connected via shared memory. The peak performance of the total system is 40 TFLOPS. As part of the development of basic software system, we are developing an operation supporting software system what is called a "center routine". The total system will be completed in the spring of 2002.

1 Introduction

The Science and Technology Agency of Japan has proposed a project to promote studies for global change prediction by an integrated three-in-one research and development approach: earth observation, basic research, and computer simulation. It goes without saying that basic process and observation studies for global change are very important. Most of these basic processes, however, are tightly coupled and form a typical complex system. A large-scale simulation in which the coupling between these basic processes are taken into consideration is the only way for a complete understanding of this kind of complicated phenomena As part of the project, we are developing an ultra-fast computer named the "Earth Simulator".

The Earth Simulator has two important targets, one is the applications to the atmospheric and oceanographic science and the other is the applications to the solid earth science. For the first applications, high resolution global, regional and local

M. Valero et al. (Eds.): ISHPC 2000, LNCS 1940, pp. 33-42, 2000.

models will be developed and for the second a global dynamic model to describe the entire solid earth as a system and a simulation model of earthquake generation process etc., will be developed.

2 Outline of the Hardware System

Taking as an example of a global AGCM (atmospheric general circulation model), here we consider the requirements for computational resources for the Earth Simulator. Present typical global AGCM uses about a 100km mesh in both longitudinal and latitudinal directions. The mesh size will be reduced to 10km in the high resolution global AGCM on the Earth Simulator. The number of layers will also be enhanced up to several to 10 times that of the present model. According to the resolution level, the time integration mesh must be reduced. Taking all these conditions into account, both the CPU and main memory of the Earth Simulator must be at least 1000 times lager than those of present computers. The effective performance of present typical computers is about 4-6 GFLOPS. Therefore, we set the sustained performance of the Earth Simulator for a high resolution global AGCM to be more than 5 TFLOPS.

Reviewing the trends of commercial parallel computers, we can consider two types of parallel architectures for the Earth Simulator; one is a distributed parallel system with cache-based microprocessors and the other is a system with vector processors. According to the performance evaluation for a well-known AGCM (CCM2), it is shown that the efficiency is less than 7% on cache-based parallel systems, where the efficiency is the ratio of the sustained performance to the theoretical peak. On the other hand, an efficiency about 30% was obtained on parallel systems with vector processors [1]. For this reason, we decided to employ a distributed parallel system with vector processors.

Another key issue for a parallel system is the interconnection network. As mentioned above, many different types of applications will run on the Earth Simulator. Judging from the flexibility of parallelism for many different types of applications, we employ a single-stage crossbar network in order to make the system completely flat.

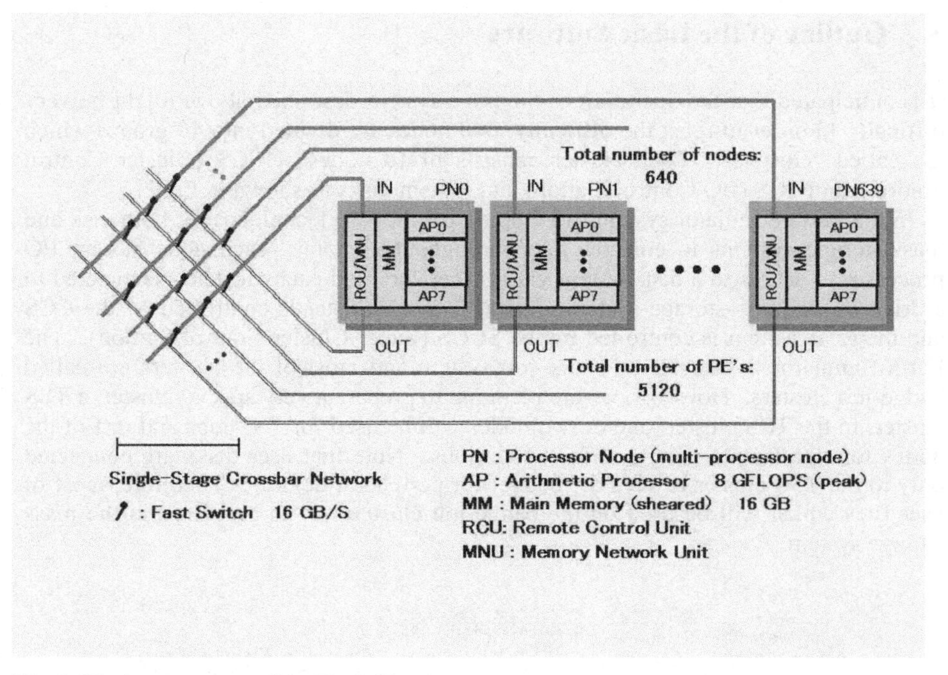

Fig. 1 Hardware system of the Earth Simulator

An outline of the hardware system of the Earth Simulator which is shown in Fig.1 can be summarized as follows:

- **Architecture: MIMD-type distributed memory parallel system consisting of computing nodes with shared memory vector type multi-processors.**
- **Performance: Assuming the efficiency 12.5%, the peak performance 40 TFLOPS (the effective performance for an AGCM is more than 5 TFLOPS).**
  - **Total number of processor nodes 640**
  - **Number of PE's for each node 8**
  - **Total number of PE's 5120**
  - **Peak performance of each PE 8 GFLOPS**
  - **Peak performance of each node 64 GFLOPS**

  - **Main memory: 10 TB (total).**
    Shared memory / node: 16 GB

  - **Interconnection network: Single-Stage Crossbar Network**

3 Outline of the Basic Software

It is anticipated that the operation of the huge system described above might be very difficult. In order to relax the difficulty, 640 nodes are divided into 40 groups which are called "clusters". Each cluster consists of 16 nodes, a CCS (Cluster Control Station), an IOCS (I/O Control Station) and system disks as shown in Fig.2.

In the Earth Simulator system, we employ parallel I/O techniques for both disk and mass-storage systems to enhance data throughputs. That is, each node has an I/O processor to access to a disk system via fast network and each cluster is connected to a drive of the mass-storage system via IOCS. Each cluster is controlled by the CCS and the total system is controlled by the SCCS (Super Cluster Control Station). The Earth Simulator is basically a badge-job system and most of the clusters are called badge-job clusters. However, we are planning to prepare a very special cluster, a TSS cluster. In the TSS cluster, one or two nodes will be used for TSS jobs and rest of the nodes for small scale (single node) badge jobs. Note that user disks are connected only to the TSS cluster to save the budget for peripheral devices. Therefore, most of user files which will be used on the badge-job clusters are to be stored in the mass storage system.

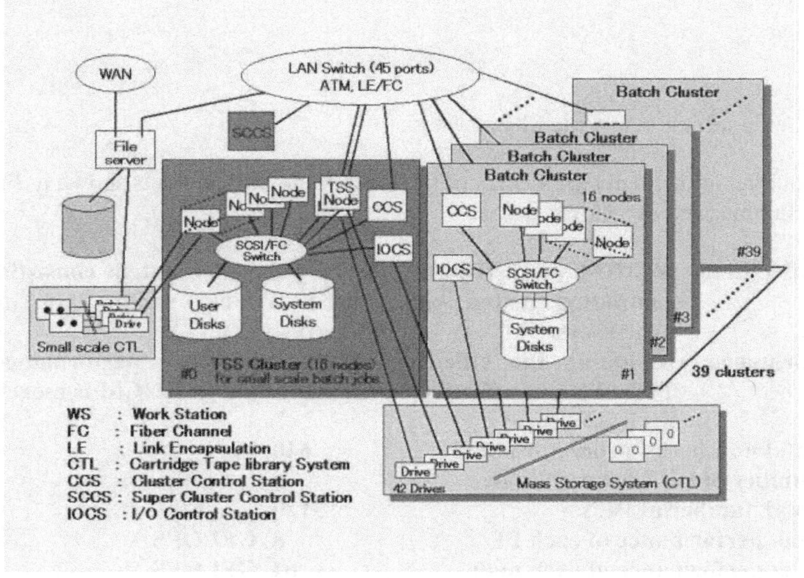

Fig. 2 Clusters and peripheral devices of the Earth Simulator

We basically use an operating system which will be developed by the vender (NEC) for their future system. However, some high level functions characteristic to the Earth Simulator will be newly developed such as:

- large scalability of distributed-memory parallel processing,
- total system control by SCCS ,
- large scale parallel I/O,
- interface for center routine (operation support software), etc..

Concerning parallel programming languages, automatic parallelization, OpenMP and micro-tasking will be prepared as standard languages for shared-memory parallelization, and both MPI2 and HPF2 for distributed memory parallelization. As described above, the Earth Simulator has a memory hierarchy. If a user wants to optimize his code on the system, he must take the memory hierarchy into consideration as shown in Fig.3.

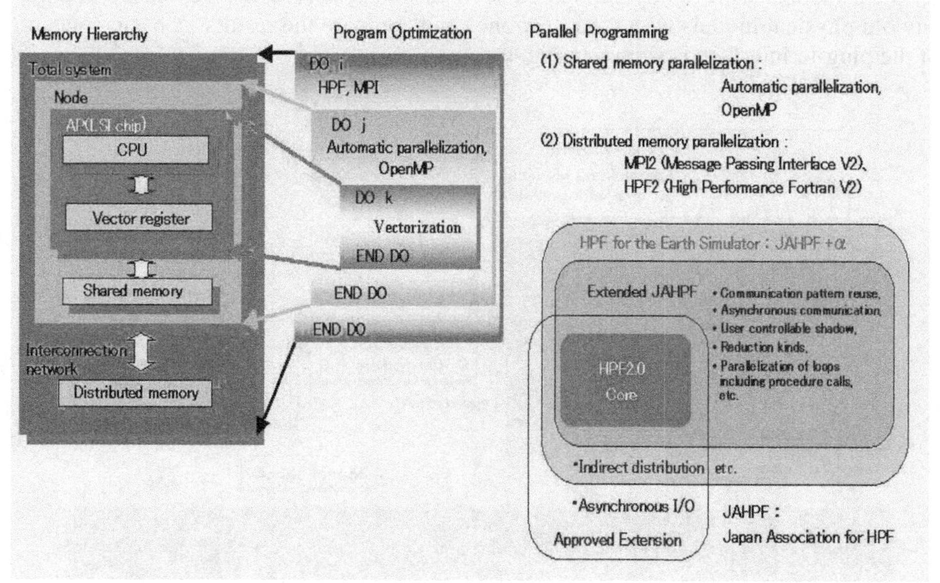

Fig. 3 Memory hierarchy and program optimization

As part of the development of basic software system, we are developing an operation supporting software system what is called a "center routine". We are going to use an archival system as a main storage of user files. It usually takes very long time to transfer user files from the mass-storage system to the system disks. Therefore, the most important function of the center routine is the optimal scheduling of not only submitted batch jobs but also user files necessary for the jobs. Because the Earth Simulator is a special machine for numerical Earth science, our job scheduler has following important features:

- higher priority to job turn-round rather than total job throughput,
- prospective reservation of batch-job nodes, Jobs are controlled by wall-clock time.
- automatic recall and migration of user files between system disks and mass-storage system.

4 Outline of the Application Software

Also, we are developing application programs optimized for the architecture of the Earth Simulator. A typical AGCM can be divided into two parts, a dynamical core and a set of physical models. We employ a standard plug-in interface to combine the physical models with the dynamical core. With this system, users can easily change any old physical models with their new ones and compare the results. A basic concept of the plug-in interface is shown in Fig.4.

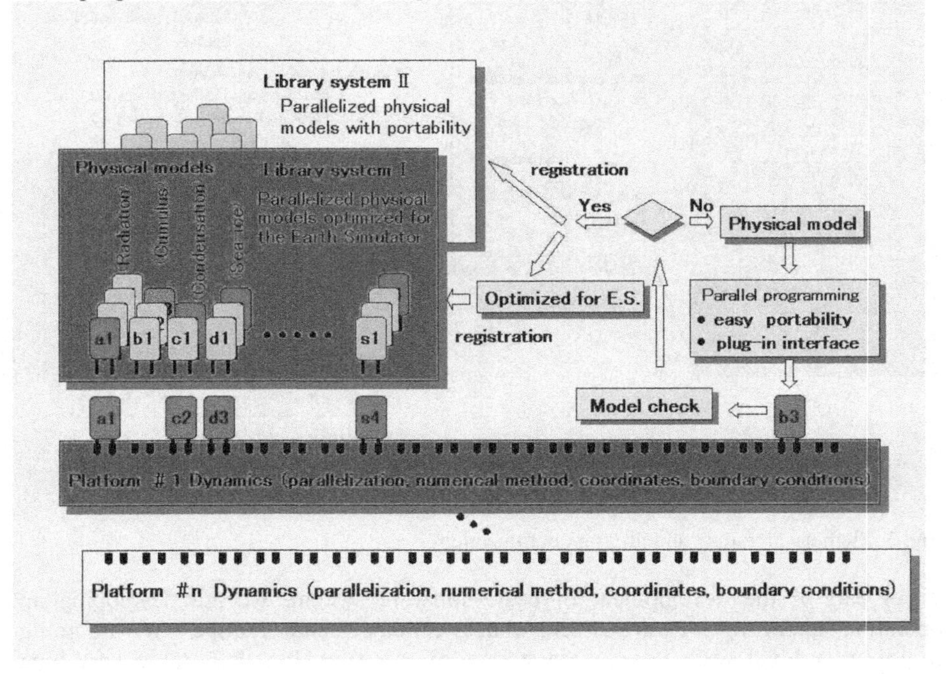

Fig. 4 Concept of plug-in interface

Taking the idea of the plug-in interface, we are developing two application software systems which will be opened and used by many users as standard models on the Earth Simulator system, an atmosphere-ocean coupled model and a large scale FEM model for solid Earth science which is called "GeoFEM". The configuration of the atmosphere-ocean coupled model is shown in Fig. 5 (a). Preliminary results on the 1-day averaged global distribution of precipitation are also shown in Fig.5 (b). The results have been obtained by averaging over the results at 4th and 5th years with an initial condition of static atmosphere. Also, the configuration of the GeoFEM and preliminary results on the crust/mantle activity in the Japan Archipelago region are shown in Fig.6 (a) and (b), respectively [2].

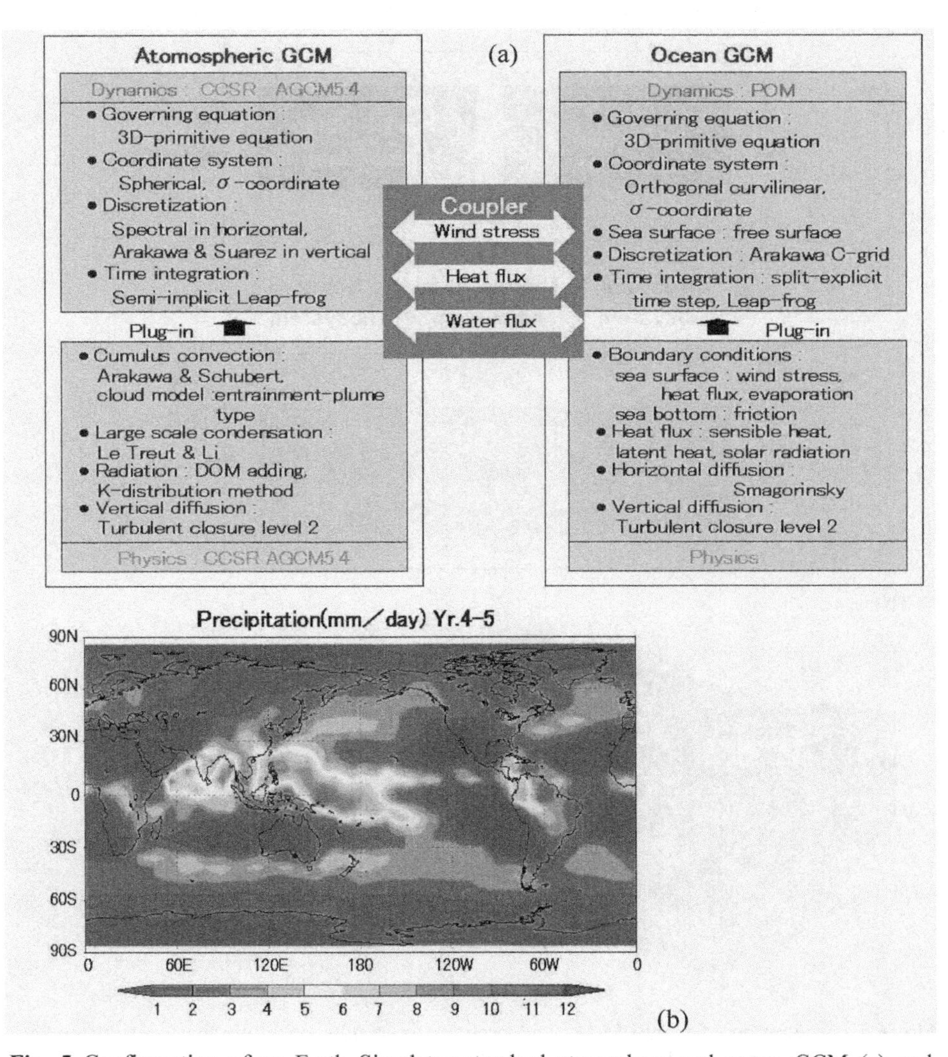

Fig. 5 Configuration of an Earth Simulator standard atmosphere and ocean GCM (a) and preliminary results on 1-day averaged global distribution of precipitation (b)

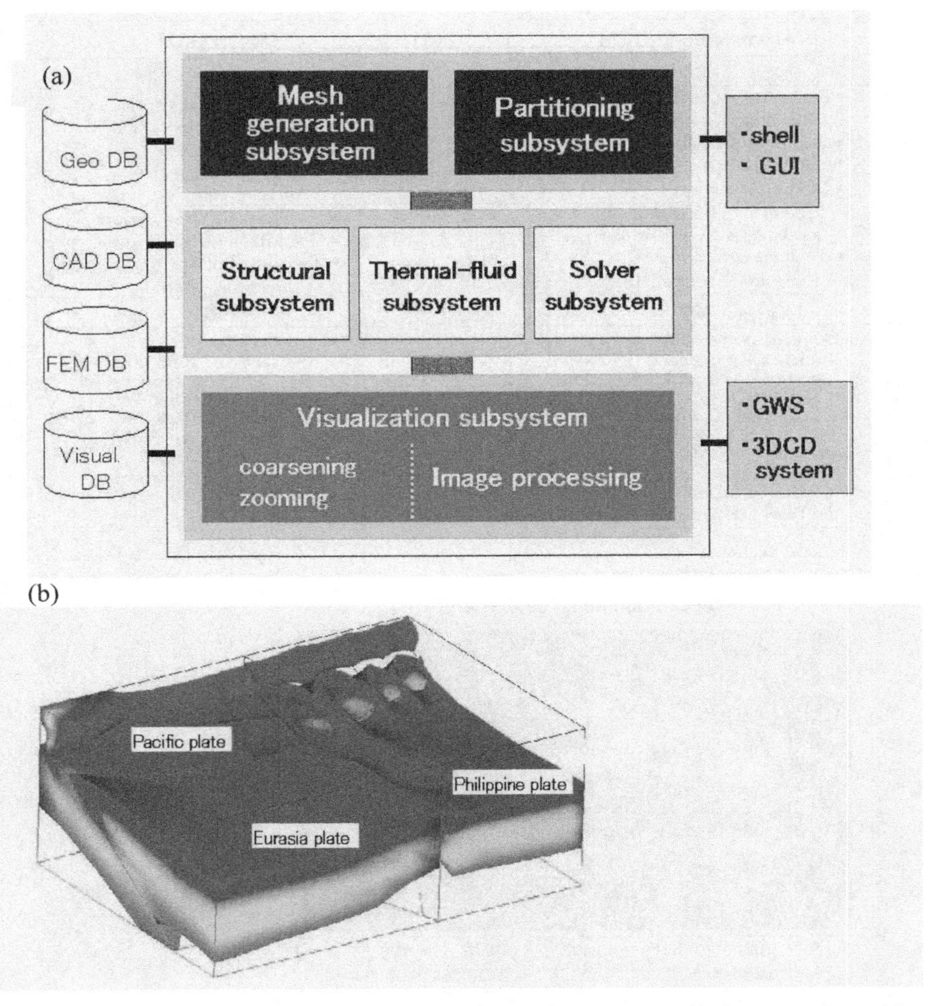

Fig. 6 Configuration of GeoFEM (a) and preliminary results on preliminary results on the crust/mantle activity in the Japan Archipelago region (b)

5 Schedule

All the design and R&D works for both hardware and basic software systems, the conceptual design, basic design, the design of parts and packaging, the R&D for parts and packaging , and the detailed design, have been completed during the last three fiscal years, FY97,98 and 99. The manufacture of the hardware system is now underway. The development of the center routine is also underway. Facilities necessary for the Earth Simulator including buildings are also under construction.

The bird's-eye view of the Earth Simulator building is shown in Fig. 7 The total system will be completed in the spring of 2002.

An outline of the schedule is shown in Table 1.

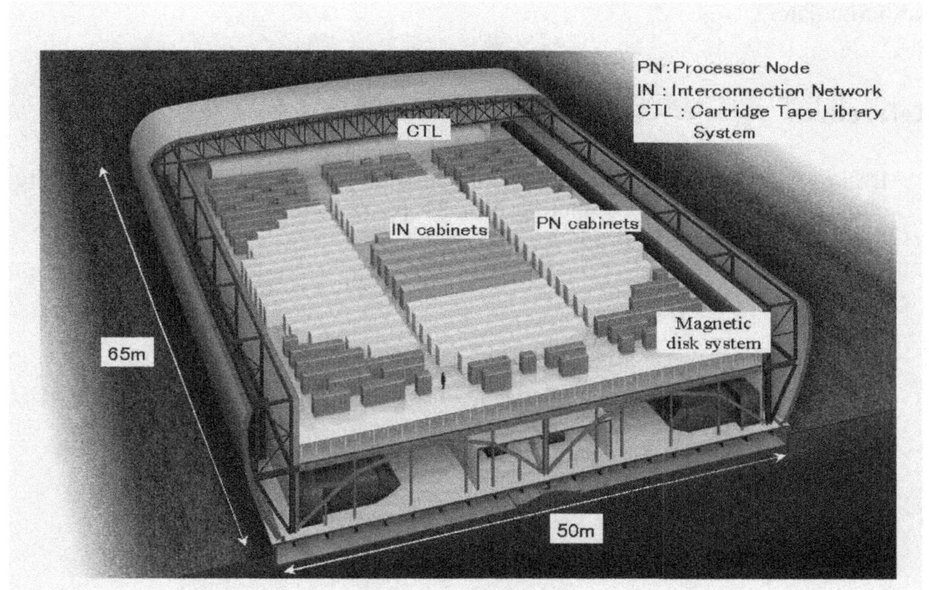

Fig. 7 Bird's-eye view of the Earth Simulator building

Table 1 Outline of schedule

	1997	1998	1999	2000	2001	2002
Hardware System						
Conceptual design						
Basic design						
Design of parts and packaging						
R&D for parts and packaging						
Detailed design						completion
Manufacture and installation						
Software System (center routine)						
Basic design						
Detailed design						
Program development & test						
Peripheral Devices						
Facilities (Buildings)						
Application Software						
Operation						

Acknowledgement

This report describes an outline of the results of design work made by NEC. The author thanks all the staffs of NEC who have made great effort for the design of the Earth Simulator.

References

1. Drake, J., et al., : Design and performance of a scalable parallel community climate model, Parallel Computing, Vol. 21, pp.1571-1591 (1995)
2. http://geofem.tokyo.rist.or.jp/

Limits of Task-based Parallelism in Irregular Applications

Barbara Kreaseck, Dean Tullsen, and Brad Calder

University of California, San Diego
La Jolla, CA 92093-0114, USA
{kreaseck, tullsen, calder}@cs.ucsd.edu

Abstract. Traditional parallel compilers do not effectively parallelize irregular applications because they contain little loop-level parallelism. We explore Speculative Task Parallelism (STP), where tasks are full procedures and entire natural loops. Through profiling and compiler analysis, we find tasks that are speculatively memory- and control-independent of their neighboring code. Via speculative futures, these tasks may be executed in parallel with preceding code when there is a high probability of independence. We estimate the amount of STP in irregular applications by measuring the number of memory-independent instructions these tasks expose. We find that 7 to 22% of dynamic instructions are within memory-independent tasks, depending on assumptions.

1 Introduction

Today's microprocessors rely heavily on instruction-level parallelism (ILP) to gain higher performance. Flow control imposes a limit to available ILP in single-threaded applications [8]. One way to overcome this limit is to find parallel tasks and employ multiple flows of control (*threads*). Task-level parallelism (TLP) arises when a task is independent of its neighboring code. We focus on finding these independent tasks and exploring the resulting performance gains.

Traditional parallel compilers exploit one variety of TLP, loop level parallelism (LLP), where loop iterations are executed in parallel. LLP can overwhelming be found in numeric, typically FORTRAN programs with regular patterns of data accesses. In contrast, general purpose integer applications, which account for the majority of codes currently run on microprocessors, exhibit little LLP as they tend to access data in irregular patterns through pointers. Without pointer disambiguation to analyze data access dependences, traditional parallel compilers cannot parallelize these irregular applications and ensure correct execution.

In this paper we explore task-level parallelism in irregular applications by focusing on *Speculative Task Parallelism* (STP), where tasks are speculatively executed in parallel under the following assumptions: 1) tasks are full procedures or entire natural loops, 2) tasks are speculatively memory-independent and control-independent, and 3) our architecture allows the parallelization of tasks via speculative futures (discussed below). Figure 1 illustrates STP, showing a dynamic instruction stream where a task Y has no memory access conflicts

M. Valero et al. (Eds.): ISHPC 2000, LNCS 1940, pp. 43-58, 2000.

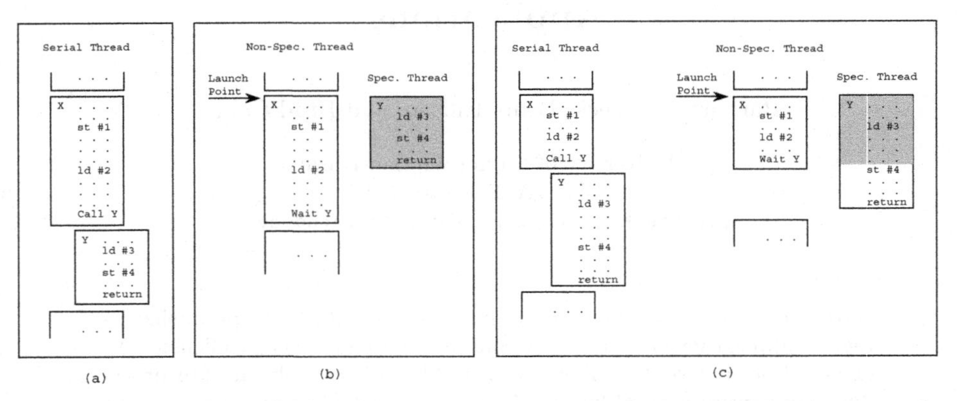

Fig. 1. *STP example: (a) shows a section of code where the task Y is known to be memory-independent of the preceding code X. (b) the shaded region shows memory- and control-independent instructions that are essentially removed from the critical path when Y is executed in parallel with X. (c) when task Y is longer than X.*

with a group of instructions, X, that precede Y. The shorter of X and Y determines the overlap of memory-independent instructions as seen in Figures 1(b) and 1(c). In the absence of any register dependences, X and Y may be executed in parallel, resulting in shorter execution time. It is hard for traditional parallel compilers of pointer-based languages to expose this parallelism.

The goals of this paper are to identify such regions within irregular applications and to find the number of instructions that may thus be removed from the critical path. This number represents the maximum possible STP. To facilitate our discussion, we offer the following definitions.

A *task*, exemplified by Y in Figure 1, is a bounded set of instructions inherent to the application. Two sections of code are *memory-independent* when neither contains a store to a memory location that the other accesses. When all load/store combinations of the type [load,store], [store,load] and [store,store] between two tasks, X and Y, access different memory locations, X and Y are said to be memory-independent. A *launch point* is the point in the code preceding a task where the task may be initiated in parallel with the preceding code. This point is determined through profiling and compiler analysis. A *launched task* is one that begins execution from an associated launch point on a different thread.

Because the biggest barrier to detecting independence in irregular codes is memory disambiguation, we identify memory-independent tasks using a profile-based approach and measure the amount of STP by estimating the amount of memory-independent instructions those tasks expose. As successive executions may differ from the profiled execution, any launched task would be inherently speculative. One way of launching a task in parallel with its preceding code is through a parallel language construct called a *future*. A future conceptually forks a thread to execute the task and identifies an area of memory in which to relay status and results. When the original thread needs the results of the futured task, it either waits on the futured task thread, or in the case that the task was never futured due to no idle threads, it executes the futured task itself.

To exploit STP, we assume a speculative machine that supports speculative futures. Such a processor could speculatively execute code in parallel when there is a high probability of independence, but no guarantee. Our work identifies launch points for this speculative machine, and estimates the parallelism available to such a machine. With varying levels of control and memory speculation, 7 to 22% of dynamic instructions are within tasks that are found to be memory-independent, on a set of irregular applications for which traditional methods of parallelization are ineffective.

In the next section we discuss related work. Section 3 contains a description of how we identify and quantify STP. Section 4 describes our experiment methodology and Section 5 continues with some results. Implementation issues are highlighted in Section 6, followed by a summary in Section 7.

2 Related Work

In order to exploit Speculative Task Parallelism, a system would minimally need to include multiple flows of control and memory disambiguation to aid in mis-speculation detection. Current proposed structures that aid in dynamic memory disambiguation are implemented in hardware alone [3] or rely upon a compiler [5, 4]. All minimally allow loads to be speculatively executed above stores and detect write-after-read violations that may result from such speculation.

Some multithreaded machines [21, 19, 2] and single-chip multiprocessors [6, 7] facilitate multiple flows of control from a single program, where flows are generated by compiler and/or dynamically. All of these architectures could exploit non-speculative TLP if the compiler exposed it, but only Hydra [6] could support STP without alteration.

Our paper examines speculatively parallel tasks in non-traditionally parallel applications. Other proposed systems, displaying a variety of characteristics, also use speculation to increase parallelism. They include Multiscalar processors [16, 12, 20], Block Structured Architecture [9], Speculative Thread-level Parallelism [15, 14], Thread-level Data Speculation [18], Dynamic Multithreading Processor [1], and Data Speculative Multithreaded hardware architecture [11, 10].

In these systems, the type of speculative tasks include fixed-size blocks [9], one or more basic blocks [16], dynamic instruction sequences [18], loop iterations [15, 11], instructions following a loop [1], or following a procedure call [1, 14]. These tasks were identified dynamically at run-time [11, 1], statically by compilers [20, 9, 14], or by hand [18]. The underlying architectures include traditional multiprocessors [15, 18], non-traditional multiprocessors [16, 9, 10], and multithreaded processors [1, 11].

Memory disambiguation and mis-speculation detection was handled by an Address Resolution Buffer [16], the Time Warp mechanism of time stamping requests to memory [9], extended cache coherence schemes [14, 18], fully associative queues [1], and iteration tables [11]. Control mis-speculation was always handled by squashing the mis-speculated task and any of its dependents. While a few handled data mis-speculations by squashing, one rolls back speculative execution to the wrong data speculation [14] and others allow selective, dependent re-execution of the wrong data speculation [9, 1].

Most systems facilitate data flow by forwarding values produced by one thread to any consuming threads [16, 9, 18, 1, 11]. A few avoid data mis-speculation through synchronization [12, 14]. Some systems enable speculation by value prediction using last-value [1, 14, 11] and stride-value predictors [14, 11].

STP identifies a source of parallelism that is complimentary to that found by most of the systems above. Armed with a speculative future mechanism, these systems may benefit from exploiting STP.

3 Finding Task-based Parallelism

We find Speculative Task Parallelism by identifying all tasks that are memory-independent of the code that precedes the task. This is done through profiling and compiler analysis, collecting data from memory access conflicts and control flow information. These conflicts determine proposed launch points that mark the memory dependences of a task. Then for each task, we traverse the control flow graph (CFG) in reverse control flow order to determine launch points based upon memory and control dependences. Finally, we estimate the parallelism expected from launching the tasks early. The following explain the details of our approach to finding STP.

Task Selection The type of task chosen for speculative execution directly affects the amount of speculative parallelism found in an application. Oplinger, et. al. [14], found that loop iterations alone were insufficient to make speculative thread-level parallelism effective for most programs. To find STP, we look at three types of tasks: *leaf procedures* (procedures that do not call any other procedure), *non-leaf procedures*, and entire *natural loops*. When profiling a combination of task types, we profile them concurrently, exposing memory-independent instructions within an environment of interacting tasks.

Although all tasks of the chosen type(s) are profiled, only those that expose at least a minimum number of memory-independent instructions are chosen to be launched early. The final task selection is made after evaluating memory and control dependences to determine actual launch points.

Memory Access Conflicts Memory access conflicts are used to determine the memory dependences of a task. They occur when two load/store instructions access the same memory region. Only a subset of memory conflicts that occur during execution are useful for calculating launch points. Useful conflicts span task boundaries and are of the form [load, store], [store, load], or [store, store]. We also disregard stores or loads due to register saves and restores across procedure

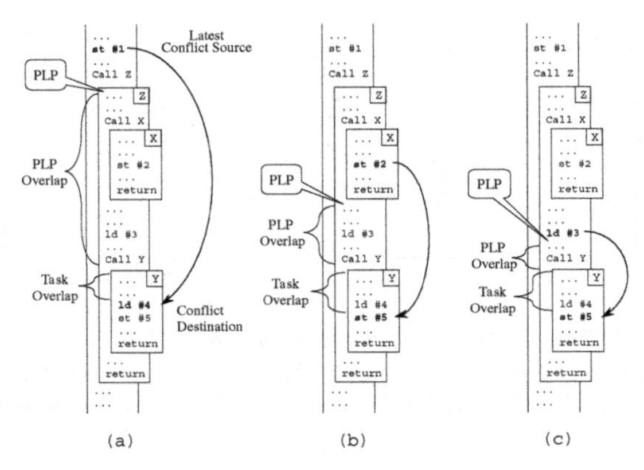

Fig. 2. *PLP Locations: conflict source, conflict destination, PLP candidate, PLP overlap, and task overlap, when the latest conflict source is (a)* before the calling routine, *(b)* within a sibling task *(c)* within the calling routine.

calls. We call the conflicting instruction preceding the task the *conflict source*, and the conflicting instruction within the task is called the *conflict destination.* Specifically, when the conflict destination is a load, the conflict source will be the last store to that memory region that occurred outside the task. When the conflict destination is a store, the conflict source will be the last load or store to that memory region that occurred outside the task.

Proposed Launch Points The memory dependences for a task are marked, via profiling, as proposed launch points (PLPs). A PLP represents the memory access conflict with the latest (closest) conflict source in the dynamic code preceding one execution of that task. Exactly one PLP is found for each dynamic task execution. In our approach, launch points for a task occur only within the task's calling region, limiting the amount and scope of executable changes that would be needed to exploit STP. Thus, PLPs must also lie within a task's calling routine.

Figure 2 contains an example that demonstrates the latest conflict sources and their associated PLPs. Task Z calls tasks X and Y. Y is the currently executing task in the example and Z is its calling routine. When the conflict source occurs before the beginning of the calling routine, as in Figure 2(a), the PLP is directly before the first instruction of the task Z. When the conflict source occurs within a sibling task or its child tasks, as in Figure 2(b), the PLP immediately follows the call to the sibling task. In Figure 2(c), the conflict source is a calling routine instruction and the PLP immediately follows the conflict source.

Two measures of memory-independence are associated with each PLP. They are the PLP overlap and the task overlap, as seen in Figure 2. The *PLP overlap* represents the number of dynamic instructions found between the PLP and the beginning of the task. The *task overlap* represents the number of dynamic instructions between the beginning of the task and the conflict destination. With PLPs determined by memory dependences that are dynamically closest to the

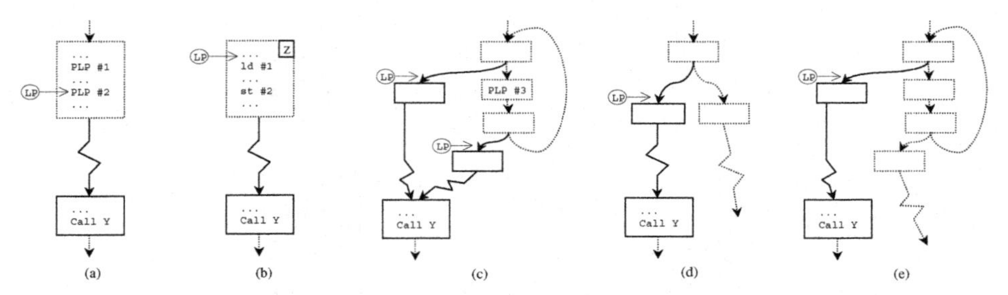

Fig. 3. *Task Launch Points: Dotted areas have not been fully visited by the back-trace for task Y. (a) CFG block contains a PLP. (b) CFG block is calling routine head. (c) Loop contains a PLP. (d) Incompletely visited CFG block. (e) Incompletely visited loop.*

task call site, and by recording the smallest task overlap, we only consider conservative, safe locations with respect to the profiling dataset.

Task Launch Points Both memory dependences and control dependences influence the placement of task launch points. Our initial approach to exposing STP determines task launch points that provide two guarantees. First, *static control dependence* is preserved: all paths from a task launch point lead to the original task call site. Second, *profiled memory dependence* is preserved: should the threaded program be executed on the profiling dataset, all instructions between the task launch point and the originally scheduled task call site will be free of memory conflicts. Variations which relax these guarantees are described in Section 5.2.

For each task, we recursively traverse the CFG in reverse control flow order starting from the original call site, navigating conditionals and loops, to identify task launch points. We use two auxiliary structures: a stalled block list to hold incompletely visited blocks, and a stalled loop list to hold incompletely visited loops. There are five conditions under which we record a task launch point. These conditions are described below. The first three will halt recursive back-tracing along the current path. As illustrated in Figure 3, we record task launch points:

a. when the current CFG block contains a PLP for that task. The task launch point is the last PLP in the block.
b. when the current CFG block is the head of the task's calling routine and contains no PLPs. The task launch point is the first instruction in the block.
c. when the current loop contains a PLP for that task. Back-tracing will only get to this point when it visits a loop, and all loop exit edges have been visited. As this loop is really a sibling of the current task, task launch points are recorded at the end of all loop exit edges.
d. for blocks that remain on the stalled block list after all recursive back-tracing has exited. A task launch point is recorded only at the end of each visited successor edge of the stalled block.
e. for loops that remain on the stalled loop list after all recursive back-tracing has exited. A task launch point is recorded only at the end of each visited loop exit edge.

Each task launch point indicates a position in the executable in which to place a task future. At each task's original call site, a check on the status of the future will indicate whether to execute the task serially or wait on the result of the future.

Parallelization Estimation We estimate the number of memory-independent instructions that would have been exposed had the tasks been executed at their launch points during the profile run. Our approach ensures that each instruction is counted as memory-independent at most once. When the potential for instruction overlap exceeds the task selection threshold the task is marked for STP. We use the total number of claimed memory-independent instructions as an estimate of the limit of STP available on our hypothetical speculative machine.

4 Methodology

To investigate STP, we used the ATOM profiling tools [17] and identified natural loops as defined by Muchnick [13]. We profiled the SPECint95 suite of benchmark programs. Each benchmark was profiled for 650 million instructions. We used the reference datasets on all benchmarks except compress. For compress, we used a smaller dataset, in order to profile a more interesting portion of the application.

We measure STP by the number of memory-independent task instructions that would overlap preceding non-task instructions should a selected task be launched (as a percentage of all dynamic instructions).

The task selection threshold comprises two values, both of which must be exceeded. For all runs, the task selection threshold was set at 25 memory-independent instructions per task execution and a total of 0.2% of instructions executed. We impose this threshold to compensate for the expected overhead of managing speculative threads and to enable allocation of limited resources to tasks exposing more STP.

Our results show a limit to STP exposed by the launched execution of memory-independent tasks. No changes, such as code motion, were made or assumed to have been made to the original benchmark codes that would heighten the amount of memory-independent instructions. Overhead due to thread creation, along with wakeup and commit, will be implementation dependent, and thus is not accounted for. Register dependences between the preceding code and the launched task were ignored. Therefore, we show an upper bound to the amount of STP in irregular applications.

5 Results

We investigated the amount of Speculative Task Parallelism under a variety of assumptions about task types, memory conflict granularity, control and memory dependences. Our starting configuration includes profiling at the page-level (that is, conflicts are memory accesses to the same page) with no explicit speculation and is thus our most conservative measurement.

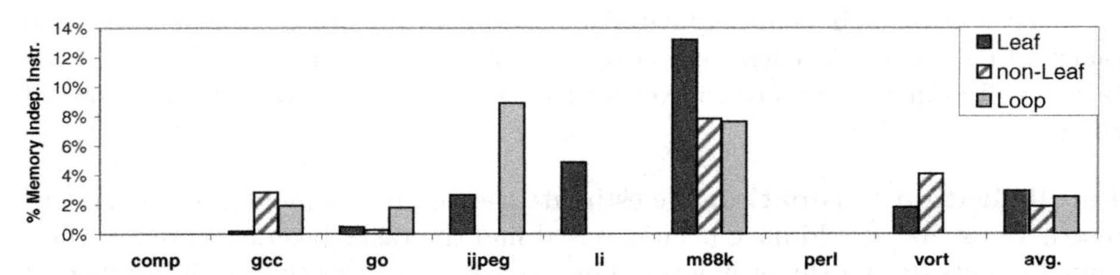

Fig. 4. *Task Types: Individual data points identify memory independent instructions as a percentage of all instructions profiled and represent our starting configuration.*

5.1 Task Type

We first look to see which task types exhibit the most STP. We then explore various explicit speculation opportunities to find additional sources of STP. Finally, we investigate any additional parallelism that might be exposed by profiling at a finer memory granularity.

We investigate leaf procedures, non-leaf procedures, and entire natural loops. We profiled these three task types to see if any one task type exhibited more STP than the others. Figure 4 shows that, on average, a total of 7.3% of the profiled instructions were identified as memory independent, with task type contributions differing by less than 1% of the profiled instructions. This strongly suggests that all task types should be considered for exploiting STP. The succeeding experiments include all three task types in their profiles.

5.2 Explicit Speculation

The starting configuration places launch points conservatively, with no explicit control or memory dependence speculation. Because launched tasks will be implicitly speculative when executed with different datasets, our hypothetical machine must already support speculation and recovery. We explore the level of STP exhibited by explicit speculation, first, by speculating on control dependence, where the launched task may not actually be needed. Next, we speculate on memory dependence, where the launched task may not always be memory-independent of the preceding code. Finally, we speculate on both memory and control dependences by exploiting the memory-independence of the instructions within the task overlap.

Our starting configuration determines task launch points that preserve **static control dependences**, such that *all paths* from the task's launch points lead to the original task call site. Thus, a task that is statically control dependent upon a condition whose outcome is constant, or almost constant, throughout the profile, will not be selected, even though launching this task would lead to almost no mis-speculations. We considered two additional control dependence schemes that would be able to exploit the memory-independence of this task.

Profiled control dependences exclude any static control dependences that are based upon branch paths that are never traversed. When task launch points

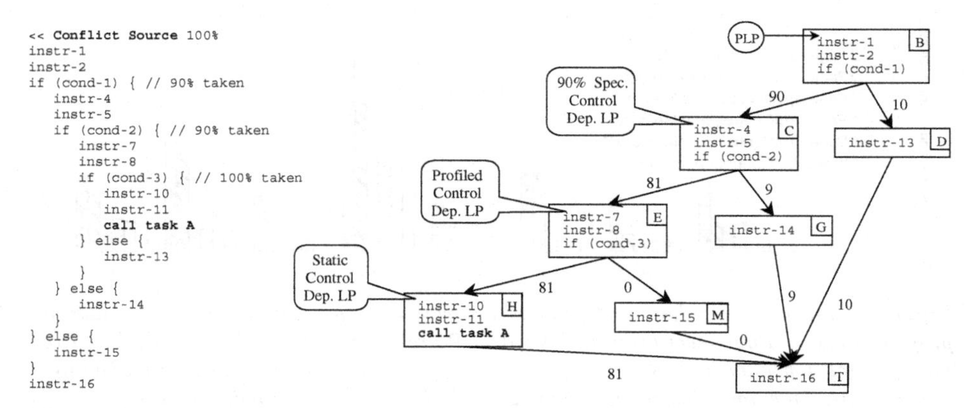

<< Conflict Source 100%
instr-1
instr-2
if (cond-1) { // 90% taken
instr-4
instr-5
if (cond-2) { // 90% taken
instr-7
instr-8
if (cond-3) { // 100% taken
instr-10
instr-11
call task A
} else {
instr-13
}
} else {
instr-14
}
} else {
instr-15
}
instr-16
PLP
instr-1 instr-2 if (cond-1) B
90% Spec. Control Dep. LP
90
10
instr-4 instr-5 if (cond-2) C
instr-13 D
Profiled Control Dep. LP
81
9
instr-7 instr-8 if (cond-3) E
instr-14 G
Static Control Dep. LP
81
0
9
10
instr-10 instr-11 call task A H
instr-15 M
0
81
instr-16 T

Fig. 5. *Control Dependence Speculation: Using the profile information that the conditions are true 90% , 90% , and 100% , respectively, profiled control dependence and speculative control dependence allow task A to be launched outside of the inner if statement. The corresponding CFG displays the launch points as placed by each type of control dependence. The edge frequencies reflect that the code was executed 100 times.*

preserve profiled control dependences, *all traversed paths* from the launch points lead to the original call site.

When task launch points preserve **speculative control dependences**, all *frequently* traversed paths from the launch points lead to the original call site. The amount of speculation is controlled by setting a minimum frequency percentage, c. For example, when c is set to 90, then at least 90% of the traversed paths from the launch points must lead to the original call site.

In Figure 5, the call statement of task A is statically control dependent on all three if-statements. The corresponding CFG in Figure 5 highlights the launch points as determined by the three control dependence options. All paths beginning with block H, all traversed paths beginning with block E, and 90% of the traversed paths beginning with block C lead to the call of task A. Therefore, the static control dependence launch point is before block H, the profiled control dependence launch point is before block E, and with c set to 90, the speculative control dependence launch point is before block C.

The price of using speculative control dependences will be the waste of resources used to speculatively initiate a launched task when the executed path does not lead to the task call site. These extra launches can be squashed at the first mis-speculated conditional.

The first three bars per benchmark in Figure 6 show the effect of control dependence speculation. The bars display static control dependence, profiled control dependence, and speculative control dependence at $c = 90$, respectively. On average, profiled control dependence exposed an additional 1.3% of dynamic instructions as memory-independent, while speculative control dependence only exposed an additional 0.6% over profiled.

The choice of using profiled or speculative control dependence will be influenced by the underlying architecture, and the degree to which speculative

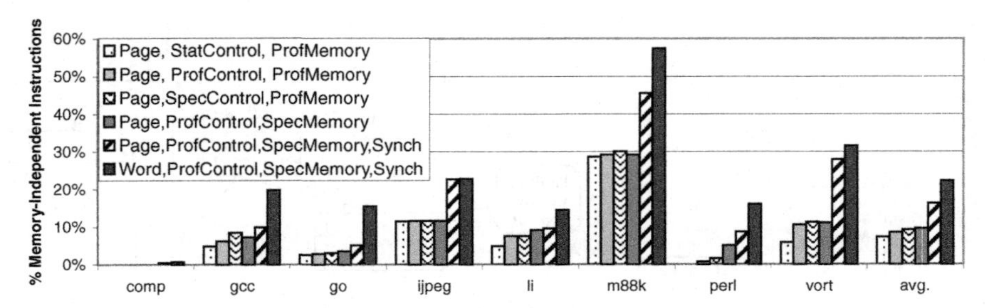

Fig. 6. *Memory-independent instructions reported as a percentage of all instructions profiled.* Page = *Page-level profiling,* Word = *Word-level profiling,* ProfControl = *Profiled control dependence,* SpecControl = *Speculative control dependence,* SpecMemory = *Speculative memory dependence,* Synch = *Early start with synchronization.*

threads compete with non-speculative for resources. Further results in this paper use profiled control dependence, due to the low gain from speculative control dependence.

Memory dependence provides another opportunity for explicit speculation. Our starting configuration determines launch points that preserve **profiled memory dependences** such that all instructions between each launch point and its original task call site are memory-independent of the task. This approach results in a conservative, but still speculative, placement of launch points.

We also consider the less conservative approach of determining task launch points by **speculative memory dependences**, which ignores profiled memory conflicts that occur infrequently. The amount of speculation is controlled by setting a minimum frequency percentage, m. For example, when m is set to 90, then at least 90% of the traversed paths from the task launch points to the original call site must be memory-independent of the task. Using speculative memory dependences is especially attractive when PLPs are far apart, and the ones nearest the task call site seldom cause a memory conflict.

We examine the effect of task launch points that preserve speculative memory dependence at $m = 90$ (the fourth bar in Figure 6). Speculative memory dependence provides small increases in parallelism. Despite the small gains, we include task launch points determined by speculative memory dependence for the remaining results.

By placing launch points (futures) at control dependences or memory dependences (PLPs), we have used the limited synchronization inherent within futures to synchronize these dependences with the beginning of the speculative task. This limits the amount of STP that we have been able to expose to the number of dynamic instructions between the launch point and the original task call site, which we call the *LP overlap*. The instructions represented by the task overlap, between the beginning of the speculative task and the earliest profiled conflict destination, are profiled memory-independent of all of the preceding code. By using explicit additional synchronization around the earliest profiled

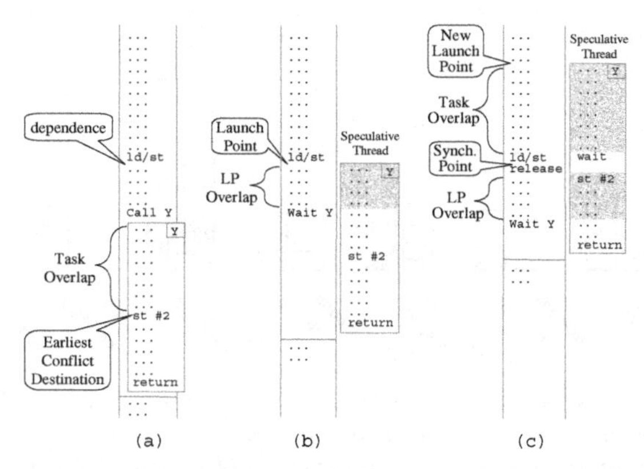

Fig. 7. *Synchronization Points: Gray areas represent memory-independent instructions. (a) a task with a large amount of task overlap on a serial thread. (b) When the dependence is used as a launch point, only LP overlap contributes to memory-independence. (c) By synchronizing the dependence with the earliest conflict destination, both LP overlap and task overlap contribute to memory-independence.*

conflict destination, early start with synchronization enables the task overlap to contribute to the number of exposed memory-independent instructions.

Early Start with Synchronization Currently, a task with a large task overlap and a small LP overlap would not be selected as memory-independent, even though a large portion of the task is memory-independent with its preceding code. By synchronizing the control or memory dependence with the earliest conflict destination, the task may be launched earlier than the dependence. Where possible, we placed the task launch point above the dependence a distance equal to the task overlap. Any control dependences between the new task launch point and the synchronization point would be handled as speculative control dependences.

Figure 7 illustrates synchronization points. When the dependence determines a launch point, in Figure 7(b), all memory-independent instructions come from the LP overlap. Figure 7(c) shows that by synchronizing the dependence with the earliest conflict destination, both the LP overlap and the task overlap contribute to the number of memory-independent instructions.

Early start shows the greatest increase in parallelism so far, exposing on the average an additional 6.6% of dynamic instructions as memory-independent (the fifth bar per benchmark of Figure 6). The big increase in parallelism came from tasks that had not previously exhibited a significant level of STP, but now are able to exceed our thresholds.

The extra parallelism exposed through early start will come at the cost of additional dynamic instructions and the cost of explicit synchronization. We did not impose any penalties to simulate those costs as they will be architecture-dependent.

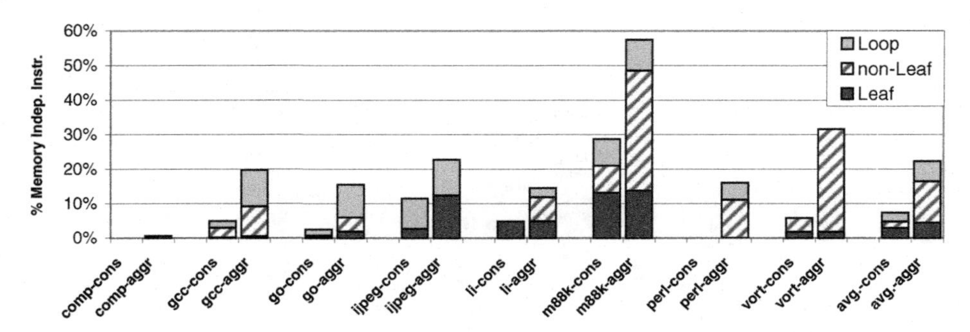

Fig. 8. *Conservative vs. Aggressive Speculation: Conservative is page-level profiling on all task types with static control dependence and profiled memory dependence. Aggressive is word-level profiling on all task types with profiled control dependence, speculative memory dependence, and early start.*

5.3 Memory Granularity

We define a memory access conflict to occur with two accesses to the same memory region. The memory granularity (the size of these regions) effects the amount of parallelism that is exposed. Reasonable granularities are full bytes, words, cache-lines, or pages. When a larger memory granularity is used, this may result in a conservative placement of launch points. The actual granularity used will depend on the granularity at which the processor can detect memory ordering violations. Managing profiled-parallel tasks whose launch points were determined with a memory granularity of a page would allow the use of existing page protection mechanisms to detect and recover from dependence violations. Thus, our starting configuration used page-level profiling. We also investigate word-level profiling.

In Figure 6, the last bar shows the results of word-level profiling on top of profiled control dependence, speculative memory dependence and early start. The average gain in memory-independence across all benchmarks was about 6% of dynamic instructions.

5.4 Experiment Summary

Figure 8 re-displays the extremes of our STP results from conservative to aggressive speculation broken down by task types. The conservative configuration includes page-level profiling on all task types with static control dependence and profiled memory dependence. The aggressive configuration comprises word-level profiling on all task types with profiled control dependence, speculative memory dependence, and early start. M88ksim showed the largest increase in the percentage of memory-independent instructions at over 28%, with vortex very close at over 25%, and the average across benchmarks at about 14%. Each of these increases in parallelism were largely seen in the non-leaf procedures. Ijpeg was the only benchmark to see a sizable increase contributed by leaf procedures. Loops accounted for increases in gcc, go, li and perl.

	Conservative			Aggressive		
	Selected Tasks	Avg Dynamic Task Length	Average Overlap	Selected Tasks	Avg Dynamic Task Length	Average Overlap
compress	0	0	0	1	282	42
gcc	22	412	93	74	363	89
go	12	546	118	49	334	76
ijpeg	5	803	300	12	996	266
li	3	50	50	9	14198	55
m88ksim	5	34	29	20	123	43
perl	0	0	0	10	515	42
vortex	14	114	45	92	215	47
average	8	245	79	33	2128	83

Table 1. *Task Statistics (Conservative vs. Aggressive)*

Table 1 displays statistics from the conservative and aggressive speculation of those tasks which exceed our thresholds. The average overlap is that part of the average task length that can be overlapped with other execution. The number of tasks selected for STP is greatly affected by aggressive speculation.

In this Section, early start with synchronization provided the highest single increase among all alternatives. Speculative systems with fast synchronization should be able to exploit STP the most effectively. Our results also indicate that a low-overhead word-level scheme to exploit STP would be profitable.

6 Implementation Issues

For our limit study of Speculative Task Parallelism, we have assumed a hypothetical speculative machine that supports speculative futures with mechanisms for resolving incorrect speculation. When implementing this machine, a number of issues need to be addressed.

Speculative Thread Management Any system that exploits STP would need to include instructions for initialization, synchronization, communication and termination of threads. As launched tasks may be speculative, any implementation would need to handle mis-speculations.

Managing speculative tasks would include detecting load/store conflicts between the preceding code and the launched task, buffering stores in the launched task, and checking for memory-independence before committing the buffered stores to memory. One conflict detection model includes tracking the load and store addresses in both the preceding code and the launched task. The amount of memory-independence accommodated by this model will be determined by the size and access of load-store address storage, and the conflict granularity.

Another conflict detection model uses a system's page-fault mechanism. When static analysis can determine the page access pattern of the preceding code, the launched task is given restricted access to those pages, while the preceding code is given access to only those pages. Any page access violation would cause the speculative task to fail.

Inter-thread Communication Any implementation that exploits STP will benefit from a system with fast communication between threads. At the minimum, inter-thread communication is needed at the end of a launched task and when the task results are used. Fast communication would be needed to enable early start with synchronization. The ability to quickly communicate a misspeculation would reduce the number of instructions that are issued but never committed. This is especially important for systems where threads compete for the same resources.

Adaptive STP We select tasks that exhibit STP based upon memory access profiling and compiler analysis. The memory access pattern from one dataset may or may not be a good predictor for another dataset. Two feedback opportunities arise that allow the execution of another data set to adapt to differences from the profiled dataset. The first would monitor the success rate of particular launch points, and suspend further launches when it fails too frequently.

The second feedback opportunity is found in continuous profiling. Rather than have a single dataset dictate the launched tasks for all subsequent runs, let datasets from all previous runs dictate the launched tasks for the current run. It is possible that the aggregate information from the preceding runs would have a better predictive relationship with future runs. Additionally, the profiled information from the current run could be used to supersede the profiled information from previous runs, with the idea that the current run may be its own best predictor. Although profiling is expensive and must be optimized, the exact cost is beyond the scope of this paper.

7 Summary

Traditional parallel compilers do not effectively parallelize irregular applications because they contain little loop-level parallelism due to ambiguous memory references. A different source of parallelism, namely Speculative Task Parallelism arises when a task (either a leaf-procedure, a non-leaf procedure or an entire loop) is control- and memory-independent of its preceding code, and thus could be executed in parallel. To exploit STP, we assume a speculative machine that supports speculative futures (a parallel programming construct that executes a task early on a different thread or processor) with mechanisms for resolving incorrect speculation when the task is not, after all, independent. This allows us to speculatively parallelize code when there is a high probability of independence, but no guarantee.

Through profiling and compiler analysis, we find memory-independent tasks that have no memory conflicts with their preceding code, and thus could be speculatively executed in parallel. We estimate the amount of STP in an irregular application by measuring the number of memory-independent instructions these tasks expose. We vary the level of control dependence and memory dependence to investigate their effect on the amount of memory-independence we found. We profile at different memory granularities and introduced synchronization to expose higher levels of memory-independence.

We find that no one task type exposes significantly more memory-independent instructions, which strongly suggests that all three task types should be profiled for STP. We also find that starting a task early with synchronization around dependences exposes the highest additional amount of memory-independent instructions, an average across the SPECint95 benchmarks of 6.6% of profiled instructions. Profiling memory conflicts at the word-level shows a similar gain in comparison to page-level profiling. Speculating beyond profiled memory and static control dependences shows the lowest gain which is modest at best. Overall, we find that 7 to 22% of instructions are within memory-independent tasks. The lower amount reflects tasks launched in parallel from the least speculative locations.

8 Acknowledgments

This work has been supported in part by DARPA grant DABT63-97-C-0028, NSF grant CCR-980869, equipment grants from Compaq Computer Corporation, and La Sierra University.

References

1. H. Akkary and M. Driscoll. A dynamic multithreading processor. In *31st International Symposium on Microarchitecture*, Dec. 1998.
2. R. Alverson, D. Callahan, D. Cummings, B. Koblenz, A. Porterfield, and B. Smith. The Tera computer system. *1990 International Conf. on Supercomputing*, June 1990.
3. M. Franklin and G. S. Sohi. ARB: A hardware mechanism for dynamic reordering of memory references. *IEEE Transactions on Computers*, May 1996.
4. D. M. Gallagher, W. Y. Chen, S. A. Mahlke, J. C. Gyllenhaal, and W. W. Hwu. Dynamic memory disambiguation using the memory conflict buffer. In *Proceedings of the 6th International Conference on Architecture Support for Programming Languages and Operating Systems*, Oct. 1994.
5. L. Gwennap. Intel discloses new IA-64 features. *Microprocessor Report*, Mar. 8 1999.
6. L. Hammond, M. Willey, and K. Olukotun. Data speculation support for a chip multiprocessor. *ACM SIGPLAN Notices*, 33(11):58–69, Nov. 1998.
7. S. Keckler, W. Dally, D. Maskit, N. Carter, A. Chang, and W. Lee. Exploiting fine-grain thread level parallelism on the MIT Multi-ALU processor. In *Proceedings of the 25th Annual International Symposium on Computer Architecture (ISCA-98)*, pages 306–317, June 1998.
8. M. S. Lam and R. P. Wilson. Limits of control flow on parallelism. In *Proceedings of the 19th Annual International Symposium on Computer Architecture (ISCA-92)*, May 1992.
9. R. H. Litton, J. A. D. McWha, M. W. Pearson, and J. G. Cleary. Block based execution and task level parallelism. In *Australasian Computer Architecture Conference*, Feb. 1998.
10. P. Marcuello and A. Gonzalez. Clustered speculative multithreaded processors. In *Proceedings of the [ACM] International Conference on Supercomputing*, June 1999.

11. P. Marcuello and A. Gonzalez. Exploiting speculative thread-level parallelism on a SMT processor. In *Proceedings of the International Conference on High Performance Computing and Networking*, April 1999.
12. A. Moshovos, S. E. Breach, T. N. Vijaykumar, and G. S. Sohi. Dynamic speculation and synchronization of data dependences. In *Proceedings of the 24th Annual International Symposium on Computer Architecture (ISCA-97)*, June 1997.
13. S. S. Muchnick. *Advanced Compiler Design and Implementation.* Morgan Kaufmann Publ., San Francisco, 1997.
14. J. T. Oplinger, D. L. Heine, and M. S. Lam. In search of speculative thread-level parallelism. In *Proceedings of the 1999 International Conference on Parallel Architectures and Compilation Techniques (PACT99)*, October 1999.
15. J. T. Oplinger, D. L. Heine, S. Liao, B. A. Nayfeh, M. S. Lam, and K. Olukotun. Software and hardware for exploiting speculative parallelism with a multiprocessor. Technical Report CSL-TR-97-715, Stanford University, Computer Systems Laboratory, 1997.
16. G. S. Sohi, S. E. Breach, and T. N. Vijaykumar. Multiscalar processors. In *Proceedings of the 22nd Annual International Symposium on Computer Architecture (ISCA-95)*, June 1995.
17. A. Srivastava and A. Eustace. ATOM: A system for building customized program analysis tools. Research Report 94.2, COMPAQ Western Research Laboratory, 1994.
18. J. G. Steffan and T. C. Mowry. The potential of using thread-level data speculation to facilitate automatic parallelization. In *Proceedings of the 4th International Symposium on High-Performance Computer Architecture*, Feb. 1998.
19. D. M. Tullsen, S. J. Eggers, and H. M. Levy. Simultaneous multithreading: Maximizing on-chip parallelism. In *Proceedings of the 22nd Annual International Symposium on Computer Architecture (ISCA-95)*, June 1995.
20. T. N. Vijaykumar and G. S. Sohi. Task selection for a multiscalar processor. In *31st International Symposium on Microarchitecture*, Dec. 1998.
21. S. Wallace, B. Calder, and D. Tullsen. Threaded multiple path execution. In *Proceedings of the 25th Annual International Symposium on Computer Architecture (ISCA-98)*, June 1998.

The Case for Speculative Multithreading on SMT Processors

Haitham Akkary, and Sébastien Hily

Intel Microprocessor Research Labs
5350 NE Elam Young Parkway
Hillsboro, OR 97124, USA
{haitham.akkary, sebastien.hily}@intel.com

Abstract. Simultaneous multithreading (SMT) processors achieve high performance by executing independent instructions from different programs simultaneously [1]. However, the SMT model doesn't help single-thread applications and performs at its full potential only when executing multithreaded applications or multiple programs. Moreover, to minimize the total execution time of one selected high priority thread, that thread has to run alone. Recently, several speculative multithreading architectures have been proposed that exploit far away instruction level parallelism in single-thread applications. In particular, the dynamic multithreading or DMT model [2] uses hardware mechanisms to fork speculative threads at procedure and loop boundaries along the execution path of a single program, and executes these threads on a multithreaded processor. In this paper, we explore the performance scope of an SMT architecture in which spare thread contexts are used to support the DMT execution of procedure and loop threads. We show two significant advantages of this approach: (1) it increases processor utilization and total execution throughput when few programs are running, and (2) it eliminates or reduces the performance degradation of one selected high priority program when running simultaneously with other programs, without reducing total SMT throughput.

1 Introduction

In a very near future, tens of millions of transistors will be available to build single-chip microprocessors. The single-thread superscalar model used in most current commercial processors may not be able to take full advantage of this opportunity. For example a straight approach consisting of increasing the size of already existing mechanisms such as memory hierarchy or branch prediction tables will not bring spectacular performance gains. Also, inherently sequential tasks such as instruction fetch and register renaming, and the high frequency of branches in many typical programs will make a direct increase of pipeline width more difficult.

Simultaneous multithreading is a promising technique to obtain more performance from a single-chip processor [1]. SMT integrates several hardware contexts to allow the concurrent execution of multiple programs. The threads share most of the processor resources and in each cycle, instructions from multiple threads can be

M. Valero et al. (Eds.): ISHPC 2000, LNCS 1940, pp. 59-72, 2000.

simultaneously issued to the functional units. The availability of independent instructions from different threads enables an SMT processor to address several bottlenecks of conventional superscalar processors: low ILP of non-scientific applications, cache miss latency, branch mispredictions, and waste of issue slots due to instruction misalignment and taken branches. On mixed applications from the Spec92 benchmarks suite, with 8 threads, a 2.5 throughput gain over a complexity-equivalent superscalar processor could be expected [3].

SMT, however, relies heavily on the availability of multiple threads in each cycle and does not improve performance when only a single-thread application is present for execution. This limitation could be overcome with parallelizing compilers that automatically create multiple threads from the same program. Unfortunately, many non-numeric programs have proven to be difficult to parallelize due to their complex control flow and memory access patterns. Recently, there have been several promising proposals for speculative multithreaded architectures [4,5,6,7,8,9,10,11,12,13]. Some of these architectures [4,5,6,7,10,11] combine thread-level data speculation hardware with special compilers to extract parallelism from general-purpose programs. Others [2,8,9,13] are completely dynamic and target existing single-thread binaries that run on current superscalar processors. These dynamic multithreading techniques may be especially interesting if the basic architecture provides the flexibility of mixing speculative and simultaneous multiple threads, providing performance gain for single-thread as well as multithreaded applications.

In particular, the Dynamic Multithreading (DMT) model [2] improves single program execution with out-of-order instruction fetch and issue, and speculative execution far ahead in a program. As on an SMT microprocessor, several hardware contexts support the simultaneous execution of several threads through a shared superscalar pipeline. Unlike traditional SMT, the threads come from a single program and are created dynamically by hardware. On a DMT processor, the program is dynamically forked at the end of a loop or at a procedure call, and a speculative thread is initiated to follow the fall-through path. By this means, [2] shows that the performance of Spec95 integer applications could be improved by a factor of up to 35%, on average.

In this paper, we explore the performance impact of adding speculative multithreading to an SMT model. We evaluate whether there is any headroom left in the SMT architecture, with its heavy resource sharing, to support the significant amount of speculative execution that can take place on a speculative multithreaded architecture [14]. We show that the benefit of speculative multithreading exceeds the disadvantage arising from misprediction and execution resource sharing, resulting in net increase in total execution throughput of the SMT processor. We also show that speculative multithreading speeds up the execution of one high priority program running with one other program simultaneously, over the non-speculative SMT processor running the same program alone. Our results are presented for a variety of instruction fetch configurations and highlight some of the tradeoffs that would face the designer of such architecture. In our study, we have used the DMT architecture, which we believe represents an extreme case in its speculation penalty, since it relies completely on hardware and requires no compiler support for thread selection and scheduling.

2 Architecture Overview

Our baseline model of processor is derived from the DMT architecture presented in [2] and illustrated in Figure 1. It supports the execution of up to 8 threads. In our case, a thread can be a conventional SMT program or can be created dynamically by hardware. Each thread has its own program counter, *trace buffer* in which speculative instructions and their results are stored, load and store queues, and return address stack. The threads share the memory hierarchy, the physical register file, the functional units, and the branch prediction tables. In Figure 1, the dark shaded boxes correspond to duplicated hardware. Depending on the configuration, the hardware corresponding to the light shaded boxes can be either duplicated or shared.

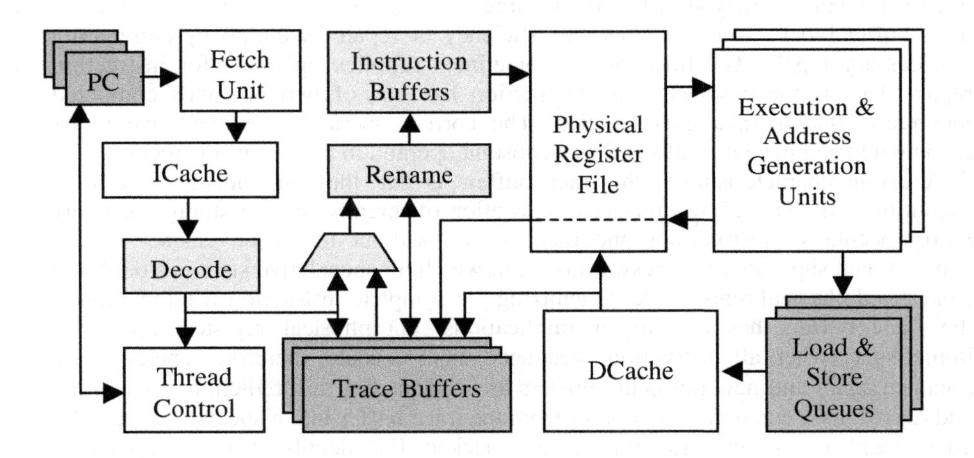

Fig. 1. Processor block diagram

In each cycle, active threads are competing for fetch and issue resources. A priority mechanism is implemented using a round-robin policy. Fetched instructions are decoded, renamed, and put into an instruction buffer where they wait for their operands and for the availability of functional units. Decoded instructions are also written into the trace buffer attached to their thread. When instructions execute, results are written back into the physical register file, as well as into the trace buffer.

A trace buffer acts as a large speculative instruction window and its primary goal is to improve the execution throughput by allowing far ahead speculation. New threads are created at loop *backward branches* and procedure calls. They are executed speculatively starting at the instructions following the procedures or the loops. In other words, two different threads pursue both the branch or call path and the fall-through path simultaneously. Threads are not necessarily created at every opportunity. A thread prediction table of 2-bit saturating counters is used to predict which opportunities are taken.

Threads may depend on register or memory values produced by other threads from the same program. Waiting for other threads to compute all the input data would seriously hamper the *run-ahead* ability of the architecture. To solve this problem, a simple data speculation mechanism is implemented: a new thread executes using the register context from the spawning thread as input. Although some of these register values may be modified later, such as *procedure return value* register, this mechanism successfully predicts most of the register input values due to the scope of the variables in the caller [2]. Loads to memory are speculatively issued as if there were no conflicting stores to the same addresses from prior threads. Memory disambiguation circuitry in the load queues, and register value matching circuitry in the trace buffer, activated before a thread is committed, are used to identify data mispredictions. A recovery sequence is initiated when a misprediction is detected.

During recovery, instructions affected by data mispredictions are sequentially fetched from the trace buffer (instead of the instruction cache), renamed, and issued again for *repair execution.* Instructions that are not affected by data mispredictions do not need repair and they are filtered out using a register scoreboard that resides in the trace buffer block. This increases the efficiency of repair execution by eliminating unnecessary repair. The trace buffer sometimes supplies operands for instruction repair. This is the case when an instruction has one of two operands computed incorrectly due to data misprediction. The correct operand does not have to be recomputed again, so it is attached as a constant operand to the repair instruction.

A useful characteristic of the trace buffers is that they sit outside the critical execution path. This allows far-ahead execution of threads and the storage of many more speculative instructions and their results without the inconvenience of the conventional superscalar processor model, in which all speculative state is stored in a centralized physical register file. Organizing the complete instruction window around the trace buffers has two major implications: (1) physical registers are freed immediately when all instructions that need them execute, therefore, operands are renamed again and new physical registers are allocated at misprediction repair time, and (2) results are committed in order from the trace buffer into a special register file, after thread inputs from prior threads are checked. Full details of the trace buffers, data speculation, and repair algorithm are described in [2].

The pipeline used in the baseline model is a generic 7-stage out-of-order pipeline that includes fetch, decode, rename, issue, register file read, execute, and retire stages.

3 Simulation Methodology

In our simulation model, the instructions can be issued to 10 functional units: 4 ALUs, 1 multiply/divide unit, 4 floating point add units and 1 floating point multiply/divide unit. Two of the four ALUs perform Load/Store address computation. All the functional units, except the divide unit, are pipelined. The execution unit latencies are listed in Table 1. All threads share 128 physical registers for renaming purposes. The trace buffers have 512 entries per thread.

A description of the configuration of the memory hierarchy chosen for our experiments is also given in Table 1. All caches are fully pipelined. The data cache and the L2 cache are lock-up free. The load latency to the first level data cache, not including the address computation, is 2 cycles. The 2 cycles account for address translation and data cache access, which happen in parallel with the address match and possibly store forwarding from the store queues. Memory latency is 30 cycles.

For branch prediction, we have used a 512-entry 4-way associative BTB and a 4K-entry PHT. The PHT is indexed using a 12-bit XOR of a subset of the instruction address and the content of a 12-bit branch history register. Each thread has a private branch history register and 32-entry return address stack.

Table 1. Instruction latencies and configuration of the cache hierarchy

Inst. type	Latency
Integer ALU	1
Load/Store	2
Multiply	3
Divide	20
FP-ALU	2
FP-multiply	4
FP-divide	12

Cache	Size	Assoc.	Block size	Latency
Inst.	32K	2-way	64 bytes	1 cycle
Data	32K	2-way	64 bytes	2 cycles
L2	1M	4-way	64 bytes	5 cycles

The performance simulator is derived from the SimpleScalar tools set [15] and supports a modified version of the MIPS-IV ISA. To run our experiments, we used 6 of the Spec95 benchmarks: four integer applications (gcc, go, vortex, and perl) and two floating-points applications (applu and fpppp). These applications were compiled using the SimpleScalar version of gcc with level-3 optimization. We have selected gcc, go, and perl because their execution throughput is low and their parallelization is difficult. Vortex and fpppp are two of the more memory-intensive applications in the spec95 benchmarks.

Our choice of the Spec95 benchmarks suite to evaluate a multithreaded architecture may seem arbitrary, since it is not clear that a practical multithreaded processor would be running combinations of workloads with similar characteristics to Spec95 [16]. Unfortunately, there is no standard set of benchmarks established for SMT based architectures. Our results should be viewed as a demonstration of the potential of the evaluated architecture model when running concurrently multiple programs, some of which may be non-numeric and difficult to parallelize using a compiler. Since the SMT model emphasizes instruction throughput, we have run all possible 1, 2, 3 and 4 combinations of the selected benchmarks, and averaged the overall throughput. Each simulation was stopped after 50 million simulation cycles. This gave us approximate run lengths between 50 million and 200 million instructions per simulation, depending on the number of benchmarks in each combination.

4 Simulation Results

In this section, we investigate the performance of 3 different architectural models. To execute several threads simultaneously, the front-end of the processor has to provide a lot of bandwidth. Current microprocessors typically allow up to 4 instructions to be fetched and issued into the scheduling window per cycle. To offer a significantly better bandwidth on a multithreaded processor, the most forward way is to increase the number of blocks that can be issued simultaneously. Using this approach, the instruction cache would be multi-ported, and the decode and rename units would be duplicated. For our baseline model, we have chosen a configuration that doubles the fetch and issue bandwidth of typical current superscalar processors for a small increase in complexity. The base configuration has an instruction cache with 2 fetch ports, and can rename and issue two blocks of 4 instructions from two different threads every cycle. As long as there is more than one active thread, the base machine can provide the peak issue bandwidth of an 8-wide processor, but with fetch and rename complexity per thread of a 4-wide processor. The performance of the base model is presented in section 4.1. We also present in section 4.2 results for two more aggressive architectures featuring one and two 8-wide fetch ports respectively. These two architectures may be more complex to build than the baseline machine because of the high frequency of branches in general purpose programs, the high number of read ports into the rename table, and the intra-block dependency check and bypass logic at the rename table outputs [17].

4.1 Baseline

Figure 2 shows the performance of two multithreaded architectures that use the 2x4 fetch/issue configuration of the base machine. The SMT key refers to a conventional simultaneous multithreading architecture with 8 thread contexts. The DSMT key refers to an SMT architecture capable of creating hardware threads, à la DMT, also with 8 thread contexts. DSMT performs consistently better than SMT, with an average increase in throughput of 5%, 8%, 13%, and 28% for 4, 3, 2 and 1 programs respectively. These results highlight the performance potential of mixing speculative multithreading with the simultaneous execution of independent software threads.

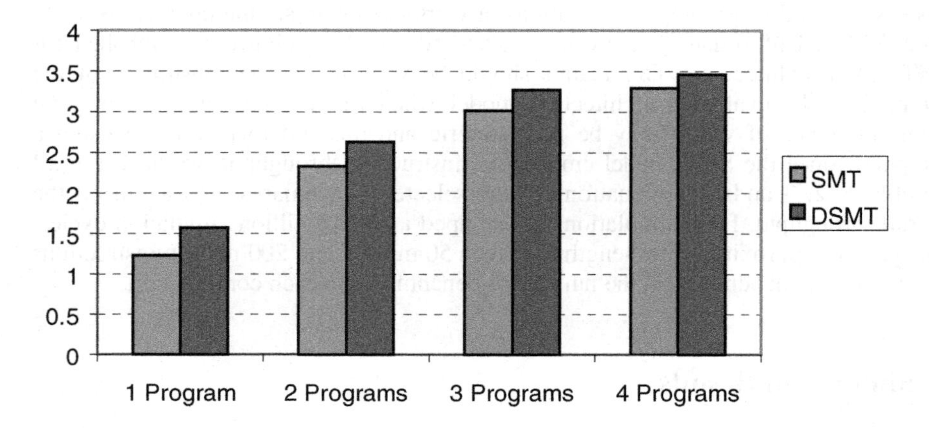

Fig. 2. Average IPC for the SMT and DSMT architectures, with two 4-wide fetch ports

It should be noted that when running one program, the SMT architecture behaves as a simple 4-wide superscalar processor, while the DSMT architecture becomes the DMT model presented in [2]. There is a possibility for an improvement on the SMT model by implementing a mechanism to predict two addresses per cycle, in order to use both fetch ports in the same cycle when there is only one active thread. One such mechanism has been proposed in [18]. The DSMT model would also benefit from that mechanism, but to a lesser degree, since the frequency of having only one thread in active fetch is limited, as shown in table 3. It should also be noted here that the results of the comparison between DSMT and SMT might vary if a different thread arbitration policy such as *Icount* [3] is used.

Table 2 gives the average number of active threads on the DSMT model (both speculative and non-speculative) when 1, 2, 3 or 4 programs are running. When only one program is present, the DSMT architecture average thread utilization is 3.6, significantly less than the machine total of 8 threads. With 4 programs, 85% of the thread contexts are utilized. Notice that the increase in throughput over SMT with 3 programs running is significant, even though there are 3 speculative hardware threads competing for resources with the 3 programs, on average. This indicates that the benefits of the speculative hardware threads out-weigh the misprediction penalties and the associated loss of fetch and execution bandwidth. Even with 4 programs, there is still a 5% gain in the average IPC. Table 3 gives the time distribution of active threads for 1, 2, 3 and 4 programs running in DSMT mode.

Table 2. Average number of DSMT active threads for various numbers of programs

1 Program	2 Programs	3 Programs	4 Programs
3.6	4.8	6	6.8

Table 3. Time distribution of the number of active threads for various numbers of simultaneous programs running in DSMT mode

Threads	*1*	*2*	*3*	*4*	*5*	*6*	*7*	*8*
1 Program	26%	15%	16%	12%	8%	7%	8%	8%
2 Programs		15%	18%	17%	14%	11%	9%	16%
3 Programs			10%	14%	16%	16%	15%	29%
4 Programs				8%	13%	16%	18%	45%

An interesting note on the DSMT model is related to the cache hierarchy design. A study presented in [19] showed that going over 4 programs on an SMT architecture with a conventional memory hierarchy would not be effective since the L2-cache bandwidth becomes a severe bottleneck. The fact that the speculative threads are very likely to belong to the execution path of the running programs is a valuable characteristic of the DSMT model. This reduces the pressure on the cache hierarchy and allows the DSMT model to utilize effectively the 8 available thread contexts.

The Asymmetric model. One of the major concerns regarding SMT architectures is that the high execution throughput is reached without direct control of the threads. This can result in an imbalance of the execution throughput of the individual threads. This phenomenon is magnified by priority mechanisms, such as *Icount* or *Brcount*, introduced in [3] to enhance the overall execution rate by giving more resources to the best performing thread. However, many users may look for low latency on one program, even when several other programs are executing in parallel. It is unrealistic to expect both the highest total throughput and the lowest latency of one high-priority program at the same time, but it may be useful to find ways to reduce the performance loss of a single high-priority program, when other lower priority threads are running simultaneously.

To address this problem, [20] presents a scheme based on a simple fetch-stage prioritization. Low-priority threads can only share the fetch bandwidth that is not used by the high-priority thread. This scheme limits substantially the latency degradation of the priority thread due to SMT execution, but at the cost of a global loss of throughput.

We have evaluated an asymmetric architecture (ASMT) in which one high priority program is favored in comparison to other *background programs*. In our simple model, the execution of one selected *foreground program* is supported by the dynamic multithreading technique, while the other lower-priority programs are executed in non-speculative mode using only the SMT technique. Figure 3 shows that the ASMT processor model exhibits only a slightly lower average performance than the DSMT model (except for one program, where the two models are equivalent) and that the average performance remains higher than the SMT model.

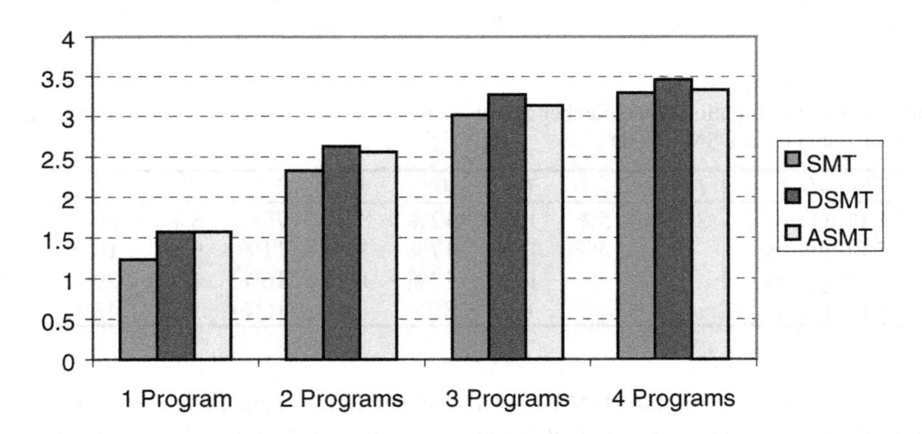

Fig. 3. Average IPC for the ASMT architecture with two 4-wide fetch ports

Even though ASMT outperforms SMT in total throughput, the high priority program performs significantly better in ASMT than in SMT or DSMT mode. Figure 4 illustrates more precisely this advantage of asymmetric multithreading execution for two different applications. Among the benchmarks we have simulated, vortex has the highest execution throughput in single thread execution mode and gcc has one of the lowest average IPC numbers. The metric displayed in Figure 4 is the average IPC exhibited by each of gcc and vortex running as high-priority program, while various combinations of the other applications in our suite are running as background threads. The performance gain brought by the ASMT processor is significant and increases as the number of simultaneous threads increases. When 4 programs are executing, the ASMT model allows gcc to exhibit an average of 18% and 21% higher throughput than the DSMT and SMT models respectively. In the case of vortex, the performance advantage given by ASMT execution reaches 23% and 31% respectively. It should be noted that with two programs running, the ASMT foreground program still outperforms the SMT architecture running the same program alone. With 3 programs running, the degradation of the foreground program is still very small compared to the single thread performance on the SMT architecture.

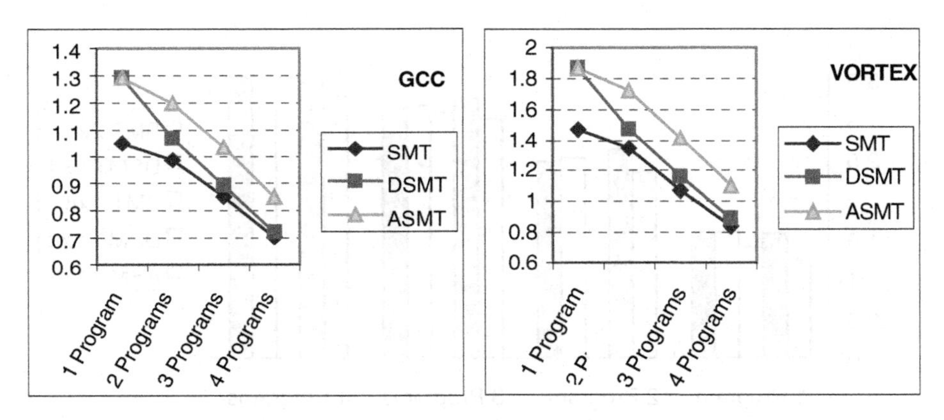

Fig. 4. Average IPC of a foreground application when running alone, or with 1, 2 or 3 other programs, for 3 different execution models with two 4-wide fetch ports.

The ASMT processor model provides an option for having high throughput without sacrificing the execution latency of one particular program. A software controllable mode could be implemented to allow switching between DSMT and ASMT execution. It is important to note that the ASMT mechanism still results in a significant loss in the foreground program performance when several other programs are active. The loss is 34% for gcc and 41% for vortex when running with 3 other programs. We are currently investigating more aggressive priority schemes and instruction schedulers to reduce this loss.

4.2 Using 8-wide issue model

In this section, we explore alternative architectures featuring different fetch and issue mechanisms. The first one is an 8-wide issue machine with a single 8-wide fetch port. It offers the same peak fetch and issue bandwidth as the 2x4 base machine studied in section 4.1. The second architecture feeds instructions into an 8-wide rename unit using two 8-wide fetch ports. It can simultaneously rename instructions from multiple threads, if available, to fill up the rename block.

Sharing a single 8-wide fetch port. In this model, the active threads share a single fetch port, on a round-robin basis. The thread accessing the instruction cache can fetch up to 8 instructions in a single cycle. This model offers the same peak fetch and issue bandwidth as the base machine, and could be equally or even more complex to implement. First, it shares all the pipeline stages. This means the rename unit must handle renaming for 8 instructions of the same thread simultaneously. Second, 8-wide fetch could be quite complex, especially if multiple branch prediction and a trace cache [21,22] were used to maximize fetch bandwidth.

Figure 5 shows the average throughput for the different multithreaded execution models that were defined in section 4.1 (SMT, DSMT, ASMT). The performance of the SMT and DSMT models on the 2x4 base machine is also shown for comparison.

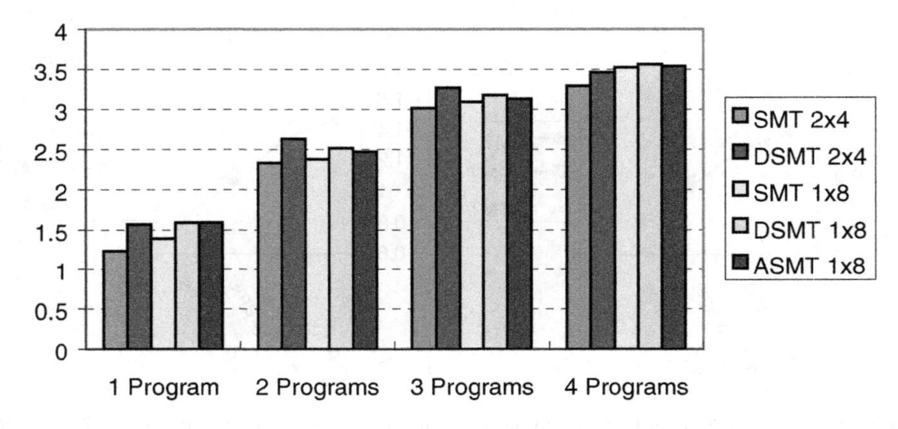

Fig. 5. Average IPC for various multithreaded architectures with one 8-wide fetch port. The performance of SMT and DSMT for two 4-wide fetch ports is given as a reference.

First, we compare the performance of the 1x8 and 2x4 SMT machines. As expected, a boost in performance can be noticed for the SMT architecture with 1 program, in comparison to the base machine. The fetch and issue bandwidth available when only one thread is active is double the bandwidth offered by the 2x4 base machine. The gain is lower when running SMT with 2, 3 and 4 programs, since the 2x4 machine has a good chance of using both fetch ports. In this case, it is only when some threads stall due to instruction cache misses that it may not be possible to utilize both fetch ports.

Second, looking closely at the SMT and DSMT performance on the 1x8 and 2x4 machines shows a very complex behavior. There are many interesting interactions occurring. On one hand, having many threads active and fetching two 4-instruction blocks every cycle helps performance by limiting instruction cache fragmentation effects (misalignment and taken branches). On the other hand, when there is only one thread fetching instructions (other threads stalling due to instruction cache misses) or when a thread resumes execution along the correct path after a branch misprediction, 8-wide fetch and issue helps feed the pipeline quickly. Speculative threads complicate things even further. On one hand, speculative threads are subject to control flow as well as data mispredictions and incorrectly predicted threads waste fetch resources. On the other hand, speculative threads in the DMT model increase fetch efficiency by allowing instructions to be fetched and executed out-of-order relative to mispredicted branches. There are probably other subtle interactions that we have not yet uncovered.

Characterizing all these interactions completely will require a lot of work, possibly on a larger set of benchmarks. From the measurements we have done, a trend seems to come out. The 1x8 wide fetch configuration seems to favor simulations with more programs (non-speculative threads) than speculative threads. In general, the 1x8 SMT performs better than 2x4 SMT, but the 2x4 DSMT performs better than 1x8 DSMT with the exception of 4-programs simulations, which have more non-speculative than speculative threads active (Table 2).

Again, the ASMT model gives better throughput than SMT, and ASMT performance is very close to DSMT performance. However, the performance of the foreground program is hardly enhanced compared to the ASMT or DSMT model. In Figure 6, one can see that the overall IPC of gcc or vortex is increased, but the difference between the asymmetric execution and the regular SMT execution is very small: in the best case 7% for 2 programs, for both gcc and vortex.

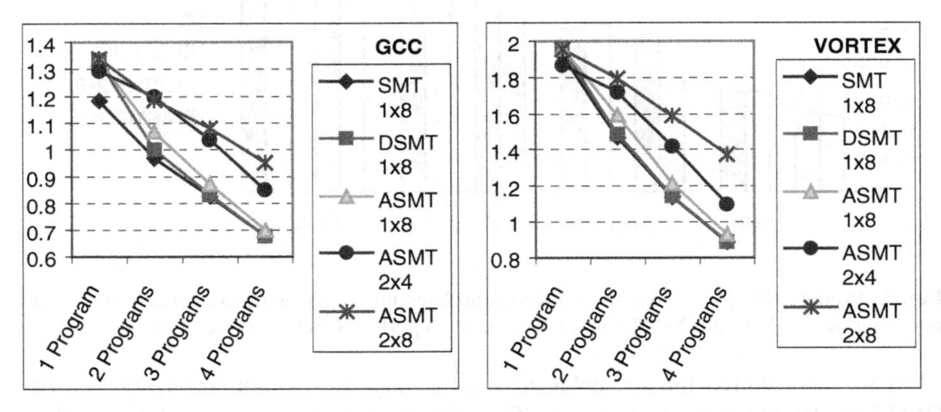

Fig. 6. Average IPC of a foreground application when running alone, or with 1,2 or 3 other programs, for 3 different multithreaded architectures.

In contrast to the 1x8 ASMT configuration, the 2x4 ASMT allows a significantly better IPC for the foreground program. The average gain in IPC brought by the two fetch ports over the single fetch port is 12%, 19% and 21% for gcc with 2, 3 and 4 programs respectively, and 8%, 17% and 21% for vortex. Therefore, in order to build a speculative SMT microprocessor in which a good balance is maintained between the performance of a selected program and the overall throughput, two narrow fetch/issue ports should be favored over one wide port.

Using two 8-wide fetch ports. Fetch bandwidth is a major bottleneck in multithreaded architectures. In this section, we look at a higher fetch bandwidth processor model in which the active threads share two 8-wide fetch ports. The processor is 8-wide issue, and in each cycle, up to 8 instructions from up to 2 threads can be fetched. As many instructions as possible are taken from the first thread and the second thread fills the remaining available issue slots. The main benefit of this new fetch mechanism is to limit block fragmentation while allowing high issue bandwidth when there is only one thread active. This front-end configuration remains reasonably balanced with the execution resources in the base model, whereas having a wider than 8-issue pipeline would require a bigger register file and more functional units.

The performance exhibited by various multithreaded architectures featuring two 8-wide fetch ports is shown in Figure 7. As expected, SMT gets a major benefit from the new mechanism when the number of programs increases. The gain in IPC averages 5%, 13% and 25% for 2, 3 and 4 programs respectively compared to the two 4-wide fetch ports model (3%, 10% and 16% when compared to the one 8-wide fetch port model).

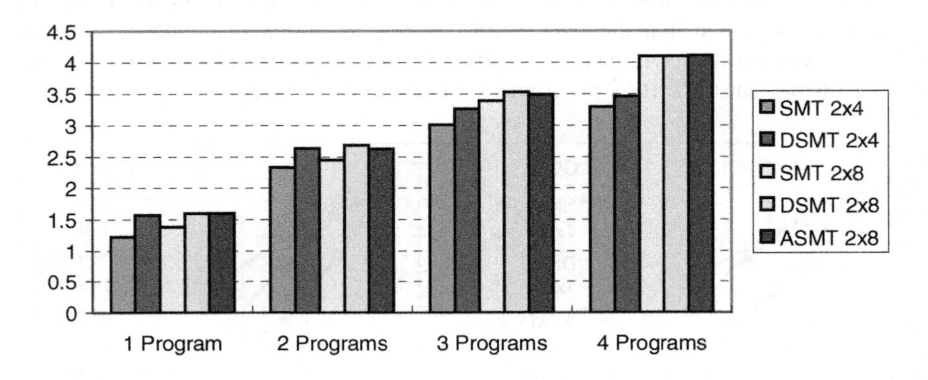

Fig. 7. Average IPC for various multithreaded architectures with two 8-wide fetch ports. The performance of SMT and DSMT for two 4-wide fetch ports is given as a reference.

Again, speculative threads enhance significantly the overall throughput of the DSMT model and up to 3 programs. The gain averages 16%, 10% and 4% for 1, 2 and 3 programs respectively. With 4 programs, the 4 spare contexts are not sufficient for efficient speculative thread execution. In general, several free contexts are needed to achieve significant performance gain with DMT execution [2]. This appears to be the case here as well for the DSMT and ASMT configurations, when multiple programs are running.

The comparison of the relative performance of DSMT with 4-wide or 8-wide fetch ports is another result that reflects the trend pointed out earlier. The gain in going wider is only 2% on average for 1 or 2 programs, but 8% and 19% for 3 and 4 programs respectively. This shows again that if we have enough resources to create more speculative than non-speculative threads, the contribution of the speculative threads to performance outweighs the benefit of having wider fetch blocks. When there is less opportunity for speculative thread execution (e.g. 3 and 4-program simulations), the benefit of having wider fetch blocks becomes significant.

The ASMT model exhibits very good overall performance, and with 4 programs, it shows the highest IPC. In fact, the ASMT architecture has 4 spare contexts, when 4 programs are running, to create speculative threads for the priority program. This is a sufficient number of thread contexts to significantly boost the performance of the high-priority foreground program. On the other hand, the high fetch bandwidth lowers the negative impact of the speculative threads on the other concurrent programs. Both observations together help explain the unexpected result that with 4 programs, the 2x8 ASMT model provides the best total throughput of any other configuration we have simulated, as well as the best high-priority program performance of all 4-program simulations, as shown in Figure 6.

5 Summary

We have evaluated an architecture that combines two approaches to multithreading: simultaneous and speculative dynamic multithreading. Simultaneous multithreading relies on sharing processor resources among software-generated threads or multiple programs to increase performance. Dynamic multithreading uses hardware mechanisms and data speculation to automatically create multiple threads from the same program. Dynamic multithreading provides the means to increase throughput on programs that have proven difficult to parallelize automatically with a compiler.

The two techniques complement each other very well. When there is one or few programs to take full advantage of the SMT processor, speculative DMT threads increase the processor utilization and throughout via far ahead execution. On the other hand, the speculative DMT model still falls short of utilizing the full capability of an SMT processor. Running one or two other programs simultaneously significantly increases the processor utilization.

The most striking example of the potential of the combined technologies is the asymmetric model with 2 programs. The foreground thread running in DMT mode can provide equal to or better performance than a conventional SMT processor running the program alone, while maintaining a significantly higher total processor throughput due to the background thread running simultaneously.

We have also presented a performance comparison of several fetch and issue organizations. We have found that, in general, for the same total fetch and issue width, multiple fetch/issue ports organization favors speculative multithreading, while a wider fetch and issue pipeline is better for a pure SMT model. Increasing fetch supply using multiple narrow fetch/issue ports on DMT processors is not only attractive for its simplicity, but it also provides superior performance as well.

References

1. D. M. Tullsen, S. J. Eggers and H. M. Levy. Simultaneous Multithreading: Maximizing on-Chip Parallelism. *The 22nd International Symposium on Computer Architecture*, June 1995.
2. H. Akkary and M.A. Driscoll. A Dynamic Multithreading Processor. *The 31st International Symposium on Microarchitecture*, November 1998.
3. D. M. Tullsen, S. J. Eggers, J. S. Emer, H. M. Levy, J. L. Lo and R. L. Stamm. Exploiting Choice: Instruction Fetch and Issue on an Implementable Simultaneous Multithreading Processor. *The 23rd International Symposium on Computer Architecture*, May 1996.
4. M. Franklin. *The Multiscalar Architecture*. Ph.D. Thesis, University of Wisconsin-Madison, November 93.

5. G. S. Sohi, S. E. Breach, and T.N. Vijaykumar. Multiscalar Processors. *The 22nd International Symposium on Computer Architecture*, June 1995.
6. J. Y. Tsai, Z. Jiang, E. Ness, and P.-C. Yew. Performance Study of a Concurrent Multithreaded Processor. *The 4th International Symposium on High-Performance Computer Architecture*, January 1998.
7. J. G. Steffan and T. C. Mowry. The Potential for Using Thread-Level Data Speculation to Facilitate Automatic Parallelization. *The 4th International Symposium on High-Performance Computer Architecture,* January 1998.
8. S. Wallace, B. Calder, and D. M. Tullsen. Threaded Multiple Path Execution. *The 25th International Symposium on Computer Architecture*, June 1998.
9. P. Marcuello, A. González, and J. Tubella. Speculative Multithreaded Processors. *International Conference on Supercomputing'98*, July 1998.
10. V. Krishnan and J. Torrellas. Hardware and Software Support for Speculative Execution of Sequential Binaries on a Chip-Multiprocessor. *International Conference on Snupercomputing'98*, July 1998.
11. L. Hammond, M. Willey, and K. Olukotun. Data Speculation support for a chip multiprocessor. *International Conference on Architectural Support for Programming Languages and Operating Systems*, October 1998.
12. M. Tremblay. Magic: Microprocessor Architecture for Java Computing. Presentation at *HotChips'99,* August 1999.
13. L. Codrescu and D. S. Wills. Architecture of the Atlas Chip-Multiprocessor: Dynamically Parallelizing Irregular Applications. *The 1999 International Conference on Computer Design, VLSI in Computers & Processors (ICCD'99),* October 1999.
14. H. Akkary. A Dynamic Multithreading Processor. Ph.D. Thesis, Portland State University, June 1998.
15. D. Burger and T. M. Austin. The SimpleScalar Tool Set, Version 2.0. *Computer Architecture News*, Vol. 25, No. 3, pp. 13-25, June 1997.
16. D. Lee, P. Crowley, J.-L. Baer, T. Anderson and B. Bershad. Execution Characteristics of Desktop Applications on Windows NT. *The 25th International Symposium on Computer Architecture*, June 1998.
17. A.S. Palacharla, N.P. Jouppi, and J.E. Smith. Complexity-Effective Superscalar Processors. *The 24th International Symposium on Computer Architecture*, June 1997.
18. A. Seznec, S. Jourdan, P. Sainrat, and P. Michaud. Multiple-Block Ahead Branch Predictors. *International Conference on Architectural Support for Programming Languages and Operating Systems*, October 1996.
19. S. Hily and A. Seznec. Standard Memory Hierarchy Does Not Fit Simultaneous Multithreading. In Proceedings of *Workshop on MultiThreaded Execution, Architecture and Compilation*, held in conjunction with *HPCA-4*, Colorado State Univ. Technical Report CS-98-102, January 1998.
20. S.E. Raasch and S.K. Reinhardt. Applications of Thread Prioritization in SMT Processors. In Proceedings of *Workshop on MultiThreaded Execution, Architecture and Compilation*, held in conjunction with *HPCA-5*, January 1999.
21. S. Patel, D. Friendly, and Y. Patt. Critical Issues Regarding the Trace Cache Fetch Mechanism. University of Michigan Technical Report CSE-TR-335-97, 1997.
22. E. Rotenberg, S. Bennett, and J. E. Smith. Trace Cache: a Low Latency Approach to High Bandwidth Instruction Fetching. *The 29th International Symposium on Microarchitecture*, December 1997.

Loop Termination Prediction

Timothy Sherwood and Brad Calder

Department of Computer Science and Engineering
University of California, San Diego
{sherwood,calder}@cs.ucsd.edu

Abstract. Deeply pipelined high performance processors require highly accurate branch prediction to drive their instruction fetch. However there remains a class of events which are not easily predictable by standard two level predictors. One such event is loop termination. In deeply nested loops, loop terminations can account for a significant amount of the mispredictions. We propose two techniques for dealing with loop terminations. A simple hardware extension to existing prediction architectures called *Loop Termination Prediction* is presented, which captures the long regular repeating patterns of loops. In addition, a software technique called *Branch Splitting* is examined, which breaks loops with iteration counts above the detection of current predictors into smaller loops that may be effectively captured. Our results show that for many programs adding a small loop termination buffer can reduce the missprediction rate by up to a difference of 2%.

1 Introduction

Branch prediction is the architectural feature which allows the front-end of the processor to continue fetching instructions in the presence of control flow changes. Branch prediction predicts the directions of branches during fetch, so the fetch engine knows which cache block to fetch from in the next cycle. When the wrong direction is predicted, the whole pipeline following the branch must be flushed, significantly impacting performance.

Accurate branch prediction uses the past history of branches to predict the future behavior of a branch. Early branch prediction architectures used an N-bit saturating counter to predict the direction of each branch [11]. Using only an N-bit counter accurately predicts branches which are biased in either a taken or not-taken direction, but looses prediction accuracy if the branch exhibits a more complex pattern. To capture these patterns local and global branch history prediction were proposed [16,14].

Local branch history can be used to accurately predict a branch by storing the last L directions for a given branch, and using this to index into a 2nd level Pattern History Table (PHT) [16,14]. The PHT is a table of N-bit saturating counters used to predict the direction of the branch. Local branch history increases prediction accuracy by capturing arbitrary patterns that are repeated within the last L times the branch was executed.

M. Valero et al. (Eds.): ISHPC 2000, LNCS 1940, pp. 73–87, 2000.

Global branch history uses correlation information between branches to accurately predict a branch's outcome. A history of the last G executed branch directions are kept track of in a global history register. The global history register is then used to index into a pattern history table of 2-bit counters to predict the branch direction.

It is hard for local and global histories to capture loop terminations of the pattern $((1)^N 0)^m$, where 1 represents a taken branch and a 0 represents a fall-through branch. These patterns cannot be captured by a local history predictor if N is larger than L (the local history size). Loop termination can only be captured using global history if N is smaller than G[1] or there is a unique branching sequence right before the loop termination. For the programs examined in this study, 43% of the executed branches are loop branches, and 7% of the executed loop branches are mispredicted on average.

In this paper we propose to use a Loop Termination Buffer (LTB) to predict branch patterns of type $((1)^N 0)^m$. The LTB keeps track of branches with this behavior, and predicts when the pattern changes (terminates). This allows us to achieve up to 100% prediction accuracy for loop branches after a short warm up period.

In addition, we examine the potential of using a software approach called Branch Splitting to correctly predict loop branches. Local branch history can only accurately predict loop terminations for branches that execute less than L times, where L is the number of bits used for the local history. For loops that have an iteration count larger than L, the loop guarding branch can be split into two or more branches all which have an interaction count less than L, as long as the product of new iteration counts equals the old iteration count. The new branches' patterns will then fit into local history and will have all of their backwards and termination behavior accurately predicted.

The rest of the paper is organized as follows. Section 2 describes Loop Termination Prediction and our Loop Termination Buffer implementation. Section 3 describes Branch Splitting. Simulation methodology is described in section 4. Section 5 evaluates the performance of Loop Termination Prediction and Branch Splitting. Section 6 describes prior research for loop termination prediction. Finally, we summarize our contributions in section 7.

2 Loop Termination Prediction

Traditionally loops are thought of as the steady state operation of execution, and at first glance loop termination seems to account for only a small portion of the total amount of branches seen. However, this is an important part of a branch prediction architecture, because a regular loop has the miss rate which is the inverse of it's iteration count. Since the branch prediction accuracy of most processors is already in the high nineties, loop branch mispredictions can account for a large fraction of the remaining branch mispredictions.

[1] Assuming that there are not any branches internal to the loop.

In this section, we propose using a very simple architecture extension which can predict how many times a loop branch will iterate to provide loop termination prediction.

2.1 Predicting Loop Termination

To allow instruction fetch to continue without stalling, each cycle a traditional branch prediction architecture predicts (1) if the current instruction fetch contains a branch, (2) the direction of the branch, and (3) provides the branch target address to be used in the I-cache fetch in the next cycle. This prediction is typically performed given only the current instruction cache fetch address each cycle.

The goal of our research is to predict when a loop branch terminates its looping behavior. Therefore, in addition to the above prediction information, during branch prediction we must (4) determine that the branch is a loop branch, and (5) predict if it has terminated or not. In order to provide this prediction information, the branch has to be labeled as a loop branch in the branch prediction architecture and the loop's trip count must be predicted.

Identifying Loop Branches. Loop branches can be identified in hardware by either having special loop branch instructions (effectively having the compiler mark a branch as loop branch), or identifying the loop branches by looking at the sign bit of their displacement.

For this study, we targeted loops with the conditional branch at the bottom of the loop as shown in the doubly nested loop code example in figure 1. The number of loop branches generated this way depends upon the compiler being used. The Compaq C and FORTRAN Alpha compilers (with -O4 optimization) compiles most `for`, `while`, and `do` loops with the conditional branch check at the bottom of the loop with a negative displacement.

For our programs, loop branches usually have a negative displacement, and we dynamically predict that a conditional branch is a loop branch if it has a negative displacement. While this is not a perfect definition of a loop conditional branch, we used it to dynamically classify in hardware which branches are loop branches, so that we may know which branches should attempt to use loop termination prediction.

Predicting Loop Trip Count and Termination. In order to correctly predict loop termination the branch predictor needs to (1) predict the loop trip count for the branch, and (2) keep track of the current loop iteration for the loop branch.

The *loop trip count* is the number of times in a row the loop branch is taken. The trip count can be a constant determined at compile-time for some loops, constant for a given run of an application but only determined at run-time for that run, or dynamically changing during a program's execution. Predicting the loop trip count can catch all three of these types of loop branches.

We use a *loop iteration* counter to record the current iteration count – the number of times the branch has been taken since it was last not-taken. The

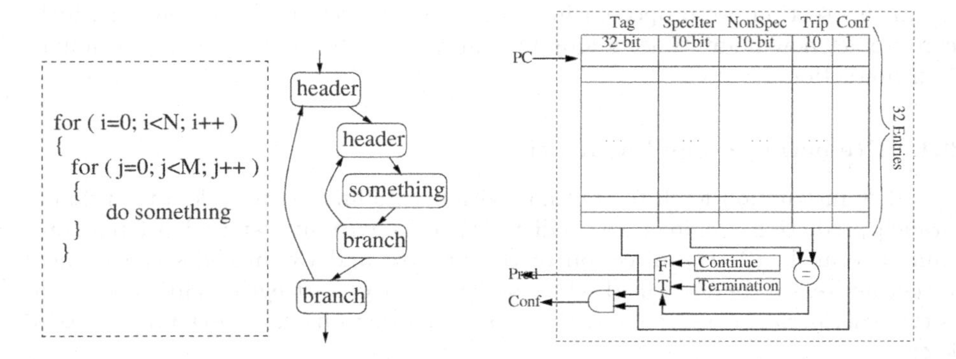

Fig. 1. Doubly nested for loop **Fig. 2.** Loop Termination Buffer

iteration counter is used to (1) predict when the loop has terminated, and (2) to initialize the loop trip count.

The loop termination prediction information described above could be stored directly into a Branch Target Buffer used for predicting target addresses during branch prediction [9,15]. Instead of doing this, in our study we examine having a small associative buffer on-the-side to accurately predict the loop termination for loop branches.

2.2 Loop Termination Buffer

The *Loop Termination Buffer* (LTB) is a small hardware structure capable of detecting the impending termination of loop-branches. The prediction mechanism takes advantage of the fact that many loops have trip counts that do not change often over the course of execution. Take for example the doubly nested loop in figure 1. If we assume a traditional predictor which has been properly warmed up, the inner loop will cause N branch misses, and the outer loop will cause one, for a total of N+1 misses. The inner branch has a highly regular pattern of being taken N-1 times and then falling through the Nth time. This last time, the loop's termination, is an event which we wish to correctly predict.

As can be seen in figure 2, the LTB has five fields: a tag field to store the PC of a branch; a *speculative* and *non-speculative* iteration count to store the number of times the branch has been taken in a row; the loop trip count field to track the number of consecutive times the loop-branch was taken before the last not-taken; and a confidence bit indicating that the same loop trip count has been seen at least twice in a row.

During Branch Prediction. To access the LTB, the fetch PC, which is used to perform the normal branch prediction, is also used to index in parallel into the LTB. If there is a tag match, the speculative iteration counter is checked against the trip count. If they are equal, the branch is a candidate for predicting termination. The confidence bit is then tested. If it is set, the loop branch is predicted to be not-taken (exiting the loop). If instead the speculative iteration counter and the trip count are not equal, then the speculative iteration count is speculatively incremented by one. We count from 0 up with the iteration counter, instead of down, so that we have the iteration count available in the counter for initialization of the trip count.

When Resolving a Branch. A branch is considered for insertion into the LTB when it completes and its branch direction is resolved. When a branch resolves and is found to not be in the LTB, it is inserted into the LTB if determined to be a loop-branch and if it is mispredicted by the default predictor. In our study, we assume backwards conditional branches are loop-branches. When inserted into the LTB, all of the LTB entries counters are initialized to zero. Since an entry is inserted into the LTB when a loop-branch is found to be mispredicted, there are no outstanding branches and this allows the speculative and non-speculative iteration counters to start out synchronized.

During resolution, a taken branch found in the LTB has its non-speculative iteration counter incremented by one.

A not-taken loop-branch found in the LTB during branch resolution updates its trip count and confidence bit. If the non-speculative iteration counter is equal to the trip count stored in the LTB, then the confidence bit is set, otherwise it is cleared. The non-speculative iteration count is then incremented by one and is copied to the trip count. The speculative iteration count is then set to the current speculative iteration count minus the non-speculative iteration count. The speculative iteration count is reset to this value because the same loop-branch may have already been fetched again before the not-taken branch resolves. Finally, the non-speculative iteration counter is reset to zero.

One reason for having two iteration counters, as shown in figure 2, is to recover the iteration counters during a branch misprediction. When a branch misprediction occurs, all of the non-speculative iteration counters copy their values into the speculative iteration counters. Therefore, this synchronizes the speculative and non-speculative counters. This approach is similar to prior architectures proposed to recover branch prediction state during a misprediction [10].

2.3 Loop Termination Predictor

Applying the loop termination buffer to branch prediction is a fairly straight forward process. When a loop branch is predicted to terminate, the branch is predicted as not-taken. Otherwise the branch is predicted taken. Although a more integrated approach of loop termination prediction into existing branch predictor architectures is possible (e.g., putting the counters into the branch target buffer), we examined using a separate buffer to concentrate on the effect of loop termination prediction.

Looking to figure 3 we can see how the branch predictors are combined at a high level. The final predictor generates a prediction for every branch, however the primary predictor will be overridden if the loop termination buffer generates a confident prediction. Recall from above that a confident prediction is generated if the branch PC is found in the LTB and it has seen the same trip count twice in a row.

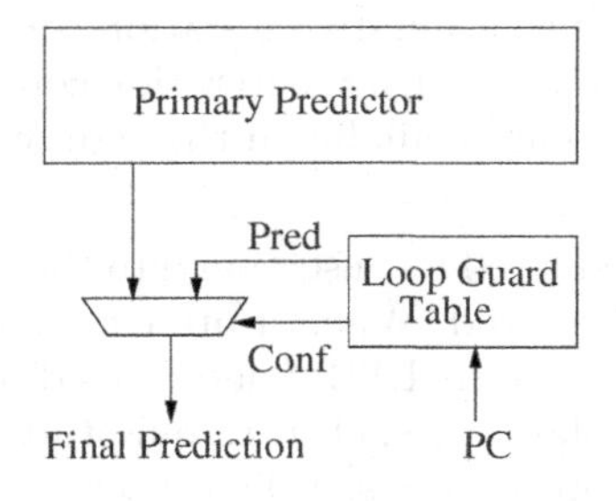

Fig. 3. Using the Loop Termination Buffer for Branch Prediction. The primary predictor is used except in the case where the loop termination predictor predicts loop exit and is highly confident

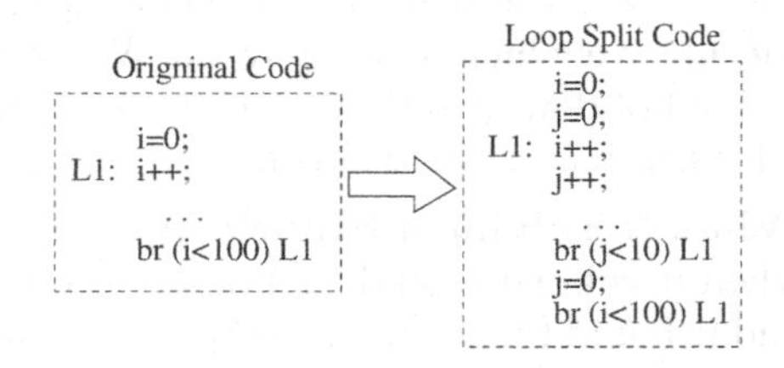

Fig. 4. Example of Branch Splitting. The branch in the original code will cause a branch misprediction 1 out 100 times the branch is executed. The transformed code can fit in the local history and will correctly predict both loop branches

3 Branch Splitting

A local branch history can only accurately predict loop terminations for branches that execute less than L times. For example, let us assume that the local history size is 12. If the predictor is predicting a loop branch whose iteration count is 100 (i.e. the pattern $(1^{99}0)^M$) as in figure 4, then the local history will not be able to predict the loop exit.

Because the local branch history can only accurately predict loop terminations for branches that execute less than L times, where L is the number of bits used for the local history, breaking the loop into multiple smaller loops will allow local history to correctly predict them. For loops that have an iteration count larger than L, the loop guarding branch can be split into two or more branches all which have an interaction count less than L, as long as the product of new iteration counts equals the old iteration count. The new branches' patterns will then fit into local history and will have all of their backwards and termination behavior accurately predicted.

Continuing with example above, let us split the branch into two branches, B1 with iteration count of 10, and B2 with iteration count of 10. This will create a doubly nested loop, whose loop body will be executed the same number of times as the original loop. Both the branch patterns will now be $(1^9 0)^M$ which can be easily captured in the local history. This is very much akin to the technique of loop tiling, except here the resource we are tiling for is the branch predictors local history.

To apply branch splitting, we need accurate knowledge of the loop bounds, in order to create new bounds with little or no cleanup code. If cleanup code is generated, then this may more than offset the branch mispredictions saved by the main loop. In addition, when applying branch splitting one needs to be aware of the increased pressure splitting the loop branch will create on the branch predictor, since it will increase the number of entries used.

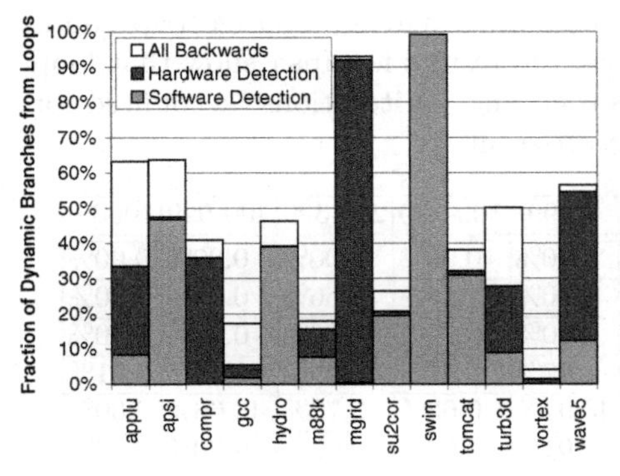

Fig. 5. Percentage of conditional branches executed which are looping branches, as detected by being backwards, or as detected by hardware or software

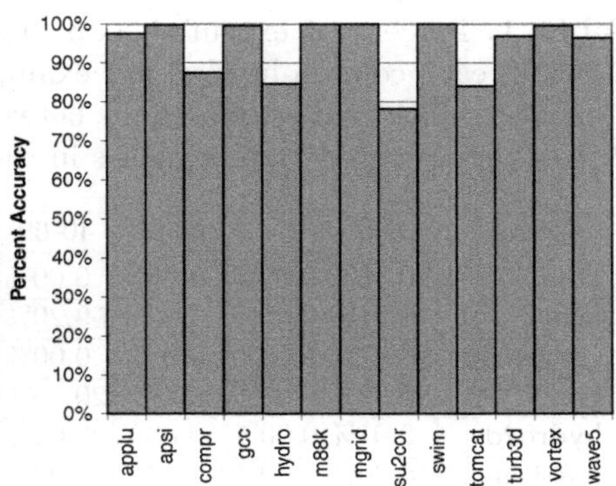

Fig. 6. Confident Prediction Accuracy. The prediction accuracy of the LTB when it returns it's prediction as confident

4 Methodology

To perform our evaluation we gathered results for 13 of the SPEC95 INT and FP benchmarks. The benchmarks were compiled on a Alpha AXP-21164 processor under OSF/1 V4.0 using the DEC C and FORTRAN compilers. All benchmarks were compiled at full optimization (`-O4 -ifo`). Since we are using full optimization, some loops in the programs were unrolled and software pipelined.

For the results in this paper we used ATOM [12] to instrument the programs and gather simulation results. The ATOM instrumentation tool has an interface that allows the elements of the program executable, such as instructions, basic blocks, and procedures, to be queried and manipulated. In particular, ATOM allows an "instrumentation" program to navigate through the basic blocks of a program executable, and collect information about registers used, opcodes, branch conditions, and perform control-flow and data-flow analysis. Programs were executed for a total of five billion instructions or program termination.

5 Results

To evaluate the benefit of loop termination prediction we examine the branch prediction performance of two predictors based on McFarling's meta predictor [8].

McFarling's meta predictor has a *meta* chooser table of 2-bit counters to choose between *bimodal* and *gshare* branch prediction. We use a bimodal table of 2-bit counters indexed by the PC to produce the bimodal prediction. In addition, we use a global history register XORed with the branch PC as an index into a table of 2-bit counters to provide the gshare prediction. The meta table is also

Table 1. Percent of executed loop branches that had an average trip count shown in each column header. For example, the results for `apsi` show that loop branches, which had a loop count between 40 and 69 iterations, accounted for 69% of the executed loop branches in the program

application	0-9	10-19	20-39	40-69	70-99	100-199	200-399	400-999	1000+
applu	70.15%	0.00%	29.85%	0.00%	0.00%	0.00%	0.00%	0.00%	0.00%
apsi	16.31%	0.00%	7.93%	59.20%	0.00%	11.88%	4.66%	0.01%	0.00%
compress	99.97%	0.00%	0.03%	0.00%	0.00%	0.00%	0.00%	0.00%	0.00%
gcc	32.39%	0.00%	23.47%	20.49%	1.97%	5.67%	5.29%	10.73%	0.01%
hydro2d	2.01%	0.00%	0.03%	0.01%	43.94%	0.63%	9.72%	43.66%	0.00%
m88ksim	18.80%	0.00%	1.14%	0.03%	0.01%	0.08%	79.92%	0.00%	0.00%
mgrid	2.17%	0.00%	11.57%	86.25%	0.00%	0.00%	0.00%	0.00%	0.00%
su2cor	22.84%	0.00%	0.53%	1.07%	1.24%	69.54%	0.00%	4.77%	0.00%
swim	0.00%	0.00%	0.00%	0.00%	0.00%	0.20%	0.00%	99.80%	0.00%
tomcatv	4.03%	0.00%	0.00%	0.00%	0.00%	45.94%	0.00%	50.03%	0.00%
turb3d	60.60%	0.00%	36.89%	2.50%	0.00%	0.00%	0.00%	0.00%	0.00%
vortex	82.61%	0.00%	1.97%	3.69%	1.64%	7.98%	1.24%	0.87%	0.00%
wave5	3.19%	0.00%	0.00%	2.63%	0.00%	32.96%	6.56%	54.66%	0.00%

indexed with the PC, and the 2-bit counter keeps track of which predictor (the bimodal predictor or gshare predictor) is more often correctly predicting the branch, and then the corresponding prediction is used.

The predictor with local history, called here the local/global chooser or LGC, is very similar to the McFarling predictor with the exception that the bimodal predictor is replaced with two tables for local history prediction. A per branch history is tracked in a first level *local* history table, which is then used to index into a second table of 2-bit counters. This extra level of indirection allows local branch history patterns to be captured. The major difference between the LGC and the branch predictor found on the Alpha 21264 [7] is that LGC uses the PC to index into the chooser, as opposed to using the global history.

5.1 Loop Characterization

We begin our analysis with a characterization of the loop behavior in the programs we examined. Figure 5 shows the percent of conditional branches executed which we classified as loop branches. These are executed branches having a negative sign displacement as described in section 2.1. The graph also shows the fraction of these executed branches that were found in our loop termination buffer (Hardware Loop Detection) during execution. Then layered on top of that is the percent of executed branches that had branch splitting applied to them. For `swim`, almost 100% of its executed branches were loop branches. These were all found in the LTB, and had branch splitting applied to them. For `applu`, 63% of its executed branches were classified as loop branches, 34% of the executed branches were found in the LTB, and 9% of the executed branches had the branch splitting optimization applied to them.

We further examine these loop branches by breaking them down in terms of their iteration count. Table 1 shows the breakdown of loop branches in terms of the iteration count represented by the range in the column headers. These numbers are calculated by taking the average trip count for each static branch, multiplying this by the number of times the branch was executed, and then dividing this by the total number of loop branches executed. For example, the results for `mgrid` show that 88% of the loop branches executed were executed in a loop which went from 0 to a loop trip count somewhere between 40 to 69.

In looking at the results and source code for `compress`, we saw that 97% of the loop branches that were executed were in a loop which iterated between 0 and 7. This is the reason for the dominating short trip counts in `compress`. The Compaq compiler did not unroll this loop, perhaps due to a rarely taken `break` statement inside the loop. For this program, local (per-branch) history will correctly predict the loop branches, because the trip count is less than the local history size (see figure 7). But for many of the other programs, local branch history can not correctly capture all of the loop termination. The local history length is not sufficient to capture a trip count in the range of 40 to 69 iterations as in `mgrid`.

5.2 Branch Splitting

To examine the potential of Branch Splitting as presented in section 3, we profile each branch over the execution of the application. We track the number of mispredictions from each branch, along with information on the number of iterations between loop exists, and the regularity of the pattern found. A backwards branch is said to be regular if the branch direction pattern is $((1)^N 0)^m$ over the *entire* execution of the program, so the iteration count was always $N + 1$. If the branch is regular and the cycle count is larger than the local history size, then we applied branch splitting to the loop branch. The light gray part of the bar in figure 5 shows the percent of executed branches splitting was applied to.

Table 2 shows the potential reduction in branch mispredictions from applying branch splitting. The first two columns are the before and after overall branch misprediction rates, and the third column is the difference between them. The next column shows the total number of static branches in the executable. The last column is the percent increase in static branches when branch splitting was used. While the number of dynamic branches should not change significantly, any extra static branches can lead to more collisions in the table. Some applications such as `apsi` do quite well, achieving an absolute decrease in misprediction rate of 1.3%.

5.3 Loop Termination Prediction

In this section, we examine the reduction in branch misprediction rate from adding our Loop Termination Buffer to the Meta predictor, represented by `Meta` and `Meta+LTP`, and to the LGC predictor, represented by `LGC` and `LGC+LTP`. For

Table 2. Effects of Branch Splitting. The first two columns are the before and after overall branch misprediction rates, and the third is the difference between them. Static branches is the total number of branches in the executable. The last column is the percent increase in static branches when branch splitting is used

app	miss rate	after splits	difference	static branches	% branch inc
applu	2.79%	2.52%	0.27%	1153	4.08%
apsi	2.64%	1.33%	1.31%	1642	4.45%
compress	10.96%	10.96%	0.00%	182	1.65%
gcc	9.07%	9.03%	0.04%	15952	0.42%
hydro2d	0.33%	0.09%	0.24%	1667	7.26%
m88ksim	4.67%	4.19%	0.48%	1017	0.10%
mgrid	1.64%	1.61%	0.03%	1172	0.34%
su2cor	4.15%	3.83%	0.32%	1717	3.38%
swim	0.20%	0.00%	0.20%	983	3.26%
tomcatv	0.99%	0.63%	0.37%	891	3.03%
turb3d	2.98%	2.65%	0.32%	1274	1.18%
vortex	1.69%	1.69%	0.00%	6259	0.18%
wave5	0.74%	0.30%	0.44%	1794	2.84%

the results gathered, each branch prediction table (Meta, Local, Bimodal) has 32K entries per table. For the loop termination predictor, we simulate adding a small 32 entry fully associative LTB with random replacement to the base predictor. The size of this predictor (for both the tag and data) is less than 256 bytes.

The confidence prediction accuracy of LTP is shown in figure 6, and is independent of the base predictor used. This is the percent of loop branches that were predicted to be accurate by the confidence counter in the LTB. Recall from section 2 that the confident bit is set when the the branch PC is found in the table and it has seen the same iteration count two times in a row. The reason that the prediction accuracy is not near 100% for all applications is that some loops have break statements, and/or they change the number of times they iterate during the execution of the program.

We next examine the effects of LTP on the prediction accuracy of loop branches and all executed branches. Figure 7 shows the branch misprediction rates for only the loop branches, and figure 8 shows the misprediction rate in terms of all executed branches. Results are shown using 32K entry tables for the base Meta and LGC predictor, and when adding our LTB to the predictor. `Compress`, as discussed above, has a small inner loop that contains a break statement and is not unrolled and pipelined by the Compaq C complier. This loop escapes the Meta prediction but is fully captured in the LTB. It is also captured by the LGC predictor in its local history. The results in figure 8 show that we reduce the overall miss rate for `mgrid` from 1.7% down to 0.1% using loop termination prediction. Figure 9 shows the code example of one of the routines

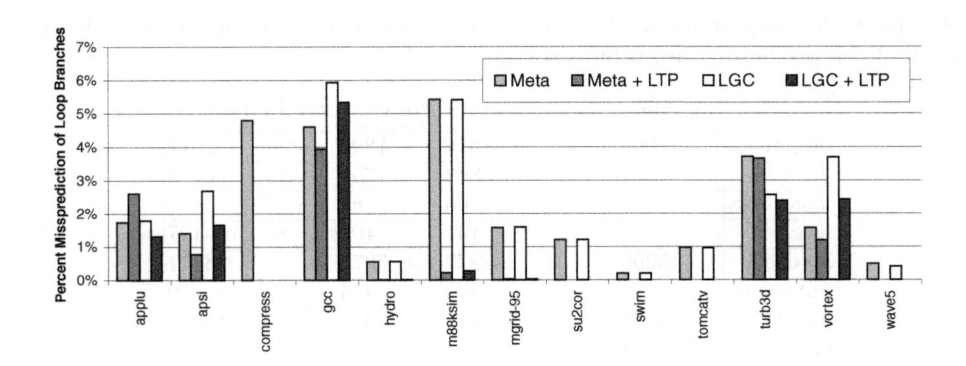

Fig. 7. Loop Misprediction Rate. The prediction accuracy of McFarling's Meta predictor and a local/global chooser with 32k entry tables, with and without loop termination prediction for only loop branches

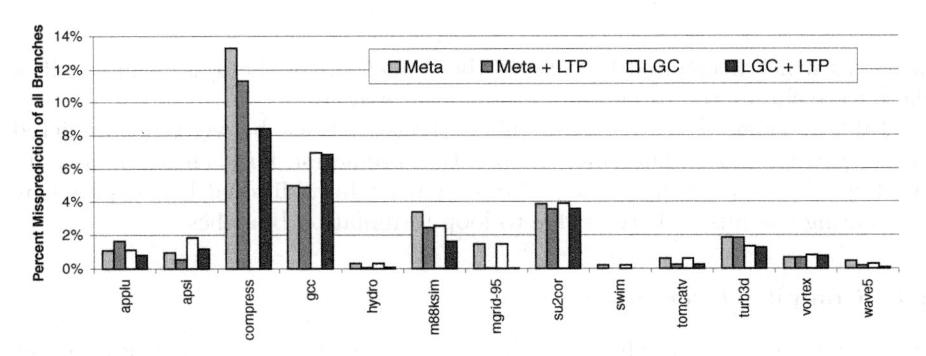

Fig. 8. Overall Misprediction Rate. The prediction accuracy of Meta and LGC with 32k entry tables, with and without loop termination prediction for all branches

```
      SUBROUTINE RESID(U,V,R,N,A)
      INTEGER N
      REAL*8 U(N,N,N),V(N,N,N),R(N,N,N),A(0:3)
      INTEGER I3, I2, I1
      DO 600 I3=2,N-1
      DO 600 I2=2,N-1
      DO 600 I1=2,N-1
 600  R(I1,I2,I3)=V(I1,I2,I3)
     >      -A(0)*( U(I1,  I2,  I3  ) )
     >      ... additional matrix computations left out of example
      CALL COMM3(R,N)
      RETURN
      END
```

Fig. 9. Looping code example from `mgrid`, which accounts for 22% of the mispredicted branches

Table 3. Average number of instructions between a mispredicted branch for each of the predictors and table sizes simulated

name	32k Meta	32k Meta+LTP	32k LGC	32k LGC+LTP
applu	1969	1310	1898	2617
apsi	3394	5863	1755	2745
compress	54	63	85	85
gcc	147	151	105	107
hydro2d	5206	26277	5071	23871
m88ksim	311	430	414	650
mgrid	4370	197808	4320	215815
su2cor	333	363	331	360
swim	25002	2721829	24994	2635741
tomcatv	3731	9356	3691	9111
turb3d	1508	1532	2102	2244
vortex	1562	1599	1306	1397
wave5	6844	16752	10421	45375

in `mgrid` which accounted for 22% of the branch mispredictions. Almost all of these were eliminated using loop termination prediction.

Table 3 shows the average number of instructions between mispredicted branches with and without loop termination prediction for each of the 3 simulated systems. The results show a large increase in additional ILP exposed by eliminating the mispredictions due to loop termination branches.

5.4 Compiler Interaction

Our loop termination prediction architecture is a highly accurate structure that rarely makes incorrect predictions. However, one fairly common loop type will cause problems. Loops with `break` or `continue` instructions. For example, a loop with a `break` statement in it will have the same trip count several times in a row, and then terminate abruptly. While these types of loops are still predicted with higher accuracy than in a traditional predictor, they prevent perfect prediction, even with confidence. Table 2 shows the results of eliminating mispredictions for only loop branches (using branch splitting) for only branches that always have the same iteration count. Figure 8 shows that we achieve a much higher reduction in misprediction rate using hardware loop termination prediction. This is because the LTB is accurately predicting loops whose iteration count change over the lifetime of the program.

A common compiler optimization to improve memory performance it to tile loops. This optimization creates smaller inner loops to traverse over the data, keeping as much data in the cache creating reuse and reducing cache misses. This creates much more loop termination behavior, which could be correctly predicted by our loop termination buffer. In addition, our branch splitting results indicate that it may be worthwhile to also take into consideration the size of the branch history registers for branch performance, when tiling a loop for memory performance. For the results gathered in this paper, the compiler did not perform any tiling.

6 Related Work

Branch prediction research has concentrated on reducing aliasing into branch prediction tables, and to increase accuracy by using chooser/meta predictors which allow branches to be predicted correctly by both local or global history, based upon confidence information [1].

6.1 Other Branch Predictors for Predicting Loop Termination

Recently, researchers have examined using prior committed values [4] or value prediction in combination with traditional branch prediction to improve branch prediction accuracy [2,3]. These predictors can accurately predict loop terminations for long loop histories, since they predict the branches based on past or predicted values. They are more general and can potentially capture more data correlation for branches besides loop termination, although at the cost of adding large buffers to store values or value differences. Our loop termination buffer approach accurately predicts loops even with just the addition of a small 32 entry prediction buffer (less than 256 bytes in size). In addition, they are fundamentally different when predicting loop terminations from LTB. While the value-based approaches predict the input operands (or their difference), the LTB predicts loop trip counts and keeps track of the current speculative loop iteration of the branch. The benefit our LTB approach is its simplicity and low cost, and it can easily be added to an existing predictor with very little overhead or increase in cycle time.

6.2 Predicting Loop Iterations for Speculative Thread Generation

Knowing the speculative loop iteration is an important area of research for speculative threaded execution. Tubella and Gonzalez [13] examined adding a *Loop Execution Table* (LET) and *Loop Iteration Table* (LIT) to a speculative threaded processor, to provide prediction information for loop iterations. The LET is used to predict the number of iterations for each loop to guide the generation of speculative threads. Whereas the LIT is used to keep track of information related to the live-in registers and memory locations from the last loop iteration.

Our loop termination buffer is very similar to the function of the loop execution table. Since their goal was to guide speculative thread generation, they did not examine using the LET for branch prediction. Our LTB has a confidence predictor, which is not in the LET, to determine when to use the loop termination prediction or the original branch predictor. In addition, the LTB keeps track of a speculative and non-speculative loop count, which is used to restore the loop count in the LTB in case of a misprediction.

6.3 Loop Counting Branch Instructions

Some architectures have special instructions for loop branches. The IBM Power PC [5] has a loop-based branch instruction that branches based on the value of a Count Register. The Intel IA-64 [6] has an identical branch called the Counted Loop Branch. For both of these architectures, when a loop-branch instruction is executed, if the count register is non-zero, then the register is decremented and the branch is taken. Otherwise, the branch is not-taken (the loop is terminated).

The PowerPC and IA-64 architectures currently do not use this loop-based branch instruction for LTP. In order to use it for loop termination prediction, an architecture would need to speculatively keep track of the loop count value. Each time the loop-branch is predicted, the speculative loop count would then be decremented. When the speculative count value reaches zero, the branch would be predicted as terminated. In actuality, a stack of speculative count values would needed, so that the count register could be saved between procedure calls. This is if the loop-branch instruction can be saved in order to allow different nested procedures to use the loop-branch register.

A stack-based speculative loop count predictor would be very similar to the Loop Termination Buffer we propose. Even so, the Loop Termination Buffer would be preferred, since it can capture more types of branches than those that are implemented using the loop-based branch instruction, and we can achieve a high degree of loop branch coverage with a LTB of size less than 256 bytes.

7 Summary

Loop terminations can constitute a large percentage of the mispredictions generated by current branch predictors. In this paper we proposed two schemes for dealing with this problem, the Loop Termination Buffer (LTB) and Branch Splitting.

The LTB is a hardware mechanism that detects and predicts branch patterns of the form $((1)^N 0)^m$. The LTB tracks branches with this behavior and then informs the predictor when it has found such a pattern. The addition of the LTB allows regular loop guarding branches to be predicted with 100% accuracy. However, we also find a significant benefit from the LTB for branches that have a loop trip count that changes over the lifetime of the program. The LTB reduced the loop mispredictions from 5.4% down to 0.2% for `m88ksim`, and aided the overall prediction accuracy by a significant amount for most programs.

In addition, we examined the potential of using a software approach called Branch Splitting to correctly predict loop branches. For loops that have iteration counts larger than may be captured by local history, the loop guarding branch may be split into two or more branches all of which have an interaction count that would be captured by local history. This approach resulted in only small reductions in misprediction rates for some programs, since we only applied it to loops that had the same iteration count for the execution of the whole program. For `apsi`, branch splitting decreased the misprediction rate from 2.6% down to 1.3%.

Acknowledgments

We would like to thank the anonymous reviewers for providing useful comments on this paper. This work was funded in part by NSF CAREER grant number CCR-9733278, by DARPA/ITO under contract number DABT63-98-C-0045, and a grant from Compaq Computer Corporation.

References

1. A. Eden and T. Mudge. The YAGS branch prediction scheme. In *31st International Symposium on Microarchitecture*, pages 69–77, December 1998. 85
2. A. Farcy, O. Temam, R. Espasa, and T. Juan. Dataflow analysis of branch mispredictions and its application to early resolution of branch outcomes. In *31st International Symposium on Microarchitecture*, December 1998. 85
3. Gonzalez and Gonzalez. Control-flow speculation thorough value prediction for superscalar processors. In *International Conference on Parallel Architectures and Compilation Techniques*, October 1999. 85
4. T. Heil, Z. Smith, and J. E. Smith. Improving branch predictors by correlating on data values. In *32nd International Symposium on Microarchitecture*, November 1999. 85
5. IBM. *The PowerPC Architecture: A Specification for a New Family of RISC Processors.* Morgan Kaufmann Publishers, 1994. 86
6. Intel. *IA-64 Application Developer's Architecture Guide.* Intel Corporation, Order Number 245188-001, 1999. 86
7. R. E. Kessler, E. J. McLellan, and D. A. Webb. The alpha 21264 microprocessor architecture. In *International Conference on Computer Design*, October 1998. 80
8. S. McFarling. Combining branch predictors. Technical Report TN-36, Digital Equipment Corporation, Western Research Lab, June 1993. 79
9. C. H. Perleberg and A. J. Smith. Branch target buffer design and optimization. *IEEE Transactions on Computers*, 42(4):396–412, 1993. 76
10. K. Skadron, P. Ahuja, M. Martonosi, and D. Clark. Improving prediction for procedure returns with return-address-stack repair mechanisms. In *Proceedings of the 31st Annual International Symposium on Microarchitecture*, pages 259–271, December 1998. 77
11. J. E. Smith. A study of branch prediction strategies. In *8th Annual International Symposium of Computer Architecture*, pages 135–148. ACM, 1981. 73
12. A. Srivastava and A. Eustace. ATOM: A system for building customized program analysis tools. In *Proceedings of the Conference on Programming Language Design and Implementation*, pages 196–205. ACM, 1994. 79
13. J. Tubella and A. Gonzalez. Control speculation in multithreaded processors through dynamic loop detection. In *4th International Symposium on High Performance Computer Architecture*, February 1998. 85
14. T. Yeh. Two-level adpative branch prediction and instruction fetch mechanisms for high performance superscalar processors. Ph.D. Dissertation, University of Michigan, 1993. 73
15. T. Yeh and Y. Patt. A comprehensive instruction fetch mechanism for a processor supporting speculative execution. In *Proceedings of the 25th Annual International Symposium on Microarchitecture*, pages 129–139, December 1992. 76
16. T.-Y. Yeh and Y. Patt. Two-level adaptive branch prediction. In *18th Annual International Symposium on Computer Architecture*, pages 51–61, May 1991. 73

Compiler–Directed Cache Assist Adaptivity*

Xiaomei Ji, Dan Nicolaescu, Alexander Veidenbaum, Alexandru Nicolau, and Rajesh Gupta

Department of Information and Computer Science, University of California Irvine
444 Computer Science, Building 302, Irvine, CA 92697–3425
{xji,dann,alexv,nicolau,rgupta}@ics.uci.edu

Abstract. The performance of a traditional cache memory hierarchy can be improved by utilizing mechanisms such as a victim cache or a stream buffer (cache assists). The amount of on–chip memory for cache assist is typically limited for technological reasons. In addition, the cache assist size is limited in order to maintain a fast access time. Performance gains from using a stream buffer or a victim cache, or a combination of the two, varies from program to program as well as within a program. Therefore, given a limited amount of cache assist memory, there is a need and a potential for "adaptivity" of the cache assists i.e., an ability to vary their relative size within the bounds of the cache assist memory size. We propose and study a compiler-driven adaptive cache assist organization and its effect on system performance. Several adaptivity mechanisms are proposed and investigated. The results show that a cache assist that is adaptive at loop level clearly improves the cache memory performance, has low overhead, and can be easily implemented.

1 Introduction

The area available for on–chip caches is limited and the size and associativity of a cache for a given processor cannot be significantly increased without causing an increase in the cycle time. A small area dedicated to a victim cache and/or a stream buffer [7] can increase the performance of the memory system while it may not be large enough to double the cache size. Victim caches eliminate conflicts and exploit temporal locality of the programs, while stream buffers exploit spatial locality because they fetch data that is likely to be accessed in the near future. We call a victim cache, a stream buffer or a combination of the two a *cache assist*.

A cache assist needs to have a high degree of associativity, and it needs to have an access time equal to that of the level of cache utilizing it, i.e. its access time is very small. This imposes a limit on the size of the cache assist memory. In [8] it is shown that for any CMOS process technology the cache size cannot be increased too much without causing an increase in cycle time and access time.

* This work was supported in part by the DARPA ITO under Grant DABT63-98-C-0045.

M. Valero et al. (Eds.): ISHPC 2000, LNCS 1940, pp. 88–104, 2000.

When both a victim cache and a stream buffer are desirable, their relative sizes have to be selected within the bounds of the (small) cache assist memory size.

Unfortunately, neither a victim cache nor a stream buffer are a panacea: in some programs a victim cache performs much better, in others a stream buffer performs much better. In this paper we show that a dynamic combination of the two improves the overall performance the most. This happens across different applications as well as within a single application.

We propose a simple system that allows the cache assist configuration to vary at run time. A set of four special instructions is used to change the functioning of the cache assist, making it work as a stream buffer, victim cache or a combination of the two. A compiler can insert these instructions in the code at points it determines suitable by either static code analysis or using profile–directed feedback.

While the hardware modifications are modest, the following questions determining the feasibility of the approach need to be answered:

1. when should the cache assist configuration be changed,
2. how often is it necessary to reconfigure,
3. what is the optimal reconfiguration policy?

On one hand it would not be feasible to change the cache assist configuration every few instructions as the overhead associated with such reconfiguration would make the approach prohibitively expensive. On the other hand if we reconfigure too infrequently, e.g. once per function call, we might miss some optimization opportunities because a function may contain a number of loops, each of them with a distinct cache behavior.

It has been shown that the majority of dynamic instructions in a program are executed in innermost loops. An inner loop is also likely to have reasonably stable spatial/temporal locality characteristics. This suggests that an inner loop may be a good place to change the organization of the cache assists and maintain the setting for the duration of such a loop. In this paper we propose and study different schemes of adapting the cache assist at loop level, trying to determine which one has better performance. We also propose other schemes using a more static assist memory partitioning and compare their performance with the loop–level adaptive cache assist configurations.

We currently use a profile–based mechanism for the control of adaptation by the compiler. Future work will study the opportunity to use compile–time analysis for making adaptivity decisions. The size of the cache assist memory is very important from both the access time and the effectiveness of adaptivity. The effect of varying the cache assist size on the miss rate of the memory system is studied as well.

2 Related Work

Victim cache [7] is a mechanism that is aimed specifically at conflict misses. It predicts that a replaced line of data will be accessed again shortly and stores the replaced data in a small fully-associative buffer on the refill path of the cache. On a cache miss, the victim cache is checked to see whether the data is present. If so, the data is copied from the victim cache to the cache.

A stream buffer [7] is a mechanism to prefetch and store data. It consists of a FIFO memory plus an address generator. On a cache miss, all stream buffers are searched in parallel to find whether the data is present. On a hit, the data is copied to the cache and the stream buffer is refilled from successive addresses in the lower memory hierarchy. On stream buffer miss, a buffer is allocated and addresses following the miss address will be prefetched into the buffer.

To the best of our knowledge there is no previous work in applying adaptivity to configure a cache assist memory. However, adaptivity has been applied in various forms. Selected examples of its use are:

Adaptive routing pioneered by ARPANET in computer networks and, more recently, applied to multiprocessor interconnection networks [1], [3] to avoid congestion and route messages faster to their destination.

Adaptive throttling for interconnection networks [3]. [16] shows that "optimal" limit varies and suggests admitting messages into the network adaptively based on current network behavior.

Adaptive cache control of coherence protocol choice were proposed and investigated in the FLASH and JUMP-1 projects [4], [11].

Adapting branch history length in branch predictors was proposed in [9] since optimal history length was shown to vary significantly among programs.

Adaptive page size has been proposed in [14] to improve the page management overhead and it is used in to reduce the TLB and memory overhead in [12].

Adaptive adjustment of data prefetch length in hardware was shown to be advantageous [2], while in [5] the prefetch lookahead distance was adjusted dynamically either purely in hardware or with compiler assistance. A cache with a fixed large cache line is used in [10] in association with a predictor to only fetch the parts of the cache line that are likely to be used.

Adaptive cache line size was shown to improve the miss rate without an appreciable increase in bandwidth in [18], [19] and [6]. A scheme for adapting the cache line size dynamically was proposed in [18]. A special adaptive controller is incorporated in the cache access controller to monitor the memory access pattern of an application and change the line size to double or half its original size at a time in order to suit the application's needs. In [18] the cache line is truly variable, whereas [19] uses a set of four predefined values for the line size. A scheme that uses two fixed sizes was proposed in [6].

A method to use compiler provided information to do software assistance for data caches was proposed in [15]. The compiler decides through static analysis when data exhibits spatial or temporal locality and generates code to attach a special spatial/temporal tag. The tag is used by the hardware when deciding if cache lines replaced from the cache should be placed in a victim cache.

3 System Organization

Figure 1 shows the components of the system being studied. It consists of a 3–level memory hierachy plus a partitionable cache assist memory that can function as either a stream buffer, a victim cache or a combination of the two. The cache assist memory consists of N cache–line sized buffers connected to L1 fill path. Separate control units utilize the allocated memory as a victim cache or as a stream buffer. A fully associative write buffer with a line size identical to the L1 line size is also used.

The L1 cache is direct mapped and the hit latency is assumed to be 1 cycle. The L1 bus transfer takes 2 cycles. L2 is a 2–way set–associative with the access latency of 15 cycles. The main memory access latency is 100 cycles.

When the processor requests data, the L1 cached is searched. On a miss the victim cache and the stream buffer are searched in parallel. If both miss the request is sent to the next level of memory, otherwise the cache assist supplies the data.

Associated with the cache assist area are configuration registers. The registers contain the size of the victim cache, the size of the stream buffer, and hit counters for both of them. The configuration for the cache assist can be changed dynamically at run time using four operations:

- shrink_stream_buffer(cache_lines_to_shrink)
- shrink_victim_cache(cache_lines_to_shrink)
- extend_stream_buffer(cache_lines_to_enlarge)
- extend_victim_cache(cache_lines_to_enlarge).

Extending the stream buffer marks the new entries as invalid, shrinking it does the same and deletes any pending requests from the "issued prefetch" queue. Shrinking and extending the victim cache sets the victim cache size register to the new value and, in the case of extending, marks the added entries as invalid.

The compiler can insert these instructions in places in the program where static anlysis or profile based feedback determine that changing the configuration and relative sizes of the cache assists will improve the performance.

4 Experimental Infrastructure

4.1 Simulator

The framework provided by the ABSS [13] simulation system is used in this study. ABSS is a simulator that runs on SUN Sparc systems and is derived from the MINT simulator [17].

The ABSS simulator consists of 5 parts: augmentor, thread management, cycle-counting libraries, user-defined simulator of the memory system and the application program.

The augmentor program (called *doctor*) parses the original application assembly code, and adds instrumentation code that sends information about the loads and stores executed by the program to the simulator.

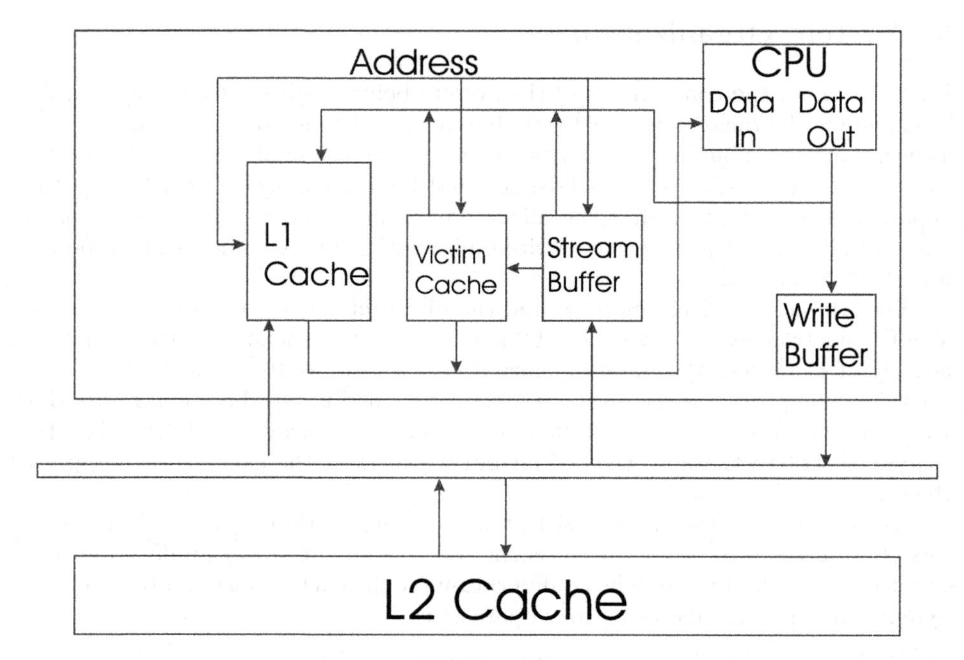

Fig. 1. System design

Our custom memory architecture simulator simulates a 3–level memory hierarchy plus a highly configurable memory cache assist with modules for modeling a stream buffer and a victim cache. The sizes of the victim cache and stream buffer are changeable at run time via commands embedded in the simulated program.

4.2 Compilation

We have used version 2.95 of the GCC compiler collection to conduct all the experiments. The compiler back–end was modified to emit special code sequences before entering a loop, or on the code path for exiting a loop. Given that the compiler back–end is common to the C and Fortran77 compiler we were able to use this instrumentation for compiling all the SPEC95 benchmarks.

The code sequences were used for adjusting the cache assist allocation, and for collecting statistics and identifying the loop (source file name and line number), and signaling to the cache simulator that a loop is being entered or exited.

In order not to modify the behavior of the program, the code sequences leave the processor in the same state as it was before the sequence in question has run. This is achieved by saving and restoring all the registers that the code sequence uses, including the flag registers. Furthermore, the loop instrumentation is done in the assembly emitting pass of the compiler (the last compilation pass), so it does not affect the code generation.

All the benchmarks where compiled using the -O2 optimization flag, the target instruction set was SPARC V8plus.

4.3 Benchmarks

The set of benchmarks shown in Table 1 was chosen because it has a good mix of both numeric and non–numeric programs, because they are fairly memory hierarchy intensive, and because SPEC95 is a standard set of benchmarks. All benchmark programs were simulated until completion.

Table 1. Benchmarks used

Benchmark	Decription	Instructions	Memory references
go	Plays the game GO	3.20e+10	7.76e+09
ijpeg	Image compression	2.70e+10	7.39e+09
perl	Perl interpreter	1.42e+10	3.42e+07
apsi	Calculates statistics on temperature	3.74e+10	1.20e+10
fpppp	Performs multi-electron derivatives	3.18e+11	1.03e+11
swim	Solves shallow water equations	3.21e+10	1.32e+10
turb3d	Simulates turbulence	1.13e+11	2.86e+10
wave	Solves Maxwell's equations	3.80e+10	1.20e+10

For some of the experiments profiling was used to select an "optimal" cache assist configuration. Profiling was performed using the SPEC training input set. The profile information was then used to run the benchmarks with the reference input set. We have verified that such profiling is accurate.

5 Performance Evaluation

To compare the relative performance of different cache assist configurations we use two main metrics: miss rate and execution time. For each experiment we gather the following kinds of data in order to evaluate the cache and cache assist performance.

- L1 and L2 miss rates
- number of hits in assist buffer
- miss rate reduction

We define the following equation to determine the overall performance improvement for the system:

$$miss\_rate\_reduction = (old\_miss\_rate - new\_miss\_rate) \\ *100.0/old\_miss\_rate \quad (1)$$

We simulate a base cache hierarchy with a 16KB direct mapped L1 cache and a 256KB 2 way set–associative L2 cache. The line size is 32 bytes for L1 and 64 bytes for L2. We will call this the *base system configuration*. Figures 2 and 3 show the L1 and L2 miss rates respectively, for the benchmarks using the base configuration. Only *swim* and *wave* have L1 miss rates that are greater than 15% and, except for *apsi*, all of them have L2 miss rates less than 3%.

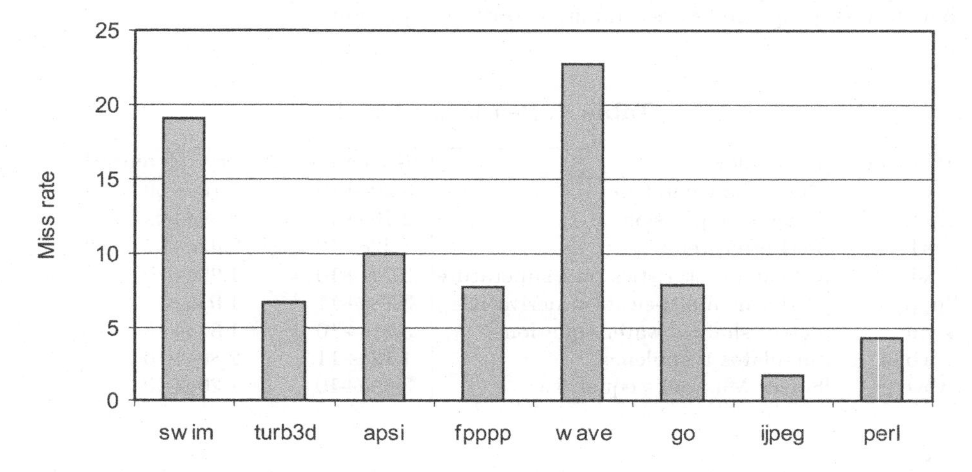

Fig. 2. L1 miss rate of 16KB direct-mapped, 32B line size cache

5.1 The Performance of Individual Cache Assists

The performance of the individual cache assists is evaluated using the base system configuration and either a 1KB victim cache or a 1KB stream buffer. Figure 4 shows the miss rate reduction for each of the assists when compared to the base configuration.

The effect varies from program to program. In *go* the stream buffer barely has an impact (under 5% miss rate reduction), but the victim cache reduces the miss rate by 50%. The same is observed for *perl* and *fpppp* where the victim cache reduces the miss rate much more than the stream buffer. The reverse is observed in the case of *turb3d* where the stream buffer reduces the miss rate by 55%, but the victim cache only reduces it by 23%. For *apsi*, *ijpeg* and *wave* the difference is not as pronounced.

The above results confirm the advantage of using a cache assist, but the type of cache assist that is most useful varies from application to application. Thus we conjecture that a system that has a cache assist that can be reconfigured between a victim cache or a stream buffer at run time on a per program basis would improve performance.

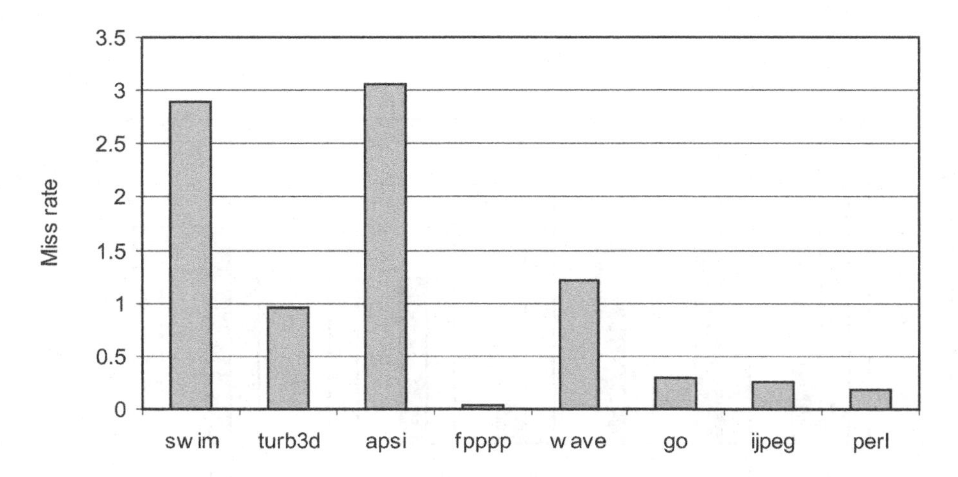

Fig. 3. L2 miss rate of 256KB, 2-way set associative, 64B line size L2 cache

The fact that memory accesses in a program very seldom follow a uniform pattern suggests that the effect of cache assists also varies within a program. To evaluate the effect of cache assists on different portions of the code we instrument and collect performance data for all the inner loops in a program. The inner loops' memory access behavior is indicative of the entire program behavior since instructions executed in the inner loops often account for more than 98% of the memory reference instructions executed by a program.

Figure 5 shows the miss rate reduction per loop for the *apsi* benchmark when using either a 1KB stream buffer or a 1KB victim cache for a given loop. For some loops the victim cache reduces the miss rate much more than the stream buffer, whereas the opposite is true for other loops.

Figure 6 shows the miss rate reduction compared to a normal cache hierarchy for different instantiations of the loop at line 276 from file jidcting.c in the *ijpeg* benchmark when using a 1KB victim cache or stream buffer. The miss rate reduction varies a lot between loop instantiations, with some instances preferring a victim cache and others preferring a stream buffer.

The miss rate reduction when using a cache assist varies widely between different loops, and between instantiations of the same loop. We can now conclude that cache assist adaptivity is not only desirable at the program level, but it should also be applied dynamically within a program.

5.2 Dynamic Combination of Cache Assist Techniques

So far we discussed using the cache assist memory either as a stream buffer or as a victim cache. Given the fact that few program exhibit pure temporal locality or spatial locality, but rather a mix of them, one can expect that using both cache assists at the same time would have a better performance. To take advantage

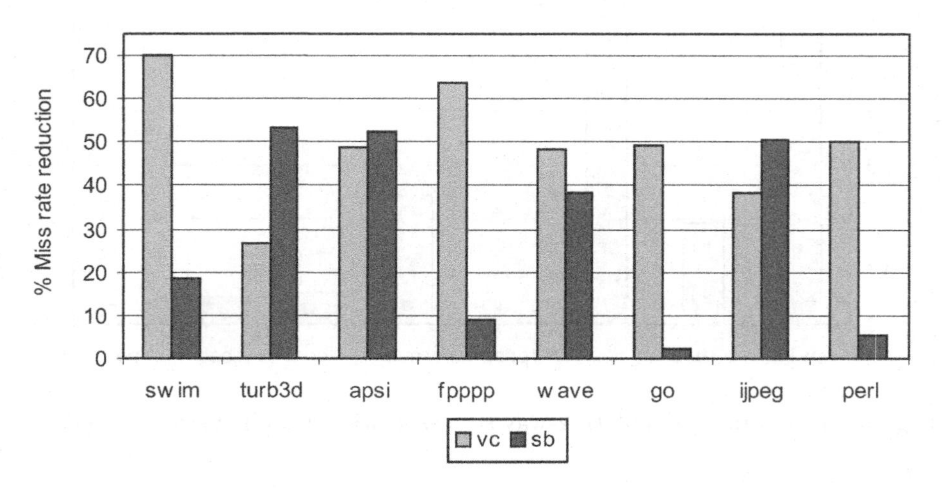

Fig. 4. Miss reduction rate for a 1KB cache assist

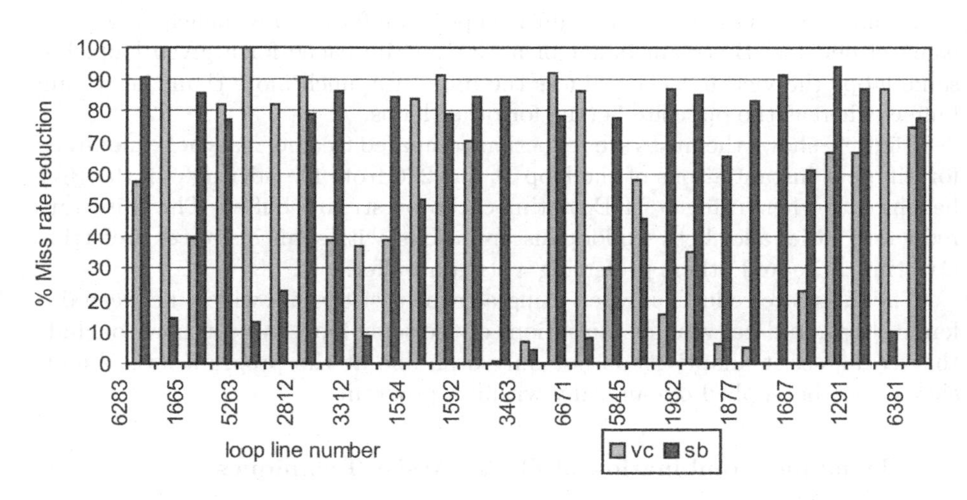

Fig. 5. Miss rate reduction per loop (a 1KB assist, the *apsi* benchmark)

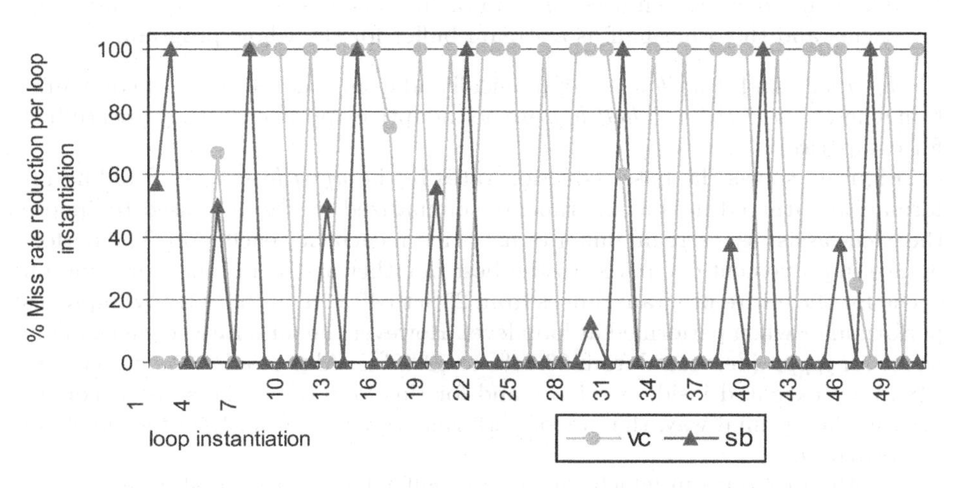

Fig. 6. Miss rate reduction per loop instantiation in *ijpeg* benchmark

of the facts presented above a program could change the cache assist structure either initially or before entering a loop so that it is either a victim cache or a stream buffer, depending on what configuration results in a lower miss rate. The question is, given limited cache assist memory, what is the best way to partition it.

To investigate different possibilities of adaptation we propose four approaches to partitioning the total (limited) cache assist space between the victim cache and the stream buffer. They are:

1. Use the entire cache assist memory either as a victim cache or a stream buffer, changing the use for each loop (the *dyna_loop* approach). The decision to use one configuration or the other is taken based on which achieves a greater miss removal rate for that loop. The miss reduction information comes from profiling.
   In the *dyna_loop* case the cache assist can be used either as a stream buffer or as a victim cache for any loop. We conjecture that splitting the cache assist and using a part of it as a stream buffer, and another as a victim cache would further improve the performance. The following three strategies use this kind of partitioning.
2. Partition the cache assist memory between the victim cache and the stream buffer in the same ratio as the miss reduction rate of the victim cache and the stream buffer for the whole program (the *part_buf* approach). The partition is fixed for the duration of the program.
3. The *dyna_buf* approach partitions the cache assist memory between the victim cache or stream buffer per inner loop, proportionally to the miss removal rate ratio of victim cache and stream buffer for that loop.

4. The *half_buf* approach uses one half of the cache assist memory as a victim cache, and the other half as a stream buffer for the whole program.

dyna_loop and *dyna_buf* are dynamically adapting the cache assist configuration whereas *part_buf* and *half_buf* are not adaptive approaches, they are studied for comparison.

Figure 7 shows the miss reduction rates in the *dyna_loop* case. Profiling information gathered in the experiments summarized in Fig. 5 is used to choose the cache assist as a stream buffer or as a victim cache for each loop. The performance improvement compared to the best of either a stream buffer or a victim cache for the entire program ranges from 25% to 49%. Thus adaptivity improves performance when performed at loop level. However the miss rate got marginally worse for *fpppp* (decreased from 63.54% to 62.27%). Almost all memory accesses (98%) are executed inside one loop, and for this loop the cache assist is configured in the optimal way, the loss of performance comes from the other loops in the program.

For the programs in which the stream buffer has a very small improvement as compared to a victim cache (*go*, *perl*) the additional miss reduction rate is minimal because any possible gain from using a stream buffer is minimal.

The results for *part_buf* appear in Figure 8. With the exception of *fpppp* all the benchmarks show gains when compared to just using victim cache or a stream buffer. *Fpppp*'s loss is determined by the fact that its most dominant loop would need a bigger victim cache than what the part_buf approach allocates. However, the degradation is again minimal, a 2% decrease in miss rate reduction.

Figure 9 shows the results for the *dyna_buf*. It improves the miss ratio by 32% for *turb3d*, 43% for *apsi*, 53% for *wave*, and 51% for *ijpeg*. All the results are better than the case of using just a stream buffer or a victim cache, except for *fpppp* (see an explanation for Fig. 7). The improvement is minor for the benchmarks that show very little improvement from using a stream buffer.

Finally, the *half_buf* approach uses one half of the assist cache memory as victim cache and the other half as stream buffer for the whole program. This is not a dynamic approach, but it is used for comparison with the *dyna_loop* and *dyna_buf* approaches. The results are shown in Fig. 10 as relative percentage improvement over the miss rate reduction for the *half_buf* approach using the formula:

$$\frac{miss\_rate\_reduction(dyna) - miss\_rate\_reduction(half)}{miss\_rate\_reduction(half)} * 100.0 \qquad (2)$$

The *half_buf* configuration marginally outperforms the *dyna_buf* configuration for *apsi* and *turb3d* It is significantly outperformed for *fpppp*, *go*, *ijpeg* and *perl* by up to 28%.

We can correlate this result with the experiments using the cache assist just as stream buffer or victim cache. It shows that the *dyna_buf* configuration outperforms the *half_buf* configuration in the cases where the victim cache performs clearly better than the stream buffer.

The *dyna_loop* configuration noticeably outperformed by *half_buf* in two cases, *apsi* and *ijpeg*, by up to 14%. It outperforms *half_buf* by 15 to 26% in 3 cases: *fpppp*, *go* and *perl*. Thus *dyna_loop* is not always a win.

The Figure 11 compares the miss rate reduction for all the techniques presented. Because in the previous paragraph we have compared the fixed size, non–reconfigurable cache assist *half_buf* with the reconfigurable approaches we are not going to repeat that comparison here. For *fpppp* the performance of using the cache assist as a victim cache is marginally better than any adaptive approach, but this does not happen for any other benchmark. The programs that show high miss reductions rates from using a victim cache, but very low from using a stream buffer (*swim*, *go*, *perl*) get only minimal benefits from any of the proposed adaptive schemes. *Dyna_buf* consistently outperforms *dyna_loop* except for *swim* and *fpppp*. It also outperforms *part_buf* with the exception of *ijpeg* where the difference is negligible. Thus, the most dynamic approach, the *dyna_buf* is the best. Adapting the cache assist configuration is most helpful in cases when both a victim cache and a stream buffer individually show noticeable improvement.

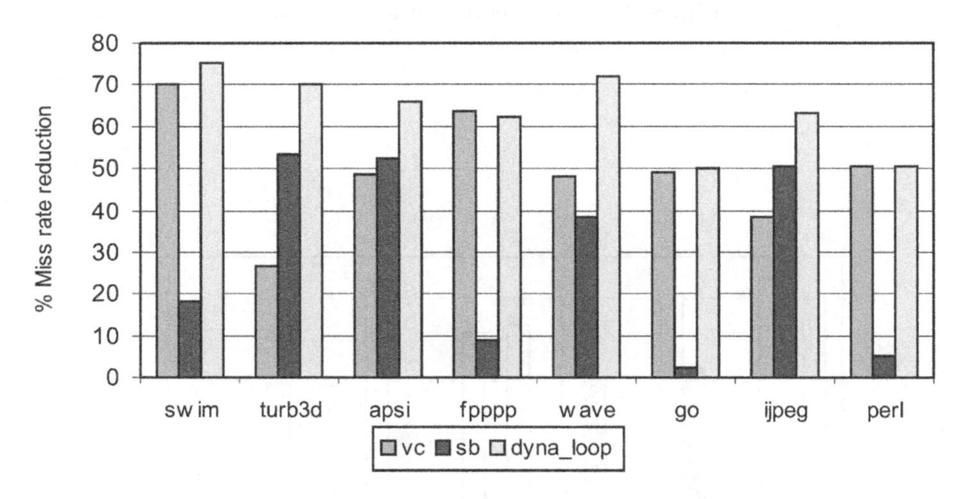

Fig. 7. Miss rate reduction for dyna_loop

5.3 The Effect of Cache Assist Buffer Size

The overall size of cache assist memory is an important parameter, the effectiveness of adaptation may depend on it. The miss reduction rate for a 256B cache assist memory is shown in Figure 12. Compared to a 1KB cache assist memory in Fig. 10 one can see that for the small cache assist the *dyna_buf* approach is a win in all but one case, while only in two cases the performance decreases

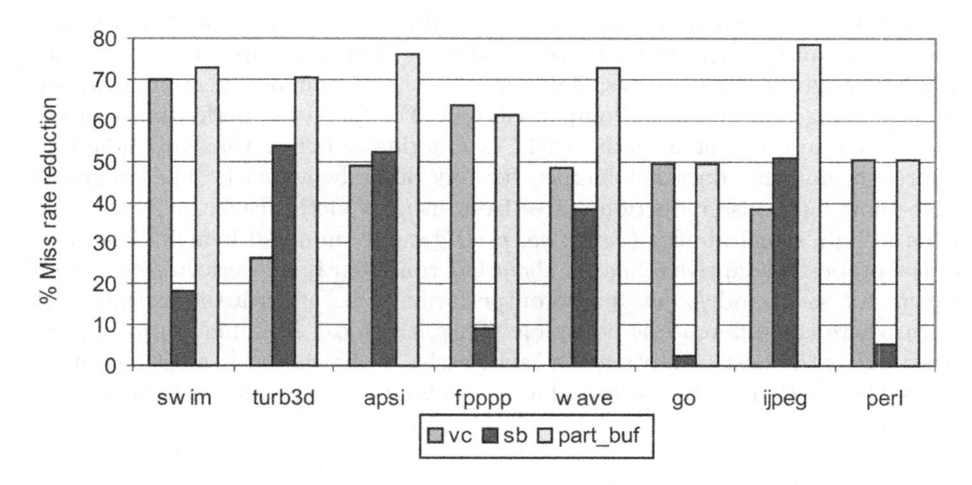

Fig. 8. Miss rate reduction for part_buf

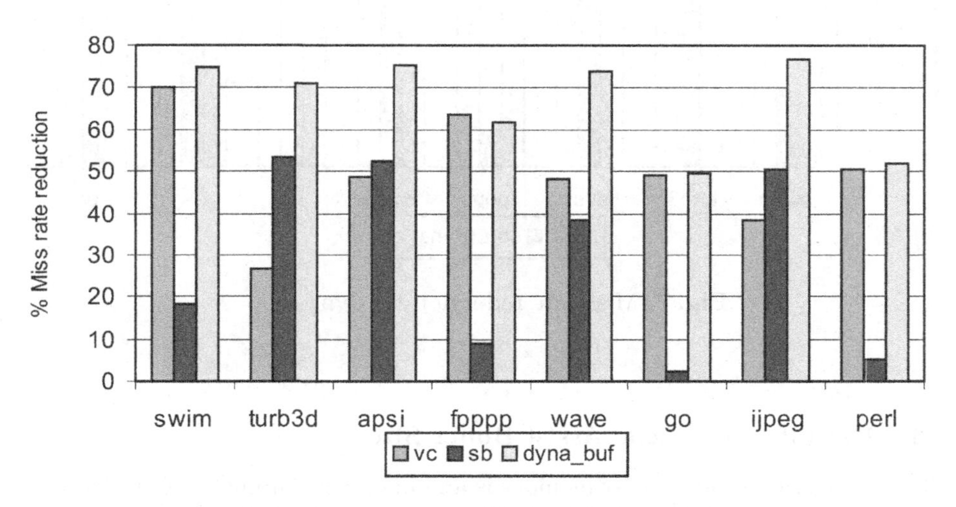

Fig. 9. Miss rate reduction for dyna_buf

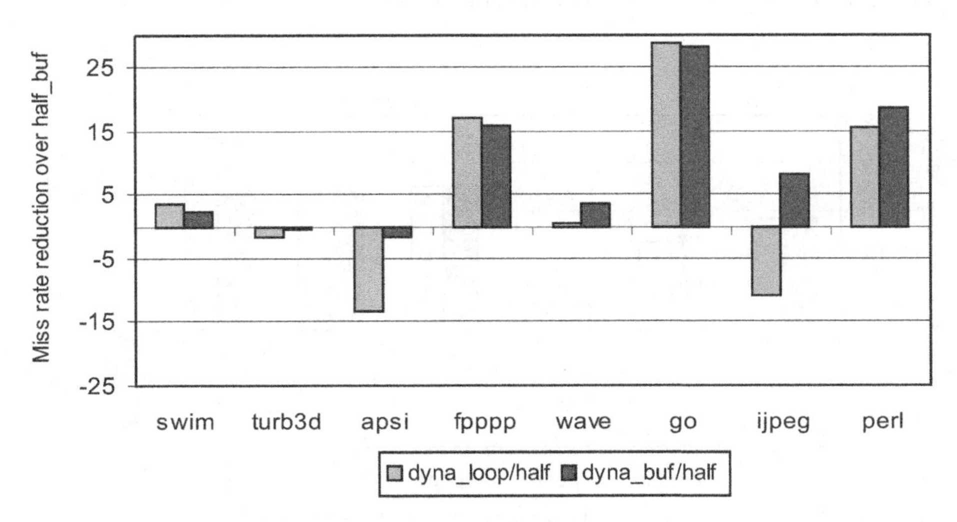

Fig. 10. Dyna_loop and dyna_buf performance relative to half_buf

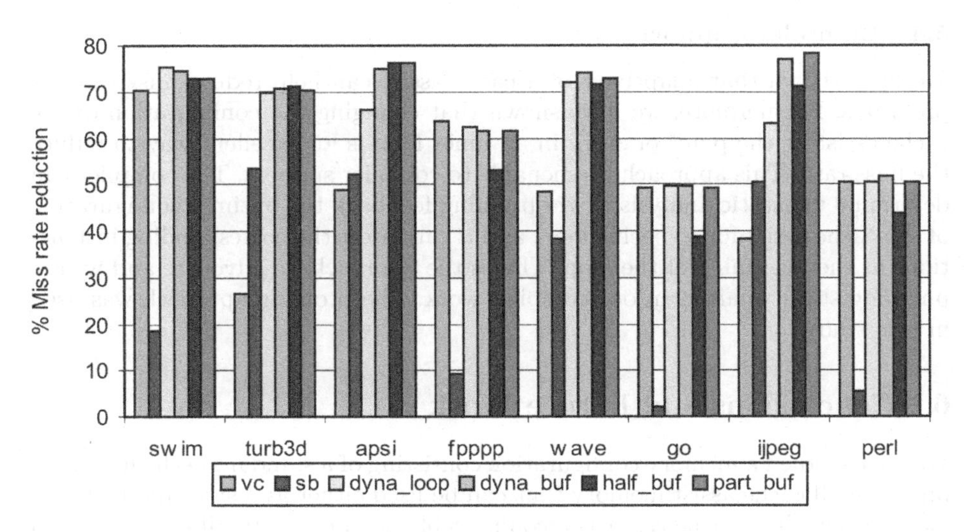

Fig. 11. Miss rate reduction for all the configurations

as compared to the *half_buf* approach. Therefore, when adaptive cache assist memory space is smaller the adaptive cache assist improves performance more than it does when the cache assist memory is larger.

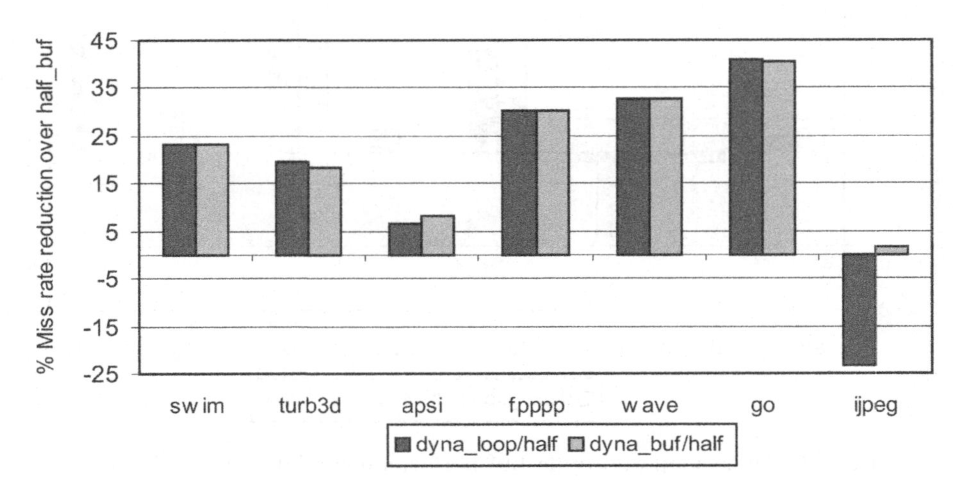

Fig. 12. Dyna_loop and dyna_buf performance relative to half_buf for a 256B cache assist

5.4 Compiler Support

We have shown that adaptivity of a cache assist can help reduce miss rates of programs. Furthermore, we have shown that changing the configuration of the cache assist at the point of entry in an inner loop is an excellent way to reduce the miss rate. This approach is amenable to compiler support. The compiler can determine via static analysis or via profiling feedback the optimal configuration of the cache assist for a specific loop, and it can insert the corresponding instructions at the beginning of the loop. This is the approach we advocate and we are pursuing static analysis in our compiler work. The profiling approach was used in this study.

6 Conclusions and Future Work

We have studied a memory configuration consisting of a standard cache hierarchy plus a small cache assist memory that can be used either as a stream buffer or a victim cache. The cache assist is reconfigurable at run time to allocate a certain fraction of memory to victim cache and/or to stream buffer.

We have shown that using a cache assist reduces the miss rate of the cache and that adapting the configuration of the cache assist reduces it even more.

Several approaches have been studied and we have concluded that an approach that reconfigures the cache assist per inner loop at run time achieves best performance. Using a 1KB adaptive assist memory, up to 50% additional miss rate reduction is achieved by the best of the proposed methods. Simple static assist memory partitioning, on the other hand can suffer up to 15% loss of performance.

References

[1] Andrew A. Chien and Jae H. Kim. Planar-adaptive routing: Low-cost adaptive networks for multiprocessors. In *Proc. 19th Annual Symposium on Computer Architecture*, pages 268–277, 1992. 90

[2] Fredrik Dahlgren, Michel Dubois, and Per Stendstrom. Fixed and adaptive sequential prefething in shared memory multiprocessors. In *Intl. Conference on Parallel Processing*, 1993. 90

[3] W. J. Dally and H. Aoki. Deadlock-free adaptive routing in multicomputer networks using virtual channels. In *IEEE Transactions on Parallel and Distributed Systems*, pages 466–475, 1993. 90

[4] Jeffrey Kuskin et al. The Stanford FLASH multiprocessor. In *Proc. 21st Annual Symposium on Computer Architecture*, pages 302–313, 1994. 90

[5] Edward H. Gornish and Alexander Veidenbaum. An integrated hardware/software data prefething scheme for shared-memory multiprocessors. In *Intl. Conference on Parallel Processing*, pages 247–254, 1994. 90

[6] Teresa L. Johnson and Wen mei Hwu. Run-time adaptive cache hierarchy management via reference analysis. In *Proceedings of the 24th Annual International Symposium on Computer Architecture*, 1997. 90

[7] Norman P. Jouppi. Improving direct-mapped cache performance by the addition of a small fully-a ssociative cache and prefecth buffer. 88, 89, 90

[8] Norman P. Jouppi and Steven J. E. Wilton. Tradeoffs in two-level on-chip caching. In *Proc. 21st Annual Symposium on Computer Architecture*, 1994. 88

[9] Toni Juan, Sanji Sanjeevan, and Juan J. Navarro. Dynamic history-length fitting: A third level of adaptivity for branch prediction. In *Proceedings of the 25th Annual International Symposium on Computer Architecture*, pages 155–166, 1998. 90

[10] Sanjeev Kumar and Christopher Wilkerson. Exploiting spatial locality in data caches using spatial footprints. In *Proceedings of the 25th Annual International Symposium on Computer Architecture*, pages 357–368, 1998. 90

[11] T. Matsumoto, K. Nishimura, T. Kudoh, K. Hiraki, H. Amano, and H. Tanaka. Distributed shared memory architecure for JUMP-1. In *Intl. Symposium on Parallel Architecures, Algorithms, and Networks*, pages 131–137, 1996. 90

[12] Ted Romer, Wayne Ohlich, Anna Karlin, and Brian Bershad. Reducing TLB and memory overhead using on-line superpage promotion. 1996. 90

[13] D. Sunada, D. Glasco, and M. Flynn. ABSS v2.0: SPARC simulator. Technical Report CSL-TR-98-755, Stanford University, 1998. 91

[14] Madhusudhan Talluri and Mark D. Hill. Surpassing the TLB performance of superpages with less operating system support. 1996. 90

[15] O. Temam and N. Drach. Software-assistance for data caches. In *Proceedings IEEE High Performance Computer Architecture*, 1995. 90

[16] Steve Turner and Alexander Veidenbaum. Scalability of the Cedar system. In *Supercomputing*, pages 247–254, 1994. 90

[17] Jack E. Veenstra and Robert J. Fowler. Mint: A front end for efficient simulation of shared-memory multiprocessors. In *Intl. Workshop on Modeling, Analysis and Simulation of Computer and Telecommunication Systems*, pages 201–207, 1994. 91

[18] Alexander V. Veidenbaum, Weiyu Tang, Rajesh Gupta, Alexandru Nicolau, and Xiaomei Ji. Adapting cache line size to application behavior. In *Proceedings ICS'99*, June 1999. 90

[19] Peter Van Vleet, Eric Anderson, Lindsay Brown, Jean-Loup Baer, and Anna Karlin. Pursuing the performance potential of dynamic cache line sizes. In *Proceedings of 1999 International Conference on Computer Design*, November 1999. 90

Skewed Data Partition and Alignment Techniques for Compiling Programs on Distributed Memory Multicomputers

Tzung-Shi Chen[1] and Chih-Yung Chang[2]

[1] Dept. of Information Management, Chang Jung University
Tainan 711, Taiwan, R.O.C.
chents@mail.cju.edu.tw

[2] Dept. of Computer and Information Science, Aletheia University
Tamsui, Taipei, Taiwan, R.O.C.
changcy@email.au.edu.tv

Abstract. Minimizing data communication over processors is the key to compile programs for distributed memory multicomputers. In this paper, we propose new data partition and alignment techniques for partitioning and aligning data arrays with a program in a way of minimizing communication over processors. We use skewed alignment instead of the dimension-ordered alignment techniques to align data arrays. By developing the skewed scheme, we can solve more complex programs with minimized data communication than that of the dimension-ordered scheme. Finally, the experimental results show that our proposed scheme has more opportunities to align data arrays such that data communications over processors can be minimized.

1 Introduction

Over the last decade, a great number of researchers paid their attention on maximizing parallelism and minimizing communication for a given program executed on a parallel machine [1,3,10,11,12,13]. Chen and Sheu [3], Lim *et al.* [10,11], Ramanujam and Sadayappan [12], and Shih, Sheu, and Huang [13] presented approaches to analyze data reference patterns on a program with structures of nested loops so that the parallelized program can be run on a parallel machine in a communication-free manner with some constraints. Furthermore, Lim *et al.* [10,11] tried to maximize parallelism and minimize communication on a scalable parallel machine by using affine transformations when a program cannot be partitioned in a communication-free manner. For a program running on a distributed memory multicomputer, it is not easy to distribute and manage partitioned data and computations over processors when affine transformation methods addressed in [10,11] are used. In general, their methods are suitable for use in shared memory or distributed shared memory multiprocessor systems. This is because affine transformation methods are not easy to handle regular data distribution, such as block or cyclic block data distribution, and to consider the situation of the workload balancing.

M. Valero et al. (Eds.): ISHPC 2000, LNCS 1940, pp. 105–119, 2000.

The distribution of data across processors is of critical importance to the efficiency of the parallel program in a distributed memory machine. For a good data distribution pattern, we should consider the case that it has to allow the workload to be evenly distributed over processors so that we can maximize parallelism and minimize communication. Recently, a number of researchers developed parallelizing compilers that take a sequential program based on automatic partitioning of data arrays and computations and generate the target parallelized program for a parallel machine [6,8,14,15]. The PARADIGM project [6,14] developed a fully automated technique for translating serial programs for efficient execution on distributed memory multicomputers. Lee [8] proposed a dynamic programming technique to efficiently solve the data redistribution problem among program segments based on the approaches proposed in [6,9]. Ayguadé, Garcia, and Kremer [2], and Tandri and Abdelrahman [15] also addressed data redistribution techniques for parallel programs with explicitly specifying **do/doall** loop constructs.

There are numerous researchers concentrated on the alignment and distribution problem [2,6,8,9,15]. Li and Chen [9] studied the problem of aligning data arrays in order to minimize communications. Meanwhile, they also showed the alignment problem is NP-complete in the number of data arrays with a loop nest. Gupta and Banerjee [6] used the constraint-based method to extend the alignment method [9] for the distributed memory multicomputers. Lee [8] solved the data redistribution problem among loop nests using a dynamic programming technique based on Gupta and Banerjee's method [6]. Previous investigations [2,15] addressed some approaches to solve the alignment problem by considering how to distribute and align data arrays as well as how to preserve parallelism on a given program.

As we know, a large number of researches as mentioned above have paid their attention in aligning multiple arrays by using array dimensions to match the other array dimensions of different arrays. In contrast to dimension-ordered data layouts, Kandemir *et al.* [7] addressed a linear algebra framework to automatically determine the optimal data layouts expressed by hyperplanes for each array referenced in a program. That is, determining skewed data layouts is important to a program for analyzing data access behavior and exploiting parallelism. The skewed data layouts are useful for banded-matrix operations such as in BLAS library [5]. In addition, most data arrays are referenced in a skewed manner after applying loop skewing or unimodular transformations [16] to the program for extracting maximum parallelism.

The main aim of this paper is to propose a framework for determining the skewed data partition and distribution as well as skewed data alignment for each data array accessed on a given source program. This makes the partitioned data and program efficient while they are distributed and performed on a multicomputer system. Here the source program is assumed to be a loop nest with explicit **doall** and **do** constructs [16]. First, we show how to identify a perfect skewed data alignment relation between data arrays and the loop nest. Based on these relations, a skewed alignment scheme is proposed to align data arrays with a loop nest program so as to minimize communication. The experimental results show that our skewed alignment scheme is more efficient than the dimension-ordered matching scheme.

The rest of this paper is organized as follows. Section 2 states the machine model and program model used here as well as the strategies of the skewed data partition and data distribution. In Section 3, we explore the alignment relations between a data array and a loop nest. Furthermore, we propose a skewed array alignment scheme to optimize the data alignment problem in Section 4. Experimental results are presented in Section 5. Conclusions are summarized in Section 6.

2 Preliminaries

2.1 Machine Model and Program Model

Here we give the abstract target machine, a q-dimensional grid of $N_1 \times N_2 \times \cdots \times N_q$ $(=N)$ processors, where q is the maximum dimensionality of any array used in the source program, and is less than or equal to the deepest level of the loop nest program appeared in the source program. A processor in the q-D (D stands for dimensional) grid is represented by the tuple $(p_1, p_2, \ldots, p_q)$, where $0 \leq p_i \leq N_i - 1$ for $1 \leq i \leq q$. Such a topology can be easily embedded into most of distributed memory machines, such as hypercubes and tori.

The program model used here is assumed to be an l-deep non-perfect loop nest containing the explicit **doall** and **do** loop constructs. Programs belonging to this loop model can be obtained via a sequence of loop transformations used in [16]. The parallel program generated from a sequential program corresponds to the SPMD (simple program-multiple-data) model, in which each processor performs the same program code but operates on distinct data items [6,8,16]. In addition, the owner-computes rule is used here. Processor that owns left-hand side variable of a statement computes the whole statement [16].

2.2 Skewed Data Partition and Distribution

In this subsection, we introduce how to express the skewed data partition and data distribution over processors. We first state the difference of data representation between traditional data space and skewed data space for a data array. Consider a 2-dimensional array $A[i, j]$ with $0 \leq i \leq 3$ and $0 \leq j \leq 3$. From the point view of traditional data space representation, the data space of array A has two axes i and j. In other words, we can say that the data space is spanned with two base vectors $(1, 0)$ and $(0, 1)$ corresponding to axes i and j, respectively. Therefore, an element $A[i, j]$ of array A can be represented by the matrix representation

$$\bar{I} = D\bar{I}, \text{where } \bar{I} = \begin{bmatrix} i \\ j \end{bmatrix} \text{ and } D = \begin{bmatrix} 1\,0 \\ 0\,1 \end{bmatrix}$$

which uses the two base vectors as the basis of the data space. Now we intend to derive a new data space representation, named skewed data space representation, used in designing our proposed skewed alignment scheme. To illustrate the idea of skewed data alignment technique, we use two new vectors $\bar{s}_1 = (1,0)$ and $\bar{s}_2 = (1,1)$, which are linearly independent, as the basis of a new data space. The transformation from the original data space to the new data space can be derived as follows. We start evaluating all components of the vectors $\bar{s}_1$ and $\bar{s}_2$. Let g_1 and g_2 be their greatest common divisors (gcd) of all components in $\bar{s}_1$ and $\bar{s}_2$, respectively. Then, we divide each component by g_1 in $\bar{s}_1$ to obtain the new vector $\bar{s}'_1$ as well as divide each component by g_2 in $\bar{s}_2$ to obtain the new vector $\bar{s}'_2$. In this example, we have, $\bar{s}_1 = \bar{s}'_1$ and $\bar{s}_2 = \bar{s}'_2$. Now we put the two new vectors $(1,0)$ and $(1,1)$ as columns to form a transformation matrix $D' = \begin{bmatrix} 1 & 1 \\ 0 & 1 \end{bmatrix}$ so that we have two corresponding new axes i' and j'. Detailed derivations are shown in below.

$$\begin{aligned} &D\bar{I} = D'\bar{I}', \text{ where } \bar{I}' = \begin{bmatrix} i' \\ j' \end{bmatrix} \\ &\Longrightarrow \begin{bmatrix} 1 & 0 \\ 0 & 1 \end{bmatrix} \begin{bmatrix} i \\ j \end{bmatrix} = \begin{bmatrix} 1 & 1 \\ 0 & 1 \end{bmatrix} \begin{bmatrix} i' \\ j' \end{bmatrix} \Longrightarrow D'^{-1}D\bar{I} = D'^{-1}D'\bar{I}' \\ &\Longrightarrow D'^{-1}D\bar{I} = I_{2\times 2}\bar{I}' \Longrightarrow \bar{I}' = D'^{-1}D\bar{I} \end{aligned}$$

where

$$D'^{-1} = \begin{bmatrix} 1 & -1 \\ 0 & 1 \end{bmatrix} \text{ (an inverse matrix of } D'\text{) and } I_{2\times 2} = \begin{bmatrix} 1 & 0 \\ 0 & 1 \end{bmatrix} \text{ (an identity matrix).}$$

That is, we have the following index transformation

$$\begin{bmatrix} i' \\ j' \end{bmatrix} = \begin{bmatrix} i - j \\ j \end{bmatrix},$$

where $-3 \le i' \le 3$ and $\max(0, -i') \le j' \le \min(3, 3 - i')$. Hence, i' is referred to as the axis of the vector $\bar{s}_1 = (1,0)$ and j' is referred to as the axis of the vector $\bar{s}_2 = (1,1)$. For example, $\begin{bmatrix} i' \\ j' \end{bmatrix} = \begin{bmatrix} 0 \\ 1 \end{bmatrix}$ while $\begin{bmatrix} i \\ j \end{bmatrix} = \begin{bmatrix} 1 \\ 1 \end{bmatrix}$. For general cases of the above transformation derivations, we can refer to reference [3] for details.

With these transformed axes of data arrays, we will discuss how to perform skewed data partition and data distribution among processors on a q-D grid below. For traditional data distribution, the k-th dimension of an n-dimensional data array A is denoted as A_k, $1 \le k \le n$. Here we use vectors to denote the dimensions of a data array. Let the original data space be spanned by the base vectors $\bar{e}_i$ for $1 \le i \le n$, where $\bar{e}_i$ is a $1 \times n$ vector whose components are set to 0 except for the i-th position set to 1. Each element of array A can be represented by such n base vectors. Thus, each array dimension $\bar{e}_k$ will be mapped to a unique dimension $map(\bar{e}_k)$ of the processor grid, where $1 \le map(\bar{e}_k) \le q$. For skewed data partition and distribution, a data array A in general can be represented by n row vectors $\bar{m}_k$, $1 \le k \le n$, where these n vectors are linearly independent. Thus, each array dimension $\bar{m}_k$ will be mapped into a unique dimension $map(\bar{m}_k)$ of the processor grid, where $1 \le map(\bar{m}_k) \le q$. Here we suppose that each array

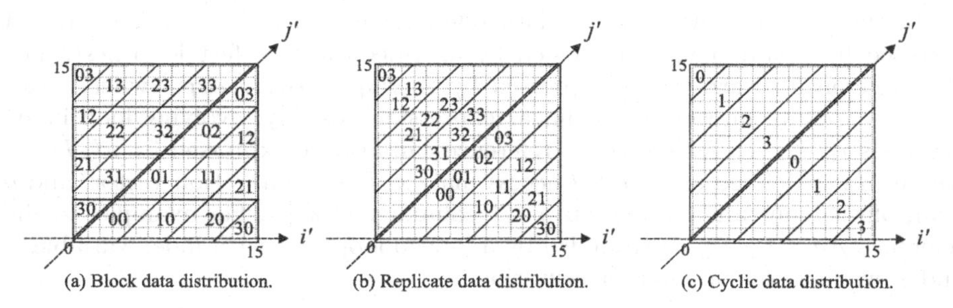

Fig. 1. Some data distribution schema for a 16 × 16 data array A

dimension $\bar{m}_k$ has the corresponding transformed index axis i'_k. The skewed data distribution for dimension $\bar{m}_k$ of a data array A is of the form

$$f_A^{\bar{m}_k}(i'_k) = \begin{cases} \lfloor \frac{d \times i'_k - offset}{block} \rfloor [\bmod N_{map(\bar{m}_k)}] & \text{if } A \text{ is distributed} \\ X & \text{if } A \text{ is replicated} \\ \text{constant or } * & \text{if } A \text{ is not distributed,} \end{cases}$$

where $d \in \{-1, 1\}$ and the part of square parentheses surrounding is optional in this expression. Symbol d stands for index increasing or decreasing along the direction $\bar{m}_k$, which depends on whether d is 1 or -1, respectively. Symbol $offset$ indicates displacement for the mapping of the dimension $\bar{m}_k$. Symbol $block$ indicates the block size for distribution in this dimension. Function $f_A^{\bar{m}_k}(i'_k)$ returns the processor index along the dimension $map(\bar{m}_k)$ of the processor grid. This distribution function extended with skewed data distribution is generalized based on the proposed method [8]. Here we show different data distributions possible for a 16 × 16 data array $A[0..15, 0..15]$ on a 4 × 4 processor grid as in Fig. 1 (a) and Fig. 1 (b) and on a four-processor multicomputer as in Fig. 1 (c). The block size in these examples is 4. Their corresponding data distribution functions are shown below.

(a) $f_A^{(1,0)}(i') = \lfloor \frac{i'}{4} \rfloor \bmod 4$
$\quad f_A^{(1,1)}(j') = \lfloor \frac{j'}{4} \rfloor \bmod 4$

(b) $f_A^{(1,0)}(i') = \lfloor \frac{i'}{4} \rfloor \bmod 4$
$\quad f_A^{(1,1)}(j') = X$

(c) $f_A^{(1,0)}(i') = \lfloor \frac{i'}{4} \rfloor \bmod 4$
$\quad f_A^{(1,1)}(j') = *$

3 Data Alignment Relations

An iteration space of an l-deep loop nest as an l-dimensional polyhedron where each point (iteration) is denoted by an $l \times 1$ column vector $\bar{I} = (i_1, i_2, \ldots, i_l)^t$, where t is denoted as the operation of matrix transpose and each i_k denotes a loop index, $1 \le k \le l$. Here i_1 is the outmost loop while i_l is the innermost loop from outer to inner. Herein, $\bar{\mathbf{0}}^l$ represents an $l \times 1$ column vector where all of each component (element) are zero.

In the following, we define a data reference for an n-dimensional array A accessed by a statement surrounded by an l-deep non-perfect loop nest. The data reference to array A is expressed by $A[exp_1, exp_2, \ldots, exp_n]$ where exp_j, $1 \leq j \leq n$ is an integer-valued linear expression possibly involving loop index variables i_1, i_2, ..., i_l. This data reference can also be expressed by $M_A \bar{I} + \bar{o}$, where M_A is defined as an $n \times l$ access matrix, $\bar{I}$ is the iteration vector, and $\bar{o}$ is an offset (constant vector), an $n \times 1$ column vector [3]. For example, to the reference $A[i-1, i+j]$ surrounded by a 2-deep loop nest with index variables i and j, we have the access representation

$$M_A \begin{bmatrix} i \\ j \end{bmatrix} + \bar{o}, \text{where } M_A = \begin{bmatrix} 1 & 0 \\ 1 & 1 \end{bmatrix} \text{ and } \bar{o} = \begin{bmatrix} -1 \\ 0 \end{bmatrix}.$$

Next, we will explore alignment relations between a certain data array surrounded by an l-deep loop nest and the k-th outermost loop, $1 \leq k \leq l$.

Definition 1: [Perfect Data Alignment]
Suppose that all the iterations on the k-th outermost loop in an l-deep loop nest access the elements of array A along a certain direction (dimension) $\bar{m}$. We call the dimension $\bar{m}$ of array A as *perfect data alignment* which is aligned with the k-th outermost loop.

□

By Definition 1, it turns out that if the property of perfect data alignment is held, no communication is incurred while distributing all iterations along the k-th outermost loop as well as distributing the elements of array A accessed by these iterations along dimension $\bar{m}$ over processors. This is because that the reference data and its corresponding computations (iterations) are distributed to the same processor. Hence, we have the following theorem based on the concept of skewed data layouts presented in [7].

Theorem 1:
Suppose we have an n-dimensional array reference R surrounded by an l-deep loop nest and there exists an $n \times l$ access matrix and an offset vector $\bar{o}$ to the array reference R. If there exists the k-th column vector $\bar{m}_k^t \neq \bar{\mathbf{0}}^l$ in M_R, $1 \leq k \leq l$, aligning the array direction $\bar{m}_k$ with the k-th outermost loop is referred to as a *perfect data alignment*.
Proof:
The detailed proof can refer to the reference in [4].

□

For example, consider the following 2-deep loop nest $L1$.

do $i = 0$ to $m-1$ /* **doall** loop */
  do $j = 0$ to $m-1$
    $B[j+1, i] = B[j, i] * A[i+j, j]$ $(L1)$
  enddo
enddo

We have the access matrix to the data reference $A[i+j,j]$

$$M_A = \begin{bmatrix} 1 & 1 \\ 0 & 1 \end{bmatrix} = \left[\bar{m}_1^t \bar{m}_2^t\right].$$

According to Theorem 1, array A along $\bar{m}_1 = (1,0)$ is perfectly aligned with the loop i. For instance, elements of array A accessed by the two consecutive iterations $\bar{I}_1 = (1,1)$ and $\bar{I}_1' = (2,1)$ to loop i have the relation of the data reference direction $\bar{m}_1^t = (1,0)^t$; that is,

$$M_A \begin{bmatrix} 2 \\ 1 \end{bmatrix} - M_A \begin{bmatrix} 1 \\ 1 \end{bmatrix} = \begin{bmatrix} 1 \\ 0 \end{bmatrix}.$$

Apparently, we do gain a good reference pattern, by distributing array A along the direction $\bar{m}_1 = (1,0)$, while distributing the iterations of loop i over processors. As a result, we are able to make a profit on reducing communication while array A is distributed along $(1,0)$ and thus is aligned with loop i. At the same time, we do also minimize data communication by distributing array A along the direction $\bar{m}_2 = (1,1)$ while the iterations of loop j is distributed over processors. In contrast to the perfect data alignment, we have the non-perfect data alignment either aligning the array A along direction $(0,1)$ with loop i or aligning the array A along direction $(0,1)$ with loop j. Applying the non-perfect data alignment to distributing data and iterations generally incurs a lot of data communication among processors.

4 Skewed Array Alignment

We define a directed graph, called *computation-communication alignment graph*, CCAG, to extract and express the characteristics of a program and data arrays accessed by the program.

Definition 2: **[Computation-Communication Alignment Graph]**

A *computation-communication alignment graph*, CCAG, is a directed graph $G = \{V, E\}$ to a program with an l-deep loop nest, where V is a set of vertices and E is a set of edges, defined below. V is composed of three different types of vertices:

(1) array dimension vertex $A_{\bar{m}_k}$ with respect to an array dimension $\bar{m}_k$ whose transpose column vector $\bar{m}_k^t \neq \bar{\mathbf{0}}^l$ is the k-th column vector of the $n \times l$ access matrix M_R for reference R of an n-dimensional array A, on a statement inside an l-deep loop nest, $1 \leq k \leq l$,

(2) loop vertex with respect to loop nest with **do** (sequential loop) construct, and

(3) loop vertex with respect to loop nest with **doall** (parallel loop) construct.

E is composed of three different types of edges:

(1) a read edge, $(A_{\bar{m}_k}$, loop $k)$, links array dimension $\bar{m}_k$ of array A to the incurred k-th outermost loop if this reference R is on the right-hand side of the statement,

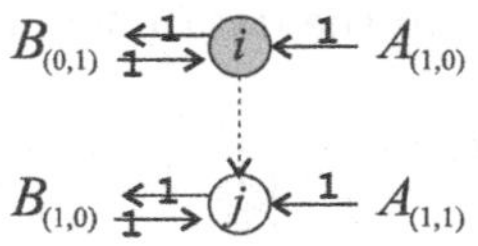

$B_{(0,1)}$ $\underset{1}{\overset{1}{\rightleftarrows}}$ (i) $\overset{1}{\leftarrow}$ $A_{(1,0)}$

$B_{(1,0)}$ $\underset{1}{\overset{1}{\rightleftarrows}}$ (j) $\overset{1}{\leftarrow}$ $A_{(1,1)}$

Fig. 2. CCAG for loop nest $L1$

(2) a write edge, (loop k, $A_{\bar{m}_k}$), links an incurred k-th outermost loop to array dimension $\bar{m}_k$ of array A if this reference R is on the left-hand side of the statement, and
(3) a **do** edge links loop nesting between two consecutive loops from outer to inner if there exists such a loop structure in a program.
There are weights on read and write edges which, repectively, indicate the number of edges ($A_{\bar{m}_k}$, loop k) and (loop k, $A_{\bar{m}_k}$).

□

Note that if there is no loop nest or more than one consecutive loop nests in a program, we assume that there exists an outermost loop surrounding the original program. According to Definition 2, if there exists such a loop structure, the root of this loop structure with loop level 1 is the outermost loop vertex. The constructed loop structure can be labeled from root down to leaves one by one numbered with loop levels 1 to l.

As in loop nest $L1$, we have a CCAG with 4 array dimension vertices, $A_{(1,0)}$, $A_{(1,1)}$ $B_{(0,1)}$, and $B_{(1,0)}$, and 2 loop vertices, where loop vertex i, indicated by the gray point, is a **doall** loop and loop vertex j, indicated by the empty point, is a **do** loop as shown in Fig. 2. An edge of solid arrow line with weight one links perfect data alignment dimension to its incurred loop, and vice versa. A **do** edge with dashed arrow line is represented by the loop structure linked from outer loop i to inner loop j.

Apparently, we have a lot of information to be extracted under the construction of a CCAG. These include what loop (iterations) can be parallelized, what array dimension needs to be aligned with some loop, and what array dimension needs to be aligned or matched with the other array dimensions. Each partitioned array dimensions can be identified while all of each dimension of the array are determined as well.

We know that the data alignment problem is an NP-complete problem [2] [9]. However, in this paper, we use a new kind of graphs to capture the characteristics of a program and data arrays as well as intent to optimize the data communication overhead. Thus, we will address a heuristic alignment scheme in the following. To the best of our knowledge, this is the first discussion to explore the skewed data alignment problem.

In the following subsections, we first address how to pick up the alignment directions from a given data array to a loop nest. Next, we shall present how to align multiple data arrays with the loop nest for minimizing data communication overhead.

4.1 Alignment for a Data Array

A heuristic algorithm to efficiently align a data array with a loop nest to minimize communication over processors based on a given CCAG is described below. Now, we discuss how to select data alignment dimensions from a given data array. We have two main phases to align multiple data references to an array: the doall-loops phase followed by the do-loops phase. Based on a constructed CCAG, we first examine the loop structure with existing **doall** loops from loop level 1 to the deepest loop level l in sequence in the doall-loops phase. That is, we give high priority to distributing the most external parallel loop from outer to inner while distributing computations over processors. This will lead to the largest granularity of this program that is distributed to processors. That is, we expect to explore the coarse-grained parallelism to a program. This consideration will be benefited to the distributed memory machines in general. In the former phase, we are concerned with minimizing communication while distributing computations along **doall** loops and distributing data arrays along the selected skewed dimensions. This is because we require to distribute the computations along **doall** loops such that these computations can be executed in a parallel manner. After that, we use the same operation as the doall-loops phase does to proceed with the processing of the **do** loops in the latter phase. In this phase, we examine the loop structure with the existing **do** loops from loop level 1 to the deepest loop level l as well. In the second phase, we are concerned with minimizing the residual communication while partitioning and distributing data arrays is performed along the selected skewed dimensions in this phase.

For each phase, we have two sub-phases; the former is for aligning write data references with the loop nest, whereas the latter is for aligning read data references with the loop nest. That is, we give higher priority to align write references with the loop nest than the read ones. It turns out that the computation distribution for this loop can meet the owner-computes rule for the generated parallel code.

The algorithm of aligning an n-dimensional data array A with l-deep loop nest is described below. Initially, let a set of alignment vectors for array A, S_A, be empty, ϕ, i.e., $S_A = \phi$.

Algorithm Skewed-Alignment(A, S_A)
Begin
Doall-loops phase:
 Alignment(Type of **doall** loop, Type of write edge, A, S_A);
 Alignment(Type of **doall** loop, Type of read edge, A, S_A);
Do-loops phase:
 Alignment(Type of **do** loop, Type of write edge, A, S_A);
 Alignment(Type of **do** loop, Type of read edge, A, S_A);
End

Procedure Alignment(*loop-type*, *edge-type*, A, S_A)
Step 1: Set p to 1.

Step 2: Examine loop level p, $1 \le p \le l$, on loop structure in CCAG of the l-deep loop nest.

Step 3: Assume there are r reference dimensions to array A connected to loop vertices of *loop-type* in loop level p. For each reference dimension $\bar{m}_k$, $1 \le k \le r$, we sum up the weights, which are on the edges connecting $\bar{m}_k$ with all of loops of *loop-type* in loop level p. We sort them according to weights and obtain the results, each aligned array dimension $\bar{m}_k$, with the weight w_k, $1 \le k \le r$, in a decreasing sequence; that is, $w_1 \ge w_2 \ge \cdots \ge w_r$. Suppose when $w_a = w_{a+1} = \cdots = w_b$, we have $w'_a \ge w'_{a+1} \ge \cdots \ge w'_b$ where w'_k is the total weight of summing up the weights, which are on the edges connecting $\bar{m}_k$ with all the loops of *loop-type* in loop level $p+1$ to loop level l, $1 \le a \le k \le b \le r$.

Step 4: We add the alignment direction one by one from $\bar{m}_1$ to $\bar{m}_r$ into the set of S_A if the vector is not the linear combination of the vectors in the predecessor set of S_A.

Step 5: We examine the set of S_A. If the cardinality of the set is not equal to n, dimensionality of array A, and $p < l$, add one to p and goto Step 2; otherwise, goto Step 6.

Step 6: Complete the construction of skewed data alignment dimensions associated with the loops of *loop-type* in the loop nest.

We know that it is possible that the construction of aligned dimension set S_A for array A is incomplete when the above algorithm is applied. That is, the cardinality of the constructed S_A is less than n, the dimensionality of array A. If the cardinality of S_A is still less that n, we add suitable base vectors to the alignment set S_A to form n vectors, linearly independent, as a complete set of dimensions for skewed data representation. This is because we have to use n linearly independent vectors to represent the data space of array A. Using these heuristic criteria of selection, the larger the sum of weights to an aligned dimension of an array, the more the number of references accessed to the aligned dimension of the array is. Therefore, we obtain the best set of aligned dimensions to an array in order to minimize communication under such a selection scheme.

4.2 Aligning Multiple Data Arrays for a Program

Assume we have K data arrays, A_1, A_2, ..., A_K, accessed by the source program. For each data array A_i, we can obtain the corresponding set S_{A_i} of selected skewed alignment vectors, $1 \le i \le K$, by applying our proposed skewed-alignment algorithm. First, we transform and reduce the CCAG into a reduced CCAG, an undirected graph, called RCCAG. Each vertex in the RCCAG is a vertex, which was obtained from the set of S_{A_i} for each array A_i. The set of edges in RCCAG is composed of edges connecting loop vertices to the array dimension vertices in S_{A_i} for each array A_i. Next, we transform the RCCAG into a CAG (component affinity graph) in [9]. Finally, we use the alignment algorithm [9], maximum-weight bipartite graph matching, to solve our perfect alignment problem to the transformed CAG.

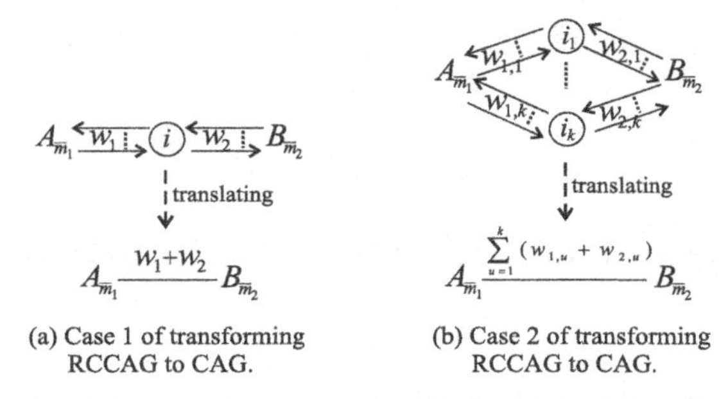

(a) Case 1 of transforming RCCAG to CAG.
(b) Case 2 of transforming RCCAG to CAG.

Fig. 3. Two cases of transforming RCCAG to CAG

In what follows, we present how to translate the RCCAG into CAG graph. The vertices in RCCAG are as the vertices in the transformed CAG. The undirected edges in the transformed CAG are generated by one of the following two cases.

Case 1: Suppose there is a weight w_1 which is the summation of weights on all edges connecting array dimension vertex $A_{\bar{m}_1}$ with a loop vertex i, whatever **do** or **doall** loops for array A. Also, suppose there is a weight w_2 which is the summation of weights on all edges connecting array dimension vertex $B_{\bar{m}_2}$ with the loop vertex i for the other array B. There exists an undirected edge, connecting $A_{\bar{m}_1}$ with $B_{\bar{m}_2}$, with having weight $w_1 + w_2$ on the edge as shown in Fig. 3 (a).

Case 2: Suppose there are weights $w_{1,1}, \ldots w_{1,k}$ which are the summation of weights on all edges connecting array dimension vertex $A_{\bar{m}_1}$ with the corresponding loop vertices $i_1, \ldots, i_k$, whatever **do** or **doall** loops for array A. Also, suppose there are weights $w_{2,1}, \ldots w_{2,k}$ which are the summation of weights on all edges connecting array dimension vertex $B_{\bar{m}_2}$ with the corresponding loop vertices $i_1, \ldots, i_k$, whatever **do** or **doall** loops for the other array B. There exists an undirected edge, connecting $A_{\bar{m}_1}$ with $B_{\bar{m}_2}$, with having weight $\sum_{u=1}^{k}(w_{1,u} + w_{2,u})$ on the edge as shown in Fig. 3 (b).

The weight on each edge in the transformed CAG is represented as the unaligned penalty. That is, the larger the weight of an edge is, the more both the array dimensions need to be aligned. Therefore, we can use the approach, maximum-weight bipartite graph matching [9], to solve our alignment problem. For a given program, we can perform this scheme to transform the data arrays with the new axes and to explore the skewed alignment relations among these data arrays.

Now we demonstrate how to align arrays with the loop nest $L1$. The relations, CCAG for $L1$, among program segment and arrays A and B are shown in Fig. 2. Via selecting the perfect data alignments for each array, we have the sets $S_A = \{\bar{m}_1^A = (1,0), \bar{m}_2^A = (1,1)\}$ and $S_B = \{\bar{m}_1^B = (0,1), \bar{m}_2^B = (1,0)\}$. Thus, the RCCAG is constructed from CCAG of loop nest $L1$ while all of directed

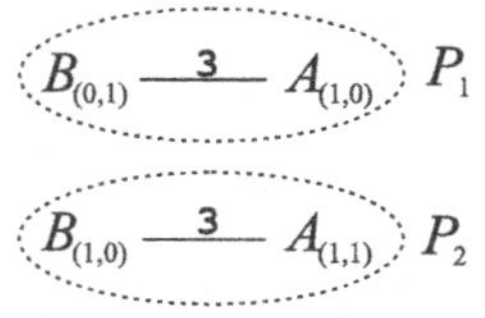

$B_{(0,1)}$ —3— $A_{(1,0)}$ P_1

$B_{(1,0)}$ —3— $A_{(1,1)}$ P_2

Fig. 4. The transformed CAG for RCCGA and alignment between arrays A and B

edges were transformed to undirected edges. Therefore, we have the transformed CAG and two matched dimension partitions P_1 and P_2 as shown in Fig. 4. Apparently, $A_{(1,0)}$ is aligned with $B_{(0,1)}$ and $A_{(1,1)}$ is aligned with $B_{(1,0)}$ when the proposed approach [9] is applied to this transformed CGA.

5 Experimental Results

Based on the skewed data alignment technique, we are going to discuss how to determine the data distribution and block size for each data array with the loop nest $L1$. This example is written by hand to the parallelized versions. Here assume N processors with 1-D grid machine used in our experiments. We first demonstrate how to determine data partition and block size with the loop nest $L1$ here. By applying our proposed scheme, $A_{(1,0)}$ is aligned with $B_{(0,1)}$ while $A_{(1,1)}$ is aligned with $B_{(1,0)}$ for $L1$. In $L1$, loop i is a **doall** loop, which can be executed in parallel. Clearly, we distribute loop i to N processors in block while loop j is sequentially executed without the need of distribution. As derived from Section 2, we have the following index transformations.

$$\begin{bmatrix} i'_A \\ j'_A \end{bmatrix} = \begin{bmatrix} i-j \\ j \end{bmatrix} \text{ for array } A[i,j],\ 0 \le i \le 2m-1 \text{ and } 0 \le j \le m-1, \text{ and}$$

$$\begin{bmatrix} i'_B \\ j'_B \end{bmatrix} = \begin{bmatrix} j \\ i \end{bmatrix} \text{ for array } B[i,j],\ 0 \le i \le m \text{ and } 0 \le j \le m-1.$$

Thus, we have the following skewed data distribution for array A as shown in Fig. 5 (a) and for array B as shown in Fig. 5 (b) with block size of $\frac{m}{N}$, provided the assumption that m is divisible by N.

$$\text{(a)}\ f_A^{(1,0)}(i'_A) = \lfloor \tfrac{i'_A}{m/N} \rfloor \bmod N \qquad \text{(b)}\ f_B^{(0,1)}(i'_B) = \lfloor \tfrac{i'_B}{m/N} \rfloor \bmod N$$
$$f_A^{(1,1)}(j'_A) = * \qquad\qquad f_B^{(1,0)}(j'_B) = *$$

With such data and computation distributions over processors, the loop nest $L1$ can be executed in parallel in a communication-free manner.

For the loop nest $L1$, the experimental results are shown in Table 1. In this experimental study, we implement two versions of parallel codes on a 32-node nCUBE-2 multicomputer: the first, called SA, designed by our proposed skewed alignment scheme and the second, called CAG, designed by the CAG's alignment

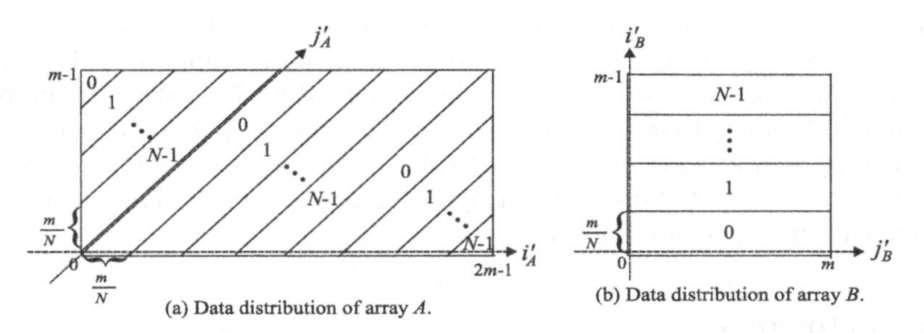

Fig. 5. Optimum skewed data partition and distribution for loop nest $L1$ on 1-D grid

scheme where $A_{(1,0)}$ is aligned with $B_{(0,1)}$ while $A_{(0,1)}$ is aligned with $B_{(1,0)}$. In Table 1, the problem size is represented by m and the number of processors is represented by N. Here, the block size of data distribution for arrays A and B is $\frac{m}{N}$. Obviously, the running performance using our proposed skewed alignment scheme is superior to the traditional dimension-ordered alignment scheme. The major reason is that the parallel version implemented by the skewed alignment scheme is running on processors in a communication-free manner for this sample example.

Table 1. Execution time in seconds for loop nest $L1$ running in parallel on a 32-node nCUBE-2 multicomputer

m	$N=2$		$N=4$		$N=8$		$N=16$		$N=32$	
	SA	CAG	SA	CAG	SA	CAG	SA	CAG	SA	CAG
2^8	0.43	1.02	0.23	1.13	0.13	1.15	0.06	1.88	0.03	2.43
2^9	0.81	1.82	0.45	1.03	0.27	1.36	0.14	1.57	0.05	1.89
2^{10}	1.65	3.85	0.87	2.91	0.47	2.46	0.28	2.97	0.16	3.08
2^{11}	3.23	8.76	1.65	5.54	0.88	5.02	0.46	4.89	0.27	4.76
2^{12}	6.43	15.87	3.23	12.11	1.65	11.89	0.87	11.03	0.46	10.55

We can see that data arrays A and B are perfectly aligned with each other. Assume we align arrays A and B using the traditional dimension-ordered alignment. That is, either $A_{(1,0)}$ is aligned with $B_{(1,0)}$ while $A_{(0,1)}$ is aligned with $B_{(0,1)}$, or $A_{(1,0)}$ is aligned with $B_{(0,1)}$ while $A_{(0,1)}$ is aligned with $B_{(1,0)}$. Both of them might incur a great amount of communication while these data arrays are distributed among processors whatever block or cyclic data distribution are used. This is because there does not exist any perfect data alignment between arrays A and B when the dimension-ordered alignment scheme is used.

In most of parallelizing compiling systems, they use some powerful compiling techniques as tools to explore loop parallelism and data locality for improving the performance of executing programs. The techniques used, in general, include loop skewing, loop interchange, loop tiling, unimodular transformation, and more [16]. Our skewed scheme can be combined with most of the powerful compiling techniques together, such as loop skewing, loop interchange, and so on. The proposed scheme can automatically detect and extract program parallelism as well as distribute data arrays over processors so that the incurred communication can be minimized.

6 Conclusions

In this paper, we proposed a heuristic alignment technique for efficiently aligning data arrays so as to minimize communication over multicomputers. We used skewed alignment instead of the traditional techniques with dimension-ordered alignment to align data arrays. With this development of skewed alignment scheme, we can solve complex programs having minimized data communication more efficiently than that of the dimension-ordered scheme. Finally, we show that our proposed scheme is outperformed over the scheme proposed by the previous work as dimension-ordered data distribution used.

Acknowledgements

This work was supported by the National Science Council of Republic of China under Grant NSC-89-2213-E-309-005.

References

1. Anderson, J. M. and Lam, M. S.: Global Optimizations for Parallelism and Locality on Scalable Parallel Machines. Proceedings of the ACM SIGPLAN'93 Conference on Programming Language Design and Implementation. Albuquerque, NM, USA (1993) 112-125 105
2. Ayguadé, E., Garcia, J., and Kremer, U.: Tools and Techniques for Automatic Data Layout: A Case Study. Parallel Computing, **24** (1998) 557-578 106, 112
3. Chen, T.-S. and Sheu, J.-P.: Communication-Free Data Allocation Techniques for Parallelizing Compilers on Multicomputers. IEEE Transactions on Parallel and Distributed Systems, **5(9)** (1994) 924-938 105, 108, 110
4. Chen, T.-S. and Chang, C.-Y.: Skewed Data Partition and Alignment Techniques for Compiling Programs on Distributed Memory Multicomputers. Technical Report. Chang Jung University, Taiwan (2000) 110
5. Dongarra, J. J., Croz, J. D., Hammarling, S., and Duff, I.: A Set of Level 3 Basic Linear Algebra Subprograms. ACM Trans. Mathematical Software, **16(1)** (1990) 1-7 106
6. Gupta, M. and Banerjee, P.: Demonstration of Automatic Data Partitioning Techniques for Parallelizing Compilers on Multicomputers. IEEE Transactions on Parallel and Distributed Systems, **3(2)** (1992) 179-193 106, 107

7. Kandemir, M., Choudhary, A., Shenoy, N., Banerjee, P., and Ramanujam, J.: A Linear Algebra Framework for Automatic Determination of Optimal Data Layouts. IEEE Transactions on Parallel and Distributed Systems, **10(2)** (1999) 115-135 106, 110
8. Lee, P.-Z.: Techniques for Compiling Parallel Programs on Distributed Memory Multicomputers. Parallel Computing, **21(12)** (1995) 1895-1923 106, 107, 109
9. Li, J. and Chen, M.: The Data Alignment Phase in Compiling Programs for Distributed-Memory Machines. Journal of Parallel and Distributed Computing, **13** (1991) 213-221 106, 112, 114, 115, 116
10. Lim, A. W. and Lam, M. S.: Maximizing Parallelism and Minimizing Synchronization with Affine Partitions. Parallel Computing, **24(3-4)** (1998) 445-475 105
11. Lim, A. W., Cheong, G. I., and Lam, M. S.: An Affine Partitioning Algorithm to Maximize Parallelism and Minimize Communication. 13th ACM International Conference on Supercomputing. Rhodes, Greece (1999) 228-237 105
12. Ramanujam, J. and Sadayappan, P.: Compile-Time Techniques for Data Distribution in Distributed Memory Machines. IEEE Transactions on Parallel and Distributed Systems, **2(4)** (1991) 472-482 105
13. Shih, K.-P., Sheu, J.-P., and Huang, C.-H.: Statement-Level Communication-Free Partitioning Techniques for Parallelizing Compilers. The Journal of Supercomputing, **15(3)** (2000) 243-269 105
14. Su, E., Lain, A., Ramaswamy, S., Palermo, D. J., Hodges IV, E. W., and Banerjee, P.: Advanced Compilation Techniques in the PARADIGM Compiler for Distributed-Memory Multicomputers. Proceedings of 1995 ACM International Conference on Supercomputing. Barcelona, Spain (1995) 424-433 106
15. Tandri, S. and Abdelrahman, T. S.: Automatic Partitioning of Data and Computations on Scalable Shared Memory Multiprocessors. Proceedings of 1997 International Conference on Parallel Processing. Bloomingdale, IL, USA (1997) 64-73 106
16. Wolfe, M.: High Performance Compilers for Parallel Computing. Addison-Wesley Inc., CA. (1996) 106, 107, 118

Processor Mechanisms for Software Shared Memory

Nicholas P. Carter[1], William J. Dally[2], Whay S. Lee[3], Stephen W. Keckler[4], and Andrew Chang[2]

[1] Coordinated Science Laboratory, University of Illinois at Urbana-Champaign
{npcarter@crhc.uiuc.edu}
[2] Computer Systems Laboratory, Stanford University
{billd@cva.stanford.edu, achang@cva.stanford.edu}
[3] Sun Microsystems
{Whay.Lee@EBay.Sun.COM}
[4] Department of Computer Sciences, The University of Texas at Austin
{skeckler@cs.utexas.edu}

Abstract. The M-Machine's combined hardware-software shared-memory system provides significantly lower remote memory latencies than software DSM systems while retaining the flexibility of software DSM. This system is based around four hardware mechanisms for shared memory: status bits on individual memory blocks, hardware translation of memory addresses to home processors, fast detection of remote accesses, and dedicated thread slots for shared-memory handlers. These mechanisms have been implemented on the MAP processor, and allow remote memory references to be completed in as little as 336 cycles at low hardware cost.

1 Introduction

Distributed Shared-Memory (DSM) systems use a variety of methods to implement shared-memory communication between processors. Some designers provide substantial hardware support for shared memory, such as hardware protocol engines [1] or dedicated co-processors [10]. Other systems rely completely on software to implement shared memory [13]. Hardware protocol engines can give very low remote memory latencies, but increase the complexity and hardware cost of the system. In addition, they restrict the system to one shared-memory protocol, limiting performance on applications whose communications patterns do not match the assumptions of the shared-memory protocol. Software-based shared-memory is very flexible, and does not increase the cost of the system, but tends to have relatively poor performance due to the overheads imposed by conventional networks and virtual memory systems. Providing co-processors to execute shared-memory handlers gives both flexibility and speed, but the hardware cost of this approach is substantial.

In this paper, we present an alternative approach to implementing DSM systems, based around four hardware mechanisms for shared memory that are integrated into the processor itself, substantially improving shared-memory performance at low hardware cost while retaining the flexibility of software-based approaches. Shared-memory protocols share several common features, which can be exploited in the design of hardware

M. Valero et al. (Eds.): ISHPC 2000, LNCS 1940, pp. 120-133, 2000.

mechanisms to accelerate shared memory. They must be able to detect references to remote memory, determine where the request for the remote memory must be sent, transfer data back to the requesting processor, and complete operations which are waiting for the data. In addition, it is desirable that an architecture support transfers of small blocks of data between processors to reduce false sharing (unnecessary remote memory operations caused by placing two or more unrelated data objects within a block of memory that the shared-memory system treats as an atomic unit), and that the architecture allow user programs to continue executing during remote memory references.

Based on these requirements, we have designed four hardware mechanisms for shared memory, which have been implemented as part of the MIT M-Machine project [5]. Block status bits on each eight-word block of memory allow individual blocks to be transferred between processors. A fast event system detects remote memory references and invokes software handlers in as little as 10 cycles. A Global Translation Lookaside Buffer caches translations between virtual addresses and their home processors. Dedicated thread slots for software handlers eliminate context switch overhead when starting handlers, and allow user programs to execute in parallel with shared-memory handlers.

Using all of our mechanisms allows system software to complete a remote memory reference in 336 cycles on the M-Machine, almost 20x faster than most software-only shared-memory systems, and only 2.5x slower than current-generation hardware shared-memory systems such as the SGI Origin 2000 [11]. On applications, we achieve a 9% performance improvement on a latency-bound FFT computation and a 30% improvement on an occupancy-bound multigrid computation, as compared to the performance of our system without these hardware mechanisms [4].

The remainder of this paper begins with a brief overview of the MIT M-Machine, followed by a description of the mechanisms and their use in implementing shared memory. We continue with an analysis of the performance impact of our mechanisms for shared memory, followed by a discussion of related work and some future research directions.

2 The M-Machine

The M-Machine Multicomputer is an experimental multicomputer that we have designed at M.I.T. and Stanford to explore architectural techniques to take advantage of improvements in silicon fabrication technology. An M-Machine consists of a two-dimensional array of processing nodes, each of which consists of a custom Multi-ALU processor (MAP) and five SDRAM chips. The MAP chip has been fabricated in a 0.5-micron process, and work is ongoing as of July, 2000 on a prototype M-Machine.

As shown in Figure 1, each MAP chip contains three processor clusters [8], two cache memory banks, and a network subsystem. Each of the clusters acts as an independent, multithreaded processor. The instruction issue logic in each cluster implements zero-overhead multithreading between the five active threads in the cluster, selecting an instruction to issue each cycle based on operand and resource availability. Threads running in the same thread slot (hardware registers which hold program state) on each of the clusters are assumed to be part of the same job, and may use the cluster switch to write into each other's register files. Memory addresses are interleaved between the two cache banks, allowing two memory operations to be completed each cycle.

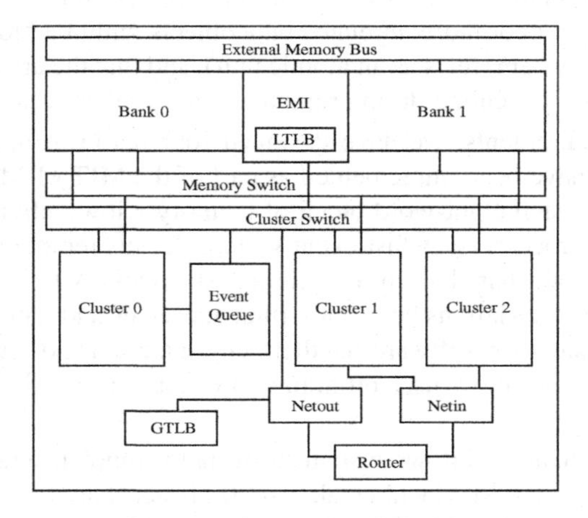

Figure 1: MAP Chip Block Diagram

The MAP chip's architecture allows it to exploit parallelism at multiple granularities. Each cluster acts as a 2- or 3-wide LIW processor[1], allowing fine-grained instruction-level parallelism to be exploited within a cluster. The MAP chip's inter-cluster communication mechanisms [9] provide low-latency communication between threads running on different clusters, making it feasible to exploit medium-grained parallelism within a MAP chip. Finally, the on-chip network hardware [12] provides fast, user-level messaging between processors in an M-Machine, allowing coarser-grained parallelism to be exploited across multiple processors.

Shared memory is implemented on the M-Machine using a combination of hardware and software. Hardware detects remote memory operations and passes them to software to resolve. Remote memory references are enqueued in software while their data is being obtained, and are then resolved by the shared-memory handlers. Completing pending operations in software is made easier by the MAP chip's *configuration space*, a mechanism which maps all of the register state of the chip into an address space that can be accessed using normal load and store operations, relying on the Guarded Pointers [3] protection scheme to prevent unauthorized programs from modifying other program's register states. This allows the system software to write the result of each pending operation directly into its original destination register, making software resolution of remote memory requests transparent to user programs.

---

1. The original design called for each cluster to have 3 ALUs: one integer, one memory, and one floating-point. Chip-space constraints forced the removal of the FP ALUs from two of the clusters during implementation.

3 Mechanisms for Shared Memory

We have designed four hardware mechanisms for software shared memory: block status bits, a fast event system, a global translation lookaside buffer, and dedicated thread slots for shared-memory handlers. Together, these mechanisms implement several important sub-tasks of most shared-memory protocols in hardware, freeing the software handlers to implement higher-level DSM policies. Block status bits allow 8-word blocks of data to be transferred between processors, reducing false sharing. The event system detects remote references and invokes software handlers to resolve them. The Global Translation Lookaside Buffer provides a flexible mapping of addresses to home processors, allowing data to be mapped for maximum locality. Dedicated thread slots for software handlers eliminate context switch overheads when invoking software handlers, reducing the remote reference time.

3.1 Block Status Bits

The block status bits allow small blocks of data to be transferred between processors by associating two bits of state with each 8-word block of data on a node. These bits encode the states invalid, read-only, read-write, and dirty, and the hardware enforces the permissions represented by these states. Block status bits eliminate much of the false sharing which occurs in software-only shared memory systems because conventional virtual memory systems are unable to record presence information on units of data smaller than a page. They are stored in the page table and copied into the local TLB (LTLB) and the cache when a block is referenced, allowing remote data to be stored at all levels in the memory hierarchy.

During a memory reference, the memory system checks the block status bits of the referenced address to determine if the operation is allowed. This check is done in parallel with hit/miss determination in the cache or LTLB and therefore does not impact memory latency. If the block status bits allow the operation, it is completed in hardware. Otherwise, the event system starts a software handler to resolve the operation. Implementing block status bits requires 1KB of SRAM in the LTLB, and 0.25 KB of SRAM in the caches.

3.2 The Event System

The event system is responsible for invoking software handlers in response to remote memory accesses and other events which require intervention by system software. When the hardware detects a situation which requires software intervention, such as a remote memory reference, it places an event record describing the situation in a 128-word hardware *event queue*. It then discards the original operation, allowing programs to continue to execute while the event is resolved. A dedicated event handler thread processes the event records and resolves events.

The head of the event queue is mapped onto a register in the event handler's register file. This speeds up queue accesses and provides a low-overhead mechanism for blocking the event handler when there are no pending events. If the event queue is empty, the register for the head of the queue is marked empty by the scoreboard logic, preventing instructions that read the register from issuing. When the event handler tries to read the head of the event queue to begin processing the next event, the instruction stalls while the queue is empty, but issues as soon as an event record is placed in the queue, allowing the event handler to respond quickly to events without consuming execution cycles that could be used by other threads in polling.

3.3 Dedicated Thread Slots for Shared-Memory Handlers

The M-Machine takes advantage of the MAP chip's multithreaded architecture to eliminate context switch overhead from software handlers by dedicating a set of thread slots to software handler threads. Since the handler threads are always resident in their thread slots, there is no need to perform a context switch when a handler executes, in contrast to single-threaded implementations of software shared memory.

Thread	Cluster 0	Cluster 1	Cluster 2
0	User	User	User
1	User	User	User
2	Exception	Exception	Exception
3	*Event Handler*	*Evict Proxy*	*Bounce Proxy*
4	LTLB Miss Handler	*Request Handler (P0 Message Handler)*	*Reply Handler (P1 Message Handler)*

Figure 2: Thread Slot Assignments on the MAP Chip

Figure 2 shows the thread slot assignments on the MAP chip when shared-memory programs are being run. Threads which are involved in implementing shared memory are shown in italics, while thread slots which include special hardware are underlined. The event and message handlers process the events which occur when a remote memory reference is made as well as the various messages which are used to implement the shared-memory protocol. In addition, two thread slots are used for *proxy* threads, which are used to break potential deadlock situations that occur when the shared-memory system needs to send three sequential messages over the MAP chip's two network priorities.

Allocating thread slots to software handlers significantly improves the M-Machine's remote access time, and simplifies the design of the software handlers because a handler is never interrupted to allow another handler to execute. The main incremental cost of dedicating thread slots to handlers is the 1.25KB of storage required to hold the register files of the dedicated threads, since the substantial hardware complexity incurred by multithreading is required by the MAP chip's base architecture.

3.4 The Global Translation Lookaside Buffer

Determining how data will be mapped across the processors in a DSM is an important part of program implementation. Mapping data for maximum locality can substantially reduce the number of remote references made by a program, and thus improve performance. However, providing flexibility in address mapping increases the latency of software shared memory, as the shared-memory handlers must translate each remote reference to find the home processor of the reference, creating another area where a small amount of hardware support can significantly improve performance.

The Global Translation Lookaside Buffer (GTLB) acts as a cache for translations between virtual addresses and their home processors, similar to the way a normal TLB caches translations between virtual and physical addresses. The format of a GTLB entry (Figure 3) allows each entry to map a variable-sized group of pages across variable-sized regions of the machine. The data mapped by a GTLB entry is specified by a base address and a size field, which specifies the number of pages in the page group. The region of the machine that the page group is mapped across is specified by its start node, the X- and Y-extents of the region (in the 2-D network), and the number of contiguous pages mapped per node. All fields except the base address and the start node are logarithmically encoded to reduce space.

Base Address	Size	Start Node	X Extent	Y Extent	Pages Per Node

Figure 3: GTLB Entry Format

The GTLB allows substantial flexibility in mapping addresses. For example, Figure 4 shows the three ways in which 16 pages of data can be mapped across a 2x2 block of processors. Note that changing the value of the start field allows the pages to be translated across the machine, facilitating space sharing of a multiprocessor.

The GTLB's entry format allows it to be implemented in very little hardware. The MAP chip implements a 4-entry GTLB due to space constraints (a 16-entry GTLB was specified in the initial design), which requires 64 bytes of content-addressable memory. For the experiments run for this paper, only two GTLB entries were required -- one to map the code segment locally on each processor, and one to map the data segment across the entire machine.

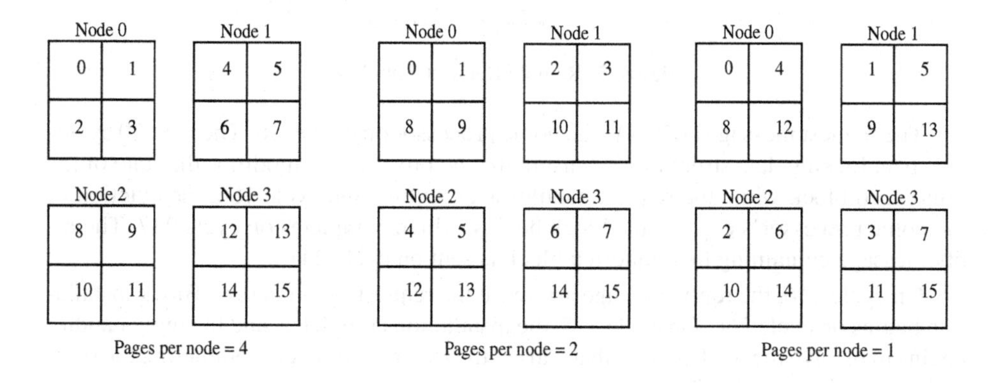

Figure 4: GTLB Mappings

4 Using the Hardware Mechanisms to Implement Shared Memory

Figure 5 shows the steps involved in completing a remote memory reference on the M-Machine when all of our hardware mechanisms are used. On cycle 1, a user program issues a load or store which references a remote address. By cycle 10, the event system has determined that the referenced block is remote and has started the event handler to resolve the event. By cycle 33, the event handler has decoded the type of the event and jumped to the correct routine to resolve it.

On cycle 49, the event handler completes the computation of the configuration space address which will be used to resolve the original load or store, and executes a **GPRB** operation to probe the GTLB for the home processor of the requested address.

In cycles 63-111, the event handler creates a record describing the remote operation and enqueues it in the software pending operation structure that records all remote memory references being completed in software. It then sends the request message to the home node and terminates. The request message must be sent after the pending operation data structure has been unlocked to avoid a potential deadlock with the reply handler.

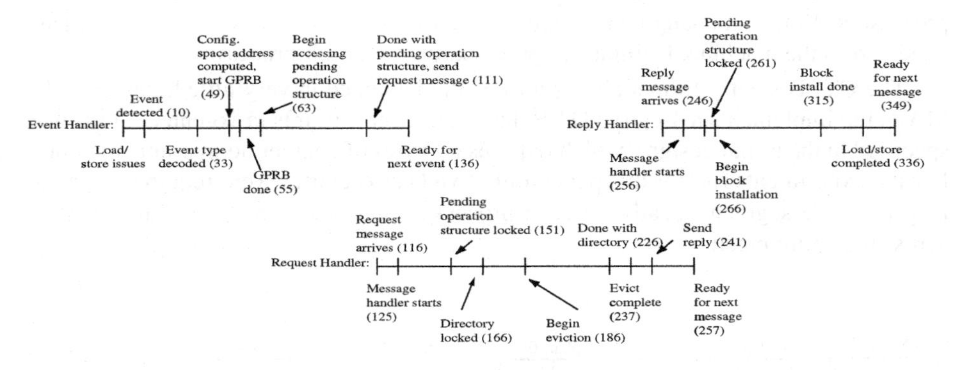

Figure 5: Remote Request Timeline

The request message arrives at the home processor on cycle 116. The next 50 cycles are spent locking data structures, to ensure that no other thread modifies the state of the referenced block while the request handler is executing. On cycle 186, the eviction of the home processor's copy of the block begins, which completes on cycle 237. The reply message containing the requested block is sent on cycle 241.

On cycle 246, the reply message arrives at the requesting processor. Block installation begins on cycle 266. On cycle 315, the installation completes, and the reply handler begins to resolve the load or store that caused the remote reference, completing the original operation on cycle 336.

5 Evaluation

A three-hop cache-coherence protocol was implemented in software on the MSIM simulator to evaluate our mechanisms for shared memory. MSIM is a C-language model of the M-Machine that gives execution times within 10% of those given by the cycle-accurate RTL model of the MAP chip on our verification test suite. The model of the MAP chip used for these experiments has a 128-entry, two-way set-associative LTLB as well as floating-point units in all three clusters, restoring features that were removed late in the implementation process due to area constraints. The shared-memory handlers use the floating-point registers as temporary storage and perform constant generation in the floating-point units, making them relevant for this study.

The handlers used for these experiments were implemented in hand-coded assembly language. Four versions of the handlers were written, one which uses all of the hardware mechanisms, one which only uses the block status bits, one which uses the block status bits and the GTLB, and one which uses the block status bits and the dedicated thread slots for software handlers. Versions which did not use the GTLB included a software address translation routine, while versions which did not use the thread slots simulated context switches when starting or exiting handlers.

5.1 Remote Access Times

Figure 6 shows the M-Machine's remote access time as a function of the set of mechanisms used. The column labelled "Full M-Machine Mechanisms" shows the remote access time when all of the hardware mechanisms are in use, measured from the cycle on which the processor issues a load to the cycle on which an instruction which uses the result of the load issues. Proceeding to the left, the columns show the remote access time if various subsets of the mechanisms for shared-memory are used. The leftmost column shows an estimate of the remote access time if none of the MAP chip's mechanisms are used, while the rightmost column shows the remote access time of an 8-processor SGI Origin 2000 [11] for comparison purposes. All of the columns which show measured data from the M-Machine have been subdivided into the time spent in the event handler on the requesting processor, the request handler on the home processor, and the reply handler on the requesting processor.

Based on block transfer programs written for the M-Machine, we estimate that using the block status bits reduces remote access time by approximately 1000 cycles when only one block of a page is required. This estimate was used to generate the leftmost column of Figure 7. If a program requires fine-grained sharing of data, this latency reduction can significantly improve performance. For programs with more coarse-grained data sharing, using software to implement shared memory allows the block size of the protocol to be increased to match the needs of the application.

If the GTLB is added to the block status bits, remote access time is reduced to 427 cycles, an improvement of 18%. Using the block status bits and the dedicated thread slots for software handlers has similar results, reducing the remote access time to 433 cycles (17% improvement). Combining the block status bits, the GTLB, and the dedicated thread slots for shared-memory handlers so that all of the M-Machine's hardware

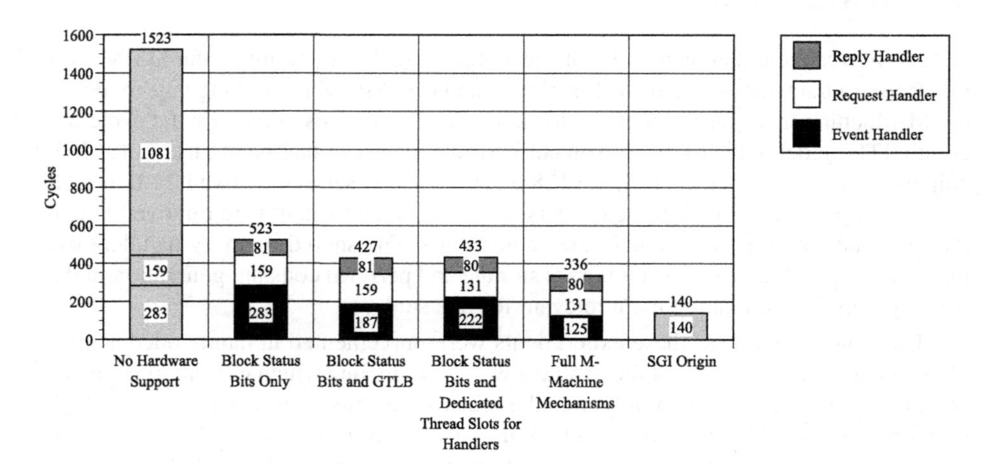

Figure 6: Remote Access Times

mechanisms are in use gives a remote access time of 336 cycles, a 35% improvement over the block status bits alone.

The mechanisms affect the execution time of different handlers in the shared-memory protocol, which has a significant impact on program performance, as will be shown later. When the GTLB is used, only the execution time of the request handler changes, since the determination of the requested address' home processor is only done once, in the request handler. On the other hand, using the dedicated thread slots affects the execution time of both the event and the request handlers, as it eliminates context switch overhead from all of the handlers. The execution times for the reply handlers shown on this graph do not change when the dedicated thread slots are used because execution time is measured from the point at which the first instruction of a handler executes and the context switch at the end of the reply handler does not affect the total latency.

Figure 7 shows how the remote access time changes when one or more invalidations are required to complete a remote reference. In the protocol implemented for these experiments, the home processor does not send an invalidation message to itself when it needs to invalidate its copy of a block, so the points shown on the graph represent 2-, 4-, and 8-way sharing of data. All versions of the protocol see an increase in remote access time of approximately 50% when an invalidation is required to complete a remote memory reference. As the number of invalidations required to complete the reference increases, all of the protocols see a linear increase in remote memory latency, because the bottleneck is the time required to process the acknowledgement message from each invalidation on the requesting processor.

Use of the GTLB produces a constant improvement in remote access time, independent of the number of invalidations, while the dedicated thread slots give an improvement which increases with the number of invalidations. Again, this is due to the fact that the determination of the home processor of an address is only done once. In contrast, the use of dedicated thread slots eliminates the context switch overhead from

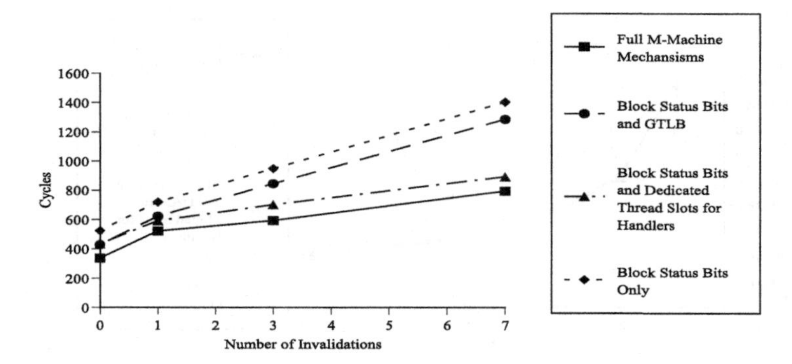

Figure 7: Remote Access Time With Invalidations

each handler. As the number of invalidations increases, the number of handlers in the critical path of the remote request increases, and therefore the performance impact of the dedicated thread slots increases with the number of invalidations.

5.2 Program Results

Two programs were simulated to evaluate the impact of the mechanisms for shared memory on program execution time: a 1,024-point FFT, and an 8x8x8 multigrid computation. These programs were based on source code provided by the Alewife group at M.I.T. [2], and were written in C with annotations for parallelism. The measurements presented here show the execution time of the parallel kernel of each application.

These programs display very different shared-memory characteristics. In FFT, remote memory references are relatively infrequent and are fairly evenly distributed across the processors. Multigrid makes many more remote references, and spends much more time waiting for memory than FFT. More importantly, multigrid's remote memory references are poorly distributed across the processors, because the destination matrix fits in a single page of memory and is thus mapped onto only one processor. Due to this difference in memory access patterns, the shared-memory performance of FFT is dominated by the latency of the shared-memory handlers, while the performance of multigrid is dominated by the occupancy of the request (priority 0 message) handler slot on the hot-spot processor.

Figure 8 shows the execution time of a 1,024-point FFT on the M-Machine. As would be expected from the low demands that this program places on the shared-memory system, overall performance is good, achieving better than 4 times speedup on eight processors even when only the block status bits are used. Adding either the GTLB or the dedicated thread slots for software handlers to the block status bits improves performance by 2-5%, depending on the number of processors. Adding both of these mechanisms gives speedups almost equal to the speedups provided by each of the mechanisms independently, reducing execution time by up to 9%. .

Figure 9 shows the execution time of an 8x8x8 multigrid computation. When just the block status bits and the GTLB are used, execution time is up to 20% greater than

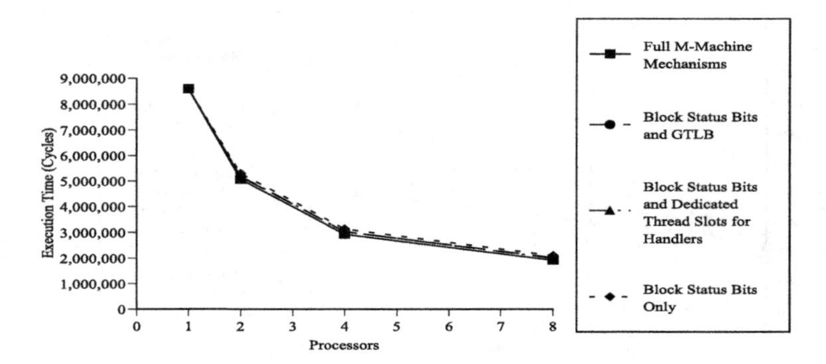

Figure 8: 1,024-Point FFT

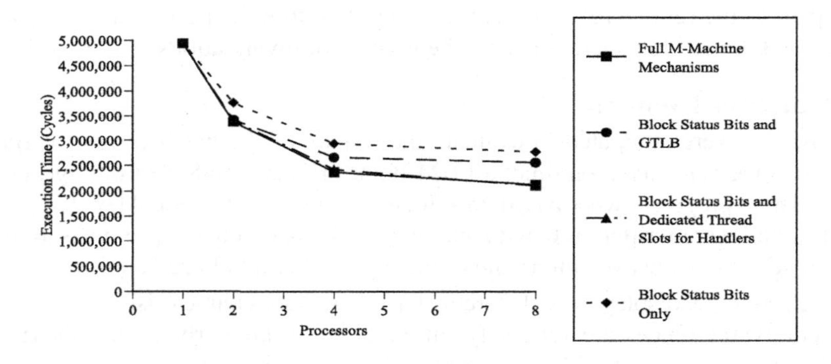

Figure 9: 8x8x8 Multigrid

when all of the M-Machine's mechanisms are in use. Using only the block status bits increases execution time by up to 30% over the full-mechanism case.

Interestingly, disabling the GTLB so that only the block status bits and dedicated thread slots are used does not significantly increase the execution time of multigrid. In fact, performing address translation in software gives a 1% performance improvement over using all of the hardware mechanisms when the program is run on eight processors. This counter-intuitive behaviour occurs because the GTLB reduces the execution time of the event handler on the requesting processor, while the dominant factor in Multigrid's performance is the occupancy of the request handler thread on the hot-spot processor, which is not affected by the use of the GTLB. In fact, using the GTLB increases the rate at which requests arrive at the hot-spot processor, increasing the number of requests which must be returned to the requesting processor because the block they request is in the process of being invalidated. These requests must be retried later, increasing the occupancy of the request handler on the hot-spot processor and decreasing overall performance.

6 Related Work

A number of systems have explored different hardware/software tradeoffs in implementing distributed shared memory. IVY [13] implements shared memory in software through user-level extensions to the virtual memory system. Shasta [16] and Blizzard-S [17] rely on the compiler to insert check code before each memory reference to determine whether the referenced data is local or remote. Blizzard-E [17] takes a different approach, modifying the error correction code (ECC) bits on blocks of data to force traps to software when remote blocks are referenced.

These systems show the advantages of the MAP chip's base architecture over commodity processors in implementing software shared memory. Blizzard-S and Blizzard-E have remote memory latencies of approximately 6000 cycles, while Shasta achieves latencies as low as 4200 cycles. In contrast, the M-Machine's remote memory latency is estimated to be approximately 1500 cycles when the event system is the only mechanism in use, giving it a 2.8-4x speed advantage over these software-only systems. This speed advantage is due to the MAP chip's event system and integrated network hardware, which reduce the time to invoke software handlers and inter-processor communication delay. Adding the block status bits, GTLB, and dedicated thread slots for software handlers to the base MAP architecture reduces the remote access time to 336 cycles, increasing the M-Machine's advantage to 12.5x-17.8x.

The Typhoon [14][15] and FLASH [10][6][7] projects explored the use of a dedicated co-processor to implement shared memory. Typhoon used a commodity processor to execute the shared-memory handlers, while FLASH relied on a custom MAGIC chip. Typhoon-0, the least-integrated of the systems studied in the Typhoon project, was implemented using FPGA technology, and achieved a remote memory latency of 1461 cycles. Two more-integrated versions of the Typhoon architecture, Typhoon-1 and Typhoon, were studied in simulation, and had remote memory latencies of 807 and 401 cycles respectively. With all of its mechanisms in use, the M-Machine has better than a 4x advantage in remote memory latency over Typhoon-0, and a 19% advantage over the full Typhoon system. Most of this advantage comes from the M-Machine's superior network subsystem and low-latency event system.

FLASH is able to complete a remote memory access in 111-145 cycles, depending on whether the remote data is cached. The ISA of the MAGIC chip is a major contributor to FLASH's low memory latency, as it contains many non-standard instructions to accelerate shared-memory protocols. In [6], the authors report that at least 38% of protocol processor issue slots contain one of these non-standard instructions, suggesting that the latency of shared-memory handlers running on MAGIC is significantly reduced by the addition of these instructions.

Comparing the M-Machine to these other systems shows the advantages of integrating hardware support for software shared memory into the processor. The M-Machine achieves significantly better remote memory latencies than the software-only shared memory systems by performing common tasks in hardware. In addition, the M-Machine has better remote memory latencies than any of the Typhoon systems, in spite of the fact that they utilize substantial custom hardware and a commodity co-processor to implement shared memory. While FLASH's remote memory latencies are significantly better than the M-Machine's, much of this is due to the MAGIC chip's optimized instruction set, creating an opportunity for future work which combines hardware support for software shared memory with an optimized instruction set.

7 Conclusion

In this paper, we have shown that adding a small set of hardware mechanisms to support software shared memory to a processor can significantly improve remote memory latency at low hardware cost. We have implemented four key mechanisms for software shared memory: block status bits to allow small blocks of data to be transferred between processors, a fast event system to detect remote memory accesses and invoke software handlers, dedicated thread slots to eliminate context switch overhead when starting handlers, and a global translation lookaside buffer to determine the home processors of remote addresses in hardware. In combination, these mechanisms allow remote memory accesses to be performed in as little as 336 cycles, significantly faster than most software-only or combined hardware/software shared memory systems. Hardware cost of these mechanisms is small -- approximately 3.5KB of storage, and some control logic.

Program-level experiments showed that the impact of our mechanisms on program execution time depends strongly on whether the dominant factor in the program's performance was the latency or the occupancy of the shared-memory handlers. On a 1,024-point FFT, in which latency was the dominant factor, using all of the mechanisms improved performance by up to 9% when compared to using only the block status bits. On an 8x8x8 multigrid, which is dominated by the occupancy rate of the request handler on the hot-spot processor, the mechanisms improved execution time by up to 30%.

The mechanisms implemented on the MAP chip substantially improve the M-Machine's shared-memory performance at a low hardware cost. However, the M-Machine's remote access time is still more than a factor of 2.5x greater than that of contemporary full-hardware shared-memory systems, suggesting that additional hardware support is required to close the performance gap between hardware- and software-based shared-memory systems.

8 References

[1] Anant Agarwal *et al.* The MIT Alewife Machine: Architecture and Performance. In *Proceedings of the 22nd International Symposium on Computer Architecture*, June 1995.

[2] Ricardo Bianchini. Application Performance on the Alewife Multiprocessor. Alewife Systems Memo #43, LCS, Massachusetts Institute of Technology, 1994.

[3] Nicholas P. Carter, Stephen W. Keckler, and William J. Dally. Hardware Support for Fast Capability-Based Addressing. In *Proceedings of the Sixth International Conference on Architectural Support for Programming Languages and Operating Systems,* October 1994.

[4] Nicholas P. Carter. Processor Mechanisms for Software Shared Memory. Ph.D. Thesis, Massachusetts Institute of Technology, 1999.

[5] Marco Fillo, Stephen W. Keckler, William J. Dally, Nicholas P. Carter, Andrew Chang, Yevgeny Gurevich, and Whay S. Lee. The M-Machine Multicomputer. In *Proceedings of the 28th International Symposium on Microarchitecture*, Ann Arbor, MI, December 1995.

[6] Mark Heinrich *et al.* The Performance Impact of Flexibility in the Stanford FLASH Multiprocessor. In *Proceedings of the Sixth International Conference on Architectural Support for Programming Languages and Operating Systems* (ASPLOS VI), October 1994

[7] John Heinlein, Kourosh Gharachorloo, Scott Dresser, and Anoop Gupta. Integration of Message Passing and Shared Memory in the Stanford FLASH Multiprocessor. In *Proceedings of the Sixth International Conference on Architectural Support for Programming Languages and Operating Systems*, October 1994

[8] Stephen W. Keckler and William J. Dally. Processor Coupling: Integrating Compile Time and Runtime Scheduling for Parallelism. In *Proceedings of the 19th International Symposium on Computer Architecture*, May 1992.

[9] Stephen W. Keckler, William J. Dally, Daniel Maskit, Nicholas P. Carter, Andrew Chang, and Whay S. Lee. Exploiting Fine-grain Thread-level Parallelism on the MIT Multi-ALU Processor. In *Proceedings of the 25th International Symposium on Computer Architecture*, June 1998.

[10] Jeffery Kuskin *et al.* The Stanford FLASH Multiprocessor. In *Proceedings of the 21st Annual Symposium on Computer Architecture,* June 1994

[11] James Laudon and Daniel Lenoski. The SGI Origin: a ccNUMA Highly Scalable Server. In *Proceedings of the 24th Annual Symposium on Computer Architecture*, June 1997.

[12] Whay S. Lee, William J. Dally, Stephen W. Keckler, Nicholas P. Carter, and Andrew Chang. Efficient, Protected Message Interface in the MIT M-Machine. In *IEEE Computer*, November 1998.

[13] Kai Li. IVY: a Shared Virtual Memory System for Parallel Computing. In *Proceedings of the International Conference on Parallel Processing*, 1988.

[14] Steven K. Reinhardt, James R. Larus, and David A. Wood. Tempest and Typhoon: User-level Shared Memory. In *Proceedings of the 21st International Symposium on Computer Architecture*, April 1994.

[15] Steven K. Reinhardt, Robert W. Pfile, and David A. Wood. Decoupled Hardware Support for Distributed Shared Memory. In *Proceedings of the 23rd Annual Symposium on Computer Architecture*, May 1996

[16] Daniel J. Scales, Kourosh Gharachorloo, and Chandramohan A. Thekkath. Shasta: a Low Overhead, Software-only Approach for Supporting Fine-grained Shared Memory. In *Proceedings of the Seventh International Conference on Architectural Support for Programming Languages and Operating Systems*, October 1996.

[17] Ioannis Schoinas *et al.* Fine-grain Access Control for Distributed Shared Memory. In *Proceedings of the Sixth International Conference on Architectural Support for Programming Languages and Operating Systems*, October 1994.

An Evaluation of Page Aggregation Technique on Different DSM Systems

Mario Donato Marino and Geraldo Lino de Campos*

Computing Engineering Department- Polytechnic School of the University of So Paulo

Abstract. The page aggregation technique consists of considering a larger granularity unit than a page, in a page-based DSM system. In this paper an initial evaluation of the influence of the page aggregation technique in the speedup of a DSM system is done, by applying it in two DSMs: JIAJIA and Nautilus. TreadMarks, a DSM well known by the scientific community, is also included in this comparison as a reference for optimal speedups. Different granularity sizes are considered in this study: 4kB, 8kB, 16kB and 32kB. The benchmarks evaluated in this study are SOR (from Rice University), LU and Water N-Squared (both from SPLASH-II). The first results show that this technique can improve the JIAJIA's speedup by 'up to 4.1% and the Nautilus's speedup by up to 37.7%.

1 Introduction

In recent years, several factors have contributed to make the network of workstations (NOW) the most used as a parallel computer:

1. the evolution of microprocessors;
2. the decrease of costs of interconnection technologies;
3. the adoption of hardware on the shelf components.

Big projects such as Beowulf[11] can be mentioned to exemplify these tendencies.

The *Distributed Shared Memory* (DSM) paradigm[8], which has been widely discussed for the last 9 years, is an abstraction of shared memory which permits viewing of a network of workstations as a shared memory parallel computer. By moving or replicating data[8], shared memory uniform accesses are done by the different nodes, implementing in this way the DSM's main aim. These movements and/or replications of data guarantee its consistency, allowing programs done by physically shared memory machines to be easily ported and developed[1], since to develop message passing programs is more difficult than to develop shared memory programs. The research of the DSM area can be resumed mainly by the development and evolution of a large number of consistency models and DSM systems. Carter [1] has classified the DSM evolution in two generations:

* Address: Av. Professor Luciano Gualberto 158, trav3, sala C2-24, prdio E. Eltrica, Cidade Universitria, CEP 05508-900, So Paulo, Brazil; phone: +55 11 818-5583; fax: +55 11 818-5294; e-mail:{mario,geraldo}@regulus.pcs.usp.br

M. Valero et al. (Eds.): ISHPC 2000, LNCS 1940, pp. 134-145, 2000.

- a big number of consistency messages and the adoption of the sequential consistency model; one can exemplify this with the Ivy [6];
- a drastic reduction of the number of consistency messages by the adoption of the release consistency model, applying techniques to reduce false sharing; several examples can be mentioned: Munin[2], Quarks[7], TreadMarks[3][19], CVM[10], Midway[9], JIAJIA[4] and Nautilus[5].

In terms of granularity, in most cases page-grained DSMs approaches were chosen instead of fine-grained ones. Also, the study of Iftode[17] showed that for several applications from SPLASH-II, page-grain DSMs perform similarly to or better than fine-grain ones, although generally higher bandwidth and message handling costs favor page-based DSM while lower latency favors fine-grained approach[17].

In page-based DSM systems, shared memory accesses are detected using virtual memory protection, thus one page is the unit of access detection and can be used as unit of transfer. Depending on the memory consistency model and the situation, also the diffs[1] are used as an unit of transfer. For example, in homeless lazy release consistency (LRC), such as TreadMarks, if the node has a dirty page, diffs are fetched from several nodes, when an invalid page is accessed. On the other hand, in a home-based DSM like JIAJIA, pages are fetched from the home nodes when a remote page fault occurs.

The unit of access detection and the unit of transfer can be increased by using a multiple of the hardware page size. In this way, if aggregation is done, false sharing is increased. Aggregation reduces the number of messages exchanged. If a processor accesses several pages successively, a single page fault request and reply can be enough, instead of multiple exchanges, which are usually required. A secondary benefit is the reduction of the number of page-faults. On the other hand, false sharing can increase the amount of data exchanged and the number of messages[16].

The main contribution of this paper is to evaluate the page aggregation technique[16], for different benchmarks, on two different DSMs: JIAJIA and Nautilus. The page aggregation technique is evaluated in both DSMs with a PC network, with a free operating system. In order to have a reference of optimal speedups, TreadMarks's speedups are included in this study. Unfortunately, the TreadMarks version used is a demo version (1.0.3), thus the source code is not available and it was not possible to evaluate it with other grain sizes (default is 4kB), in order to compare it with JIAJIA and Nautilus.

The evaluation comparison is done by applying some different benchmarks: LU (kernel from SPLASH-II)[15], SOR (from Rice University) and Water N-Squared (from SPLASH-II). The environment of the comparison is a 8PC network interconnected by a fast-Ethernet shared media. The operating system used in each PC is Linux (2.x). This study is a preliminary evaluation of this technique and four aggregation sizes are used: 4kB, 8kB, 16kB and 32kB.

In section 2, a brief description of Nautilus is given. In section 3, the page aggregation method and its consequences are explained. In sections 4 and 5,

[1] diffs: codification of the modifications suffered by a page during a critical section.

some benchmarks are evaluated showing the application of the page aggregation technique. Section 6 concludes this work.

2 Nautilus

The main motivation of the new software DSM Nautilus is to develop a DSM with a simple consistency memory model, in order to provide good speedups, and compatible with TreadMarks and JIAJIA. This idea is very similar to the ideas utilized by JIAJIA, mentioned in the studies of Hu[4] and Eskicioglu[12], but Nautilus makes use of some other techniques, which distinguishes it from JIAJIA. These techniques will be mentioned below. In order to be portable, it was developed as a runtime library like TreadMarks, CVM and JIAJIA, because there is no need to change the operating system kernel[2].

To summarize the Nautilus features: i) scope consistency only sending consistency messages to the owner of the pages and invalidating pages in the acquire primitive; ii) multiple writer protocols; iii) multi-threaded DSM; iv) no use of SIGIO signals(which notice the arrival of a network message); v) minimization of diffs creation; vi) primitives compatible with TreadMarks, Quarks and JIAJIA; vii) network of PCs; viii) operating under Linux 2.x; ix) UDP protocols.

Nautilus is a page-based DSM, like TreadMarks and JIAJIA. In this scheme, pages are replicated through the several nodes of the net, allowing multiple reads and writes[8], thus improving speedups. By adopting the multiple writer protocols proposed by Carter[2], false sharing is reduced. The mechanism of coherence adopted is write invalidation[8], because several studies [2][3][4][12] show that this type of mechanism provides better speedups for general applications. Nautilus uses scope consistency model, which also reduces the false sharing effect[14]. In Nautilus, this consistency model is implemented through a locked-based protocol[13].

Nautilus is the first multi-threaded DSM system implemented on top of a free Unix platform that uses the scope consistency model because:

1)there are versions of TreadMarks implemented with threads, but it does not use scope consistency memory model;

2)JIAJIA is a DSM system based on scope consistency, but it is not implemented using threads.

3)CVM[10] is a multi-threaded DSM system, but it uses lazy release consistency and at the moment, it does not have a Linux-based version.

4)Brazos[18] is a multi-threaded DSM and it uses scope consistency, but it's implemented on a Windows NT platform.

Nautilus manages the shared memory using a home-based scheme, but with a directory structure of all pages instead of only a structure of the relevant pages (cached), used by JIAJIA. Also, a different memory organization from JIAJIA, explained in item 2.1.

To improve the speedup of the applications submitted, Nautilus uses two techniques: i)multi-threaded implementation; ii) diffs of pages that were written by the owner are not created.

The multi-threaded implementation of Nautilus permits:

1) minimization of context switch: Nautilus's threads are used, for example, to help the reply a request for a page, to apply a diff and not to run a user program, as in Brazos;

2) no use of SIGIO signals: most page-based DSM systems created until present implemented on top of a Unix platform uses SIGIO signals to activate a handler to take care of the arrival of messages which come from the network. One of the threads remains blocked trying to read messages from the net. While blocked, it remains slept and thus is a non consuming CPU. This technique decreases the overhead of the DSM and allows as much CPU time as possible to the user program. Thus, Nautilus is the first scope consistency DSM system of the second generation which does not use a SIGIO signal in its implementation.

On the same way that TreadMarks and JIAJIA do, Nautilus is also concerned with network protocols. So, it also uses UDP protocol to minimize overheads.

Nautilus also deals with the compatibility of primitives. Its primitives are simple and totally compatible with TreadMarks, JIAJIA and Quarks; as a result there is no need for code rearrangements. One example of this compatibility is that in this study, LU and SOR are converted from JIAJIA and SOR from TreadMarks, basically changing the name of the primitives.

Like TreadMarks and JIAJIA, Nautilus is also concerned with synchronization messages. To minimize the number of messages, the synchronization messages would carry consistency information, minimizing the emission of the latter.

Nautilus follows the lock-based protocol proposed by JIAJIA[12], because of its simplicity, thus minimizing overheads. Resuming this protocol, the home nodes of the pages always contain a valid page, and the diffs corresponding to the remote cached copies of the pages are sent to the home nodes. A list with the pages to be invalidated in the node is attached to the acquire lock message.

JIAJIA[3] only contains information of the relevant pages, the cached copies of the pages, because it argues that it *reduces the space overhead the system*[4]. On the other hand, Nautilus maintains a local directory structure for all pages, since it does not occupy a relevant space and does not increase the overhead of the system. Differently, this helps increasing the speedup of the system.

Following JIAJIA [3][12] concept, in Nautilus, the owner nodes of the pages do not need to send the diffs to other nodes, according to the scope consistency model. So, diffs of pages written by the owner are not created, which is believed to be more efficient than the lazy diff creation of TreadMarks.

The implementation of the state diagram of the page transitions is done in Unix using the *mprotect()* primitive. With this primitive, pages can be in read-only(RO), invalid (INV) or read-write (RW) states, thus pages can have their states changed easily.

3 Page Aggregation

In terms of implementation, following the other DSMs directions, in Nautilus there is a handler responsible for requesting a page from a remote node when a segmentation fault occurs. When a page is accessed and it's in the INV state, a SIGSEVG signal is generated and the respective handler, as it was said before, requests the page from the home node. When the page arrives, the primitive *mprotect()* changes the state from INV to RO.

When the page is written, another SIGSEGV signal is generated and the primitive *mprotect()* changes the state of the page from RO to RW. After the generation of the diffs, also with the *mprotect()* primitive, pages go to RO state again. And, when the write-notices arrive, indicating the pages are modified by other nodes, pages go to INV state.

The primitive *mprotect()* permits the consideration of a granularity multiple of a page, thus giving the same permission for a region multiple of a page. Thus, this fact gives the condition to modify more than one page at the same time, which is named page aggregation technique.

The study [16] says that if aggregation is done, false sharing is increased and aggregation reduces the number of messages exchanged. Also, the processor accesses several pages successively, a single page fault request and reply can be enough, instead of multiple exchanges of requests and replies, which are usually required. The study [16] also shows that there is a reduction of the number of page-faults, but false sharing can increase the amount of data exchanged and the number of messages.

By changing the page size default (4kB) to, for example, 8kB using the *mprotect()* primitive, it's possible to evaluate the effects of the incremented size in page fault reduction in the speedups.

4 Experimental Platform and Applications

Here, the experimental platform and the applications are detailed.

4.1 Experimental Platform

The results reported here are collected on a 8 PC network. Each node (PC) is equipped with a K6 - 233 MHz (AMD)processor, 64 MB of memory and a fast ethernet card (100 Mbits/s) . The nodes are interconnected with a hub. In order to measure the speedups, the network above was completely isolated from any other external networks. Each PC runs Linux Red Hat 6.0. The experiments are executed with no other user process.

In this study, **four** sizes are considered for page size: 4kB, which is the default (memory hardware), 8kB, 16kB and 32kB.

4.2 Applications

The test suite includes three programs: LU (from SPLASH-II[15]), SOR (from Rice University) and Water N-Squared (from SPLASH-II). SPLASH-II is a collection of parallel applications implemented to evaluate and design shared memory multiprocessors.

"The LU kernel from SPLASH II factors a dense matrix into the product of a lower triangular and upper triangular matrix. The NxN matrix is divided into an nxn array of bxb blocks (N = n*b) to exploit temporal locality on sub-matrix elements. The matrix is factored as an array of blocks, allowing blocks to be allocated contiguously and entirely in the local memory of processors that own them. LU is a kernel from SPLASH2 benchmarks that has a rate computation/communication $O(N^3)/O(N^2)$, which increases with the problem size N. The nodes frequently synchronize in each step of computation and none of the phases are fully parallelized[4].

Water is an N-body molecular simulation program that evaluates forces and potentials in a system of water molecules in the liquid state using a brute force method with a cutoff radius. Water simulates the state of the molecules in steps. Both intra- and inter-molecular potentials are computed in each step. The most computation- and communication-intensive part of the program is the inter-molecular force computation phase, where each processor computes and updates the forces between each of its molecules and each of the n/2 following molecules in a wrap-around fashion[12].

SOR from Rice University solves partial differential equations (Laplace equations) with an Over-Relaxation method. There are two arrays, black and red array allocated in shared memory. Each element from the red array is computed as an aritmethic mean from the black array and, each element from the black array is computed as an aritmethic mean from the red array. Communication occurs across the boundary rows on a barrier. The SOR from Rice University solves Laplace partial equations. For a number of iterations it has two barriers each iteration and communication occurs across boundary rows on a barrier. The communication does not increase with the number of processors and the relation communication/computation reduces as the size of the problem increases[4].

5 Result Analysis

Before presenting the results and their analysis, it is necessary to emphasize that the execution time for number of nodes = 1 in all evaluated benchmarks is obtained from the sequential version of the benchmarks without any DSM primitive. So, the primitive used to allocate memory to obtain the sequential time (t(1) and number of nodes = 1) is **malloc()**, default primitive of C programming.

In order to have an accurate, homogeneous and fair comparison, the same programs are executed using TreadMarks (version 1.0.3), JIAJIA (version 2.1) and Nautilus (version 0.0.1).

Table 1 shows some features and results of the benchmarks: sequential time (t(1)), 8-processor parallel run time(t(8)), speedup for 8 nodes(Sp8), remote get page request counts per node (gp) and number of local SIGSEGV per node (SG). The sequential time t(1) was obtained from the sequential program without DSM primitives, as has already been mentioned.

For table 1 and for the graphics below, there are several extensions: "J4k" means JIAJIA using 4kB page size, "J8k" means JIAJIA using 8kB page size,

app	LU	Water	SOR
t(1)	350.90	2983.00	29.10
t(8).Tmk	55.45	403.20	8.66
t(8).J4k	62.81	429.82	21.22
t(8).J8k	61.24	432.59	14.44
t(8).J16k	60.24	440.03	20.87
t(8).J32k	60.63	452.98	30.69
t(8).N4k	55.17	426.88	7.67
t(8).N8k	55.56	422.96	6.30
t(8).N16k	58.03	428.99	5.60
t(8).N32k	60.20	437.06	5.56
Sp8.Tmk	6.33	7.40	3.36
Sp8.J4k	5.59	6.94	1.37
Sp8.J8k	5.73	6.90	2.02
Sp8.J16k	5.82	6.78	1.39
Sp8.J32k	5.79	6.59	0.95
Sp8.N4k	6.36	6.99	3.79
Sp8.N8k	6.32	7.05	4.62
Sp8.N16k	6.05	6.95	5.20
Sp8.N32k	5.83	6.82	5.22
SG.J4k	87	106	112
SG.J8k	44	78	56
SG.J16k	22	63	56
SG.J32k	11	56	56
SG.N4k	7980	851	12425
SG.N8k	5029	602	7912
SG.N16k	3030	532	3990
SG.N32k	1650	398	2010
gp.J4k	3542	1921	893
gp.J8k	2220	1206	461
gp.J16k	1212	849	461
gp.J32k	663	632	461
gp.N4k	1528	445	118
gp.N8k	1232	312	72
gp.N16k	940	281	51
gp.N32k	540	195	42

Table 1. table comparing TreadMarks, JIAJIA and Nautilus for page sizes: 4k, 8k, 16k and 32k

"N16k" means Nautilus using 16kB page size and "N32k" means Nautilus using 32kB page size.

There are some constraints with the TreadMarks version (1.0.3) used:

i) the applications were executed and the speedups measured using *Nautilus* running on up to **8 nodes;**

ii)**bigger input sizes:** the shared memory size is limited in this version;

iii)the source of this demo version is not available, thus it was neither possible to evaluate TreadMarks with other page size nor to measure the parameters gp (get page request) and SG (local SIGSEGV).

5.1 LU

By looking at table 1 and by applying the aggregation technique, i.e., by increasing the page size from 4kB to 32kB, JIAJIA's speedup was improved by up to 4.1% due to the reduction of the number of SIGSEGVs (SG.J* rows) from about 39.04% up to 49.4% and the reduction of the number of gp (gp.J* rows) from about 37.32% up to 45.4%. For the page size of 32kB, the JIAJIA's speedup is lower than for the page size of 16kB due to the different data distribution resulting from the application of the page aggregation technique. For page size of 4kB to 16kB, it is possible to notice by looking at figure 1 or by observing the Sp.J* rows of table 1, that the increase of the speedups of JIAJIA grows with the increase of the page size.

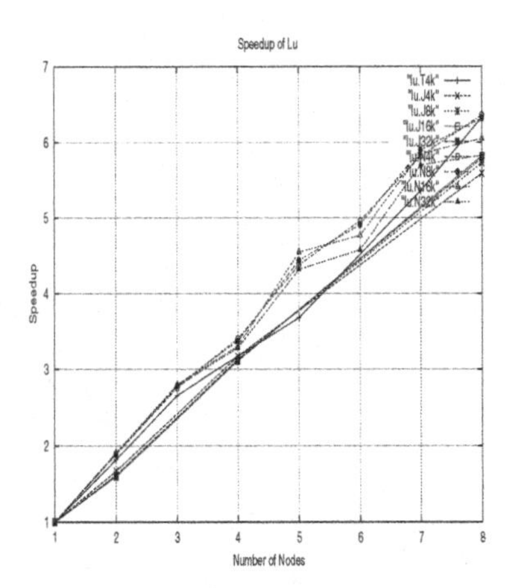

Fig. 1. speedups of LU: **N=1792**

For 8 nodes, from table 1, a reduction of 8.33% of Nautilus's speedup is observed when the page aggregation technique is applied (increasing page size from 4kB to 32kB). Although the number of SIGSEGVs and the number of get page requests decreases by up to 45.4% and 42.6% respectively, as can be observed from 1, the employment of the page aggregation technique changes the data distribution. This new data distribution changes the home nodes, resulting in a distribution that is not as adequate as the initial one (4kB), decreasing the speedups of Nautilus. The speedups of Nautilus with this technique applied can also be observed in figure 1.

By comparing TreadMarks with JIAJIA, from table 1 and from figure 1, it can be noticed that TreadMarks is faster than JIAJIA by up to 13.24% for a page size of 4kB, and 8.76% faster for a page size of 16kB. Generically without the page aggregation technique (4kB's page size), the lazy release consistency model and the lower number of diffs sent are responsible for the better speedups of TreadMarks over JIAJIA for the LU benchmark.

By comparing Nautilus with TreadMarks, for a page size of 4kB, i.e. without the page aggregation technique, Nautilus is 0.47% faster than TreadMarks, due to better data distribution, the avoidance of SIGIO signals and multi-threading. By applying the page aggregation technique, the speedup of Nautilus decreases, and TreadMarks becomes up to 8.58% faster than Nautilus, when the latter uses pages of 32kB (N32k).

Comparing JIAJIA with Nautilus, it is possible to notice from table 1 and from figure 1 that Nautilus is faster than JIAJIA by up to 13.77%. Although the number of SIGSEGVs of JIAJIA is two orders of magnitude lower than Nautilus, as can be noticed from table 1, the number of get page requests is up to 50.0% lower for Nautilus, thus improving the data locality and thus, its speedup. The multi-threading and the avoidance of SIGIO signals helps to improve the performance of Nautilus .

5.2 Water

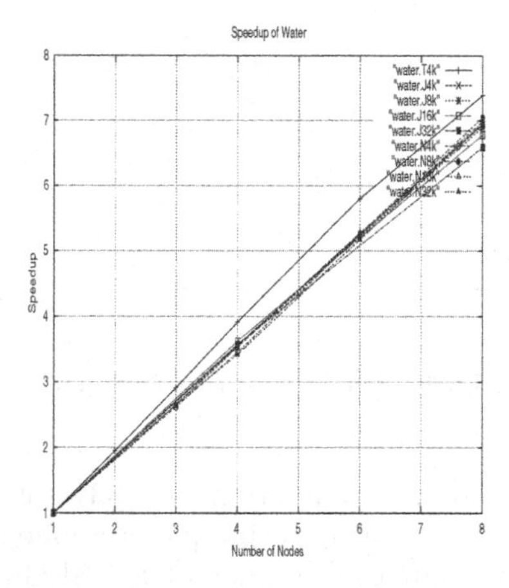

Fig. 2. speedups of Water: **1728 molecules** and **25 steps**

By looking at table 1 and by applying the aggregation technique increasing the page size from 4kB to 32kB, there is a reduction of about 5.0% in JIAJI-A's speedup. Still observing this table, there is a reduction of the number of SIGSEGVs (SG.J* rows) by up to 26.4% and a reduction of the number of gp (gp.J* rows) by up to 37.2%. The employment of page aggregation technique changes the data distribution and in this new data distribution, the home nodes are changed, resulting in a distribution that is not as adequate as the initial (4kB).

For 8 nodes, from table 1, it can be observed that Nautilus's speedup is increased by up to 1.0%, when the page aggregation technique is applied (increasing page size from 4kB to 32kB). Also, the number of SIGSEGVs and the number of get page requests decreased by up to 29.3% and 30.6% respectively. The problem with this benchmark is the high level of synchronization presented, which dominates its behavior.

For Water, TreadMarks is up 6.6% to faster than JIAJIA with a page size of 4k (J4k), and 12.29% faster than JIAJIA using the 32kB page size (J32k), since the page aggregation technique changed the JIAJIA data distribution. Generically comparing TreadMarks with JIAJIA, without page aggregation technique, TreadMarks is faster than JIAJIA due to the LRC model adopted by TreadMarks and the high number of synchronization messages of Water.

Comparing TreadMarks with Nautilus, TreadMarks is 4.96% faster than Nautilus using a page size of 8kB (N8k), and 8.50% faster than Nautilus using the 32kB page size (N32k). Generically for Water, without the page aggregation technique (page size of 4kB), TreadMarks is up to 5.86% faster than Nautilus due to the lazy consistency model and high synchronization of the Water benchmark. Also, Nautilus's semaphore implementation is still under development.

Comparing JIAJIA with Nautilus, it is possible to notice from table 1 and from figure 2 that Nautilus is faster than JIAJIA by up 1.65%. Although the number of SIGSEGVs of JIAJIA is one order of magnitude lower than Nautilus, as can be noticed from table 1, the number of get page requests is up to 76.8% lower for Nautilus, thus a lower number of pages transfered improvies the data locality and so, its speedup. And with the avoidance of SIGIO signals, the multithreading helps to improve its speedup.

5.3 SOR

As can be noticed from table 1, the speedups of JIAJIA are very unusual. Therefore, any related speedups are not considered for SOR analysis.

By observing table 1 and figure 3, for SOR benchmark, the page aggregation technique decreased the number of SIGSEGVS by up to 49.6%, and also the number of pages requested by up to 39.0%. These reductions justify the increase of the speedups of 37.7% for 8 nodes, for Nautilus DSM.

Comparing Nautilus with TreadMarks, without the page aggregation technique applied, Nautilus is up to 12.80% faster. By applying this technique, Nautilus becomes up to 55.36% faster than TreadMarks, the best known reference of the DSM area.

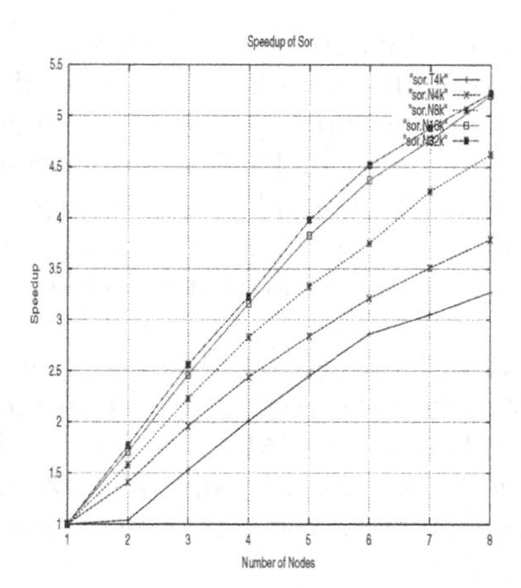

Fig. 3. speedups of SOR: **1792x1792**

Generically for this benchmark, without the page aggregation technique applied, the excellent speedup of Nautilus over TreadMarks can be justified by the data distribution (choice of the page owners) adopted by itself improving the matrix data locality (minimizing the number of messages through the net) and giving a lower cold start up time to distribute shared data. Also the avoidance of SIGIO signals and the multi-threading help to improve the SOR speedup.

6 Conclusion

In this paper the page aggregation technique was evaluated for two different DSMs: JIAJIA and Nautilus. The main contribution was to evaluate the application of this technique and its effects on the speedup of these DSMs. Also, JIAJIA, TreadMarks and Nautilus were evaluated and compared.

It was shown that the page aggregation technique has improved Nautilus speedups by up to 37.7% for the SOR benchmark and improved JIAJIA speedup up to 4.61 % for the LU benchmark, reducing the number of page faults and the number of SIGSEGVs. For Water, the improvement of the technique applied was up to 1.0%, due to the high synchronization and the access method of this benchmark. In addition, the speedup of the three DSMs, JIAJIA, TreadMarks and Nautilus, without the page aggregation technique are compared, but not as the main goal of the paper. In a future study, other benchmarks from SPLASH-2 and NAS benchmark will be evaluated.

And the intention is to acquire a complete version of TreadMarks, including its sources, to evaluate and compare it with JIAJIA and Nautilus for several grain sizes.

References

1. Carter J. B., Khandekar D., Kamb L., *Distributed Shared Memory: Where We are and Where we Should Headed*, Computer Systems Laboratory, University of Utah, 1995.
2. Carter J. B., *Efficient Distributed Shared Memory Based on Multi-protocol Release Consistency*, PHD Thesis, Rice University, Houston, Texas, September, 1993.
3. Keleher P. , *Lazy Release Consistency for Distributed Shared Memory*, PHD Thesis, University of Rochester, Texas, Houston, January 1995.
4. Hu W., Shi W., Tang Z., *JIAJIA: An SVM System Based on a new Cache Coherence Protocol,* technical report no. 980001, Center of High Performance Computing , Institute of Computing Technology, Chinese Academy of Sciences, January, 1998.
5. Marino M. D., Campos G. L., Sato L. M.; *An Evaluation of the Speedup of Nautilus DSM System* published at IASTED PDCS99.
6. Li K, *Shared Virtual Memory on Loosely Coupled Multiprocessors*, PHD Thesis, Yale University, 1986.
7. Swanson M., Stoller L., Carter J., *Making Distributed Shared Memory Simple, Yet Efficient*, Computer Systems Laboratory, University of Utah, technical report , 1998.
8. Stum M. , Zhou S. , *Algorithms Implementing Distributed Shared Memory*, University of Toronto, IEEE Computer v.23 , n.5 , pp.54-64 , May 1990.
9. Bershad B. N. , Zekauskas M. J. , SawDon W. A. , *The Midway Distributed Shared Memory System* , COMPCOM 1993.
10. Keleher P., *The Relative Importance of Concurrent Writers and Weak Consistency Models*, in Proceedings of the 16th International Conference on Distributed Computing Systems (ICDCS-16), pp. 91-98, May 1996.
11. Becker D., Merkey P.; *Beowulf: Harnessing the Power of Parallelism in a Pile-of-PCs*, Proceedings, IEEE Aerospace, 1997.
12. Eskicioglu, M.S., Marsland T.A., Hu W, Shi W.; *Evaluation of the JIAJIA DSM System on High Performance Computer Architectures*, Proceeding of the Hawai'i International Conference on System Sciences, Maui, Hawaii, January, 1999.
13. Hu W. , Shi W., Tang Z.; *A lock-based cache coherence protocol for scope consistency*, Journal of Computer Science and Technology, 13(2):97-109, March, 1998.
14. Iftode L., Singh J.P., Li K; *Scope Consistency: A bridge between release consistency and entry consistency.* Proceedings of the 8th ACM Annual Symposium on Parallel Algorithms and Architectures (SPAA'96), pp. 277-287, June, 1996.
15. Woo S., Ohara M., Torrie E., Singh J.P., Gupta A.; *The SPLASH-2 programs: Characterization and methodological considerations. In Proceedings of the 22th Annual Symposium on Computer* Architecture, pages 24-36, June, 1995.
16. Amza C., Cox A. L., Dwarkadas S., Jin L. J., Rajamani K., Zwaenepoel W., *Adaptive Protocols for Software Distributed Shared Memory*, Proceedings of IEEE, Special Issue on Distributed Shared Memory, pp. 467-475, March 1999.
17. Iftode L., Singh J. P.; *Shared Virtual Memory: Progress and Challenges*; Proceedings of the IEEE, Vol 87, No. 3, March 1999, 1999.
18. Speight E., Bennett J. K., *Brazos: A third generation DSM system*, In Proceedings of the 1997 USENIX Windows/NT Workshop, pp. 95-106, August, 1997.
19. Keleher P., Update Protocols and Iterative Scientific Applications, In The 12th International Parallel Processing Symposium, March 1998.

Nanothreads vs. Fibers for the Support of Fine Grain Parallelism on Windows NT/2000 Platforms

Vasileios K. Barekas, Panagiotis E. Hadjidoukas,
Eleftherios D. Polychronopoulos, and Theodore S. Papatheodorou

High Performance Information Systems Laboratory
Department of Computer Engineering and Informatics, University of Patras
Rio 26500, Patras, Grecce
{bkb,peh,edp,tsp}@hpclab.ceid.upatras.gr
http://www.hpclab.ceid.upatras.gr

Abstract. Support for parallel programming is very essential for the efficient utilization of modern multiprocessor systems. This paper focuses on the implementation of multithreaded runtime libraries used for the fine-grain parallelization of applications on the Windows 2000 operating system. We have implemented and introduce two runtime libraries. The first one is based on standard Windows user-level fibers, while the second is based on nanothreads. Both follow the Nanothreads Programming Model. A systematic evaluation comparing both implementations has also been conducted in three levels: the user-level thread packages, the runtime libraries and the applications level. The results demonstrate that nanothreads outperform the Windows fibers. The performance gains of the thread creation and context switching mechanisms are reflected on both runtime libraries. Experiments with fine-grain applications demonstrate up to 40% higher speedup in the case of nanothreads compared to that of fibers.

1 Introduction

During the last few years, there have been significant technological advances in the area of workstations and servers. These systems are based on low-cost multiprocessor configurations running conventional operating systems, like Windows NT. Although the performance of these systems is comparable to that of other more expensive small-scale Unix-based multiprocessors, the software used is inadequate to utilize the existing hardware efficiently. Parallel processing on these systems is in a primitive stage, due to the lack of appropriate tools for the efficient implementation of parallel applications. The parallelization of a sequential application requires the explicit use and knowledge of the underlying thread architecture. Furthermore, the user himself must detect the potential parallelism. As we show in this paper, the existing support provided by the Windows is inadequate for the efficient implementation of a wide range of parallel applications.

M. Valero et al. (Eds.): ISHPC 2000, LNCS 1940, pp. 146–159, 2000.

On the other hand, most of the high-end multiprocessors running Unix-like operating system provide adequate and convenient tools to the user in order to build parallel applications, which can be executed with the lowest possible overhead. Such advanced tools consist of an automatic parallelizing compiler, an optimized multi-threading runtime library and the appropriate operating system support [1,3]. Directives are inserted manually or automatically at compile time into the sequential code to specify the existing parallelism [11]. The modified code is analyzed and the directives are interpreted into appropriate runtime API functions. The final code is compiled and linked with the runtime library to produce the parallelized executable.

In this paper, we present the implementation of two multithreaded runtime libraries, both based on user-level threads running on Windows 2000. These libraries are specially designed to provide the user with the necessary support for the efficient parallelization of applications. The first library, called FibRT (Fibers RunTime), uses the standard Windows fibers, while the second one, called NTLib (NanoThreads Library for NT), uses nanothreads, a custom user-level threads package. The NTLib runtime library was ported to Windows 2000 operating system from its original implementation on the IRIX operating system [1]. Both FibRT and NTLib libraries are implemented according to the Nanothreads Programming Model [12]. These libraries export the same API to the user, providing the same functionality. For the rest of this paper, the term runtime library will refer to both FibRT and NTLib, unless otherwise specified. We compare both implementations in terms of runtime overhead, and the performance gains by using them in the parallelization and execution of real applications.

The rest of this paper is organized as follows; Section 2 provides the necessary background. In Section 3, we present the two runtime libraries introduced in this paper, together with the necessary details of our implementations. Performance study and experimental results are presented in Section 4. In Section 5, we present related work; finally, we summarize in Section 6.

2 Background

In this section we outline the multithreading support provided in the Windows 2000 operating system, for both kernel-level and user-level threads, along with the Nanothreads Programming Model.

2.1 Windows Multithreaded Architecture

The Windows 2000 operating system supports multiple kernel-level threads, through a powerful thread management API [14]. These threads are the operating system's smallest kernel-level objects of execution and processes may consist of one or more threads. Each thread can create other threads that share the same address space and system resources having however, independent execution stack and thread specific data. Kernel-level threads are scheduled on a system wide basis by the kernel in order to be executed on a processor. It is through threads that Windows allows programmers to exploit the benefits of concurrency and parallelism. Since these threads are kernel-level, their state resides in the operating system kernel, which is responsible for their scheduling.

Besides kernel-level threads, Windows also provides user-level threads, called fibers. These are the smallest user-level objects of execution. Fibers run in the context of the threads that schedule them and are unknown to the operating system [7]. Each thread can schedule multiple fibers. However, a fiber does not have all the same state information associated with it as that associated with a thread. The only state information maintained for a fiber is its stack, a subset of its registers, and the fiber data provided during its creation. The saved registers are the set of registers typically preserved across a function call. A fiber is scheduled when switching to it from another fiber. The operating system still schedules threads to run. When a thread running fibers is preempted, its currently running fiber is also preempted. Figure 1 illustrates the overall Windows multithreaded architecture.

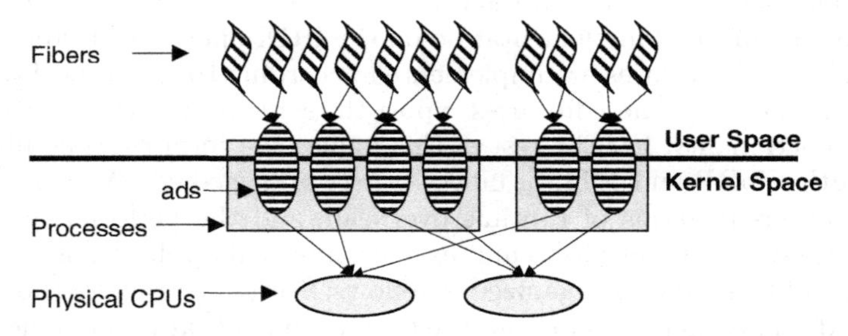

Fig. 1. Threads architecture in the Windows 2000 operating system, fibers reside completely in the user space

The efficient utilization of a multiprocessor system requires multiple kernel-level threads of control to be active at any time. The presence of more than one processors, means the simultaneous execution of corresponding number of threads. Although Windows thread API provides extensive functionality, kernel-level threads overhead makes them insufficient for fine-grain parallelization of applications. An application that uses hundreds of ready-to-execute threads reserves a significant part of the process address space. Furthermore, a large number of context switches occurs resulting in excessive scheduling overhead.

On the other hand, fiber management occurs entirely in user space and consequently their overhead is significantly lower than that of kernel-level threads. The cost of suspending a kernel-level thread is an order of magnitude more than that of a fiber's switching, which is performed in user-space. Similarly, there is an order of magnitude difference in the cost of creation between fibers and kernel-level threads. The application programmer is responsible for the management of fibers such as allocating memory, scheduling them on kernel threads and preempting them. This means that the user has to manage the scheduling of fibers, while the scheduling of threads is controlled entirely by the operating system. Thus, fibers programming becomes more difficult. Obviously, an appropriate runtime system that provides support for programming with fibers can take advantage of all of their potential benefits.

2.2 The Nanothreads Programming Model

The Nanothreads Programming Model (NPM) [12] exploits multiple levels of loop and functional parallelism. The integrated compilation and execution environment consists of a parallelizing compiler, a multithreaded runtime library and the appropriate operating system support. According to this model, applications are decomposed into fine-grain tasks and executed in a dynamic multiprogrammed environment. The parallelizing compiler analyzes the source program to produce an intermediate representation, called the Hierarchical Task Graph (HTG). The HTG is an directed acyclic graph with various hierarchy levels, which are determined by the nesting of loops and correspond to different levels of exploitable parallelism. Nodes at the same level of the hierarchy are connected with directed arcs that represent data or control dependencies between them. They can also have several input and output dependencies. Any arc between two nodes implies that these nodes have to be executed sequentially. A node is instantiated with a task that uses a user-level thread to execute its associated code. The task is not ready to be executed, as its dependencies may be unresolved. When all its dependencies have been satisfied, the task becomes ready for execution.

The runtime library controls the creation of tasks, ensuring that the generated parallelism matches the number of processors allocated to the application by the operating system . The overhead of the runtime library is low enough to make the management of parallelism affordable. In other words, the runtime library implements dynamic program adaptability to the available resources by adjusting the granularity of the generated parallelism. The runtime library environment co-operates with the operating system, which distributes physical processors among the running applications. The operating system provides virtual processors to applications, as the kernel abstraction of physical processors on which applications can execute. Virtual processors provide user-level contexts for the execution of the tasks' user-level threads. The main objective of the NPM is that both application scheduling (user-level threads to virtual processors mapping) at the user-level, and virtual processor scheduling (virtual to physical processors mapping) at the kernel-level, must be tightly coordinated in order to achieve high performance.

3 Runtime Libraries Implementation

In this section we present a description of our runtime libraries implementation. These libraries are primarily designed to provide support for the parallel execution of tasks at the backend of a parallelizing compiler. The compiler takes as input either sequential code or code annotated with special directives that indicate the presence of code that can be executed in parallel by multiple processors. The latter case is the most frequently technique used today after the adoption of the OpenMP standard [11]. The compiler translates the annotated code into parallel code that takes advantage of the runtime libraries. Beside the compiler, our runtime libraries can be used by a programmer directly, for the development of applications that use the exported API. For the rest of this section, the term user will refer to either the parallelizing compiler or the application programmer.

As stated before, we have implemented two multithreaded runtime libraries for the Windows NT/2000 environment; the NTLib library that uses custom user-level threads, named nanothreads, and the FibRT library that uses the standard Windows user-level fibers. The custom user-level threads used by the NTLib are based on the QuickThreads package [5], which provides similar functionality for non-preemptive thread management, with that provided for fibers by the Windows API. Additionally, the QuickThreads package enables the passing of multiple arguments to the thread function. Both libraries have been designed to provide the user with a suitable interface for the exploitation of application parallelism. Their light-weight user-level threads support fine-grain parallelism, and the programming model used allows the exploitation of multiple levels of parallelism.

The currently exported API implements the Nanothreads programming model, which has been described in Section 2.2. Other programming models such as the fork-join, which is required for the OpenMP standard, can be implemented easily using the existing infrastructure. Both NTLib and FibRT runtime libraries export the same API to the user, providing the same functionality. This API is responsible for the task management, the handling of the ready task queues, the control of the dependencies between the tasks and the initialization of the environment. The implementation details are described in the rest of this section. A detailed description of the exported API can be found in [2].

Task Management. A task is the fundamental object that the runtime libraries manage. Tasks are blocks of application code, that can be executed in parallel. The responsibility of our runtime libraries is to instantiate them using user-level threads. Each task has to execute some work that is represented by a user supplied function, which can take multiple arguments. Task management implementation is different in the two runtime libraries because they use different user-level thread packages for the instantiation of tasks. On both libraries, each task is represented by a compact structure, called task structure, which describes the work that a task will execute, its input and output dependencies, the virtual processor where the task runs on and a pointer to the associated user-level thread stack. In the NTLib, the creation of a task involves the creation of a nanothread, which will execute the task's work. The nanothread's stack is initialized with the work information (user function and its arguments). In order to execute the user function we just switch to the task's user-level thread.

On the contrary, in the FibRT we create a fiber for each task. In this case, the work information resides in the task structure and its pointer refers to the associated fiber. This fiber is created in order to execute a helper function in which we pass as argument a pointer to the task structure. This function pushes the user's function arguments in the stack and calls the user function. When switching to the associated fiber, the helper function is executed, having as result the user function to be executed. This way the user can directly execute a function with multiple arguments through a fiber, bypassing the fibers disadvantage of taking only one argument for the user function.

The exported API provides functions for the creation of tasks; the user specifies the function, its arguments and task's input and output dependencies. The task creation procedure involves the initialization of the task structure and the associated user-level thread (nanothread or fiber). In the NTLib we must allocate space for both the task structure and the nanothread's stack, while in the FibRT we allocate space only for the task structure and then we call the Win32 API function `CreateFiber` to create and initialize a new fiber.

Optionally, we can reuse already allocated space; this is another distinguishing point between the two libraries. In the NTLib, we maintain a global queue called reuse queue, where we insert the nanothread stacks of the already executed tasks. When we need to create a nanothread for a new task, we extract an already allocated task structure and nanothread's stack and finally we reinitialize them with the new task information. If we cannot find an already executed nanothread in this queue to reuse, we allocate space for a new task structure and the nanothread's stack. In the FibRT we reuse only task structures, not the fibers themselves, due to the limitations of the fibers' implementation. The space allocated for each fiber is released after the termination of its execution.

Queue Management. Both libraries maintain ready queues, where the ready for execution tasks are inserted. A task is ready to be executed when all its input dependencies have been satisfied from its ancestors in the HTG graph. There is one global queue where all the processors have equal access and per-processor local queues, where only the owner processor has access. Although this configuration is very flexible and preserves affinity for the task scheduling, we optionally allow a processor with an empty local queue to steal work from another processor's local queue to maintain load balancing and better system utilization.

Dependency Management. Inside the task structure, we keep information for both input and output dependencies. Input dependencies are represented using a counter in the structure, while output dependencies are maintained by keeping a pointer for each successor task. When a task finishes its execution, it satisfies one input dependency on each one of its successors. Every time a task creates a subtask, additional care must be taken to preserve the input dependencies of the creator task. For this reason, we must increase by one the creator task input dependencies and declare it as the successor of the subtask. This way, we can maintain multiple levels of nested tasks according to the HTG structure, making our runtime libraries capable to exploit multiple levels of parallelism.

Runtime Initialization. The runtime libraries environment has to be initialized before the user calls any of the libraries routines. An initialization routine is provided for the instantiation of the environment and is the first library function that an application executes. This routine takes as arguments some user parameters such as the maximum number of processors that the application will run on, the thread scheduling policy that will be used, etc. This routine initializes the ready queue environment, and creates one kernel-level thread for each processor that the application will run on. These kernel-level threads play the role of virtual processors, which will execute the application created tasks, and are bound to specific processors. Additionally, in the FibRT library the created kernel-level threads have to initialize their internal structures in order to support fibers, so we call the `ConvertThreadToFiber` Win32 API function.

Task Scheduling. Each virtual processor enters a task dispatching routine and searches for ready tasks to execute, either after its creation, or when the execution of a task has finished. To maintain the affinity across tasks, the next task is selected from the set of the satisfied successors of the previously ran task. If there are more than one satisfied successors, the first is selected and the remainders are inserted into the local ready queue. If there are no satisfied successors, the virtual processor will search for the next task, first in its local ready queue and then in the global ready queue. Selecting the next task using this order maximizes the exploited task affinity. A virtual processor that cannot find any task with the above method can search in another processor's local queue [13]. Using this technique, we maximize load balancing across the executing processors.

4 Experimental Evaluation

This section reports a systematic evaluation of the implementation of both runtime libraries. In addition, we present the performance of native Windows kernel-level threads, in order to show their inefficiency compared to that of user-level threads. More specifically, the measurements were conducted in three levels: the user-level threads packages, the runtime libraries, and manually parallelized applications built using the two runtime libraries. In subsection, 4.1 an evaluation of the user-level threads primitives cost is presented. Subsection 4.2 reports the performance of the multithreaded runtime libraries. Finally, in subsection 4.3 we use applications parallelized with our libraries to measure the performance delivered to the final user.

All the experiments were conducted on a Compaq Proliant 5500 4-processor 200MHz Pentium Pro system, running Windows 2000 Advanced Server, equipped with 512 MB of main memory. Both runtime libraries were developed using the Microsoft Visual C++ compiler. Time measurements were collected using the Pentium Pro processor's time-stamp counter register.

4.1 Evaluating User-Level Threads

In this section, we measure the cost of the thread primitives for both the Windows fibers and nanothreads implementation. These primitives include thread creation and thread context switching. Additionally, we present the cost of the corresponding kernel-level threads primitives. These experiments were conducted using only one processor, by setting the process affinity mask to this processor.

First, we use a simple microbenchmark to measure the creation time of suspended kernel-level threads, fibers and nanothreads. Figure 2.a illustrates the results for various numbers of threads. The measured times are given in clock cycles and presented in logarithmic scale.

The time required for the creation of a number of nanothreads is almost half the time needed to create the same number of fibers. This difference between nanothreads and fibers is due to the smaller stack which is allocated during the nanothreads initialization and the more independent nature of the nanothreads versus that of fibers, which depend on their creator kernel-level thread. During this experiment we didn't use the stack reuse mechanism of the nanothreads package. As expected, the creation time for both user-level thread packages is almost an order of magnitude less than that of Windows kernel-level threads. This is due to the heavy nature of the kernel level threads, which require during their creation the initialization of both their user and kernel context, and the internal kernel structures.

In the second experiment, we evaluate the cost of the context switching for the three thread classes. For this reason, we use a microbenchmark to create a number of suspended kernel-level threads, fibers, and nanothreads, which execute an empty function. For user-level threads, we measure the time it takes to execute all threads, using peer-to-peer scheduling. In the case of kernel-level threads, the measurement includes the time to resume the initially suspended threads until they finish their execution. In fact, the benchmark execution time indicates the cost of the context switching to a thread that has not run before in the system.

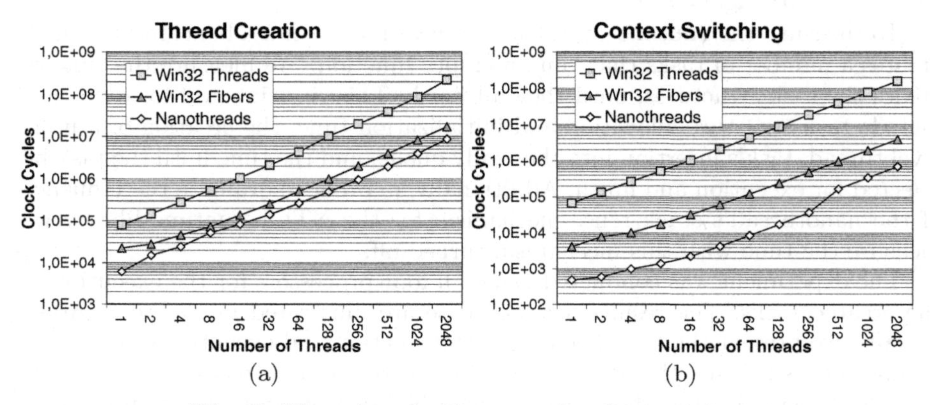

Fig. 2. Threads primitives overhead comparison

Our measurement includes the overhead to read the new thread's state from memory; this means that several cache misses and even page faults can occur. In this case, the cost differs from the pure context switching cost, which is measured using only two threads, switching to each other several times. This implies that both thread states reside in the cache memory. Our experiment represents a more realistic measurement, since it refers to the context switching cost which occurs during the execution of real applications. Figure 2.b illustrates the measured time from the microbenchmark execution, for each case.

The measured time for the nanothread case is almost an order of magnitude less than the corresponding time for fibers, and more than two orders of magnitude less than kernel-level threads' time. These results are very interesting and show that the nanothreads context switching mechanism is faster compared to that of fibers. Although both user-level thread packages preserve almost the same state between context switching, we measured a substantial difference in their costs. This difference is due to the more compact implementation of the nanothreads package which uses its stack to save its state registers, while the fibers use a separate context structure. Additionally, fibers preserve some state about exception handling. The large difference between the two user-level threads packages and the kernel-level threads case is because of the full processor context that is preserved between consecutive kernel-level threads context switching.

Both experiments demonstrate that nanothreads provide significantly less creation and context switching costs than the standard Windows fibers. Their low overhead makes them suitable for the fine-grain parallelization of applications. Furthermore the above experiments show the inefficiency of kernel-level thread for their use in fine-grain parallel applications.

4.2 Evaluating the Runtime Libraries

In this section, we evaluate the runtime overhead that NTLib and FibRT libraries impose in the creation and execution of tasks. In fact, we want to see what are the advantages of using the faster primitives of the nanothreads in a complex runtime library, against fibers.

To measure the runtime overhead we have built a microbenchmark, which creates a number of tasks that run an empty function. We measure the time to create the task structures, until they all finish. This time interval for each task include task structure allocation and initialization time, the creation of a user-level thread, task's insertion into the ready queue, and the time until the task be selected for execution and finish. Additionally, in order to evaluate the influence of the nanothread's stack reuse mechanism for the NTLib runtime library we measure the time with this mechanism turned off.

The experiment was conducted using all four processors for the execution of the created tasks. The results for each case are illustrated in Figure 3. Additionally, we present the time needed for the creation and execution of the same number of tasks instantiated using Windows kernel-level threads.

The measured times show clearly that the creation and execution of tasks in NTLib runtime library is much faster than both FibRT runtime library and

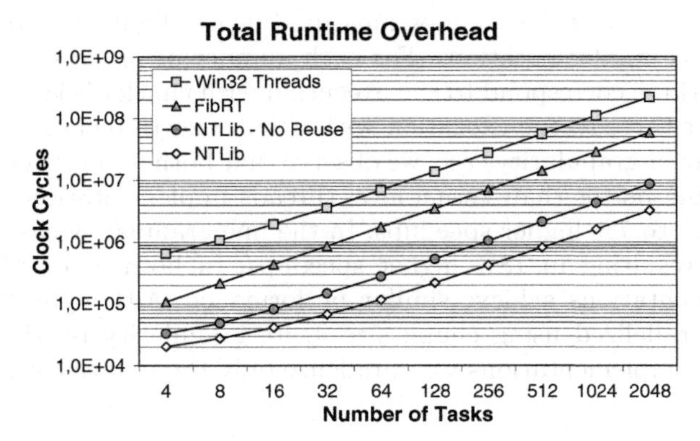

Fig. 3. Runtime library overhead for the creation and execution of a number of tasks

kernel-level threads, in all cases. In addition, the reuse mechanism in the NTLib library lowers the total runtime overhead by more than 50%. Although this mechanism improves only the task creation procedure, by allowing a new task to be created using an already allocated nanothread's stack, its influence is very important in the total execution time of the microbenchmark. The runtime overhead of the NTLib library is more than an order of magnitude lower than that of FibRT library and almost two orders of magnitude less than that of kernel-level threads. These results were expected due to the lower costs of the nanothreads primitives, measured in the previous section.

4.3 Performance Evaluation of Applications

In this section, we investigate whether the difference in the performance between the NTLib runtime library and the FibRT, is reflected on the execution of parallel applications. We are interested in the fine-grain parallelism because of its advantages over coarse-grain parallelism. For our experiments, we select three well-known applications in order to measure the performance of our runtime libraries. These applications have been parallelized using the runtime libraries, with variable granularity. The parallelization was made by hand since currently the source code of a parallelizing compiler is not available. Additionally, we have implemented these applications using Windows kernel-level threads just to illustrate their unsuitability for fine-grain parallelism. These applications are CMM, which performs complex matrix multiplication, BlockLU, which decomposes a dense matrix using blocking, and Raytrace, which renders a three-dimensional scene. The CMM is an application that has been previously used for the evaluation of the Nanothreads Model [9,13] while the other two applications come from the Splash-2 benchmark suite [16].

To examine the behavior of CMM under several levels of granularity we use matrices of size 192x192 with variable chunk size for the outer-most loop. In

Figure 4.a, we present the speedups measured using chunk size equal to 4 and 16, for our three implementations. For each number of processors we show six bars; the first three correspond to the execution with chuck of size 4 (fine-grain), while the other three to the execution with chunk size 16 (coarser-grain).

For the coarser granularity case, we observe that both user-level thread implementations scale better than kernel-level threads implementation. Particularly, they achieve up to 7% higher speedups. In the fine granularity case, these gains are increased resulting in 18% higher speedups. In both cases, the user-level thread implementations achieve similar performance. Although the CMM application is parallelized using chunk size 4, its granularity is still very coarse because in our implementations we parallelize only the outermost loop.

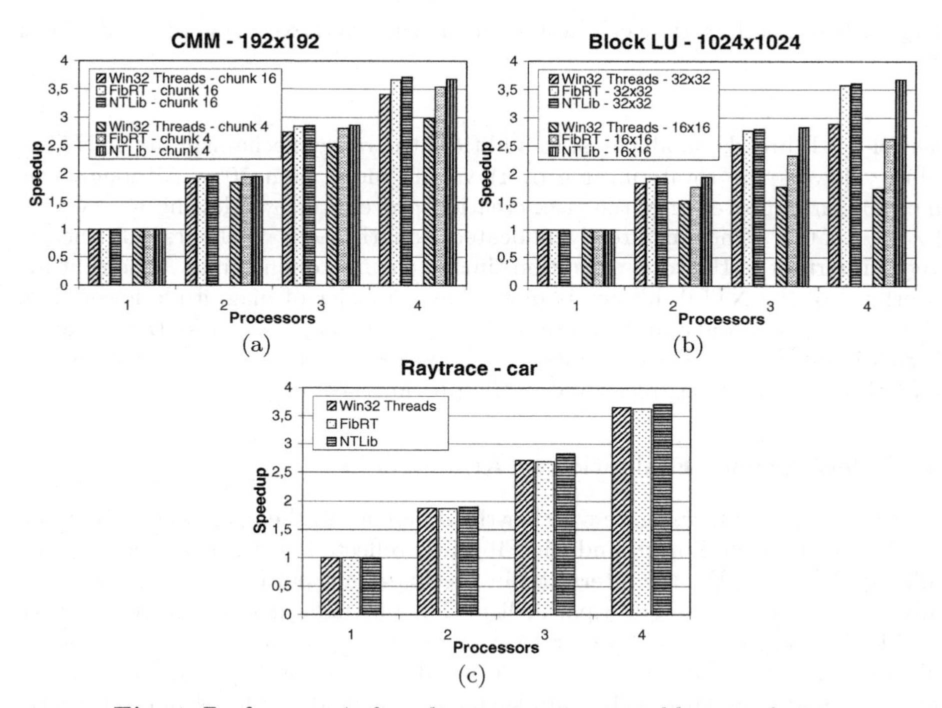

(a) (b) (c)

Fig. 4. Performance of applications using variable granularity

The second application, BlockLU divides the matrix into blocks. We use one task to represent the computation required for each block, so using different block sizes we can observe the performance of the application for different granularity. We execute BlockLU for matrix of size 1024x1024, using block sizes 32x32 and 16x16, to express the coarse-grain and the fine-grain case respectively. Speedups for each case are illustrated in Figure 4.b using the same form as Figure 4.a.

The speedup measurements for BlockLU exhibit larger divergences than for CMM, due to its finer granularity. During the execution of BlockLU, for a matrix of size 1024x1024 and block size 16x16, several thousands of tasks are created. For the coarse-grain case, we observe that the speedups of both runtime library implementations are up to 24% higher, than those of the kernel-level thread implementation. In the fine-grain case, the kernel-level threads implementation does not scale for more than three processors. On the contrary, both user-level implementations scale well; the NTLib implementation achieves 40% better speedup than the FibRT implementation, mainly due to its lower runtime overhead.

Our last application, Raytrace, differs from the other applications due to its coarser granularity nature. Although Raytrace creates several tasks, the work associated to each one task is quite large, in terms of execution time. Consequently, the task management overhead can be considered negligible compared to the total execution time. We execute Raytrace with the car scene as an input. In Figure 4.c, the speedups of the Raytrace execution are illustrated. As we can see, while all the implementations scale well, the NTLib implementation achieves approximately 3% better speedup that the other implementations.

Summarizing, the experiments make clear that for the exploitation of the fine-grain parallelism we must minimize the runtime overhead. Kernel-level threads are inefficient for that purpose due to their high overhead. Runtime libraries based on user-level thread packages provide adequate support, and low overhead for the efficient exploitation of the fine-grain parallelism. Furthermore, as shown by our application experiments the NTLib runtime library is more suitable than the corresponding FibRT.

5 Related Work

Although the Windows thread API provides the infrastructure for high performance multithreaded programming, only few runtime systems have been developed until recently. The runtime systems presented below, concern only the platform of Windows NT/2000. Visual KAP [6] from Kuck and Associates Inc. is a commercial optimizing preprocessor for Windows NT/95 that provides a multiprocessing runtime library. Another system for structured high-performance multithreaded programming has been presented in [15]. It is based on the Sthreads library on top of the support that Windows NT provides. In both cases, there is no support for multilevel parallelism and wherever there are nested parallel loops, only the outermost is parallelized. Furthermore, they are based on Windows kernel-level threads only, which according to our experiments in Section 4 are inappropriate for the exploitation of fine-grain parallelism.

The Illinois-Intel Multithreading Library [4] supports various types of parallelism and extends the degree of available support for multithreading by providing the capability to express nested loop, co-begin/co-end and DAG parallelisms. IML does support multiple levels of general, unstructured parallelism. IML uses Windows NT's fibers as user-level threads, against our NTLib which defines its own threads with the advantages of lower overhead for creation and context switching and as a result can support fine-grain parallelism. However, the most important difference is that our system provides, according to the defined programming model, the necessary functionality for multiprogramming support with arbitrary forms of parallelism.

Although in Unix platforms there is a variety of custom-based user-level thread packages, this is not the case for the Windows NT platforms. The only one user-level thread package that we found is that of Cthreads [8]. However, there is no information given about the implementation and technical details of this porting. The development on Cthreads itself continues only to the extent that it supports ongoing projects.

6 Conclusions and Future Work

This paper presented the development of two multithreaded runtime libraries NTLib and FibRT, based on nanothreads and fibers respectively. The overhead of both runtime libraries is low enough to make the management of fine-grain parallelism affordable. The systematic comparison of the two libraries showed that nanothreads are more efficient lightweight user-level threads than fibers. Experiments with applications and application benchmarks indicate that NTLib provides more efficient parallelization and better scalability than FibRT. The main objective of our system is the effective integration of fine-grain parallelism exploitation and multiprogramming. Our future work is concentrated on the implementation of a kernel interface that provides a lightweight communication path between active user applications and the Windows operating system. This interface will support requests of resources from the user-level execution environment and will inform it of actual resource allocation and availability. The implementation of the kernel interface relies on shared memory as the communication mechanism between the kernel and the application and vice versa. This mechanism and the scheduling policies are the major part of a series of modifications performed to Windows kernel and have already been implemented in other operating systems [9,10]. According to the above, we are in progress of implementing the necessary mechanism for integrating the kernel interface for multiprogramming support to our runtime system in the context of Windows 2000.

Acknowledgements

We acknowledge the contribution of all the research staff of the Nanos project, at UPC, University of Patras and CNR on the Nanos research infrastructure that led to this research and the results reported in our paper. This work was partially supported by G.S.R.T. research program 99EΔ-566.

References

1. E. Ayguadé, M. Furnari, M. Giordano, H-C. Hoppe, J. Labarta, X. Martorell, N. Navarro, D. Nikolopoulos, T. Papatheodorou and E. Polychronopoulos, *Nano-Threads Programming Model Specification*, ESPRIT Project No. 21907, Deliverable M1.D1, 1997. 147

2. V. Barekas, P. Hadjidoukas, E. Polychronopoulos and T. Papatheodorou, *Supporting Fine-Grain Parallelism on Windows NT Platforms*, Technical Report 012000, High Performance Information Systems Laboratory, University of Patras, January 2000. 150
3. M. Giordano and M. Mango Furnari, *A Graphical Parallelizing Environment for User-Compiler Interaction*, In Proc. of the 1999 International Conference on Supercomputing, Rhodes, Greece, June 1999. 147
4. M. Girkar, M. Haghighat, P. Grey, H. Saito, N. Stavrakos, and C. Polychronopoulos, *Intel-Illinois Multi-threading Library: Multithreaded support for IA-based multiprocessors systems*, Intel Technology Journal, 1998. 157
5. D. Keppel, *Tools and Techniques for Building Fast Portable Thread Packages*, Technical Report UW-CSE-93-05-06, University of Washington at Seattle, June 1993. 150
6. Kuck and Associates, Inc, *Visual KAP for OpenMP User's Manual Version 3.6*, available at `http://www.kai.com`. 157
7. Microsoft Corp., *Microsoft Developer Network Library*, available at `http://msdn.microsoft.com/library`. 148
8. B. Mukherjee, G. Eisenhauer and K. Ghosh, *A Machine Independent Interface for Lightweight Threads*, Operating Systems Review of the ACM Special Inter-est Group in Operating Systems, pages 33-47, January 1994. 158
9. D. Nikolopoulos, E. Polychronopoulos and T. Papatheodorou, *Fine-Grain and Multiprogramming-Conscious Nanothreading with the Solaris Operating System*, In Proc. of the 1999 International Conference on Parallel and Distributed Processing Techniques and Applications (PDPTA '99), CSREA Press, Vol. IV, pp. 1797-1803, Las Vegas, Nevada, June 28 - July 1 1999. 155, 158
10. D. Nikolopoulos, E. Polychronopoulos, T. Papatheodorou, C. Antonopoulos, I. Venetis and P. Hadjidoukas, *Achieving Multiprogramming Scalability of Parallel Programs on Intel SMP Platforms: Nanothreading in the Linux Kernel*, In Proc. of the Parallel Computing'99 Conference (ParCo'99), Delft, The Netherlands, August 1999. 158
11. OpenMP Architecture Review Board, *OpenMP C and C++ Application Program Interface Ver. 1.0*, Octomber 1998, available at `http://www.openmp.org`. 147, 150
12. C. Polychronopoulos, N. Bitar and S. Cleiman, *Nanothreads: A User-Level Threads Architecture*, Technical Report 1295, University of Illinois at Urbana-Champaign, 1993. 147, 149
13. E. Polychronopoulos and T. Papatheodorou, *Scheduling User-Level Threads on Distributed Shared Memory Multiprocessors*, In Proc. of the 5th EuroPar Conference (EuroPar '99), Toulouse, France, September 1999. 152, 155
14. J. Ritcher, *Advanced Windows: The Professional Developer 's Guide to the Win32 API for Windows NT 4.0 and Windows 95*, Microsoft Press, 1995. 147
15. J. Thornley, K. Mani Chandy and H. Ishii, *A System for Structured High-Performance Multithreaded Programming in Windows NT*, In Proc. of the 2nd USENIX Windows NT Symposium, pp. 67-76, Seattle, Washing-ton, August 1998. 157
16. S. Woo, M. Ohara, E. Torrie, J. P. Singh and A. Gupta, *The SPLASH-2 programs: Characterization and Methodological Monsiderations*, In Proc. of the 22th International Symposium on Computer Architecture, Santa Margherita Ligure, Italy, June 1995. 155

Partitioned Parallel Radix Sort*

Shin-Jae Lee[1], Minsoo Jeon[1], Andrew Sohn[2], and Dongseung Kim[1]

[1] Dept. Electrical Engineering, Korea University
Seoul, 136-701 Korea
dkim@classic.korea.ac.kr
[2] Computer & Info. Science Dept., New Jersey Institute of Technology
Newark, NJ 07102-1982, USA
sohn@cis.njit.edu

Abstract. *Load balanced parallel radix sort* solved the load imbalance problem present in parallel radix sort. Redistributing the keys in each round of radix, each processor has exactly the same number of keys, thereby reducing the overall sorting time. Load balanced radix sort is currently known the fastest internal sorting method for distributed-memory multiprocessors. However, as the computation time is balanced, the communication time emerges as the bottleneck of the overall sorting performance due to key redistribution. We present in this report a new parallel radix sorter that solves the communication problem of balanced radix sort, called *partitioned parallel radix sort*. The new method reduces the communication time by eliminating the redistribution steps. The keys are first sorted in a top-down fashion (left-to-right as opposed to right-to-left) by using some *most* significant bits. Once the keys are localized to each processor, the rest of sorting is confined within each processor, hence eliminating the need for global redistribution of keys. It enables well balanced communication and computation across processors. The proposed method has been implemented in three different distributed-memory platforms, including IBM SP2, CRAY T3E, and PC Cluster. Experimental results with various key distributions indicate that partitioned parallel radix sort indeed shows significant improvements over balanced radix sort. IBM SP2 shows 13% to 30% improvement while Cray/SGI T3E does 20% to 100% in execution time. PC cluster shows over 2.5 fold improvement in execution time.

1 Introduction

Sorting is one of the fundamental problems in computer science. Its use can be found essentially almost everywhere, be it scientific computation or non-numeric computation [8,9]. Sorting of a certain number of keys has been used in benchmarking various parallel computers or judging the specific algorithm performance when it is experimented on the same parallel machine. Serial sorts often need $O(N \log N)$ time, and the time becomes significant as the number of

* This work is partially supported by STEPI grant no. 97-NF-03-04-A-01, KRF grant no. 985-0900-003-2, and NSF grant no. INT-9722545.

M. Valero et al. (Eds.): ISHPC 2000, LNCS 1940, pp. 160–171, 2000.

keys becomes large. Because of its importance, numerous parallel sorting algorithms have been developed to reduce the overall sorting time, including bitonic sort [1], sample sort [4,5], and column sort [7]. In general, parallel sorts consist of multiple rounds of serial sort, called *local sort*, performed in each processor in parallel, followed by moving keys among processors, called *redistribution step* [6]. Local sort and data redistribution may be intermixed and iterated a few times depending on the algorithms used. The time spent in local sort depends on the number of keys. Parallel sort time is the sum of the times of local sort and the times for data redistribution in all rounds. To make the sort fast, it is important to distribute the keys as evenly as possible throughout the rounds of sort, since the execution time is dependent on the most heavily loaded processor in each round [5,11]. If a parallel sort has made its work load balanced perfectly in each round, there would be no further improvement of the time spent in that part. However, the communication time varies depending on the data redistribution schemes (e.g. all-to-all, one-to-many, many-to-one), the amount of data and its frequency of communication (e.g. many short messages, or a few long messages), and network topologies (hypercube, mesh, fat-tree) [8,3]. It was reported that for a large number of keys, the communication times occupy a great portion of the sorting time[3,12]. *Load balanced parallel radix sort* [11] or LBR, reduces the execution time by perfectly balancing the load among processors in every round. *Partitioned parallel radix sort* or PPR, proposed in this paper, further improves the performance by reducing the multiple rounds of data redistribution to *one*. While partitioned radix sort may introduce slight load imbalance among processors due to its not-so-perfect key distribution, the overall performance gain can be of particular significance since it substantially reduces the overall communication time. It is precisely the purpose of this report to introduce this new algorithm that features balanced computation and balanced communication.

The paper is organized as follows. Section 2 briefly explains balanced parallel radix sort and identifies its deficiency in terms of communication. Section 3 presents a new partitioned parallel radix sort. Section 4 lists the experimental results of the algorithm on three different distributed-memory parallel machines including SP2, T3E and PC cluster. The last section concludes this report.

2 Parallel Radix Sort

Radix sort is a simple yet very efficient sorting method that outperforms many well known comparison-based algorithms. Suppose N keys are evenly distributed to P processors initially such that there are $n = \frac{N}{P}$ keys per processor. When sort completes, we expect that all keys are ordered according to the rank of processors $P_0, P_1, \cdots, P_{P-1}$, besides keys in each processor have also been sorted. Serial radix sort is implemented in two different ways: *radix exchange sort* and *straight radix sort* [10]. Since parallel radix sorts are typically derived from a serial radix sort, we first briefly describe them, and ideas of parallelization will be given. We define some symbols used later in this paper as listed below:

- b is the number of bits of an integer key such that an integer key is represented as $(i_{b-1}i_{b-2}\cdots i_1 i_0)$.
- g is the group of bits used at each round of scanning.
- r is the number of rounds each key goes through, also the number of rounds in radix sort.

Radix exchange sort generates and maintains an *ordered queue*. Initially, it reads the g *least significant bits* (in other words, $i_{g-1}i_{g-2}\cdots i_1 i_0$) of each key, and stores it in a new queue at the location determined by the g bits. If all keys are examined and placed in the new queue, the round completes. Keys are ordered according to their least significant g bits. The following rounds use the next g least significant bits $(i_{2g-1}i_{2g-2}\cdots i_{g+1}i_g)$ to order them in a new queue. Keys move back and forth during the round. In the following rounds, the same operations are done as before. After r rounds (where $r = \lceil \frac{b}{g} \rceil$), all bits are scanned, and the sort is complete. LBR parallelizes the sort by repeating the following process a given number of times: it builds an ordered queue globally by using the least-significant g bits of keys, then divides it into P *equal sized* segments, allocates them to all processors. In other words, a globally ordered queue is created, then divided equally, and each is assigned to one processor. In the next round a new ordered queue is again generated using the next g bits by P processors together, then divided equally, and each is distributed. Load of processors is always perfectly balanced in this scheme because of the same number of keys given in each processor. LBR is reported to outperform fastest parallel sorts by up to 100% in execution time [11]. LBR, however, requires data redistribution across processors in *every round*, thus, it consumes a considerable amount of time in communication.

Straight radix sort initially uses $M = 2^g$ buckets instead of the ordered queues. It first *bucket–sorts* [10] keys using the g *most significant bits* $(i_{b-1}i_{b-2}\cdots i_{b-g})$ of each key. Bucket–sort puts keys into buckets whose index corresponds to the g bits. Thus, keys with the same g bits gather in the same bucket. Similarly in the second round, keys in each bucket are bucket-sorted again using the next g most significant bits $(i_{b-(g+1)}i_{b-(g+2)}\cdots i_{b-2g})$, generating M new subbuckets per bucket. The remaining rounds are done in the same manner. In this scheme keys never leave the bucket where they have been placed in a previous round. One significant problem in the scheme is that the number of overall buckets(subbuckets) explodes quickly, and many buckets with few keys waste a lot of resource (memory) if not carefully implemented.

In parallel implentation, the first round is done exactly the same as the serial straight radix sort. Then, each processor is assigned and will be in charge of a few consecutive buckets obtained in the first round. Now, buckets are exchanged among processors according to their index, thus keys with the same g most significant bits are collected from all processors into one. In the remaining $r - 1$ rounds, bucket sorts continues locally by using $b - g$ bits without data exchange. If keys are evenly distributed among buckets, each processor will hold M/P buckets in average. However, it is possible that some processors may be allocated with buckets with a lot of keys while others have few, depending on the distribution characteristics of the keys. This static/naive partitioning of keys may cause severe load imbalance among processors. PPR solves this problem as described in the next section.

3 Partitioned Parallel Radix Sort

Assume that we use only $M = 2^g$ buckets per processor throughout the sort. B_{ij} represents bucket-j in processor P_i. PPR consists of two phases: *local sort* and *key partitioning*. PPR needs $r = \lceil \frac{b}{g} \rceil$ rounds in all. Details are as follows:

I. Key Partitioning

Each processor bucket-sorts its keys using the g most significant bits. From now on, the left-most g bits of a key is represented by *the most significant digit* (*MSD*). Each key is placed into an appropriate bucket, thus, processor P_k stores a key to bucket B_{kj}, where j corresponds to the MSD of the key. At the end of the bucket sort, all keys have been partially sorted internally with respect to their MSDs, i.e. the first bucket includes the smallest keys, the second the next smallest, $\cdots$, and the last the largest. Then, processors collect the key counts of their buckets together to find a global key distribution map as follows. An illustratation is given in Figure 1.

For all $j = 0, 1, \cdots, M-1$, key counts of B_{kj} are added up to get G_j, a global count of keys in all buckets B_{kj} across processors (k=0,1, $\cdots$, $P-1$). Then prefix sums of global key counts of G_js are computed.

Let's consider hypothetical buckets (called *global buckets*) GB_js which are a collection of jth buckets of B_{kj} from all processors P_0, P_1, $\cdots$, P_{P-1}. Then G_j corresponds to the key count of bucket GB_j. Taking into account the prefix sums and the average number of keys ($n = N/P$), global buckets are to be divided into P groups, each having one or more consecutive buckets, in such a way that the key counts of each group become as equal as possible. The first group consists of first few buckets $GB_0, GB_1, \cdots GB_{k-1}$ whose counts add up to approximately n keys, the second $GB_k, GB_{k+1}, \cdots GB_l$ again to have approximately n keys, etc. jth group of buckets is now allocated P_j, which becomes the *owner* of the buckets.

Now all processors send their buckets of keys to their owners simultaneously except those whose owner is itself. After this movement, keys are sorted partially across processors, since any key in GB_i is smaller than any key in GB_j for $i < j$. Note that keys have not been sorted locally yet.

II. Local Sort

Keys in each processor are now sorted locally at a time by all processors, to make all N keys in order. Serial radix exchange sort is performed at first with the rightmost g bits, then, with the next rightmost g bits, $\cdots$, until all $b - g$ bits are used up. Only $b - g$ bits are examined because the left most g bits have already been used in Phase I. Phase II needs $\lceil (b - g)/g \rceil$ rounds.

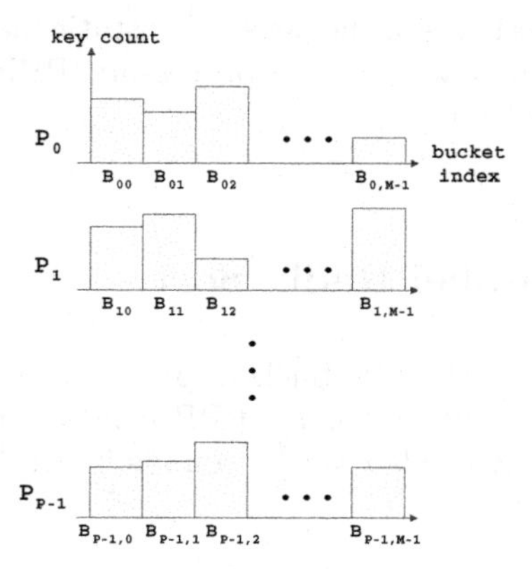

(a) Bucket counts in individul processors

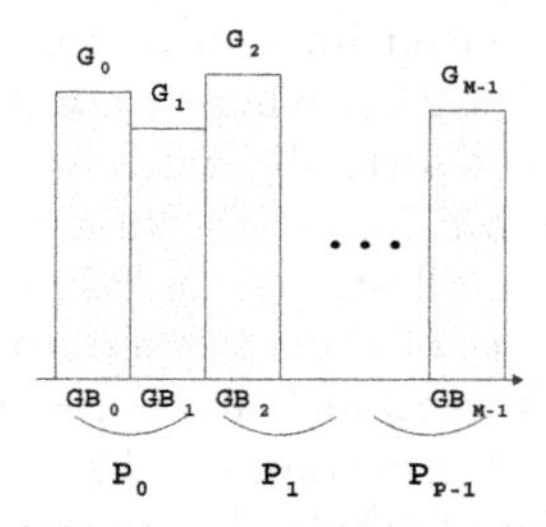

(b) Global bucket count and the partitioning map

Fig. 1. Local and global key count maps and bucket partitioning

The performance of PPR relies on how evenly the keys are distributed in the first phase. It is not very likely that each processor gets exactly the same number of keys. Refinement of the partitioning of keys can be made in Phase I by further dividing the buckets that lie in the partition boundary and have more counts than needed for even partitioning. However, splitting a bucket and allocating to two neighboring processors could not produce the desired sorted output by Phase II, since keys having the same MSD would stay in different processors. Thus, we avoid splitting buckets further, and keys should be distributed to processors *bucket by bucket*, on which our refinement method is based, as explained below.

PPR resembles *sample sort* [4,5] in the aspect of data partitioning and local sort. In sample sort, after keys have moved according to the *splitters (pivots)* to each processor, they are partially ordered across processors, thus further movement of keys across processors is not needed. One significant difference is in that the global key distribution statistic in sample sort is not known until keys actually move to designated processors, while in PPR it is known *before* the costly data movement. Thus, it is possible to adjust the partitioning before the actual key movement. If current partitioning is not likely to giving satisfactory balance in work load, PPR increases g so that the keys in the boundary buckets can spread out further into a larger number of *sub*-buckets to produce an evener partition. For example, if g is increased by 2 (bits), the keys in each boundary bucket are splitted into four buckets, enabling finer partitioning. The process repeats until a satisfactory partitioning is obtained.

4 Experiments and Discussion

PPR is experimented on three different parallel machines: IBM SP2, Cray T3E, and PC cluster. PC cluster is a set of 16 personal computers with 300MHz Pentium-II CPUs interconnected by a 100Mbps fast ethernet switch. T3E is the fastest machine among them, as long as the computational speed is concerned. As inputs of sort, N/P keys are synthetically generated in each processor with the distribution characteristics called *uniform, gauss,* and *stagger* [11]. Uniform creates keys with uniform distribution. Gauss forms keys with Gaussian distribution. Stagger produces specially distributed keys as described in [4]. We run the programs onto up to 64 processors, each with maximum of 64M keys. Keys are 32-bit integers for SP2 and PC cluster, and 64-bit integers for T3E. Code is written in C with MPI communication library [13]. Among many experiments we have performed, only a few representative results are shown here.

We first verify that PPR can reduce the communication time while it tolerates load imbalance. We expect the communication time be cut down to 1/4 and 1/8 at maximum with $g = 4, 8$, compared to LBR for sorts of 32-bit and 64-bit integer keys, respectively. As seen in Figures 2-3, there is a great reduction in communication times: they are now about 1/4 for 32–bit keys in SP2, and around 1/6 for 64 bit integers in T3E.

The load imbalance among processors is shown in Figure 4. It is the greatest for the case of Gauss, with maximum difference of 5.2% against the perfect balanced case, which proves it is not so severe as to significantly impair the overall performance of PPR. Improved performance of PPR over LBR can be observed in Figures 5 &6 for SP2, and Figures 7 & 8 for T3E.

We have found that in T3E the communication portion in sorting time is greater than SP2. In addition, since the keys are 64-bit integers in T3E, more improvement of PPR over LBR is expected due to larger r because we save $r - 1$ rounds of interprocessor communication. More enhancement on T3E can be observed in Figures 7 & 8 compared to Figures 5 & 6, respectively. In PC cluster, the network is so slow that the two parallel sorts are *slower* than the uniprocessor sort for the cases of $P \geq 8$ as shown in Figures 9-10. Nevertheless, PPR delivers remarkable performance over LBR since the communication time dominates the computation time. Table 1 lists the performance.

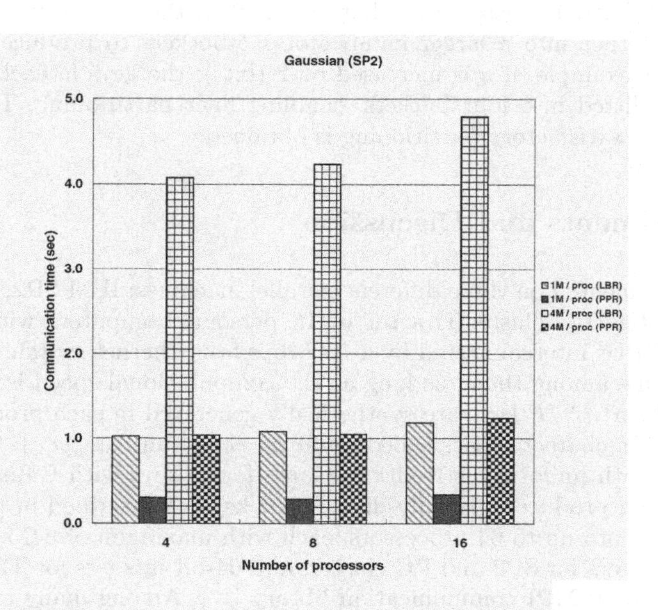

Fig. 2. Comparison of communication times of PPR and LBR on SP2 with gaussian distribution

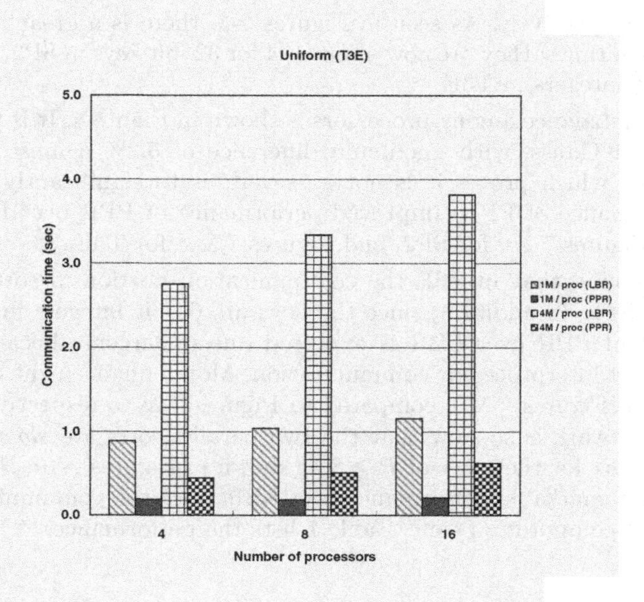

Fig. 3. Comparison of communication times on T3E with uniform distribution

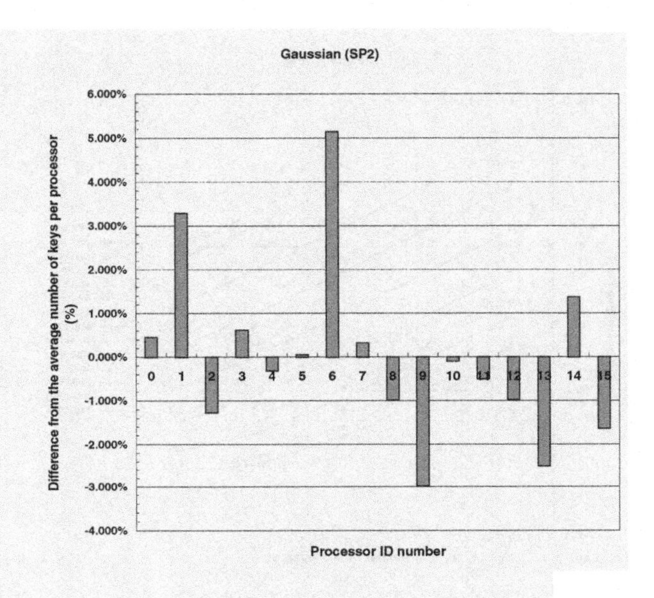

Fig. 4. Percentage deviation of work load from perfect balance on SP2 with gaussian distribution

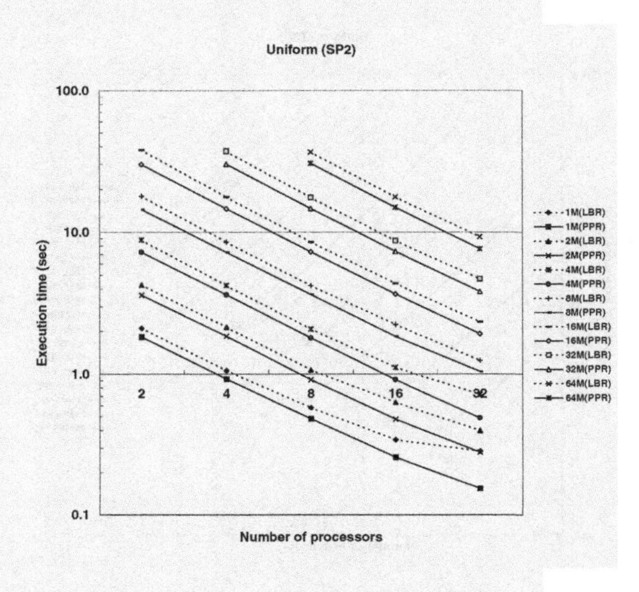

Fig. 5. Execution times on SP2 with uniform distribution

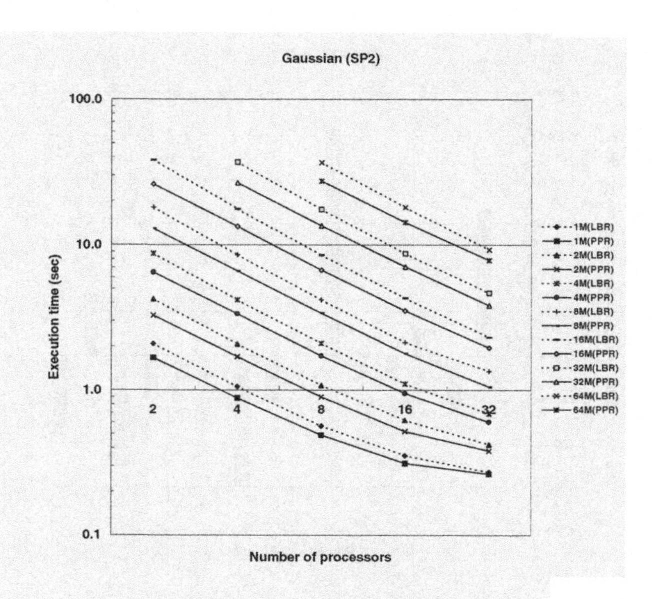

Fig. 6. Execution times on SP2 with gaussian distribution

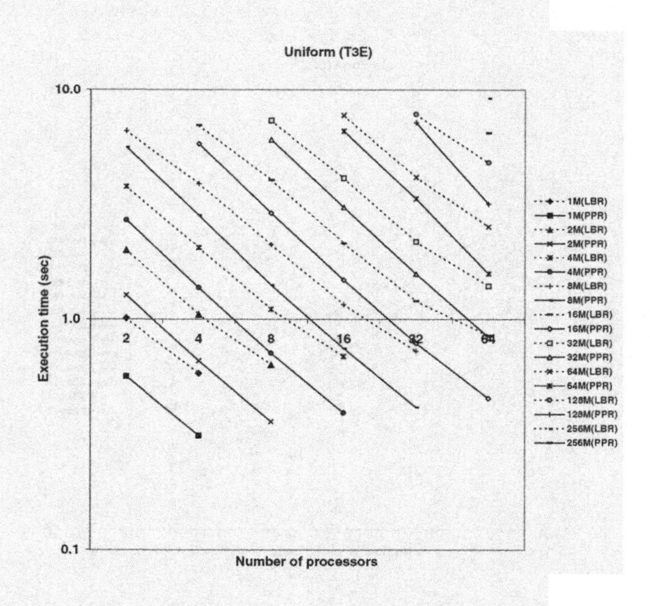

Fig. 7. Execution times on T3E with uniform distribution

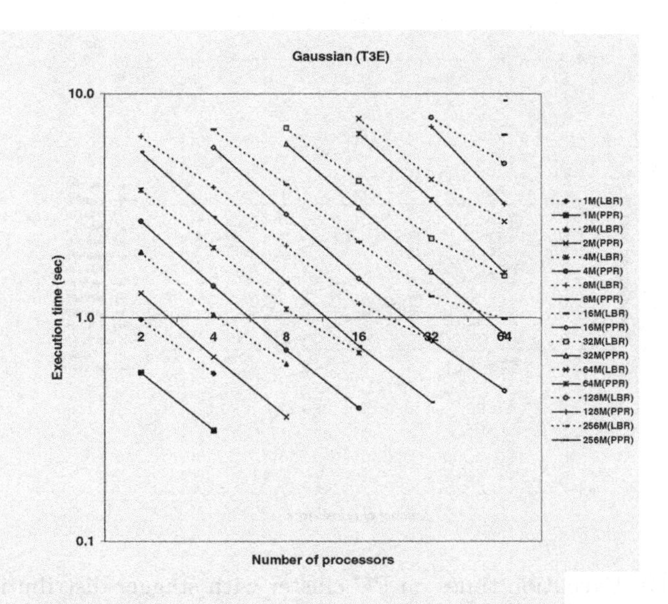

Fig. 8. Execution times on T3E with gaussian distribution

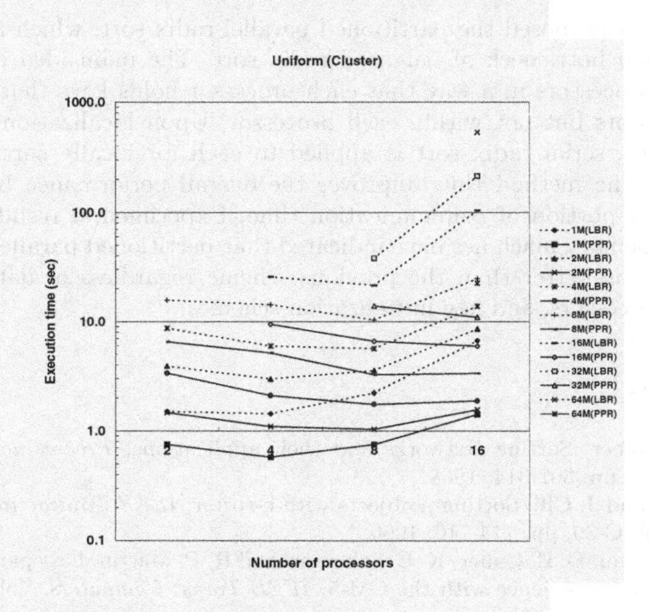

Fig. 9. Execution times on PC cluster with uniform distribution

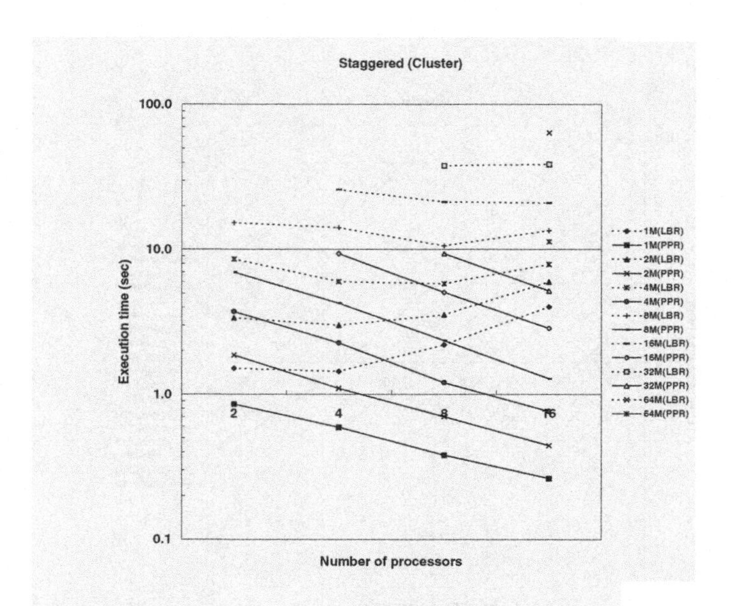

Fig. 10. Execution times on PC cluster with stagger distribution

5 Conclusion

This paper has proposed the partitioned parallel radix sort, which reduces the communication bottleneck of balanced radix sort. The main idea is to divide the keys to processors in a way that each processor holds keys that are sorted across processors but not within each processor. Upon localization of keys to each processor, serial radix sort is applied to each for locally sorting the assigned keys. The method thus improves the overall performance by reducing the significant portion of communication time. Experimental results on three distributed-memory machines have indicated that partitioned parallel radix sort always performs better than the previous scheme regardless of data size, the number of processors, and key initialization schemes.

References

1. M. E. Batcher, Sorting Networks and their applications, *Proceedings of AFIPS Conference*, pp. 307-314, 1968. 161
2. R. Beigel and J. Gill, Sorting n objects with k-sorter, *IEEE Transactions on Computers*, vol. C-39, pp. 714-716, 1990.
3. A. C. Dusseau, D. E. Culler, K. E. Schauser, and R. P. Martin, Fast parallel sorting under LogP: experience with the CM-5, *IEEE Trans. Computers*, Vol. 7(8), Aug. 1996. 161
4. D. R. Helman, D. A. Bader, and J. JaJa, Parallel algorithms for personalized communication and sorting with an experimental study, *Procs. ACM Symposium*

Table 1. Execution time of sorts on 4–processor PC cluster

Keys	LBR	PPR	speedup
uniform			
1M	1.434	0.577	2.485
2M	2.967	1.094	2.712
4M	5.974	2.090	2.858
8M	14.30	5.166	2.768
16M	25.660	9.480	2.706
stagger			
1M	1.442	0.593	2.432
2M	3.008	1.101	2.732
4M	6.034	2.268	2.660
8M	14.129	4.279	3.302
16M	26.007	9.342	2.784

on Parallel Algorithms and Architectures, Padua, Italy, pp. 211-220, June 1996. 161, 164, 165

5. J. S. Huang and Y. C. Chow, Parallel Sorting and Data Partitioning by Sampling, *Procs. the 7th Computer Software and Applications Conference*, pp. 627-631, November 1983. 161, 164
6. J. JaJa, *Introduction to Parallel Algorithms*, Addison-Wesley, 1992. 161
7. F. T. Leighton, Tight Bounds on the Complexity of Parallel Sorting, *IEEE Transactions on Computers*, C-34: pp. 344-354, 1985. 161
8. F. T. Leighton, *Introduction to Parallel Algorithms and Architectures: Arrays, Trees, Hypercubes*, Addison-Wesley, Morgan Kauffman, 1992. 160, 161
9. W. A. Martin, Sorting, *ACM Computing Surveys*, Vol. 3(4), p.p. 147-174, 1971. 160
10. Sedgewick, *Algorithms*, Wiley, 1990. 161, 162
11. A. Sohn and Y. Kodama, Load balanced parallel radix sort, *Procs. 12th ACM Int'l Conf. Supercomputing*, Melbourne, Australia, July 14-17, 1998. 161, 162, 165
12. A. Sohn, Y. Kodama, M. Sato, H. Sakane, H. Yamada, S. Sakai, Y. Yamaguchi, Identifying the capability of overlapping computation with communication, *Procs. ACM/IEEE Parallel Architecture and Compilation Techniques*, Boston, MA, Oct. 1996. 161
13. Message Passing Interface Forum, MPI: A Message-Passing Interface Standard. Technical report, University of Tennessee, Knoxville, TN, June 1995. 165

Transonic Wing Shape Optimization Based on Evolutionary Algorithms

Shigeru Obayashi,[1] Akira Oyama[2] and Takashi Nakamura[3]

[1] Tohoku University, Department of Aeronautics and Space Engineering,
Sendai, 980-8579, Japan.
obayashi@ieee.org
[2] Tohoku University, Currently, NASA Glenn Research Center, Cleveland, OH, USA.
[3] National Aerospace Laboratory, Chofu, Tokyo, 182-8522, Japan.
nakamura@nal.go.jp

Abstract. A practical three-dimensional shape optimization for aerodynamic design of a transonic wing has been performed using Evolutionary Algorithms (EAs). Because EAs coupled with aerodynamic function evaluations require enormous computational time, Numerical Wind Tunnel (NWT) located at National Aerospace Laboratory in Japan has been utilized based on the simple *master-slave* concept. Parallel processing makes EAs a very promising approach for practical aerodynamic design.

1 Introduction

Most of commercial aircrafts today cruise at transonic speeds. During the long duration of cruise, engine thrust is applied to maintain aircraft speed against aerodynamic drag. Since a large part of their maximum takeoff weights is occupied by the fuel weight, the objective of an aerodynamic design optimization of a transonic wing is, in principle, minimization of drag.

Unfortunately, drag minimization has many tradeoffs. There is a tradeoff between drag and lift because one of the drag component called induced drag increases in proportion to the square of the lift. A wing that achieves no induced drag would have no lift. Another tradeoff lies between aerodynamic drag and wing structure weight. An increase in the wing thickness allows the same bending moment to be carried with reduced skin thickness with an accompanying reduction in weight. On the other hand, it will lead to an increase in another component of the drag called wave drag. Therefore, the aerodynamic design of a transonic wing is a challenging problem.

Furthermore, optimization of a transonic wing design is difficult due to the followings. First, aerodynamic performance of a wing is very sensitive to its shape. Very precise definition of the shape is needed and thus its definition usually requires more than 100 design variables. Second, function evaluations are very expensive. An aerodynamic evaluation using a high fidelity model such as the Navier-Stokes equations usually requires 60-90 minutes of CPU time on a vector computer.

M. Valero et al. (Eds.): ISHPC 2000, LNCS 1940, pp. 172-181, 2000.

Among optimization algorithms, Gradient-based Methods (GMs) are well-known algorithms, which probe the optimum by calculating local gradient information. Although GMs are generally superior to other optimization algorithms in efficiency, the optimum obtained from these methods may not be a global one, especially in the aerodynamic optimization problem.

On the other hand, Evolutionary Algorithms, in particular, Genetic Algorithms (GAs) are known to be robust methods modeled on the mechanism of the natural evolution. GAs have capability of finding a global optimum because they don't use any derivative information and they search from multiple design points. Therefore, GAs are a promising approach to aerodynamic optimizations.

Finding a global optimum in the continuous domain is however challenging for GAs. In traditional GAs, binary representation has been used for chromosomes, which evenly discretizes a real design space. Since binary substrings representing each parameter with a desired precision are concatenated to form a chromosome for GAs, the resulting chromosome encoding a large number of design variables for real-world problems would result in a string length too long. In addition, there is discrepancy between the binary representation space and the actual problem space. For example, two points close to each other in the real space might be far away in the binary-represented space. It is still an open question to construct an efficient crossover operator that suits to such a modified problem space.

A simple solution to these problems is the use of floating-point representation of parameters as a chromosome [1]. In these real-coded GAs, a chromosome is coded as a finite-length string of the real numbers corresponding to the design variables. The floating-point representation is robust, accurate, and efficient because it is conceptually closest to the real design space, and moreover, the string length reduces to the number of design variables. It has been reported that the real-coded GAs outperformed binary-coded GAs in many design problems [2]. However, even the real-coded GAs would lead to premature convergence when applied to aerodynamic shape designs with a large number of design variables.

To apply GAs to practical, large-scale engineering problems, the idea of dynamic coding, in particular Adaptive Range GAs [3,4], is incorporated with the used of the floating-point representation. The objective of the present work is to apply the resulting approach to a practical transonic wing design and to demonstrate the feasibility of the present approach.

2 Adaptive Range Genetic Algorithms

To treat a large search space with GAs more efficiently, sophisticated approaches have been proposed, referred to as dynamic coding, which dynamically alters the coarseness of the search space. In [5], Krishnakumar et al. presented Stochastic Genetic Algorithms (Stochastic GAs) to solve problems with a large number of real design parameters efficiently. Stochastic GAs have been successfully applied to Integrated Flight Propulsion Controller designs [5] and air combat tactics optimization [6]. As they mentioned, the Stochastic GAs bridge the gap between ES and GAs to handle large design problems.

Adaptive Range Genetic Algorithms (ARGAs) proposed by Arakawa and Hagiwara [3] are a quite new approach, also using dynamic coding for binary-coded GAs to treat continuous design space. The essence of their idea is to adapt the population toward promising regions during the optimization process, which enables efficient and robust search in good precision while keeping the string length small. Moreover, ARGAs eliminate a need of prior definition of search boundaries since ARGAs distribute solution candidates according to the normal distributions of the design variables in the present population. In [4], ARGAs have been applied to pressure vessel designs and outperformed other optimization algorithms.

Since the ideas of the Stochastic GAs and the use of the floating point representation are incompatible, ARGAs for floating point representation are developed. The real-coded ARGAs are expected to possess both advantages of the binary-coded ARGAs and the floating point representation to overcome the problems of having a large search space that requires continuous sampling.

2.1 ARGAs for Binary Representation

When conventional binary-coded GAs are applied to real-number optimization problems, discrete values of real design variables p_i are given by evenly discretizing prior-defined search regions for each design variable [$p_{i,min}$, $p_{i,max}$] according to the length of the binary substring $b_{i,l}$ as

$$p_i = (p_{i,\max} - p_{i,\min})\frac{c_i}{2^{sl}-1} + p_{i,\min} \quad (1)$$

where sl represents string length and

$$c_i = \sum_{l=1}^{sl}(b_{i,l} \cdot 2^{l-1}).$$

In binary-coded ARGAs, decoding rules for the offspring are given by the following normal distributions,

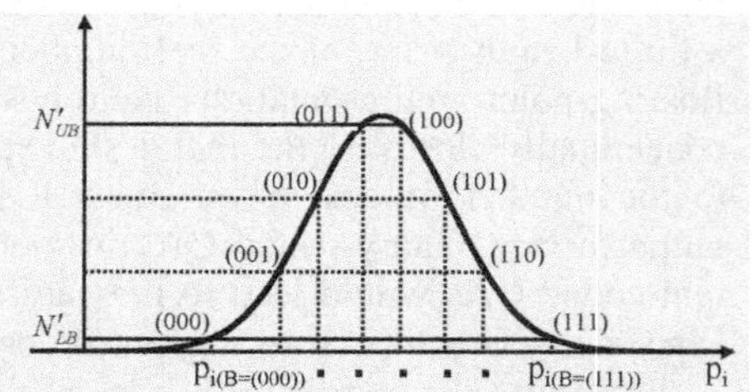

Fig. 1 Decoding for original ARGAs

$$N'(\mu_i,\sigma_i^2)(p_i) = \sqrt{2\pi}\sigma_i \cdot N(\mu_i,\sigma_i^2)(p_i) = \exp(-\frac{(p_i-\mu_i)^2}{2\sigma_i^2}) \quad (2)$$

where the average μ_i and the standard deviation σ_i of each design variable are determined by the population statistics. Those values are recomputed in every generation. Then, mapping from a binary string into a real number is given so that the region between N'_{UB} and N'_{LB} in Fig. 1 is divided into equal size regions according to the binary bit size as

$$p_i = \begin{cases} \mu_i - \sqrt{-2\sigma_i^2 \cdot \ln(N'_{LB} + (N'_{UB} - N'_{LB})\frac{c_i}{2^{sl-1}-1})} & for \quad c_i \le 2^{sl-1}-1 \\ \mu_i + \sqrt{-2\sigma_i^2 \cdot \ln(N'_{UB} - (N'_{UB} - N'_{LB})\frac{c_i - 2^{sl-1}}{2^{sl-1}-1})} & for \quad c_i \ge 2^{sl-1} \end{cases} \quad (3)$$

where N'_{UB} and N'_{LB} are additional system parameters defined in [0,1]. In the ARGAs, genes of design candidates represent relative locations in the updated range of the design space. Therefore, the offspring are supposed to represent likely a range of an optimal value of design variables.

Although the original ARGAs have been successfully applied to real parameter optimizations, there is still room for improvements. The first one is how to select the system parameters N'_{UB} and N'_{LB} on which robustness and efficiency of ARGAs largely depend. The second one is the use of constant intervals even near the center of the normal distributions. The last one is that since genes represent relative locations, the offsprings become constantly away from the centers of the normal distributions when the distributions are updated. Therefore, the actual population statistics does not coincide with the updated population statistics.

2.2 ARGAs for Floating-Point Representation

In real-coded Gas, real values of design variable are directly encoded as a real string r_i, $p_i=r_i$ where $p_{i,\min} \le r_i \le p_{i,\max}$.
Otherwise, sometimes normalized values of the design variables are used as

$$p_i = (p_{i,\max} - p_{i,\min}) \cdot r_i + p_{i,\min} \qquad (4)$$

where $0 \le r_i \le 1$.

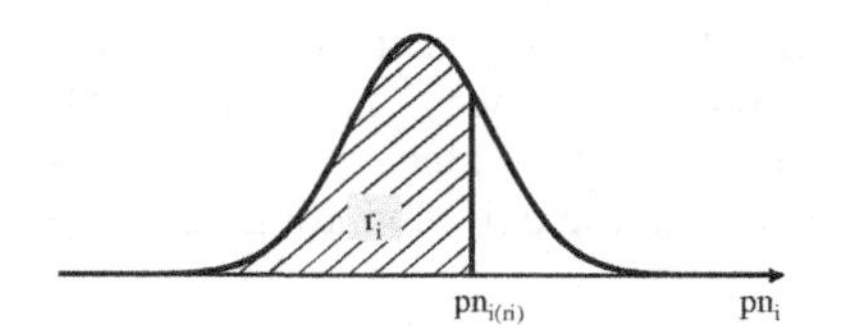

Fig. 2 Decoding for real-coded ARGAs

To employ floating-point representation for ARGAs, the real values of design variables p_i are rewritten here by the real numbers r_i defined in (0,1) so that integral of the probability distribution of the normal distribution from $-\infty$ to pn_i is equal to r_i as

$$p_i = \sigma_i \cdot pn_i + \mu_i \qquad (5)$$

$$r_i = \int_{-\infty}^{pn_i} N(0,1)(z)dz \qquad (6)$$

where the average μ_i and the standard deviation σ_i of each design variable are calculated by sampling the top half of the previous population so that the present population distributes in the hopeful search regions. Schematic view of this coding is illustrated in Fig. 2. It should be noted that the real-coded ARGAs resolve drawbacks of the original ARGAs; no need for selecting N'_{UB} and N'_{LB} as well as arbitrary resolution near the average. Updating μ_i and σ_i every generation, however, results in inconsistency between the actual and updated population statistics in the next generation because the selection operator picks up the genes that correspond to the promising region according to the old population

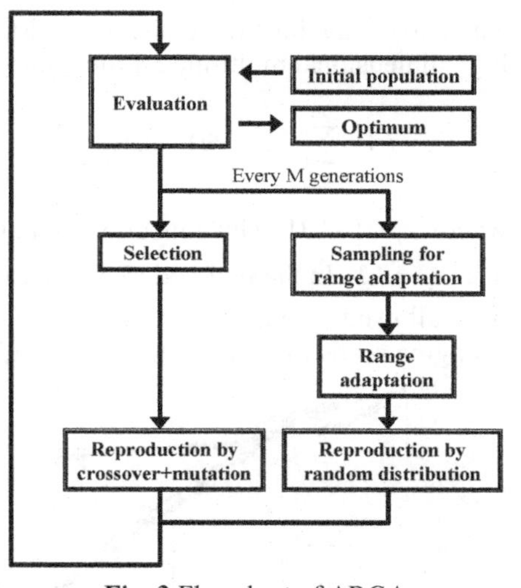

Fig. 3 Flowchart of ARGA

statistics. To prevent this inconsistency, the present ARGAs update μ and σ every M (M>1) generations and then the population is reinitialized. Flowchart of the present ARGA is shown in Fig. 3.

To improve robustness of the present ARGAs further, relaxation factors, ω_μ and ω_σ are introduced to update the average and standard deviation as

$$\mu_{new} = \mu_{present} + \omega_\mu(\mu_{sampling} - \mu_{present}) \tag{7}$$

$$\sigma_{new} = \sigma_{present} + \omega_\sigma(\sigma_{sampling} - \sigma_{present}) \tag{8}$$

where $\mu_{sampling}$ and $\sigma_{sampling}$ are determined by sampling the top half of the population. Here, ω_μ, ω_σ and M are set to 1, 0.5 and 4, respectively. They are determined by parametric studies using some simple test functions.

In this study, design variables are encoded in a finite-length string of real numbers. Fitness of a design candidate is determined by ist rank among the population based on ist objective function value and then selection is performed by the stochastic universal sampling [7] coupled with the elitist strategy. Ranking selection is adopted since it maintains sufficient selection pressure throughout the optimization. One-point crossover is always applied to real-number strings of the selected design candidates. Mutation takes place at a probability of 0.1, and then a uniform random disturbance is added to the corresponding gene in the amount up to 0.1.

2.3 Test Problem Using a Multi-modal Function

To demonstrate how the real-coded ARGA works, it was applied to minimization of a high dimensional multi-modal function:

$$F1 = \sum_{i=1}^{20} (x_i^2 + 5(1 - \cos(x_i \cdot \pi))) \tag{9}$$

where $x_i \in [-3,3]$. This function has a global minimum at $x_i=0$ and two local optima near $x_i = \pm 2$. In the real-coded ARGA, x_i correspond to p_i in eq.(5). 150 generations were allowed with a population size of 300. Five trials were run for each GA changing seeds for random numbers to give different initial populations. Figure 4

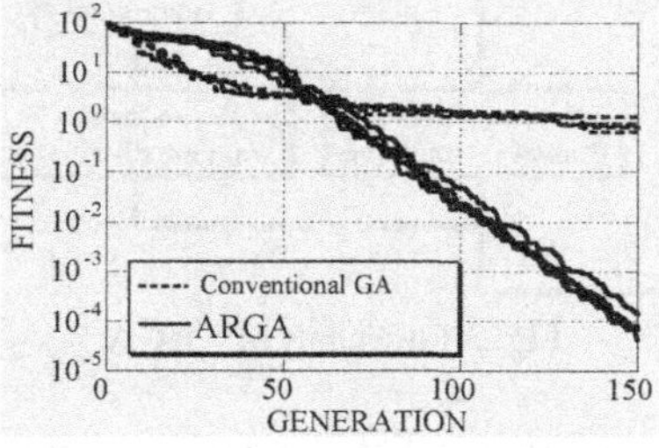

Fig. 4 Comparison of convergence histories

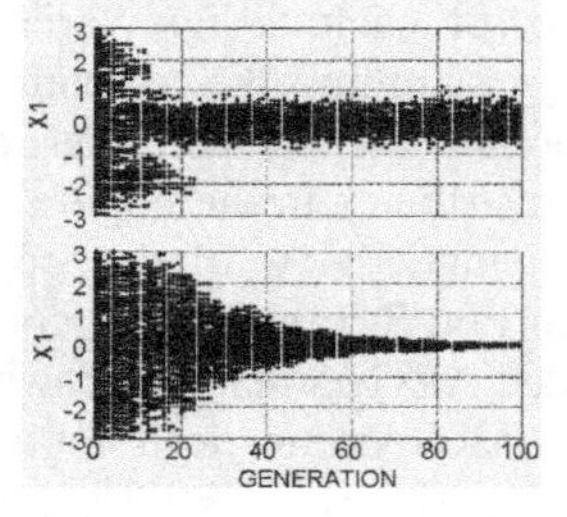

Fig. 5 Comparison of convergence histories of x_1 between GA (above) and ARGA (below)

compares the performances of the conventional GA and the ARGA. Figure 5 plots all x_1's from the temporary solutions, which helps to understand why the ARGA works better than the conventional GA. This figure shows that the ARGA maintains gene diversity longer than the conventional GA in the initial phase and then adapts to their search space to the local region near the optimal. While the initial gene diversity contributes to the ARGA' s robustness, the adaptive feature of the ARGA improves their local search capability. The ARGA also showed its advantages over a real-coded GA on dynamic control problem and aerodynamic airfoil shape optimization [8].

3 Aerodynamic Design of a Transonic Wing

A wide range of approximations can represent the flow physics. Among them, the Navier-Stokes equations provide the state-of-the-aft of aerodynamic performance evaluation for engineering purposes. Although the three-dimensional Navier-Stokes calculation requires large computer resources to estimate wing performances within a reasonable time, it is necessary because a flow around a wing involves significant viscous effects, such as potential boundary-layer separations and shock wave/boundary layer interactions in the transonic regime. Here, a three-dimensional Reynolds-averaged Navier-Stokes solver [9] is used to guarantee an accurate model of the flow field and to demonstrate the feasibility of the present algorithm.

The objective of the present wing design problem is maximization of lift-to-drag ratio *L/D* at the transonic cruise design point, maintaining the minimum wing thickness required for structural integrity against the bending moment due to the lift distribution. The cruising Mach number is set to 0.8. The Reynolds number based on the chord length at the wing root is assumed to 10^7.

In the present optimization, a planform shape of generic transport was selected as the test configuration (Fig. 6). Wing profiles of design candidates are generated by the PARSEC airfoils as briefly described in the next section. The PARSEC parameters and the sectional angle of attack (in other words, root incident angle and twist angle) are given at seven spanwise sections, of which spanwise locations are also treated as design variables except for the wing root and tip locations. The PARSEC parameters are rearranged from root to tip according to the airfoil thickness so that the resulting wings always have maximum thickness at the wing root. The twist angle parameter is also rearranged into numerical order from tip to root. The wing surface is then interpolated in spanwise direction by using the second-order Spline interpolation.

In total, 87 parameters determine a wing geometry. Parameter ranges of the design space are shown in Table 1. It should be noted that in ARGAs, user-defined design space is used just to seed the initial population. ARGA can promote the search space outside of the initially defined design space.

To estimate the required thickness distribution to stand the bending moment due to the lift distribution, the wing is modeled by a thin walled box-beam as shown in Fig. 6. The constraint for wing thickness t_1 is specified by using the minimum thickness t_{min} calculated from the wing box sustaining the aerodynamic bending moment M as,

$$t_1 > \frac{M}{\sigma_{ultimate} \cdot c \cdot t_2} = t_{\min} \tag{10}$$

where following assumptions are made: the thickness of the skin panels are 2.5[cm] and its ultimate normal stress $\sigma_{ultimate}$ is 39[ksi]. The length of the chord at wing root *c* and maximum wingspan *b/2* are 10[m] and 18.8[m], respectively (for the derivation of Eq. (10), see [10] for example).

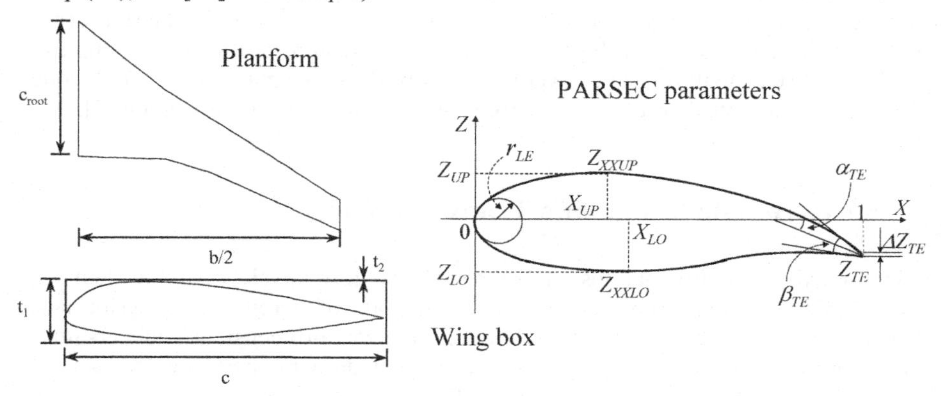

Fig. 6 Wing geometry definition. Planform shape is frozen during the optimization. Wing box is used to estimate its structural strength. PARSEC parameters are the design variables for airfoil shapes defined at seven spanwise sections

Table 1 Parameter ranges of the design space. PARSEC is determied by leading-edge radius (r_{LE}), upper and lower crest locations including curvatures (X_{UP}, Z_{UP}, Z_{XXUP} X_{LO}, Z_{LO}, Z_{XXLO}), trailing-edge ordinate (Z_{TE}) and thickness (ΔZ_{TE}) and direction and wedge angles (α_{TE}, β_{TE})

parameters	r_{LE}	Z_{TE}	α_{TE}	β_{TE}	X_{UP}	Z_{UP}	Z_{XXUP}	X_{LO}	Z_{LO}	Z_{XXLO}	twist angle
Upper bound	0.030	0.01	-3.0	8.0	0.7	0.18	0.0	0.6	0.02	0.9	7 deg
Lower bound	0.002	-0.01	-13.0	4.0	0.3	0.08	-0.3	0.2	-0.04	0.3	-1 deg

3.1 PARSEC Airfoils

An airfoil family “PARSEC” has been recently proposed to parameterize an airfoil shape [11]. A remarkable point is that this technique has been developed aiming to control important aerodynamic features effectively by selecting the design parameters based on the knowledge of transonic flows around an airfoil.

Similar to 4-digit NACA series airfoils, the PARSEC parameterizes upper and lower airfoil surfaces using polynomials in coordinates *X*, *Z* as, $Z=\sum_{n=1}^{6} a_n \cdot X^{n-1/2}$ where a_n are real coefficients. Instead of taking these coefficients as design parameters, the PARSEC airfoils are defined by basic geometric parameters: leading-edge radius (r_{LE}), upper and lower crest locations including curvatures (X_{UP}, Z_{UP}, Z_{XXUP} X_{LO}, Z_{LO}, Z_{XXLO}), trailing-edge ordinate (Z_{TE}), thickness (ΔZ_{TE}) and direction and wedge angles

(α_{TE}, β_{TE}) as shown in Fig. 6. These parameters can be expressed by the original coefficients a_n by solving simple simultaneous equations. Eleven design parameters are required for the PARSEC airfoils to define an airfoil shape in total. In the present case, the trailing-edge thickness is frozen to 0. Therefore, ten design variables are used to give each spanwise section of the wing.

3.2 Optimization Using Real-Coded ARGA

Because the objective function distribution of the present optimization is likely to be more complex than the above test function minimization, the relaxation factor ω_σ is now set to 0.3. The structured coding coupled with one-point crossover proposed in [12] is also incorporated. The present ARGA adopts the elitist strategy where the best and the second best individuals in each generation are transferred into the next generation without any crossover or mutation. The parental selection consists of the stochastic universal sampling and the ranking method using Michalewicz's nonlinear function. Mutation takes place at a probability of 10% and then adds a random disturbance to the corresponding gene in the amount up to ± 10% of each parameter range in Table 1. The population size is kept at 64 and the maximum number of generations is set to 65 (based on the CPU time allowed). The initial population is generated randomly over the entire design space.

The main concern related to the use of GAs coupled with a three-dimensional Navier-Stokes solver for aerodynamic designs is the computational cost required. In the present case, each CFD evaluation takes about 100 min. of CPU time even on a vector computer. Because the present optimization evaluates 64 x 65 = 4160 design candidates, sequential evolutions would take almost 7000 h (more than nine months!).

Fortunately, parallel vector computers are now available at several institutions and universities. In addition, GAs are intrinsically parallel algorithms and can be easily parallelized. One of such computers is *Numerical Wind Tunnel* (*NWT*) located at National Aerospace Laboratory in Japan. NWT is a MIMD parallel computer with 166 vector-processing elements (PEs) and its total peak performance and the total main memory capacity are about 280 GFLOPS and 45GB, respectively. For more detail, see [13]. In the present optimization, evaluation process at each generation was parallelized using the master-slave concept. This made the corresponding turnaround time almost 1/64 because the CPU time used for GA operators are negligible.

To handle the structural constraint with the single-objective GA, the constrained optimization problem was transformed into an unconstrained problem as

$$fitness\ function = \begin{cases} 100 + L/D & if \quad t \geq t_{\min} \\ (100 + L/D) \cdot \exp(t - t_{\min}) & otherwise \end{cases} \tag{11}$$

where t and t_{min} are thickness and minimum thickness at the span station of the maximum local stress. The exponential term penalizes the infeasible solutions by reducing the fitness function value. Because some design candidates can have negative L/D, the summation of 100 and L/D is used.

3.3 Results

The optimization history of the present ARGA is shown in Fig. 7 in terms of L/D. During the initial phase of the optimization, some members had a strong shock wave or failed to satisfy the structural constraint. However they were weeded out from the population because of the resultant penalties to the fitness function. The final design has L/D of 18.91 (C_L = 0.26213 and C_D = 0.01386) satisfying the given structural constraint. Turnaround time of this optimization was about 108 h on NWT.

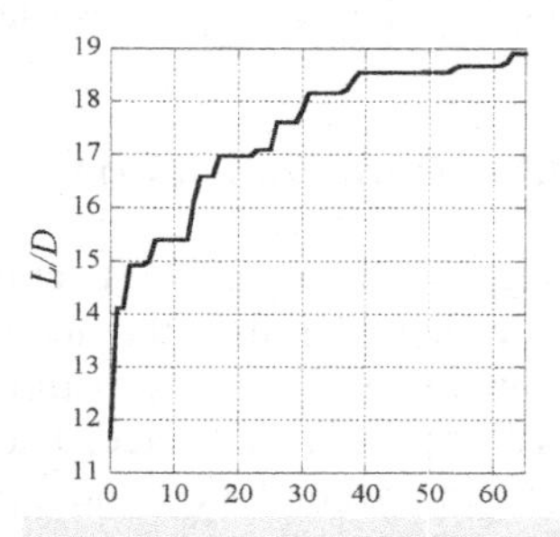

Fig. 7 Optimization history

To examine whether the present optimal design is close to a global optimum, we have checked it against analytically and empirically established design guidelines. In aerodynamics, spanwise lift distribution should be elliptic to minimize the induced drag. However, the structural constraint leads to a tradeoff between induced drag and wave drag. This enforces the spanwise lift distribution to be linear rather than elliptic. The present solution does have a linear distribution. To produce this distribution, a wing is usually twisted in about five degrees. The present wing is twisted in six degrees.

Figure 8 shows the designed airfoil sections and the corresponding pressure distributions at the 0, 33, and 66% spanwise locations. In the pressure distributions, neither any strong shock wave nor any flow separation is found. This ensures that the present wing has very little wave drag and pressure drag. At 33 and 66% spanwise locations, the rooftop, front-loading and rear loading patterns are observed, which are typical for the supercritical airfoils [14] used for advanced transport today. The corresponding airfoil shapes are indeed similar to supercritical airfoils. Overall, these detailed observations of the design confirm that the present design is very close to a global optimum expected by the present knowledge in aerodynamics.

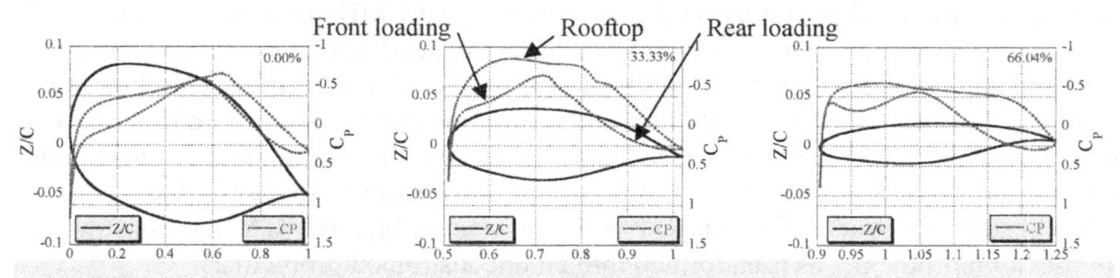

Fig. 8 Designed airfoil sections and corresponding pressure distributions

4 Summary

To develop GAs applicable to practical aerodynamic shape designs, the real-coded ARGAs have been developed by incorporating the idea of the binary-coded ARGAs with the use of the floating-point representation. The real-coded ARGA has been applied to a practical aerodynamic design optimization of a transonic wing shape for generic transport as well as a simple test case. The test case result confirms the present GA outperforms the conventional GA.

Aerodynamic optimization was performed with 87 real-number design variables by using the Navier-Stokes code. The realistic structural constraint was imposed. The resulting wing appears very similar to advanced wing designs based on supercritical airfoils. The straight span load distribution of the resulting design represents a compromised design between minimizations of induced drag and wave drag. The designed wing also has a fully attached flow and the allowable minimum thickness so that pressure drag and wave drag are minimized under the present structural constraint. These results confirm the feasibility of the present approach for future applications.

References

1. Michalewicz, Z., *Genetic Algorithms + Data Structures = Evolution Programs,* third revised edition, Springer-Verlag, (1996).
2. Janikow, C. Z. and Michalewicz, Z., An Experimental Comparison of Binary and Floating Point Representations in Genetic Algorithms, Proc. of the 4th Intl. Conference on Genetic Algorithms, (1991), pp.31-36.
3. Arakawa, M. and Hagiwara, I., Development of Adaptive Real Range (ARRange) Genetic Algorithms, *JSME Intl. J.*, Series C, Vol. 41, No. 4 (1998), pp. 969-977.
4. Arakawa, M. and Hagiwara, I., Nonlinear Integer, Discrete and Continuous Optimization Using Adaptive Range Genetic Algorithms, Proc. of 1997 ASME Design Engineering Technical Conferences, (1997).
5. Krishnakumar, K., Swaminathan, R., Garg, S. and Narayanaswamy, S., Solving Large Parameter Optimization Problems Using Genetic Algorithms, Proc. of the Guidance, Navigation, and Control Conference, (1995), pp.449-460.
6. Mulgund, S., Harper, K., Krishnakumar, K. and Zacharias. G., Air Combat Tactics Optimization Using Stochastic Genetic Algorithms, Proc. of 1998 IEEE Intl. Conference on Systems, Man, and Cybernetics, (1998), pp.3136-3141.
7. Baker, J. E., Reducing Bias and Inefficiency in the Selection Algorithm, Proc. of the 2nd Intl. Conference on Genetic Algorithms, (1987), pp.14-21.
8. Oyama, A., Obayashi, S. and Nakahashi, K., Wing Design Using Real-Coded Adaptive Range Genetic Algorithm, Proc. of 1999 IEEE Intl. Conference on Systems, Man, and Cybernetics [CD-ROM], (1999).
9. Obayashi, S. and Guruswamy, G. P., "Convergence Acceleration of an Aeroelastic Navier-Stokes Solver," *AIAA Journal*, Vol. 33, No. 6, 1995, pp.1134-1141.
10. Case, J., Chilver, A. H. and Ross, C. T. F., *Strength of Materials & Structures with an Introduction to Finite Element Methods*, 3rd Edn., Edward Arnold, London, 1993.
11. Sobieczky, H, Parametric Airfoils and Wings, *Recent Development of Aerodynamic Design Methodologies –Inverse Design and Optimization –*, Friedr. Vieweg & Sohn Verlagsgesellschaft mbH, Braunschweig/Wiesbaden, (1999), pp.72-74.
12. Oyama, A., Obayashi, S., Nakahashi, K. and Hirose, N., Fractional Factorial Design of Genetic Coding for Aerodynamic Optimization, AIAA Paper 99-3298, (1999).
13. Nakamura, T., Iwamiya, T., Yoshida, M., Matsuo, Y. and Fukuda, M., Simulation of the 3 Dimensional Cascade Flow with Numerical Wind Tunnel (NWT), Proc. of the 1996 ACM/IEEE Supercomputing Conference [CD-ROM], (1996).
14. Harris, C. D., NASA Supercritical Airfoils, NASA TP 2969, (1990).

A Common CFD Platform UPACS

Hiroyuki Yamazaki[1], Shunji Enomoto[2], and Kazuomi Yamamoto[3]

[1] Computational Science Division, National Aerospace Laboratory,
7-44-1 Chofu, Tokyo, Japan
yamazaki@nal.go.jp
[2] Aeroengines Division, National Aerospace Laboratory,
7-44-1 Chofu, Tokyo, Japan
eno@nal.go.jp
[3] Aeroengines Division, National Aerospace Laboratory,
7-44-1 Chofu, Tokyo, Japan
kazuomi@nal.go.jp

Abstract. NAL(National Aerospace Laboratory) is developing a common parallel CFD Platform UPACS(Unified Platform for Aerospace Computational Simulation), for the purpose of the efficient CFD programming and the aggregation of the CFD technology of NAL. UPACS is coded based on three-dimensional Navier-Stokes equations and supports multi-block grids. It is parallelized by MPI message passing library. In this paper the concept and the structure of UPACS is described and its computational performance is evaluated on NAL NWT(Numerical Wind Tunnel) supercomputer.

1 Introduction

UPACS (Unified Platform for Aerospace Computational Simulation) is introduced, which NAL is developing for the efficient CFD programming and the aggregation of CFD technology of NAL. The parallel and vector performance of the current version is evaluated on NAL NWT vector parallel supercomputer with simple test cases.

The progress in CFD and parallel computers during 1990s enabled massive numerical simulations of flow around realistic complicated aircraft configurations, direct optimization of aerodynamic design including structure analysis or heat transfer, complicated unsteady flow in jet engines, and so on. However, the increased complexity in computer programs due to the adaptation to complex configurations, the parallel computation, and the multi-physics couplings brought various problems. One of them is a difficulty in sharing analysis codes and know-how among CFD researchers, because they tend to make their own variations of the programs for their own purposes. The parallel programming is another problem that it made the program much more complicated and the portability was sometimes lost.

M. Valero et al. (Eds.): ISHPC 2000, LNCS 1940, pp. 182-190, 2000.

In order to overcome the difficulties in such complicated programming and to accelerate the code development, NAL has started a pilot project UPACS (Unified Platform for Aerospace Computer Simulation) in 1998. UPACS is expected to free CFD researchers from the parallel programming, but the biggest advantage would be that both CFD users and code developers can easily share various simulation codes through the UPACS common base.

2 Concept of the UPACS Code

The following design concepts have been determined.
(1) Multi-block methods: Considering the adaptation to complex configurations, multi-block structured grid methods has been chosen as the first step. The multi-block approach is easily applied to parallel computing. The extension to unstructured grid methods, and the overset grid method are also under consideration.
(2) Separation of multi-block multi-processor procedures from solver routines: The parallel computation and multi-block data control processes are clearly separated from the CFD solver modules so that the solver can be modified without considering the parallel process and the multi-block data handling as if the solver module itself is only for single block problems.
(3) Portability: The parallelization based on domain-decomposition using the message passing interface, MPI, is used to minimize the dependency on hardware architectures.
(4) Structure and capsulation: The hierarchical structure of program and data are clearly defined and the modules are encapsulated so that the code can be shared and modified more easily among CFD researchers and developers.

UPACS is written in Fortran 90 and runs on workstations and NWT. Since the original version of the program is not vectorized by the compiler of NWT, the NWT vector version of the code is provided. The difference of the vector version differs little from the original one. It is added the compiler directives and the inner DO loops are vectorized.

3 Basic Structure of the Code

3.1 Hierarchical Structure

One of the key features in the UPACS code design is the hierarchical structure (Fig.1). This structure consists of three layers;

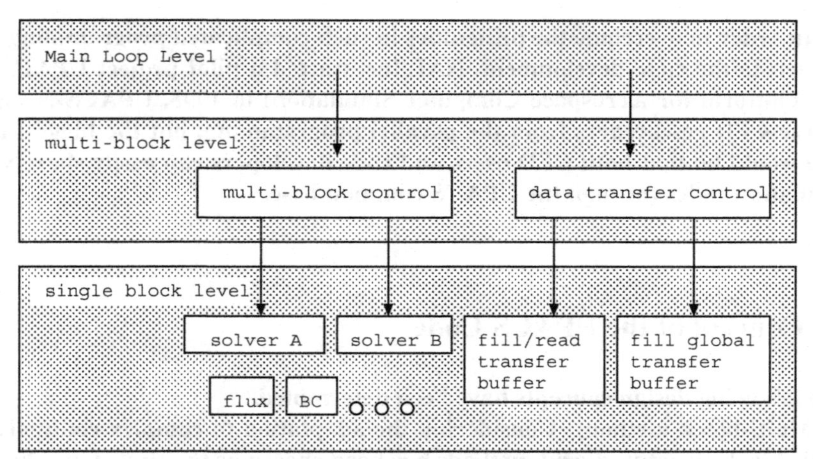

Fig.1. Hierarchical structure of UPACS

(1) Bottom layer: Single-block level

Subroutines in this layer calculate inside the block and its boundary conditions. Data of the neighboring blocks are already prepared as boundary values by the intermediate layer. Thus the routines in this layer can be easily programmed and replaced with other numerical solvers.

(2) Intermediate layer: Multi-block level

The data transfer between blocks and the distribution of blocks to processors are handled in this layer. These procedures are independent of calculations inside each

block therefore the intermediate layer can be generalized and used like libraries. Users can freely modify solvers in the bottom layer without touching the parallel procedure which are controlled in the intermediate layer.

(3) Top layer: Main loop level

The top layer is the main routine which determines the framework of iteration algorithm. It is dependent on the solution method or numerical models and users would want to modify the main routine. Some of the calculation procedures related to parallel processing would also be defined in this layer.

The flow of the program is described as follows:

Initial process: reading grids and calculation of metrics are performed. A grid is extended to outside in two layers of cells. This extra cells are used by communication of values with neighboring grids, or for wall or entrance/exit boundary.

Iteration: mainly two subroutine, 'updateGhostPhys' and 'timeIntegration' are called here. UpdateGhostPhys subroutine updates the extended cells of grids. At the wall boundary, values are set at the cells which is calculated for the boundary to satisfy the boundary condition.

3.2 Calculation

UPACS is programmed to calculate based on three-dimensional Navier-Stokes equations.

The equations are discretized by the finite volume method. A cell is a hexahedron which has a finite volume. The physical variable Q is defined at the center of a cell (cell-centered grid).

The convective term is calculated with Roe[2] scheme or AUSMDV[3] scheme. The numerical flux on the cell boundary is evaluated with MUSCL scheme.

The viscous term is calculated with second-order central difference discretization.

3.3 Parallelization and Boundary Treatment

The information of the connected grid and boundary conditions are unified by the concept of "window". A window is a two-dimensional region which covers the surfaces of a grid. This information is given to the solver code by a text file which describes the geometry of windows on the grid and the type of windows (boundary or connected region to other grid).

The values at the boundary connecting to a neighboring grid, are set by communication. The timeIntegration subroutine calculates values of the next time step inside the grids.

The basic communication strategy is described as follows:
(1) A grid sends data to the connected grid asynchronously.
(2) Receives data from the connected grid. Transforms and sets the received data to the extended area of the grid. The transformation of the data is the receiver's task. It is necessary because a grid is not generally connected in same i,j,k-direction with the neighbouring grids.
(3) Waits for the completion of asynchronous send. This is necessary because the data to be sent asynchronously must be kept unchanged until finishing to send.

Data in the face, the edge and the corner cells are transferred respectively(Fig.2). The edge cells and the corner cells are extrapolated with the values of the face cells and the edge cells before being set the received data. These are usually overwritten by the received data.

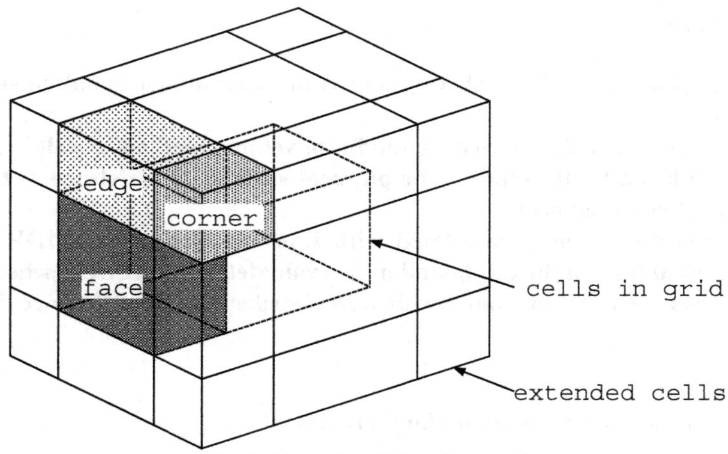

Fig.2 Transfered data in a grid

3.4 Current Status of Development

The UPACS code is currently under development through discussions on the detailed design and validation of the CFD solver and the multi-block multi-processor procedures. The figure 3 and 4 show the example of inviscid calculations around the experimental SST(supersonic transport) with 105 blocks.

Because of the intensive development of UPACS, the modification to provide vectorization version is currently intended to be little and the code is vectorized in one direction.

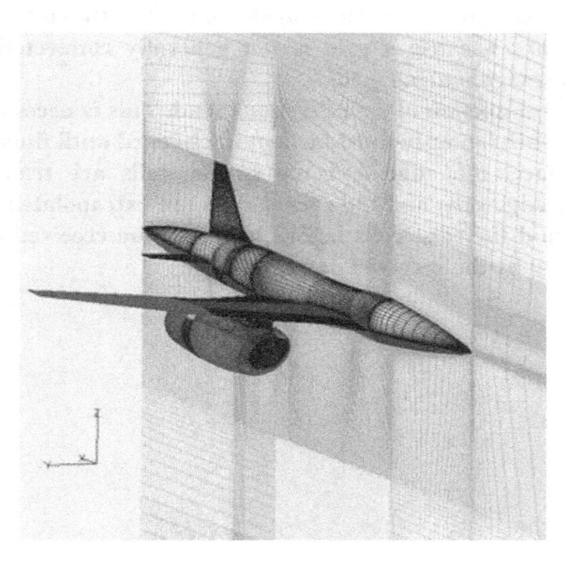

Fig .3 Grids of the Experimental SST Model

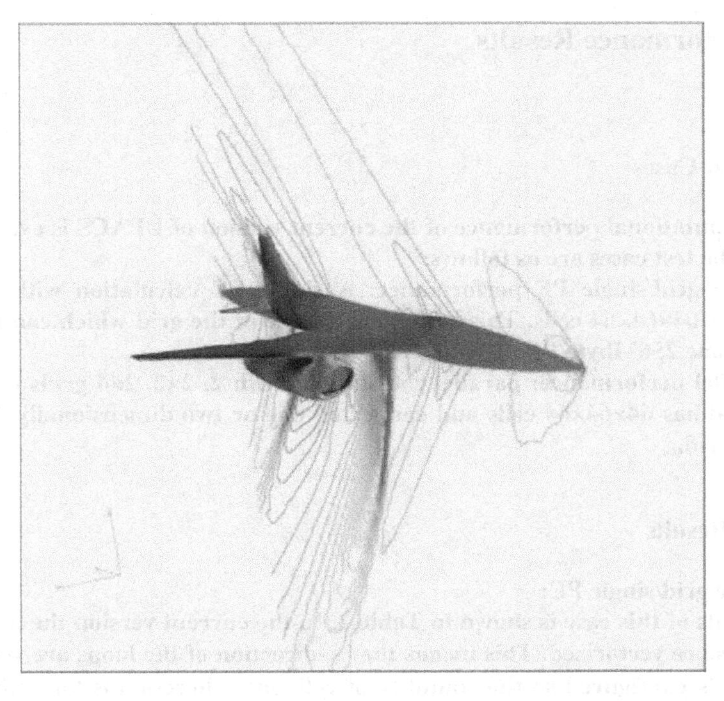

Fig.4 Calculation Results of Experimental SST model

4 Specification of NWT

NWT (Numerical Wind Tunnel) is a parallel computer consisting of 166PEs (processor elements) connected with a cross-bar network. Each PE is a vector processor with pipelines of multiplicity 8. The add, multiply, load and store pipelines can be operated simultaneously. A scalar instruction is a Long Instruction Word(LIW) which issues 3 scalar instructions or 2 floating point instructions at once. Each PE has a main memory of 256MB exept for 4Pes with 1GB. The cross-bar network exclusively connects any pair of PEs without any influence by other PEs. Total peak performance is 280GFLOPS and the total capacity of the main memory is 44.5 GB.

5 Performance Results

5.1 Test Cases

The computational performance of the current version of UPACS is evaluated on NWT. The test cases are as follows:
(1) single grid/single PE performance: A single PE calculation with one grid which has 64×64×64 cells. This is the largest size of the grid which can be calculated by one 256Mbyte PE of NWT.
(2) Parallel performance: parallel calculations with 2, 2×2, 2×4 grids connected. Each grid has 64×64×64 cells and connected one or two dimensionally. Each PE has one grid.

5.2 Results

(1) Single grid/single PE:
The results of this case is shown in Table 1. In the current version the most inner DO loops are vectorized. This means the -direction of the loops are vectorized. If a grid is configured so that number of cells in -direction is long, the vector performance is improved (240×30×30 case).

Table 1: Performance of single PE cases on NWT

Grid size	Equation	Time[s] (one iteration)	Performance [MFLOPS]
64×64×64	Euler	1.452	280.0
64×64×64	Navier-Stokes	2.573	317.6
240×30×30	Navier-Stokes	1.595	426.2

(2) Parallel performance:
Table 2 shows the parallel performance, based on Navier-Stokes equations. The size of one grid is constant (64×64×64) and this represents scalarbility performance. The performance ratio is the parallel performance relative to the single PE.

The calculation times of cell values are almost constant, and the difference of the iteration times between cases come from the diffence of the transfer times in the updateGhostPhys routine. The transfer time is affected by the number of the surfaces connected to other grids. In the 4×4 case, the inner four grids take more time than the outer grids. The inner grids have four surfaces to send or receive data, while the outer ones have two or three. In the 4×8 and 8×8 cases, grids have at most four surfaces to communicate, and the transfer time is almost same as 4×4 cases.

Table 2 parallel performance on NWT

Number of PEs	Time[s] (one iteration)	Time[s] UpdateGhostPhys	Performance ratio
1	2.573	0.063	1
2×1	2.792	0.288	1.84
2×2	2.836	0.325	3.63
2×4	2.870	0.359	7.17
4×4	3.091	0.581	13.32
4×8	3.099	0.589	26.57
8×8	3.105	0.595	53.03

Table 3 shows the parallel performance on Sun workstation Enterprise 4500. The calculation is based on Navier-Stokes equations. The size of one grid is constant (64×64×64). The original code is tested, and the performance improvement effort for the workstation is not held.

Table 3. Parallel performance on Enterprise 4500

Number of PEs	Time[s] (one iteration)	Time[s] UpdateGhostPhys	Performance ratio
1	157.36	0.339	1
2×1	159.14	1.17	1.98
2×2	159.28	1.451	3.95
2×4	159.39	1.762	7.90

The difference of iteration times between cases comes from the UpdateGhostPhys routine which sets values into the extended cells of grids due to the boundary conditions or data transfer (if parallel).

The time charts about transferring data of 2×2 PE case are shown in Fig. 5 created by VAMPIR analyzing tool. In the chart, the light colored bands show times spent in the user codes, the dark ones in MPI libraries. The oblique lines represent actual communication by MPI library. This figure shows most of time is spent in the transformation of received data and extrapolation processes. These processes are not currently modified to improve performance.

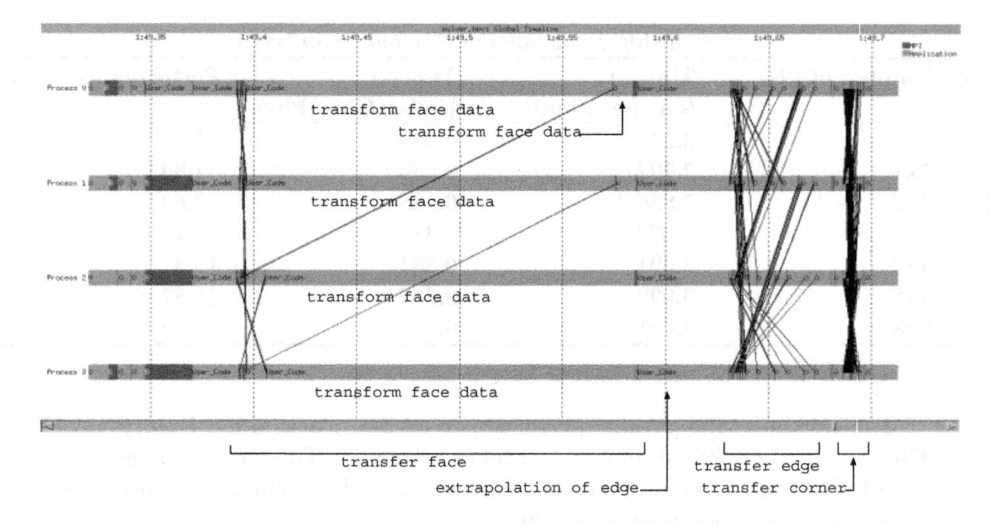

Fig. 5 Tranfer Times of 2×2 PE cases on NWT

6 Conclusion

A common CFD platform UPACS is introduced, which is under development by NAL. The concept of UPACS is described. It supports the multi-block grids, and uses the message passing library MPI for portability. The basic structure of the code is explained. The code has a structure of three layers to achieve the concept of multi-block, capsulation and so on.

The parallel and vector performance of the code is evaluated on NWT, with simple cases. The results show good performance, though there are rooms for further performance improvements.

7 References

1. Ryoji Takagi, "Development of a CFD Code for a Flow around a Complex Configuraton", National Aerospace Laboratory technical report TR-1375, Oct., 1998.

2. P.L.Roe, "Approximate Riemann Solvers, Parameter Vectors, and Difference Schemes", J.Comput. Physics. Vol43, pp.357-372,1981.

3. Y.Wada and M.S.Liou, "A Flux Splitting Scheme with High-Resolution and Robustness for Discontinuities", AIAA Paper 94-0083, 1994.

On Performance Modeling for HPF Applications with ASL*

Thomas Fahringer[1], Michael Gerndt[2],
Graham Riley[3], and Jesper Larsson Träff[4]

[1] Institute for Software Science, University of Vienna
tf@par.univie.ac.at
[2] Central Institute for Applied Mathematics, Research Centre Juelich
m.gerndt@fz-juelich.de
[3] Department of Computer Science, University of Manchester
griley@cs.man.ac.uk
[4] C&C Research Laboratories, NEC Europe Ltd.
traff@ccrl-nece.technopark.gmd.de

Abstract. Multiprocessor systems are increasingly being used to handle large-scale scientific applications that demand high-performance. However, performance analysis is not as mature for multiprocessor systems as for uniprocessor systems, and improved ways of automatic performance analysis are needed to reduce the cost and complexity of developing distributed/parallel applications.
Performance analysis is commonly a cyclic process of measuring and analyzing performance data, identifying and possibly eliminating performance bottlenecks in slow progression. Currently this process is controlled manually by the programmer. We believe that the implicit knowledge applied in this cyclic process should be formalized in order to provide automatic performance analysis for a wider class of programming paradigms and target architectures. This article describes the performance property specification language (ASL) developed in the APART Esprit IV working group which allows specifying performance-related data by an object-oriented model and performance properties by functions and constraints defined over performance-related data. Performance problems and bottlenecks can then be identified based on user- or tool-defined thresholds. In order to demonstrate the usefulness of ASL we apply it to HPF (High Performance Fortran) by successfully formalizing several HPF performance properties.

1 Introduction

Although rapid advances in processor design and communication infrastructure are bringing teraflops performance within grasp, the software infrastructure for multiprocessor systems simply has not kept pace. The lack of useful, accurate

* The ESPRIT IV *Working Group on Automatic Performance Analysis: Resources and Tools* is funded under Contract No. 29488

M. Valero et al. (Eds.): ISHPC 2000, LNCS 1940, pp. 191–204, 2000.

performance analysis is particularly distressing, since performance is a key issue of multiprocessor systems.

Despite the existence of a large number of tools assisting the programmer in performance experimentation, it is still the programmer's responsibility to take most strategic decisions. Many performance tools are platform and language dependent, cannot correlate performance data gathered at a lower level with higher-level programming paradigms, focus only on specific program and machine behavior, and do not provide sufficient support to infer important performance properties.

In this article we describe a novel approach to formalize performance bottlenecks and the data required in detecting those bottlenecks with the aim to support automatic performance analysis for a wider class of programming paradigms and architectures. This research is done as part of APART Esprit IV *Working Group on Automatic Performance Analysis: Resources and Tools* [2]. In the remainder of this article we use the following terminology:

Performance-related Data: Performance-related data defines information that can be used to describe performance properties of a program. There are two classes of performance related data. First, static data specifies information that can be determined without executing a program on a target machine. Examples include program regions, control and data flow information, and predicted performance data. Second, dynamic performance-related data describes the dynamic behavior of a program during execution on a target machine. This includes timing events, performance summaries and metrics, etc.

Performance Property: A performance property (e.g. load imbalance, communication, cache misses, etc.) characterizes a specific performance behavior of a program and can be checked by a set of *conditions*. Conditions are associated with a *confidence value* (between 0 and 1) indicating the degree of confidence about the existence of a performance property. A confidence value 0 means that the condition is most likely never true whereas a confidence value 1 implies that the condition presumably always holds. In addition, for every performance property a *severity figure* is provided that specifies the importance of the property. The higher this figure the more important or severe a performance property is. The severity can be used to concentrate first on the most severe performance property during the performance tuning process. Performance properties, confidence, and severity are defined over performance-related data.

Performance Problem: A performance property is a performance problem, iff its severity is greater than a user- or tool-defined threshold.

Performance Bottleneck: A program can have one or several performance bottlenecks which are characterized by having the highest severity figure. If these bottlenecks are not a performance problem, then the program's performance is acceptable and does not need any further tuning.

For example, a code region may be examined for the existence of a communication performance property. The condition for this property holds, if any process executing this region invokes communication (communication time is larger than zero). The confidence value is 1 because measured communication time represents a proof for this property. In contrast, if prediction would be used to reflect communication, then the confidence value is likely to be less than 1. The severity is given by the percentage of communication time relative to the execution time of the entire program. If the severity is above a user or tool defined threshold, then the communication performance property defines a performance problem. If this performance problem is the most severe one, then it denotes the performance bottleneck of a program. If there are several performance properties with identical highest severity figure, then all of them are performance bottlenecks. Commonly, a programmer may try to eliminate or at least to alleviate the bottlenecks before examining any other performance problems.

In this paper we introduce the APART Specification Language (ASL) which allows the description of performance-related data through the provision of an object-oriented specification model and which supports the definition of performance properties in a novel formal notation. Our object-oriented specification model is used to declare – without the need to compute – performance information. It is similar to Java, uses only single inheritance and does not require methods. A novel syntax has been introduced to specify performance properties. The objective of ASL is to support performance modeling for a variety of programming paradigms including message passing, shared memory, and distributed memory programming. In this paper we demonstrate the usefulness of ASL for HPF ([8] - High Performance Fortran) by successfully formalizing several HPF performance properties. ASL has also been applied to OpenMP and message passing programs details of which can be found in [4].

The organization of this article is as follows. Related work is presented in Section 2. ASL constructs for specifying performance-related data are presented in Section 3. This includes the base classes which are programming paradigm independent and classes to specify performance-related data that are specific to HPF programs. The syntax for the specification of performance properties is described in Section 4. Examples for HPF property specifications including HPF code excerpts are presented in Section 5. Conclusions and Future work are discussed in Section 6.

2 Related Work

The use of specification languages in the context of automatic performance analysis tools is a new approach. Paradyn [9] performs an automatic online analysis and is based on dynamic monitoring. While the underlying metrics can be defined via the *Metric Description Language* (MDL) [12], the set of searched bottlenecks is fixed. It includes *CPUbound*, *ExcessiveSyncWaitingTime*, *ExcessiveIOBlockingTime*, and *TooManySmallIOOps*.

A rule-based specification of performance bottlenecks and of the analysis process was developed for the performance analysis tool OPAL [7] in the SVM-Fortran project. The rule base consists of a set of parameterized hypothesis with

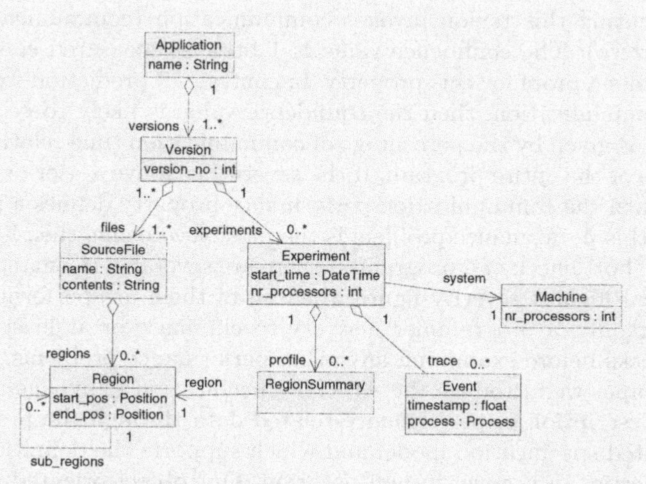

Fig. 1. Base classes of performance-related data models

proof rules and refinement rules. The proof rules determine whether a hypothesis is valid based on the measured performance data. The refinement rules specify which new hypotheses are generated from a proven hypothesis.

Another approach is to define a performance bottleneck as an event pattern in program traces. EARL [15] describes event patterns in a more procedural fashion as scripts in a high-level event trace analysis language which is implemented as an extension of common scripting languages like Tcl, Perl or Python.

The language presented here served as a starting point for the definition of the ASL but is too limited. MDL does not allow to access static program information and to integrate information for multiple performance tools. It is specially designed for Paradyn. The design of OPAL focuses more on the formalization of the analysis process and EDL and EARL are limited to pattern matching in performance traces.

Some other performance analysis and optimization tools apply automatic techniques without being based on a special bottleneck specification, such as KAPPA-PI [1], FINESSE [11], and the online tuning system Autopilot [13].

3 Performance-Related Data Specification

In the following we describe a library of classes that represent static and dynamic information for performance bottleneck analysis. We distinguish between two sets of classes. First, the set of base classes which is independent of any programming paradigm, and second, programming paradigm dependent classes which are shown for HPF.

3.1 Standard Class Library

Note that we expect most data models described with this language will have a similar overall structure. This similarity is captured in the base classes. Future data models can build specialized classes in form of subclasses.

Figure 1 shows the UML [14] representation of the base classes which are programming paradigm independent. The translation of the UML diagrams into the specified syntax is straightforward. Initially, there is an application for which performance analysis has to be done. Every application has a name and may possibly have a number of implementations, each with a unique version number. Versions may differ with respect to their source files and experiments. Every source file (the contents of which is stored in a generic string) has one or several static code regions each of which is uniquely specified by *start_pos* (position where region begins in the source file) and *end_pos* (position where region ends in the source file). A position in a region is defined by a line and column number with respect to the given source file. Experiments – denoting the second attribute of a version – are described by the time (*start_time*) when the experiment started and the number of processors (*nr_processors*) that were available to execute the version. Furthermore, an experiment is also associated with a static description of the machine (e.g. number of processors available) that is used for the experiment. Every experiment includes also dynamic data, i.e. a set of region summaries (*profile*) and a set of events (*trace*). The class *RegionSummary* describes performance information across all processes employed for the experiment. Region summaries are associated with the appropriate region. The class *Event* represents information about individual events occurring at runtime, such as sending a message to another process. Each event has a *time_stamp* attribute determining when the event occurred and a *process* attribute determining in which process the event occurred.

3.2 HPF Class Hierarchy

High Performance Fortran (HPF) [8] defines a set of language extensions to Fortran 90/95 to facilitate efficient parallel programming of scalable parallel architectures. The main concept of HPF relies on data distribution. The programmer writes a sequential program and specifies how the data space of a program should be distributed by adding data distribution directives to the declarations of arrays. A compiler then translates a program into an efficient parallel SPMD target program using explicit message passing on distributed memory machines. More details about HPF can be found in [8]. Figure 2 shows the classes that model static information for HPF programs [8]. Class *HPFRegion* is a subclass of *Region* (see Figure 1) and contains the following attributes (Figure 3) representing static performance-related information:

- *dirs* describes HPF directives such as PROCESSORS, DISTRIBUTED, etc.
- *deps* specifies data dependences implied by code regions. A data dependence is described by the source, destination, type (e.g. true-, anti- and output-dependence), direction, distance, and level (loop-carried or -independent).

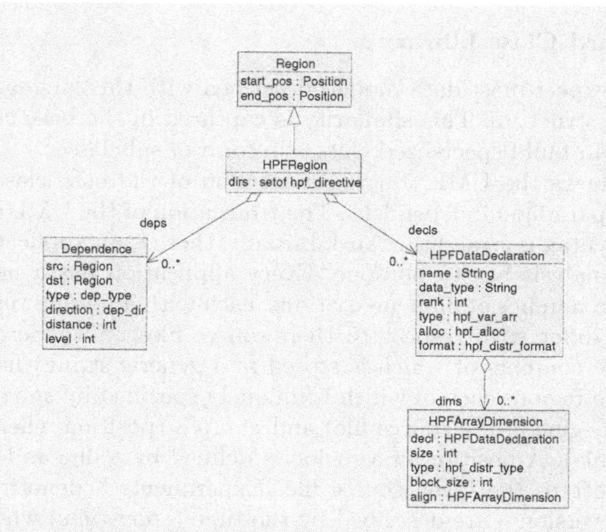

Fig. 2. Static performance-related information for *HPFRegions*

- *decls* specifies HPF data declarations for scalars and arrays. Data declarations are described by attributes *name* (name of a data object), *data_type* (e.g. integer), *rank* (dimensionality of data), *type* (data is an array or a scalar), *alloc* (data has been declared DYNAMIC or STATIC), and *format* (distributed, replicated, prescriptive, etc.). In case of an array, additional information about every dimension is provided which includes the size of the dimension, distribution type (e.g. BLOCK or CYCLIC), block size in case of BLOCK distribution, and alignment information.

Figure 3 displays several subclasses which extend *HPFRegion* by following code regions: *HPFProcedure*, *HPFLoop*, *HPFIfBlock*, *HPFBasicBlock*, *HPFProcedureCall*, and *HPFArrayAssignment*. Class *HPFLoop* can be further specified by attribute ltype which indicates the type of loop, for instance, DO loop, DO INDEPENDENT loop, etc. A DO INDEPENDENT loop relates to HPF's parallel loop which asserts that the loop iterations can be executed in any order. HPF array assignment cover both Fortran77 and Fortran90 array operations.

Figure 4 shows the HPF class library for dynamic information. Class *HPFRegionSummary* extends class *RegionSummary* (see Figure 1) and comprises three attributes: *processes* specifies the set of processes executing a region, *sums* reflects performance summary information across all processes executing the region, and *proc_sums* indicates performance summary information for a region with respect to individual processes.

Class *HPFSummary* contains several performance attributes which are average values across all processes with respect to a specific region:

- *nr_executions*: number of times the region has been executed
- *duration*: time spent in executing the region

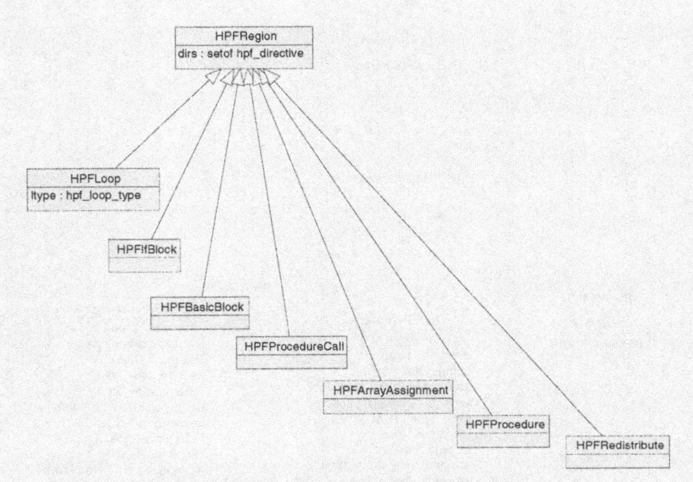

Fig. 3. Subclasses of *HPFRegion*

- *comm_time*: communication time
- *dep_comm_time*: communication time caused by data dependences
- *align_comm_time*: communication time caused by data alignment
- *sync_time*: synchronization time
- *idle_time*: idle time
- *io_time*: input/output time
- *compiler_ovh_time*: compiler overhead time
- *inspector_time*: time spent in inspector/executor phase (compiler inserted code to handle irregular problems)
- *expl_redistr_time*: time spent in explicit redistribution of data caused by HPF REDISTRIBUTE directive.
- *impl_redistr_time*: time spent in implicit redistribution of data at procedure boundaries.
- *nr_cache_misses*: number of cache misses.

Compiler overhead time amounts for extra costs implied by compiler-inserted statements in a parallel program. For instance, in order to enforce the owner-computes paradigm (write operations can only be executed by the owning process) additional IF-conditions may be inserted in a parallel program. Executing these IF-conditions is considered to be compiler overhead.

Class *HPFProcessSummary* contains all attributes of class *HPFSummary* with respect to a specific process (identified by attribute *process*). Note that dynamic performance-related information can both be measured [9,10,3] or predicted [6]. Some of this information however require a close interaction between performance measurement/prediction tools and compilers [6].

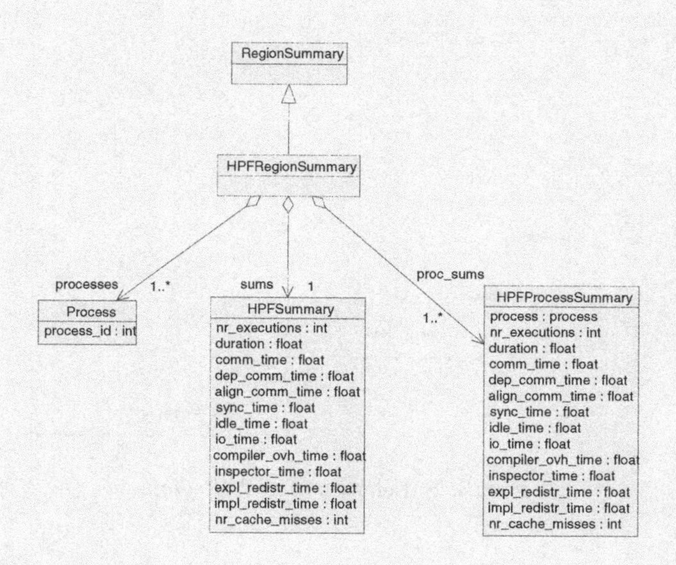

Fig. 4. Dynamic performance-related information (summaries) in the HPF data model

4 Performance Property Specification

A performance property (e.g. load imbalance, synchronization, cache misses, redundant computations, etc.) characterizes a specific performance behavior of a program. The ASL property specification (Figure 5) syntax defines the name of the property, its context via a list of parameters, and the condition, confidence, and severity expressions. The property specification is based on a set of parameters. These parameters specify the property's context and parameterize the expressions. The context specifies the environment in which the property is evaluated, e.g. the program region and the test run. The condition specification consists of a list of conditions. A condition is a predicate that can be prefixed by a condition identifier. The identifiers have to be unique with respect to the property since the confidence and severity specifications may refer to the conditions by using the condition identifiers. The confidence specification is an expression that computes the maximum of a list of confidence values. Each confidence value is computed as an arithmetic expression in an interval between zero and one. The expression may be guarded by a condition identifier introduced in the condition specification. The condition identifier represents the value of the condition. The severity specification has the same structure as the confidence specification. It computes the maximum of the individual severity expressions of the conditions. The severity specification will typically be based on a parameter specifying the rank basis. If, for example, a representative test run of the application has been monitored, the time spent in remote accesses may be compared to the total execution time. If, instead, a short test run is the basis for performance evaluation since the application has a cyclic behavior, the remote access overhead may be compared to the execution time of the shortened loop.

5 HPF Performance Properties

This section demonstrates the ASL constructs for specifying performance properties in the context of HPF. Some global definitions are presented first which are then used in the definitions of a number of HPF properties.

```
property       is PROPERTY pp-name '(' arg-list ')' '{'
                  [LET def * IN]
                       pp-condition
                       pp-confidence
                       pp-severity
                  '};'
arg            is type ident
pp-condition   is CONDITION ':' conditions ';'
conditions     is condition
               or condition OR conditions
condition      is ['(' cond-id ')' ]bool-expr
pp-confidence  is CONFIDENCE ':' MAX '(' confidence-list ')' ';'
               or CONFIDENCE ':' confidence ';'
confidence     is ['(' cond-id ')' '->' ] arith-expr
pp-severity    is SEVERITY ':' MAX '(' severity-list ')' ';'
               or SEVERITY ':' severity ';'
severity       is ['(' cond-id ')' '->' ] arith-expr
```

Fig. 5. ASL property specification syntax

5.1 Global Definitions

In most property specifications it is necessary to access the summary data of a given region for a given experiment. Therefore, we define the *summary* function that returns the appropriate *HPFRegionSummary* object. It is based on the set operation *UNIQUE* that arbitrarily selects one element from the set argument which has cardinality one due to the design of the data model.

```
HPFRegionSummary summary(Region r, Experiment e)=
                     UNIQUE({s IN e.profile WITH s.region==r});
```

The *duration* function returns the execution time of a region which is given by the arithmetic mean across all processes that execute the region.

```
float duration(Region r, Experiment e)=summary(r,e).sums.duration;
```

For all HPF performance properties the severity is computed by relating some aspect of the execution time to the duration of a given *rank_basis* region (for instance, execution time of the entire program) in the experiment.

5.2 Property Specifications

The *parallel_costs* property determines whether the total parallel cost in the execution of a parallel program is non-zero. The parallel costs of a region can be subdivided into four categories: communication, synchronization, compiler overhead, and input/output time.

A region accounts for this property if *cost_sum* is greater than 0. If parallel costs can be measured then the confidence for this condition is one. In case of predicting parallel costs, the confidence may be less than one. The severity of this property is the fraction of the time spent for parallel costs compared to the duration of rank basis, typically the duration of the main program. Note, that *comm_time*, *sync_time*, *compiler_ovh_time*, *io_time*, and *duration* are summary figures across all processes executing the region.

```
Property parallel_costs(HPFRegion r, Experiment e, Region rank_basis) {
LET
  float cost_sum = summary(r,e).sums.comm_time +
                   summary(r,e).sums.sync_time +
                   summary(r,e).sums.compiler_ovh_time +
                   summary(r,e).sums.io_time;
IN
  CONDITION:   cost_sum>0;
  CONFIDENCE:  1;
  SEVERITY:    cost_sum/duration(rank_basis,e);
}
```

The severity of this property is larger than the severity of the individual properties for each of the categories. This may lead to the selection of the parallel cost property as a performance problem according to the predefined severity threshold while the individual properties, i.e. *comm_time*, *sync_time*, *compiler_ovh_time*, and *io_time*, may not be referred to as performance problems. This property determines whether a region implies communication.

```
Property communication_costs (HPFRegion r, Experiment e,
                              Region rank_basis) {
LET
  float comm_time = summary(r,e).sums.comm_time;
IN
  CONDITION:   comm_time > 0;
  CONFIDENCE:  1;
  SEVERITY:    comm_time / duration(rank_basis,e);
}
```

The condition and severity of property *communication_costs* is based on *comm_time* which is the arithmetic mean across all processes executing a region. The severity is the communication time divided by the execution time of the rank basis. A performance property for input/output overhead can be described very similar to the communication cost property.

```
Property redist_costs (HPFRegion r,Experiment e,Region rank_basis) {
LET
redist1 = summary(r,e).sums.impl_redistr_time;
redist2 = summary(r,e).sums.expl_redistr_time;
IN
CONDITION:
  (Cond1)
          ((typeof(summary(r,e).region)==HPFProcedure)
              AND
            (EXISTS dec IN summary(r,e).region.decls
             SUCH THAT
                dec.format==PRESCRIPTIVE OR dec.format==TRANSCRIPTIVE)
              AND
            (redist1 > 0))
        OR
  (Cond2)
         ((typeof(summary(r,e).region)==HPFRedistribute)
             AND
          (EXISTS dec IN summary(r,e).region.decls
           SUCH THAT dec.alloc == DYNAMIC)
             AND
          (redist2 > 0))
CONFIDENCE:  MAX((Cond1)->0.8, (Cond2)->1.0);
SEVERITY:    MAX((Cond1)->redist1 / duration(rank_basis,e),
                  (Cond2)->redist2 / duration(rank_basis,e));
}
```

Property *redist_costs* specifies the time spent in redistributing data inside (redistribution of dummy arrays and dynamic arrays) or outside of procedures (redistribution of dynamic arrays). We include two different conditions for this property to hold each of which is associated with its own confidence value. The first condition ensures that a region is a procedure. If the procedure has prescriptive or transcriptive mapping and *impl_redistr_time* is non-zero then redistribution may occur at the procedure boundary. The second condition covers the case where DYNAMIC arrays are redistributed based on the HPF `REDISTRIBUTE` directive which causes *expl_redistr_time* to be non-zero. A close integration between performance measurement tool and HPF compiler is needed in order to associate measured performance data with the location in the HPF program that causes redistribution. Commonly, the more conditions are included for a property the more options are available to determine whether a property holds and the more refined information about the cause of the property is supplied. For instance, if prediction fails, then monitoring could be employed to prove a certain performance property. For property *redist_costs*, we include two conditions with more detailed information than if only a single value for all redistribution costs would be provided.

The confidence of property *redist_costs* is the maximum confidence value across both conditions as specified above. Note that for this property we assume that measuring implict redistribution (condition 1) may be less precise than timings of explicit redistribution (condition 2) due to a lack of compiler information. The severity is the maximum across the time spent in, respectively, implicit and explicit redistribution of data relative to the execution time of the region selected as the rank basis.

Property *uneven_work_distribution* specifies how even the computations of a parallel program have been distributed across all processes executing a region.

```
Property uneven_work_distribution (HPFRegion r, Experiment seq,
                                   Experiment par, Region rank_basis) {
LET
int nr_processes = COUNT(procs WHERE procs IN summary(r,par).processes);
float opt_duration = duration(r,seq)/nr_processes;

float deviation = SQRT(SUM (EXP(proc_sums.duration - proc_sums.comm_time
      - proc_sums.sync_time - proc_sums.idle_time -
      proc_sums.compiler_ovh_time - proc_sums.io_time - opt_duration, 2)
                  WHERE proc_sums IN summary(r,par).proc_sums ))
IN
  CONDITION: (deviation /  opt_duration) > uneven_threshold;
  CONFIDENCE:  1;
  SEVERITY:  summary(r,par).sums.duration / duration(rank_basis,par);
}
```

The standard deviation of the computational costs (excluding communication, synchronization, idle time, compiler overhead, and input/output time) of every process with respect to the optimal duration (sequential execution time divided by number of processes) is computed. The condition is then given as the variation coefficient normalized with the optimal duration compared against a threshold. The severity is defined as the execution time divided by the rank basis.

The data distribution directives of an HPF program have a decisive impact on the resulting quality of the work distribution. For instance, the nested DO loop in the following code excerpt implies a triangular loop iteration space.

```
!HPF$ DISTRIBUTE A(BLOCK,*)
DO i=1,n
   DO j=i,n
      A(i,j) = A(i,j)-A(i,j-1)*A(i,i)
   END DO
END DO
```

Distributing array A column-wise causes a very uneven distribution of the loop iterations across a set of processes executing this loop. If array A would be distributed CYCLIC (for instance, CYCLIC in the first and replicated in the second dimension), then the corresponding work distribution can be nearly optimal depending on how many processes are involved in executing this loop. A more detailed technical report with many more examples about performance modeling for HPF applications based on ASL can be found in [5].

6 Conclusions and Future Work

In this article we describe a novel approach to the formalization of performance problems and the data required to detect them with the future aim of supporting automatic performance analysis for a variety of programming paradigms and architectures. We present the APART Specification Language (ASL) developed as part of the APART Esprit IV Working Group on *Automatic Performance Analysis: Resources and Tools.* This language allows the description of performance-related data through the provision of an object-oriented specification model and supports definition of performance properties in a novel formal notation. Performance-related data can either be static (gathered at compile-time, e.g. code regions, control and data flow information, predicted performance data, etc.) or dynamic (gathered at run-time, e.g. timing events, performance summaries, etc.) and is used as a basis for describing performance properties. A performance property (e.g. uneven work distribution, communication, redistribution costs, etc.) characterizes a specific type of performance behavior which may be present in a program. A set of conditions defined over performance-related data verify the existence of properties during (the execution of) a program. We applied ASL to HPF by successfully formalizing several HPF performance properties. ASL has also been used to formalize a large variety of MPI and OpenMP performance properties which is described in [4].

Two extensions to the current language design will be investigated in the future: First, the language will be be extended by templates which facilitate specification of similar performance properties both within and across program paradigms. Second, meta-properties may be useful as well. For example, communication can be proven based on summary information, i.e. communication exists if the sum of the communication time in a region over all processes is greater than zero. Meta-properties can be useful to evaluate other properties based on region instances instead of region summaries.

ASL should be the basis of a common interface for a variety of performance tools that provide performance-related data. Based on this interface we plan to develop a system that provides automatic performance analysis for a variety of programming paradigms and target architectures.

References

1. E. Luque A. Espinosa, T. Margalef. Automatic Performance Evaluation of Parallel Programs. In *IEEE Proc. of the 6th Euromicro Workshop on Parallel and Distributed Processing*. IEEE Computer Society Press, January 1998. 194
2. APART – Esprit Working Group on Automatic Performance Analysis: Resources and Tools, Feb. 1999. 192
3. T. Fahringer, P. Blaha, A. Hössinger, J. Luitz, E. Mehofer, H. Moritsch, and B. Scholz. Development and Performance Analysis of Real-World Applications for Distributed and Parallel Architecture. AURORA Technical Report TR1999-16, http://www.vcpc.univie.ac.at/aurora/publications/, University of Vienna, August 1999. 197
4. T. Fahringer, M. Gerndt, G. Riley, and J. Träff. Knowledge Specification for Automatic Performance Analysis. APART Technical Report, Workpackage 2, Identification and Formalization of Knowledge, Technical Report FZJ-ZAM-IB-9918, Research Centre Jülich, Zentralinstitut für Angewandte Mathematik (ZMG), Jülich, Germany, November 1999. 193, 203
5. T. Fahringer, M. Gerndt, G. Riley, and J. Träff. On Performance Modeling for HPF Applications with ASL. Aurora Technical Report AuR_00-10, University of Vienna, http://www.vcpc.univie.ac.at/aurora/publications/, University of Vienna, Institute for Software Science, Vienna, Austria, August 2000. 202
6. T. Fahringer and A. Požgaj. P^3T+: A Performance Estimator for Distributed and Parallel Programs. *Scientific Programming, IOS Press, The Netherlands*, accepted for publication, to appear in 2000. 197

7. M. Gerndt, A. Krumme, and S. Özmen. Performance analysis for svm-fortran with opal. In *Int. Conference on Parallel and Distributed Processing Techniques and Applications (PDPTA'95), Athens, Georgia*, pages 561–570, 1995. 193
8. High Performance Fortran Forum, High Performance Fortran Language Specification. Version 2.0.δ, Technical Report, Rice University, Houston, TX, Jannuary 1997. 193, 195
9. B. Miller, M. Callaghan, J. Cargille, J. Hollingsworth, R. Irvin, K. Karavanic, K. Kunchithapadam, and T. Newhall. The Paradyn Parallel Performance Measurement Tool. *IEEE Computer*, 28(11):37 – 46, November 1995. 193, 197
10. B. Mohr, D. Brown, and A. Malony. TAU: A portable parallel program analysis environment for pC++. In *CONPAR*, Linz, Austria, 94. 197
11. N. Mukherjee, G. D. Riley, and J. R. Gurd. Finesse: A prototype feedback-guided performance enhancement system. In *Euromicro Workshop on Parallel and Distributed Processing PDP'2000, Rhodes, Greed.* IEEE Computer Society, January 2000. 194
12. Paradyn Project: Paradyn Parallel Performance Tools: User's Guide, 1998. 193
13. R. Ribler, J. Vetter, H. Simitci, and D. A. Reed. Autopilot: Adaptive control of distributed applications. In *Proc. of the 7th IEEE Symposium on High-Performance Distributed Computing.* IEEE Computer Society, 1998. 194
14. James Rumbaugh, Ivar Jacobson, and Grady Booch. *The Unified Modeling Language Refrence Manual.* Object Technology Series. Addison Wesley Longman, Reading, Mass., 1999. 195
15. F. Wolf and B. Mohr. EARL - A Programmable and Extensible Toolkit for Analyzing Event Traces of Message Passing Programs. In *Proc. of 7th International Conference, HPCN Europe 1999*, pages 503–512, Amsterdam, The Netherlands, April 1999. 194

A "Generalized k-Tree-Based Model to Sub-system Allocation" for Partitionable Multi-dimensional Mesh-Connected Architectures

Jeeraporn Srisawat[1] and Nikitas A. Alexandridis[2]

[1] King Mongkut's Institute of Technology
Ladkrabang, Bangkok 10520, THAILAND
ksjeerap@kmitl.ac.th
[2] The George Washington University,
Washington, DC 20052, U.S.A.
alexan@seas.gwu.edu

Abstract. This paper presents a new processor allocation approach called "a generalized k-Tree-based model" to perform dynamic sub-system allocation/deallocation decision for partitionable multi-dimensional mesh-connected architectures. Time complexity of our generalized k-tree-based sub-system allocation algorithm is $O(k^4 2^k (N_A+N_F)+k^2 2^{2k})$ for the partitionable k-D meshes and $O(N_A+N_F)$ for the partitionable 2-D meshes, where N_A is the maximum number of allocated tasks, N_F is the corresponding number of free sub-meshes, N is the system size, and $N_A+N_F \leq N$. Most existing processor allocation strategies have been proposed for the partitionable 2-D meshes with various degrees of time complexity and system performance. In order to evaluate the system performance, the generalized k-Tree-based model was developed and by simulation studies the results of applying our k-Tree-based approach for the partitionable 2-D meshes were presented and compared to existing 2-D mesh-based strategies. Our results showed that the k-Tree-based approach (when it was applied for the partitionable 2-D meshes) yielded the comparable system performance to those recently 2-D mesh-based strategies.

1 Introduction

A multi-dimensional (k-D) mesh-connected parallel architecture is a useful network type of parallel systems for high performance parallel applications that may require different degrees of processor interconnection (such as 1-D text processing, 2-D image processing, 3-D graphic processing, etc.). The partitionable k-D mesh system is provided (at run time) for executing various independent applications (or tasks) in parallel. Examples of prototypes and commercial parallel systems (which support a multi-user environment) include the Intel Touchstone system [5], the Intel Paragon XP/S [6], the Intel/Sandia ASCI System [10], etc. In the partitionable parallel systems, a number of independent smaller tasks (from the same or different applications) come in, each requiring at run time a separate sub-system (or partition) to execute. In order to provide appropriate free sub-systems for

M. Valero et al. (Eds.): ISHPC 2000, LNCS 1940, pp. 205-217, 2000.

new tasks, a special designed operating system (known as the processor allocator) has to dynamically partition the computer system to allocate a sub-system for each incoming task, as well as to deallocate a sub-system and recombine partitions as soon as they become available when a task completes. For the partitionable 2-D mesh-connected systems, a number of processor allocation/deallocation (decision) techniques have been proposed in the past such as bit-map approaches [11], [16] and non-bit-map approaches [1], [2], [3], [4], [7], [8], [9], [12], [13], [14], [15]. Among those existing 2-D mesh-based strategies, time complexities of recently sub-system allocation methods (that yield the comparable system performance) are $O(N_a^3)$ of the Busy List (BL) [3], $O(N_a\sqrt{N})$ of the Quick Allocation (QA) [15], and $O(N_f^2)$ for the 2-D meshes and $O(N_f^k)$ for the k-D meshes of the Free Sub-List (FSL) [7], where N is the system size, N_a is the number of allocated tasks ($N_a \leq N$), and N_f is the number of free sub-systems ($N_f \leq N$).

In this paper, we propose the designing of "a generalized k-tree-based model" in order to perform dynamic sub-system allocation/deallocation decision for partitionable multi-dimensional mesh-connected architectures. Our generalized model includes a k-Tree system state representation and a number of generalized algorithms (such as the network partitioning algorithm, the sub-system combining algorithm, the best-fit heuristic for allocation decision, the sub-system allocation/deallocation decision algorithm) Time complexity of the generalized k-Tree-based approach for the partitionable k-D meshes is $O(k^4 2^k(N_A+N_F)+k^2 2^{2k})$, where N_A is the maximum number of allocated tasks, N_F is the corresponding number of free sub-meshes, N is the system size, and $N_A+N_F \leq N$. Therefore, time complexity of our approach for the partitionable 2-D meshes is only $O(N_A+N_F)$. For the performance evaluation (by simulation study), the generalized k-Tree-based model was developed and the system performance of our application for the partitionable 2-D meshes was presented and compared to the recently 2-D mesh-based strategies such as the Busy List (BL) [3], the Free-Sub List (FSL) [7], and the Quick Allocation (QA) [15]. Our simulation results showed that the k-Tree-based approach yielded the comparable system performance to those recently 2-D mesh-based strategies.

In the next section, we present the generalized k-Tree-based model to perform processor allocation/deallocation decision for the partitionable k-D mesh parallel machines as well as the corresponding time complexity. Section 3 presents the evaluated system performance of the generalized k-Tree-based model. Then, the performance results (by simulation study) for the 2-D mesh networks are presented and compared to existing 2-D mesh-based strategies. Finally, conclusions are discussed in Section 4.

2 k-Tree-Based Model to Sub-System Allocation for k-D Meshes

In this paper, "the generalized k-Tree-based model" is proposed to perform sub-system allocation/deallocation decision for the partitionable multi-dimensional (k-D) mesh-connected architectures. In the generalized k-Tree-based model, we use a data structure, called a "k-Tree" to represent the system states of the partitionable k-D mesh-connected systems. Our generalized k-Tree-based model includes a k-Tree system states representation (Section 2.1) and a number of generalized algorithms for each sub-system allocation/deallocation decision such as the network partitioning algorithm (Section 2.2), the sub-system combining algorithm (Section 2.3), the best-fit heuristic for allocation decision (Section 2.4), and the sub-system allocation/deallocation decision algorithm (Section 2.5).

2.1 k-Dimensional Mesh Systems and k-Tree System Stage Representation

DEFINITION 1: A *k-Dimensional (k-D) Mesh Network* (of size $N = n_1 \times n_2 \times \ldots \times n_k$) is defined as a network graph $G(V, E) = G_1 \times G_2 \times \ldots \times G_k$ [or a product of k linear arrays $G_i(V_i, E_i)$ of n_i nodes; $V_i = \{1, 2, \ldots, n_i\}$ and $E_i = \{<j, j+1> \mid j = 1, 2, \ldots, n_i-1\}$], where $V = \{\alpha = (a_1, a_2, \ldots, a_k)$ be an address of any PE (processing element) in $G \mid a_1 \in V_1, a_2 \in V_2, \ldots, a_k \in V_k\}$ and $E = \{<\alpha, \beta>$ be a link between any two PEs in G, $\alpha = (a_1, a_2, \ldots, a_k)$ and $\beta = (b_1, b_2, \ldots, b_k) \mid$ there exists an i such that $<a_i, b_i> \in E_i$, where $i = 1, 2, \ldots, k.\}$.

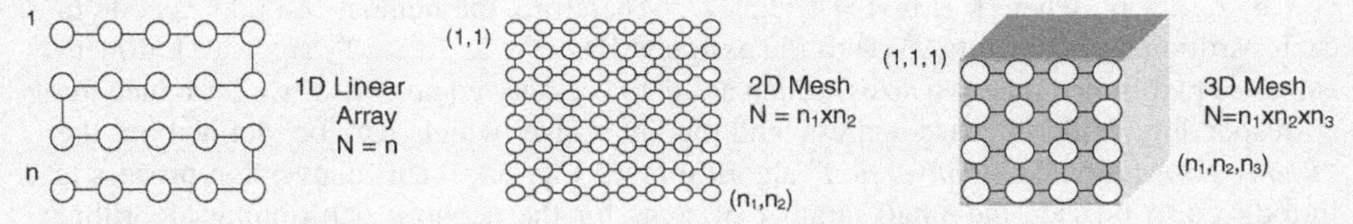

The "*k-Tree*" data structure is used to represent system states (or store allocation information) of the partitionable k-D mesh-connected parallel system, where k is a number of dimensions of the multi-dimensional (k-D) mesh-connected system. This is a special k-Tree since sub-systems' sizes after partitioning are not necessary equal, unlike a balance k-Tree of the same sub-system sizes after partitioning. In this paper, a "system" refers to a given partitionable k-D system, represented as a k-product network $G_1 \times G_2 \times \ldots \times G_k$ of size $N = n_1 \times n_2 \times \ldots \times n_k$ and a "sub-system" refers to a smaller k-D system of size $N' = n_1' \times n_2' \times \ldots \times n_k'$, where $n_i' \leq n_i$ and $i = 1, 2, \ldots, k$, which is provided for incoming task(s).

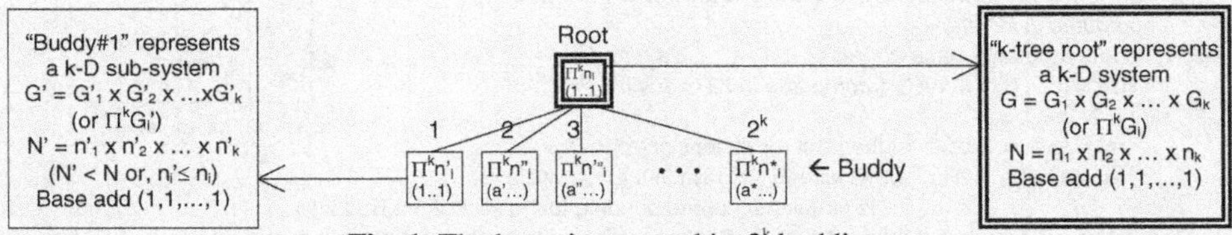

Fig. 1. The k-tree's root and its 2^k buddies.

In our k-Tree-based approach, the number of nodes in the k-Tree is dynamic, corresponding to the number of tasks allocated. At the start (see Fig. 1.), the k-Tree consists of only one node (called the root), used to store the system information (i.e., a size, a base-address, a status, etc.) of the initial system (which is the k-D network $G_1 \times G_2 \times \ldots \times G_k$ (or $\Pi^k G_i$) of size $N = n_1 \times \ldots \times n_k$ (or $\Pi^k n_i$) with a base address $\alpha = (1, 1, \ldots, 1)$, where $i = 1, 2, \ldots, k$. During execution (run) time when many jobs (or tasks) are allocated, each leaf node (representing a sub-system) may be available or free (status = 0) or unavailable or busy (status = 1) and each internal node is partially available (status = x). In order to allocate an incoming task, each larger free node in the k-Tree can be dynamically created and partitioned into a number of children/node, called buddies. Assume that the incoming tasks always request the same value "k" as provided by the system. Therefore, the number of buddies = 2^k (described later in Section 2.2).

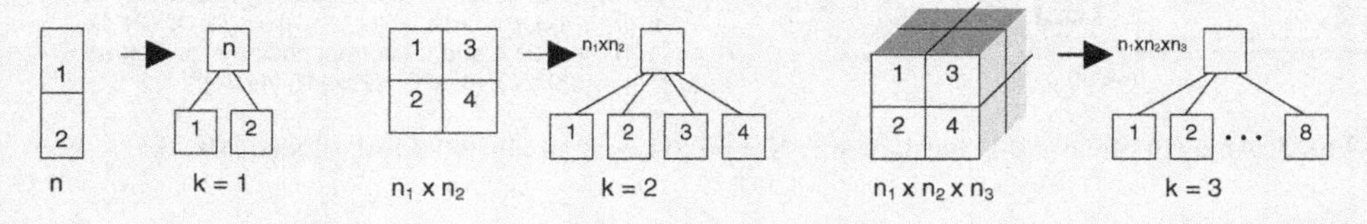

2.2 Network Partitioning Algorithm

The network partitioning is a partitioning process that partitions all k dimensions (or k linear array networks) of the k-D mesh system (of size $N = n_1 \times n_2 \times \ldots \times n_k$) into smaller sub-systems and allocates one for the request (k-D mesh of size $p_1 \times p_2 \times \ldots \times p_k$, where $p_i \le n_i$, $i = 1, 2, \ldots, k$). Under this network partitioning, the relationship of the corresponding buddy's ID, buddy's base-address, and buddy's size are defined as follows:

Let a requested network is defined as a network graph $G' = G_1' \times G_2' \times \ldots \times G_k'$(of size $p_1 \times p_2 \times \ldots \times p_k$, where $p_i \le n_i$, $i = 1, 2,\ldots,k$). Therefore, the number of buddies/node in each partitioning is equal to 2^k (with the assigned IDs = 1, 2, 3, …, 2^k) since all k dimensions are partitioned into two sizes, which are not necessary equal. Next, we introduce the corresponding buddy's base-address and buddy's size which can be defined in the "*Buddy-ID-Address-Size conversion*" algorithm (see Fig. 2.). This conversion process is introduced to provide the small number of steps for the network partitioning algorithm (described next) and the sub-system combining algorithm (described later in Section 2.3). This process (i.e., identifying #buddies = 2^k, their base-addresses and sizes) is computed in $k2^k$ steps and hence the total time complexity is $O(k2^k)$.

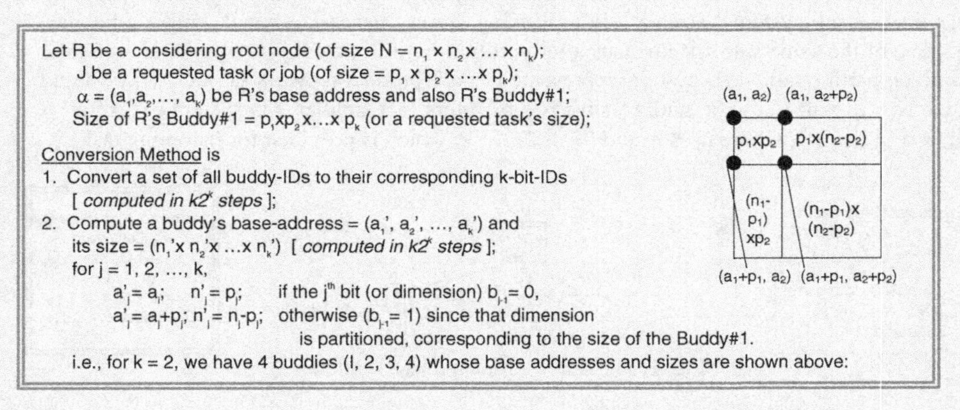
Let R be a considering root node (of size $N = n_1 \times n_2 \times \ldots \times n_k$);
J be a requested task or job (of size = $p_1 \times p_2 \times \ldots \times p_k$);
$\alpha = (a_1, a_2, \ldots, a_k)$ be R's base address and also R's Buddy#1;
Size of R's Buddy#1 = $p_1 \times p_2 \times \ldots \times p_k$ (or a requested task's size);

<u>Conversion Method</u> is
1. Convert a set of all buddy-IDs to their corresponding k-bit-IDs
   [*computed in $k2^k$ steps*];
2. Compute a buddy's base-address = $(a_1', a_2', \ldots, a_k')$ and its size = $(n_1' \times n_2' \times \ldots \times n_k')$ [*computed in $k2^k$ steps*];
   for j = 1, 2, …, k,
   $a_j' = a_j$; $n_j' = p_j$; if the jth bit (or dimension) $b_{j-1} = 0$,
   $a_j' = a_j + p_j$; $n_j' = n_j - p_j$; otherwise ($b_{j-1} = 1$) since that dimension is partitioned, corresponding to the size of the Buddy#1.

i.e., for k = 2, we have 4 buddies (l, 2, 3, 4) whose base addresses and sizes are shown above:

Fig. 2. The "Buddy-ID-Address-Size conversion" algorithm for the network partitioning.

For example (see Fig. 3.), consider a 2-D mesh system (of size $N = n_1 \times n_2 = 64 \times 64$) which is stored in the k-Tree's root with a base address $\alpha = (a_1, a_2) = (1, 1)$, where k = 2. Suppose the first incoming task requests a sub-system of size 20x20. For this task (20x20), the root (2-D mesh system of size 64x64), is partitioned into $2^k = 4$ buddies. Let's assume that the request will be allocated to the sub-Buddy#1 (at level 2).

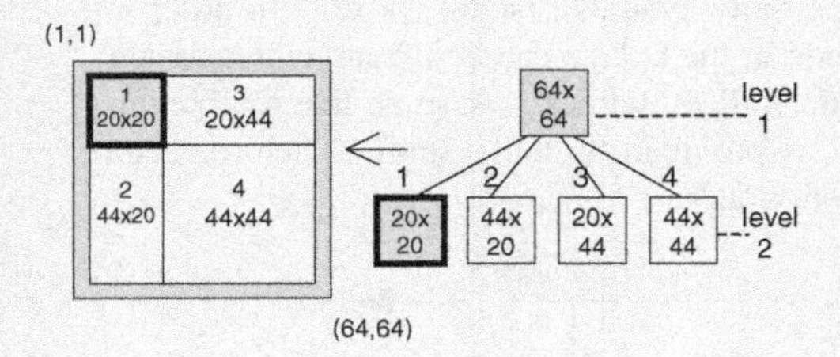

By applying "Buddy-ID-Address-Size conversion" algorithm (in Fig. 2.), the sub-Buddy#1's base-address is $(a_1, a_2) = (1, 1)$ and its size is $(p_1 \times p_2) = (20\times20)$. First the set of bit-address of each buddy i, {i / i =1, 2, 3, 4} is $\{(b_1\ b_0) \mid b_i = 0 \text{ or } 1\}$ = {00, 01, 10, 11}, and hence its corresponding base-address is $\{(a_1', a_2') \mid a_j' = a_j + (p_j * b_{j-1})\}$ = {(1,1), (21,1), (1,21), (21,21)} and its corresponding size is $\{(n_1', n_2')\}$ = {(20x20), (44x20), (20x44), (44x44)}.

Fig. 3. An example of a 2-D mesh of size N = 64x64 with an allocated task of size 20x20.

2.3 Sub-System Combining Algorithm

The sub-system combining is applied during processor allocation or deallocation. First, the "*Combinations of* 2^j *Adjacent Buddies*" algorithm is introduced (see Fig. 4.) in order to combine 2^j buddies (where j = 1, 2, ..., k-1) into the larger free sub-systems. This algorithm is computed in $O(k2^{k-j})$ time for each j and hence $O(k2^k)$ time for all values of the variable j, where j = 1, 2, 3, ..., k-1 (since $k2^1 + k2^2 + k2^3 + \ldots + k2^{k-1} = 2k[2^{k-1}-1]$).

Let R be a root node (at level L-1) that has all buddies = 2^k at level L.

<u>$k(2^{k-j})$ Combinations</u> are described (to combine 2^j free buddies or the jth dimension, j=1, 2,..., k-1) as follows:
For each j,

1. Create a set (2^{k-j} elements) of (k-j)-bit binary strings [*computed in* $(k-j)2^{k-j}$ *steps*];
2. Append these 2^{k-j} elements with j *'s to create a set of initial (adjacent) k-bit ternary strings [*computed in* $k2^{k-j}$ *steps*];
3. Shift left (k-1) times for each ternary string ($T = t_{k-1} \ldots t_1t_0$) of all 2^{k-j} strings, where $t_i \in$ {0,1,*} and then $k(2^{k-j})$ combinations are obtained [*computed in* $k2^{k-j}$ *steps*]. Note: Each ternary string T compounds of (k-j) 0's or 1's and j *'s.

Fig. 4. The "Combinations of 2^j adjacent buddies" algorithm for the sub-system combining.

For example, consider a (64x64)-mesh system with 4-buddy partitioning, where k = 2 and $2^k = 4$ (see Fig. 3.). In order to combine one of k dimensions or 2^j Buddies (such as j = 1), there are $k2^{k-j}$ (= $2x2^{2-1}$ = 4) possible combinations, recognized as follows: First, a 4-element set of "a compound of 2^j adjacent buddies" is computed as a set of binary strings {0, 1}. Next, each of these two elements is appended with j (or 1) * to create the corresponding set of ternary string {0*, 1*}. Then, each ternary string is shifted left (k-1) times to create a 4-element set of ternary string {0*, *0, 1*, *1} which represents a set of combinable strings (i.e., a ternary string 1* is interpreted as a set of 2-adjacent Buddies {1<u>0</u>, 1<u>1</u>} (or {Buddy#3, Buddy#4}) for a combined system's size = 64x44.)

Next, we classify the k-Tree-based sub-system combining algorithms into three groups, which will be applied later in the allocation and deallocation procedures (Section 2.5):

<u>ALGORITHM C.1</u> "*Combine All Buddies*": This combining procedure is used to combine all free 2^k buddies (at level L) into a larger free k-Tree's node at level L-1. This combining process is computed in $O(2^k)$ time and it is applied after finishing the deallocation of any finished task in order to maintain the minimum number of nodes and the maximum free nodes' sizes in the k-Tree as much as possible.

<u>ALGORITHM C.2</u> "*Combine Some Buddies*" (or called "Buddy-Buddy combining"): This combining procedure is used to combine a number (i.e., 2, 4, ..., or 2^{k-1}or j = 1, 2, ..., k-1) of adjacent free buddies (of the same root sub-tree) at level L into a larger free sub-system in order to allocate for an incoming task (whose requested a size which is larger than that of each of 2^k buddies but is less than or equal to that of the combined sub-system.) After applying the algorithm (in Fig. 4.: see also the following figures for the combining of 2-D meshes), for a ternary string T, the combined sub-system's size and its base-address are computed by using the "2^j *Combinable Buddy-ID-Address-Size conversion*" algorithm (see Fig. 5.) which can be computed in ($3k2^j + k$) steps. Therefore, time complexity of this combining for each j in order to combine all possible $\sum_{j=1}^{k-1} k2^{k-j}$ combined sub-systems (from 2^j adjacent free buddies) is $O(k^2 2^k)$ and hence $O(k^3 2^k)$ for all value of j, where j = 1, 2, ..., k-1.

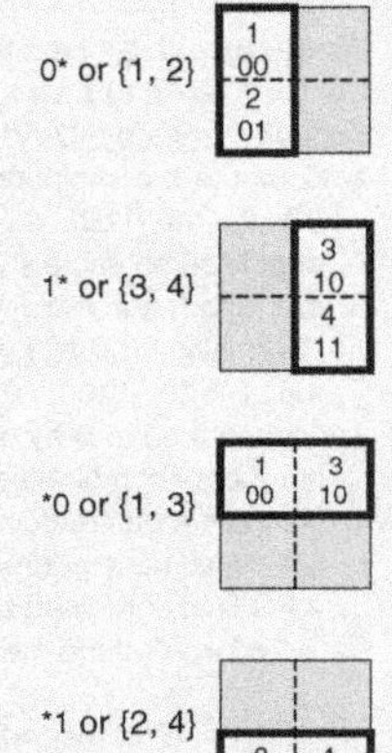

Let R be a considering root node (of size $N = n_1 x\, n_2 x \ldots x\, n_k$ & base address $\alpha = (a_1, a_2, \ldots, a_k)$);
T be a ternary string $(t_{k-1} \ldots t_1 t_0)$, from Algo. (in Fig. 4.) i.e., T = 1* or {Buddy#3, Buddy#4} of the 2D meshes;
Buddy-ID-Address-Size Conversion Method is

1. Convert the ternary string $T=(t_{k-1} \ldots t_1 t_0)$ to a set (2^j elements) combinable binary-IDs $(b_{k-1} \ldots b_1 b_0)$ [i.e., T = 1* → {10, 11} = {Buddy#3, Buddy#4}] [*computed in* $(k2^j + k)$ *steps*] by
   1.1 create 2^j binaries $(r_{j-1} \ldots r_1 r_0)$ and replace them in the location(s) of j "*"s in T;
   1.2 convert binary-IDs to a set of integer-IDs ($1 \le ID \le 2^k$, where 1 = 0…00, 2 = 0…01, 3 = 0…10, … , 2^k = 1…11);
2. Compute a base-address & a size of each combined sub-system of 2^j sub-systems [*computed in* $2k2^j$ *steps*] by
   2.1 a base address = the base address of the minimum buddy-ID;
   2.2 a combined size = $(n_1' x\, n_2' x \ldots x\, n_k')$, for $\forall i$, i = 1, 2, …, k;
   $n_i' = n_i$ (of the min buddy-ID) + n_i (of the max buddy-ID) if (b_{i-1} (of the min buddy-ID) $\neq b_{i-1}$ (of the max buddy-ID));
   $n_i' = n_i$ (of the min buddy-ID) otherwise.

Fig. 5. The "2^j Combinable Buddy-ID-Address-Size conversion" algorithm (Algorithm C.2).

ALGORITHM C.3 "*Combine a Buddy and Corresponding Sub-Buddies*" (or called "Buddy-SubBuddy combining") and "*Combine Some SubBuddies*" (or called "SubBuddy-SubBuddy combining"): These combining procedures are used to combine some free buddies (at level L) and their corresponding sub-buddy nodes (at Level L+1) to yield more sub-system recognition (from any partitioning size) than those obtained from the combining in Algorithm C.2. Then, we introduce the "Buddy-SubBuddy combining" algorithm (see Fig. 6.) for recognizing $k2^k$ combined sub-systems of a free buddy at level L and its adjacent free nodes (or sub-buddies) at level L+1 and also the "SubBuddy-SubBuddy combining" algorithm (see Fig. 7.) for recognizing $k2^{k-1}$ combined sub-systems of some adjacent free nodes at level L+1 [see also following figures for all possible combining for the 2-D meshes]. Each of these conversion processes can be computed in $O(k^2 2^k)$ time for each combined node (or string) and $O(k^2 2^{2k})$ time for all possible 2^k strings.

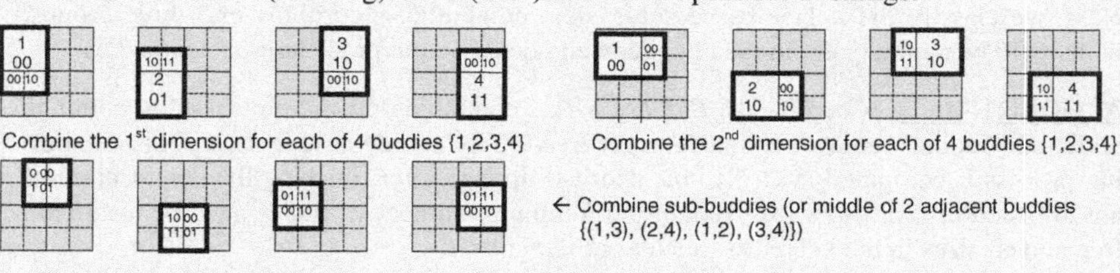

Combine the 1st dimension for each of 4 buddies {1,2,3,4}
Combine the 2nd dimension for each of 4 buddies {1,2,3,4}
← Combine sub-buddies (or middle of 2 adjacent buddies {(1,3), (2,4), (1,2), (3,4)})

Let $B = (b_{k-1} \ldots b_1 b_0)$ be a free buddy (i.e., 00 or {Buddy#1} of the 2D meshes);
$T = (t_{k-1} \ldots t_1 t_0)$ be a ternary string (one of k dimensions to be combined);
Buddy&SubBuddy-ID-Address-Size Conversion Method is

1. Compute k combinable buddies $C = (c_{k-1} \ldots c_1 c_0)$ of B from all possible k dimensions [*computed in* k^2 *steps*]; (i.e., for the jth dim, if j = 1, then T = 0* →C = 01 or {Buddy#2} of the 2D meshes)
2. Identify combinable sub-buddies (for each C) as a set of ternary string $S = (s_{k-1} \ldots s_1 s_0)$ by replacing (k-j) 0's/1's in T with *; and replacing j *'s in T with j b's [*computed in* k^2 *steps for k C's*].
   [Formally, for $\forall$ bit i = 0,1,2, .., k-1, $c_i = t$; s_i = *; (if t_i = 0/1) or $c_i = b_i'$ $s_i = b_i$; (if t_i = *)]
   (i.e., S = *0 → {00, 10} or {SubBuddy#1, SubBuddy#3} of the 2D meshes)
3. Convert the set S by using "2^j combinable-Buddy-ID-Address-Size conversion" algorithm (in Fig. 5.), including a number of combined sub-systems $(n_1' x n_2' \ldots x n_k')$ and their base addresses [*computed in* $(k-j)(3k2^{k-j}+k)$ *steps*];
4. Compute a base-address & size of each combined S from 2 Buddies in the set [*computed in* k^2 *steps for k C's*]:
   4.1 a final base address = the base address of the min buddy-ID (of the combined S);
   4.2 a final combined size = $(n_1'' x\, n_2'' \ldots x\, n_k'')$, for $\forall i$, i = 1, 2, …, k;
   $n_i'' = n_i'$(of the min buddy-ID) + n_i' (of the max buddy-ID) if (b_{i-1} (of the min buddy-ID) $\neq b_{i-1}$ (of the max buddy-ID));
   $n_i'' = n_i'$(of the min buddy-ID) otherwise.

Fig. 6. The "Buddy&SubBuddy-ID-Address-Size conversion" algorithm (Algorithm C.3-1).

Let T = (t_{k-1} ... $t_1 t_0$) be a ternary string of a pair of combinable 2 buddies
<u>SubBuddy&SubBuddy-ID-Address-Size Conversion Method</u> is (in this case j = 1)

1. Convert T to a set of (2 elements) buddy-ID {B_1,B_2/B_i= (bi_{k-1} ... $bi_1 bi_0$) } of 2 partially free buddies at level L by creating 2^j binary-numbers (r_{j-1}...$r_1 r_0$); replacing them into j'*s of (t_{k-1} ... $t_1 t_0$) [*computed in k2 steps*].
2. Identify combinable sub-buddies of each of 2 buddies as a set of ternary string S (s_{k-1} ... $s_1 s_0$) [*computed in k^2 steps*]: for ∀ bit i = 0, 1, 2, ..., k-1, $b1_i = t_i$; $b2_i = t_i$; and $s1_i$ = *; $s2_i$ = *; if t_i = 0 or 1,
   or $b1_i = 0_i$; $b2_i = 1_i$; and $s1_i = b1_i'$; $s2_i = b2_i'$; if t_i = *
3. Similar to that of algorithm in Fig. 6.
4. Similar to that of algorithm in Fig. 6.

Fig. 7. The "SubBuddy&SubBuddy-ID-Address-Size conversion" algorithm (Algorithm C.3-2)

2.4 Best-Fit Heuristic for Allocation Decision

In this sub-section, we present *a generalized best-fit heuristic* for the partitionable k-D mesh-connected systems. The best-fit heuristic is to find the best free sub-system (of all possible available sub-systems) for an incoming task by introducing a number of criteria that tend to cause the minimum system fragmentation(s), which are:

<u>Best-Fit Criteria</u>:

1. Find all free sub-systems that can preserve the "maximum free size" as possible [see *AlgorithmA.1 in O(k) time*].
2. If there are many candidates (sizes ≥ the request) that have the same property in (1), then the candidate that gives the "minimum different size factor (diffSF)" is selected [see *AlgorithmA.2 in O(k^2) time*].
3. If there are many candidates (sizes ≥ the request) that have the same property in (1) & (2), then the "smallest size" candidate that yields the "minimum combining factor (CF)" [*see Algorithm A.3 in O(k) time*] is selected. Otherwise, select by random.
4. After searching on all nodes in the k-Tree,
   - If the best free sub-system is "equal to" the request, then it is directly allocated to the request.
   - Otherwise (it is "larger than" the request), it is partitioned and one of its buddies which yields properties similar to that given in Step 1 – Step 3 plus being the "best buddy location" or providing the "minimum modified CF (MCF)" will be selected [see *Algorithm A.4 in O($k2^k$) time*].

<u>Note:</u> Criteria 1-3 are applied for every free node or combined sub-system; however, a criterion 4 is computed only once for the best free sub-system, obtained from Steps 1-3.

For example, given a 64x64-system with two tasks (20x20, 22x15) allocated (see Fig. 8). Suppose there are two new incoming tasks (22x5 and 15x10).

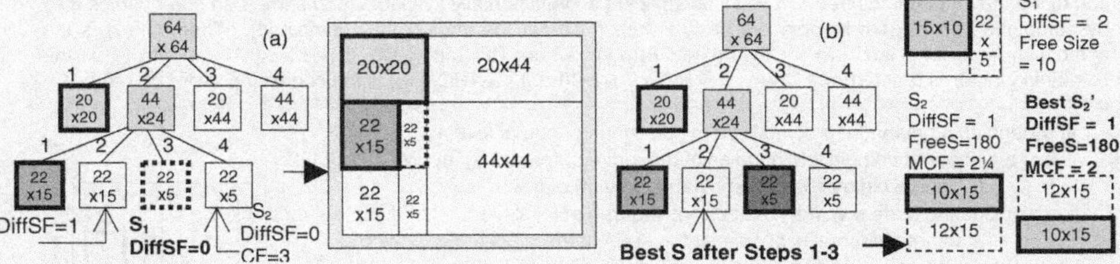

For the new task (22x5), searching process starts from the root and then visits all nodes in the k-Tree (in order to find the best free S). After applying Steps 1-3 of the best-fit heuristic, the best free node that can accommodate the request is S_1(22, 5) (with min diffSF, min CF) at level 3 (see Fig. 8.a). Therefore, it is allocated to the requested task 22x5. For the next task (15x10) (see Fig. 8.b), after applying the best-fit criteria (Steps 1-3), the best sub-system is S(22, 15) at level 3 since it can preserve the maximum free size (64x44). Since the sub-system S(22, 15) is larger than the request 15x10, Step 4 is applied that is the S(22, 15) is partitioned. After partitioning, the rotated size 10x15 (S_2) provides the better best-fit value (diffSF = 1, free size = 180) than that of the regular size 15x10 (S_1) (diffSF = 2, free size = 110). Finally, among S_2 (the top buddy) and S_2' (the bottom buddy), the S_2'(10,15) is selected and allocated to the request task (15x10) since it yields the minimum modified CF (MCF) (see AlgorithmA.4).

Fig. 8. An example of the "best fit" k-Tree-based allocation.

ALGORITHM A.1: "*Overlap Status*" of two available sub-systems S_1 (the maximum free size) and S_2 (any free sub-system) is identified as follows: If these two sub-systems (S_i, i = 1, 2) in the k-Tree have base-address $\alpha_i = (a_{i1}, a_{i2}, \ldots, a_{ik})$, last-cover address $\beta_i = (b_{i1}, b_{i2}, \ldots, b_{ik})$, and size $|S_i| = (n_{i1} \times n_{i2} \times \ldots \times n_{ik})$, where $b_{ij} = a_{ij} + n_{ij} - 1$, i = 1, 2, and j = 1,2,..., k, then Case 1: the sub-system S_2 is a *subset* of the sub-system S_1 if ($a_{2j} \geq a_{1j}$ and $b_{2j} \leq b_{1j}$) for $\forall$j, j =1,2,..., k. Case 2: the sub-systems S_1 and S_2 are *disjoint* either if (a_{1j}-$a_{2j} \geq n_{2j}$) or if (a_{2j}-$a_{1j} \geq n_{1j}$) for $\exists$j, j = 1, 2, ..., k. Otherwise (neither Case 1 nor Case 2) the sub-system S_1 *intersects* the sub-system S_2. Each of these three statuses (subset, disjoint, intersect) can be computed in O(k) time.

For instance, the following figure illustrates various overlap statuses between S_1 "the maximum free size" [n_1 = 6 x 6 at <(1, 5), (6, 10)>] and other free sub-systems such as S_2 [n_2 = 4 x 4 at <(5,1), (8,4)>], S'_2 [n'_2 = 2 x 4 at <(5, 7), (6, 10)>], and S''_2 [n''_2 = 4 x 2 at <(5, 5), (8, 6)>].

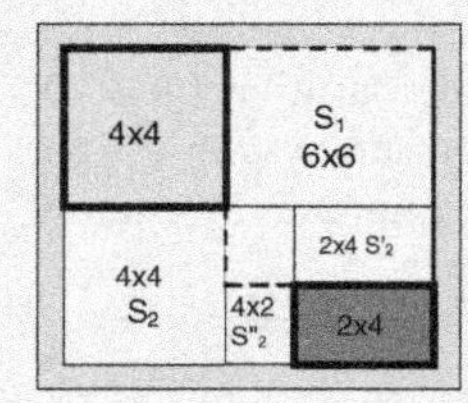

S_1 and S_2 are *disjoint* since for i = 2 there exists (a_{12} - a_{22}) = (5 – 1 = 4) $\geq n_{22}$ (= 4). S'_2 is *subset* of S_1 since S'_2 = <(5,7), (6, 10)> $\subset$ S_1 = <(1, 5), (6, 10)> (or $\forall$i = 1, 2; $a_{2i} \geq a_{1i}$; and $b_{2i} \leq b_{1i}$). S''_2 *intersects* to S_1 since they are not disjoint and also neither is a subset of the other.

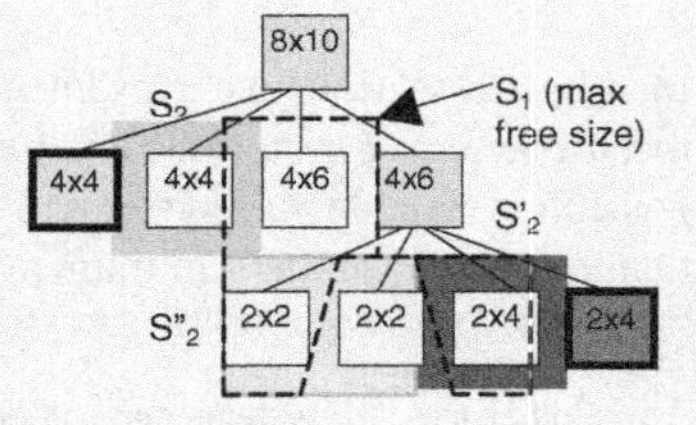

ALGORITHM A.2: "*Task Rotation*" in this paper is a process of shifting a given task size k-1 times to find the suitable location of the free sub-system for the requested (k-D) task of size $p_1 \times p_2 \times \ldots \times p_k$. Thus, there are k possible rotated sizes that can be allocated for the task: ($p_1 \times p_2 \times \ldots \times p_k$), ($p_2 \times p_3 \times \ldots \times p_k \times p_1$), ..., and ($p_k \times p_1 \times \ldots \times p_{k-1}$). For the partitioning of each rotated task size against a given free S, the different size factor ($0 \leq$ diffSF $\leq k$) and the maximum (remaining) free size (FS) (0 < |FS after partition| < |FS before partition|) are computed in O(k) time and hence in O(k^2) time for all k possible rotated sizes. See an example of the task rotation in Fig. 8.b that is the rotated S_2(10x15) is selected rather than the regular S_1(15x10)).

ALGORITHM A.3: "*Combining Factor* (CF)" of any free sub-system S (at level L) is computed from its adjacent neighbor nodes as a summation of the probability of combining (PC) of each of 2^k combinable nodes of the same root sub-tree.

In this study, for an adjacent side we define PC=0 if that particular combined side is one of 2^k system boundaries (since that side cannot be combined), PC=¼ if its adjacent node of that particular side is busy (since it can be combined after it becomes free); PC=½ if its adjacent node is partially available (and some free sub-buddies may be combined); or PC=1 if its adjacent node is free (or it can be immediately combined). Then, CF(α) is the combining factor of α is CF(α) = $CF_1(\alpha,\beta) + CF_2(\alpha,\gamma)$, where $CF_1(\alpha,\beta)=PC(\alpha,\beta_1)+PC(\alpha,\beta_2)+\ldots+PC(\alpha,\beta_k)$ is the combining factor of α at L-1 and $CF_2(\alpha, \gamma) = PC(\alpha,\gamma_1)+PC(\alpha,\gamma_2) +\ldots+PC(\alpha,\gamma_k)$ is the combining factor of α at L-2:

Let α denotes a binary-ID ($b_{k-1} \ldots b_1b_0$) of a considering node at level L
k is a number of combinable buddies of the root sub-tree (of α) at level L-1 or L-2.
$\beta_1,\beta_2,\ldots,\beta_k$ denote binary-IDs of combinable node(s) of the considering node α with the same root sub-tree at level L-1
$\gamma_1, \gamma_2,\ldots,\gamma_k$ denote binary-IDs of adjacent node(s) of the considering node α with the same root sub-tree at level L-2

Identify adjacent nodes by using the following rules:

1) For a k-Tree node, each β_i or γ_i is identified by negating the ith bit of that node (α) which are $\beta_1 = (b_{k-1} \ldots b_1b_0')$, $\beta_2 = (b_{k-1} \ldots b_1'b_0)$, ..., and $\beta_k = (b_{k-1}' \ldots b_1b_0)$). Therefore, for a root(α) = ($r_{k-1} \ldots r_1r_0$), its combinable buddies are $\gamma_1 = (r_{k-1} \ldots r_1r_0')$, $\gamma_2 = (r_{k-1} \ldots r_1'r_0)$, ..., and $\gamma_k = (r_{k-1}' \ldots r_1r_0)$, respectively (See the following example).
2) For a combined sub-system S = ($t_{k-1} \ldots t_1t_0$) of 2^j free (or partially free) buddies, $1 \leq j < k$, (or 2^j combined nodes) and β_i = negate the ith 0/1 (or non *) bit, represented S as a binary number, where i=1, 2,..., j.

Given an 8x8-mesh. Let $\alpha = b_1b_0 = 11$ (or 4), residing at level 3 and 2 adjacent buddies of α are β_1 = 1<u>0</u> (or 3) ; β_2 = <u>0</u>1 (or 2). Assume the root of node α at level 2 = 10 (or 3) and then 2 adjacent nodes of the root (α) are γ_1 =1<u>1</u> (or 4) and γ_2 = <u>0</u>0 (or 1).

ALGORITHM A.4: "*The Best Buddy* (*or Best Sub-partition*)" is applied after partitioning (for Step 4 in the best-fit heuristic). In this case, assume the best free sub-system from Steps 1-3 is the node S, whose size is larger than the requested task. Then, the node S will be partitioned into 2^k buddies and the best sub-partition (or one of 2^k buddies will be allocated to the request. Let $\beta_i = (\beta_{i1}, \beta_{i2}, ..., \beta_{ik})$ be a set of combinable buddies of the considering node (α_i), where i = 1, 2, ..., 2^k. Since the combining factor (see Algo. A.3) CF(α, β) is the same for all α_i, i = 1, 2, ..., 2^k, the modified combining factor MCF (α_i, β_{ij}) for each α_i is computed as MCF (α_i, β_{ij}) = Σ_1^k PC (α_i, β_{ij}) in O($k2^k$) time and the one (yielding min MCF) is selected as the best buddy.

See an example of finding the best buddy node in Fig. 8.b: the MCF of S_2 (10x15) = 1+1+¼+0 = 2¼ (since its four combinable boundaries are two free node, one busy node, and one system boundary) and the MCF of S'_2 = 1+1+0+0 = 2 (since its four combinable boundaries are two free nodes and two system boundaries); and hence the S'_2 yields the minimum MCF and it is selected.

2.5 Allocation/Deallocation Decision Algorithm

In "*the best-fit k-Tree-based processor allocation*" procedure, the searching starts from the k-Tree's root and goes to the left most (leaf) node. If that node is free and its size can accommodate the request, then its best-fit value (see Section 2.4) is computed. Then, the best sub-system is updated if the new free sub-system yields the better best-fit value (since it tends to cause the minimum system fragmentation). The above process is repeated for the next node in the k-Tree (if there exists). After all nodes (including each leaf node and each internal node (for a number of combined sub-systems: see Section 2.3)) are visited, the final process is applied, which is either 1) to allocate the best sub-system directly to the request (if its size is equal to that of the request) or 2) to partition the corresponding node (see Section 2.2) for the request (since its size is larger than that of the request).

Finally, whenever a task is finished, "*the k-Tree-based processor deallocation*" procedure is applied by searching for the location of the finished sub-system starts from the k-Tree's root and goes to the subset path until reaching the leaf node that stores information of the finished task. After finding the corresponding k-Tree's node of the finished task, its status is updated (or removed from the k-tree). Finally, the combining process is recursively applied from the finished node(s) to the root (if it is possible).

Note: the expand-node-size function (in the allocation process) will be applied if a combined sub-system is selected as the best sub-system for the current incoming task in order to limit the number of nodes in the k-Tree and provide the same methodology to update and partition as a regular (free) leaf node. Then, other corresponding nodes in the combined sub-system have to be updated as busy. Finally (in the deallocation process), these corresponding busy nodes will be free whenever that expanded node is free (in the resume-node-size function).

For example, assume that in the current system, there are three tasks allocated: Buddy#1 and Buddy#2 at level 2, Buddy#1 at level 3 and suppose that a new incoming task requests a sub-system of size 5x6 (which is larger than each buddy node but less than a combined sub-system of size 6x6).

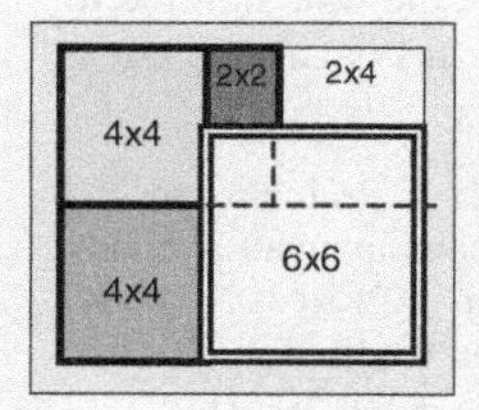

Therefore, the expanded node processing of a combined (6x6) sub-system (stored in the Buddy#4 (at level 2) of the root) is applied before partitioning. Note: the old information of this node is also stored (in dashed node) for resuming later when it is finished.

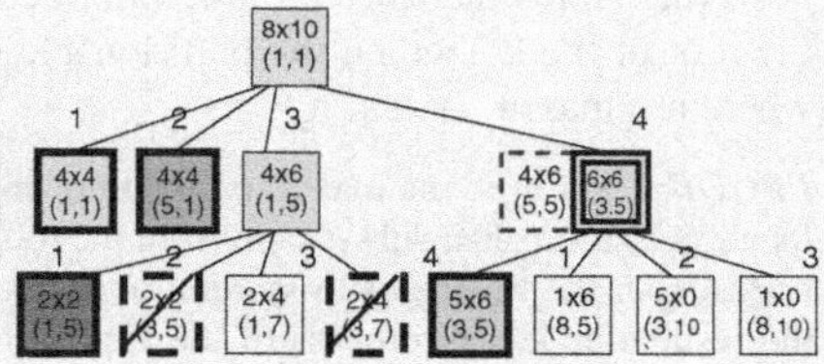

2.6 Time Complexity Analysis

Let N be the system size ($N = n_1 \times n_2 \times \ldots \times n_k$), N_A be the maximum number of allocated tasks ($N_A \leq N$), N_F be the corresponding number of free nodes in the k-Tree ($N_A + N_F \leq N$), and M be the maximum number of nodes in the k-Tree (where M = external (leaf) nodes + internal (non-leaf) nodes < 2N).

THEOREM 1: Time complexity of the k-Tree-based allocation to find the best free sub-system for each incoming task on a k-D mesh (of size $N= n_1 \times n_2 \times \ldots \times n_k$) is $O(k^4 2^k (N_A+N_F)+k^2 2^{2k})$.

PROOF: In the allocation algorithm (see Section 2.5), a number of recursive iterations of the DFS (depth first search) are at most a number of nodes in the k-Tree and only nodes whose sizes are larger than (or equal to) the request are visited. In this (non-bit map) approach, the number of nodes in the k-Tree is proportional to the number of allocated tasks or busy nodes (N_A) and the number of free nodes (N_F) in the k-Tree, where $N_A + N_F \leq N$ and $N_F \leq (2^k-1)N_A$. Since the number of external (or leaf) nodes in the k-Tree are at most $N_A + N_F \leq N$ and the number of internal nodes are at most (#leaf nodes-1) divided by (2^k-1); therefore the total number of nodes in the k-Tree is at most M nodes, where $M = (N_A+N_F) + (N_A+N_F-1) / (2^k-1)$. For each (free) leaf node (of N_F nodes), the best-fit value is computed in $O(k^2)$ time and hence $O(k^2 N_F)$ for all leaf nodes. For each internal node (of $(N_A+N_F-1)/(2^k-1)$ nodes), the best-fit value is computed in $O(k^2)$ time for each node and hence $O(k^4 2^{2k})$ time for $k^3 2^k + k^2 2^{2k}$ combined sub-systems, as summarized in Table 1. (Note that the maximum free size (in Step 0) is computed only once for each task by using DFS in $O(k^3 2^k (N_A+N_F))$ time.) Finally, after finding the best free sub-system (Step 3), if its size is equal to the request, then it is directly allocated to the request. Otherwise, the network partitioning and the best sub-partition will be applied, which can be computed in $O(k2^k)$ time. Then, the corresponding node(s) in the k-Tree is updated in O(k) time. Note: for the combined sub-system that is larger than the request, before partitioning, expand-node-size process is applied in $O(k^2 2^{2k})$ time. Thus, total time complexity to visit all nodes in the k-Tree is approximately $[N_A+kN_F+k^3 2^{2k}(N_A+N_F-1)/(2^k-1)] + [N_A+k^2 N_F] + [(k^4 2^{2k}+k^3 2^k)(N_A+N_F-1)/(2^k-1)] + [k^2 2^{2k}+2k2^k+k] = O(k^4 2^k(N_A+N_F) + k^2 2^{2k})$, where $N_A+N_F \leq N$.

Table 1. Time complexity of the k-Tree-based approach for the k-D meshes.

Functions in each k-Tree's node for sub-system Allocation (see Section 2.5)	Time complexity
(0) Before searching to find the best free sub-system	
- Compute a maximum free size (from all M nodes in the k-Tree) Busy Leaf + Free Leaf + Internal Node operations [$\cong N_A+kN_F+k^3 2^{2k}(N_A+N_F-1)/(2^k-1)$]	$O(k^3 2^k(N_A + N_F))$
(1) Leaf node operation (for N_F nodes) : - Compute best-fit value $O(k^2)$ [$\cong N_A + k^2 N_F$]	$O(N_A + k^2 N_F)$
(2) Internal node operation (for $(N_A+N_F -1)/(b-1)$ nodes) [$\cong (k^4 2^{2k}+ k^3 2^k)(N_A+N_F -1)/(2^k-1)$]	$O(k^4 2^k(N_A+ N_F))$
Sub-system combining:Algo.C.2 $O(k^3 2^k)$ and Algo.C.3 (next level) $O(k^2 2^{2k})$)	
- Compute best-fit value (of all combined subsystems) $O(k^4 2^{2k})$	
(3) Partitioning after finding the best free node [$\cong k^2 2^{2k}+ 2k^2 2^k+ k$]	$O(k^2 2^{2k})$
- Expanding node (for a combined sub-system) $O(k^2 2^{2k})$	
- Best sub-partition $O(k2^k)$	
- Network partitioning $O(k2^k)$	
- Allocate (update k-Tree) $O(k)$	
Total time (for M nodes) = (0) + (1) + (2) + (3)	$O(k^4 2^k(N_A+ N_F) + k^2 2^{2k})$

Note:Our k-Tree-based model, when applied to the 2-D/3-D meshes, provides a linear time complexity $O(N_A+N_F)$

THEOREM 2: Time complexity of the k-Tree-based deallocation to free the particular k-Tree node that stores the finished task and to combine the free buddy nodes of the root sub-tree to the root of the k-Tree on the partitionable k-D mesh ($N = n_1 \times n_2 \times \ldots \times n_k$) is $O(n2^k+ k^2 2^{2k})$, where $n = \max(n_1, n_2, \ldots, n_k)$.

PROOF: Let n be the maximum depth of the k-Tree ($n = \max(n_1, n_2, \ldots, n_k)$). Searching for the location of a finished sub-system from the root is at most $n(2^k)$ steps. Then, combining all 2^k buddy nodes from the finished sub-system to the root (if it is possible) takes another $n(2^k)$ steps. Finally, the resume-node-size process of the expand-node-size process (if any) may be required in $O(k^2 2^{2k})$ time. Therefore, total time complexity of the k-Tree-based deallocation is $O(n2^k + k^2 2^{2k} < M)$.

3 Performance of the k-Tree-Based Approach on the 2-D Meshes

In order to evaluate the system performance, the generalized k-Tree-based approach was developed. By simulation study, a number of experiments are performed to investigate the effect of applying the "k-tree-based model" for performing processor allocation/deallocation for the partitionable 2-D meshes and compare to recently 2-D mesh-based strategies ([3], [15], [7]). The investigated system performance includes system utilization, system fragmentation, average allocation time, etc. For each experiment, a number of simulation time units are iterated around 5,000-50,000 time units and a number of incoming tasks are generated approximately 1,000-10,000 tasks, according to the setting of the system size parameter, the task size (i.e., row, column) parameter and the task size's distribution. For each evaluated result, a number of different data sets are generated and the algorithm is repeated until an average system performance does not change (or at least 100 iterations). Experimental results of applying the k-Tree-based strategy are represented for both *static system performance* (with concerning processor allocation for incoming tasks (or jobs) only (or it is assumed that no task finishes during the considering time)) and *dynamic system performance* (with taking into account of deallocation for some finished tasks). In this study, in order to set the same incoming tasks and environment to all strategies for the comparison purpose, the static system performance is concerned (i.e., when we measure the system utilization and system fragmentation); otherwise the dynamic system performance is concerned. In each experiment, two task-size distributions are considered: the Uniform distribution $U(\alpha, \beta)$ and the Normal distribution $N(\mu, \sigma)$. For each of these distributions, the system sizes (N = RxC) are varied and the task sizes [1x1 - RxC] are generated, where $\alpha = 1$, $\beta = \max(R, C)$ for the Uniform distribution $U(\alpha, \beta)$ and $\mu=\sigma=\max(R, C)/2$ for the Normal distribution $N(\mu, \sigma)$. Other parameters are fixed such as task arrival rate ~ Poisson (λ) (or inter-arrival time ~ Exp($1/\lambda=5$)), and service time ~ Exp($\mu=10$), etc.

In Experiment 1, we investigated "*the effect of system sizes to the system utilization* (U_{sys}) *and the system fragmentation* (F_{sys})". In this experiment, the system sizes (N = RxC) were varied and the task sizes (1x1 - RxC) were generated and fixed. In Table 2 (the system utilization result), for all test cases the k-Tree strategy performed ~60% system utilization which was comparable to those of the recently 2-D mesh-based strategies (i.e., for the uniform distribution, the FSL, the BL, and the QA strategies yielded ~56%, ~57%, and ~56% system utilization, etc.). For the system fragmentation, the k-Tree approach and these existing strategies also performed the same results for the system fragmentation since there was no internal system fragmentation ($F_{sys} = 1-U_{sys}$). In Experiment 2, we investigated "*the effect of task sizes to the system utilization and the system fragmentation*". In this case, the system size was fixed (N = 512x512) and the task sizes were generated and varied. In Fig. 9. (the system fragmentation result), for all test cases the k-Tree strategy performed the comparable system fragmentation to the FSL, BL, and QA strategies, which were ~30%, ~40%, and ~41% system fragmentation for task sizes [1x1-250], [1x1-350x350], and [1x1-512x512], respectively. For the system utilization, the k-Tree approach and these existing 2-D mesh-based strategies also performed the same results since $U_{sys} = 1-F_{sys}$ (or no effect of the internal system fragmentation). In Experiment 3, we investigated "*the effect of allocation time*" of the k-Tree and existing 2-D mesh-based strategies when the system sizes were increased. In Fig. 10. (the average allocation time), our k-Tree approach yielded the improved average allocation time, compared to the existing strategies for all tested cases, except when the system size was small (N = 64x64). In this case, the average allocation time of the k-Tree, FSL, and BL strategies were approximately constant since they depended on the number of allocated tasks (N_a). However, the average allocation time of the QA strategy was increase linearly which depended on the number of allocated tasks (N_a) and the system size (N). In Experiment 4, we investigated "*the effect of allocation and deallocation time*" when the system sizes were increased. In Fig. 11.. (the average allocation and deallocation time), our k-Tree approach yielded the improved average allocation and deallocation time, compared to the existing strategies when system sizes were increased. In this case, the average allocation and deallocation time of the k-Tree was approximately constant while those of existing 2-D mesh-based strategies were increased linearly.

Table 2. Effect of "the system sizes" to the system utilization (%).

Task size Distributions	System sizes (N = RxC)	k-Tree	Free Sub-List (FSL)	Busy List (BL)	Quick Allocation (QA)
Uniform (α, β)	64 x 64	60.97	56.06	57.27	55.64
α=1, β=min(R,C)	128 x 128	60.11	56.98	56.67	57.84
For [1x1 – RxC]	256 x 256	59.77	55.77	54.61	55.87
	512 x 512	58.86	56.11	55.91	56.53
Normal (μ, σ^2)	64 x 64	61.18	61.42	58.46	57.53
μ = min(R,C)/2	128 x 128	59.16	56.89	58.02	54.79
For [1x1 – RxC]	256 x 256	60.10	58.02	55.12	55.09
	512 x 512	61.28	57.04	57.22	55.89

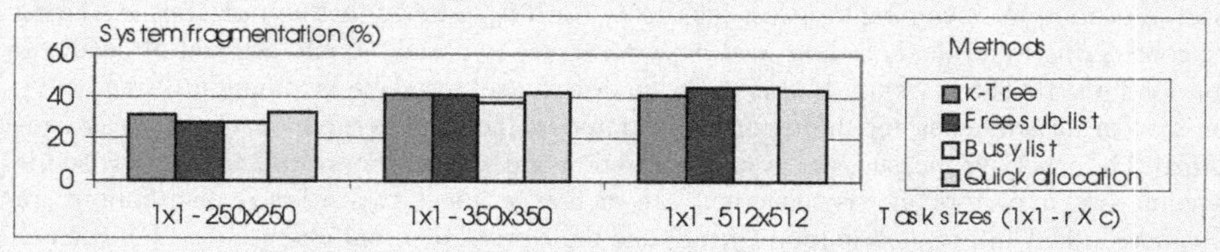

Fig. 9. Effect of "the task sizes" to the system fragmentation (%).

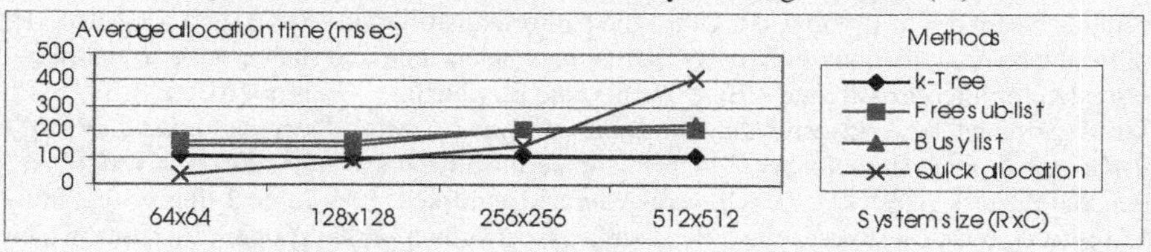

Fig. 10. Effect of "the system sizes" to the average allocation time.

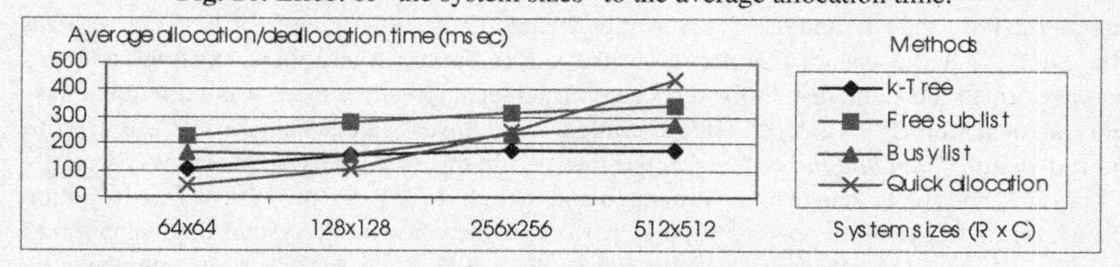

Fig. 11. Effect of "the system sizes" to the average allocation and deallocation time.

4 Conclusion

In this paper, we present the design and the development of the "generalized k-tree-based sub-mesh allocation" model for the partitionable multi-dimensional mesh-connected systems. Time complexity of the k-Tree-based approach is $O(N_A + N_F)$ for the partitionable 2-D and 3-D meshes and $O(k^4 2^k (N_A+N_F)+k^2 2^{2k})$ for the partitionable k-D meshes, where N_A is the maximum number of allocated tasks and N_F is the corresponding number of free sub-meshes. By simulation studies, a number of experiments were performed to investigate the system performance of applying the k-Tree-based model for the partitionable 2-D meshes. In the experimental results, the system performance (i.e., the system utilization and system fragmentation) of the k-Tree-based model for the partitionable 2-D meshes was comparable to existing 2-D mesh-based strategies. In addition, the k-tree-based approach yielded the improved allocation/deallocation decision time (i.e., the average allocation time, the average allocation and deallocation time), compared to those 2-D mesh-based strategies when the system sizes (N) are very large.

5 References

[1] Chuang, P.J., Tzeng, N.F.: An Efficient Submesh Allocation Strategy for Mesh Computer Systems. In Procs. of Int'l Conf. on Distributed Computing Systems, May (1991) 256-263

[2] Das Sharma, D., Pradhan, D. K.: A Fast and Efficient Strategy for Submesh Allocation in Mesh-Connected Parallel Computers. In Procs. of the 5th IEEE Symp. on Parallel and Distributed Processing (1993) 682-689

[3] Das Sharma, D., Pradhan, D. K.: Submesh Allocation in Mesh Multicomputers Using Busy-List: A Best-Fit Approach with Complete Recognition Capability. Journal of Parallel and Distributed Computing, Vol. 36 (1996) 106-118

[4] Ding, J., Bhuyan, L.N.: An Adaptive Submesh Allocation Strategy for Two-Dimensional Mesh Connected Systems. In Procs. of Int'l Conf. on Parallel Processing, Vol. II (1993) 193-200

[5] Intel: A Touchstone DELTA System Description. Supercomputer Systems Division, Intel Corporation, Beaverton, OR 97006 (1991)

[6] Intel: Paragon XP/S Overview. Supercomputer Systems Division, Intel Corporation, Beaverton, OR 97006 (1991)

[7] Kim, G., Yoon, H.: On Submesh Allocation for Mesh Multicomputers: A Best-Fit Allocation and a Virtual Submesh Allocation for Faulty Meshes. IEEE Transactions on Parallel and Distributed Systems, Vol. 9(2) (1998) 175-185

[8] Li, K., Cheng, K.H.: Job Scheduling in a Partitionable Mesh Using a Two-Dimensional Buddy System Partitioning Scheme. IEEE Transactions on Parallel and Distributed Systems, Vol. 2(4) (1991) 413-422

[9] Liu, T., et. al.: A Submesh Allocation Scheme for Mesh-Connected Multiprocessor Systems. In Proceedings of 1995 Int'l Conf. on Parallel Processing, Vol. II (1995) 159-163

[10] Mattson, et al.: Intel/Sandia ASCI system. In Procs. of Int'l Parallel Processing Symposium (1996)

[11] Mohapatra, P.: Processor Allocation Using Partitioning in Mesh Connected Parallel Computers.: Journal of Parallel and Distributed Computing, Vol. 39 (1996) 181-190

[12] Srisawat, J., Alexandridis, N.A.: Efficient Processor Allocation Scheme with Task Embedding for Partitionable Mesh Architectures. In Procs. of Int'l Conf. on Computer Applications in Industry and Engineering, Las Vegas, November (1998) 305-308

[13] Srisawat, J., Alexandridis, N.A.: Reducing System Fragmentation in Dynamically Partitionable Mesh-Connected Architectures. In Procs. of Int'l Conf. on Parallel and Distributed Computing and Networks, Australia, December (1998) 241-244

[14] Srisawat, J., Alexandridis, N.A.: A New Quad-Tree-Based Sub-System Allocation Technique for Mesh-Connected Parallel Machines. In Procs. of the 13th ACM-SIGARCH Int'l Conf. on Supercomputing, Greece, June (1999) 60-67

[15] Yoo, S., et. al.: An Efficient Task Allocation Scheme for 2D Mesh Architectures. IEEE Transactions on Parallel and Distributed systems, Vol. 8(9) (1997) 934-942

[16] Zhu, Y.: Efficient Processor Allocation Strategies for Mesh-Connected Parallel Computers. Journal of Parallel and Distributed Computing, Vol.16 (1992) 328-337

An Analytic Model for Communication Latency in Wormhole-Switched *k*-Ary *n*-Cube Interconnection Networks with Digit-Reversal Traffic

H. Sarbazi-Azad[1], L. M. Mackenzie[1], M. Ould-Khaoua[2]

[1] Department of Computing Science, University of Glasgow, Glasgow G12 8QQ, U.K.
Email: {hsa,lewis}@dcs.gla.ac.uk

[2] Department of Computer Science, University of Strathclyde, Glasgow G1 1XH, U.K.
Email: mohamed@cs.strath.ac.uk

Abstract. Several analytical models of fully adaptive routing in wormhole-routed *k*-ary *n*-cubes under the uniform traffic pattern have recently been proposed in the literature. This paper describes the first analytical model of fully adaptive routing in *k*-ary *n*-cubes in the presence of non-uniform traffic generated by the digit-reversal permutation, which is an important communication operation found in many matrix computation problems. Results obtained through simulation experiments confirm that the model predicts message latency with a good degree of accuracy under different working conditions.

1 Introduction

Most current multicomputers, e.g. Cray T3E [2], and Cray T3D [14], employ *k*-ary *n*-cube interconnection network for low-latency and high-bandwidth inter-processor communication. The *k*-ary *n*-cube has an *n*-dimensional grid structure with *k* nodes in each dimension such that every node is connected to its neighbouring nodes in each dimension by direct channels. The two most popular instances of *k*-ary *n*-cubes are the hypercube (where k=2) and the 2- and 3-dimensional torus (where n = 2 and 3) [10].

Current routers significantly reduce message latency by using *wormhole* switching (also widely-known as wormhole routing [17]). Wormhole routing divides a message into elementary units called *flits*, each of a few bytes for transmission and flow control, and advances each flit as soon as it arrives at a node. The *header* flit (containing routing information) governs the route and the remaining data flits follow it in a pipelined fashion. Moreover, throughput in wormhole routed networks can be increased by organizing the flit buffers associated with a given physical channel into several virtual channels [7].

Most interconnection networks, including *k*-ary *n*-cubes, provide multiple paths for routing messages between a given pair of nodes. *Deterministic* routing, where messages with the same source and destination addresses always take the same network path, has been popular because it requires a simple deadlock-avoidance algorithm, resulting in a simple router implementation [10]. However, messages cannot use alternative paths to avoid blocked channels. *Fully-adaptive* routing overcomes this limitation by enabling messages to explore all alternative paths. Several authors like Duato [9], Lin *et al* [16], and Su and Shin [24] have proposed fully-adaptive routing algorithms, which can achieve deadlock-freedom with only one extra virtual channel compared to deterministic routing.

Analytical models of deterministic routing in common wormhole-routed networks, including the *k*-ary *n*-cube, have been widely reported in the literature [1, 5,

M. Valero et al. (Eds.): ISHPC 2000, LNCS 1940, pp. 218-229, 2000.

6, 8]. Several researchers have recently proposed analytical models of fully-adaptive routing under the uniform traffic pattern [3, 20, 18].

A number of studies [10] have revealed that the performance advantages of adaptive routing over deterministic routing are more noticeable when the traffic is non-uniform. Many real-world parallel applications in science and engineering exhibit these kinds of traffic patterns [13, 23]. For instance, computing multi-dimensional FFTs, finite elements, matrix problems and divide and conquer strategies exhibit regular communication patterns [12], which are highly non-uniform as they put uneven bandwidth requirement on network channels. Permutations patterns, such as digit-reversal, matrix-transpose, shuffle, exchange, butterfly and vector-reversal are examples of regular patterns that generate typical non-uniform traffic in the network (see [12, 23] for more details on these permutations). To the best of our knowledge, there has been hardly any attempt to propose an analytical model of adaptive routing in wormhole-routed networks under the non-uniform traffic patterns including permutation traffic patterns. In fact, most existing studies have resorted to simulation to evaluate the performance of adaptive routing under such traffic conditions [8, 16, 19, 24]. In an effort to fill this gap, this paper proposes the first analytical model for computing message latency in the k-ary n-cube with fully adaptive routing in the presence of digit-reversal permutation traffic, which is one of the most important permutations found in typical parallel applications, such as matrix problems and radix-k FFT computation [12, 13]. The model is developed for Duato's fully adaptive routing algorithm [9], but the modelling approach can equally be used for the other routing algorithms described in [16, 24].

2 The k-Ary n-Cube and Duato's Routing Algorithm

The unidirectional k-ary n-cube where k is referred to as the *radix* and n as the *dimension*, has $N=k^n$ identical nodes, arranged in n dimensions, with k nodes per dimension. Each node can be identified by an n-digit radix k address $(a_1, a_2, \ldots, a_n)$. The ith digit of the address vector, a_i, represents the node position in the ith dimension. There is a link from node $(a_1, a_2, \ldots, a_n)$ to node $(b_1, b_2, \ldots, b_n)$ if and only if there exists an i, $(1 \le i \le n)$, such that $a_i = (b_i + 1)$ $mod\ k$ and $a_j = b_j$ for $(1 \le j \le n$; $i \ne j)$. Each node consists of a processing element (PE) and a router, as illustrated in Fig.1. The PE contains a processor and some local memory. The router has $(n+1)$ input and $(n+1)$ output channels. A node is connected to its neighbouring nodes through n inputs and n output channels in a unidirectional k-ary n-cube. The injection channel is used by the PE to inject messages to the network (via router) and messages at the destination eject the network via the ejection channel. Each physical channel is associated with some, say V,

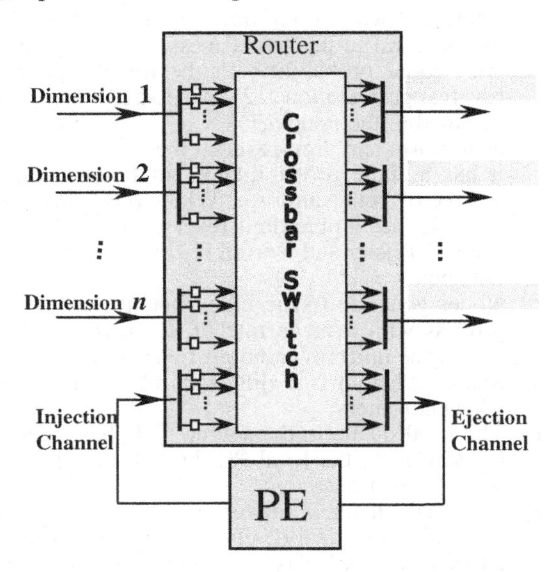

Fig.1- The node structure in the k-ary n-cube.

virtual channels. Each virtual channel has its own flit queue, but shares the bandwidth of the physical channel with other virtual channels in a time-multiplexed fashion [7]. The router contains flit buffers for any incoming virtual channel. An $(n+1)V$-way crossbar switch directs message flits from any input virtual channel to any output virtual channel. Such a switch can simultaneously connect multiple input to multiple output virtual channels while there is no conflicts [7, 10].

The high cost of adaptivity [4] has motivated researchers to develop adaptive routing algorithms that require a fewer number of virtual channels. Several authors have proposed routing algorithms that exhibit a trade-off between adaptivity and the number of virtual channels required to ensure deadlock-freedom [10]. For instance, *planar* adaptive routing [10] and the turn model [11] are partially adaptive.

Authors in [9, 16, 24] have proposed fully-adaptive routing algorithms, which can achieve deadlock freedom using a minimal number of virtual channels. Their proposed algorithms require only one extra virtual channel per physical channel, compared to deterministic routing, allowing for an efficient router implementation. For instance, Duato's algorithm divides the virtual channels into two classes: *a* and *b*. At each routing step, a message visits adaptively any available virtual channel from class *a*. If all the virtual channels belonging to class *a* are busy, it visits a virtual channel from class *b* using deterministic routing. Duato's algorithm requires at least three virtual channels per physical channel to ensure deadlock-freedom where the class *a* contains one virtual channel and class *b* owns two virtual channels. When there are more than three virtual channels, network performance is maximised when the extra virtual channels are added to class *a* [9]. Thus, when V virtual channels are used per physical channel in a k-ary n-cube, the best performance is achieved when class *a* and *b* contain (V-2) and 2 virtual channels respectively.

3 The Analytical Model

The model uses assumptions which are commonly used in the literature [1, 5, 7, 8, 18, 20, 21].

a) There are two types of traffic in the network: "uniform" and "digit-reversal". In the uniform traffic pattern, a message is destined to any other nodes in the network with equal probability. In the traffic pattern generated according to the digit-reversal permutation [12], a message generated in the source node $x = x_1 x_2 \cdots x_n$ is destined to the node $d(x) = x_n x_{n-1} \cdots x_1$. Let us refer to these two types of messages as uniform and digit-reversal messages respectively. When a message is generated it has a finite probability α of being a digit-reversal message and probability $(1-\alpha)$ of being uniform. When α =0 the traffic pattern is purely uniform while $\alpha = 1$ defines a pure digit-reversal traffic. A similar traffic model has already been used by Pfister and Norton [19] to generate non-uniform traffic patterns containing hot spots.
b) Nodes generate traffic independently of each other, and which follows a Poisson process with a mean rate of λ_g messages/cycle. Therefore, the mean generation rate of the uniform and digit-reversal traffics are respectively $(1-\alpha)\lambda_g$ and $\alpha\lambda_g$.
c) Message length is B flits, each of which is transmitted in one cycle between two adjacent routers.
d) The local queue in the source node has infinite capacity. Moreover, messages are transferred to the local PE through the ejection channel as soon as they arrive at their destinations.
e) V virtual channels are used per physical channel. With Duato's fully adaptive routing algorithm [9], class *a* contains (V-2) virtual channels which are crossed adaptively and class *b* contains two virtual channels which are crossed

deterministically (e.g. in an increasing order of dimensions). Let the virtual channels belonging to class a and b be called the adaptive and deterministic virtual channels respectively. When there is more than one adaptive virtual channel available a message chooses one at random. To simplify the model derivation, no distinction is made between the deterministic and adaptive virtual channels when computing the different virtual channels occupancy probabilities [3, 20, 18].

The mean message latency is composed of the mean network latency, $\overline{S}$, that is the time to cross the network, and the mean waiting time seen by a message in the source node, $\overline{W_s}$. However, to capture the effects of virtual channels multiplexing, the mean message latency is scaled by a factor, $\overline{V}$, representing the average degree of virtual channels multiplexing, that takes place at a physical channel. Therefore, the mean message latency can be approximated by

$$Latency = (\overline{S} + \overline{W_s})\overline{V} \tag{1}$$

Examining the address patterns generated by digit-reversal permutations reveals that we need to consider two cases where n is even and odd separately for computing the different quantities, $\overline{S}$, $\overline{W_s}$, and $\overline{V}$. This is because when n is even all network channels receives both uniform and digit-reversal traffics. However, when n is odd not all channels receives both types of messages. While the channels at the center dimension (i.e., $(n+1)/2$) receive uniform messages only, channels at the other dimension receive the uniform as well as digit-reversal messages.

3.1 Outline of the Model When n Is Even

Given that a uniform message can make between 1 and $n(k$-1) hops (i.e., the network diameter), the average number of hops that a uniform message makes across the network, L_u, is given by

$$L_u = \sum_{i=1}^{n(k-1)} iP_{u_i} \tag{2}$$

where P_{u_i} is the probability that a uniform message makes i hops to reach its destination. The average number of hops that a uniform message makes in each dimension can therefore be given by [1]

$$k_u = d_u / n = (k-1)/2 \tag{3}$$

In [18, 21], the number of way that i hops can be distributed among the dimensions of a k-ary n-cube with at most h hops in each dimension is derived to be

$$D_0^h(i,n) = \sum_{l=0}^{n} (-1)^l \binom{n}{l} \binom{i-lh+n-l-1}{n-1} \tag{4}$$

Thus, the probability that a uniform message generated in a given source node, makes i hops to reach its destination, P_{u_i}, can be written as

$$P_{u_i} = D_0^{k-1}(i,n)/(N-1) = \sum_{l=0}^{n} (-1)^l \binom{n}{l} \binom{i-lk+n-1}{n-1} \Big/ (N-1) \tag{5}$$

Let us now calculate the average number of hops that a digit-reversal message makes across the network, L_d. The number of possible combinations, where the address patterns $x_1 x_2 \cdots x_n$ and $x'_1 x'_2 \cdots x'_n$ differ in exactly i digits (i= 0, 1,..., n), is given by [22]

$$n_{d_i} = \begin{cases} \binom{n/2}{i/2}(k-1)^{\frac{i}{2}} k^{\frac{n}{2}} & \text{if } i \text{ is even} \\ 0 & \text{otherwise} \end{cases} \tag{6}$$

Hence, the probability that the source and destination addresses for a digit-reversal message differ in i digits, P_{d_i}, can therefore be written as

$$P_{d_i} = \frac{n_{d_i}}{N - n_{d0}} = \begin{cases} \binom{n/2}{i/2}\frac{(k-1)^{i/2}}{k^{n/2}-1} & \text{if } i \text{ is even} \\ 0 & \text{otherwise} \end{cases} \tag{7}$$

Let us assume that the i-th digit (i= 0, 1,..., n) in the source address, x_i, is different from that of the destination address, i.e. x'_i. Considering all possible values that x_i and x'_i may take (i.e. $0 \le x_i, x'_i < k$), the average difference between x_i and x'_i, which is the average number of hops that a digit-reversal message makes in the i-th dimension, is given by [1]

$$k_d = (k-1)/2 \tag{8}$$

Thus, the average number of hops that a digit-reversal message makes across the network is given by

$$L_d = \sum_{i=1}^{n} i k_d P_{d_i} \tag{9}$$

Examining the traffic generated by the digit-reversal permutation shows that a fraction of n_{d0}/N of the network nodes send only uniform messages and the remaining fraction $(1-n_{d0}/N)$ send a combination of uniform (with probability $1-\alpha$) and digit-reversal (with probability α) messages. Using the above equations, the average number of hops, $\overline{L}$, that a message makes in the network is derived to be

$$\overline{L} = (n_{d0}/N)L_u + [1 - n_{d0}/N](\alpha L_d + (1-\alpha)L_u) = \zeta_d L_d + \zeta_u L_u \tag{10}$$

where the uniform and digit-reversal messages contribute with the following weights

$$\zeta_d = \alpha(1 - n_{d0}/N) \tag{11}$$

$$\zeta_u = n_{d0}/N + (1-\alpha)(1-n_{d0}/N) \tag{12}$$

Adaptive routing enables a message to explore any available channel that brings it closer to its destinations, resulting in an approximately even traffic rate on network channels. Since a message makes, on average, $\overline{L}$ hops in the network the total traffic existing in the network at a given time is $N\overline{L}\lambda_g$. Given that a router in the k-ary n-cube has n output channels the rate of messages arriving at each channel, λ_c, can be written as [1, 6]

$$\lambda_c = \overline{L}\lambda_g / n \tag{13}$$

The uniform and digit-reversal messages see different network latencies as they cross different number of channels to reach their destinations. If S_u and S_d denote the mean network latency for uniform and digit-reversal messages, respectively, the mean network latency taking into account both types of messages is given by

$$\overline{S} = \zeta_d S_d + \zeta_u S_u \tag{14}$$

Averaging over all possible cases for a digit-reversal message, gives the mean network latency for digit-reversal messages, S_d, as

$$S_d = \sum_{i=1}^{n} P_{d_i} S_{d_i} \tag{15}$$

where S_{d_i} is the network latency for a digit-reversal message whose source and destination addresses differ in i digits. As a uniform message makes, on average, d_u hops across the network the mean network latency for a uniform message, S_u, is given by

$$S_u = B + L_u + \sum_{j=1}^{d_u} P_{block_{u_j}} w_t \tag{16}$$

The term $B + L_u$ in the above equation accounts for the message transmission time while the $P_{block_{u_j}} w_t$ accounts for the delay due to blocking at the j-th hop channel $(1 \le j \le L_u)$ along the message path. $P_{block_{u_j}}$ is the probability of blocking when the uniform message arrives at the j-th hop channel and w_t is the mean waiting time to acquire a virtual channel when the message is blocked. Similarly, the network latency for a digit-reversal message whose source and destination address patterns are different in i digits, S_{d_i}, is given by

$$S_{d_i} = B + ik_d + \sum_{j=1}^{ik_d} P_{block_{d_{i,j}}} w_t \tag{17}$$

where $P_{block_{d_{i,j}}}$ is the probability of blocking when the digit-reversal message arrives at the j-th hop channel. A message (either a uniform or digit-reversal message) is blocked at the j-th hop channel when all the adaptive virtual channels of the remaining dimensions to be visited and also the deterministic virtual channel of the lowest dimension to be visited are busy. Given that blocking has occurred a message has to wait for the deterministic virtual channel at the lowest dimension. To compute the probability of blocking for a uniform message, $P_{block_{d_j}}$, let us consider a uniform message making L_u hops across the network (k_u hops in each of n dimensions) that has arrived at the j-th hop channel along its path. Such a message may already have passed up to $(j-1)/k_u$ dimensions. If l $(0 \le l \le (j-1)/k_u)$ dimensions are passed, then the message still has to cross the remaining $(n\text{-}l)$ dimensions. Therefore, the probability of blocking can be expressed as

$$P_{block_{u_{l,j}}} = P_{pass_{u_{l,j}}} P_a^{n-l-1} P_{a\&d} \tag{18}$$

where $P_{pass_{u_{l,j}}}$ is the probability that l dimensions are passed at the j-th hop channel, P_a is the probability that all adaptive virtual channels at a physical channel are busy, and $P_{a\&d}$ is the probability that all adaptive and deterministic virtual channels at a physical channel are busy. Since l can be 0, 1, …, or $(j-1)/k_u$, the probability of blocking at the j-th hop channel is given by

$$P_{block_{u_j}} = \sum_{l=0}^{(j-1)/k_u} P_{block_{u_{l,j}}} \tag{19}$$

The probability that l dimensions are passed at the j-th hop channel, $P_{passu_{l,j}}$, can be computed as follows. The number of combinations that l particular dimensions are passed is $D_0^{k_u-1}(j-lk_u,n-l)$. These l dimensions can be chosen from n dimensions in $\binom{n}{l}$ ways resulting in a total of $\binom{n}{l}D_0^{k_u-1}(j-lk_u,n-l)$ combinations that l dimensions may be passed. Dividing this by the total number of combinations that j hops can be made over n dimensions gives the probability that a uniform message has passed l dimensions at its j-th hop as

$$P_{passu_{l,i}} = \binom{n}{l}D_0^{k_u-1}(j-lk_u,n-l)/D_0^{k_u-1}(j,n) \quad (20)$$

Let us now consider a digit-reversal message that passes i dimensions to reach its destination, i.e. a digit-reversal message whose source and destination addresses differ in i digits. Such a message makes ik_d hops over i dimensions. Adopting the same approach taken above for calculating P_{blocku_j} for uniform messages, we can derive $P_{blockd_{i,j}}$ as

$$P_{blockd_{i,j}} = \sum_{l=0}^{(j-1)/k_d} P_{blockd_{l,i,j}} \quad (21)$$

$$P_{blockd_{l,i,j}} = P_{passd_{l,i,j}} P_a^{i-l} P_{a\&d}, \qquad (0 \le l \le (j-1)/k_d) \quad (22)$$

$$P_{passd_{l,i,j}} = \binom{i}{l}D_0^{k_d-1}(j-lk_d,i-l)/D_0^{k_d}(j,i) \quad (23)$$

The calculation of the probabilities P_a and $P_{a\&d}$ have already been outlined in [18]. If P_v $(0 \le v \le V)$ denotes the probability that v virtual channels are busy at a physical channel (P_v is calculated in the next sub-section), P_a and $P_{a\&d}$ are given in terms of P_v as (see [18] for a more detailed derivation of the probabilities P_a and $P_{a\&d}$)

$$P_a = P_V + 2P_{V-1}/\binom{V}{V-1} + P_{V-2}/\binom{V}{V-2} \quad (24)$$

$$P_{a\&d} = P_V + 2P_{V-1}/\binom{V}{V-1} \quad (25)$$

To determine the mean waiting time, w_t, to acquire a virtual channel when a message is blocked, a physical channel is treated as an M/G/1 queue. The mean arrival rate to the channel is λ_c (given by equation 13) and the mean service seen by a message at a given channel can be approximated by $\overline{S}$ (given by equation 14). Using results from queuing theory, the mean waiting time in the event of blocking can be expressed as [15]

$$w_t = \frac{\rho \overline{S}\left(1+C_{\overline{S}}^2\right)}{2(1-\rho)} \quad (26)$$

$$\rho = \lambda_c \overline{S} \quad (27)$$

$$C_{\overline{S}}^2 = \sigma_{\overline{S}}^2 / \overline{S}^2 \quad (28)$$

where $\sigma_{\overline{S}}^2$ is the variance of the service time distribution. Since the minimum service time is equal to the message length B, following a suggestion proposed in [8], the variance of the service time distribution can be approximated as

$$\sigma_{\overline{S}}^2 = (\overline{S} - B)^2 \tag{29}$$

As a result, the mean waiting time, w_t, to acquire a virtual channel when a message is blocked becomes

$$w_t = \lambda_c \overline{S}^2 \left[1 + \left(\overline{S} - B\right)^2 / \overline{S}^2\right] / \left[2(1 - \lambda_c \overline{S})\right] \tag{30}$$

The probability, P_v, that v adaptive virtual channels are busy at a physical channel, can be determined using a Markov chain with V+1 states. State π_v, $(0 \le v \le V)$, corresponds to v virtual channels being busy. The transition rate out of state π_v to state π_{v+1} is the traffic rate λ_c (given by equation 13) while the rate out of state π_v to state π_{v-1} is approximated by $1/\overline{S}$ ($\overline{S}$ is given by equation 14). The transition rate out of state π_V are reduced by λ_c to account for the arrival of messages while a channel is in this state. The probability P_v can be computed using the steady-state equations as

$$P_v = \begin{cases} \lambda_c \overline{S} - (\lambda_c \overline{S})^v & 0 \le v < V \\ (\lambda_c \overline{S})^V & v = V \end{cases} \tag{31}$$

In virtual channel flow control, multiple virtual channels share the bandwidth of a physical channel in a time-multiplexed manner. The average degree of multiplexing of virtual channels in the network is given by [7]

$$\overline{V} = \sum_{i=0}^{V} i^2 P_i \Big/ \sum_{i=0}^{V} i P_i \tag{32}$$

The calculation of the mean waiting time, $\overline{W_s}$, at the local queue in the source node is calculated in the same manner used for calculating the mean waiting time at a given network channel (equation 30). The local queue is treated as an M/G/1 queue with an arrival rate of λ_g / V (recalling that a message in the source node can enter the network through any of the V virtual channels), a service time of $\overline{S}$, and thus a mean waiting time of [15]

$$\overline{W_s} = \frac{\lambda_g}{V} \overline{S}^2 \left[1 + \left(\overline{S} - B\right)^2 / \overline{S}^2\right] \Big/ \left[2(1 - \frac{\lambda_g}{V} \overline{S})\right] \tag{33}$$

Examining the above equations reveals that there are several inter-dependencies between the different variables of the model. For instance, Equations (14) and (15) reveal that $\overline{S}$ is a function of S_u and S_{d_i} while equations (16), (17) and (30) show that S_u and S_{d_i} are functions of $\overline{S}$. Given that closed-form solutions to such inter-dependencies are very difficult to determine the different variables of the model are computed using iterative techniques.

3.2 Outline of the Model When *n* Is Odd

As stated above, when n is odd, channels belonging to dimension $(n+1)/2$ receive uniform messages only. The traffic due to digit-reversal messages falls only on the channels belonging to dimensions 1, 2, …, $(n-1)/2$, $(n+3)/2$, …, n. Let us refer to dimension $(n+1)/2$ as the "centre-dimension" and the channels belonging to this dimension as the "centre-channels". Similarly, let us refer to other dimensions as "other-dimensions" and their associated channels as the "other-channels". When n is odd, equations (6) and (18) will change (see [22] for more details) to

$$n_{d_i} = \begin{cases} \binom{(n-1)/2}{i/2}(k-1)^{i/2} k^{(n+1)/2} & \text{if } i \text{ is even} \\ 0 & \text{otherwise} \end{cases} \tag{34}$$

$$P_{block_{ul,i}} = P_{pass_{ul,i}} \Bigg[\left(\binom{n-1}{l-1} \Big/ \binom{n}{l} \right) P_{a_{other}}^{n-l-1} P_{a\&d_{other}} + \left(1 - \binom{n-1}{l-1} \Big/ \binom{n}{l} \right) \left(\frac{1}{(n-l)} P_{a_{other}}^{n-l-1} P_{a\&d_{center}} + \frac{(n-l-1)}{(n-l)} P_{a_{other}}^{n-l-2} P_{a_{center}} P_{a\&d_{other}} \right) \Bigg] \tag{35}$$

where

$$P_{a_{centre}} = P_{V_{centre}} + 2P_{V-1_{centre}} / \binom{V}{V-1} + P_{V-2_{centre}} / \binom{V}{V-2} \tag{36}$$

$$P_{a\&d_{centre}} = P_{V_{centre}} + 2P_{V-1_{centre}} / \binom{V}{V-1} \tag{37}$$

$$P_{a_{other}} = P_{V_{other}} + 2P_{V-1_{other}} / \binom{V}{V-1} + P_{V-2_{other}} / \binom{V}{V-2} \tag{38}$$

$$P_{a\&d_{other}} = P_{V_{other}} + 2P_{V-1_{other}} / \binom{V}{V-1} \tag{39}$$

While all channels receive uniform traffic, only the channels belonging to other-dimensions receive digit-reversal traffic. Therefore, when n is odd the traffic rate arriving at each centre-channel, $\lambda_{c_{centre}}$ and other-channel, $\lambda_{c_{other}}$, can be expressed as

$$\lambda_{c_{center}} = \zeta_u L_u \lambda_g / n \tag{40}$$

$$\lambda_{c_{other}} = \lambda_{c_{center}} + \zeta_d L_d \lambda_g / (n-1) \tag{41}$$

The mean waiting time for a centre-channel and an other-channel, $w_{t_{centre}}$ and $w_{t_{other}}$, can be expressed as

$$w_{t_{centre}} = \lambda_{c_{centre}} S_u^2 \left[1 + (S_u - B)^2 / S_u^2\right] / \left[2(1 - \lambda_{c_{centre}} S_u)\right] \tag{42}$$

$$w_{t_{other}} = \lambda_{c_{other}} S_{t_{other}}^2 \left[1 + (S_{t_{other}} - B)^2 / S_{t_{other}}^2\right] / \left[2(1 - \lambda_{c_{other}} S_{t_{other}})\right] \tag{43}$$

where S_u (given by equation 16) and $S_{t_{other}}$ (calculated below) are approximated values for service time of a centre-channel and an other-channel, respectively. Therefore, the mean waiting time for a channel considering both types of channels would be

$$w_t = \tfrac{1}{n} w_{t_{centre}} + \left[1 - \tfrac{1}{n}\right] w_{t_{other}} \tag{44}$$

The mean service time for an other-channel, $S_{t_{other}}$, can be approximated as

$$S_{t_{other}} = \frac{\lambda_{c_{centre}}}{\lambda_{c_{other}}} S_u + \left[1 - \frac{\lambda_{c_{centre}}}{\lambda_{c_{other}}}\right] S_d \tag{45}$$

Adapting the approach used to calculate P_v when n is even, we can write the expression of the probability of having v busy virtual channels at a centre-channel, $P_{v_{centre}}$, and at an other-channel, $P_{v_{other}}$, as

$$P_{v_{centre}} = \begin{cases} \lambda_{c_{centre}} S_u - (\lambda_{c_{centre}} S_u)^v & 0 \le v < V \\ (\lambda_{c_{centre}} S_u)^v & v = V \end{cases} \tag{46}$$

$$P_{v_{other}} = \begin{cases} \lambda_{c_{other}} S_{t_{other}} - (\lambda_{c_{other}} S_{t_{other}})^v & 0 \le v < V \\ (\lambda_{c_{other}} S_{t_{other}})^v & v = V \end{cases} \tag{47}$$

The average degree of multiplexing of virtual channels belonging to a centre-channel and an other-channel in the network, and the total average multiplexing degree of virtual channels in the network, are given by

$$\overline{V}_{centre} = \sum_{i=0}^{V} i^2 P_{i_{centre}} \Big/ \sum_{i=0}^{V} i P_{i_{centre}} \tag{48}$$

$$\overline{V}_{other} = \sum_{i=0}^{V} i^2 P_{i_{other}} \Big/ \sum_{i=0}^{V} i P_{i_{other}} \tag{49}$$

$$\overline{V} = \overline{V}_{centre} / n + (1 - 1/n) \overline{V}_{other} \tag{50}$$

3.3 Simulation Experiments

The above model has been validated through a discrete-event simulator that mimics the behaviour of Duato's fully-adaptive routing at the flit level in k-ary n-cubes. In each simulation experiment, a total number of 100K messages are delivered. Statistics gathering was inhibited for the first 10K messages to avoid distortions due to the initial startup conditions. The mean message latency is defined as the mean amount of time from the generation of a message until the last data flit of the message reaches the local PE at the destination node. Numerous experiments have been performed for several combinations of network sizes, message lengths, digit-reversal traffic fractions, and number of virtual channels to validate the model. However, for the sake of specific illustration, Fig.2 depicts latency results predicted by the proposed models plotted against those provided by the simulator for a 8-ary 2-cube ($N=8^2$) and 8-ary 3-cube ($N=8^3$) respectively and for different message lengths, B = 32 and 64 flits. Moreover, the number of virtual channels per physical channel was set to V=3 and 5 and the fraction of digit-reversal messages was assumed to be α = 0.1 and 0.7. The horizontal axis in the figures shows the traffic generation rate at each node (λ_g) while the vertical axis shows the mean message latency. The figures reveal that in all cases, the analytical model predicts the mean message latency with a good degree of accuracy in the steady state regions. However, some discrepancies around the saturation point are apparent. However, the simplicity of the model makes it a practical evaluation tool that can be used to gain insight into the performance behavior of fully adaptive routing in the k-ary n-cube interconnection network.

It is worth noting that latency results for different values of α reveal that digit-reversal traffic has a little impact on the mean message latency since adaptive routing is able to exploit alternative paths of the k-ary n-cube to route blocked messages, and as a result it manages to distribute the traffic load approximately evenly among the network channels.

4 Conclusion

This paper has presented an analytical model to compute message latency in wormhole-switched k-ary n-cubes with fully adaptive routing in the presence of traffic generated by digit-reversal permutations, which are used in many parallel applications (e.g., matrix problems and radix-k FFT computation). Although the model uses Duato's fully adaptive routing algorithm, it can equally be adapted for other adaptive routing algorithms proposed in [16, 24]. To our best knowledge, this is the first model proposed in the literature that considers non-uniform traffic generated by digit-reversal permutation in wormhole-routed *k*-ary *n*-cubes. Simulation experiments have revealed that the latency results predicted by the analytical model are in good agreement with those obtained through simulation experiments.

References

[1] Agarwal, A.: Limits on interconnection network performance. *IEEE TPDS* **2**(**4**) (1991) 398-412.

[2] Anderson, E., Brooks, J., Grass, C., Scott, S.: Performance of the Cray T3E multiprocessor. *Proc. Supercomputing Conference* (1997).

[3] Boura, Y., Das, C.R., T.M. Jacob: A performance model for adaptive routing in hypercubes. *Proc. Int. Workshop Parallel Processing* (1994) 11-16.

[4] Chien, A.: A cost and performance model for *k*-ary *n*-cube wormhole routers. *IEEE TPDS* **9**(**2**) (1998)150-162.

[5] Ciciani, B., Colajanni, M., Paolucci, C.: An accurate model for the performance analysis of deterministic wormhole routing. *Proc. 11th IPPS* (1997) 353-359.

[6] Dally, W.J.: Performance analysis of *k*-ary *n*-cubes interconnection networks. *IEEE TC* **C-39**(**6**) (1990) 775-785.

[7] Dally, W.J.: Virtual channel flow control. *IEEE TPDS* **3**(**2**) (1992)194-205.

[8] Draper, J.T., Ghosh, J.: A Comprehensive analytical model for wormhole routing in multicomputer systems. *J. Parallel & Distributed Computing* **32** (1994) 202-214.

[9] Duato, J.: A new theory of deadlock-free routing in wormhole networks. *IEEE TPDS* **4**(**12**) (1993) 1320-1331.

[10] Duato, J., Yalamanchili, S., Ni, L.: *Interconnection networks: An engineering approach.* IEEE Computer Society Press (1997).

[11] Glass, C.J., Ni, L.M.: The turn model for adaptive routing. *Proc. 19th Annual Symp. Computer Architecture* (1992) 268-277.

[12] Grammatikakis, M., Hsu, D.F., Kratzel, M., Sibeyn, J.: Packet routing in fixed-connection networks: A survey. *J. Parallel & Distributed Computing* **54**(1998) 77-132.

[13] Hwang, K.: Advanced Computer Architecture: Parallelism, Scalability, Programmability. McGraw-Hill Computer Science Series (1993).

[14] Kessler, R.E., Schwarzmeier, J.L.: Cray T3D: A new dimension for Cray Research. *in CompCon* (1993) 176-182.

[15] Kleinrock, L.: Queueing Systems. Vol. 1, John Wiley, New York (1975).

[16] Lin, X., Mckinley, P.K., Lin, L.M.: The message flow model for routing in wormhole-routed networks. *Proc. ICPP* (1993) 294-297.

[17] Ni, L., McKinley, P.K.: A survey of wormhole routing techniques in direct networks. *IEEE Computer* **25**(**2**) (1993) 62-76.

[18] Ould-Khaoua, M.: A performance model for Duato's adaptive routing algorithm in *k*-ary *n*-cubes. *IEEE TC* **48**(**12**) (1999) 1-8.

[19] Pfister, G.J., Norton, V.A.: Hotspot contention and combining in multistage interconnection networks. *IEEE TC* **34**(**10**) (1985) 943-948.

[20] Sarbazi-Azad, H., Ould-Khaoua, M., Mackenzie, L.M.: Performance Analysis of *k*-ary *n*-cubes with Fully Adaptive Routing. *Proc. ICPADS'2000* (2000), Iwate, Japan.

[21] Sarbazi-Azad, H., Ould-Khaoua, M., Mackenzie, L.M.: An analytical model of fully-adaptive wormhole-routed *k*-ary *n*-cubes in the presence of hot-spot traffic. *Proc IPDPS'2000* (2000) 605-610.

[22] Sarbazi-Azad, H., Ould-Khaoua, M., Mackenzie, L.M.:Modelling adaptive routing in wormhole-routed *k*-ary *n*-cubes in the presence of digit-reversal traffic.Technical report, Department of Computing Science, Glasgow University, U.K. (2000).

[23] Seigel, H.J.: Interconnection networks for large-scale parallel Processing. McGraw-Hill (1990).

[24] Su, C., Shin, K.G.: Adaptive deadlock-free routing in multicomputers using one extra channel. *Proc. ICPP* (1993) 175-182.

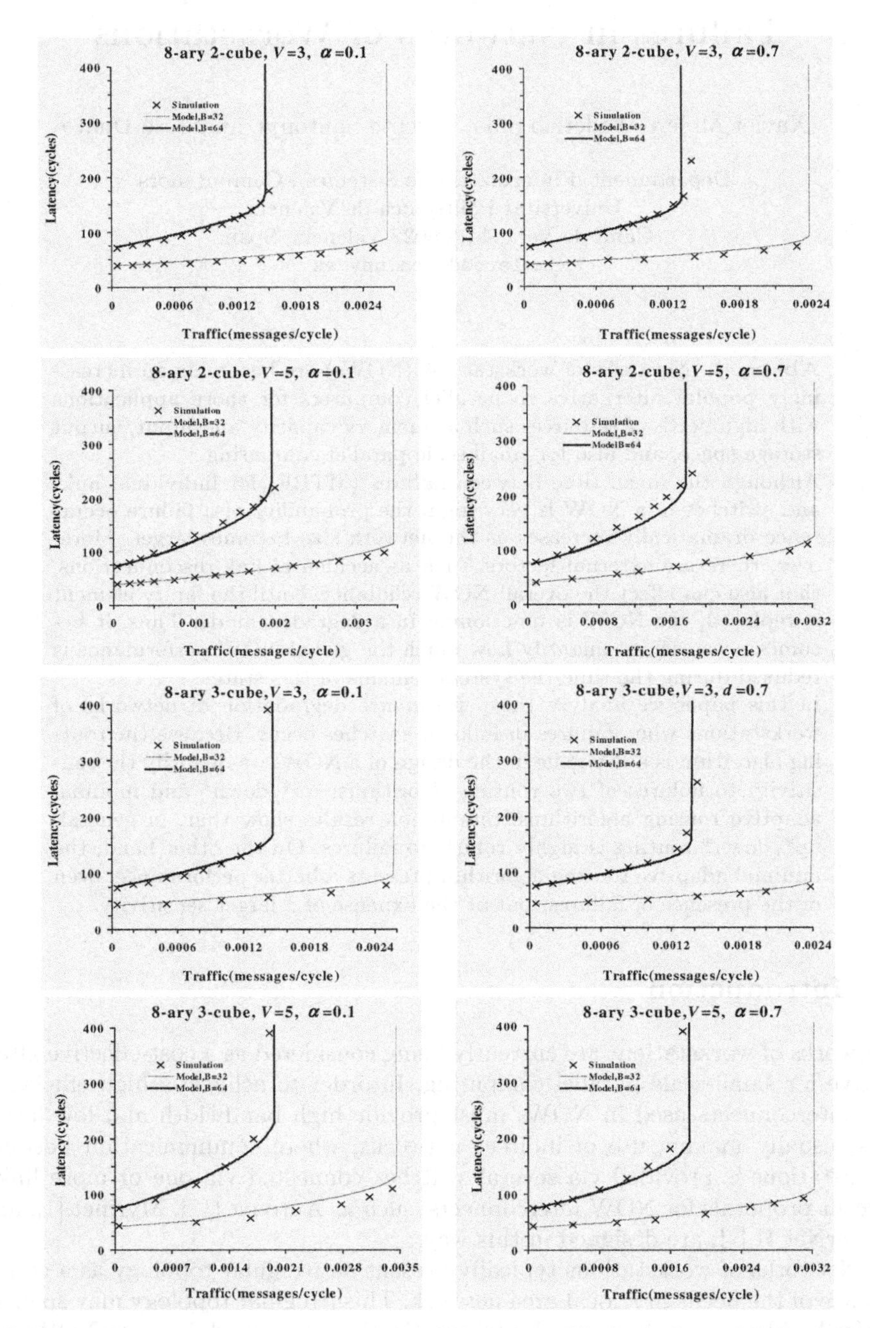

Fig.2- Average latency versus message generation traffic in an 8-ary 2-cube and an 8-ary 3-cube for V=3 and 5 virtual channels per physical channel, message length B=32 and 64 flits, and digit-reversal traffic portions α=0.1 and 0.7.

Performance Sensitivity of Routing Algorithms to Failures in Networks of Workstations

Xavier Molero, Federico Silla, Vicente Santonja, and José Duato

Departament d'Informàtica de Sistemes i Computadors
Universitat Politècnica de València
Camí de Vera, 14. 46022 València, Spain
jmolero@disca.upv.es

Abstract. Networks of workstations (NOWs) are becoming an increasingly popular alternative to parallel computers for those applications with high needs of resources such as memory capacity and input/output storage space, and also for small scale parallel computing.
Although the mean time between failures (MTBF) for individual links and switches in a NOW is very high, the probability of a failure occurrence dramatically increases as the network size becomes larger. Moreover, there are external factors, such as accidental link disconnections, that also can affect the overall NOW reliability. Until the faulty element is replaced, the NOW is functioning in a degraded mode. Thus, it becomes necessary to quantify how much the global NOW performance is reduced during the time the system remains in this state.
In this paper we analyze the performance degradation of networks of workstations when failures in links or switches occur. Because the routing algorithm is a key issue in the design of a NOW, we quantify the sensitivity to failures of two routing algorithms: up*/down* and minimal adaptive routing algorithms. Simulation results show that, in general, up*/down* routing is highly robust to failures. On the other hand, the minimal adaptive routing algorithm presents a better performance, even in the presence of failures, but at the expense of a larger sensitivity.

1 Introduction

Networks of workstations are currently being considered as a cost-effective alternative for small-scale parallel computing. In order to achieve a high efficiency, the interconnects used in NOWs must provide high bandwidth and low latencies, usually making use of indirect networks, where communication between workstations is provided via several switches connected via one or more links. Recent proposals for NOW interconnects, such as Autonet [12], Myrinet [1], and ServerNet II [7], are designed in this way.

Networks of workstations typically present an irregular topology as a consequence of the needs in a local area network. This irregular topology may span an entire building, or even several buildings. On the other hand, because NOWs use switch-based interconnects, messages must be routed through several switches

M. Valero et al. (Eds.): ISHPC 2000, LNCS 1940, pp. 230–242, 2000.

until they reach the destination. Routing algorithms consist of the rules that messages must follow to go from a source to a destination workstation.

The design and evaluation of some routing algorithms in networks of workstations with irregular topologies has been previously done in [13]. In this study the topology was supposed to be failure free, that is, the possibility that switches or/and links could fail was not considered. In fact, switches and links have very high values for the mean time between failures (MTBF) parameter. For example, the commercial M2LM-SW16 Myrinet switch with 16 ports has a calculated MTBF of approximately 5×10^5 hours (about fifty-seven years) [10]. However, the probability of a switch failure increases linearly with the number of switches in the network. Therefore, although the value of reliability is very high for the Myrinet switch, if network size is large enough, the probability of failure is much higher. For example, a 64-switch network built with the previous commercial switch will have a global MTBF of $5 \times 10^5/64 \approx 7813$ hours (about one year). On the other hand, in large networks spanning one or several buildings, faulty switches are not the only possible source of failures. Switches may be unintentionally turned off, or even the power supply may fail. Also, it is possible that some links may be incorrectly installed, may suffer accidental disconnections, and even they can be affected by electromagnetic interferences (EMI).

Managing switch and/or link failures has been analyzed in the context of direct networks with regular topologies [2,8]. In these networks a fault-tolerant routing algorithm is needed in order to bypass the faulty region. These algorithms are usually topology dependent. In the case for networks of workstations, where generic routing algorithms must be used due to the topology irregularity, managing failures may be faced in a more general way. In these networks, every time a new workstation is connected or disconnected to/from the network, it is necessary to run a reconfiguration process in order to update routing tables. Also, every time a switch is attached/unattached to the network, routing tables must be updated. Therefore, managing failures in irregular networks may be seen as another instance of the general reconfiguration process. Nevertheless, performance of the reconfigured network should be analyzed.

When a link or a switch fails, network topology changes and thus a reconfiguration process starts in order to compute the routing rules for the new topology. During this process, a reconfiguration algorithm is used in order to make the network operational as soon as possible [12,1,3]. Because messages will be discarded during the reconfiguration phase, it is important that the entire process is as fast as possible. However, after the reconfiguration phase, the network will have a lower aggregate bandwidth than before the failure. Also, connectivity between workstations will be degraded.

In this paper we will focus on analyzing the impact on network performance when functioning in a degraded mode once the network has been reconfigured after a failure. A study about how much network performance is degraded would be useful to know how user applications see the failure impact and how much they are affected by switch or link failures. Because a faulty link or switch modifies network topology, it is important to know how much partial modifications in the topology will affect the performance of the routing algorithm used by messages. In this way, we have evaluated the sensitivity to failures of two different routing algorithms: up*/down* routing [12], and minimal adaptive routing [13].

The rest of the paper is organized as follows. Section 2 briefly introduces networks of workstations. In Section 3 the routing algorithms evaluated are described. Section 4 carries out a performance evaluation. Finally, Section 5 summarizes the conclusions from this paper.

2 Networks of Workstations

Networks of workstations are usually arranged as switch-based networks with irregular topology. In these networks each switch is shared by several workstations, which are connected to the switch through some of its ports. The remaining ports of the switch are either left open or connected to other switches to provide connectivity between the workstations. Links in a NOW are typically bidirectional full-duplex, and multiple links between two switches are allowed. Figure 1(a) shows a typical network of workstations. It is assumed in this figure that switches have eight ports and each workstation has a single port.

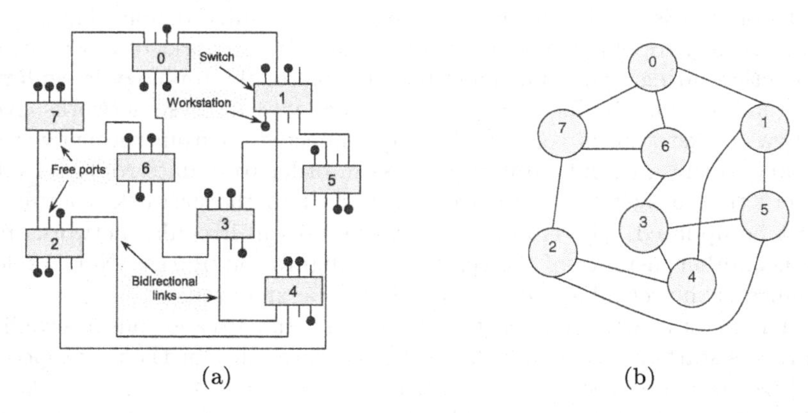

Fig. 1. A NOW with irregular topology and its corresponding graph

A routing algorithm must determine the path to be followed by messages. Several deadlock-free routing schemes have been proposed for irregular networks [12,11]. Moreover, a general methodology for the design of adaptive routing algorithms for irregular networks has been recently proposed in [13].

Finally, once a message reaches a switch directly connected to its destination workstation, it can be delivered as soon as the corresponding link becomes free. Thus, we are going to focus on routing messages between switches, modeling the interconnection network I by a multigraph $I = G(N, C)$, where N is the set of switches, and C is the set of bidirectional links between switches. Figure 1(b) shows the graph for the irregular network in Figure 1(a).

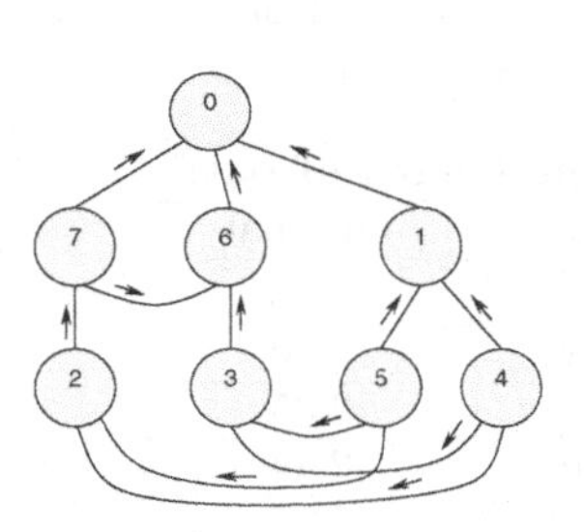

Fig. 2. NOW without link nor switch failures

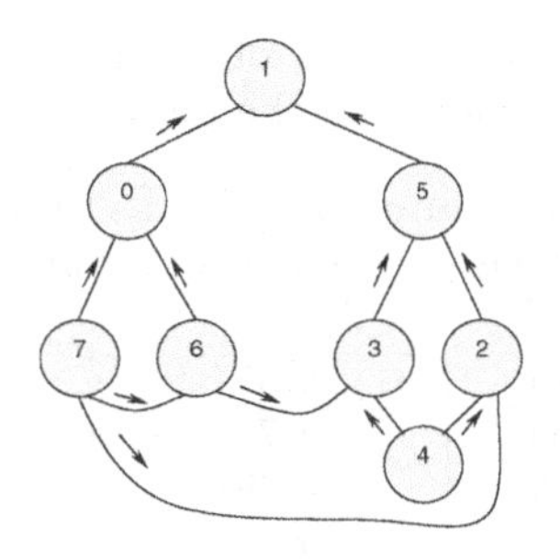

Fig. 3. NOW after the first link (4–1) failure

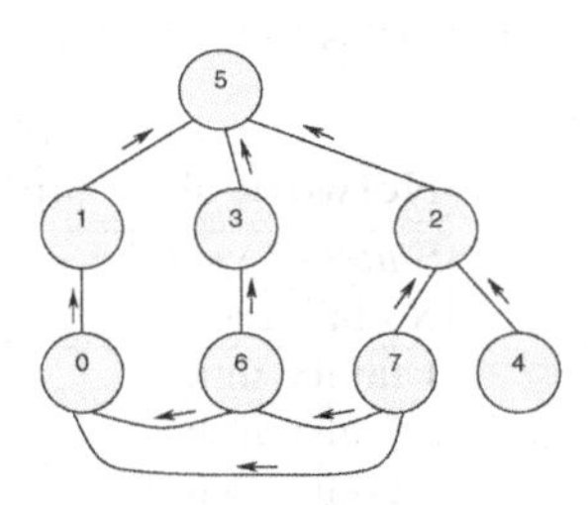

Fig. 4. NOW after the second link (3–4) failure

3 Routing Algorithms

In this section, the two routing algorithms analyzed in the performance evaluation are briefly described. For more details, see [12] and [13], respectively.

3.1 Up*/down* Routing

The up*/down* routing scheme provides partially adaptive routing in irregular networks. It is based on table lookup, and thus routing tables must be filled before messages can be routed. In order to fill these tables, a breadth-first spanning tree on the graph G of the network is first computed. The root switch in the tree is the switch which distance to the rest of switches in the network is minimum. Routing is based on an assignment of direction to all the operational links in the network. In particular, the "up" end of each link is defined as: (1) the end whose switch is closer to the root in the spanning tree; or (2) the end whose switch has the lower identifier, if both ends are at switches at the same tree level. After this assignment, each cycle in the network has at least one link in the "up" direction and one link in the "down" direction. To avoid deadlocks, routing is based on the following up*/down* rule: a message cannot traverse a link in the "up" direction after having traversed a link in the "down" direction.

Although up*/down* routing provides some adaptivity, it is not always able to provide a minimal path between every pair of workstations due to the restriction imposed by the up*/down* rule. As network size increases, this effect becomes more important.

Figure 2 shows the example irregular network depicted in Figure 1. The root for the corresponding breadth-first spanning tree is switch 0. The assignment of "up" direction to the links in the network is illustrated. The "down" direction is along the reverse direction of the link. Figure 3 shows the reconfigured network after the failure of the link connecting switches 4 and 1. In this configuration, the root of the tree moves from switch 0 to switch 1. Figure 4 shows a new reconfigured network after the failure of a second link. In this case, the faulty

Table 1. Average number and length of paths between workstations for the three irregular topologies employed in the performance evaluation

	Average number of paths		Average length of paths	
Network size	up*/down*	minimal	up*/down*	minimal
Small (8 switches)				
No failures	1.13	1.31	2.19	2.00
1 faulty link	1.65	1.71	2.41	2.35
2 faulty links	1.28	1.39	2.50	2.44
1 faulty switch	1.50	1.58	2.25	2.17
2 faulty switches	1.63	1.88	2.25	2.25
Medium (32 switches)				
No failures	1.24	1.48	4.61	3.77
1 faulty link	1.21	1.45	4.64	3.84
2 faulty links	1.34	1.55	4.84	4.24
1 faulty switch	1.38	1.51	4.68	4.01
2 faulty switches	1.41	1.38	4.77	4.02
Large (64 switches)				
No failures	1.19	1.40	5.86	4.55
1 faulty link	1.19	1.39	5.88	4.59
2 faulty links	1.19	1.40	5.91	4.64
3 faulty links	1.18	1.40	5.95	4.66
1 faulty switch	1.17	1.40	5.97	4.65
2 faulty switches	1.26	1.41	6.05	4.74

link was the one connecting switches 3 and 4. Note that the root of the tree has changed again. In this case it has moved to switch 5. It must be pointed out that, even if the root does not move to another switch after a failure, routing tables should be computed again.

Table 1 numerically shows the effect of the number of faulty links or switches on the number of paths between workstations and their lengths when up*/down* routing is used in the topology of Figure 1. It can be seen that the first link failure increases the average number of up*/down* paths from 1.13 to 1.65. This increment in the mean number of paths may seem to be contradictory, since one could expect it to decrement instead of increment. Actually, the reason is that the mean number of paths has been increased, but at the cost of increasing their average lengths. As can be seen in the table, after the first link failure, the average path length increases from 2.19 to 2.41. The second link failure also increases the average number of up*/down* paths from 1.13 to 1.28, but it increases their average length from 2.19 to 2.50. A similar effect occurs when switch failures are considered.

When network size increases, one can see that the effect of failures is similar: the mean number of paths changes (increases or decreases depending on each particular topology) but the average length of the paths always increases.

3.2 Adaptive Routing

A design methodology for adaptive routing algorithms for irregular networks has been proposed in [13]. The aim of this methodology is providing minimal routing between every pair of workstations, as well as increasing adaptivity. This design methodology can be applied to any deadlock-free routing algorithm. In particular, the second routing scheme evaluated in this paper is the result of applying this methodology to the up*/down* routing algorithm. Concisely, physical channels in the network are split into two virtual channels, called the *original* channel and the *new* channel. Newly injected messages can only leave the source switch using new channels belonging to minimal paths. When a message arrives at a switch through a new channel, the routing function gives a higher priority to the new channels belonging to minimal paths. If all of them are busy, then the up*/down* routing algorithm is used, selecting an original channel belonging to a minimal path (if any). If none of the original channels supplied provides minimal routing, then the one that provides the shortest path will be used. Once a message reserves an original channel, it will be routed using only original channels according to the up*/down* routing function until it is delivered.

Table 1 shows how link and switch failures affect the number of minimal paths between workstations for this adaptive routing algorithm when used in the network of Figure 1. Note that in this case the table reflects the minimal topological distances, because when the minimal adaptive routing algorithm is used, messages can follow non-minimal paths if they are routed along original channels. It can be seen that the first link failure increases the average number of minimal paths from 1.31 to 1.71, but the increment of their average length ranges from 2.00 to 2.35. The second link failure increases the average number of paths from 1.31 to 1.39, also increasing their average length from 2.00 to 2.44. Results when switches fail are also displayed. It can be observed that the effect of switch or link failures is similar to the case for the up*/down* routing scheme, but the mean number of minimal paths remains higher than the mean number of up*/down* paths, and also their lengths is noticeably smaller, especially for large networks.

In general, the impact of failures on topology is more significant as network size decreases. For example, regarding average path length, which is directly related to message latency, the absolute increase is lower for the large network. Specifically, for 2 faulty links, the absolute increment when using up*/down* routing is 0.31, 0.23, and 0.05, for small, medium, and large networks, respectively. In the case for minimal adaptive routing, these increments are 0.44, 0.47, and 0.09, respectively. If we look at the relative degradation, the increase is even less dramatic.

4 Performance Evaluation

This section evaluates the impact of link and switch failures in network performance for the two routing algorithms previously presented. Before the analysis of simulation results, we will describe the models used for switches and networks, the traffic pattern generation, and the output variables used in the performance evaluation.

4.1 Switch Model

Each switch has a routing control unit that selects the output channel for a message as a function of its destination workstation, the input channel, and the output channel status. Table look-up routing is used. The routing control unit can only process one message header at a time. It is assigned to waiting messages in a demand-slotted round-robin fashion. When a message gets the routing control unit but it cannot be routed because all the alternative output channels are busy, it must wait in the input buffer until its next turn. A crossbar inside the switch allows simultaneous multiple message traversal. It is configured by the routing control unit each time a successful route is established. We have assumed that it takes one clock cycle to compute the routing algorithm. Also, it takes one clock cycle to transmit one flit across the internal crossbar. On the other hand, data are injected into the physical channel, which is pipelined, at a maximum rate of one flit per cycle. Link propagation delay has been assumed to be 4 clock cycles.

Input buffer size is 32 flits. Since the minimal adaptive routing algorithm presented before makes use of two virtual channels, the number of virtual channels used by the up*/down* routing scheme has been set to two in order to compare both routing algorithms in the same conditions.

4.2 Network Model

Network topology is completely irregular and has been generated randomly. However, for the sake of simplicity, we imposed three restrictions to the topologies that can be generated. First, we assumed that there are exactly 4 workstations connected to each switch. Also, two neighboring switches are connected by a single link. Finally, all the switches in the network have the same size. We assumed 8-port switches: we will suppose that one port remains free, thus leaving 3 ports available to connect to other switches. The location of link and switch failures is chosen in a random way. The only restriction is that the resulting topology after the failures must be connected.

We have evaluated networks with a size of 8 (small), 32 (medium), and 64 (large) switches (32, 128, and 256 workstations, respectively). In order to study how much the variability in network topology affects its performance, five different irregular topologies have been generated, in a completely random way, for each of the three sizes. For each of the topologies, we ran simulations evaluating them under the (exactly) same workload. However, although we have studied five different topologies for each network size, only results for the most representative of them are presented.

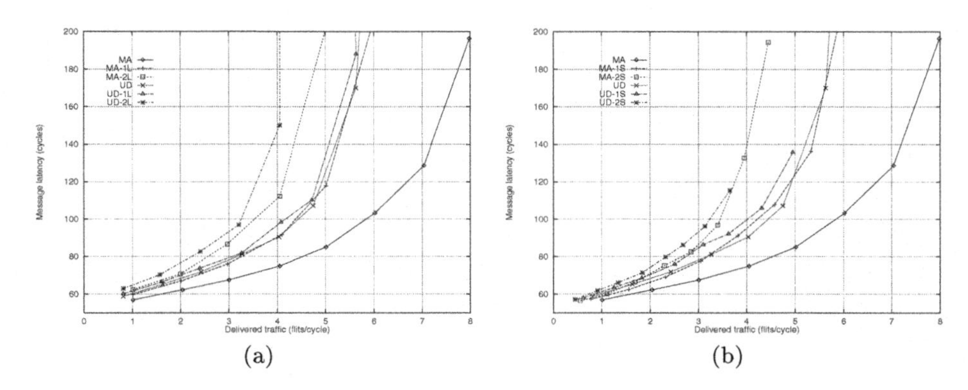

(a) (b)

Fig. 5. Message latency versus traffic for a small network

4.3 Message Generation and Performance Variables

We considered that message generation rate is exponentially distributed and it is the same for all the workstations. Also, we have assumed that message destination is randomly chosen among all the workstations in the network. With respect to message length, 16-flit and 32-flit messages have been considered. Conclusions in both cases are very similar, so we will present results concerning only to the last case.

The most important performance measures are latency and throughput. Message latency lasts since the message is introduced in the network until the last flit is received at the destination workstation, and it is measured in clock cycles. Traffic is the flit reception rate measured in flits per cycle.

4.4 Simulation Results

Here we present the analysis of results obtained by simulation. The NOW simulator [9] has been implemented in the CSIM language [4]. We will refer to the up*/down* and minimal adaptive routing algorithms without failures as UD and MA, respectively. A suffix like nL will indicate n link failures, whereas mS will indicate m switch failures.

Small Size Networks This kind of networks consists of 8 switches (32 workstations) and 12 links. The first faulty link represents about the 8.5% of the total number of links, and the first faulty switch represents exactly the 12.5% of the total number of switches in the network. A second failure in a link or in a switch will affect about the 9%, and 14%, respectively. Thus, failures in links or switches highly affect the network topology, due to its small size.

As can be seen in Figure 5, the UD routing algorithm is not greatly affected by the first faulty link or switch, whereas the second one decreases throughput approximately in a 30%. On the other hand, the MA routing algorithm is highly

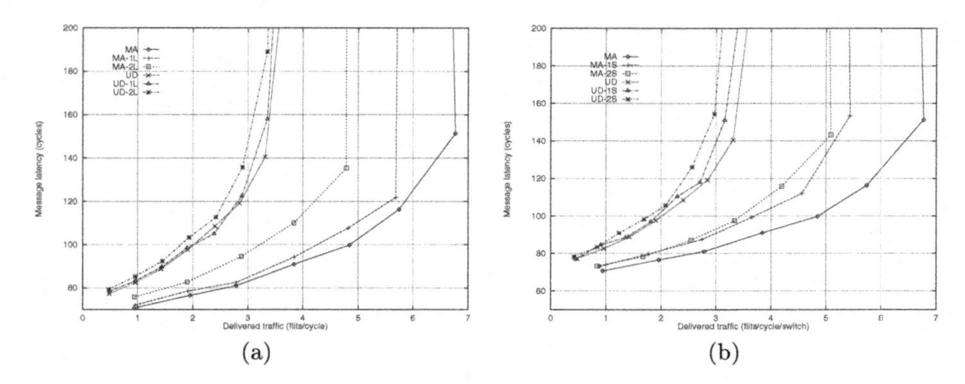

Fig. 6. Message latency versus traffic for a medium network

affected by the first failure, both in a link or in a switch, then increasing network latency and decreasing the delivered traffic. With a faulty link or switch, the maximum throughput of the network is decreased a 25%. The second failure significantly affects network performance for both routing algorithms. In particular, the second faulty link affects as much as the second failure in a switch. This is due to the small size of the network: paths that communicate workstations are quite similar for both routing algorithms in the absence of failures, both for the mean number of paths and for their average length. When failures occur, the adaptivity of the MA routing algorithm decreases drastically, while the average length of the paths increases to values similar to those achieved by the UD scheme. This causes that the performance of the MA algorithm is more affected by failures than the one for the UD scheme. In all the analyzed topologies, MA-1L routing scheme always performs slightly better than the UD routing algorithm. When a switch failure occurs, MA-1S routing scheme and the UD routing algorithm become similar. In any case, the negative effect of failures increases with high loads, and for small size networks highly depends on the underlying topology.

Medium Size Networks These networks consist of 32 switches (128 workstations) and 43 links. Thus, a faulty link represents about the 2.5% of the total number of links, and a faulty switch represents exactly the 3.125% of the amount of switches; therefore, a failure in a link or in a switch would not affect the topology as much as in the small size networks. The influence of a second failure in a link or in a switch will not be so important in these networks because there are several paths connecting the workstations.

In Figure 6 it can be appreciated that the MA routing algorithm is more sensitive to failures than the UD routing scheme: while the latter only presents a slight increment in network latency, the former highly increases latency and decreases maximum throughput. The reason for the MA routing scheme to be more sensitive to failures than the UD algorithm is that as links and switches

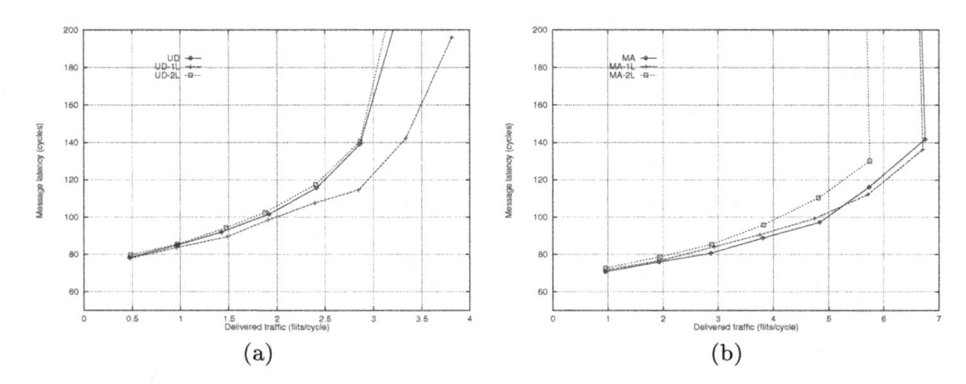

(a) (b)

Fig. 7. Effect of link failures for another medium size network

fail, minimal paths increase their lengths, as can be seen in Table 1, and therefore the performance of the MA algorithm becomes much worse. This effect is not so noticeable in the UD scheme because paths are usually non-minimal.

For this network size, two different performance zones clearly appear: one for the UD routing algorithm and another one for the MA routing algorithm. The MA routing algorithm always offers, even with faulty links or switches, better performance than the UD routing algorithm. This is because the network size allows the MA routing algorithm to take advantage of its degree of adaptivity, even when the number of minimal paths is reduced by failures. As can be expected, the first switch failure has greater impact on network performance than the first link failure, although this effect is more important in the MA routing algorithm.

Finally, as mentioned before, once the network has been reconfigured after the failure of links or switches, the new routing tables have nothing to do with the previous ones. In fact, they may distribute better message traffic. This is the case for one of the studied topologies, where the UD routing algorithm obtains some benefit of faulty links, as can be seen in Figure 7. In this case, after the first link failure, both latency and throughput are improved because the resultant topology balances traffic better. The MA routing algorithm also experiments this improvement near network saturation. In this case, for low and medium network loads, there is a slight increment in latency, because minimal paths are longer due to the faulty link. However, for high network loads the performance obtained with one faulty link is better because messages make use more intensively of escape paths, which in this case offer better performance.

Large Size Networks Large networks consist of 64 switches (256 workstations) and 96 links. In this case, topology is not as much affected by failures as in the case for networks of small and medium sizes. Due to the high number of links in the network, we have also considered the failure of a third link.

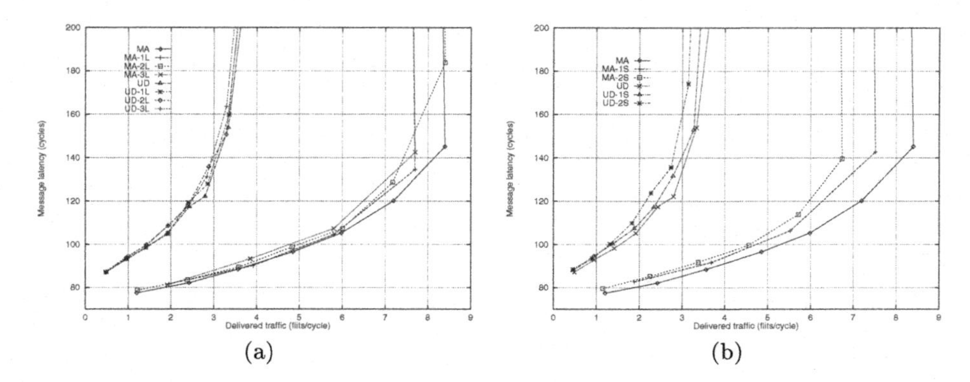

Fig. 8. Message latency versus traffic for a large network

Figure 8 shows that the MA routing algorithm is more sensitive to failures than the UD routing algorithm, although not as much as in medium size networks, because in these large networks a link or a switch represents a lower percentage of the whole network. The UD routing scheme only increases message latency, whereas the MA routing algorithm both increases latency and reduces the maximum delivered traffic. In both routing algorithms, the sensitivity to switch failures is higher than to link failures because a greater number of minimal paths are lost. In general, the first link failure slightly affects network performance, because of network size: it still provides enough alternative paths to absorb the network traffic. The worse case occurs when there are two faulty switches.

5 Conclusions

In this paper, the sensitivity to failures in links and switches of two routing algorithms for NOWs has been analyzed. The evaluation study was performed on networks with 8, 32, and 64 switches, in order to consider from small size to large size networks. Network performance degrades in the presence of failures, but this degradation highly depends on the routing algorithm and the network size.

In general, the performance of the up*/down* routing algorithm is slightly decreased by failures, only resulting in a low increment in latency. The minimal adaptive routing algorithm is more sensitive to failures, leading to an increment in latency, and a decrement of the maximum traffic accepted by the network. More precisely, and taking into account the size of the network, this study has provided the following main insights:

- For small networks, both routing algorithms are affected in a similar manner, mainly due to the few number of paths between switches. A great performance decrease is caused by the first failure in a link or in a switch, with a degree of degradation quite similar in both kinds of failures. The minimal adaptive routing algorithm performs better than the up*/down* routing scheme, even under the presence of a faulty link. In any case, the underlying topology highly affects the conclusions on network performance.

- When network size increases, the effect of failures on network performance decreases, because there are more links, and thus more paths exist between switches. Also, faulty links and switches represent a small percentage of the network. The existence of a greater number of paths can be used by the minimal adaptive routing algorithm to take advantage of its adaptivity. Independently from network topology, up*/down* routing performance is mainly limited by the early saturation of the links near the root switch. Moreover, even in the case of failures, the upper bound of network performance remains limited by the saturation of these links. On the other hand, the minimal adaptive routing algorithm is highly sensible to failures because the number of minimal paths is reduced, whereas their length is increased.
- In some of the analyzed topologies, the up*/down* routing algorithm has been benefited by a failure. The minimal adaptive routing algorithm can also take benefit because it uses the previous routing algorithm for the escape paths. Thus, the design of the topology is a very important issue when the up*/down* routing algorithm is used.

References

1. N. J. Boden, et al., "Myrinet - A gigabit per second local area network," in *IEEE Micro*, pp. 29–36, February 1995. 230, 231
2. R. V. Boppana and S. Chalasani, "Fault-tolerant wormhole routing algorithms for mesh networks," in *IEEE Trans. on Computers*, vol. 44, no. 7, pp 848–864, July 1995. 231
3. R. Casado, et al., "Performance evaluation of Dynamic reconfiguration in high-speed local area networks," in *Proc. of 6th Int. Symp. on High-Perf. Comp. Arch.*, 2000. 231
4. CSIM18 Simulation Engine (C version). Mesquite Software, Inc. 237
5. W. J. Dally and H. Aoki, "Deadlock-free adaptive routing in multicomputer networks using virtual channels", IEEE TPDS, vol. 4, no.4, pp. 466-477, April 1993.
6. J. Duato, S. Yalamanchili, L. Ni. *Interconnection Networks. An Engineering Approach.* IEEE Computer Society. 1997.
7. D. Garcia and W. Watson. "ServerNet II", in *Proc. of the 1997 Parallel Computing, Routing and Communication Workshop*. June, 1997. 230
8. P. T. Gaughan and S. Yalamanchili, "A Family of Fault-Tolerant Routing Protocols for Direct Multiprocessor Networks", IEEE Trans. on Parallel and Distributed Systems, vol. 6, no. 5, pp. 482-497, May 1995. 231
9. X. Molero et al., "Modeling and simulation of a network of workstations with wormhole switching", in *Proc. of the 33rd Annual Simulation Symposium*, April 2000. 237
10. M2LM-SW16 Myrinet switch. Available at `http://www.myri.com`, July 1999. 231
11. W. Qiao and L. M. Ni, "Adaptive routing in irregular networks using cut-through switches," in *Proc. of the 1996 Int. Conf. on Parallel Processing*, August 1996. 232

12. M. D. Schroeder et al., "Autonet: A high-speed, self-configuring local area network using point-to-point links," Technical Report SRC 59, DEC, April 1990. 230, 231, 232, 233
13. F. Silla and J. Duato, "Improving the efficiency of adaptive routing in networks with irregular topology," in *Proc. of the 1997 Int. Conf. on High Perf. Comp.*, 1997. 231, 232, 233, 235

Decentralized Load Balancing in Multi-node Broadcast Schemes for Hypercubes

Satoshi Fujita[1] and Yuji Kashima[2]

[1] Faculty of Engineering, Hiroshima University
Higashi-Hiroshima, 739-8527, Japan
[2] Oki Electric Industry Co. Ltd.
{fujita,kashima}@se.hiroshima-u.ac.jp

Abstract. In this paper, we study the multi-node broadcast problem in hypercubes, that is an asynchronous and repetitive version of all-to-all broadcast problem. Suppose that nodes of a hypercube asynchronously broadcast a piece of information. They can asynchronously initiate their broadcasts while other broadcasts are in process. The multi-node broadcast problem is the problem of completing each of those broadcasts quickly. We propose several decentralized schemes for solving the problem. The effectiveness of the schemes is demonstrated by simulation.

Keywords: Parallel processing, broadcast, load balancing, hypercube, decentralized communication scheme.

1 Introduction

Quick completion of broadcast significantly improves the overall performance of many important applications, such as numerical computation and combinatorial optimization. This is a main motivation of the study of broadcast problem, that has been investigated extensively during the past three decades [2,3]. Unfortunately however, most of previous studies on efficient broadcast schemes have been carried out under "strong" assumptions on the timing of broadcast initiation to reduce the difficulty of analyses, although in most of real applications, broadcasts are initiated asynchronously and repeatedly.

We challenge to solve this problem in general. Suppose that nodes asynchronously and repeatedly initiate their broadcasts to distribute a piece of information called *token* to every other node. The **multi-node broadcast problem** (MBP, in short) is the problem of designing an algorithm for quickly completing each of those broadcasts. In this paper, we consider MBP in hypercube. Note that hypercube is one of the most popular network topologies for parallel computers [5], and design and analysis of efficient routing algorithms on hypercubes is one of the most attractive issues to be studied [4].

Intuitively speaking, a multi-node broadcast scheme determines the route of each token, in a static or dynamic manner. A key point in designing efficient multi-node broadcast schemes is how to balance the load of communication links

M. Valero et al. (Eds.): ISHPC 2000, LNCS 1940, pp. 243–251, 2000.

that may vary dynamically depending on the timing of several broadcast initiations. On the other hand, it is also true that the collection of global information about the network traffic will increase the overall cost of the resultant scheme. Hence in order to complete each broadcast as quickly as possible, we should develop a multi-node broadcast scheme that can adapt itself to the change of traffic load by using informations that is "locally" observable by each node.

The remainder of this paper is organized as follows. Section 2 introduces some basic definitions and notation. A simple single-node broadcast scheme on hypercubes is also given. In Section 3, several multi-node broadcast schemes are proposed. The effectiveness of the schemes is demonstrated by simulation in Section 4. Section 5 concludes the paper.

2 Preliminaries

2.1 Model

Let $Q_n = (V, E)$ be an undirected binary n-cube, where V is the set of nodes representing processors, and E is the set of edges representing bidirectional communication links between processors. Each node in V corresponds to a binary string of length n, i.e., $V = \{0,1\}^n$, and two nodes $u, v \in V$ are connected by an edge in E iff u and v differ in exactly one bit. If u and v differ in the ith bit, then edge $\{u, v\} \in E$ is said to be of dimension i, and denote it as $v = \oplus_i u$.

Nodes in V communicate to each other by passing **tokens** through the communication network. When a node initiates a (multi-node) broadcast scheme, it computes information called *routing information* which specifies the destinations and routes the token will flow, and attaches it to the token. Then the token is sent out to the first node of each route specified in the routing information. Upon receiving a token, a node looks up its routing information, and forwards the token to the successor on each path specified in the routing information.

Each edge is assumed to be full-duplex, and conflicts of tokens at each edge are resolved by a *conflict resolution strategy* (CRS, for short) specified as a part of multi-node broadcast scheme. We further assume that, for simplicity, the local computation time at nodes is negligible. Nodes, therefore, can send at most one token to each of their adjacent nodes, and simultaneously, can receive at most one token from each of them in a unit time, called **step**. Note that tokens sent out to different adjacent nodes in a step can be different.

Finally, the **broadcast time** of a token is defined to be the maximum elapsed time necessary to distribute the token to every other node.

2.2 Single-Node Broadcast Scheme SIMPLE

In the multi-node broadcasting, the broadcast time of each token depends on the broadcast processes of other tokens. In [1], we have shown that if K ($\leq |V|$) nodes simultaneously initiate their broadcasts, it takes at least $\max\{n, \lceil (K-1)/n \rceil\}$ steps to complete all of the K broadcasts in Q_n (hereafter we call it Remark

1). In this subsection, we introduce a single-node broadcast scheme that will be used as a building block in later sections. In the scheme, called SIMPLE, the routing information for a token is given in the form of a permutation over $\{1, 2, \ldots, n\}$. A **permutation** π over $\{1, 2, \ldots, n\}$ is described as $(i_1, i_2, \ldots, i_n)$, where $i_j \in \{1, 2, \ldots, n\}$ for $1 \leq j \leq n$, and $i_j \neq i_k$ if $j \neq k$. For $1 \leq j \leq n$, denote i_j by $\pi(j)$. Let S_n be the set of all permutations over $\{1, 2, \ldots, n\}$.

In scheme SIMPLE, each broadcast initiator u first selects a permutation π from S_n, and attaches it to the token as the routing information. Then the token is broadcast according to the following two rules.

scheme SIMPLE

Rule 1: Broadcast initiator u sends a copy of the token to every neighbor $\oplus_{\pi(1)} u, \oplus_{\pi(2)} u, \ldots, \oplus_{\pi(n)} u$ simultaneously, and terminates.
Rule 2: If a node v receives a token from node $\oplus_{\pi(i)} v$ for $1 \leq i < n$, then v sends a copy of the token to each of neighbors $\oplus_{\pi(i+1)} v, \oplus_{\pi(i+2)} v, \ldots, \oplus_{\pi(n)} v$ simultaneously, and terminates. □

For any given permutation π ($\in S_n$), scheme SIMPLE completes a single broadcast in n steps, which is time optimal [1]. In scheme SIMPLE, for any $1 \leq i < j \leq n$, no edge of dimension $\pi(j)$ is followed by an edge of dimension $\pi(i)$ on any delivery path. Since each delivery path contains at most one edge of dimension i for each $1 \leq i \leq n$, permutation π completely determines the order of edges that occur in each delivery path.

2.3 Conflict Resolution Strategies

CRS plays an important role in designing efficient multi-node broadcast schemes. Let $t(u)$ be a token waiting to pass through an edge e of dimension i, and π the permutation attached to $t(u)$. The rank of token $t(u)$ at edge e is an integer j such that $i = \pi(j)$. The *furthest-destination first-serve* (FDFS) rule is a CRS that selects a token with the *smallest rank* among all conflicting tokens, where a tie is broken by the first-come first-serve (FCFS) rule. In order to minimize the maximum broadcast time, in the following, we often introduce the notion of *deadline* to the FDFS rule. Let τ be a natural number representing the deadline. In the deadlined FDFS, we modify the original FDFS in such a way that if it spends τ time units after entering a waiting queue, then the token is given the highest priority among others in the queue until it is taken out of the waiting queue.

3 Load Balancing Schemes

3.1 Naive Oblivious Schemes

The simplest way for realizing a multi-node broadcast is to invoke SIMPLE for a *fixed* permutation π regardless of the broadcast initiator, and to resolve conflicts by the FDFS rule. We call this naive scheme NAIVE. Although it achieves the

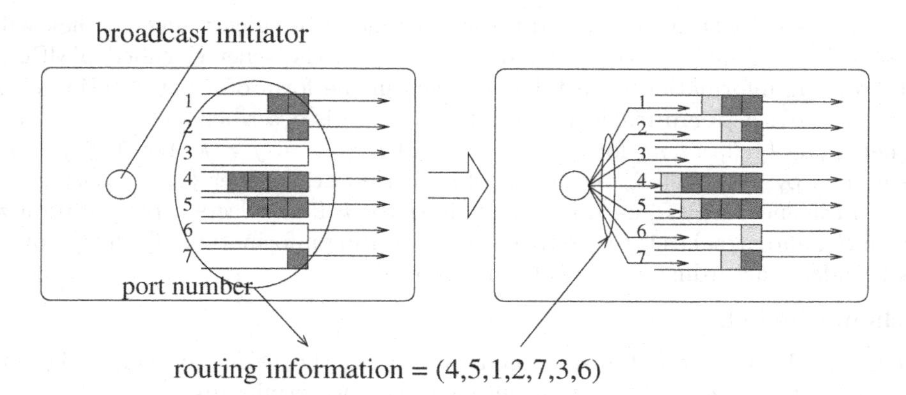

Fig. 1. Initiation of a broadcast according to scheme SORT

lower bound on the broadcast time in the best case, in the worst case, it requires a time factor of $n/2$ from a lower bound [1]. A simple (but powerful) way for balancing the traffic load in NAIVE is to introduce a randomization. More concretely, we may modify NAIVE in such a way that the attached permutation is randomly and independently selected from set S_n for each broadcast initiator. In the following, we call this randomized scheme RANDOM.

3.2 An Adaptive Scheme Based on Local Information

In [1], we proposed a non-oblivious multi-node broadcast scheme SORT. In the scheme, the selection of a permutation is based on the length of the waiting queues incident on the initiator. More precisely, the initiator generates a permutation by "sorting" the dimensions in the nonincreasing order of the length of the outgoing waiting queues. Note that the queue length of an outgoing edge is "locally" observable by the initiator and it assumes no global information about the network traffic. A formal description of the scheme is as follows (Figure 1 illustrates a broadcast initiation according to SORT):

scheme SORT

1) **Preprocessing:** Let $M = \{1, 2, \ldots, n\}$ be the set of indices. Let $\ell(i)$ denote the queue length of the outgoing edge of dimension i that is incident on the initiator. We first sort set M in the nonincreasing order of $\ell(i)$. Let $(i_1, i_2, \ldots, i_n)$ be the resulting list; i.e., $i_j \neq i_k$ for $j \neq k$, and $\ell(i_j) \geq \ell(i_{j+1})$ for $1 \leq j \leq n-1$.
2) **Broadcast tokens:** Attach permutation $(i_1, i_2, \ldots, i_n)$ to the token as the routing information, and initiate the broadcast according to scheme SIMPLE.
3) **Conflict resolution strategy:** Conflicts of tokens are resolved by the deadlined FDFS rule. □

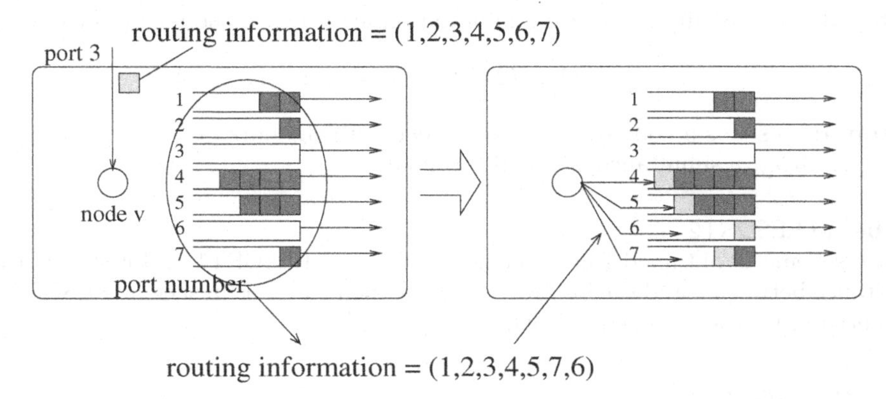

Fig. 2. Forwarding of token in scheme RESORT1

3.3 Introduce More Adaptiveness

In our preliminary experiments [1], it was shown that SORT is more robust than RANDOM when there is a spatial imbalance of broadcast initiators. Although it seems to be a natural consequence since SORT is non-oblivious and can use more information than RANDOM, from the view point of "decentralized" load balancing, it is very interesting since in SORT, each initiator only collects "local" information before determining the routing information.

However, it should be too optimistic to assume that the information local to the broadcast initiator *precisely* reflects the global load distribution. Hence, it is a natural idea for improving the performance of SORT to allow adaptive changes of attached permutation according to the local informations that can be observed by the intermediate nodes on the routing path. In the following, we propose two ways for such an extension and evaluate their effectiveness by simulations: The first way is to change the attached permutation at *every* node on the delivery path. We call it RESORT1 (Figure 2 illustrates forwarding of token in scheme RESORT1). The second way is to allow adaptive change of permutations at most once on any delivery path. We call the resulting scheme RESORT2. Formal descriptions of those two schemes are given as follows.

scheme RESORT1
The scheme is obtained from scheme SORT by modifying Rule 2 of the underlying single-node broadcast scheme SIMPLE, as follows:

Step 1: Suppose that a node v receives a token from node $\oplus_{\pi(i)} v$ for $1 \le i < n$, where π is the permutation attached to the token as the routing information.

Step 2: Let $M = \{\pi(i+1), \pi(i+2), \dots, \pi(n)\}$ be the set of indices that have not been used in the delivery path from the initiator to v, and should be used in the remaining paths. Sort set M in the nonincreasing order of $\ell(i)$, where $\ell(i)$ is the queue length of the outgoing edge of dimension i that is incident on v. Let $(i_{i+1}, i_{i+2}, \dots, i_n)$ be the resulting list.

Step 3: We modify the routing information attached to the received token to

$$\pi' = (\pi(1), \pi(2), \ldots, \pi(i), i_{i+1}, i_{i+2}, \ldots, i_n).$$

Step 4: v sends a copy of the token to each of neighbors $\oplus_{\pi'(i+1)}v, \oplus_{\pi'(i+2)}v, \ldots, \oplus_{\pi'(n)}v$ simultaneously, and terminates. □

scheme RESORT2
This scheme is obtained from schemes SORT and RESORT1 as follows: If the distance between v and the initiator is $\lfloor n/2 \rfloor$, then v acts as in scheme RESORT1; otherwise, it acts as in scheme SORT. □

4 Simulation

We conducted simulations in order to evaluate the effectiveness of the proposed schemes. All of the following results assume $n = 9$; i.e., a binary n-cube with $2^9 = 512$ nodes is considered. Every node repeats a broadcast 30 times, and each node can initiate a new broadcast only after completing the previous broadcast initiated by the node. The time interval between two consecutive broadcasts is randomly selected from set $\{0, 1, \ldots, 5\}$; i.e., at least 0 step and at most 5 steps.

We conducted experiments under the following three patterns of initiator distribution: i.e., all of the $|V|$ nodes broadcast (Pattern 1); $|V|/2$ nodes with prefix 0 broadcast (Pattern 2); and $|V|/2$ nodes with prefix 00 or 11 broadcast (Pattern 3). In the next subsection, we demonstrate that if the spatial imbalance of broadcast initiators is large enough, then the proposed adaptive schemes could give a better performance in terms of the average broadcast time than RANDOM. The result also shows that the introduction of a further adaptiveness to SORT could not significantly improve the performance of the scheme.

4.1 Results

Figures 3 and 4 show average broadcast times for each initiation pattern. In each figure, broadcast time is averaged over all initiators and ten broadcasts; e.g., the value of the vertical axis at "10 iterations" is the average broadcast time over the first 10 broadcasts of all initiators ("20 iterations" corresponds to the next 10 broadcasts and "30 iterations" corresponds to the last 10 broadcasts).

By the figures, we have the following observations: In Pattern 1, RANDOM and SORT exhibit almost the same performance, and the performance is degraded by allowing more adaptiveness; e.g., RESORT1 takes 25 % more broadcast time than SORT, and RESORT2 takes 5 % more broadcast time than SORT (see Figure 3). In Pattern 2, the performance of RANDOM becomes apparently worse than the other schemes; e.g., it takes 40 % more broadcast time than SORT. On the other hand, the comparison of three adaptive schemes implies that even in such an "imbalanced" situation, RESORT1 and RESORT2 cannot beat SORT (see Figure 4 (a)). In Pattern 3, RANDOM beats RESORT1, and three schemes

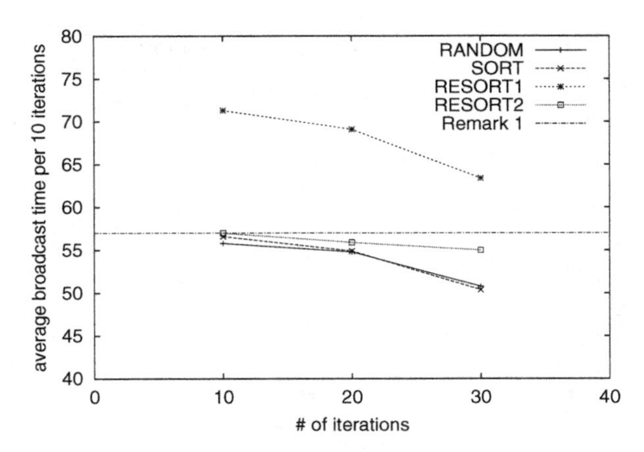

Fig. 3. Uniform distribution of broadcast initiators (Pattern 1)

except for RESORT1 exhibit almost the same performance. In fact, RESORT1 takes 20 % more broadcast time than the other three schemes (see Figure 4 (b)).

In order to find an explanation of the above phenomena, we observed difference of permutations generated by different schemes, in more detail. Note that in Pattern 2, the edges of dimension 9 becomes a bottleneck if it is used as the latter (or, the last) elements in the generated permutation. Hence in order to avoid such a possible bottleneck, the scheme should use those edges as the former elements. In fact, the percentage of dimension 9 edges as the first element in the generated permutations decreases from 38 % to 27 % by using RESORT2 instead of SORT, and to 21 % by using RESORT1. In other words, the force to move dimension 9 to the former elements (in the generated permutation) becomes weak by introducing more adaptiveness to the broadcast scheme.

On the other hand, in Pattern 3, we may avoid bottlenecks caused by two dimensions instead of one (note that dimensions 8 and 9 play the same role in Pattern 3); i.e., the spatial imbalance of broadcast initiators in Pattern 3 is smaller than that in Pattern 2. In fact, the percentage of permutations is almost equal, in contrast to Pattern 2.

5 Concluding Remarks

In this paper, we proposed several decentralized schemes for solving the multi-node broadcast problem. The effectiveness of the schemes is demonstrated by simulation. As a result, we show that the multi-node broadcast scheme in which each broadcast initiator adapts itself to the traffic load by using a locally observable information exhibits the best performance, and the force to balance the load becomes weak by introducing more adaptiveness to the scheme.

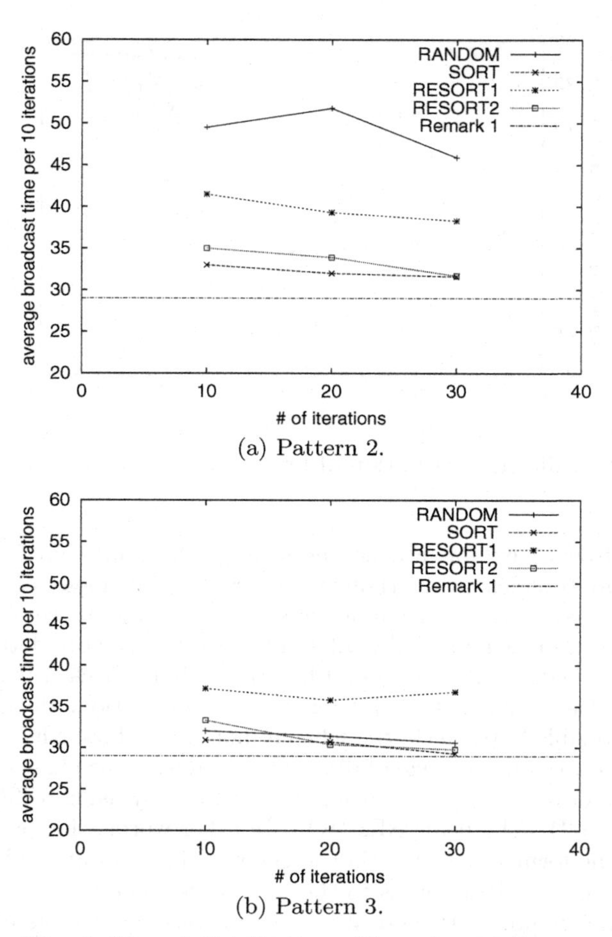

(b) Pattern 3.

Fig. 4. Biased distribution of broadcast initiators

Acknowledgments

This research was partially supported by the Ministry of Education, Science and Culture of Japan (Grant #11780228), and Research for the Future Program from Japan Society for the Promotion of Science (JSPS): Software for Distributed and Parallel Supercomputing (JSPS-RFTF96P00505).

References

1. Y. Hayakawa, S. Fujita, and M. Yamashita, A Decentralized Scheme for Multi-Node Broadcasting on Hypercubes. Proc I-SPAN'97, IEEE, pp.487–493, 1997. 244, 245, 246, 247

2. S. M. Hedetniemi, S. T. Hedetniemi, and A. L. Liestman. A survey of gossiping and broadcasting in communication networks. *Networks*, 18:319–349, 1988. 243
3. J. Hromkovič, R. Klasing, B. Monien, and R. Peine. Dissemination of information in interconnection networks (Broadcasting and Gossiping). in F. Hsu and D.-Z. Du (eds.) Combinatorial Network Theory, Science Press & AMS. 243
4. F. Thomson Leighton. *Introduction to Parallel Algorithms and Architectures: Arrays, Trees, Hypercubes.* Morgan Kaufmann Publishers, Inc., 1992. 243
5. A. Trew and G. Wilson, editors. *Past, Present, Parallel: A Survey of Available Parallel Computer Systems*, chapter 4, pages 125–147. Springer-Verlag, 1991. 243

Design and Implementation of an Efficient Thread Partitioning Algorithm

José Nelson Amaral, Guang Gao, Erturk Dogan Kocalar,
Patrick O'Neill, Xinan Tang

Computer Architecture and Parallel Systems Laboratory,
University of Delaware, Newark, DE, USA, http://www.capsl.udel.edu
Dep. of Comp. Science, Univ. of Alberta, Canada, http://www.cs.ualberta.ca

Abstract. The development of fine-grain multi-threaded program execution models has created an interesting challenge: how to partition a program into threads that can exploit machine parallelism, achieve latency tolerance, and maintain reasonable locality of reference? A successful algorithm must produce a thread partition that best utilizes multiple execution units on a single processing node and handles long and unpredictable latencies.
In this paper, we introduce a new thread partitioning algorithm that can meet the above challenge for a range of machine architecture models. A quantitative affinity heuristic is introduced to guide the placement of operations into threads. This heuristic addresses the trade-off between exploiting parallelism and preserving locality. The algorithm is surprisingly simple due to the use of a time-ordered event list to account for the multiple execution unit activities. We have implemented the proposed algorithm and our experiments, performed on a wide range of examples, have demonstrated its efficiency and effectiveness.

1 Introduction

This paper is a contribution to the development of high-performance computer systems based on a fine-grain multi-threaded program execution and architecture model. A key to the success of multi-threading is the development of compilation methods that can efficiently exploit fine-grain parallelism in application programs and match them with the parallelism of the underlying hardware architecture. In particular, partitioning programs into fine-grain threads is a new challenge that is not dealt with in conventional compiler code generation and optimization.

Our thread partitioning algorithm was developed for the Efficient Architecture for Running THreads (EARTH), a multi-threaded execution and architecture model [8,4]. Under the EARTH model, a thread becomes *enabled* for execution if and only if it has received signals from all the split-phase operations that it depends on. Furthermore, threads are non-preemptive: once a thread is scheduled for execution, it holds the execution unit until its completion. Therefore whenever an operation may involve long and/or unpredictable latencies,

M. Valero et al. (Eds.): ISHPC 2000, LNCS 1940, pp. 252-259, 2000.

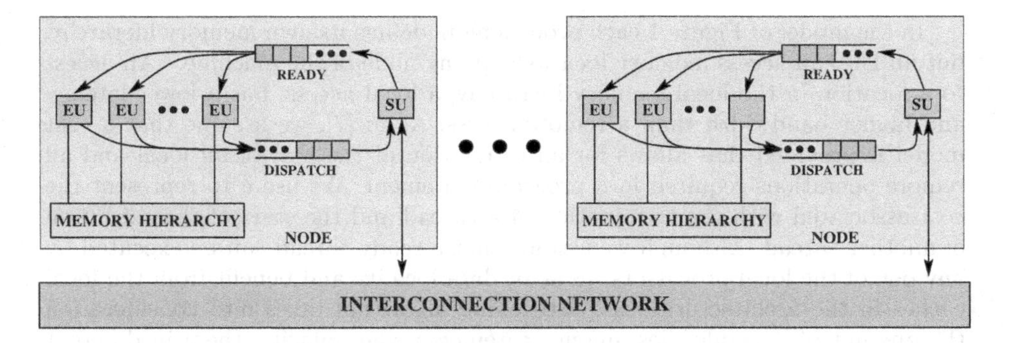

Fig. 1. Architecture abstract model.

the role of a compiler (or programmer) is to make the operation "split-phase". We call this requirement the *split-phase constraint*, and assume that such constraints are explicitly identified and presented to the thread partitioner. The thread partitioning problem studied in this paper can be stated as follows.

> *Given a machine model M and a weighed data dependence graph G with some nodes labeled as split-phase nodes, partition G into threads such that the total execution time of G is minimized subject to the split-phase constraints.*

The main contribution of this paper is the development of an heuristic thread partitioning algorithm suitable for a machine model that allows multiple execution units in each processing node.[1] Unlike previous related thread partitioning algorithms, ours faces a new challenge: the existence of more than one thread execution unit per node implies a trade-off between the need to generate enough parallel threads per node to utilize these execution units, and the need to assign related operations to the same thread to enhance locality of access.

2 Machine Model

Our architecture model is presented in Figure 1. Each processing node has N execution units (EU) and one synchronization unit (SU). Both the EU and the SU perform the functions specified in the EARTH Virtual Machine [8]. Threads that are ready for execution are placed in the ready queue that is serviced by the EUs. When an active thread performs a synchronization operation or requests a long latency data transfer, the request for such a service is placed in the dispatch queue. The SU is responsible for the communication with all other processing nodes and for the synchronization between threads within the node.

[1] Notice that the algorithm presented in [7] collapses as much local computation as possible into a single thread, thus making it inadequate for the machine model studied in this paper that has multiple execution units per processing node.

In the model of Figure 1 each processing node has its own memory hierarchy, but an EU can access memory locations in any node of the machine.[2] An access to a location in the local memory hierarchy, a local access, has a lower latency and higher bandwidth than a remote access. As in [7], we assume that a cost model is provided that allows for an estimation of the cost of all local and all remote operations required in a program statement. We use δ to represent the cost associated with the termination of a thread and the start of the execution of another thread. Although we assume that a ready thread can be executed by any one of the local processors, to favor data locality and benefit from the local caches in the architecture, the partitioning algorithm takes into consideration the amount of dependencies among statements when placing them in different threads

3 Thread Partition Cost Model

We assume that a program is written in a sequential language augmented with high level-parallel constructs, and that the data has been partitioned among the memory modules in the processing nodes of the machine. Therefore, given a program statement we can determine whether the statement is *local* or *remote.* We also assume that the program has been translated into a Data Dependence Graph (DDG).

Thus the program is represented by a graph $G(V, E)$ where each node in V represents a collection of program statements. A node can be a *simple node* such as an assignment statement or a *compound node* such as a loop. If the execution of the program statements requires accesses to a remote memory module, the node is a *remote node* otherwise the node is a *local node*. Each edge $(v_i,\ v_j)$ in E represents a data dependency from v_i to v_j. An edge departing from a remote node is a *remote edge*, and an edge departing from a local node is a *local edge.* Like in [7] we represent E by an adjacency matrix C defined as:

$$C_{i,j} = \begin{cases} 1 \text{ if } (v_i, v_j) \in E \\ 0 \text{ otherwise} \end{cases}$$

The partitioning algorithm is based on a cost model that associates a *local cost* c_i^L and a *remote cost* c_i^R to each node v_i that represent the number of cycles that the nodes spends in the EU and the number of cycles that elapses between the request and the completion of a split-phase transaction started by the node. One of the main goals of the partitioning algorithm is to decide when it is advantageous to dispatch the request, terminate the current thread and to start the execution of a new thread when the remote operation is completed. For some interconnection networks a model to predict the network performance will be necessary because the remote cost is affected by the load in the network. For the experiments with our partitioning algorithm, we assume that this cost

[2] If the underline machine does not allow direct accesses to remote memory, the EARTH systems emulates a global address space.

is bounded by a constant. The cost model presented in this paper can be constructed for any machine through profiling experiments and examination of the machine specification for the number of cycles required to execute each class of instruction. The network latency and bandwidth can also be readily measured. For our experiments we use the values obtained by Theobald for the EARTH-MANNA implementation [8].

4 Problem Formulation

Assume that at runtime threads are selected for execution from a single queue of ready threads using an efficient scheduler. Given a DDG G with each node $v_i \in G$ annotated with its local cost c_i^L and its remote cost c_i^R, and a constant thread switching cost δ, the **thread partitioning problem** is the problem of finding a thread partition P that meets two goals: (1) minimizes the total execution time, (2) maximizes the affinity between nodes assigned to the same thread. The **affinity** of a node v_i of G to a thread T_k of P, $A(v_i, T_k)$, is given by the ratio between the number of dependences between nodes of T_k and v_i, and the total number of incoming edges of v_i.

$$A(v_i, T_k) = \frac{\sum_{v_j \in T_k} C_{ji}}{\sum_j C_{ji}}$$

Observe that if all the incoming edges of v_i are from nodes in T_k, then $A(v_i, T_k) = 1$. On the other hand, if none of the incoming edges of v_i are from nodes in T_k, then $A(v_i, T_k) = 0$.

Goals (1) and (2) should be pursued under the constraint that nodes connected by a remote edge are assigned to different threads. Goal (1) is the principal goal of the partition algorithm, while goal (2) is necessary to favor locality of access because the abstract model assumes that all processors have equal probability of fetching a thread from the ready queue.

The thread partitioning algorithm uses the affinity function to decide in which thread to place a node of the DDG. The algorithm keeps a partial schedule of the threads already formed and searches into this schedule for the best place to insert a node from the DDG. To minimize the searching time, an event list is used to store the starting and finishing time of each thread. A detailed description of the thread partitioning algorithm including an example is presented in [1].

5 Experimental Results

We use the Thread Partition Test Bed presented in [7] to generate random DDGs to test the partition algorithm. We vary several properties of the DDGs generated, including the number of nodes, the average number of outgoing edges from a node, the percentage of remote nodes in the graph and the distribution of local and remote costs in the nodes.

The distribution of execution costs for the nodes is as follows: A local node can be of three types: 1) Local I/O (20%), 2) Local function call (10%), 3) Other(math,etc.) (70%). A local I/O is assigned a cost of 10 cycles. A local function call is assigned 10 cycles and a node of other type has 3 cycles. Remote nodes can be of type: 1) Remote I/O (80%), 2) Remote function call (20%). A remote I/O is assumed to take 300 cycles, and the costs of remote function calls are uniformly distributed between 400 cycles and 4000 cycles. This distribution of node types and the degree of the DDG generated are based on static profiling of EARTH-C benchmarks [7, 3]

5.1 Summary of Main Experimental Results

The main results of our experiments can be summarized as follow.

Absolute Efficiency Our new partition algorithm is very efficient for a wide range of DDGs on all the machine models except the EARTH-Dual model where only a single EU is available per node and there is a high cost associated with thread switching. The algorithm performs remarkably well when the machine model has multiple execution units – e.g. for SMP and SCMP models – the average absolute efficiency is above 99%.

Effectiveness of Search Heuristics The use of an event list and of a time line schedule results in an effective search for the placement of a new node (see [1] for details).

Latency Tolerance Capacity The algorithm is robust to variations in latency. In our experiments a remote operation latency varied between 400 and 4000 cycles. As shown in Table 1 the partition algorithm produces thread partitions that are able to tolerate these varied latencies.

5.2 Machine Architectures

We define four different machine architectures for our experiments. MANNA is a multiprocessor machine with 40 processors distributed in 20 processing nodes interconnected by a crossbar switch. MANNA is the first platform in which the EARTH model was implemented [5].

EARTH-MANNA-DUAL: An implementation of the EARTH architecture on the MANNA machine. The second processor is used for the SU function. Therefore, according to our model, this is a machine with a single EU per node. The thread switching cost in this machine is $\delta = 36$ cycles (see measurements reported in [8]).

EARTH-MANNA-SPN: The two processors in the machine are used to implement the SU functions. Therefore, this machine has two EUs per processing node. The thread switching cost is $\delta = 16$ (see measurements reported in [8]).

EARTH-SU: This is the EARTH architecture with a custom hardware SU. We consider a machine with a single execution unit per processing node and with a thread switching cost $\delta = 2$ (see measurements reported in [8]).

SMP: This is an Symmetric Multi-Processor machine with 4 processors per node. In such a machine we expect the thread switching cost to be similar to the one for the MANNA-SPN, therefore we will use $\delta = 16$.

Single-Chip Multi-threaded Processor (SCMP) In this case we consider a hypothetical machine that has multiple functional units with multi-threading support. We assume a machine with 8 EUs and with a thread switching cost of $\delta = 10$.

5.3 Measuring Absolute Efficiency

An optimal partition cannot result in an execution time that is shorter than the critical path of the program or shorter than the total amount of work to be performed divided by the number of EUs available. Thus, a lower bound for the execution time is given by:

$$T_{lowest} = \max(T_{crit}, \frac{\sum_{i \in V} c_i^L}{N})$$

where T_{crit}, is the length of the critical path, the sum of the local cost c_i^L of all nodes is the total amount of work to be performed by the program, and N is the number of EUs in the machine. We define the ***absolute efficiency*** as the ratio:

$$E = T_{lowest}/T_{end}$$

where T_{end} is the execution time for the program with the thread partition produced by our algorithm running under an efficient FIFO scheduler. $E = 100\%$ means that the partition algorithm found a partition that can result in the optimal execution time.

To measure the efficiency of the algorithm, we varied the percentage of remote nodes in the randomly generated DDG from 25% to 75%, and the size of the graph from 10 nodes to 1000 nodes. Each graph has three times as many edges as the number of nodes. Then we generated twenty distinct random DDGs and applied the partition algorithm to each one of them. We computed the average execution time for each run, compared it with T_lowest, and present the average efficiency in Table 1. The algorithm did remarkably well for machines with four or eight execution units per processing node (SMP and SCMP). The algorithm also did quite well both for the EARTH-SU that has a single EU and a very low thread switching cost and for the EARTH-MANNA-SPN that has two EUs per processing node. The results for graphs with a large number of nodes for the EARTH-MANNA-DUAL are not as good. This should not come as a surprise because this architecture has a single EU and very large thread switching costs.

6 Related Work

The thread partition problem for multi-threaded architectures is similar to the task partitioning and scheduling problem [6, 9]. In both problems a program has

Machine	% Remote Edges	Nodes						
		10	20	50	100	200	500	1000
EARTH-MANNA-DUAL	25%	87.6	91.8	90.0	83.6	82.3	77.0	78.1
	50%	96.3	95.8	96.8	93.7	83.8	71.6	71.2
	75%	96.5	98.7	98.9	98.2	95.6	69.5	66.0
EARTH-MANNA-SPN	25%	95.2	98.2	95.9	94.2	91.5	87.1	88.6
	50%	95.7	97.6	98.7	97.7	97.8	88.5	83.7
	75%	99.3	99.8	99.9	99.8	99.8	99.3	80.3
EARTH-SU	25%	95.3	98.3	92.4	93.2	96.7	99.9	100.0
	50%	95.9	95.4	99.2	97.0	92.0	92.8	98.1
	75%	98.4	99.3	98.4	97.8	98.5	90.3	94.1
SMP	25%	99.2	99.7	99.8	99.9	99.3	99.9	99.9
	50%	99.7	99.8	99.9	99.9	99.9	99.9	100.0
	75%	99.8	99.9	99.9	99.9	99.9	100.0	100.0
SCMP	25%	99.2	99.8	99.8	99.9	99.3	99.9	99.9
	50%	99.7	99.8	99.9	99.9	99.9	99.9	100.0
	75%	99.7	99.9	99.9	99.9	99.9	100.0	100.0

Table 1. Average partition algorithm efficiency.

to be divided into smaller pieces with respect to some constraints (dependencies). The focus of task partition is to allocate N tasks onto M processors in order to reduce the total execution time. This problem is often represented as a graph partition problem in which nodes denote tasks and edges represent two types of constraints: *precedent constraint*, i.e., one task must complete before another task can start; and *communication constraint*, i.e., data must be exchanged between two tasks. When the communication constraint is taken into consideration the task partitioning problem is an NP-complete problem [2]. In this case, the optimization goal is often reduced to minimize the total communication [6]. For further discussion of related work we refer to [7] and [1].

7 Conclusion

We designed, implemented, and evaluated an efficient, effective, and robust algorithm to partition a program into threads for the case in which multiple execution units are available in each processing node of a parallel architecture. The algorithm is efficient because it generates a partition that results in an execution time that is very close to the best possible execution time determined by the length of the critical path and the total amount of computation existing in the program. The algorithm is robust because it worked efficiently for a varied set of architectures and a wide range of latencies between processing node. The algorithm is effective because it employs a data structure associated with a searching algorithm that reduce the time complexity of the algorithms. On our experimental framework we tested the algorithm with several thousand data dependency graphs with up to a thousand nodes and several thousand connections.

Acknowledgements

We would like to thank Andres Marquez, Arthour Stoutchinin, Gagan Agarwal, Kevin Theobald, and Mark Butala for productive discussions and valuable help. We acknowledge the support of DARPA, NSA and NASA through a subcontract with JPL/Caltech. The current EARTH research is partly funded by the NSF.

References

1. J. N. Amaral, G. R. Gao, E. D. Kocalar, P. O'Neill, and X. Tang. Design and implementation of an efficient thread partitioning algorithm. Technical report, University of Delaware, Newark, DE, July 1999. CAPSL Technical Memo 30.
2. Michael R. Garey and David S. Johnson. *Computers and Intractability: A Guide to the Theory of NP-Completeness.* W. H. Freemann and Co., New York, New York, 1979.
3. Laurie J. Hendren, Xinan Tang, Yingchun Zhu, Guang R. Gao, Xun Xue, Haiying Cai, and Pierre Ouellet. Compiling C for the EARTH multithreaded architecture. In *Proceedings of the 1996 Conference on Parallel Architectures and Compilation Techniques (PACT '96)*, pages 12–23, Boston, Massachusetts, October 20–23, 1996. IEEE Computer Society Press.
4. Herbert H. J. Hum, Olivier Maquelin, Kevin B. Theobald, Xinmin Tian, Guang R. Gao, and Laurie J. Hendren. A study of the EARTH-MANNA multithreaded system. *International Journal of Parallel Programming*, 24(4):319–347, August 1996.
5. Olivier Maquelin, Guang R. Gao, Herbert H. J. Hum, Kevin B. Theobald, and Xin-Min Tian. Polling Watchdog: Combining polling and interrupts for efficient message handling. In *Proceedings of the 23rd Annual International Symposium on Computer Architecture*, pages 178–188, Philadelphia, Pennsylvania, May 22–24, 1996. ACM SIGARCH and IEEE Computer Society. *Computer Architecture News*, 24(2), May 1996.
6. Vivek Sarkar. *Partitioning and Scheduling Parallel Programs for Multiprocessors.* Research Monographs in Parallel and Distributed Computing. Pitman, London and The MIT Press, Cambridge, Massachusetts, 1989. Revised version of the author's Ph.D. dissertation (Stanford University, April 1987).
7. Xinan Tang, Jian Wang, Kevin B. Theobald, and Guang R. Gao. Thread partitioning and scheduling based on cost model. In *Proceedings of the 9th Annual ACM Symposium on Parallel Algorithms and Architectures*, pages 272–281, Newport, Rhode Island, June 22–25, 1997. SIGACT/SIGARCH and EATCS.
8. Kevin Bryan Theobald. *EARTH: An Efficient Architecture for Running Threads.* PhD thesis, McGill University, Montréal, Québec, May 1999.
9. T. Yang and A. Gerasoulis. List scheduling with and without communication delay. *Parallel Computing*, 19:1321–1344, 1993.

A Flexible Routing Scheme for Networks of Workstations *

José Carlos Sancho, Antonio Robles and José Duato

Departamento de Informática de Sistemas y Computadores
Universidad Politécnica de Valencia
P.O.B. 22012,46071 - Valencia, SPAIN
E-mail: {jcsancho,arobles,jduato}@gap.upv.es
Phone: +34 963877573; FAX: +34 963877579

Abstract. NOWs are arranged as a switch-based network which allows the layout of both regular and irregular topologies. However, the irregular pattern interconnect makes routing and deadlock avoidance quite complicated. Current proposals use the $up^*/down^*$ routing algorithm to remove cyclic dependencies between channels and avoid deadlock. Recently, a simple and effective methodology to compute $up^*/down^*$ routing tables has been proposed by us. The resulting routing algorithm is very effective in irregular topologies. However, its behavior is very poor in regular networks with orthogonal dimensions. Therefore, we propose a more flexible routing scheme that is effective in both regular and irregular topologies. Unlike $up^*/down^*$ routing algorithms, the proposed routing algorithm breaks cycles at different nodes for each direction in the cycle, thus providing better traffic balancing than that provided by $up^*/down^*$ routing algorithms. Evaluation results modeling a Myrinet network show that the new routing algorithm increases throughput with respect to the original $up^*/down^*$ routing algorithm by a factor of up to 3.5 for regular networks, also maintaining the performance of the improved $up^*/down^*$ routing scheme proposed in [7] when applied to irregular networks.

Keywords: Networks of workstations, regular and irregular topologies, routing algorithms, deadlock avoidance.

1 Introduction

NOWs are arranged as a switch-based network which provides the wiring flexibility, scalability, and incremental expansion capability required in this environment. In order to achieve high bandwidth and low latencies, NOWs are often connected using gigabit local area network technologies. There are recent proposals for NOW interconnects like Autonet [9], Myrinet [1], Servernet II [4], and Gigabit Ethernet [10].

Switch-based network allows the layout of both regular and irregular topologies. However, regular networks are often used when performance is the primary concern [6]. On the other hand, the irregular pattern interconnect makes routing and deadlock avoidance quite complicated. Current proposals use the $up^*/down^*$ routing algorithm. Others routing schemes have been proposed for NOWs, like adaptive-trail routing [5],

* This work was supported by the Spanish CICYT under Grant TIC97-0897-C04-01.

M. Valero et al. (Eds.): ISHPC 2000, LNCS 1940, pp. 260-267, 2000.

minimal adaptive routing [11], and smart-routing [2]. However these routing algorithms have not been considered due to their limited applicability or their high computational cost.

In *up*/down** routing [9], a breadth-first search spanning tree (BFS) is computed. This algorithm is quite simple, and has the property that all the switches in the network will eventually agree on a unique spanning tree. A direction ("up" or "down") is assigned to each network link, based on the position in the spanning tree, and messages are routed through sequences of "up" or "down" channels. The "down"/"up" transition is forbidden. As a consequence, in most cases, *up*/down** routing does not always supply minimal paths between non-adjacent switches, becoming more frequent as network size increases. Recently, it has been proved that making a different assignment of direction to links may lead to a significant increase in the number of minimal paths followed by messages. The methodology to compute routing tables is based on obtaining a depth-first search spanning tree (DFS) instead of a BFS spanning tree [7]. This methodology is very efficient in irregular networks. However in regular networks such as meshes or hypercubes, its behavior is noticeably worse than that of the dimension-order routing algorithm (DOR).

In this paper, we propose a more flexible routing scheme that is effective in both regular and irregular network topologies. It is based on computing a DFS spanning tree to break cyclic dependencies, like the algorithm proposed in [7]. However, unlike in *up*/down** routing schemes, the removal of cyclic channel dependencies can be done separately for each direction in each cycle, thus allowing us to achieve better traffic balancing in most network topologies.

The rest of the paper is organized as follows. In Section 2, the *up*/down** routing scheme and the methodologies to compute its routing tables are described. In Section 3, a flexible routing scheme that is effective in both regular and irregular networks is proposed. Section 4 shows performance evaluation results for the new routing algorithm. Finally, in Section 5 some conclusions are drawn.

2 Up*/Down* Routing

Up/down** routing is the most popular routing scheme currently used in commercial networks, such as Myrinet [1], valid for networks with regular or irregular topology. In order to compute *up*/down** routing tables, different methodologies can be applied. These methodologies are based on an assignment of direction ("up" or "down") to the operational links in the network by building a spanning tree. These methodologies differ in the type of spanning tree to be built. One methodology is based on a BFS spanning tree, such as it was proposed in Autonet [9], whereas another methodology is based on a DFS spanning tree, as it was recently proposed in [7].

In networks without virtual channels, the only practical way of avoiding deadlock consists of restricting routing in such a way that cyclic channel dependencies [1] are avoided [3]. To avoid deadlocks while still allowing all links to be used, *up*/down** routing uses the following rule: a legal route must traverse zero or more links in "up" direction followed by zero or more links in "down" direction. Thus, cyclic channel

[1] There is a channel dependency from a channel c_i to a channel c_j if a message can hold c_i and request c_j. In other words, the routing algorithm allows the use of c_j after reserving c_i. It will be represented as $c_i \rightarrow c_j$

dependencies are avoided by imposing routing restrictions, because a message cannot traverse a link along the "up" direction after having traversed one in the "down" direction. Next, we describe how to compute both a BFS and DFS spanning tree, and how to assign a direction to the links.

2.1 Computing a BFS Spanning Tree

First, to compute a BFS spanning tree, a switch must be chosen as the root. Starting from the root, the rest of the switches in the network are arranged on a single spanning tree [9]. Then, an assignment of direction ("up" or "down") to links is performed. The "up" end of each link is defined as: 1) the end whose switch is closer to the root in the spanning tree; 2) the end whose switch has the lower identifier, if both ends are at switches at the same tree level. The result of this assignment is that each cycle in the network has at least one link in the "up" direction and one link in the "down" direction.

2.2 Computing a DFS Spanning Tree

Like in the BFS spanning tree, an initial switch must be chosen as the root before starting the computation of a DFS spanning tree. The selection of the root is made by using heuristic rules. The rest of the switches are added following a recursive procedure.

Unlike in the BFS spanning tree, adding switches to build the path is made by using heuristic rules. We apply the heuristic rule recently proposed in [8]. Starting from the root switch, the switch with a higher number of links connecting to switches that already belong to the tree is selected as the next switch in the tree. In case of tie, the switch with higher average topological distance to the rest of the switches will be selected first.

According to [8] the root switch is selected after computing all the DFS spanning trees and selecting one of them based on two behavioral routing metrics: (1) the average number of links in the shortest routing paths between hosts over all pairs of hosts, referred to as *average distance*; and (2) the maximum number of routing paths crossing through any network channel, referred to as *crossing paths*. We first compute the metrics for each DFS spanning tree obtained by selecting the root among all the switches in the network. Finally, the switch selected as the root will be the one whose DFS spanning tree provides the lower value for the *crossing paths* metric. In case of tie, the switch with the lower value for the *average distance* metric will be selected. Therefore, the selected switch will be the one that allows more messages to follow minimal paths and provides better traffic balancing.

Next, before assigning direction to links, switches in the network must be labeled with positive integer numbers. A different label is assigned to each switch. When assigning directions to links, the "up" end of each link is defined as the end whose switch has the higher label.

3 Flexible Routing Scheme

Our interest on a flexible routing scheme that is effective in both regular and irregular networks is due to the following reasons: (1) in some cases, for performance reasons, it may be advisable to use regular networks; (2) in spite of using regular networks, links may fail or components may be added/removed in a NOW environment. Thus

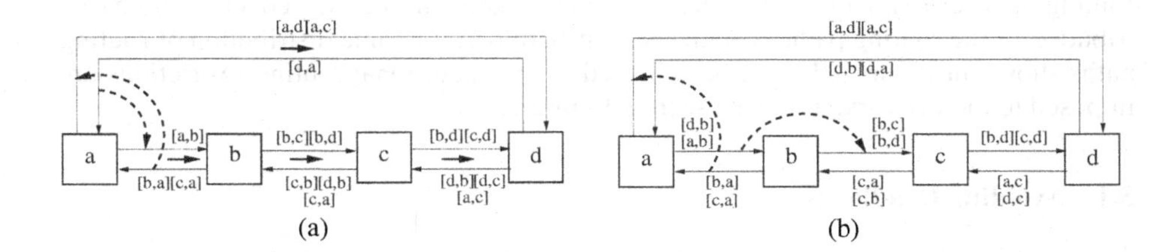

Fig. 1. (a) Link direction assignment and cyclic channel dependency removed by using the *up*/down** routing scheme, and (b) independent removal of cyclic channel dependencies for the two directions in the cycle.

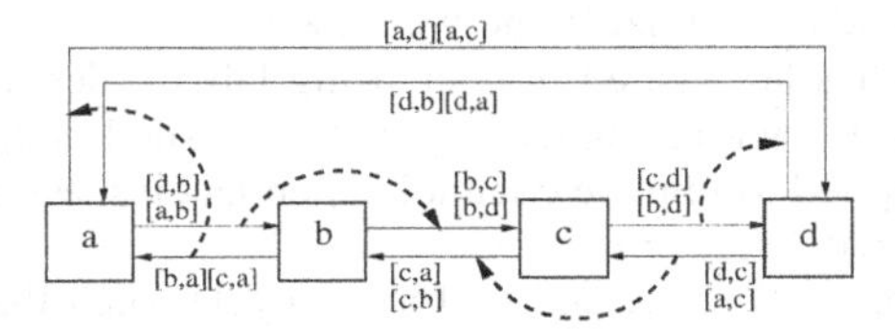

Fig. 2. Balancing paths by removing additional channel dependencies.

recomending the use of generic routing algorithms suitable for the resulting topology. (3) often, switch arrangement exhibits a certain degree of regularity or hierarchy, but this may not be enough to allow specific routing algorithms for regular networks to be used.

As shown in Section 2, *up*/down** routing algorithms remove cyclic channel dependencies by breaking cycles at the same node for both directions of each cycle. That is, if the dependency between the channels c_i and c_j is removed, then the channel dependency between the channels c_j and c_i will be removed too. We have observed that this constraint may cause an uneven distribution of traffic in some network topologies.

To ilustrate this idea, consider the simple 4-switch network depicted in Figure 1(a). Solid arrows represent the "up" direction assigned to each link by the *up*/down** routing algorithm. Also, removed channel dependencies are shown in dashed arrows. Each routing path crossing a channel is represented by $[x, y]$, where x and y represent the source and destination switches of the routing path, respectively. Every routing path is computed by selecting a single path between every pair of switches, thus minimizing the number of routing paths crossing each channel. We can observe that the *up*/down** routing algorithm unevenly distributes the routing paths among the channels, since there are some channels crossed by 3 routing paths, whereas other channels are crossed by a single routing path.

As can be observed, *up*/down** routing imposes hard constraints to remove channel dependencies. However, by removing channel dependencies independently for each direction in each cycle a more even distribution of traffic can be achieved. In Figure 1(b) we can observe that the number of dependencies removed from the network is the same as the number of dependencies removed in Figure 1(a). However, the routing restrictions are independent for the two directions of the cycle. Also, we can observe that the new removal of cyclic channel dependencies decreases the maximum number of routing paths crossing every channel in the network down to 2. Moreover, an even distribution of the routing paths is achieved. Furthermore, the same distribution of routing paths shown in Figure 1(b) can be achieved even if additional routing restrictions are imposed to the network, as can be seen in Figure 2.

3.1 Avoiding Deadlocks

To ease the identification of cyclic channel dependencies in the network graph, once a DFS spanning tree has been computed, a different label is assigned to each switch, as described in Section 2.2. The label assigned to each switch allows us to detect cycles in the network graph, because every cycle will have a single switch a that satisfies the condition $L(b) > L(a) < L(c)$, where $L(x)$ is a function that returns the label assigned to switch x, and b, a, and c are adjacent network switches.

Next, all the switches are visited in decreasing label order to break cycles in the network graph. The switch with a higher label will be visited first. At every switch, the following rules are applied to remove channel dependencies, assuming that the switch currently visited is v_i and $L(v_i) = i$:

(R_1) if the switches v_i, v_k, v_r, and v_j form a cycle, such that $i > k$ and $i > j$, the channel dependencies[2] $c_{k,i} \to c_{i,j}$, $c_{i,k} \to c_{k,r}$, $c_{j,r} \to c_{r,k}$, and $c_{r,j} \to c_{j,i}$ are removed, as can be seen in Figure 3(a).

(R_2) if there exist switches v_j and v_k, such that they are adjacent to switch v_i, where $k > i < j$ and $k < j$, the dependency $c_{k,i} \to c_{i,j}$ is removed, as can be seen in Figure 3(b).

(R_3) if there exist switches v_j and v_k, such that they are adjacent to switch v_i, where $k > i < j$, $k < j$, and all of them form part of the same cycle, the dependency $c_{j,i} \to c_{i,k}$ is removed, as can be seen in Figure 3(c).

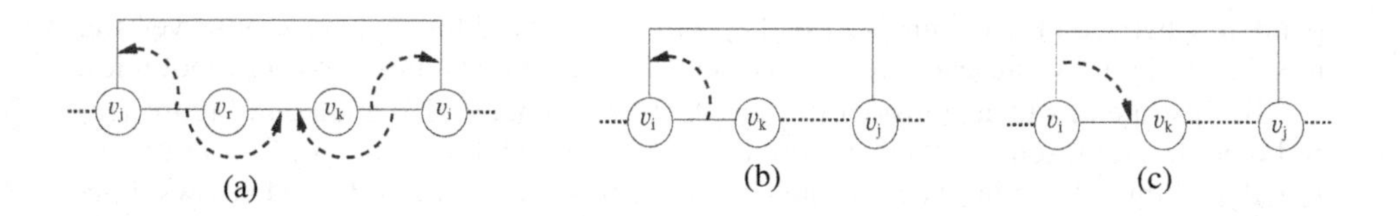

Fig. 3. Channel dependencies removed by applying (a) rule (R_1), (b) rule (R_2), and (c) rule (R_3).

The new routing scheme is deadlock-free, since the rules (R_2) and (R_3) guarantee that all the cycles in the network graph are broken. Note that some cycles in the network graph could be broken by the rule (R_1). However the aim of this rule is to remove additional channel dependencies in order to achieve better traffic balancing.

Nevertheless, notice that the network may become disconnected after applying rule (R_1). This is because, unlike rules (R_2) and (R_3), rule (R_1) may remove channel dependencies between channels belonging to the DFS spanning tree. Therefore, after applying

[2] The unidirectional channel that links the switch v_i to the switch v_j is denoted as $c_{i,j}$. Similarly, the unidirectional channel that links the switch v_j to the switch v_i is denoted as $c_{j,i}$.

rules (R_1), (R_2), and (R_3) to every switch in the network, we must apply the following rule to guarantee that the network remains connected:

(R_4) if there is no routing path to reach switch v_j from switch v_i, and: (a) $i > j$, restore the dependencies $c_{k+1,k} \rightarrow c_{k,k-1}$, $\forall k : i - 1 > k > j + 1$. (b) $i < j$, restore the dependencies $c_{k+1,k} \rightarrow c_{k,k-1}$, $\forall k : j + 1 < k < m - 1$, where m is the label associated to switch v_m, which in turn is adjacent to switch v_n, $i \leq n < j < m$. Note that the restored dependencies are always established between channels belonging to the DFS spanning tree.

Once all the switches in the network have been visited, the routing tables will be filled with all the shortest paths between every pair of switches.

4 Performance Evaluation

In this section, we evaluate by simulation the performance of the new routing algorithm proposed in Section 3 (FX_DFS). For comparison purposes, we have also evaluated the $up^*/down^*$ routing algorithms based on both BFS and DFS spanning tree, as described in Section 2 (UD_BFS and UD_DFS, respectively). Also, dimension-order routing (XY) has been evaluated for regular networks, such as meshes. In order to obtain realistic simulation results, we have used timing parameters for the switches taken from a commercial network. We have selected Myrinet because it is becoming increasingly popular due to having very good performance/cost ratio.

4.1 Network and Switch Model

Irregular network topologies have been generated randomly. Network sizes of 8, 16, 32, and 64 switches have been evaluated. We have generated ten different topologies for each network size analyzed. The maximum variation in throughput improvement of FX_DFS routing with respect to UD_DFS routing is not larger than 10%. Results plotted in this paper correspond to the topologies that achieve the average behavior for each network size. Also, we have generated regular networks, like 2-D meshes and 2-D tori, with network sizes of 4×4 and 8×8 switches. For space reasons only some significantly results are plotted.

For both regular and irregular networks, we assume that every switch in the network has 8 ports, using 4 ports to connect to workstations and leaving 4 ports to connect to other switches. For message length, 32-flit and 512-flit messages were considered. A uniform message destination distribution has been used.

The path followed by each message is obtained using table-lookup at the source host, very much like in Myrinet networks. Therefore, deterministic source routing is assumed. Wormhole switching is used. Flits are one byte wide and the physical channel is one flit wide.

4.2 Simulation Results

Figures 4(a) and 4(b) show the average message latency versus accepted traffic for both 8×8 tori and mesh, respectively. As can be seen, for 512-flit messages FX_DFS achieves a noticeable improvement in both latency and throughput with respect to

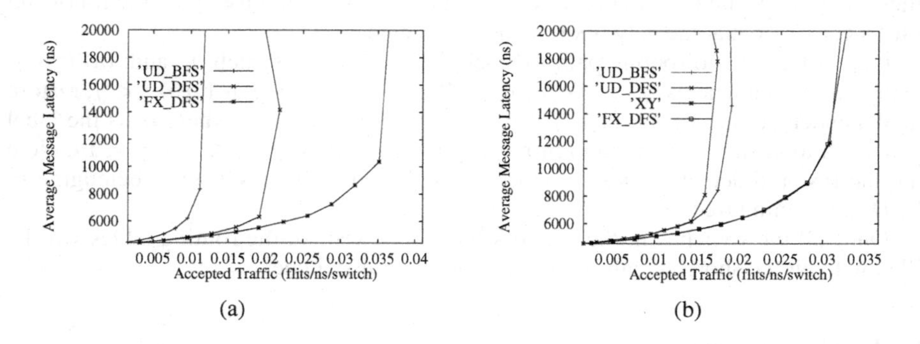

Fig. 4. Average message latency vs. accepted traffic. 8×8 (a) torus and (b) mesh. Uniform distribution. Message length is 512 flits.

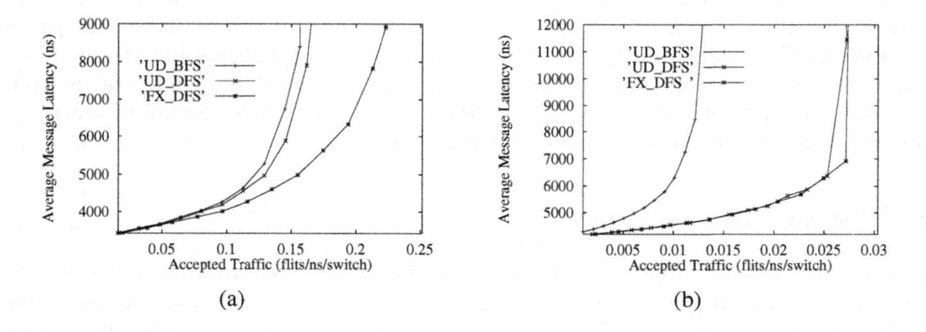

Fig. 5. Average message latency vs. accepted traffic. Irregular topology of (a) 8 and (b) 64 switches. Uniform distribution. Message length is 512 flits.

UD_BFS. Throughput improvement achieves a factor of 3.5 for a 8×8 torus. Moreover, FX_DFS achieves the same behavior than the XY for 2D meshes and increase significantly the performance with respect to UD_BFS and UD_DFS. In a 2-D torus and mesh networks, all of the evaluated routing strategies allow all messages to be routed through minimal paths. Therefore, the difference in performance between them can only be due to the differences in traffic balancing achieved by them. We can observe that better traffic balancing can be obtained by using $up^*/down^*$ routing based on a DFS spanning tree. Moreover, traffic balancing in the network can be improved by removing channel dependencies as proposed in Section 3 (FX_DFS).

On the other hand, the poor behavior of UD_BFS is due to the fact that it tends to concentrate traffic in the channels close to the root switch of the BFS spanning tree. Therefore, these channels become the bottleneck of the network, leading to saturation at relatively low traffic, especially in large networks.

On the other hand, Figures 5(a) and 5(b) show the average message latency versus accepted traffic for irregular networks with 8 and 64 switches, respectively. Message size is 512 flits. In general, we can observe that the behavior of FX_DFS is almost identical to that of UD_DFS for large networks. As shown in [8], the improvement in performance of UD_DFS with respect to UD_BFS is mainly due to the fact that UD_DFS allows most messages to follow minimal paths, especially in large networks. However, for very small networks, FX_DFS significantly outperforms UD_DFS. The improvement in throughput of FX_DFS with respect to UD_DFS is about 20%. This is only due to the fact that FX_DFS achieves better traffic balancing. Note that UD_DFS hardly improves performance with respect to UD_BFS, because in small networks it is likely that messages follow minimal paths.

5 Conclusions

In this paper, we have proposed a new routing scheme for NOWs that, unlike $up^*/down^*$ routing strategies, is effective in both regular and irregular network topologies. Like the improved $up^*/down^*$ routing scheme recently proposed in [7], the new routing algorithm is also based on computing a DFS spanning tree to break cyclic dependencies. However, the removal of channel dependencies is performed in a more flexible way than when applied to $up^*/down^*$ routing, allowing us to achieve better traffic balancing in most network topologies. Moreover, the new routing algorithm does not require new resources to be added to the network. Only routing tables need to be updated.

Evaluation results modeling a Myrinet network under uniform traffic show that the new routing algorithm significantly outperforms $up^*/down^*$ routing strategies when it is applied to regular and small irregular networks. In particular, the proposed routing scheme improves throughput with respect to the original $up^*/down^*$ routing scheme by a factor of up to 3.5 for 2-D tori. Moreover, for 2-D meshes, the new routing algorithm exhibits the same behavior as the dimension-order routing algorithm (DOR), improving throughput with respect to the original $up^*/down^*$ routing strategy.

References

1. N. J. Boden et al., Myrinet - A gigabit per second local area network, *IEEE Micro*, vol. 15, Feb. 1995.
2. L. Cherkasova, V. Kotov, and T. Rockicki, Fibre channel fabrics: Evaluation and design, *29th Hawaii International Conference on System Sciences*, Feb. 1995.
3. W. J. Dally and C. L. Seitz, Deadlock-free message routing in multiprocessors interconnection networks, *IEEE Transactions on Computers*, vol. C-36, no. 5, pp. 547-553, May. 1987.
4. D. García and W. Watson, Servernet II, in *Proceedings of the 1997 Parallel Computer, Routing, and Communication Workshop*, June 1997.
5. W. Qiao and L. M. Ni., Adaptive routing in irregular networks using cut-trough switches, in *Proc. of the 1996 Int. Conf. on Parallel Processing*, Aug. 1996.
6. R. Riesen et al., "CPLANT", in *in Proc. of the Second Extreme Linux Workshop*, June 1999.
7. J.C. Sancho and A. Robles, Improving the Up*/Down* Routing Scheme for Networks of Workstations Accepted for publishing in *Euro-Par 2000*, Aug. 2000.
8. J.C. Sancho, A. Robles, and J. Duato, New Methodology to Compute Deadlock-Free Routing Tables for Irregular Networks, in *in Proc. of CANPC'2000*, Jan. 2000.
9. M. D. Schroeder et al., Autonet: A high-speed, self-configuring local area network using point-to-point links, *SRC research report 59*, DEC, Apr. 1990.
10. R. Sheifert, *Gigabit Ethernet*, Addison-Wesley, Apr. 1998.
11. F. Silla and J. Duato, Improving the Efficiency of Adaptive Routing in Networks with Irregular Topology, in *1997 Int. Conference on High Performance Computing*, Dec. 1997.

Java Bytecode Optimization with Advanced Instruction Folding Mechanism

Austin Kim, Morris Chang

Dept. of Computer Science, Illinois Institute of Technology, 10 West 31st st., Chicago, IL. 60616
austin.kim@lucent.com, chang@charlie.iit.edu

Abstract. The execution performance of Java has been a problem since it was introduced world wide. As one of the solutions, a bytecode instruction folding process for Java processors was developed in a PicoJava model and a Producer, Operator and Consumer (POC) model. Although the instruction folding process in these models saved extra stack operations, it could not handle certain types of instruction sequences. In this paper, a new instruction folding scheme based on a new, advanced POC model is proposed and demonstrates improvement in bytecode execution. The proposed POC model is able to detect and fold all possible instruction sequence types, including a sequence that is separated by other bytecode instructions. SPEC JMV98 benchmark results show that the proposed POC model-based folder can save more than 90% of folding operations. In addition, a design of the proposed POC model-based folding process in hardware is much smaller and more efficient than traditional folding mechanisms. In this research, the proposed instruction folding technique can eliminate most of the stack operations and the use of a physical operand stack, and can thereby achieve the performance of high-end RISC processors.

1. Introduction

It has already been more than four years since Java was introduced and widely adopted in the Internet arena as well as in consumer electronics and communication systems. Java started with good initiatives, and it was well equipped with many promising features for its future, but it now faces a few well-known problems, especially in performance.

There have been several attempts to implement the Java Virtual Machine (JVM) in a hardware chip. PicoJava [1] and JEM1 [2], for instance, have been released as commercial products. A smaller-scale version of JVM in a Field Programmable Gate Array (FPGA) chip has been introduced for research purposes [3]. These hardware approaches, with direct implementation of JVM, have clearly improved Java performance, since there is no bytecode interpretation process involved. Although there are many aspects of bytecode optimization as proposed in [4], the optimization of stack operation is the most practical approach of all.

The underlying operation of Java bytecode instructions is purely stack oriented. Due

M. Valero et al. (Eds.): ISHPC 2000, LNCS 1940, pp. 268-275, 2000.

to inefficiencies of index addressing on the operand stack, the Java processors execute 15% to 30% more instructions, compared to register-based processors [5]. The problem can be eliminated by a process called "instruction folding." The instruction folding process combines several stack manipulation instructions into a single register-based instruction, resulting in less CPU cycles for the same amount of work.

In this paper, a new instruction folding algorithm based on an advanced Producer, Operator and Consumer (POC) model is proposed. New and improved features of the model will be introduced and explained in detail. Also, the hardware implementation of the folding unit will be illustrated, and its performance, compared to traditional folding units, will be shown in the results section.

SPEC JMV98 benchmark programs [6] were selected to collect bytecode data to analyze the folding behaviors. In general, over 90% of the folding operations can be eliminated by a new POC model-based folding technique, as shown later in the experimental result section.

2. Related Work

The instruction folding technique was first introduced in the PicoJava I architecture [7] in 1996. According to the simulation results, the PicoJava I folding operation could fold up to 60% of the stack operations [8]. Research shown in [4] applied similar folding techniques as PicoJava II, and illustrated more advanced folding groups and checking patterns. Finally, a POC model technique based on the characteristics of bytecode instructions has been developed and demonstrated in [9].

3. Traditional Instruction Folding Techniques

The folding techniques demonstrated by [4] and [5] are based on the instruction groups and the group sequence patterns. In PicoJava II, six groups of bytecode instructions are identified: Non Foldable (NF), Local Variable (LV), Memory (MEM), One Operand (BG1), Two Operands (BG2), and Two Operands with MEM (OP). The decode unit issues 74-bit micro-operations to the execution engine through the operand fetch unit. These bytecode instructions are converted into micro-operations by the decode unit, which then looks for folding patterns up to four consecutive instructions long. Once a pattern is found, the corresponding micro-operation replaces the original instructions. However, folding will not work if the operands are too far from the top of the stack, which results in low foldability of the folding unit [5].

A folding technique introduced by [9] categorized all the bytecode instructions into POC types and subtypes. The sequence of an instruction stream is scanned through a highly sophisticated folding-rule checker to determine its foldability. The folding-rule checker unit generates four output status signals: Serial Instruction (SI), Foldable Instruction (FI), Continuing state (C), and Ending state (E). The sequences of instruction types that the checker receives are combinations of producer-type instructions (P), four different types of operators (O_E, O_B, O_C, and O_T), and consumer-types of instructions. This folding technique uses a 4-foldable strategy to simplify the decoder. The hardware structure of the folding unit is scalable (two to n foldable), and each unit consists of 26 logic gates and 3 logic levels. Therefore, to build 4-foldable logic circuitry, a designer needs at least 96 logic gates, which would take 12 logic levels in the worst-case timing path. Although the POC model can fold more instructions than PicoJava, the folding unit gets more complicated and can't handle many types of instruction sequences.

Currently, all of the folding algorithms described above detect and fold a stream of instructions passively. This frequently results in losing many foldable instructions, which is due to the un-optimized instruction sequence rather than a deficiency in the folding logic. A new proposed POC model-based folding mechanism can manipulate the instruction sequences properly to maximize the folding logic algorithm with less hardware and simpler design.

4. New Advanced Folding Model

4.1. Characteristics of Instruction Sequences

In order to develop an instruction folding mechanism, it would be essential to understand the nature of instruction sequences and their foldable types. It would be an ideal situation if a foldable instruction sequence is followed by another foldable sequence. However, many times foldable instructions are separated by other bytecode instructions. Five distinct types of relationships between a foldable instruction group and its adjacent instructions have been found, as shown in Figure 1. The first type, namely *Normal Sequence*, is a sequence of two normal, consecutive-foldable groups. Most existing instruction folding units can fold this sequence type without any difficulties.

Types I, II, III, and *IV* show special types of sequences that ordinary folding algorithms couldn't detect or fold completely. These extra types show typical cases, which indicate that perfectly foldable instructions can be separated by a complete folding-instruction group or by non-foldable instructions. The foldability would be largely increased by finding all foldable patterns, even if the bytecodes in the foldable patterns are separated by other bytecodes.

In *Type I*, a load instruction, "iload_2," in group B, is separated by group A. The "iload_2" must be saved in an instruction queue and retrieved later to complete the folding process of group B. After execution of group A, the broken instructions in group B are then combined with the queued instruction and executed.

Type II shows a little bit more complicated instruction sequence. Notice that the value in location 3 is pushed onto a stack by "iload_3" in group B, and the same location 3 is written back by "istore_3" in group A during a normal stack operation. There exists a single dependency between group A and B. In this case, group B should be executed first, followed by the execution of group A to maintain a correct value in location 3.

Type III is an example of a double dependency between groups A and B. Location 3 is read by group C and written back by group A, while location 8 is read by group A and written back by group D. In this case, folding C with group D can't give correct values in both locations 3 and 8. Therefore, instruction C, "iload_3," and group D should be executed separately for correct operation.

The last case, *Type IV*, is a variation of *Type I*. An outcome of group A is folded with a previously queued instruction, "iload_3," to form a new group B. Then, group B subsequently forms another foldable group C, and so on. The results of these folding types are pushed onto the stack in normal stack operations. However, in this research, these arithmetic results are kept in temporary registers or accumulators to avoid extra stack operations.

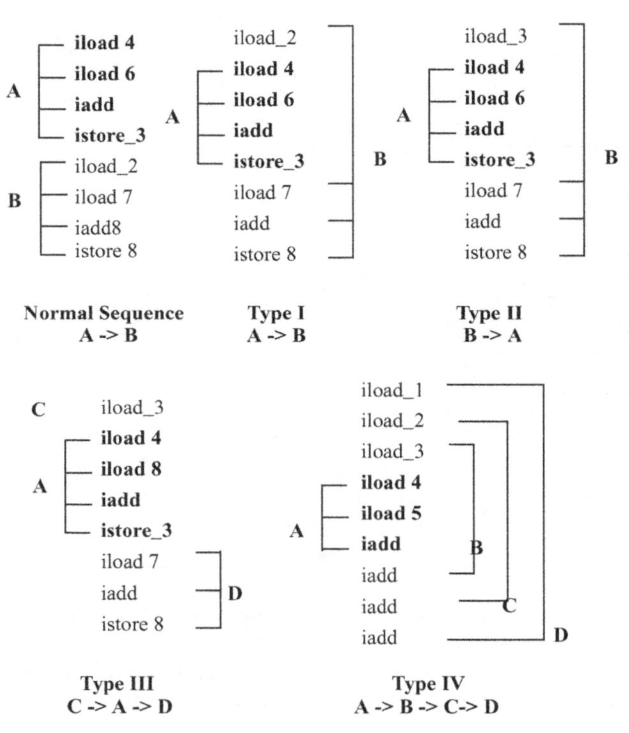

Fig. 1. Instruction Folding Sequence Types

4.2. New Advanced POC Models

A new POC model uses three major stack instruction types in the folding process, Producers, Operators, and Consumers, same as the previous POC model. These types are assigned to each instruction based on the bytecode instruction characteristics. Unlike the old POC model, the proposed POC model introduces new operator types, which are Producible Operator (Op) and Consumable Operator (Oc). Results of bytecode operations always become either producible or consumable types. These two new Operator types replace old Operators, O_E, O_B, O_C, and O_T.

This new POC model-based folding technique aggressively scans an instruction stream and attempts to locate the sequence type, as previously shown in Figure 1. If the sequence type is found, the folding unit combines foldable instructions and executes the sequence on the fly. Otherwise, the unit marks the instructions as a broken sequence in an instruction queue until they are popped and combined with the subsequent pieces of the sequence. Figure 2 illustrates a complete flow of this folding process. The bold-lined boxes and transitions indicate the portion of the proposed POC algorithm, which is not found in the traditional folding algorithms. Operations of *Type I* and *IV* sequences are shown in this flow chart. A bold-oval state in Figure 2 represents a dependency checker used for *Type II* and *III* sequences.

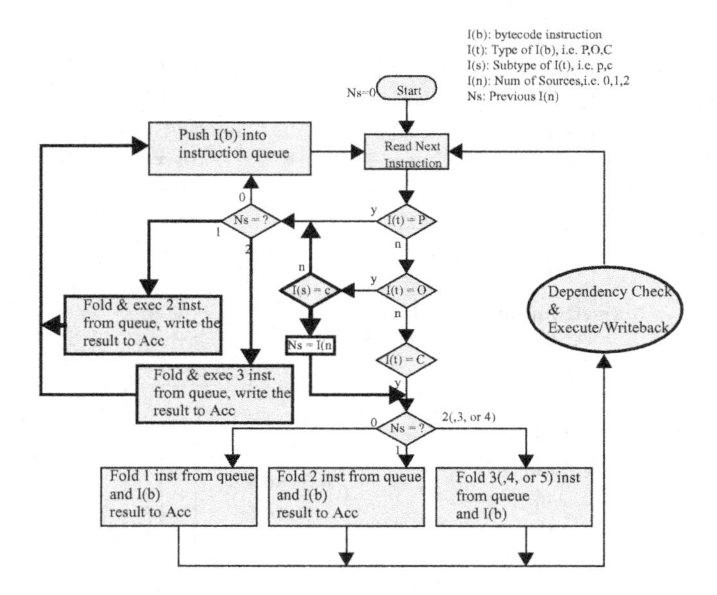

Fig. 2. New POC Folding Model Algorithm

As it will be shown in the results section, *Type II* and *Type III* rarely exist in normal bytecode sequences according to the benchmark testing statistics, which might be due to the enhancement of the Java compiler's code optimization. However, there are a number of Java bytecode assemblers, e.g. Jasmin [11], and recent custom-generated class files that produce un-optimized bytecode structures. The dependency checker algorithm has been designed and presented for these cases, and can be implemented in the folding unit optionally.

5. Experimental Results And Analysis

Data collection is made from a simple Java program and SPEC JMV98 benchmark programs [6] using trace-driven analysis of run-time bytecode instructions. A simple Java program, "addsum2," executes a loop to add integers from zero to ten. The SPEC benchmark programs selected for this experiment are "db", "compress", and "javac." The "db" program performs multiple database functions on a memory-resident database with 1.2 million bytecodes executed. The "compress" program compresses and decompresses zip files in a high-compression ratio with 3.6 million bytecodes. Finally, "javac" is the Java compiler from the JDK 1.0.2, which executes about 6.9 million bytecodes.

The proposed POC model-based algorithm composes instruction patterns based on combinations of the instruction types, i.e. Producer (P), Producible Operator (Op), Consumable Operator (Oc), and Consumer (C).

The new POC model-based instruction folder is designed to detect the sequence types as shown in Figure 1 in Section 4. The sequence types and the distributions in the benchmark programs are shown in Table I. The first numbers in the table are numbers of the folding types detected in the particular benchmark applications, and the numbers in parentheses are their percentages. Notice that *Types II* and *III* rarely appear in the programs. Implementation of these types is not necessary for normal JDK-compiled Java programs. Therefore, hardware implementation of the dependency checker can be optionally omitted from the instruction folding design. The table also indicates that 87% to 93% of foldable instructions are found across the applications. This research mainly focuses on instruction sequences that are foldable.

TABLE I Distribution of Sequence Types in Benchmarks

	ADDSUM2	DB	COMPRESS	JAVAC
NORMAL	70,404 (44.6%)	443,216 (36.5%)	442,810 (12.5%)	2,226,298 (32.3%)
TYPE I	18,390 (11.7%)	207,836 (17.1%)	1,099,156 (30.9%)	963,736 (14.0%)
TYPE II	16 (0%)	61 (0%)	176 (0%)	212 (0%)
TYPE III	0 (0%)	11 (0%)	39 (0%)	76 (0%)
TYPE IV	47,499 (30.1%)	455,310 (37.5%)	1,749,735 (49.2%)	2,922,121 (42.4%)
NON-FOLD	21,436 (13.6%)	108,998 (9.0%)	264,113 (7.4%)	785,272 (11.4%)
TOTAL	157,745	1,215,432	3,556,029	6,897,715

Unlike the traditional models, by detecting and folding a broken sequence, the new POC model-based folder is able to find more foldable instructions than the traditional folding mechanisms. Thus, the foldability improved dramatically in the proposed POC model-base folding mechanism. Figure 3 is a graphical representation of foldability of bytecode instructions between the new POC model, the old POC model, and PicoJava-based folding mechanisms for the benchmark programs. As shown in the data, the new POC model-based folder can consistently save the folding operations more than 90% throughout the applications. If the "invoke*" instructions are handled by a microcode operation, the foldability will be close to 100%. However, the old POC model and PicoJava folding achieved, at best, only 70% and 50% savings, respectively. Unlike the new POC model-base technique, the foldabilities of the old POC and PicoJava vary depending on the applications, e.g., very low foldabilities in "compress."

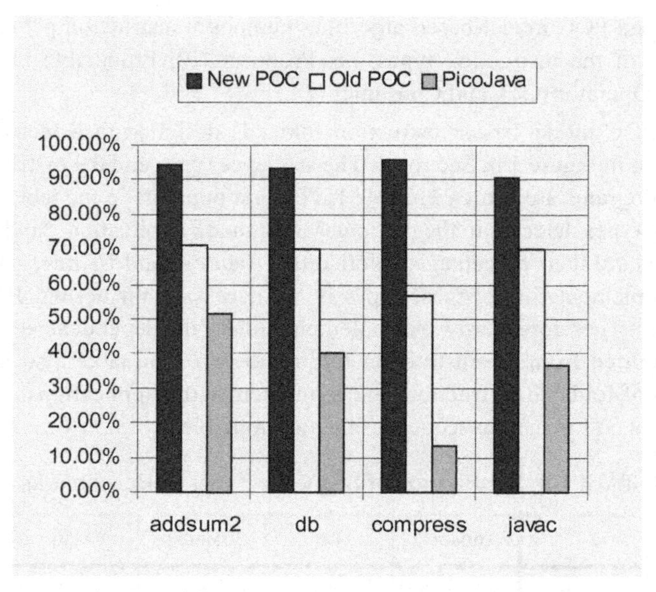

Fig. 3. Folding Operations Saved By Folding Tech-

6. Implementation of Instruction Folding

Designing a complete instruction folding unit is beyond the scope of this research. The research, however, focuses on the implementation of instruction decoding and pattern recognition units. Since the new POC model uses only four Operator types, a design of the type decoder is easier and simpler than the previous folding mechanisms.

A folding pattern recognizer is designed based on the six different folding groups described in the previous section. The instruction decoding logic and the foldability checker are mapped to generic gates. The unit can decode up to six contiguous bytecode instructions to determine the foldability. This unit is also able to detect all possible instruction folding patterns for most of the applications.

According to the hardware synthesis data, total logic resources used for the six foldable instruction pattern unit are 25 primitive gates with six logic levels. This shows considerable improvement over the old POC model's folding unit, which requires 96 logic gates in 12 logic levels for only four foldable instruction patterns.

Alternatively, the unit can be much more simplified, in case a designer wants to design a four-foldable unit by sacrificing instruction foldability. The foldable unit only requires 14 logic gates with four logic levels.

7. Future Work

This research presents a theoretical approach to folding mechanisms and partial implementations of the major units. However, complete hardware implementation of entire folding units, including instruction queues and register units, is needed to measure and understand the overall performance enhancements to Java processors. An effort to port the new POC model-based folding structure to an existing Java FPGA processor is currently in progress.

In addition, the proposed new POC model-based folding technique might be a good reusable software asset to be applied to Java compilers or Java class loaders. A research project that implements this POC model in a Java static class loader is now near completion.

8. Conclusion

In order to overcome an inherent performance problem with Java, an enhanced instruction decoding and folding process needs to be developed for Java hardware processors. In this paper, a new approach has been proposed by introducing an advanced POC model-based instruction folding technique.

The characteristics of the bytecode sequences and the patterns have been clearly verified and analyzed. Various types of bytecode sequence patterns have been developed and studied. Combining broken bytecode sequences dramatically improves the foldability. Through a series of experiments using SPEC JVM98 benchmark programs, a proposed POC model-based folder was shown to yield a very high foldability that consistently achieves 90% across the applications, while the previous POC model and PicoJava folding achieved only as much as 70% and 50% savings, respectively. Also, the proposed instruction folder is simpler and more efficient to implement in hardware. Currently, an effort to apply this new technique to an existing Java processor is under way.

References

[1] T. Halfhill, "Java Chips Boost Applet Speed," *Byte*, pp.25, May 1996

[2] A. Wolfe, "First Java-specific chip takes wing," *EETimes,* Sep. 1997. (http://eet.com/news/97/973news/java.html)

[3] A. Kim and J. M. Chang, "Designing a Java Microprocessor Core using FPGA Technology," *Proceedings of IEEE International ASIC Conference*, pp. 13-17, Sep. 1998

[4] A. Kim, Q. Yang and J. M. Chang, "Static Java Class Loader for Embedded Java Architecture," *Proceedings of CASES99 Conference*, pp. 129-133, Sep. 1998

[5] H. McGhan and J. M. O'Connor, "PicoJava: A Direct Execution Engine For Java Bytecode," *Computer*, pp. 22-30, Oct. 1998

[6] SPEC JVM98 Benchmarks, "Release version 1.03," *www.spec.org/osg/jvm98*, 1998

[7] Sun Microsystems, "PicoJava I Microprocessor Core Architecture," *White Paper*, Nov. 1996

[8] J. M. O'Connor and M. Tremblay, "PicoJava I: the Java Virtual Machine in Hardware," *IEEE Micro*, pp 45-53, 1997

[9] L. Ton, L. Chang, M. Kao, H. Tseng, S. Shang, R. Ma, D. Wang, and C. Chung, "Instruction Folding in Java Processor," *Proceedings of the International Conference on Parallel and Distributed Systems - ICPADS,* 1997, pp. 138-143

[10]N. Vijaykrishnan, N. Ranganathan and R. Gadekarla, "Object-Oriented Architectural Support for a Java Processor," *Proceedings of ECOOP'98*, July 1998, pp. 330-354

[11]J. Meyer and T. Downing, "Java Virtual Machine," *O'Reilly & Associates*, 1997

Performance Evaluation of a Java Based Chat System

Fabian Breg, Mike Lew, and Harry A. G. Wijshoff

Leiden Institute of Advanced Computer Science, Leiden University
Niels Bohrweg 1, 2333 CA Leiden, the Netherlands
{breg,mlew,harryw}@liacs.nl

Abstract. This article describes the LuChat chat system, which is a specific instance of a web based collaborative white-board style of application. To increase portability of our system among the most popular browsers, we implemented our own (subset of) the Java RMI framework, which requires no installation effort on the client side and still runs in both Netscape and Internet Explorer. The performance of our RMI system is comparable to the Java RMI version that comes with Netscape's Java virtual machine. The response time of our complete system under light load is under 30 ms, with the two most popular browsers having a response time of under 15 ms. Under normal use, our system will scale to a high number of clients.

1 Introduction

The Java programming language [5] has evolved from a language mainly used to liven up internet pages to an all-purpose programming language. We feel, however, that Java's main attraction lies in its inherent secure distributed nature, making it the language of choice for a wide variety of web based applications. The language offers powerful communication primitives like Remote Method Invocation RMI [11,9], while the virtual machine itself allows dynamic distribution of object code in a secure way.

In this paper, we introduce the LuChat chat system, which is an example of a collaborative application. The LuChat system consists of a chat server, implemented as a Java application, and a chat client, which is implemented as a Java applet, which runs in the most popular browsers without requiring any installation effort on the client side. We believe that the latter is a requirement essential for broad acceptance of any web based application.

We will discuss the design and implementation of the LuChat system in Section 2. This section will also introduce LuchatRMI, which is a replacement for Java RMI, developed to increase the portability of our system to different web browsers. We will devote the largest part of this article to the assessment of the performance of our system.

M. Valero et al. (Eds.): ISHPC 2000, LNCS 1940, pp. 276–283, 2000.

2 Design and Implementation

The LuChat system is a web based client-server application. The Chat Server and all the Chat Rooms run in one multithreaded process. The client components run as an applet inside a web browser on the client host. The chat server maintains the list of chat rooms, while each chat room manages activities taking place in these rooms.

To implement communication in the LuChat system, we chose to use Java RMI, since it provides a simple communication model. RMI allows communication between Java objects located in different Java Virtual Machines in a seamless way. Using Java RMI in a web based application does create a number of problems, however.

The first problem encountered in communication between server and client applets is the tight security restrictions imposed on applets running in web browsers. Applets are only allowed to initiate communication with the host that they are downloaded from. To deal with this problem without requiring any effort on the client side, all communication must be initiated by the clients, to the server. This makes it impossible to implement an active broadcast from the server to the clients in an efficient way, for instance by using a spanning tree algorithm.

We chose to have clients continuously poll the chat room for new events in a separate thread. Since the server now has to communicate every event to every client individually, the server becomes a bottleneck. To alleviate this problem, we try to minimize communication, by combining multiple events in a message sent to the clients.

A more severe problem is the poor support for Java RMI in certain web browsers. There do exist a number of alternative RMI implementations. KaRMI [8] was developed to be a more efficient alternative to Java RMI. Together with their improved object serialization [7], KaRMI outperforms Java RMI. NinjaRMI [6] was developed as part of the Ninja project and aimed to improve the performance of Java RMI as well as to extend its capabilities. NexusRMI [3] is an implementation of RMI built on top of NexusJava [4], which allows interoperability [2] with HPC++ [1]. Manta [10] implements a very efficient RMI by compiling the Java code to native code. None of these alternative facilitates distribution of the RMI runtime with the applet code.

Therefore, we implemented our own (subset of) RMI, called LuchatRMI. The LuchatRMI runtime system can be sent to the clients with the applet code, increasing the download size of our chat system with 6.5 Kb. LuchatRMI is built directly on top of Java TCP/IP sockets. The latency of a method invocation without parameters and return value in LuchatRMI is 6.5 ms, against 1.9 ms for standard JavaRMI. In this case, each method invocation sets up a new connection between stub and skeleton. If way change the stub and skeleton to reuse a previous connection, the latency for LuchatRMI drops to 0.6 ms. Figure 1 compares the throughput for different data sizes. The sizes specifies the size of the parameter as well as the size of the return value. Again, we see that reusing connections significantly improves performance.

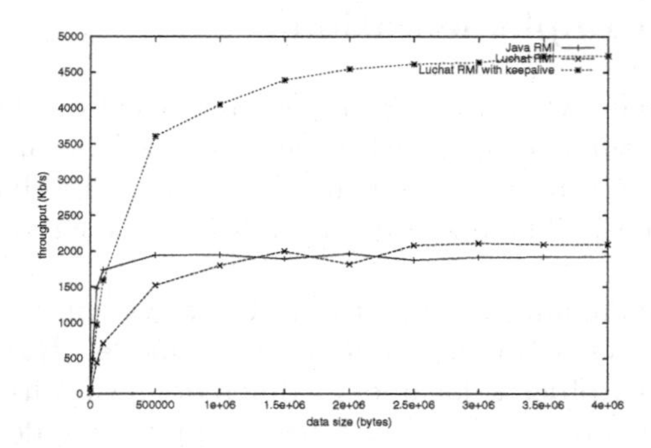

Fig. 1. Throughput comparison

3 Performance

We conducted our evaluation of the chat system performance of a cluster of Pentium II workstations running Windows NT 4. The server is running on a Pentium II machine with 128 Mb of memory, running Solaris 7 Intel Edition. Clients and server are connected with a 10 Mbits/sec Ethernet. Unless indicated otherwise in the text, we ran the clients in the Java Virtual Machine incorporated in the Netscape 4 browser.

The performance is obtained by using robot chat applets, that send 'say' events at random intervals (5-10 seconds to reflect a normal interactive chatter). We report the averages over an interval of 25 events. Each experiment was run five times, in order to identify deviations in the measurements not caused by the application. Each time, the average of five designated clients are reported. These five clients each run on their own host. The rest of the clients, up to a maximum of 75, are distributed among 15 host machines, whereby we made sure that all machines were equally loaded.

3.1 Java RMI vs. LuChat RMI Performance

Figure 2a compares the response times observed by LuChat clients under varying load using LuChat RMI and Java RMI. In this case, all clients are connected to the same room.

With a low number of clients we obtain a 10 ms respond time. As can be seen, the response time under higher load is still such that it will hardly be noticed (40 ms).

We see that Java RMI is a bit slower under light load (20 ms) and equals LuChat RMI performance under higher load. With a load of 80 clients, however, the implementation using Java RMI loses approximately 10% of the chat events, whereas the implementation using LuChat RMI loses less than 1% of the chat events.

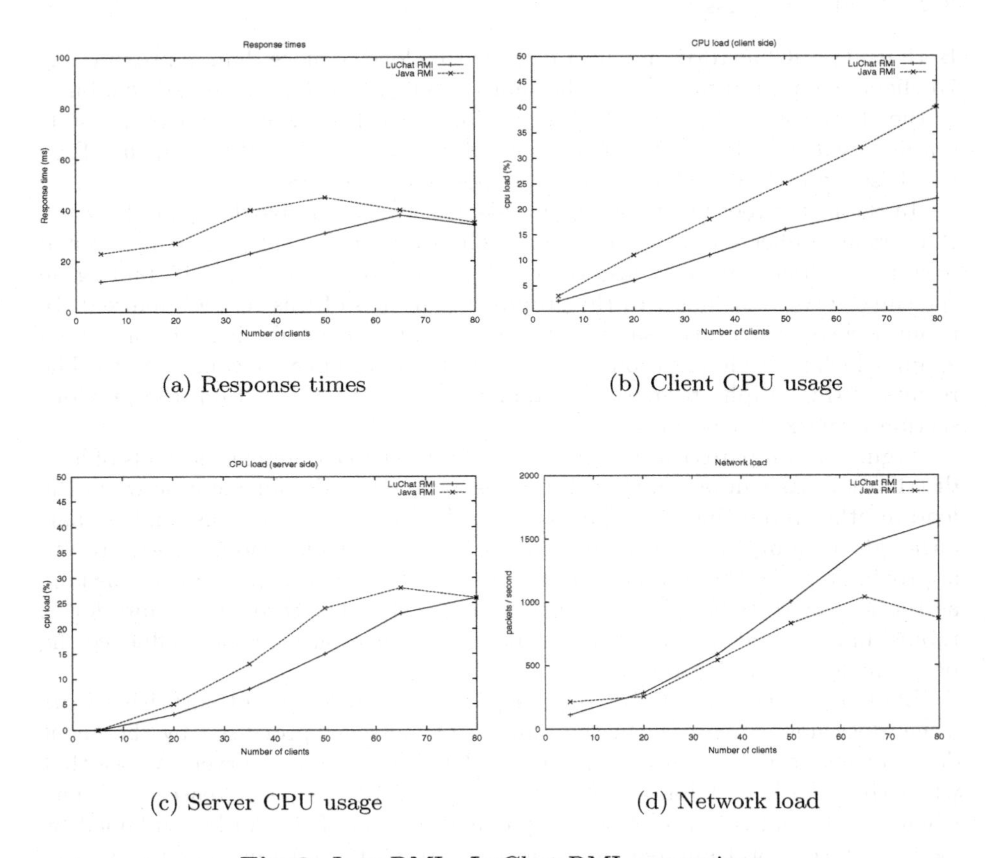

(a) Response times

(b) Client CPU usage

(c) Server CPU usage

(d) Network load

Fig. 2. Java RMI - LuChat RMI comparison

To assess what causes the difference in performance, we compare the cpu usage at both the client and the server side. Figure 2b compares the cpu usage on the client side for LuChat and Java RMI.

As can be seen, the client cpu usage when using Java RMI is a factor 1.5 higher than when using LuChat RMI. Also, on the server side, the cpu usage on high loads is higher when using Java RMI than when using LuChat RMI, as can be seen from Figure 2c. Remember that the implementation using Java RMI has a high event loss at 80 clients, which can also cause the server cpu usage to be lower, since it does not broadcast all events.

We also measured the network load using the standard 'snoop' utility from UNIX. This utility captures all incoming TCP packets and counts them. Figure 2d compares the network load when using LuChat RMI and Java RMI.

We see that the network load when using Java RMI is lower for higher loads than the network load for LuChat RMI. These graphs, however, depict packets per second and not bytes per second.

3.2 Scalability Issues

Both using LuChat RMI and Java RMI, with a load of more than approximately 75 clients connected to one and the same room, the performance and reliability of the chat system degrades. In this section, we will show experimental results obtained with similar load, but distributed differently. We will use LuChat RMI for these experiments, since that is the most portable version.

In the next experiments, we again take the average of five designated clients, that are all connected to the same room. The additional clients, again distributed over 15 machines, are now also distributed over 15 chat rooms (all running on the same server machine). In the graphs, the results of this test are marked '16 rooms'. Next, we run the same test, but put all additional clients in one room again, which is a different room from the one the initial five are connected to. The results of this graph are marked '2 rooms'. The original test from the previous section is marked '1 room'.

Figure 3a compares the response times for the three different methods of load distribution. As can be concluded, the response times do not increase when the load in other rooms increase. This is a significant result, since this implies that, although the number of clients connected to one particular room is restricted to approximately 75, the complete chat system scales to many more clients, as long as the load is distributed over multiple rooms. Furthermore, the results for '2 rooms' implies that a heavy load in one particular room does not influence the performance of another room.

Looking at the client CPU usage depicted in Figure 3b, not surprisingly, its usage does not increase when the number of messages processed by the client does not increase. Figure 3c compares the CPU usage at the server. We see that when the load is distributed of 16 rooms, the CPU usage increases less than when a high load is imposed on one particular room. This can be explained by looking at the total number of client requests that the server has to process. In all cases, the same number of events are generated by the clients. However, when the load is distributed over 16 rooms, each event has to be sent to only five clients, instead of to all clients. As can be expected, this is also reflected by the network load, as shown in Figure 3d.

Considering that in a real chat session, having a such a high load as we used in these test in one room, makes chatting clearly intractable, these results are certainly encouraging.

3.3 Platform Comparison

In this section, we will evaluate the Java performance on different operating system and Java virtual machine combinations. We will evaluate the performance of the chat system using our base setup where all clients are connected to one and the same room. Again, five clients each running alone on a certain host, will provide the actual results. Only the version with LuChat RMI will be used, since that is the most portable version. The numbers for Netscape on Windows are repeated here for comparison.

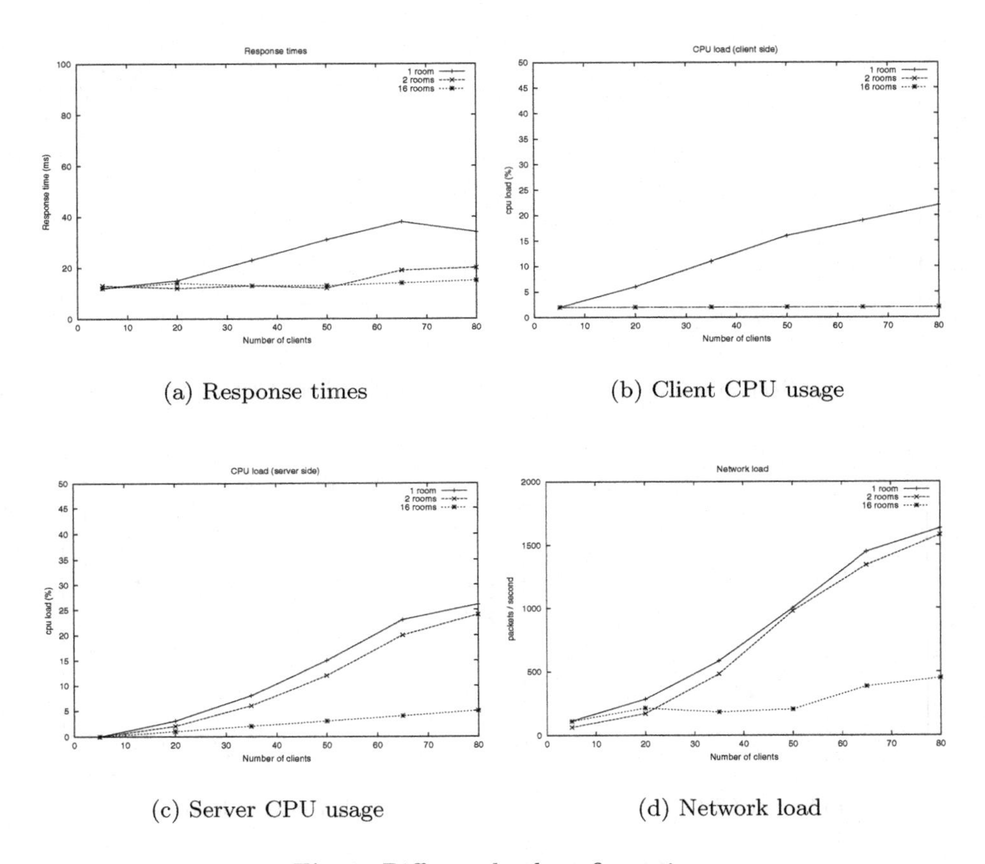

(a) Response times

(b) Client CPU usage

(c) Server CPU usage

(d) Network load

Fig. 3. Different load configurations

The first platform we will compare with is Internet Explorer 5 running on Windows NT. The next platform will be Linux, using the virtual machine that comes with Netscape 4. Linux quickly gains popularity as an alternative for Microsoft Windows, so it is interesting to see how it performs. The last platform will be Sun's own Java virtual machine that comes with the JDK 1.2. From this distribution, we use the appletviewer to run our tests.

The performance of the chat system on each of these platforms is depicted in Figure 4. As can be seen, Internet Explorer on Windows outperforms all other platforms. With Netscape on Linux we experience the highest response times, although the performance would still be acceptable for real world usage. Sun's appletviewer loses a significant amount of events (15%-20%) with a load of 65 clients, which explains why the observed response times improve. Because of the high message loss percentage, we did not bother measuring beyond 65 clients.

The amount of CPU usage at the client side could explain the difference in response times. The CPU usage on Netscape / Linux is twice as high as the CPU

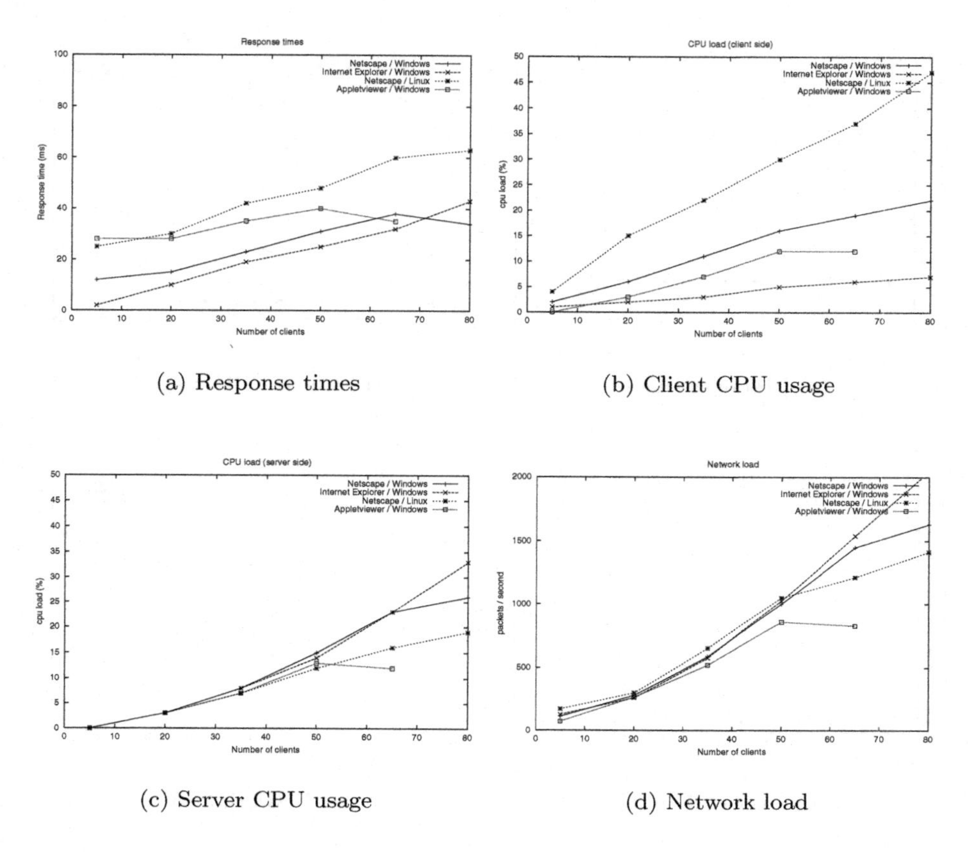

(a) Response times

(b) Client CPU usage

(c) Server CPU usage

(d) Network load

Fig. 4. Platform comparison

usage on Netscape / Windows. Internet Explorer is the least CPU demanding platform, which could explain its good performance.

The difference in CPU usage at the server could be a cause of clients not being able to keep pace. When the server detects a slow client, it will allow it to catch up by not providing all events, which also means less work for the server. Note that the difference becomes significant only under high load. The same goes for network usage. Less events sent to clients means less network usage.

4 Conclusions

In this article, we introduced our LuChat system, which is a specific instance of a collaborative world wide web based application. Since these types of applications become more important, we believe it is useful to asses the overall performance of such a system.

We also introduced the LuChat RMI system, which we developed to broaden browser support. The main attraction of our LuChat RMI system is that its runtime can be transmitted along with the application, and therefore requires no installation on the client side, making it an attractive alternative RMI system for web based applications, that need to run on a broad range of web browsers. This system is not a complete RMI implementation, but it does support basic RMI applications.

The performance of our LuChat RMI system is competitive to the RMI implementation distributed with Netscape. The performance of the overall system, at least on a Local Area Network is acceptable. We still need to assess its performance in real usage. We did, however, use the system with a very small number of users divided between the USA and Netherlands, and there the performance was acceptable.

References

1. P. Beckman, D. Gannon, and E. Johnson. Portable Parallel Programming in HPC++. 1996. 277
2. F. Breg, S. Diwan, J. Villacis, J. Balasubramanian, E. Akman, and D. Gannon. Java RMI Performance and Object Model Interoperability: Experiments with Java/HPC++. In *ACM 1998 Workshop on Java for High-Performance Network Computing*, pages 91–100, Feb 1998. 277
3. F. Breg and D Gannon. A Customizable Implementation of RMI for High Performance Computing. In *Proc. of Workshop on Java for Parallel and Distributed Computing of IPPS/SPDP99*, pages 733–747, Apr 1999. 277
4. I. Foster, G.K. Thiruvathukal, and S. Tuecke. Technologies for ubiquitous supercomputing: a Java interface to the Nexus communication system. *Concurrency: Practice and Experience*, 9(6):465–475, jun 1997. 277
5. J. Gosling, B. Joy, and G. Steele. *The Java Language Specification.* The Java Series. Addison-Wesley Developers Press, 1996. 276
6. S.D. Gribble, M. Welsh, E.A. Brewer, and D. Culler. The MultiSpace: an Evolutionary Platform for Infrastructural Services. In *Proceedings of the 1999 Usenix Annual Technical Conference*, june 1999. 277
7. B. Haumacher and M. Philippsen. More efficient object serialization. In *International Workshop on Java for Parallel and Distributed Computing*, Apr 1999. 277
8. C. Nester, M. Philippsen, and B. Haumacher. A More Efficient RMI for Java. In *ACM 1999 Java Grande Conference*, pages 153–159, Jun 1999. 277
9. Sun Microsystems. *Java(TM) Remote Method Invocation Specification*, oct 1997. revision 1.42 jdk1.2Beta1. 276
10. R. Veldema, R. van Nieuwpoort, J. Maassen, H.E. Bal, and A. Plaat. Efficient Remote Method Invocation. Technical Report IR-450, Vrije Universiteit, Amsterdam, sep 1998. 277
11. A. Wollrath, J. Waldo, and R. Riggs. Java-Centric Distributed Computing. *IEEE Micro*, 17(3):44–53, may/jun 1997. 276

Multi-node Broadcasting in All-Ported 3-D Wormhole-Routed Torus Using Aggregation-then-Distribution Strategy*

Yuh-Shyan Chen[1], Che-Yi Chen[1], and Yu-Chee Tseng[2]

[1]Department of Statistic, National Taipei University
Taipei, 10433, Taiwan, R.O.C.
chenys@chu.edu.tw
[2]Department of Computer Science and Information Engineering
National Central University,
Chungli, Taiwan, R.O.C.
yctseng@csie.ncu.edu.tw

Abstract. In this paper, we investigate the *multi-node broadcasting* problem in a all-ported 3-D wormhole-routed torus. The main technique used in this paper is based on a proposed *aggregation-then-distribution* strategy. Extensive simulations are conducted to evaluate the multi-broadcasting algorithm.

1 Introduction

A massively parallel computer (MPC) consists of a large number of identical processing elements interconnected by a network. One basic communication operation in such a machine is broadcasting. Two commonly discussed instances are: *one-to-all* broadcast and *all-to-all* broadcast, where one or all nodes need to broadcast messages to the rest of the nodes [1]. A more complicated instance is the *many-to-all* (or *multi-node*) broadcast, where an unknown number of nodes located in unknown positions each intending to perform a broadcast operation.

Saad and Schultz [3] [4] initially defined this problem and proposed a simple routing algorithm for hypercubes. A distributed approach to improve the load imbalance problem was presented by Tseng [6] for hypercubes and star graphs. However, their approach are attempted to reduce the node contention problem, which is not a congestion-free result.

This paper addresses the multi-node broadcasting problem in wormhole-routed 3-D tori. Our approach is based on an proposed *aggregation-then- distribution* strategy. The major work of this paper is to present how to develop a multi-node broadcasting using *aggregation-then-distribution* strategy in wormhole-routed 3-D tori. Given a multi-node broadcast problem with an unknown number of s source nodes located on unknown positions in an torus each intending

* This work was supported by the National Science Council, R.O.C., under Contract NSC88-2213-E-216-011.

M. Valero et al. (Eds.): ISHPC 2000, LNCS 1940, pp. 284-291, 2000.

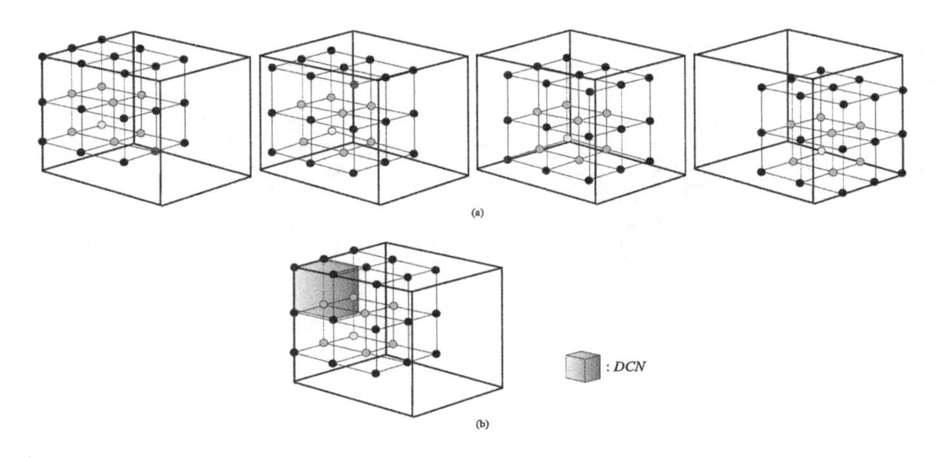

Fig. 1. (a) An example of $DDNs$, (b) DCN in a 3D torus.

to broadcast an m-byte message, our approach can solve it efficiently in time $O(\max(\lceil\log_7 n\rceil, h)T_s + \max(\lceil\log_7 \lceil\frac{n}{h}\rceil\rceil\frac{n}{h^2}m, nm)T_c)$, where h is the number of independent subnetworks. It is shown that this number has outperformed the aforementioned congestion-free scheme using edge-disjoint-spanning-trees.

2 Basic Idea

In this paper, we consider G as a 3-D torus $T_{n_1 \times n_2 \times n_3}$ with $n_1 \times n_2 \times n_3$ nodes. In 3-D tori each node is denoted as $P_{i,j,k}$, $1 \leq i \leq n_1$, $1 \leq j \leq n_2$, $1 \leq k \leq n_3$, and P_{i_1,i_2,i_3} has an edge connected $P_{(i_1 \pm 1) \bmod n_1, i_2, i_3}$ along dimension one, an edge to $P_{i_1,(i_2 \pm 1) \bmod n_2, n_3}$ along dimension two and an edge to $P_{i_1,i_2,(i_3 \pm 1) \bmod n_3}$ along dimension three. The *wormhole routing* model is assumed [2]. Under such a model, the time required to deliver a packet of L bytes from a source node to a destination node can be formulated as $T_s + LT_c$, where T_s is the start-up time and T_c represents the transmission time In addition, we adopt the *all-port* model and the *dimension-ordered routing* [6].

2.1 Network Partitioning Scheme on 3-D Torus

Consider a 3-D torus $T_{n_1 \times n_2 \times n_3}$. Suppose that there exists an integer h such that n_1, n_2, and n_3 are divisible by h. We define $h \times h$ *data-distribution network* $DDN_{u,v}$ =($V_{u,v}$,C_u)=DDN_i, u, v=0..$h-1$ and $i = u * h + v$, as follows:

$$V_{u,v} = \begin{array}{l} \{P_{i,j,l} | i = ah + ((u+v) \bmod h), j = bh + v, l = ch + u, \\ \text{for all } a = 0..\lceil\frac{n_1}{h}\rceil - 1, b = 0..\lceil\frac{n_2}{h}\rceil - 1, c = 0..\lceil\frac{n_3}{h}\rceil - 1\} \end{array}$$
$$C_{u,v} = \{\text{all channels at } x\text{-axis } ah + u + v \text{ , } y\text{-axis } bh + v \text{ and } z\text{-axis } ch + u\}$$

Each *DDN* is a *dilation-h* 3-D torus of size $\lceil\frac{n_1}{h}\rceil \times \lceil\frac{n_2}{h}\rceil \times \lceil\frac{n_3}{h}\rceil$, such that each edge is dilated by a path of h edges. Fig. 1 illustrates an example where the block

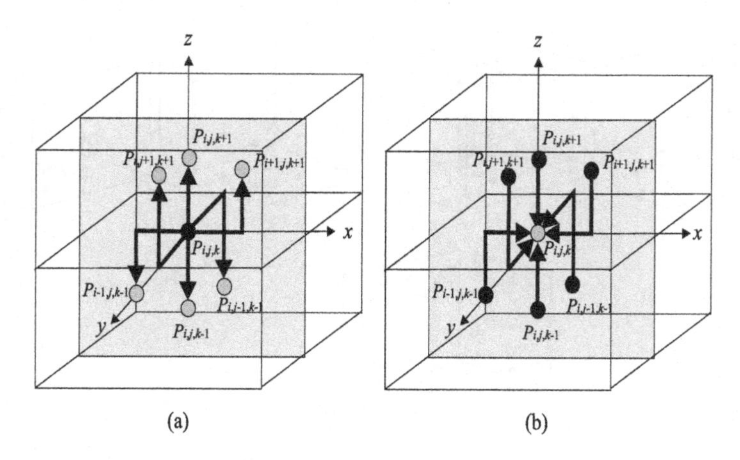

Fig. 2. An example of (a) routing matrix and (b) collecting routing matrix.

nodes denote *DDN* and the gray zone represents *DCN*. The 3-D torus $T_{n_1 \times n_2 \times n_3}$ is partitioned into $\frac{n_1 \times n_2 \times n_3}{h^3}$ *data collecting network* $DCN_k = (V_{a,b,c}, C_{a,b,c})$, $a = 0..n_1 - 1$, $b = 0..n_2 - 1$, $c = 0..n_3 - 1$, and $1 \leq k \leq \frac{n_1 \times n_2 \times n_3}{h^3}$, as follows:

$$V_{a,b,c} = \{P_{i,j,l} | i = ah + x, j = bh + y, l = ch + z, \text{ for all } x, y, z = 0..h - 1\}$$
$$C_{a,b,c} = \{\text{the set of edge induced by } V_{a,b,c} \text{ in } T_{n_1 \times n_2 \times n_3}\}.$$

2.2 Algebraic Notation

In the following, we adopt an algebraic notation to represent our routing algorithm. The torus of size n is an undirected graph. Each node is denoted as $P_{x_1,x_2,\ldots,x_k}$, $0 \leq x_i \leq n$, $1 \leq i \leq k$. Our routing algorithms is based on the concept of "span of vector spaces" in linear algebra. Conveniently, the i-th positive (resp., negative) elementary vector is denoted as $\overrightarrow{e}_i$ (resp., $\overrightarrow{e}_{-i}$) of Z^k, $i = 1..k$. We may rewrite $\overrightarrow{e}_{i_1} + \overrightarrow{e}_{i_2}$ as $\overrightarrow{e}_{i_1,i_2}$, $\overrightarrow{e}_{i_1} - \overrightarrow{e}_{i_2}$ $(=\overrightarrow{e}_{i_1} + \overrightarrow{e}_{-i_2})$ as $\overrightarrow{e}_{i_1,-i_2}$, and $\overrightarrow{e}_{i_1}$ +...+ $\overrightarrow{e}_{i_m}$ as $\overrightarrow{e}_{i_1,\ldots,i_m}$. For instance, $\overrightarrow{e}_{1,3}$=$\overrightarrow{e}_1 + \overrightarrow{e}_3$ and $\overrightarrow{e}_{1,-3}$=$\overrightarrow{e}_1 - \overrightarrow{e}_3$.

Lemma 1 *In Z^k, given a node x, an q-tuple of vectors B=($\overrightarrow{b}_1$, $\overrightarrow{b}_2$, ..., $\overrightarrow{b}_q$), and q-tuple of integer N=(n_1,n_2, ..., n_q), the span of x by vectors B and distances N is defined as*

$$SPAN(x, B, N) = \{x + \sum_{i=1}^{q} a_i \overrightarrow{b}_i \,|0 \leq a_i \leq n_i\}.$$

A 3-D tori is viewed as $SPAN(P_{0,0,0}, (\overrightarrow{e}_1, \overrightarrow{e}_2, \overrightarrow{e}_3), (n, n, n))$. We introduce two matrices: delivery routing and distance matrices. A *delivery routing matrix* $R = [r_{i,j}]_{3\times3}$ is a matrix with entries $-1, 0, 1$ such that each row indicates a message delivery; if $r_{i,j} = 1$ (resp. -1), the corresponding message will travel along the positive (resp. negative) direction of dimension j; if $r_{i,j} = 0$, the message will not travel along dimension j. A *distance matrix* $D = [d_{i,j}]_{3\times3}$ is an integer diagonal matrix (all non-diagonal elements are 0); $d_{i,i}$ represents the distance to be traveled by the i-th message described in R along each dimension.

For instance, the six message deliveries in Fig. 2(a) have three directions and thus can be represented by a delivery routing matrix:

$$R = \begin{bmatrix} e_{1,3} \\ e_{2,3} \\ e_3 \end{bmatrix} = \begin{bmatrix} 1\,0\,1 \\ 0\,1\,1 \\ 0\,0\,1 \end{bmatrix}.$$

In general, the node $p_{i,j,k}$ sends M to six nodes $p_{i+\alpha_1,j,k+\alpha_1}$, $p_{i,j+\alpha_2,k+\alpha_2}$, $p_{i,j,k+\alpha_3}$, $p_{i+\alpha_{-1},j,k+\alpha_{-1}}$, $p_{i,j+\alpha_{-2},k+\alpha_{-2}}$, $p_{i,j,k+\alpha_{-3}}$, (note that $\alpha_{\pm i} \approx \pm\frac{it}{7}$, where t is block size, see Section 3 for detail for deriving t). So we can use two distance matrices:

$$D^+ = \begin{bmatrix} \alpha_1 & 0 & 0 \\ 0 & \alpha_2 & 0 \\ 0 & 0 & \alpha_3 \end{bmatrix} \text{ and } D^- = \begin{bmatrix} \alpha_{-1} & 0 & 0 \\ 0 & \alpha_{-2} & 0 \\ 0 & 0 & \alpha_{-3} \end{bmatrix}$$

and the 6 message deliveries in Fig. 2(a) is represented by matrix multiplication:

$$D^+ \times R = \begin{bmatrix} \alpha_1 & 0 & \alpha_1 \\ 0 & \alpha_2 & \alpha_2 \\ 0 & 0 & \alpha_3 \end{bmatrix} \text{ and } D^- \times R = \begin{bmatrix} \alpha_{-1} & 0 & \alpha_{-1} \\ 0 & \alpha_{-2} & \alpha_{-2} \\ 0 & 0 & \alpha_{-3} \end{bmatrix}.$$

Further we define a similar routing matrix, namely as collecting routing matrix C. A *collecting routing matrix* $C = [c_{i,j}]_{3\times3}$ is a matrix with entries $-1, 0, 1$ such that each row indicates a path of collected message; if $c_{i,j} = 1$ (resp. -1), the corresponding message will be collected from neighboring node along the positive (resp. negative) direction of dimension j; if $c_{i,j} = 0$, the message will not be collected from neighbor along dimension j. For instance as shown in Fig. 2(b), given a collecting routing matrix

$$C = \begin{bmatrix} e_{1,3} \\ e_{2,3} \\ e_3 \end{bmatrix} = \begin{bmatrix} 1\,0\,1 \\ 0\,1\,1 \\ 0\,0\,1 \end{bmatrix},$$

then matrix multiplication

$$D^+ \times C = \begin{bmatrix} 1\,0\,1 \\ 0\,1\,1 \\ 0\,0\,1 \end{bmatrix} \text{ and } D^- \times C = \begin{bmatrix} -1 & 0 & -1 \\ 0 & -1 & -1 \\ 0 & 0 & -1 \end{bmatrix}.$$

3 Multi-Node Broadcasting in 3-D Torus

3.1 Aggregation Phase

Step 1: Diagonal-Based Data-Aggregation Operation The main function of data-aggregation operation is to regularize the communication pattern before the multi-node broadcasting. Let a 3-D torus be represented as $SPAN(P_{0,0,0}, (\vec{e}_{1,3}, \vec{e}_{1,2}, \vec{e}_1), (n,n,n))$. Each DCN is viewed as $SPAN(P_{x,y,z}, (\vec{e}_{1,3}, \vec{e}_{1,2}, \vec{e}_1), (h,h,h))$, where $0 \leq x = ih, y = jh, z = kh < n$. The data-aggregation operation is to aggregate all possible messages to a special plan in each DCN in which the plan is denoted as diagonal plan represented by $SPAN(P_{x,y,z}, (\vec{e}_{1,3}, \vec{e}_{1,2}), (h,h))$. All nodes are aggregated messages into diagonal plane $SPAN(P_{x,y,z}, (\vec{e}_{1,3}, \vec{e}_{1,2}), (h,h))$. This operation is represented by

$$C = \begin{bmatrix} 1\,0\,0 \\ 0\,1\,0 \\ 0\,0\,1 \end{bmatrix}, \; D^+ = \begin{bmatrix} t & 0 & 0 \\ 0 & 2t & 0 \\ 0 & 0 & 3t \end{bmatrix}, \; D^- = \begin{bmatrix} -t & 0 & 0 \\ 0 & -2t & 0 \\ 0 & 0 & -3t \end{bmatrix},$$

and

$$D^+ \times C = \begin{bmatrix} t & 0 & 0 \\ 0 & 2t & 0 \\ 0 & 0 & 3t \end{bmatrix} \text{ and } D^- \times C = \begin{bmatrix} -t & 0 & 0 \\ 0 & -2t & 0 \\ 0 & 0 & -3t \end{bmatrix}.$$

Lemma 2 *Diagonal-based data-aggregation operation can be recursively performed on a* $T_{n\times n\times n}$ *in* $\lceil \log_7 h \rceil T_s + \sum_{i=0}^{\lceil \log_7 h \rceil} 7^i m T_c = \lceil \log_7 h \rceil T_s + \frac{7^{\lceil \log_7 h \rceil} - 1}{6} m T_c$.

Step 2: Balancing-Load Operation After applying data-aggregation operation, each DDN_0, DDN_1,..., and DDN_{h^2-1} has different amount of messages. This is load imbalance, a data tuning procedure is presented for load balancing. This operation is divided into *prefix-sum* and *data-tuning* procedures.

Prefix-Sum Procedure: After data-aggregation operation that all source nodes' messages are aggregated to regular positions, which in diagonal plane $SPAN$ $(P_{x,y,z}, (where 0 \leq x = ih, y = jh, z = kh < n$. All those planes constitute a special cube $SPAN(P_{0,0,0}, (\vec{e}_{1,3}, \vec{e}_{1,2}, \vec{e}_1), (n, n, \lceil \frac{n}{h} \rceil))$. Our diagonal-based recursive prefix-sum procedure is to calculate prefix-sum value for each keeping-message node in $SPAN(P_{0,0,0}, (\vec{e}_{1,3}, \vec{e}_{1,2}, \vec{e}_1), (n, n, \lceil \frac{n}{h} \rceil))$. The diagonal-based prefix-sum procedure is divided into forward and backward stages. In forward stage, information of number of messages is aggregated from cube to a plane, from plane to a line, and from line to one node. After the forward stage, total number of whole source messages is kept in one node. In backward stage, partial prefix-sum value is return from node to a line, from line to plane, from plane to cube. Herein we omit the detail operations since the work is trivial.

Data Tuning Procedure: Assume that a node x is located in $DDN_{i,j}$, with a destination list. The information of destination list is to indicate that node x should move message to which neighboring nodes. To satisfying the following purpose, for node x, if $(k,l) \in$ destination list, one message from $DDN_{i,j}$ (node x) is moved to $DDN_{k,l}$. Every node x performs the following operation in parallel. (1) **Finding a destination list:** Having a prefix-sum value α and number of keeping-message β, then destination list is $F = \{\alpha \bmod h^2, (\alpha+1) \bmod h^2, ..., (\alpha+\beta) \bmod h^2\}$ if number of *DDN*s is h^2. Two communication steps are needed if intend to moving data from $DDN_{i,j}$ to $DDN_{k,l}$. Note that F' is a sequence of pairs which is constructed as follows. For every $t \in F$, let $(i = t \bmod h, j = t/h) \in F'$, where i, j indicate the offset value of row and column tuning actions in data tuning operation. (2) **Data tuning operation:** The data tuning operation is divided into row tuning and column tuning actions which is formally described below.

T1. Row tuning action ($DDN_{i,j} \rightarrow DDN_{k,j}$): An extra alignment operation is executed due to the dimension-order routing. If $|i-k| \leq 3$, then we allow $DDN_{i,j} \rightarrow DDN_{i\pm1,j}$, $DDN_{i\pm2,j}$, and $DDN_{i\pm3,j}$ within two communication steps. For each node in diagonal plane of $DDN_{i,j}$, we first align $DDN_{i\pm1,j}$ along dimension-X with distance ±1, $DDN_{i\pm2,j}$ along dimension-Y with distance ±2, and $DDN_{i\pm3,j}$ along dimension-Z with distance ±3 to six meta-nodes. Every node $P_{x,y,z}$ in diagonal plane $DDN_{i,j}$ distributes its messages to six nodes $P_{x-1,y-1,z}$, $P_{x+1,y+1,z}$, $P_{x,y-2,z}$, $P_{x,y+2,z}$, $P_{x,y,z-3}$, and $P_{x,y,z+3}$, which is represented by

$$R = \begin{bmatrix} 1\,1\,0 \\ 0\,2\,0 \\ 0\,0\,3 \end{bmatrix}, \text{ where } D^+ = \begin{bmatrix} 1\,0\,0 \\ 0\,1\,0 \\ 0\,0\,1 \end{bmatrix} \text{ and } D^- = \begin{bmatrix} -1 & 0 & 0 \\ 0 & -1 & 0 \\ 0 & 0 & -1 \end{bmatrix}.$$

T2: Column tuning action ($DDN_{k,j} \rightarrow DDN_{k,l}$): This action can be represented by

$$R = \begin{bmatrix} 1\,0\,1 \\ 0\,0\,2 \\ 0\,0\,0 \end{bmatrix}, \text{ where } D^+ = \begin{bmatrix} 1\,0\,0 \\ 0\,1\,0 \\ 0\,0\,1 \end{bmatrix} \text{ and } D^- = \begin{bmatrix} -1 & 0 & 0 \\ 0 & -1 & 0 \\ 0 & 0 & -1 \end{bmatrix}.$$

3.2 Distribution Phase

Step 1: Alignment Operation (1) **Alignment to diagonal plane:** All possible message are aligned into the diagonal plane. This task can be easily achieved by performing the diagonal-based data aggregation operation as introduced in Section 3.1, which takes time $\lceil \log_7(\lceil \frac{n}{h} \rceil) \rceil (T_s + \widetilde{m} T_c)$, where $\widetilde{m} = \frac{s}{h^2} m$. (2) **All-to-all broadcasting procedure on diagonal plane:** This procedure is to collect messages of each node in the diagonal plane $SPAN(P_{x,y,z}, (\overrightarrow{e}_{1,3}, \overrightarrow{e}_{1,2}), (\lceil \frac{n}{h} \rceil, \lceil \frac{n}{h} \rceil))$ from other nodes located in the same diagonal plane $SPAN(P_{x,y,z}, (\overrightarrow{e}_{1,3}, \overrightarrow{e}_{1,2}), (\lceil \frac{n}{h} \rceil, \lceil \frac{n}{h} \rceil))$. The plane can be viewed as $\lceil \frac{n}{h} \rceil$ rows or $\lceil \frac{n}{h} \rceil$ columns. Two broadcasting operations are needed. Basically, this work is the row and column tuning operations with different distance matrices D^+ and D^-, which is same as the row tuning (**T1**) operation and the column tuning (**T2**) operation.

Step 2: Broadcast Operation Now every node in diagonal plane of each *DDN* contains same broadcast messages. The next step is to perform a well-known result, the diagonal broadcast scheme in 3-D torus [7], on each *DDN* in parallel. The diagonal plane $SPAN(P_{x,y,z}, (\vec{e}_{1,3}, \vec{e}_{1,2}), (\lceil \frac{n}{h} \rceil, \lceil \frac{n}{h} \rceil))$ has partial source messages, and the broadcasting is based on a recursively sending messages from a diagonal plane to six planes. We mention that the operation is executed in time $\lceil \log_7 \lceil \frac{n}{h} \rceil \rceil (T_s + \widetilde{m} T_c)$, where $\widetilde{m} = \frac{s}{h^2} m$.

Step 3: Data Collection Operation Each *data collecting network* (which is $h \times h \times h$ mesh), each diagonal plane received messages M_0, M_1, ..., and M_{h^2-1}. Each received message containing the whole messages of one *DDN*. These messages should be propagated to every node of the *DCN*. This is implemented in three stages: *row broadcasting* followed by *column and horizontal broadcasting*: (1) In the *row broadcasting* stage, we use a recursive scheme. Node located in diagonal plane send messages to two nodes with distance $\pm \frac{1}{3} h$ and recursively propagate the message. This take $\lceil \log_3 h \rceil$ communication phases (2) In the *horizontal broadcasting,* every node collects the partial messages from the row broadcasting stage. The messages are belong its column nodes, every node concurrently send separate message to other nodes with pipelined scheme. A logical (directed) ring is embedded on each column of the *DCN*. The gives a dilation-2 embedding. With this embedding, every node then pipelines propagate its own message following ring. Finally, we have the following result.

Theorem 1 *The multi-node broadcasting algorithm with aggregation-then- distribution strategy can be done in a $T_{n \times n \times n}$ torus within*

$$O(\max(\lceil \log_7 n \rceil, h) T_s + \max(\left\lceil \log_7 \left\lceil \frac{n}{h} \right\rceil \right\rceil \frac{s}{h^2} m, sm) T_c).$$

4 Performance Comparison

We mainly compared our scheme against the multiple-spanning-tree scheme [5] under various situations. The parameter used in our simulations are listed below: (1) the torus size is $16 \times 16 \times 16$, (2) startup time $T_s = 30\mu$sec and $T_c = 1\mu$sec, (3) dilation $h = 7$ or 14, (4) the message size is ranging from $2k$ to $10k$. Below, we show our simulation result from several prospects. (A) *Effects of Number of Sources:* Fig. 3 shows the multi-node broadcast latency when $T_s = 30\mu$sec and $T_c = 1\mu$sec at various number of sources. Our scheme when $h = 7$ incurs higher latency than that of multiple-spanning-tree scheme, while our scheme when $h = 14$ has lower latency than that of multiple-spanning-tree scheme. It reflects the fact that our scheme has better performance than multiple-spanning-tree scheme at various number of sources. (B) *Effects of Number of h:* The value of h reflects the number of subnetworks, and thus the level of communication parallelism. So a larger h generally delivers better performance. Fig. 3 also compares multi-node broadcast latency when $h = 7$ and 14. Observe that our scheme has lower latency when $h = 14$ than our scheme $h = 7$. This is verified that high level of communication parallelism is, the better performance will be.

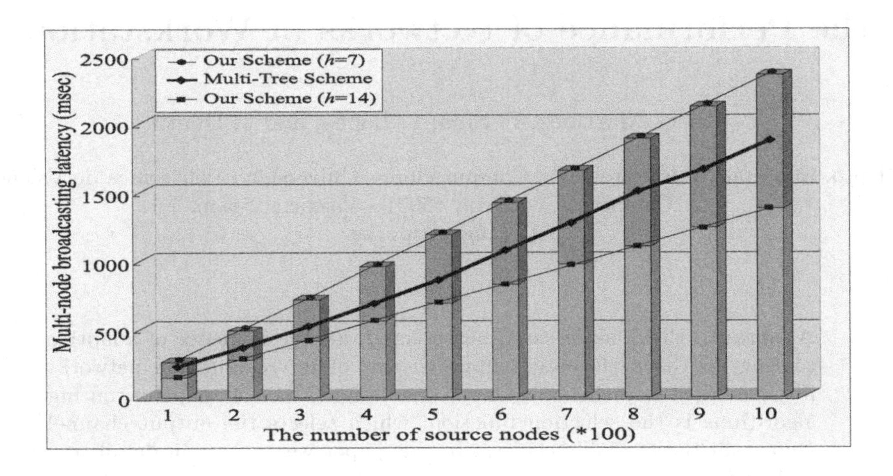

Fig. 3. Multi-Node Broadcast latency in a $16 \times 16 \times 16$ torus at various number of source nodes.

References

1. Y. S. Chen, T. Y. Juang, and E. H. Tseng. Efficient broadcasting in an arrangement graph using multiple spanning trees. *IEICE Transactions on Fundamentals of Electronic, Communications, and Computer Science*, E83-A(1):139–149, January 2000.
2. F. T. Leighton. *Introduction to Parallel Algorithms and Architectures: Arrays-Trees-Hypercubes.* Morgan Kaufmann Publishers, San Mateo, California, 1992.
3. Y. Saad and M. Schultz. Data communication in hypercubes. *Journal of Parallel and Distributed Computing*, 6(1):115–135, February 1989.
4. Y. Saad and M. Schultz. Data communication in parallel architectures. *Parallel Computing*, 11:131–150, 1989.
5. George D. Stamoulis and John N. Tsitsiklis. An efficient algorithm for multiple simultaneous broadcasts in the hypercube. *Information Processing Letter*, 46:219–224, July 1989.
6. Y. C. Tseng. Multi-node broadcasting in hypercubes and star graph. *Journal of Information Science and Engineering*, 14(4):809–820, 1998.
7. S. Y. Wang and Y. C. Tseng. Algebraic foundations and broadcasting algorithms for wormhole-routed all-port tori. *To appear in IEEE Transactions on Computers*, 2000.

On the Influence of the Selection Function on the Performance of Networks of Workstations*

J. C. Martínez, F. Silla, P. López, and J. Duato

Dpto. Informática de Sistemas y Computadores, Universidad Politécnica de Valencia
Camino de Vera, 14. 46071 - Valencia, Spain.
jc@gap.upv.es

Abstract. Previous research has pointed out the influence of adaptive routing on the performance improvement of interconnection networks for clusters of workstations. One of the design issues of adaptive routing algorithms is the selection function, which selects the output channel among all the available choices. In this paper we analyze in detail several selection functions in order to evaluate their influence on network performance. Simulation results show that network throughput may be increased up to 10%. When network is close to saturation, improvements in latency up to 40% may be achieved.

1 Introduction

Networks of workstations (NOWs) are usually interconnected by an irregular topology, which makes routing and deadlock avoidance quite complicated. Deadlocks can be avoided by removing cyclic dependencies between channels [6]. As a consequence, many messages are routed following non-minimal paths, therefore increasing message latency and wasting resources. A more efficient approach consists of allowing the existence of cyclic dependencies between channels while providing some escape paths (as dedicated virtual channels) to avoid deadlock [3,7].

Virtual channels are not only useful for designing deadlock-free routing algorithms. They may also be used in wormhole networks to increase link utilization [2]. On the other hand, virtual channels also enable the use of adaptive routing algorithms. Adaptivity makes possible to have several outgoing ports for each destination, being necessary to perform a selection among all the feasible outgoing ports. Therefore, we may divide the task of routing a message into two different phases [3]. In the first one, a routing algorithm provides a set of suitable outgoing ports to reach the message destination. In the second phase, one of the outgoing ports provided is selected according to some criterion. This second phase is performed by the selection function.

Previous work [7] has pointed out the great influence that routing algorithms have on network performance. However, the influence of selection functions on performance has not been analyzed. In this paper, we take such a challenge,

* This work was supported by the Spanish CICYT under Grant TIC97–0897–C04–01

M. Valero et al. (Eds.): ISHPC 2000, LNCS 1940, pp. 292–299, 2000.

evaluating the performance of different selection functions specially designed to be used in conjunction to the adaptive routing algorithm proposed in [7].

The following sections are organized as follows. Section 2 provides some background on routing in networks with irregular topology. In Section 3 we will describe the selection functions later evaluated in Section 4. Finally, some conclusions are drawn.

2 Routing in NOWs

Several deadlock-free routing schemes have been proposed for irregular networks [6,1,5,7]. Our analysis is centered on the Minimal Adaptive (MA) routing scheme [7]. The MA routing algorithm splits each physical channel into two virtual channels, called "original" and "new" channels, respectively. Original channels are used following the up*/down* routing algorithm used in Autonet [6]. A newly injected message can only leave the source switch using new channels belonging to minimal paths. When a message arrives at a switch through a new channel, the routing function returns a new channel belonging to a minimal path, if available. If all of them are busy, then the up*/down* routing algorithm is used, selecting an original channel belonging to a minimal path or to the shortest path if a minimal path is not available. To ensure deadlock freedom, once a message reserves an original channel, it will be routed using only original channels according to the up*/down* routing function until delivered.

Note that links may be split into more than two virtual channels. In this case, all of them except one would be used as new channels, while the other one would be the original channel. This increases adaptivity and also the number of messages that follow minimal paths.

3 Selection Functions

As mentioned above, two different phases occur when routing a message: the routing function and the selection function. The selection function selects one virtual channel from the set provided by the routing function. Usually, this selection takes into account the state (free or busy) of output virtual channels. In this case, two are the incoming parameters of the selection function: (i) the set of virtual channels provided by the routing algorithm as suitable outgoing switch ports for the message and (ii) the set of free output virtual channels. Using this information, the selection function chooses the best outgoing port for the message, according to some criterion. It can also decide to block the message until a better choice can be done.

In the MA routing algorithm (see Section 2), since new channels offer more freedom to route messages, selection functions should assign more priority to new channels than to original channels. However, several possibilities still exist, as we propose next.

The Static Priority (SP) selection function cyclically distributes priorities among the virtual channels of different physical channels. Figure 1 shows these

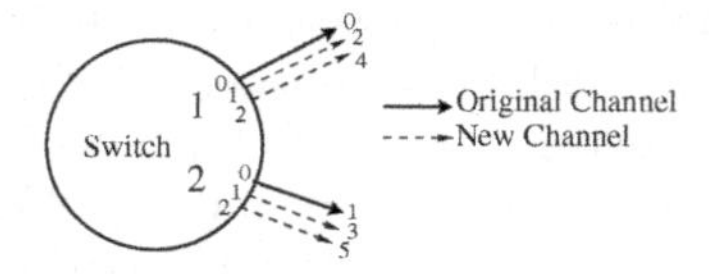

Fig. 1. Virtual channel priorities for the SP selection function

priorities for a switch with two links and three virtual channels per link. Virtual channel identifiers are inside the circle, next to the border. Big numbers inside the circle refer to physical channel identifiers. Numbers outside the switch indicate the priority assigned to that virtual channel. As can be seen, the lower priorities are assigned to original channels. Note that new channels are always assigned higher priorities than original ones. The aim of this selection function is to balance virtual channel multiplexing of physical channels, while being easily implementable in hardware.

The least recently used (LRU) selection function returns the least recently used new channel that is free, if any. If no new virtual channel is free, then it selects the least recently used free original channel. To implement this function, it is necessary to use a register of $log_2(N * M)$ bits per virtual channel, where N is the number of physical links and M is the number of virtual channels per physical link. Initially, this register is set to zero. When a message releases a virtual channel, all registers whose value is less or equal than the value of the virtual channel register increase their value by one and the register of the released virtual channel is set to zero. In this way, the higher register value a virtual channel has, the higher priority it is assigned. If several registers have the same value, a static priority is used (the same as SP).

The least frequently used (LFU) selection function returns the least frequently used new virtual channel from the set of free feasible output virtual channels. If there is no feasible new channel, then the selection function returns the least frequently used original virtual channel. In order to implement this selection function, it is necessary to log all the history of the channel in order to calculate channel utilization, being necessary a register per virtual channel to store the activity information. This register should be extremely large if we try to accurately calculate channel utilization, noticeably increasing hardware complexity. In order to reduce register size, we may calculate the channel utilization only for the last N clock cycles. In this case, we need a N-bit shift register and also an additional counter of log_2N bits per virtual channel. Initially, both the register and the counter are set to zero. The register is shifted every cycle. If the virtual channel transmits a flit in that clock cycle, a one is inserted into the shift register, and the counter is increased by one. In case the link remains idle during that clock cycle, a zero is inserted into the shift register and the counter is not increased. On the other hand, every clock cycle a one is shifted out the register, the counter is decremented in one unit. Hence, the counter value is the number of ones in the shift register, and thus, the channel utilization in the last

N clock cycles. Using this information, the channel with the lower counter value takes the higher priority.

The minimal multiplexation (MM) selection function (the one proposed in [8]) tries to minimize the number of messages being multiplexed onto a physical link at the same time that maximizes adaptivity. This selection function will select from the set of free feasible output virtual channels, the one belonging to the physical link having the highest number of free virtual channels. If two physical links have the same amount of free virtual channels, this function will select the one with the higher number of free new virtual channels. As in previous selection functions, new virtual channels have always higher priority than original virtual channels. To implement the minimal multiplexation selection function, some hardware is required to perform the necessary comparisons.

The random (RAND) selection function consists of randomly selecting one channel from the set of free feasible outgoing new virtual channels. If there is no free new channel then it randomly chooses an original channel. In this case, it is necessary to implement a random function by hardware.

Finally, we have evaluated the influence of delaying the routing decision in order to select a better channel when all of the feasible new virtual channels are busy. Original channels provide less adaptivity and usually longer paths than new channels. The use of non-minimal paths makes messages to use more resources than necessary, increasing the probability of blocking another messages and also decreasing performance. Selection functions presented above try to minimize the use of original channels by assigning them the lowest priorities. However, when no new virtual channel is free, messages must leave the switch through one of the original channels. It is possible to minimize even more the use of original channels by stopping the message at the current switch instead of allowing it to be immediately routed through an original channel even if it is available. The message would wait for a new channel becoming free. However, in order to avoid deadlock, messages must not wait indefinitely, being necessary the use of a time threshold. In this way, the message will be allowed to leave the switch using an original channel only if it is waiting for longer than the threshold.

We propose the use of two different thresholds. The first one is a simple timeout. When a message arrives at a switch, a counter attached to the input virtual channel is triggered. If the message has not being successfully routed when the counter exceeds the threshold, then the message is allowed to leave the switch through a suitable original channel. Note that the optimum value of this threshold depends on message length.

The second one is based on monitoring the activity of the requested outgoing channels. Only when all of the feasible outgoing virtual channels for that message are busy, and none of them is transmitting flits for a time that exceeds a threshold, the message is allowed to leave the switch through an original channel. This idea is based on the deadlock detection mechanism proposed in [4]. The advantage of this mechanism with respect to the use of timeouts is that the optimal value of the threshold should be less dependent on message length. Note that both mechanisms are applicable to any selection function.

4 Performance Evaluation

In this section, we evaluate by simulation the performance of the proposed selection functions. We will refer to them as the acronym used when presented. In addition, if thresholds are used, we will append a suffix to the selection function identifier. For the first case (simple timeouts), we will use the suffix Th_n, being n the threshold measured as multiples of the message length, in clock cycles. For the second case (deadlock detection), we will use the suffix Det_n, being n measured in clock cycles.

4.1 Network Model

Network topology is irregular and has been generated randomly, imposing three restrictions: (i) there are 4 workstations connected to each switch, (ii) two neighboring switches are connected by a single link and (iii) all the switches have the same size (8-port switches). We have evaluated networks with a size ranging from 16 switches (64 workstations) to 64 switches (256 workstations). For the sake of brevity, we will only show results for 64 switches. Wormhole switching is used.

Each switch has a routing control unit which applies the MA routing strategy. A crossbar inside the switch allows multiple messages traversing it simultaneously. We assumed that it takes one clock cycle to compute the routing algorithm, or to transmit one flit across a crossbar. Link propagation delay has been assumed to be 4 cycles. Links are pipelined. Data are injected into the link at a rate of one flit per cycle.

The MA routing algorithm has been evaluated with two, three, and four virtual channels per link. For the sake of shortness, we present our results only for four virtual channels. For two virtual channels, differences among choices are not significant, and for three and four virtual channels results are very similar.

Message generation rate is constant and the same for all workstations. We have evaluated the full range of traffic, from low load to saturation. Message destination is randomly chosen among all the workstations in the network. For message length, 16-flit and 64-flit messages were considered.

4.2 Simulation Results

Figure 2 shows the average message latency versus traffic for the evaluated selection functions. As can be seen, MM achieves lower latency than the rest of selection functions. Compared with the worst selection function, MM improves latency about 15% for short messages and 20% for long messages when the network is near saturation. The performance of the SP selection function is very close to MM, because both algorithms distribute the outgoing messages between all the possible virtual channels. Hence, both selection functions make physical link to be more evenly multiplexed. Moreover, they try to minimize link

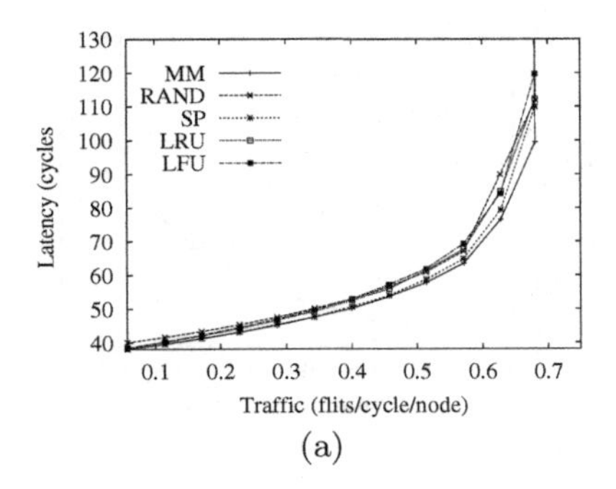

(a)

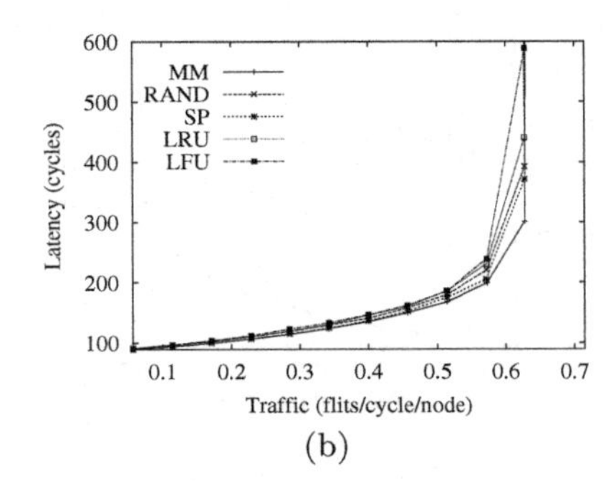

(b)

Fig. 2. Average message latency versus traffic for the analyzed selection functions. Network size is 64 switches. Message length is 16 (a) and 64 (b) flits

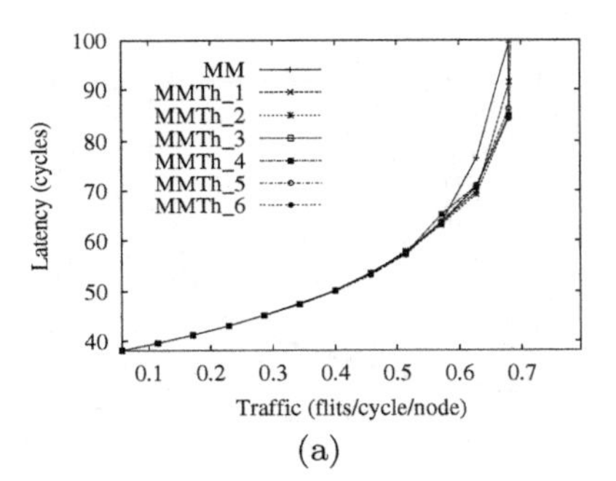

(a)

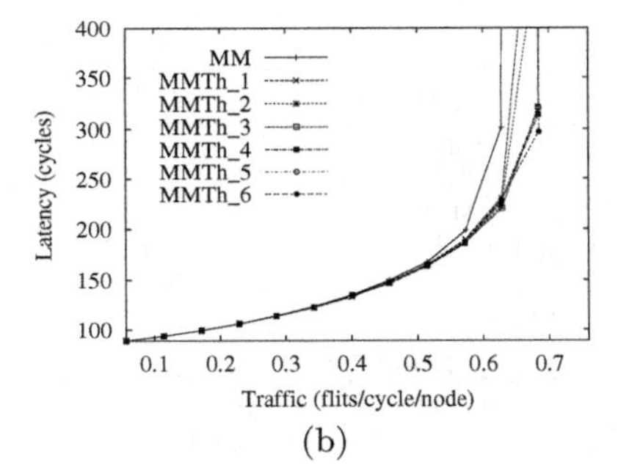

(b)

Fig. 3. Comparison of different thresholds with MM. Average message latency versus traffic for a network with 64 switches. Simple timeouts are used. Message length is 16 flits in (a) and 64 flits in (b)

multiplexing, thus allowing messages to advance faster. As a consequence, messages release resources faster, then decreasing the probability of blocking other messages. Therefore, the use of non-minimal paths is also decreased.

Let's analyze the use of time thresholds in SP and MM. Figures 3-a and 3-b show some results for MM using simple timeouts for short and long messages, respectively. A threshold equal to zero is equivalent to use the basic selection function. As can be seen, using any threshold for short messages provides similar improvements. For long messages, the performance improvement is higher. Thresholds ranging from 3 to 6 obtain the best results. The ability of the selection function to use the original channels only when necessary is what improves performance. Figures 4-a and 4-b show the results when using the deadlock detection mechanism to allow the use of original channels. In this case, a threshold equal to 0 is not the same as the basic selection function. Although the performance improvement is similar to the one obtained when using simple timeouts,

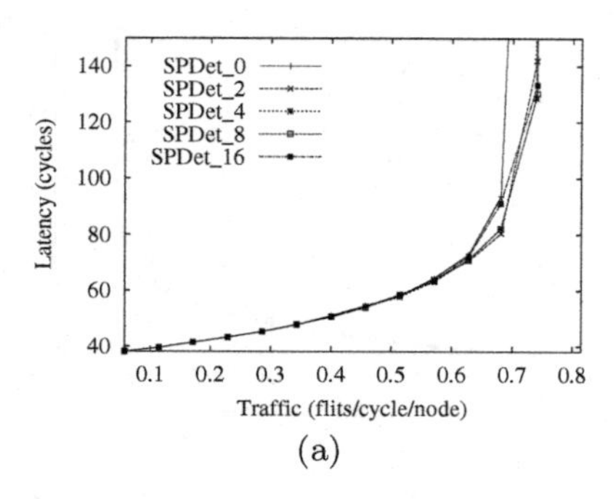

(a)

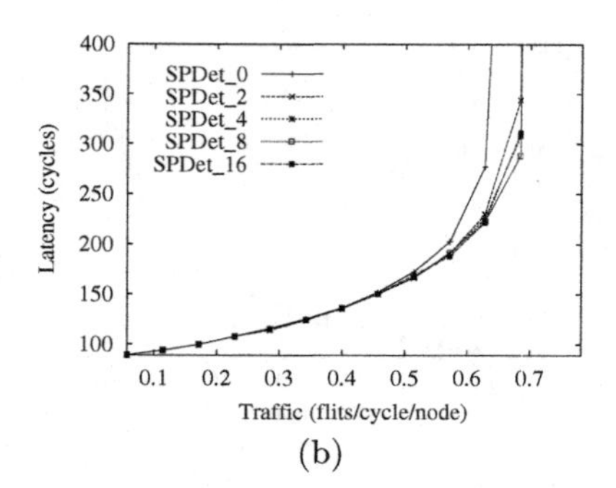

(b)

Fig. 4. Comparison of different thresholds with SP. Average message latency versus traffic for a network with 64 switches. The deadlock detection based mechanism is used. Message length is 16 flits in (a) and 64 flits in (b)

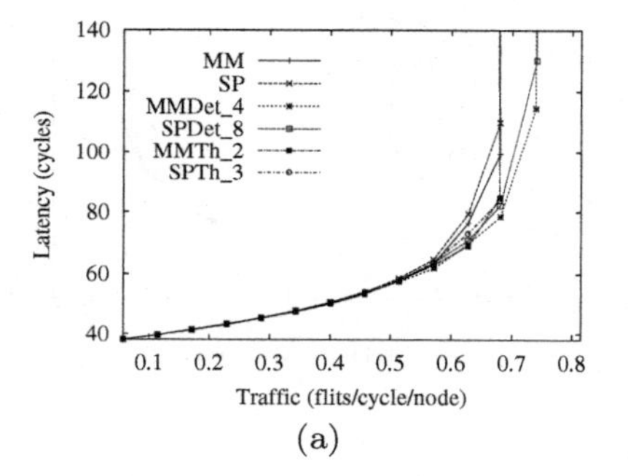

(a)

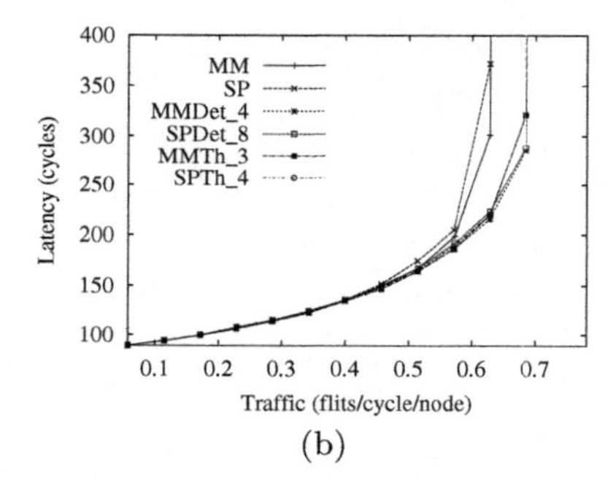

(b)

Fig. 5. Comparison of best results of MM and SP. Average message latency versus traffic for a network with 64 switches. Message length is 16 (a) and 64 (b) flits

the best threshold is the same for both message lengths, which simplifies network tuning.

In Figure 5 we compare the best combinations of thresholds and selection functions in order to determine the selection function that achieves the highest performance. For short messages (Figure 5-a), the best option is MMDet_4. This selection function reduces latency by 20% and increases throughput by 9% with respect to the basic MM function. Similar results are obtained for SPDet_8. These results are better than the ones obtained by the timeout-based selection functions. The reason is that the deadlock detection mechanism is more selective than timeouts for detecting network congestion. Thus, the use of original channels is more restricted than when the timeout-based selection functions are used. For long messages (Figure 5-b), both using a simple timeout and using the deadlock detection mechanism obtain similar improvements with respect to the basic functions. MMDet_4 and SPDet_8 present slightly better results than the selection functions based on timeouts. MMDet_4 improves latency about 30% and throughput about 9% with respect to MM, and message latency using SPDet_8 is about 40% lower than the basic SP, improving throughput also by 9%.

5 Conclusions

In this paper we have evaluated the influence on network performance of the selection function executed at each switch in order to select one of the output virtual channels provided by the routing algorithm. To do so, we have compared several selection functions. Results obtained show variations in latency about 20%, depending on the basic selection function implemented. In addition, we have extended the two best selection functions, MM and SP, in order to use time thresholds. Two alternatives, one based on simple timeouts and another based on a deadlock detection mechanism, have been analyzed. Finally, we have compared the best performance evaluation results, obtaining the best selection functions. Using a deadlock detection based mechanism improves network performance with respect to using a simple timeout, especially for short messages. On the other hand, the former can be tuned independently of message length. Results show that latency is reduced at the same time that throughput is increased about 10% for both MM and SP.

References

1. N. J. Boden, D. Cohen, R. E. Felderman, A. E. Kulawik, C. L. Seitz, J. Seizovic and W. Su, "Myrinet - A gigabit per second local area network," *IEEE Micro*, pp. 29–36, February 1995. 293
2. W. J. Dally, "Virtual-channel flow control," *IEEE Trans. on Parallel and Distributed Systems*, vol. 3, no. 2, pp. 194–205, March 1992. 292
3. J. Duato, "A new theory of deadlock-free adaptive routing in wormhole networks," *IEEE Trans. on Parallel and Distributed Systems*, vol. 4, no. 12, pp. 1320–1331, December 1993. 292
4. J. M. Martínez, P. López, J. Duato and T. M. Pinkston, "Software-Based Deadlock Recovery Technique for True Fully Adaptive," in *Proc. of the 1997 Int. Conf. on Parallel Processing*, pp. 182–189, August 1997. 295
5. W. Qiao and L. M. Ni, "Adaptive routing in irregular networks using cut-through switches," in *Proc. of the 1996 Int. Conf. on Parallel Processing*, August 1996. 293
6. M. D. Schroeder et al., "Autonet: A high-speed, self-configuring local area network using point-to-point links," Tech. Report SRC research report 59, DEC, April 1990. 292, 293
7. F. Silla and J. Duato, "Improving the Efficiency of Adaptive Routing in Networks with Irregular Topology," in *Proc. of the 1997 Int. Conf. on High Performance Computing*, December 1997. 292, 293
8. F. Silla and J. Duato, "On the Use of Virtual Channels in Networks of Workstations with Irregular Topology," in *Proc. of the 1997 Parallel Computing, Routing, and Communication Workshop*, June 1997. 295

Combining In-Transit Buffers with Optimized Routing Schemes to Boost the Performance of Networks with Source Routing*

Jose Flich[1], Pedro López[1], Manuel. P. Malumbres[1], José Duato[1], and Tom Rokicki[2]

[1] Dpto. Informática de Sistemas y Computadores, Universidad Politécnica de Valencia, Spain
jflich@gap.upv.es
[2] Instantis, Inc.
Menlo Park, California, USA
rokicki@instantis.com

Abstract. In previous papers we proposed the ITB mechanism to improve the performance of up*/down* routing in irregular networks with source routing. With this mechanism, both minimal routing and a better use of network links are guaranteed, resulting on an overall network performance improvement. In this paper, we show that the ITB mechanism can be used with any source routing scheme in the COW environment. In particular, we apply ITBs to DFS and Smart routing algorithms, which provide better routes than up*/down* routing. Results show that ITB strongly improves DFS (by 63%, for 64-switch networks) and Smart throughput (23%, for 32-switch networks).

1 Introduction

Clusters of workstations (COWs) are currently being considered as a cost-effective alternative for small-scale parallel computing. In these networks, topology is usually fixed by the location constraints of the computers, making it irregular. On the other hand, source routing is often used as an alternative to distributed routing, because non-routing switches are simpler and faster.

Up*/down* [6] is one of the best known routing algorithms for irregular networks. It is based on an assignment of direction labels ("up" or "down") to links. To eliminate deadlocks a route must traverse zero or more "up" links followed by zero or more "down" links. While up*/down* routing is simple, it concentrates traffic near the "root" switch and uses a large number of non-minimal paths.

Other routing algorithms like Smart [2] and DFS [5] achieve better performance than up*/down*. Smart first computes all possible paths for every source-destination pair, building the channel dependence graph (CDG). Then, it uses

* This work was supported by the Spanish CICYT under Grant TIC97–0897–C04–01 and by Generalitat Valenciana under Grant GV98-15-50.

M. Valero et al. (Eds.): ISHPC 2000, LNCS 1940, pp. 300–309, 2000.

an iterative process to remove dependencies in the CDG taking into account a heuristic cost function. Although Smart routing distributes traffic better than other approaches, it has the drawback of its high computation overhead. DFS computes a depth-first spanning tree with no cycles. Then, it adds the remaining channels to provide minimal paths, breaking cycles by restricting routing. A heuristic is also used to reduce routing restrictions.

These routing strategies remove cycles by restricting routing. As a consequence, many of the allowed paths are not minimal, increasing both latency and contention in the network. Also, forbidding some paths may result in an unbalanced network traffic distribution, which leads to a rapid saturation. In this paper, we propose the use of a mechanism that removes channel dependences without restricting routing. This mechanism has been first proposed in [3] to improve up*/down* routing, but it can be applied to any routing algorithm. In this paper we will apply it to improved routing schemes (Smart and DFS).

The rest of the paper is organized as follows. Section 2 summarizes how the mechanism works and its application to some optimized routing strategies. In Section 3, evaluation results for different networks and traffic load conditions are presented, analyzing the benefits of using our mechanism combined with previous routing proposals. Finally, in Section 4 some conclusions are drawn.

2 Applying the ITB Mechanism to Remove Channel Dependences

We will firstly summarize the basic idea of the mechanism. The paths between source-destination pairs are computed following any given rule and the corresponding CDG is obtained. Then, the cycles in the CDG are broken by splitting some paths into sub-paths. To do so, an intermediate host inside the path is selected and used as an in-transit buffer, (ITB); at this host, packets are ejected from the network as if it were their destination. The mechanism is cut-through. Therefore, packets are re-injected into the network as soon as possible to reach their final destination. Notice that the dependences between the input and output switch channels are completely removed because in the case of network contention, packets will be completely ejected from the network at the intermediate host. The CDG is made acyclic by repeating this process until no cycles are found. Notice that more than one intermediate host may be needed.

On the other hand, ejecting and re-injecting packets at some hosts also improves performance by reducing network contention. Packets that are ejected free the channels they have reserved, thus allowing other packets requiring these channels to advance through the network (otherwise, they would become blocked). Therefore, adding some extra ITBs at some hosts may help in improving performance. Hence, the goal of the ITB mechanism is not only to provide minimal paths by breaking some dependences but also to improve performance by reducing network contention. However, ejecting and re-injecting packets at some intermediate hosts also increases the latency of these packets and requires some additional resources in both network (links) and network interface cards (memory pools and DMA engines).

If the rules used to build the paths between source-destination pairs lead to an unbalanced traffic distribution, then adding more ITBs than the ones strictly needed will help. This is the case for up*/down*, because this routing tends to saturate the area near the root switch. Thus, there is a trade-off between using the minimum number of ITBs that guarantees deadlock-free minimal routing and using more than these to improve network throughput. Therefore, when we apply the ITB mechanism to up*/down*, we will use these two approaches. In the first case, we will place the minimum number of ITBs that guarantees deadlock-free minimal routing. Thus, given a source-destination pair, we will compute all minimal paths. If there is a valid minimal up*/down* path it will be chosen. Otherwise, a minimal path with ITBs will be used. In the second approach, we will use more ITBs than strictly needed to guarantee deadlock-free minimal routing. In particular, we will randomly choose one minimal path. If the selected path complies with the up*/down* rule, it is used without modification. Otherwise, ITBs are inserted even if there exist valid minimal up*/down* paths between the same source-destination pair.

In the case of DFS, we will use ITBs in the same way as in the second approach used for up*/down* but verifying if the paths comply with the DFS rule. However, for Smart routing, we will use a different approach. We first compute the paths between source-destination pairs that better balance network traffic. Notice that the obtained routes are not the same that Smart computes, because it computes both balanced and deadlock-free routes whereas we compute only balanced routes. For this reason, we will refer to these routes as "balanced" rather than "smart". Then, we compute the CDG and place ITBs to convert it into an acyclic one. On the other hand, since computing balanced routes alone is easier than computing both balanced and deadlock-free routes, the computational cost of the resulting routing algorithm is lower than the one of Smart routing.

3 Performance Evaluation

3.1 Network Model and Network Load

The network topologies we consider are irregular and have been generated randomly, imposing three restrictions: (i) all the switches have the same size (8 ports), (ii) there are 4 hosts connected to each switch and (iii) two neighboring switches are connected by a single link. We have analyzed networks with 16, 32, and 64 switches (64, 128, and 256 hosts, respectively).

Links, switches, and interface cards are modeled based on the Myrinet network [1]. Concerning links, we assume Myrinet short LAN cables [4] (10 meters long, 160 MB/s, 4.92 ns/m). Flits are one byte wide. Physical links are one flit wide. Transmission of data across channels is pipelined [7] with a rate of one flit every 6.25 ns and a maximum of 8 flits on the link at a given time. A hardware "stop and go" flow control protocol [1] is used to prevent packet loss. The slack buffer size in Myrinet is fixed at 80 bytes. Stop and go marks are fixed at 56 bytes and 40 bytes, respectively.

Each switch has a simple routing control unit that removes the first flit of the header and uses it to select the output link. The first flit latency is 150 ns through the switch. After that, the switch is able to transfer flits at the link rate. Each output port can process only one packet header at a time. A crossbar inside the switch allows multiple packets to traverse it simultaneously.

Each Myrinet network interface card has a routing table with one entry for every possible destination of messages. The tables are filled according to the routing scheme used.

In the case of using ITBs, the incoming packet must be recognized as in-transit and the transmission DMA must be re-programmed. We have used a delay of 275 ns (44 bytes received) to detect an in-transit packet, and 200 ns (32 additional bytes received) to program the DMA to re-inject the packet. These timings have been taken on a real Myrinet network. Also, the total capacity of the in-transit buffers has been set to 512KB at each interface card.

In order to evaluate different workloads, we use different message destination distributions to generate network traffic: *Uniform* (the destination is chosen randomly with the same probability for all the hosts), *Bit-reversal* (the destination is computed by reversing the bits of the source host id.), *Local* (destinations are, at most, 5 switches away from the source host, and are randomly computed), *Hot-spot* (a percentage of traffic (20%, 15%, and 5% for 16, 32, and 64-switch networks, respectively) is sent to one randomly chosen host and the rest of the traffic randomly among all hosts) and a *Combined* distribution, which mixes the previous ones. In the later case, each host will generate messages using each distribution with the same probability.

Packet generation rate is constant and the same for all the hosts. Although we use different message sizes (32, 512, and 1K bytes), for the sake of brevity results will be shown only for 512-byte messages.

3.2 Simulation Results

First, we analyze the behavior of the routing algorithms without using in-transit buffers. Results for up*/down*, DFS and the Smart routing algorithms will be referred to as UD, DFS and SMART, respectively. Then, we evaluate the use of in-transit buffers over up*/down* and DFS routing. For up*/down* routing, we analyze the two approaches mentioned above: using the minimum number of ITBs needed to guarantee deadlock-free minimal routing (UD_MITB), and using more ITBs (UD_ITB). For DFS routing, we only use the second approach, which will be referred to as DFS_ITB. Finally, we evaluate the use of in-transit buffers over balanced but deadlocking routes supplied by the Smart routing algorithm. This routing will be referred to as B_ITB (B from "balanced").

For each network size analyzed, we show the increase in throughput when using the in-transit buffer mechanism with respect to the original routing algorithms. Minimum, maximum, and average results for 10 random topologies are

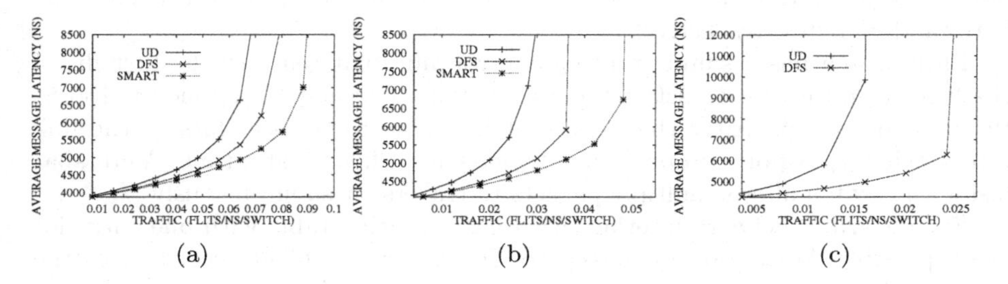

Fig. 1. Average message latency vs. accepted traffic. Message length is 512 bytes. Uniform distribution. Network size is (a) 16 switches, (b) 32 switches, and (c) 64 switches

shown. In addition, we will plot the average message latency versus the accepted traffic for selected topologies.

Routing Algorithms without ITBs Figure 1 shows the results for the uniform distribution of message destinations for selected topologies of 16, 32, and 64 switches, respectively. SMART routing is not shown for the 64-switch network due to its high computation time.

As was expected, the best routing algorithm is SMART. It achieves the highest network throughput for all the topologies we could evaluate. In particular, it increases throughput over UD and DFS routing by factors up to 1.77 and 1.28, respectively.

The performance improvement achieved by SMART is due to its better traffic balancing. Figure 3.a shows the utilization of links connecting switches for the 32-switch network. Links are sorted by utilization. Traffic is 0.03 flits/ns/switch. For this traffic value, UD routing is reaching saturation. When using UD routing, half the links are poorly used (52% of links with a link utilization lower than 10%) and a few links highly used (only 11% of links with a link utilization higher than 30%), some of them being over-used (3 links with a link utilization higher than 50%). Traffic is clearly unbalanced among all the links. DFS routing reduces this un-balancing and has 31% of links with link utilization lower than 10% and 9% of links with link utilization higher than 30%. The best traffic balancing is achieved by SMART routing. For the same traffic value, links are highly balanced, link utilization ranging from 7.76% to 20.26% (76% of links with a link utilization between 15% and 20%). As traffic is better balanced, more traffic can be handled by the SMART routing and, therefore, higher throughput is achieved.

Routing algorithms with ITBs Figure 2 shows the performance results obtained by the UD_MITB, UD_ITB, DFS_ITB and B_ITB routing algorithms for the uniform distribution of message destinations for selected 16, 32, and 64-switch networks, respectively. Table 1 shows the average results for 30 different

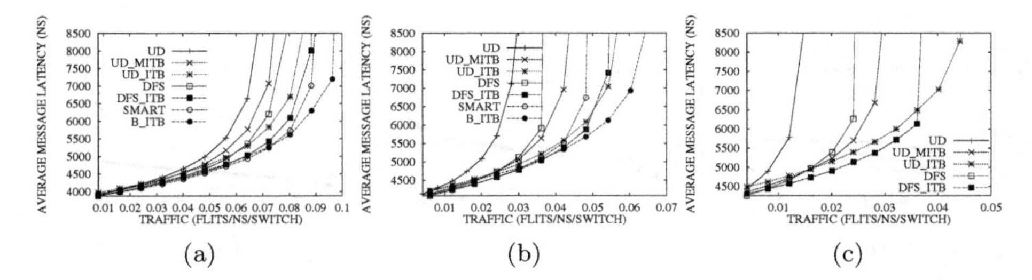

Fig. 2. Average message latency vs. accepted traffic. UD, DFS, SMART, UD_MITB, UD_ITB, DFS_ITB, and B_ITB routing. Message length is 512 bytes. Uniform distribution. Network size is (a) 16 switches, (b) 32 switches, and (c) 64 switches

Table 1. Factor of throughput increase when using in-transit buffers on the UD, DFS, and SMART routing. Uniform distribution. Message size is 512 bytes

	UD_MITB vs UD			UD_ITB vs UD			DFS_ITB vs DFS			B_ITB vs SMART		
Sw	Min	Max	Avg	Min	Max	Avg	Min	Max	Avg	Min	Max	Avg
16	1.00	1.29	1.13	1.00	1.57	1.29	1.01	1.20	1.12	1.00	1.16	1.07
32	1.16	1.72	1.46	1.50	2.14	1.88	1.25	1.56	1.41	1.11	1.33	1.23
64	1.60	2.25	1.91	2.20	3.00	2.57	1.50	1.85	1.63	N/A	N/A	N/A

topologies. For 64-switch networks, Smart routes were not available due to its high computation time.

Let us first comment on the influence of in-transit buffers on up*/down* and DFS. As can be seen, the ITB mechanism always improves network throughput over both original routing algorithms. Moreover, as network size increases, more benefits are obtained. In particular, UD_MITB improves over UD by factors of 1.12, 1.50, and 2.00 for 16, 32, and 64-switch networks, respectively. However, when more ITBs are used, more benefits are obtained. In particular, UD_ITB improves over UD by factors of 1.22, 2.14, and 2.75 for the 16, 32, and 64-switch networks, respectively. Concerning DFS, DFS_ITB routing improves network throughput over DFS by factors of 1.10, 1.39, and 1.54 for the same network sizes.

Notice that UD_ITB and DFS_ITB achieve roughly the same network throughput. These routing algorithms use the same minimal paths and the main difference between them is where the in-transit buffers are allocated and how many in-transit buffers are needed. Also, the DFS_ITB routing exhibits lower average latency than UD_ITB. This is because DFS routing is less restrictive than UD routing, and therefore, DFS_ITB needs fewer ITBs on average than UD_ITB. When using DFS_ITB routing in the 64-switch network, messages use 0.3 ITBs on average, while the average number of ITBs per message is 0.55 in UD_ITB. This also explains the higher throughput achieved by UD_ITB since messages using ITBs are removed from the network, thus reducing congestion.

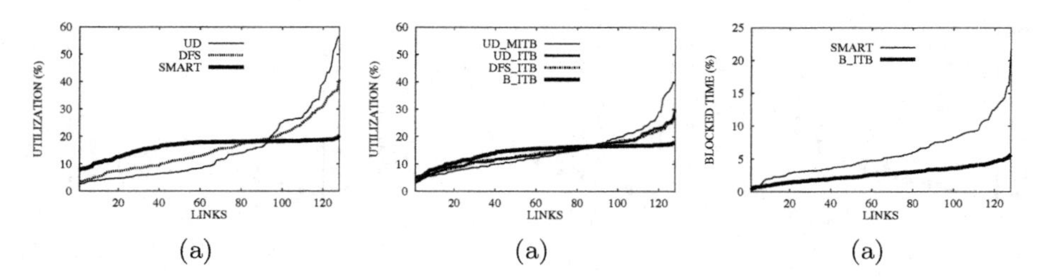

(a) (a) (a)

Fig. 3. Link utilization and link blocked time. Network size is 32 switches. Message size is 512 bytes. Uniform distribution. (a) Link utilization for original routings. (b) Link utilization for routings with ITBs. (c) Blocked time for SMART and B_ITB routing. Traffic is (a, b) 0.03 and (c) 0.05 flits/ns/switch

On the other hand, as network size increases, network throughput increases with respect to routing algorithms that do not use ITBs. UD and DFS routing are computed from a spanning tree and one of the main drawbacks of such an approach is that, as network size increases, a smaller percentage of minimal paths can be used. For the 16-switch network, 89% of the routes computed by UD are minimal. However, for 32 and 64-switch networks, the percentage of minimal routes goes down to 71% and 61%, respectively. When DFS routing is used, something similar occurs. There are 94%, 81%, and 70% of minimal routes for the 16, 32, and 64-switch networks, respectively. When using in-transit buffers, all the computed routes are minimal.

Another drawback of routing algorithms computed from spanning trees is unbalanced traffic. As network size increases, routing algorithms tend to overuse some links (links near the root switch) and this leads to unbalanced traffic. As in-transit buffers allow the use of alternative routes, network traffic is not forced to pass through the root switch (in the spanning tree), thus improving network performance. Figure 3.b shows the link utilization for UD_MITB, UD_ITB, DFS_ITB, and B_ITB routing, respectively, for the 32-switch network. Network traffic is 0.03 flits/ns/switch (where UD routing saturates). We observe that UD_MITB routing achieves a better traffic balancing than UD (see Figure 3.a). Only 33% of links have a link utilization lower than 10% and only 10% of links are used more than 30% of time. However, as this algorithm uses ITBs only when necessary to ensure deadlock-free minimal routing, a high percentage of routes are still valid minimal up*/down* paths, and therefore, part of the traffic is still forced to cross the root switch. UD_MITB traffic balance is improved by UD_ITB and DFS_ITB. With the UD_ITB routing, all links have a utilization lower than 30% and only 20% of links are used less than 10% of time. DFS_ITB routing shows roughly the same traffic balance.

Let us analyze now in-transit buffers with Smart routing. Smart routing is not based on spanning trees. Moreover, its main goal is to balance network traffic. In fact, we have already seen the good traffic balancing achieved by this routing algorithm (see Figure 3.a). Therefore, it seems that in-transit buffers will have little to offer to Smart routing. However, we observe in Table 1 that for Smart routing, the in-transit buffer mechanism also increases network throughput (except for one 16-switch network where it obtains the same network throughput). For a 32-switch network, B_ITB routing increases network throughput by a factor of 1.33.

In order to analyze the reasons for this improvement, Figure 3.b shows traffic balancing among all the links for B_ITB routing at 0.03 flits/ns/switch. As can be seen, it is very similar to the ones obtained by Smart (see Figure 3.a). The reason is that SMART routing is quite good in balancing traffic among all the links, and therefore, the in-transit buffer mechanism does not improve network throughput by balancing traffic even more.

To fully understand the better performance achieved by B_ITB routing we focus now on network contention. For this reason, we plot the link blocked time for both routing algorithms. Blocked time is the percentage of time that the link stops transmission due to flow control. This is a direct measure of network contention. Figure 3.c shows the link blocked time for a 32-switch network when using SMART and B_ITB routing. Traffic is near 0.05 flits/ns/switch. We observe that Smart routing has some links blocked more than 10% of time and some particular links being blocked more than 20% of time. On the other hand, when using in-transit buffers, blocked time is kept lower than 5% for all the links for the same traffic point.

In order to analyze the overhead introduced by ITBs, Table 2 shows the latency penalty introduced by in-transit buffers for very low traffic (the worst case). We show results for 512 and 32-byte messages. For 512-byte messages we observe that, on average, the in-transit buffer mechanism slightly increases average message latency. This increase is never higher than 5%. The latency increase is only noticeable for short messages (32 bytes). In this case, the maximum latency increase ranges from 16.66% to 22.09% for UD_ITB. The explanation is simple. The ITBs only increase the latency components that depend on the number of hops. Therefore, short messages suffer a higher penalty in latency. Additionally, the latency penalty depends on the number of ITBs needed to guarantee deadlock freedom. This is also shown in Table 2 where average latency penalty is lower when using ITBs with Smart, DFS or the minimum number of ITBs with UD (UD_MITB). Finally, the latency overhead incurred by ITBs is partially offset by the on-average shorter paths allowed by the mechanism.

Table 3 shows the factor of throughput increase for the hot-spot, bit-reversal, local, and combined traffic patterns. We observe that the in-transit buffer mechanism always increases, on average, network throughput of UD and DFS routing. In particular, when the combined traffic pattern is used, UD_ITB improves over UD by factors of 1.26, 1.65, and 2.31 for 16, 32, and 64-switch networks, respectively. Also, DFS_ITB improves over DFS by factors of 1.14, 1.35, and 1.56 for 16, 32, and 64-switch networks, respectively. Finally, B_ITB increases, on average, network throughput by a factor of 1.14 for 32-switch networks.

We conclude that, by using in-transit buffers on all the routing schemes analyzed, network throughput is increased. As network size increases, higher

Table 2. Percentage of message latency increase for very low traffic when using in-transit buffers on UD, DFS, and SMART routing. Uniform distribution

		UD_MITB vs UD			UD_ITB vs UD			DFS_ITB vs DFS			B_ITB vs SMART		
Msg. size	Sw	Min	Max	Avg	Min	Max	Avg	Min	Max	Avg	Min	Max	Avg
512	16	-0.24	0.76	0.20	1.01	2.63	1.69	0.22	0.74	0.57	-0.20	2.22	1.64
512	32	0.26	2.42	1.24	2.31	4.20	3.33	0.90	1.08	0.93	1.78	2.95	2.34
512	64	-3.85	-1.03	-2.22	-1.23	1.46	0.31	0.67	1.27	1.02	N/A	N/A	N/A
32	16	0.80	3.41	2.29	8.52	16.66	10.52	1.56	5.50	3.49	-0.35	10.18	7.77
32	32	2.33	7.57	5.32	13.16	18.07	16.44	6.09	7.28	6.59	9.26	13.28	11.00
32	64	1.40	5.97	3.64	11.69	22.09	16.87	6.44	8.56	7.64	N/A	N/A	N/A

Table 3. Factor of throughput increase when using in-transit buffers on the UD, DFS, and SMART routing for different traffic patterns. Message size is 512 bytes

		UD_MITB vs UD			UD_ITB vs UD			DFS_ITB vs DFS			B_ITB vs SMART		
Distrib.	Sw	Min	Max	Avg	Min	Max	Avg	Min	Max	Avg	Min	Max	Avg
Hot-spot	16	0.99	1.17	1.04	0.99	1.21	1.10	1.00	1.17	1.05	0.85	1.17	0.96
Hot-spot	32	1.00	1.40	1.18	1.00	1.39	1.18	0.98	1.17	1.03	1.00	1.00	1.00
Hot-spot	64	1.60	2.08	1.71	1.66	2.57	2.03	1.21	1.49	1.35	N/A	N/A	N/A
Bit-rev.	16	0.94	1.44	1.16	0.87	1.81	1.17	0.79	1.27	1.03	0.73	1.13	0.93
Bit-rev.	32	1.12	2.00	1.59	1.56	2.57	1.87	1.20	1.99	1.51	0.99	1.45	1.21
Bit-rev.	64	1.74	2.99	2.05	2.21	3.50	2.76	1.46	2.20	1.78	N/A	N/A	N/A
Local	16	0.97	1.26	1.08	1.02	1.56	1.24	1.00	1.30	1.17	1.00	1.17	1.10
Local	32	1.00	1.40	1.16	1.12	1.60	1.44	1.15	1.45	1.29	1.10	1.29	1.17
Local	64	1.00	1.20	1.07	1.40	1.57	1.49	1.13	1.33	1.24	N/A	N/A	N/A
Combined	16	1.00	1.45	1.15	1.00	1.56	1.26	0.98	1.28	1.14	1.00	1.17	1.06
Combined	32	1.12	1.57	1.31	1.31	1.86	1.65	1.20	1.50	1.35	1.04	1.27	1.14
Combined	64	1.48	2.00	1.74	1.82	2.65	2.31	1.43	1.80	1.56	N/A	N/A	N/A

improvements are obtained. In-transit buffers avoid congestion near the root switch (in the tree-based schemes), always provide deadlock-free minimal paths and balance network traffic. On the other hand, average message latency is slightly increased, but this increase is only noticeable for short messages and small networks.

4 Conclusions

In previous papers, we proposed the ITB mechanism to improve network performance in networks with source routing and up*/down* routing. Although the mechanism was primarily intended for breaking cyclic dependences between channels that may result in a deadlock, we have found that it also serves as a mechanism to reduce network contention and better balance network traffic. Moreover, it can be applied to any source routing algorithm.

In this paper we apply the ITB mechanism to up*/down*, DFS, and Smart routing schemes, analyzing its behavior in detail using up to 30 randomly generated topologies, different traffic patterns (uniform, bit-reversal, local, hot-spot, and combined), and network sizes (16, 32, and 64 switches). Network design parameters were obtained from a real Myrinet network.

Results show that, the in-transit buffer mechanism improves network performance for all the studied source routing algorithms. Up*/down* routing is significantly improved due to the many routing restrictions that it imposes and the unbalanced traffic nature of the spanning trees. Better source routing algorithms, like DFS and Smart, are also improved by the ITB mechanism. Finally, we have observed that as more ITBs are added to the network, throughput increases but the latency also increases due to the small penalty of using in-transit buffers. Therefore, there is a trade-off between network throughput and message latency. Thus, network designers have to decide on the appropriate number of ITBs depending on the application requirements.

As for future work, we plan to implement the proposed mechanism on an actual Myrinet network in order to confirm the obtained simulation results. Also, we are working on techniques that reduce ITB overhead.

References

1. N. J. Boden, D. Cohen, R. E. Felderman, A. E. Kulawik, C. L. Seitz, J. Seizovic and W. Su, "Myrinet - A gigabit per second local area network," *IEEE Micro*, pp. 29–36, February 1995. 302
2. L. Cherkasova, V. Kotov, and T. Rokicki, "Fibre channel fabrics: Evaluation and design," in *29th Int. Conf. on System Sciences*, Feb. 1995. 300
3. J. Flich, M. P. Malumbres, P. López, and J. Duato, "Improving Routing Performance in Myrinet Networks," in *Int. Parallel and Distributed Processing Symp.*, May 2000. 301
4. Myrinet, 'M2-CB-35 LAN cables, http://www.myri.com/myrinet/product_list.html' 302
5. J. C. Sancho, A. Robles, and J. Duato, "New Methodology to Compute Deadlock-Free Routing Tables for Irregular Networks," in *Workshop on Communications and Architectural Support for Network-based Parallel Computing*, January, 2000. 300
6. M. D. Schroeder et al., "Autonet: A high-speed, self-configuring local area network using point-to-point links," Tech. Rep. SRC research report 59, DEC, April 1990. 300
7. S. L. Scott and J. R. Goodman, "The Impact of Pipelined Channels on k-ary n-Cube Networks," in *IEEE Trans. on Parallel and Distributed Systems*, vol. 5, no. 1, pp. 2–16, January 1994. 302

A Comparison of Locality-Based and Recency-Based Replacement Policies

Hans Vandierendonck and Koen De Bosschere

Ghent University
Sint-Pietersnieuwstraat 41, B-9000 Gent, Belgium,
{hvdieren,kdb}@elis.rug.ac.be,
http://elis.rug.ac.be/~{hvdieren,kdb}/

Abstract. Caches do not grow in size at the speed of main memory or raw processor performance. Therefore, optimal use of the limited cache resources is of paramount importance to obtain a good system performance. Instead of a recency-based replacement policy (such as, e.g., LRU), we can also make use of a locality-based policy, based on the temporal reuse of data.
These replacement policies have usually been constructed to operate in a cache with multiple modules, some of them dedicated to data showing high temporal reuse, and some of them dedicated to data showing low temporal reuse.
In this paper, we show how locality-based replacement policies can be adapted to operate in set-associative and skewed-associative [8] caches.
In order to understand the benefits of locality-based replacement policies, they are compared to recency-based replacement policies, something that has not been done before.

1 Introduction

Trends in microprocessor development indicate that microprocessors gain in speed much faster than main memory. This discrepancy is called the memory gap. The memory gap can be hidden using multiple levels of cache memories. But even then, the delays introduced by the caches and main memory are becoming so large, that the memory hierarchy remains a bottleneck in processor performance.

An important part of the cache design is the replacement policy, which decides what data may be evicted from the cache. A recent approach to better replacement policies is using locality properties of the memory reference stream. In studies of such replacement policies, cache organisations consisting of multiple cache modules are used. Each module is a conventional cache and is dedicated to data with a specific locality type. A typical organisation is a direct mapped cache dedicated to data exhibiting temporal locality combined with a smaller and fully associative cache for data with non-temporal or highly spatial locality.

Such a cache organisation poses serious design problems. Since data can be found in either module, a multiplexer is needed to select the data from one of the

M. Valero et al. (Eds.): ISHPC 2000, LNCS 1940, pp. 310–318, 2000.

modules, increasing the cache lookup time. Furthermore, a direct mapped cache has an inherently shorter access time than a fully associative cache. It is not always possible to find two modules with the same access time. This results in an unbalanced design with one module in the critical path. Because of these difficulties, we propose to use locality-sensitive replacement policies in simple cache organisations, including the set-associative and the skewed-associative cache. However, applying locality-sensitivity to set-associative caches is not straightforward, because the operation of these replacement policies is closely interwoven with the organisation of a multi-module cache. The locality type of a block is derived from the module it is stored in. Therefore, we propose to label the blocks with their locality type, so that the replacement policy can make use of this information.

This paper is organised as follows. In section 2, we describe the various cache organisations. Section 3 describes replacement policies and their extension to set-associative and skewed-associative caches. In section 4 we present simulation results. Section 6 discusses related work and section 7 summarises the main conclusions.

2 Cache Organisations

The most wide-spread cache organisation is that of a set-associative cache. In a set-associative cache, memory blocks are mapped to cache sets by extracting bits from the block number (Figure 1(a)). An n-way set-associative cache can contain n blocks from the same set. If n is 1, the cache is called direct mapped. When there is only one set in the cache, the cache is called fully associative.

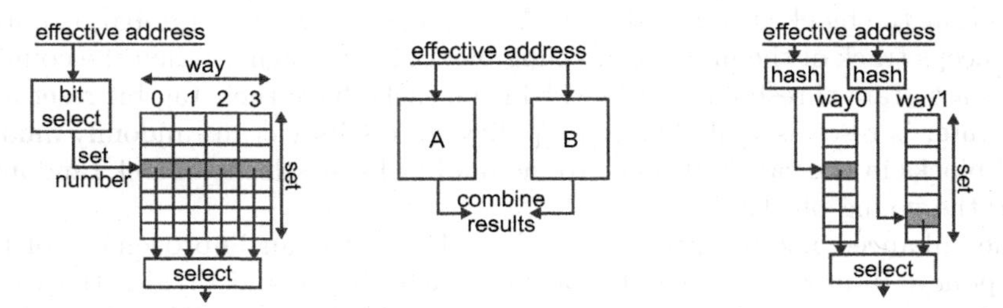

(a) Set-associative cache (b) Multi-module cache (c) Skewed-associative cache

Fig. 1. Three cache organisations

In a multi-module cache, multiple cache modules operate in parallel (Figure 1(b)). Each cache module can be thought of as a conventional cache, possibly having different associativity, block size, etc. A memory request is sent to all cache modules simultaneously, so extra combining logic is needed to obtain the result from the correct module. If data can only be cached in one module, then the combining logic consists of a multiplexer which selects the data from the module with a cache hit.

We study multi-module caches with two modules having the same block size. In the remainder of this paper, we call the cache modules A and B. The sets in module A and module B are called A-sets and B-sets, respectively.

A skewed-associative cache is a multi-bank cache. Each bank is indexed by a different set index function [8]. Furthermore, the index functions are designed in such a way that blocks are dispersed over the cache. When two blocks map to the same frame in one bank, then they do not necessarily map to the same frames in the other banks. An n-way skewed-associative cache can be modelled as a multi-module cache, where each module is direct mapped and corresponds to a bank of the skewed-associative cache.

3 Replacement Policies

In this section we describe the recency-based replacement policies and the locality-based replacement policies.

3.1 Recency-Based Replacement Policies

The *least recently used* (LRU) policy is commonly used in set-associative caches. Since the complexity of the LRU algorithm scales as the square of the number of blocks involved [11], it would be impractical to implement it for multi-module caches. Other recency-based algorithms are thus needed. We discuss the *not recently used* and the *enhanced not recently used* policies, which have been constructed for skewed-associative caches [8,9,10].

In the not recently used policy (NRU) a one bit tag is associated with every block in the cache. The tag bit is asserted every time the block is accessed and it signals that the block is young. The NRU policy also requires a (global) counter, which keeps track of the number of young blocks in the cache. When the counter reaches a certain threshold,[1] all blocks in the cache have their tag bit reset and the counter is reset as well. The NRU policy selects its victim randomly among all old blocks in the cache. If there are no old blocks, it selects a block randomly among the young blocks.

The enhanced not recently used policy (ENRU) is an improvement of the NRU policy. It uses two tag bits for each cache block and divides the cache blocks into three categories: very young, young and old. The ENRU policy selects a victim block at random, first among the old blocks, then the young blocks and, if necessary, among the very young blocks.

3.2 Locality-Based Replacement Policies

Several locality-sensitive replacement policies have been proposed in the literature. We focus on the *non-temporal streaming cache* and the *allocation by conflict*

[1] It is reported that a good threshold is half the size of the cache, expressed as a number of blocks [9].

policy. In these cache organisations, each module is dedicated to data exposing a specific type of locality. Hence, the replacement policy can be decomposed into two consecutive steps: (1) determining the locality type of the data and thus the module and (2) selecting a victim from the module e.g., using LRU.

The non-temporal streaming (NTS) cache [5], consists of three units: a temporal module A, a non-temporal module B and a locality detection unit. When a block is placed in the cache, it is placed in either module A or module B, depending on the locality properties reported by the locality detection unit. Blocks exposing mainly temporal locality are placed in module A, the other blocks in module B.

A block exposes temporal locality when at least one word in the block is used at least twice between loading the block and evicting it from the cache. Each block in the cache has one locality bit associated with it and each word in the cache has a reference bit. When a block is loaded into the cache, the block's locality bit is set to zero, indicating non-temporal use, and the reference bit of the requested word is set to one. If the reference bit of the requested word is one on a cache hit, then the locality bit is also set to one, indicating temporal use.

The detection unit is a small fully-associative cache, where the locality bits of evicted blocks are saved. In our implementation, only non-temporal blocks are stored in the detection unit's cache, since missing blocks have temporal locality.

In the *allocation by conflict* (ABC) policy [13], blocks are locked in a direct mapped cache until the number of misses exceeds the number of references to this block. The ABC policy adds a conflict bit to each block in module A. The conflict bit is set to zero when a reference is made to the associated block. It is set to one on each cache miss which maps into the same A-set. In each module, there is a candidate block for replacement, selected by LRU policies. One of these blocks is then selected using the conflict bit of the block in module A.

3.3 Locality-Sensitive Replacement Policies for Skewed- and Set-Associative Caches

A locality-sensitive replacement policy has to know the locality properties of the data in the cache, so we label each block with its type of locality. We use the not recently used (NRU) policy to account for some aging effect. The policies work similar to NRU. They define several categories of blocks and search them in their listed order.

The temporality-based NRU policy (TNRU) is a combination of NRU and NTS. The locality properties of the data are defined and detected in exactly the same way as in NTS. The recency ordering between cache blocks is maintained using the NRU policy. The TNRU policy can distinguish between four categories of blocks: old and non-temporal, old and temporal, young and non-temporal and young and temporal. The temporal properties of the block are decided using the locality information at the time of loading the block in the cache.

The second replacement policy we propose is the conflict based NRU policy (CNRU), based on the ABC policy. Each block in the cache has a conflict bit, which is managed in the same way as in ABC. The CNRU policy distinguishes between four categories of cache blocks, namely those that are old and not proven, young and not proven, old and proven and young and proven. A block is proven when its conflict bit is one.

4 Experimental Evaluation

We evaluated the performance of the replacement policies in three cache organisations: a multi-module cache, a skewed-associative cache and a set-associative cache. All caches have 32 byte blocks and use demand fetching. The skewed-associative and set-associative cache are both 8 kB large and have associativity two. The skewed-associative cache uses the index functions defined in [10].

The multi-module cache was chosen such that the modules have approximately the same cycle time. We used the cacti model [14] to obtain access times of several cache organisations in a $0.13\mu m$ technology. A fully associative 1 kB module has a 1.3 ns cycle time. To match this cycle time, the other module should be either 8 (1.23 ns) or 16 kB (1.32 ns) large and 2-way set-associative. Another possibility is to use a very large direct mapped cache, e.g. a 64 kB cache (1.36 ns). We choose to combine an 8 kB 2-way set-associative cache with a 1 kB fully associative cache.

Table 1. Miss ratios of the LRU policy in the different cache organisations

SPECfp	MM	SA	SK	SPECint	MM	SA	SK
applu	0.071	0.074	0.074	compress	0.045	0.051	0.049
apsi	0.048	0.085	0.040	gcc	0.044	0.063	0.048
fpppp	0.012	0.024	0.023	go	0.010	0.032	0.020
hydro2d	0.179	0.178	0.180	ijpeg	0.021	0.043	0.024
mgrid	0.054	0.054	0.055	li	0.038	0.045	0.041
su2cor	0.049	0.051	0.051	m88ksim	0.011	0.021	0.013
swim	0.080	0.609	0.091	perl	0.015	0.035	0.027
tomcatv	0.153	0.453	0.151	vortex	0.022	0.045	0.028
turb3d	0.061	0.083	0.050				
wave5	0.164	0.316	0.167				

We collected traces of all SPEC95 benchmarks using ATOM [12]. The number of memory references in each trace was limited to 300 million, taken from the middle of the program.

Miss ratios are used as the performance measure. However, since miss ratios vary greatly from program to program, the miss ratios are divided by the miss ratio of the LRU policy in the same cache organisation. Table 1 contains the miss ratios of the LRU policies in the different cache organisations, for reference. MM is the multi-module cache, SA is the set-associative cache and SK is the skewed-associative cache.

5 Discussion of the Results

For the multi-module cache, the most eye-catching result is that the locality-based replacement policies have very bad performance for some benchmarks (Figure 2). In one case, the miss ratio is increased with 175% (ABC for the benchmark tomcatv). In other cases, the miss ratio of a benchmark can be increased by as much as 5 or 10%. The replacement policies usually have a miss ratio that is worse than that of the LRU policy. For the SPECfp benchmarks, the ENRU and CNRU policies perform the best. These policies also closely follow the miss ratio of the LRU policy.

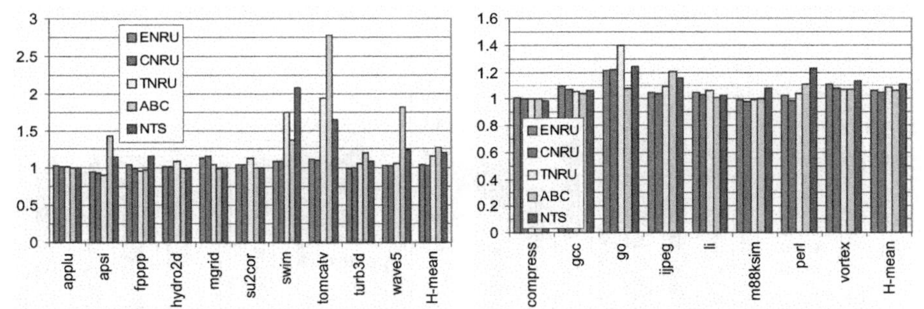

Fig. 2. Relative performance of replacement policies in the 8 kB two-way set-associative and 1 kB fully-associative multi-module cache for SPECfp (left) and SPECint (right)

For the SPECint benchmarks, the CNRU and ABC benchmarks provide the best results. In contrast to what happens for the SPECfp benchmarks, all replacement policies sometimes perform 10 to 20% worse than the LRU policy. However, this unexpected behaviour is not as bad as it is for the SPECfp benchmarks. Furthermore, the CNRU policy generally works better than the ABC policy, on which it is based. The same relation holds between TNRU and NTS.

Figure 3 shows the results for the set-associative cache. The ENRU policy has about the same performance as the LRU policy, while it has a larger cost. For some benchmarks, the TNRU policy provides a big improvement with respect to the LRU policy (e.g. 13% for fpppp and 8.6% for perl). On the average, TNRU performs about 1% better than LRU for SPECint and SPECfp.

In the skewed-associative cache, the ENRU policy usually performs worse than the LRU policy, except for fpppp (Figure 4). This was also reported in [10]. The locality-sensitive policies perform really well for the benchmark fpppp and they also work better than the ENRU policy for most benchmarks. Overall, the miss ratio of TNRU is about 2% lower than that of ENRU, both for SPECint and SPECfp. However, neither CNRU nor TNRU perform better on average than the LRU policy.

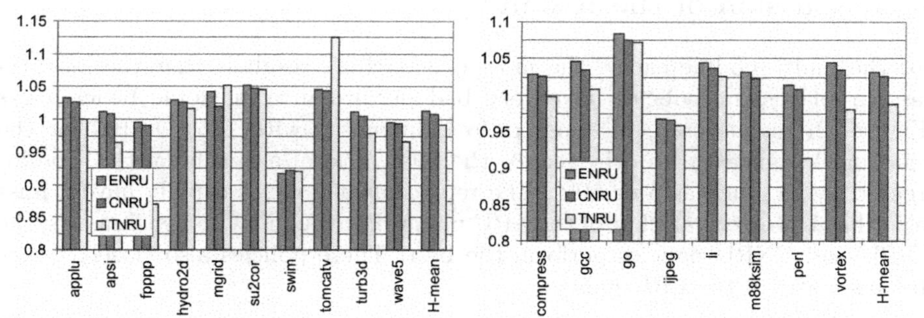

Fig. 3. Relative performance of replacement policies in the 8 kB two-way set-associative cache, for SPECfp (left) and SPECint (right)

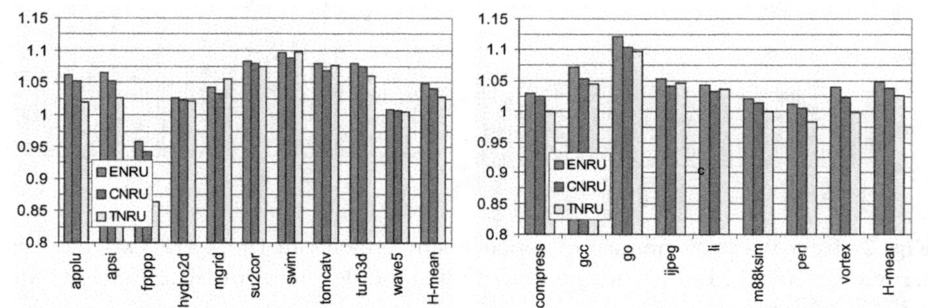

Fig. 4. Relative performance of replacement policies in the 8 kB skewed-associative cache, for SPECfp (left) and SPECint (right)

6 Related Work

Many different locality-sensitive replacement policies and accompanying cache organisations have been proposed. The NTS cache was introduced in [4] and was slightly changed in [5]. The main difference between these two versions is the way the locality properties of evicted blocks are remembered. Our implementation is based on [5].

The dual data cache [2] dedicates one module to data with high temporal locality, while the other module caches data with high spatial locality. The locality properties are detected by treating all fetched data as vectors. The stride and vector length is measured and is used to define three types of locality: non-vector data, short vectors and self-interfering vectors. The latter type is not cached at all. The speedup of the dual data cache is largely due to selective caching [2]. Alternatively, a compiler can detect the stride and vector length, as well as self- and group-reuse [6,7].

Several processors have implemented multi-module caches. The data cache of the HP PA-RISC 7200 consists of a large direct mapped cache and a small fully associative cache [1]. The purpose of the fully associative cache is to decrease the number of conflicts in the direct mapped cache.

Another approach is taken in the UltraSPARC III, which has a multi-module L1 and L2 data caches [3]. These caches are managed by splitting the reference stream, not on the basis of locality properties, but on the origin of the transfers between the caches.

7 Conclusions

We discussed the problems associated with applying locality-sensitive replacement policies to set-associative and skewed-associative caches. We extended two replacement policies from literature by labelling each block with its locality type.

We compared the locality-sensitive replacement policies to recency based replacement policies like LRU and ENRU. Overall, we find that the locality-sensitive replacement policies have approximately the same performance as recency based policies. Recency-based replacement policies can manage a multi-module cache as good as or better than locality-sensitive policies. Furthermore, the locality-based policies are mostly suited for the SPECfp benchmarks although they show very poor behaviour for some benchmarks. In contrast, the recency-based policies are more well-behaved.

For set-associative caches, the locality-based replacement policy TNRU decreases the miss ratio slightly over LRU (with about 1%). In a skewed-associative cache, the TNRU policy provides a 2% improvement over the ENRU policy.

Acknowledgements

The authors thank Lieven Eeckhout for his helpful comments in proof-reading this manuscript. Hans Vandierendonck is supported by a grant from the Flemish Institute for the Promotion of the Scientific-Technological Research in the Industry (IWT). Koen De Bosschere is research associate with the Fund for Scientific Research-Flanders.

References

1. Kenneth K. Chan, Cyrus C. Hay, John R. Keller, Gordon P. Kurpanek, Francis X. Schumacher, and Jason Zheng. Design of the HP PA 7200 CPU. *Hewlett-Packard Journal*, 47(1), February 1996. 317
2. A. Gonzalez, C. Aliagas, and M. Valero. A data cache with multiple caching strategies tuned to different types of locality. In *ICS'95. Proceedings of the 9th ACM International Conference on Supercomputing*, pages 338–347, 1995. 316
3. T. Horel and G. Lauterbach. UltraSPARC-III: Designing third-generation 64-bit performance. *IEEE Micro*, 19(3):73–85, May 1999. 317

4. J. A. Rivers and E. S. Davidson. Reducing conflicts in direct-mapped caches with a temporality-based design. In *Proceedings of the 1996 International Conference on Parallel Processing*, volume 1, pages 154–163, August 1996. 316
5. J. A. Rivers, E. S. Tam, G. S. Tyson, E. S. Davidson, and M. Farrens. Uitilizing reuse information in data cache management. In *ICS'98. Proceedings of the 1998 International Conference on Supercomputing*, pages 449–456, 1998. 313, 316
6. F. Jesús Sánchez, Antonio González, and Mateo Valero. Software management of selective and dual data caches. *IEEE Technical Committee on Computer Architecture Newsletter*, pages 3–10, March 1997. 316
7. Jesús Sánchez, , and Antonio González. A locality sensitive multi-module cache with explicit management. In *ICS'99. Proceedings of the 1999 International Conference on Supercomputing*, pages 51–59, Rhodes, Greece, June 1999. 316
8. A. Seznec. A case for two-way skewed associative caches. In *Proceedings of the 20th Annual International Symposium on Computer Architecture*, pages 169–178, May 1993. 310, 312
9. A. Seznec. A new case for skewed-associativity. Technical Report PI-1114, IRISA, July 1997. 312
10. A. Seznec and Francois Bodin. Skewed-associative caches. In *PARLE'93: Parallel Architectures and Programming Languages Europe*, pages 305–316, Munich, Germany, June 1993. 312, 314, 315
11. Alan Jay Smith. Cache memories. *ACM Computing Surveys*, 14(3):473–530, September 1982. 312
12. Amitabh Srivastava and Alan Eustace. ATOM: A system for building customized program analysis tools. Technical Report 94/2, Western Research Laboratory, March 1994. 314
13. Edward S. Tam. *Improving Cache Performance Via Active Management.* PhD thesis, University of Michigan, 1999. 313
14. Steven J. E. Wilton and Norman E. Jouppi. An enhanced access and cycle time model for on-chip caches. Technical Report 93/5, Western Research Laboratory, July 1994. 314

The Filter Data Cache: A Tour Management Comparison with Related Split Data Cache Schemes Sensitive to Data Localities*

Julio Sahuquillo[1], Ana Pont[1], and Veljko Milutinovic[2]

[1] Dept. de Informática de Sistemas y Computadores,
Universidad Politécnica de Valencia
Cno. de Vera s/n, 46071 Valencia, Spain
{jsahuqui,apont}@disca.upv.es
[2] Dept. of Computer Engineering, School of Electrical Engineering,
University of Belgrade
POB 35-54, 11120 Belgrade, Serbia, Yugoslavia
VM@etf.bg.ac.yu

Abstract. Recent cache research has mainly focussed on how to split the first-level data cache. This paper concentrates on the redesign of the filter data cache scheme, presenting some improvements on the first version of the scheme. A performance study compares these proposals with other organizations that split caches according to the criterion of data localities. The new filter cache schemes exhibit better performances than the other compared solutions. An 18 KB organization offers a block management capacity equivalent to a conventional 28 KB cache.

1 Introduction

Recent research [1-12] has focused on optimizing the first-level (L1) data cache organization in order to increase the L1 hit ratio and reduce this critical time. Usually, the proposed models classify the data lines in two independent sets; according to a predefined characteristic exhibited by the data. To improve performance, both types of data are then cached and treated separately in caches with independent organizations. For this purpose, the L1 cache is usually split into two parallel caches also called subcaches because both make up the first level and each caches one type of predefined data line. The main advantage of having two independent subcaches is that it is possible to tune each specific organization (cache size, associativity, and block size), and replacement algorithm, according to the characteristics of the data. The criterion of data locality prevails among the schemes that split the data cache. More information on this subject can be found in references [17,18,19].

* This work has been partially supported by Research Grant GV98-14-47

M. Valero et al. (Eds.): ISHPC 2000, LNCS 1940, pp. 319-327, 2000.

The filter data cache [6] is a scheme that splits the first level data cache into two subcaches, the smaller cache filters the most strongly referenced blocks, and is connected by a unidirectional data path to the larger cache. The datapath is used to move blocks to the larger subcache when a block is replaced from the smaller subcache.

This paper presents new approaches for the filter data cache scheme, and compares performance with two recent schemes [3, 4].

2 Existing Solutions

Two earlier proposals for handling spatial and temporal localities in separate caches were introduced in [1, 2]; however, both localities can appear together, or not appear at all. In the STS scheme [3] the cache is split by giving priority to temporal locality, and the lines exhibiting some temporal locality are cached together in a large organization, while the lines that do not exhibit temporal locality are cached separately. In [4] the other extreme is performed by giving priority to the spatial locality. In [5] the first level data cache is split into three organizations, one caches the lines exhibiting temporal and spatial localities together, another caches lines exhibiting only spatial locality, and the final organization only caches lines showing temporal locality.

Schemes not designed to exploit data localities in independent caches have also been proposed. The Assist cache [7] tries to reduce the conflict misses by adding a small fully associative cache. The Victim cache [8] tries to retain the most recent conflict lines in a small cache between the first level and the second level of the memory hierarchy. The Allocation By Conflict scheme [9] tries to take replacement decisions based on the behavior of the conflict block allocated in the "main subcache". To avoid introducing pollution into the cache, some schemes propose bypassing the cache lines that are infrequently referenced [10,11,13]. Other schemes propose caching them in a small bypass buffer [11]. In [12], the data cache is split according to the type of data scalar or array.

In those schemes managing reuse information, a line in the first level cache uses a hardware mechanism to gather information about the behavior of a block in cache (current information). When the line is replaced from the first level cache, this information is flushed to the L2 cache (or to another structure in the first level), and then used when the block is again referenced (reuse information) to decide in which first level cache the line must be placed. In general, the schemes have two caches at the first level, the larger one, or "main cache", and a smaller cache that usually works as an assistant to improve performance. The reuse information is reset (lost) when a line is removed from the second level cache. In addition, some schemes introduce a datapath connecting both first-level caches.

2.1 Non Temporal Streaming Cache (NTS)

In the NTS cache [3] proposed by Rivers et al. the data is dynamically tagged as temporal or non-temporal. The model shows a large temporal cache placed in parallel with a small non-temporal cache. Each line in the temporal cache has a reference bit array attached, in addition to a non-temporal (NT) bit. When a block is placed in the temporal cache, each bit in the reference bit array is reset; and the NT bit is set. When a hit occurs in this cache, the bit associated with the accessed word is set. If the bit was already set (meaning that the word had already been accessed), the NT bit is reset to indicate the line showing temporal behavior. When a line is removed from the first level cache its NT bit flushes to the second level cache. If the line is referenced again, this bit is checked to decide where it must be placed.

2.2 The Split Spatial/Non-Spatial (SS/NS)

The SS/NS cache was proposed by Prvulovic et al. [4]. This scheme makes a division between spatial and non-spatial data lines, giving priority to lines exhibiting spatial locality. The model introduces a large spatial cache in parallel with a non-spatial cache that is four times smaller. The spatial cache exploits both types of spatial locality (only spatial, or both spatial and temporal). Line size in the non-spatial cache is just one word; thus, only temporal locality can be exploited in this cache. In the spatial cache, the line size is larger (four words).

The spatial cache uses a prefetch mechanism to assist this type of locality. A hardware mechanism is introduced to recompose lines in the non-spatial cache and move them (by a unidirectional data path) to the spatial cache. It uses a reference bit array similar to the one incorporated in the NTS scheme to tag lines, which are tagged as spatial if more than two bits are set; otherwise, they are tagged as non-spatial.

2.3 The Filter Data Cache

The filter data cache was introduced in a previous paper [6]. The model presents a very small direct-mapped "filter" cache in addition to a large "main cache" in the first level. The scheme tries to identify the most heavily referenced lines and places them together in the small "filter cache".

Each cache line has a 4-bit attached counter showing the number of times that each cache line is referenced. When the access results in a hit in any subcache, the counter is increased. If the access results in a miss in both subcaches, the counter of the referenced line is compared with the counter of the conflict line in the filter cache to decide in which subcache the referenced line will be placed. If the counter of the referenced line is less than the conflict in the filter cache, then the miss line is placed in the main cache. Otherwise, the model assumes that the miss line is more likely to be referenced again than the conflict line, and so it is placed in the filter cache. As the lines in the filter cache have shown a high frequency reference, when they evict the cache they move by using a unidirectional datapath to the main cache to spend more time in the first-level. The counter value (four bits) is the only information flushed to the L2 cache when lines are evicted from the first level cache.

In this work, we add associativity to the small filter cache. So, when a conflict occurs, the line with the lower counter in the set will be compared against the referenced block counter. The result of the comparison decides in which subcache the referenced block must be placed. To avoid blocks with low, or non-temporal locality, moving to the filter cache, a minimum counter threshold is established. A threshold greater than zero means that the defect cache is the "main cache". From time to time, each line counter is shifted to ensure that lines with good temporal locality during just a phase of their execution remain in the filter cache when their temporality drops.

If an application has many blocks showing high temporal locality, then the blocks will be quickly replaced from the filter cache ending up in the main cache. In such specific situations, the effectiveness of the filter is very poor. To improve performance in such cases, a bi-directional datapath connecting the first-level caches is introduced to allow line swapping. The result of adding this feature is a different model called the Filter Cache Swap. For a performance evaluation study we consider two approaches, one of them incorporating the swapping mechanism.

3 Assumptions and Conditions of the Analysis

All organizations in the cache hierarchy are two-way set associative with a 256 bits (eight 32-bit words) block size; the "main cache" capacity is 16 KB and the "small cache" capacity is 2 KB. Due to traditional mapping function restrictions, cache capacities must be a power of two. Thus, if we wish to compare performances of the proposed organizations having 18 KB (16KB plus 2 KB) capacities with the conventional ones, we have several options. One commonly adopted solution is to compare performances against a 16 KB conventional cache which has similar capacity [9,16]. In this paper, we estimate the theoretically equivalent capacity of a conventional cache offering the same performance. To do this, we assume that performances between 16 KB and 32 KB behave linearly, and we estimate the point at which the performance would be. Thus, we compare performances among the splitting data cache schemes and present results with a conventional 16 KB cache used as baseline scheme, as well as a larger 32 KB conventional organization. The L2 cache is 256 KB. A data bus with a line-size width is assumed between the first level and the second level cache. All caches are two-way set associative.

4 Simulation Analysis

4.1 Experimental Framework and Benchmarks

Performance results have been obtained using the execution driven simulator LIMES [15] and several suites of benchmarks. We selected five SPLASH-2 benchmarks (FFT, Radix, FMM, LU and Barnes), the compress benchmark from the SPEC suite, and the two benchmarks (MM and Jacobi) discussed in [15]. The problem size in all the selected benchmarks exceeded 60M of memory references, except the benchmark Compress that was run using training data inputs.

The Tour of a Line is defined [16] as the interval of time from when the line is placed in the first level cache until the line is evicted from that level. The number of tours and their lengths, measured in mean number of accesses that hit the line while it is in a tour are used to evaluate the effectiveness of cache data management.

4.2 Hit Ratio and Tour Analysis

Table 1 shows the L1 miss ratio of the split data cache schemes, the 16 KB baseline conventional cache and the larger 32 KB cache. We are not interesting in comparing performances of the data splitting cache schemes with the performance offered by the larger classic cache, near twice their capacity. We only wish to show that performances obtained by the splitting schemes are closer to the large traditional cache than to the small, and similarly sized, traditional cache.

All the data splitting cache models improve the miss ratio of the conventional 16 KB cache, except the SS/NS model in the FMM benchmark. The filter scheme gives the best performance among the splitting data cache models, except for the Jacobi application where the NTS model performs better. This result is explained by the fact that while the data lines in this application exhibit high temporal locality; they also have bad tour persistence. In other words, some lines that are tagged as non-temporal at the end of a tour, start their next tour in the non-temporal cache, and then unexpectedly exhibit temporal locality. Consequently, despite this undesirable behavior, the NTS scheme makes excellent use of both subcaches in Jacobi.

Table 1. Miss ratio (%) of the schemes with an 8 words block size

Bencmark	16 KB	FILTER	FILTERswap	NTS	SS/NS	32 KB
Barnes	2.42	1.65	1.76	1.86	2.11	0.98
FMM	0.79	0.63	0.65	0.71	1.18	0.51
LU	1.23	0.93	1.08	1.19	1.23	1.15
FFT	3.61	3.59	3.60	3.52	3.60	3.17
RADIX	1.29	1.29	1.29	1.29	1.29	1.29
MM	10.1	10.1	9.94	10.1	10.07	10.1
Jacobi	18.7	13.3	13.54	4.40	18.73	18.7
Compress	3.12	2.91	2.92	3.01	3.03	2.17
Average	*3.23*	*3.01*	*3.03*	*3.10*	*3.22*	*2.76*

The results also show that better miss ratios are achieved in the filter scheme if the swapping mechanism is disabled, with the single exception of the MM application. The differences in the Radix kernel among the schemes are negligible because of the high data localities exhibited by these benchmarks. The miss ratio of the filter scheme is not only the best among the splitting models, but it also achieves a better miss ratio in LU and Jacobi than the larger conventional cache with nearly twice its capacity.

The lower the number of tours, the better is the effectiveness of a cache scheme. Table 2 shows the number of tours in the 16 KB conventional cache, and the percentage of reduction in tours offered by the other schemes. A negative value means that the number of tours increases with respect to the 16KB conventional cache.

The results show that the filter scheme is the 18 KB organization that most reduces the number of tours, showing significant improvements in four of the eight benchmarks used (Barnes, FMM, LU, and Compress). Its reduction sometimes doubles that obtained by the NTS and SS/NS schemes; furthermore, in some cases (LU and Jacobi) its reduction is better than the large classic cache. The only exception appears in Jacobi, where the NTS scheme shows better results than the filter scheme.

From data in Table 2 we estimated the theoretically equivalent conventional cache capacity as explained before, and show these results in Table 3. The filter cache scheme with 18 KB offers tour reductions equivalent to those offered by a theoretically conventional cache with a capacity of 28 KB. The filter swap and the NTS also offer good results equivalent to a conventional cache with a capacity of 24 KB. Poorer performances are offered by the SS/NS; only just reaching the equivalent of a 19 KB conventional cache. We have omitted the Jacobi values to estimate the average because it presents very large values showing an unusual behavior.

Table 2. Tours in the 16 KB classic cache and reduction in tours (%) offered by the other schemes

	# Tours	% Reduction in Tours				
Benchmark	16 KB	FILTER	FILTER swap	NTS	SS/NS	32 KB
Barnes	2015920	31.93	27.09	23.01	13.00	59.46
FMM	887092	20.46	17.06	10.17	-49.49	35.28
LU	2353847	7.74	4.45	3.48	0.23	3.75
FFT	1433600	2.13	-2.94	2.34	0.13	12.27
RADIX	516246	0.22	-0.07	0.20	0.30	0.26
MM	17072741	0.55	0.57	0.57	0.55	0.58
Jacobi	16967214	29.15	27.77	76.54	0.10	0.15
Compress	270003	6.61	6.46	3.46	2.89	30.55
Average	*5189583*	*12.35*	*10.05*	*14.97*	*-4.04*	*17.79*

Table 3. Theoretic Equivalents Conventional Caches Capacities

Benchmark	FILTER	Filter swap	NTS	SS/NS
Barnes	25	23	22	19
FMM	25	24	21	-6
LU	49	35	31	17
FFT	19	16	19	16
RADIX	30	16	28	35
MM	31	32	32	31
Jacobi	3093	2947	8095	27
Compress	19	19	18	18
Average	*28*	*24*	*24*	*19*

5 Hardware Cost

To calculate the hardware cost in bits in the first level of the cache hierarchy, all organizations are assumed to be two-way set associative with a line size of 32 bytes.

The hardware cost of the filter cache schemes cache is approximately 14% greater than that incurred by the 16 Kbyte conventional caches. On the other hand, the cost incurred by the 32 Kbyte cache is about 74 % greater than the filter schemes. In summary, the filter data caches with only 576 lines (512 plus 64) improve performances over organizations with the same capacity; and even sometimes surpass the performances offered by organizations using 1024 lines.

6 Conclusions

In this paper two new improvements of the filter data cache scheme have been presented. We have evaluated the performances of these schemes and compared them with two other schemes that split the cache according to the criterion of the data locality (the STS and SS/NS) and recently appeared in the literature.

In this initial study, we chose hit ratio and tour management as performance indexes that are independent of the memory access time. The unidirectional datapath between the first-level caches is used to prolong the time that a heavily referenced line spends at that level. In this sense, the results show that the proposed schemes offer better hit ratio and tour management than those splitting the cache according to the criterion of the data localities. The filter scheme offers tour management that equals a theoretically equivalent conventional cache with a capacity of 28 KB. In some cases, a cache with just 576 blocks (18 KB) improves the tour management offered by the classic cache with 1024 blocks (32 KB).

Data localities are continuously changing, so lines sometimes offer a bad tour persistency; and this reduces the efficacy of those schemes that split caches according to the criterion of data localities. Their performances consequently drop. This problem does not appear in filter schemes that place most referenced lines in a small cache. In these schemes, a bad persistency implies that lines exhibiting bad tour persistency will be quickly replaced from the filter cache. In addition, we shift the counters to ensure that lines which dynamically change their localities do not continue residing in the filter cache.

Acknowledgments

We would like to thank Prof. Milo Tomasevic, Prof. Igor Tartalja, Mr. Igor Ikodinovic, and the anonymous referees for their helpful comments. We would also like to thank Dr. Alexander Milenkovic for his help in the use of the benchmarks he developed.

References

1. A. González, Carlos Aliaga, and M. Valero, "A Data Cache with Multiple Caching Strategies Tuned to Different Types of Locality," Proceedings of the ACM International Conference on Supercomputing, Barcelona, Spain 1995, pp. 338-347.
2. V. Milutinovic, B. Markovic, M. Tomasevic, and M. Tremblay, "The Split Temporal/Spatial Cache: Initial Performance Analysis," Proceedings of the SCIzzL-5, Santa Clara, California, USA, March 1996, pp. 63-69.
3. J.A. Rivers and E.S. Davidson, "Reducing Conflicts in Direct-Mapped Caches with a Temporality-Based Design," Proceedings of the 1966 ICPP, August 1996, pp. 151-160.
4. M. Prvulovic, D. Marinov, Z. Dimitrijevic and V. Milutinovic, "The Split Spatial/Non Spatial Cache: A Performance and Complexity Analysis," IEEE TCCA Newsletter, July 1999, pp. 8-17.
5. J. Sánchez and A. González, "A Locality Sensitive Multi-Module Cache with Explicit Management," Proceedings of the ACM International Conference on Supercomputing, Rhodes, Greece, June 1999.
6. J. Sahuquillo and A. Pont, "The Filter Cache: A Run-Time Cache Management Approach," Proceedings of the 25th Euromicro Conference, Milan, Italy, September 1999, pp. 424-431.
7. K.K. Chan, C.C. Hay, J.R. Keller, G.P. Kurpanek, F.X. Schumacher, J. Zheng, "Design of the HP PA 7200 CPU," Hewlett-Packard Journal, February 1996, pp. 1-12.
8. N. Jouppi, "Improving Direct-Mapped Cache Performance by the Addition of a Small Fully-Associative Cache and Prefetch Buffers," Proceedings of the ISCA-17, June 1990, pp.364-373.
9. E.S.Tam, "Improving Cache Performance via Active Management," Ph.D. dissertation, University of Michigan, June 1999.
10. G. Tyson, M. Farrens, J. Matthews, and A.R. Pleszkun, "A Modified Approach to Data Cache Management," Proceedings of Micro-28, December 1995, pp. 93-103.
11. T. Johnson and W.W. Whu, "Run-time Adaptative Cache Hierarchy Management via Reference Analysis," Proceedings of the ISCA-24, June 1997, pp. 315-326.
12. M. Tomasko, S. Hadjiyiannis, and W. A. Najjar, "Experimental Evaluation of Array Caches," IEEE TCCA Newsletter, March 1997, pp. 11-16.
13. T. Johnson, D. A. Connors, M.C. Merten, and W.W. Whu, "Run-time Cache Bypassing," IEEE Transactions on Computers, vol. 48, no. 12, December 1999, pp. 1338-1354.
14. I. Ikodinovic, D. Magdic, A. Milenkovic, and V. Milutinovic, "Limes: A Multiprocessor Simulation Environment for PC Platforms," Proceedings of the Third International Conference on Parallel Processing and Applied Mathematics, Poland, 1999, pp. 398-412.
15. A. Milenkovic, "Cache Injection in Bus-Based Shared Memory Multiprocessors," Ph.D dissertation, University of Belgrade, 1999.
16. E.S.Tam, J.A. Rivers, V. Srinivasan, G.S. Tyson, and E.S. Davidson, "Active

Management of Data Caches by Exploiting Reuse Information," IEEE Transactions on Computers, vol. 48, no. 11, November 1999, pp. 1244-1259.
17. V. Milutinovic, and M. Valero,"Cache Memory and Related Problems: Enhancing and Exploiting the Locality," IEEE Transactions on Computers, vol. 48, no. 2, February 1999, pp. 97-99.
18. V. Milutinovic, and P. Strenstrom,"Scanning the Special Issue on Distributed Shared Memory Systems," Proceedings of the IEEE, vol. 87, no. 3, March 1999, pp. 399-403.
19. V. Milutinovic, "Microprocessor and Multimicroprocessor Systems," Wiley, 2000.

Global Magneto-Hydrodynamic Simulations of Differentially Rotating Accretion Disk by Astrophysical Rotational Plasma Simulator

Mami Machida[1], Ryoji Matsumoto[2], Shigeki Miyaji[1], Kenji E. Nakamura[3], and Hideaki Tonooka[3]

[1] Graduate School of Science and Technology, Chiba University
1-33 Yayoi-Cho, Inage-ku, Chiba 263-8522, Japan
machida@c.chiba-u.ac.jp
http://www.c.chiba-u.ac.jp/aplab/
[2] Department of Physics, Faculty of Science, Chiba University
1-33 Yayoi-Cho, Inage-ku, Chiba 263-8522, Japan
[3] Japan Science and Technology Coorporation

Abstract. We present numerical results of three-dimensional global magneto-hydrodynamic (MHD) simulations achieved on Astrophysical Rotating Plasma Simulator (ARPS) developed at Chiba University. We simulate the time evolution of differentially rotating disks by using a parallelized three-dimensional MHD code. Typical number of grid points is $(N_r, N_\phi, N_z) = (200, 64, 240)$ in a cylindrical coordinate system. We found that when the initial magnetic field is toroidal and relatively strong, the system approaches a quasi-steady state with $\beta = P_{\mathrm{gas}}/P_{\mathrm{mag}} \sim 5$. When the disk is threaded by vertical magnetic fields, magnetically driven collimated jet emanates from the surface of the disk. Fully vector-parallelized global simulations with ARPS enable us to study non-local effects such as magnetic pinch, saturation of non-linear growth of instability, and deformation of the global structure.

1 Introduction

Numerical simulation is a fundamental tool to investigate active phenomena in astrophysical objects because they occur under extreme conditions which laboratory experiments can not mimic. Rotation and magnetic fields often play essential roles in such phenomena

When matter with angular momentum infalls from the interstellar medium or from the companion star, it spirals in and forms a rotating disk around the gravitating object. The matter inside the disk gradually accretes by losing angular momentum. Such a disk is called an accretion disk. In conventional theories of accretion disks (see, e.g., Shakura and Sunyaev 1973), the turbulent viscosity and the Maxwell stress exerted by turbulent magnetic fields is postulated to transport angular momentum outward and to drive accretion.

The gravitational energy released through the accreting process is believed to be the origin of activities in X-ray binaries, dwarf novae, and active galactic

M. Valero et al. (Eds.): ISHPC 2000, LNCS 1940, pp. 328–335, 2000.

nuclei. When a rotating disk is threaded by large-scale poloidal magnetic fields, centrifugal force and magnetic pressure drive outflows (see e.g., Blandford and Payne 1982 ; Uchida and Shibata 1985).

Since three dimensional effects are essential in dynamo process and turbulence in accretion disks, we need to carry out 3D simulations.

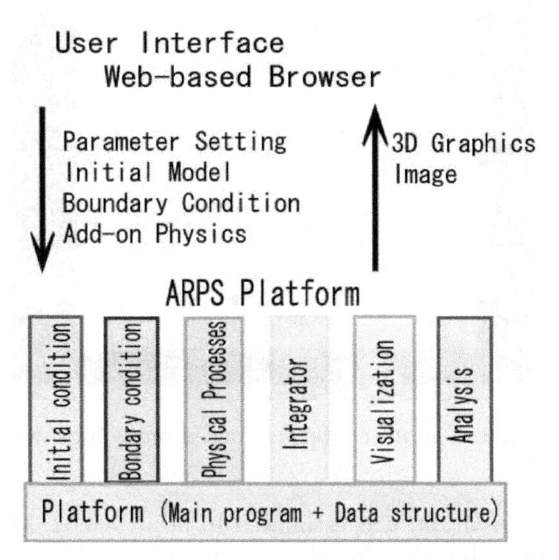

Fig. 1. The conceptual design of astrophysical rotating plasma simulator

For such needs of astrophysical community, we are developing an astrophysical rotating plasma simulator (ARPS) by which we can carry out global three-dimensional magneto-hydrodynamic (MHD) simulations of rotating plasma. Fig.1 shows the concept of ARPS. It consists of modules of initial model set-up, mesh generator, time integrator by finite differencing (engine), add-on sub-modules which incorporate various physics (e.g., resistivity, radiative cooling, heat conduction, self-gravity), and visualizer. Web based user-interface (Fig.2) enable users to set up initial model, and boundary conditions, and to monitor the progress of their simulation. The sub-modules which incorporate various physics share the same data structure and can be plugged into the platform of the simulator. Each module is parallelized by using the MPI library.

2 Importance of Global Simulations for the Study of Accretion Disks

Conventionally, the viscosity inside the accretion disk is assumed with the phenomenological parameter α, i.e., the off-diagonal component of the stress tensor $t_{r\phi}$ is assumed to be proportional to the pressure ($t_{r\phi} = -\alpha P$). By comparing

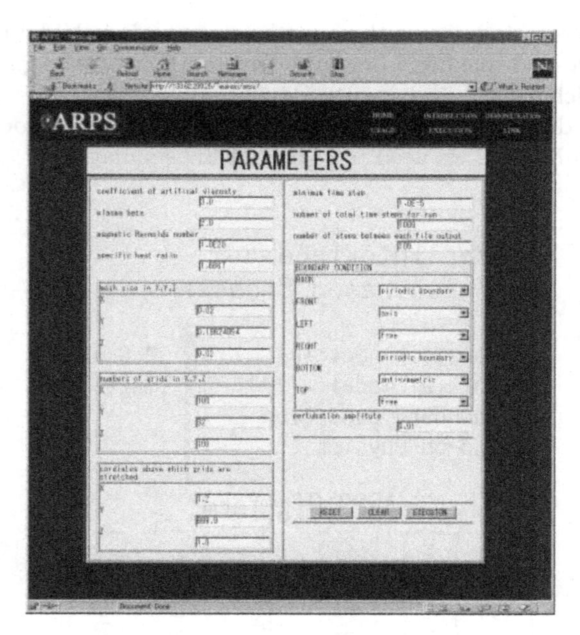

Fig. 2. The user interface based on web browser

the theory and observation, α is estimated to be $\alpha \sim 0.02$. Molecular viscosity cannot afford with such high value of α (Cannizzo et al. 1988). Balbus and Hawley (1991) pointed out the importance of local magneto-rotational (or Balbus & Hawley) instability in accretion disks which generates turbulence in accretion disks and enhances angular momentum transport rate. When a differentially rotating plasma is threaded by magnetic fields (Fig.3), instability grows if the radial force created by transporting angular momentum outward is larger than the restoring magnetic tension. This magneto-rotational instability grows even when the magnetic field is very weak. The maximum growth rate is the order of the angular velocity of the disk. The azimuthal magnetic fields also subject to the non-axisymmetric magneto-rotational instability (Balbus and Hawley 1992).

Three-dimensional local MHD simulations of the nonlinear growth of the magneto rotational instability have been carried out by several authors (Hawley et al. 1995, Matsumoto and Tajima 1995, and Brandenburg et al. 1995). It turned out, however, that the growth rate and the saturation level of the instability depend on the size of simulation box and global structure of the magnetic field. Therefore, global simulation is required in order to evaluate actual value of α. Moreover, global simulation is essential to simulate global phenomena such as jet formation and collimation.

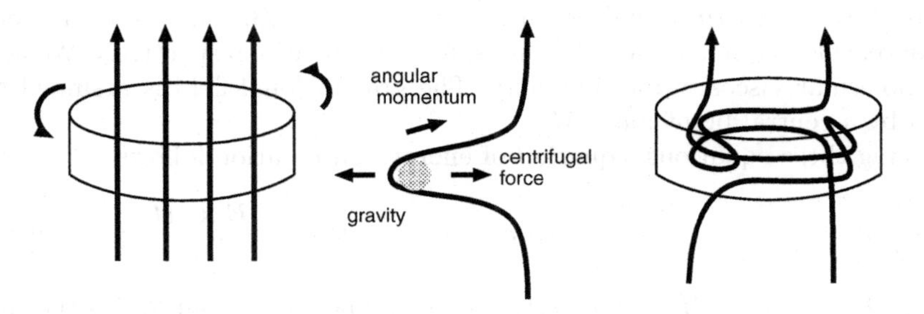

Fig. 3. The mechanism of magneto-rotational instability

3 Three-Dimensional MHD Simulations of Accretion Disks

As an initial model for the disk, we assume that a rotating polytropic ($\ln P \propto \ln \rho$) torus with constant angular momentum distribution $L = L_0$ is threaded by toroidal magnetic fields and assume that the torus is embedded in a spherical, non-rotating isothermal halo (see Fig. 4). In a cylindrical coordinate system (r, ϕ, z), the dynamical equilibrium of the torus is obtained by assuming

$$\Psi_0 = -\frac{1}{R} + \frac{L^2}{2r^2} + \frac{1}{\gamma - 1} v_s^2 + \frac{\gamma}{2(\gamma - 1)} v_A^2 = constant \ ,$$

where R is the distance from the center, v_s^2 is the square of the sound speed, γ is the specific heat ratio. We take the radius where rotation velocity is equal to the Keplerian rotation velocity ($v = v_{K0}$) as the reference radius r_0. The constant Ψ_0 is given at the reference radius, i.e., $\Psi_0 = \Psi(r_0, 0)$ (Okada et al. 1989). We also normalize the density and other variables at this radius, i.e., $r_0 = v_{K0} = \rho_0 = 1$ at $r = r_0$. A model parameter of the torus is $E_{th} = (v_{s0}/v_{K0})^2/\gamma$ where v_{s0} is the sound speed at the reference radius. We take $E_{th} = 0.05$. The halo parameters are $E_h = (v_{sh}/v_{K0})^2/\gamma$ and ρ_h/ρ_0 where v_{sh} and ρ_h are the sound speed and the density in the halo at $(r, z) = (0, r_0)$, respectively. We adopt $E_h = 1.0$ and $\rho_h/\rho_0 = 10^{-3}$.

We solved the ideal MHD equations in a cylindrical coordinate by using a modified Lax-Wendroff scheme (Rubin and Burstein 1967) with artificial viscosity (Richtmyer and Morton 1967). The ideal MHD equations incorporated in ARPS are as follows:

$$\frac{\partial \rho}{\partial t} + \nabla(\rho \, \boldsymbol{v}) = 0 \, , \tag{1}$$

$$\frac{\partial \, \boldsymbol{v}}{\partial t} + \boldsymbol{v} \cdot \nabla \, \boldsymbol{v} = -\frac{1}{\rho} \nabla P + \frac{1}{4\pi\rho} (\nabla \times \, \boldsymbol{B}) \times \, \boldsymbol{B} - \nabla \psi, \tag{2}$$

$$\frac{\partial \, \boldsymbol{B}}{\partial t} = \nabla \times (\, \boldsymbol{v} \times \, \boldsymbol{B}), \tag{3}$$

$$\frac{d}{dt} \left(\frac{P}{\rho^\gamma} \right) = 0, \tag{4}$$

where ψ is the gravitational potential, and $\rho, P, \boldsymbol{v}, \boldsymbol{B}$, and γ are the density, pressure, velocity, magnetic fields and specific heat ratio, respectively. We neglect the molecular viscosity for simplicity. The gravitational field is assumed to be given by a central point mass M.

Using these equations, equation of energy conservation follows

$$\frac{\partial}{\partial t}\left(\frac{1}{2}\rho v^2 + \frac{P}{\gamma - 1} + \frac{B^2}{8\pi}\right) + \nabla \cdot \left(\rho\, \boldsymbol{v}\frac{v^2}{2} + \frac{\gamma}{\gamma - 1}P\, \boldsymbol{v} + \frac{\boldsymbol{E} \times \boldsymbol{B}}{4\pi}\right) = -\rho\, \boldsymbol{v}\psi, \quad (5)$$

where $\boldsymbol{E} = -\,\boldsymbol{v} \times \boldsymbol{B}$ is the electric vector. The time evolution of the disk is computed by solving equations (1), (2), (3), and (5).

The number of grid points used in the simulations is $(N_r, N_\phi, N_z) = (200, 64, 240)$. We simulated only the upper half space ($z \geq 0$) and assumed that for the equatorial plane ρ, v_r, v_ϕ, B_r, B_ϕ, and P are symmetric and v_z and B_z are antisymmetric. The outer boundaries at $r = r_{\max}$ and $z = z_{\max}$ are free boundaries where waves can transmit. In order to avoid the singularity at $R = 0$, we softened gravitational potential near the gravitating center ($R \leq 0.2r_0$).

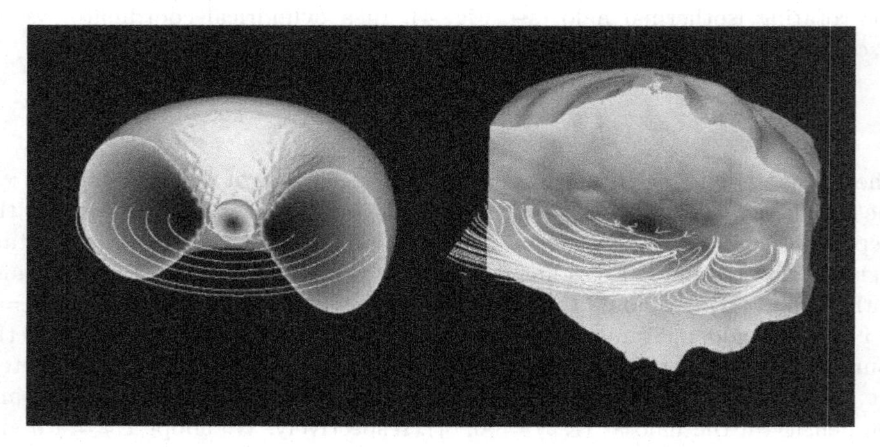

Fig. 4. (left) Initial model of a torus threaded by toroidal magnetic fields ($\beta_0 = 1$). The radius at which the gravitation force equal to the centrifugal force in the torus is $r = r_0$. (right) Density distribution and magnetic structure at $t = 6.2t_0$. Solid curves show magnetic field lines at the equatorial plane and the gray scale shows the density distribution (box size is $9r_0 \times 9r_0 \times 9r_0$)

The initial Lorentz force in the torus is assumed to be in equilibrium with the gravitational force, centrifugal force, and gas pressure gradient. The magnetic pressure is assumed to be equal to the gas pressure, $\beta_0 = P_{\rm gas}/P_{\rm mag} = 1$ at $r = r_0$. Figure 4 shows the initial model of this simulation. The solid curves show magnetic field lines and grey scale shows density distribution.

We added 1 percent ($0.01v_\phi$) random perturbation on azimuthal velocity at $t = 0$ and followed the evolution (Machida et al. 2000). The revolution time is measured by the rotation period at the reference radius ($t_0 = 2\pi r_0 / v_{\mathrm{K}0}$). When the magnetorotational instability grows after several revolutions, magnetic turbulence developing in the disk tangles magnetic field lines (Fig. 5). As the angular momentum is efficiently transported outward, the torus becomes flattened.

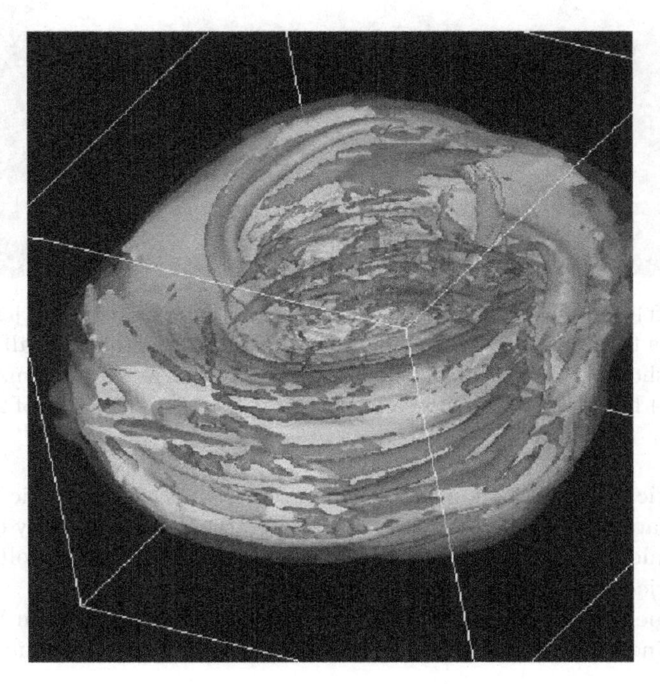

Fig. 5. Global Structure of the magnetic field (box size is $9r_0 \times 9r_0 \times 9r_0$). Isosurface of β is shown by grey scale. The dark grey region shows the isosurface of $\beta = 0.1$. It is easily recognized that magnetic loops are floated up from the disk

Figure 5 shows the magnetic structure of the accretion disk after 6.2 revolutions. The dark grey surface region is where magnetic field is largely enhanced ($\beta \leq 0.1$). As like the Solar corona, magnetic loops buoyantly rise from the disk. We can observe magnetic loops elongated in the azimuthal direction.

When an accretion disk is threaded by large scale open magnetic fields, magnetically driven bipolar jet emanates from the disk (Uchida and Shibata 1985).

Figures 6 shows a result of 3D MHD simulation of a torus initially threaded by global vertical magnetic fields. The model parameters are the same as those in the toroidal field model, though the initial plasma β at $(r, z) = (0, r_0)$ is $\beta = 2$. Small amplitude non-axisymmetric perturbations are imposed for azimuthal velocity. Within one rotation period, magnetically driven jet emanates from the torus. Although helical structure appears in the jet owing to the growth of non-axisymmetric instabilities in the disk, jets are not disrupted by the non-

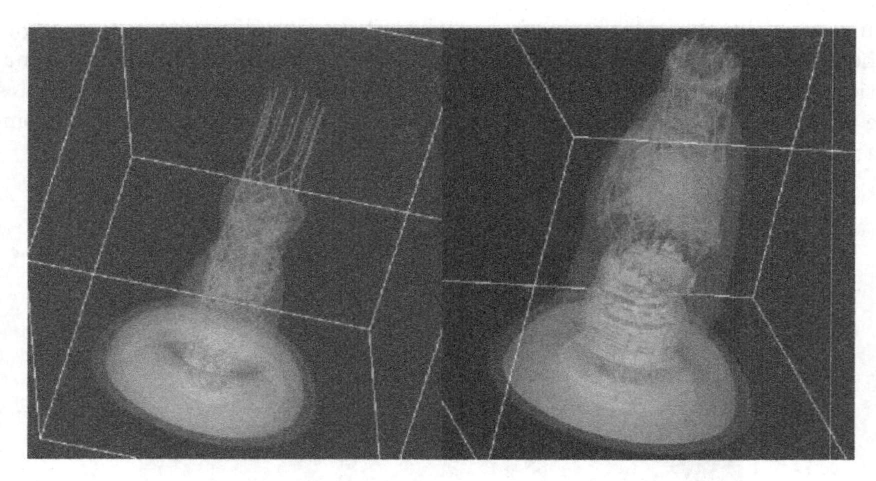

Fig. 6. (left) Numerical results of 3D global MHD simulations of jet formation from a torus threaded by vertical magnetic fields at the stage of half revolution. Grey scale shows the gradient of density distribution. Curves show magnetic field lines. (right) Density distribution and magnetic fields at the stage of 2 revolution

axisymmetric instability. In order to demonstrate that the magnetic field is wiggled on surface of sharp density gradient, the gradient of the density distribution in logarithmic scale is shown by grey scale in figures 6. After 2 revolutions, non-axisymmetric structure appears in the jet.

The numerical results can be compared with the high resolution VLBI (Very Long Baseline Interferometry) observations of jets in active galactic nuclei.

4 Summary

Global three-dimensional simulations of accretion disks reproduced various phenomena in accretion disks such as efficient angular momentum transport and jet formation. When the torus is threaded by weak magnetic fields, magnetic fields are amplified by magneto-rotational instability. After the amplification of magnetic energy saturates when $\beta \sim 10$, the system approaches a quasi-steady state. Numerically obtained value of the angular momentum transport parameter α ($\alpha \sim 0.01 - 0.1$) is consistent with observations. Inside the disk, filamentally shaped, locally magnetic pressure dominated regions appear. Magnetic energy release in strongly magnetized regions can explain violent X-ray time variabilities characteristic of black hole candidates.

The global simulation of the accretion disks needs 200,000 time steps for 13 revolutions at $r = r_0$ and it takes about 24 hours on 15 CPUs of VPP300/16R at NAOJ. For visualization analyses of data, we store the numerical data at every 2,000 time steps. The total data size is 10.8 Gbyte per one model simulation.

Three dimensional analyses are done with 3D graphic software AVS which is added on visual interface of ARPS.

The efficiency of vectoralization of ARPS is already achieved 98% and the parallel performance increases almost linearly with the number of CPUs and attains 90% performance when 15CPUs of VPP300 at National Astronomical Observatory are used (13.4 times faster than 1CPU).

The ARPS successfully demonstrated its capability to investigate accretion disks by direct numerical simulations using user-friendly interface. With this simulator, one can simulate various astrophysical objects such as X-ray binaries, Quasars, Active Galactic Nuclei, etc.

Works are in progress to install more powerful numerical engine and enable remote operation through network.

Acknowledgements

We thank Dr. K. Shibata and Dr. M. R. Hayashi for discussions. Numerical computations were carried out on Fujitsu VPP300/16R at NAOJ. This work is supported in part by the Advanced Computational Science and Technology by Japan Science and Technology Corporation (ACT-JST).

References

1. N. I. Shakura and R. A. Sunyaev, Astronomy and Astrophysics, **24** (1973) 337.
2. R. D. Blandford and D. G. Payne, Mon. Not. Roy. Astron. S oc., **199** (1982) 883.
3. Y. Uchida and K. Shibata, Publ. Astron. Soc. Japan, **37** (1985) 515.
4. J. K. Cannizzo, A. W. Shafer, and J. C. Wheeler, Astrophys. J., **333** (1988) 227.
5. S. A. Balbus and J. F. Hawley, Astrophys. J., **376** (1991), 214.
6. S. A. Balbus and J. F. Hawley, Astrophys. J., **400** (1992), 610.
7. J. F. Hawley, C. F. Gammie, and S. A. Balbus, Astrophys. J., **440** (1995) 742.
8. R. Matsumoto and T. Tajima, Astrophys. J., **445** (1995) 767.
9. A. Brandenburg, A. Nordlund, R. Stein, and U. Torkelsson, Astrophys. J., **446** (1995) 741.
10. M. Machida, M. R. Hayashi, and R. Matsumoto, Astrophys. J., **532** (2000) L67.
11. R. Okada, J. Fukue, and R. Matsumoto, Publ. Astron. Soc. Japan, **41** (1989) 41, 133.
12. R. O. Richtmyer and K. W. Morton, Differential Methods for Initial Value Problems (2nd ed., New York: Wiley 1967).
13. E. Rubin and S. Z. Burstein, J. Comput. Phys., **2** (1967) 178.
14. K. Shibata and Y. Uchida, Publ. Astron. Soc. Japan, **38** (1986) 631.

Exploring Multi-level Parallelism in Cellular Automata Networks

Claudia Roberta Calidonna*, Claudia Di Napoli, Maurizio Giordano, and Mario Mango Furnari

Istituto di Cibernetica C.N.R.,
Via Toiano, 6 I-80072 Arco Felice, Naples, ITALY
{C.Calidonna, C.DiNapoli, M.Giordano, M.MangoFurnari}@cib.na.cnr.it
http://www.cib.na.cnr.it

Abstract. Usually physical systems are characterized by different coupled parameters accounting for the interaction of their different components. The *Cellular Automata Network* (*CAN*) model [1] allows to represent each component of a physical system in terms of *Cellular Automata* (*CA*) [9], and the interaction among these components in terms of CA networks. In this paper we report our experimentations in exploiting two different kinds of parallelism offered by the CAN model using policies for network restructuring and thread assignment. At this purpose we used a prototype graphic tool (CANviz) designed to let the user experimenting heuristics to efficiently exploit two–level parallelism in CAN applications.

1 Introduction

Complex systems that evolve according to local interactions of their constituent parts can be simulated through *Cellular Automata* (*CA*) programming [9]. Applications of CA are very broad, ranging from the simulation of fluid dynamics, physical, chemical, and geological processes.

In order to capture microscopic and macroscopic aspects involved in physical phenomena simulation in a uniform representation, we used the *Cellular Automata Network* (*CAN*) model [1]. CAN model allows to represent each component of a physical system in terms of cellular automata, and the interactions among these components in terms of CA networks. CAN model offers potentially two different kinds of parallelism: *data parallelism*, coming from the local interactions of each cell composing the cellular automaton only with its neighborhood, and *control parallelism* coming from the CA network execution model.

In this paper we report our experience in exploiting both levels of parallelism and how developers can drive multi–level parallelism exploitation. In section 2 the CAN model is discussed; section 3 describes the extensions of PECANS environment [3] used to build applications in *CAN Language* (*CANL*) [2]; section 4 reports our experimental results obtained exploiting parallelism on a target parallel architecture. Finally some conclusions and future works are reported.

* Partly supported by CNR Project: "Sviluppo di una Modellistica Sperimentale Spazio-Temporale di Processi Evolutivi dell'Ambiente per la Mitigazione dei Rischi".

M. Valero et al. (Eds.): ISHPC 2000, LNCS 1940, pp. 336-343, 2000.

2 The Cellular Automata Network Model

By their nature as systems with discrete space dimensions and discrete time evolution, CA are useful for simulating spatio–temporal phenomena providing a framework for a large class of discrete models with homogeneous interactions.

A *Cellular Automaton* [9] consists of a regular discrete *lattice of cells*, with a discrete variable at each cell assuming a finite *set of states*; cells are updated synchronously in discrete time steps according to a local, identical interaction rule. According to this rule, referred to as the cellular automaton *transition function*, the state of each cell evolves in time and its evolution depends on the state, at the previous time step, both of the cell itself and of a finite number of neighbor cells. The *neighborhood* of a cell consists of the surrounding adjacent cells according to a specified topology.

The CA model we used in our work, called the *Cellular Automata Network* (*CAN*) model [1], extends the model above mentioned with the *network of cellular automata* abstraction. The CAN model can be applied when the construction of complex physical phenomenon models can be obtained by means of a reduction process in which the main model components are identified through an abstraction mechanism and interactions among components can be identified. So the CAN model provides the possibility to simulate a two–level evolutionary process in which the local cellular interaction rules evolve together with cellular automata connections. In this way, global information–processing capabilities, that are not explicitly represented in the network elementary components or in their interconnections, can be obtained.

In CAN model an automaton is denoted by a name, and its behavior is described by a set of *properties*, a *transition function*, and a *neighborhood type*. A property can correspond either to a physical property of the system to be simulated, such as temperature, volume and so on, or to some other feature of the system such as the probability of a particle to move and so on. In any case according to the standard CA model, each property corresponds to a computational grid. In this schema a cell of an automaton is considered as a functional composition among the cells of the automaton properties. A necessary requirement is that the cells of property grids must be in correspondence among them.

According to CAN model, when the physical system components are partially coupled it is necessary to introduce a *network of cellular automata*, i.e. to define a set of automata specifying a *dependence relation* among them as follows:

Definition 1. *Let A and B be two automata, if one or more properties of the automata A are used inside the transition function of the automata B, then we say that B* depends on *A, i.e. $A\delta B$.*

The CA network dependence relations can be represented by a direct acyclic graph, called the *CAN dependence graph*, whose nodes represent the cellular automata of the network and arcs represent the dependence relation between two nodes. The CA network dependence relations impose "precedences" among the execution of the network automata, so a precedence relation graph can be obtained from the dependence relation graph. In fact, let be $A = \{a_1, \ldots, a_n\}$ the set of cellular automata composing a network, we define on this set the "precedence" relation $\prec$, that is $a_i \prec a_j$ if and only if a_i must be completed before a_j can start execution; then, if $a_i \nprec a_j$ and $a_j \nprec a_i$, a_i and a_j can be concurrently executed. The couple $(A, \prec)$ is a partially ordered set.

In our computational model, we have that a_i has to be executed before a_j, i.e. $a_i \prec a_j$, if and only if there is a set of automata $\{a_1, \ldots, a_n\}$ such that $a_1 = a_i$, $a_n = a_j$, and $\forall l \in \{1, \ldots, n\} : a_l \delta a_{l+1}$.

3 The Extended PECANS Environment

Applications designed according to the CAN model are written using the *Cellular Automata Network Language* (*CANL*) [2], specifically designed to express the CAN model components. The language provides a set of primitives to define both a cellular automaton, together with all its features, and a network of cellular automata explicitly declaring the dependence relations that occur in the network.

We used the PECANS environment [3] to write CANL applications and execute them on specific architectures that can be sequential or parallel. In PECANS a CANL program is written, it is cross–compiled in the C language and then linked to the run–time environment of the architecture where the CANL application must be executed. The comparison of PECANS with other programming environments is reported in [11].

In the activity of computer simulation a very crucial requirement regards performances. CA approach proved to be a good candidate to meet this requirement since it allows to exploit the *data parallelism* intrinsic to the CA programming model coming from the local interactions of each cell composing the cellular automaton only with its neighborhood. So, the standard CA programming model maps quite naturally a SIMD execution model. Another source of parallelism in CAN applications is the *control parallelism* deriving from concurrently executing automata among which precedence relations either do not occur.

We extended PECANS with a graphic tool (CANVIZ) for the visualization of the CAN program structure and the parallel code generation control . CANVIZ is implemented in TCL/TK [10, 7] using a library, named TCLDOT [5], that adds graph manipulation facility to TCL/TK.

CANVIZ visualizes the CANL program code and the dependence graph representation of the CA network, showing the correspondence between CANL statements and CA network nodes and arcs. The main purpose of CANVIZ is to let the user drive the amount of parallelism to be exploited in CANL programs, when multi–level of parallelism is available. In fact the tool allows to graphically perform a *network restructuring policy* guaranteeing that the obtained precedence relation graph preserve the program correctness.

CANVIZ highlights nodes of the CA network which are candidate for data parallelism exploitation allowing the user to enable or disable it. Control parallelism is visible to the user in the precedence graph structure since all the branches in the graph represent tasks that can be concurrently executed, while

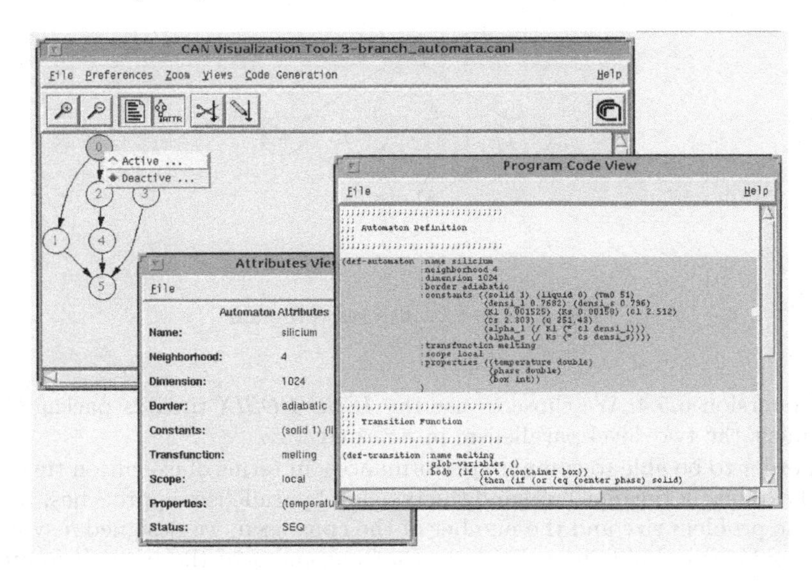

Fig. 1. The CAN Visualization Tool

branches reaching a node represent a control synchronization point. Finally, an additional feature of CANviz allows to associate weights to the cellular automata, to be taken into account when deciding the amount of parallelism to be exploited in case of multi–level parallelism exploitation.

4 Experimental Results for Multi–Level Parallelism

CA networks can be executed on a parallel computer exploiting either the data parallelism or both control and data parallelism resulting in a multi–level parallelism application. It is clear that policies are necessary to decide the amount of parallelism to be spawned for each level in order to have a better exploitation of the available parallel computer resources [8].

In our implementation only the outer level of control parallelism can be exploited: branches belonging to paths of the network deriving from previous branches are serialized in the execution. Therefore, once network execution forks on a branch, only data parallelism can be exploited.

In order to gain more information on the real advantages of using the multi–level parallelism potentially offered by the CAN model, we made experiments using the SGI Origin2000 multiprocessor computer [6]. It is a cache–coherent NUMA multiprocessor with 6 dual–processor basic node board. Each node is equipped with two 195 MHZ MIPS R10000 processors and it accommodates a 64 KBytes primary cache and a 4 MBytes secondary cache per processor. Each node has 4 GBytes of DRAM memory. The SGI Origin2000 uses the IRIX operating

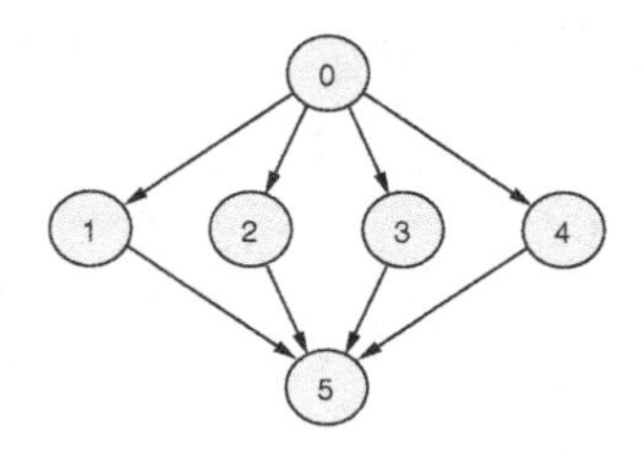

Fig. 2. The 4-branch test program

system version 6.5.4. We chose to use the *IEEE POSIX* threads package [4] to implement the two–level parallelism in a nested way.

In order to be able to compare performances, in terms of execution times, obtained exploiting the one–level and the two–level parallelism approaches, varying both the problem size and the number of the cpus used, we designed a synthetic CAN application built in such a way that automata of the network represent computational tasks with equal workloads.

In our experiments we deal with both *balanced* and *unbalanced* CA networks, where unbalancing occurs when the branches of the network have different *costs* in terms of execution time. So unbalanced networks can result when branches contain a different number of automata, and/or automata with different workloads, and/or serialized automata (i.e. automata for which no data parallelism is possible due to the use of global variables). In our test application, unbalancing derives only from the different number of automata on network branches.

The objective of our experimentations was to find out some preliminary results to individuate policies to decide which parallel execution to adopt according to the network configuration and the parallel machine resources.

The experiments were organized in two sets. For the first set, we compared speedups obtained exploiting multi-level parallelism versus one–level data parallelism approach. Therefore, we chose an application consisting of a balanced network of 6 automata whose precedence relations are shown in figure 2.

The second set of experiments was conceived to study the effects on performances of multi–level parallelization of unbalanced CA networks. In this case the application consists of the same set of automata as the previous ones, but with different dependence relations that generate unbalancing as shown in figure 3(a). Note that, from an execution time point of view, the sequential and the data parallel versions of the latter network are the same as the previous case.

In the precedence relation graph of this network, it is possible to add some precedence relations (*network restructuring policy*) without affecting the program correctness due to the precedence relation transitivity. In fact, introducing unnecessary precedence relations in our example, as shown in figure 3(c), a different synchronization point in the network parallel execution results.

In all experiments we measured application speedups obtained by the ratio of the sequential execution time with the CAN multithreaded execution time.

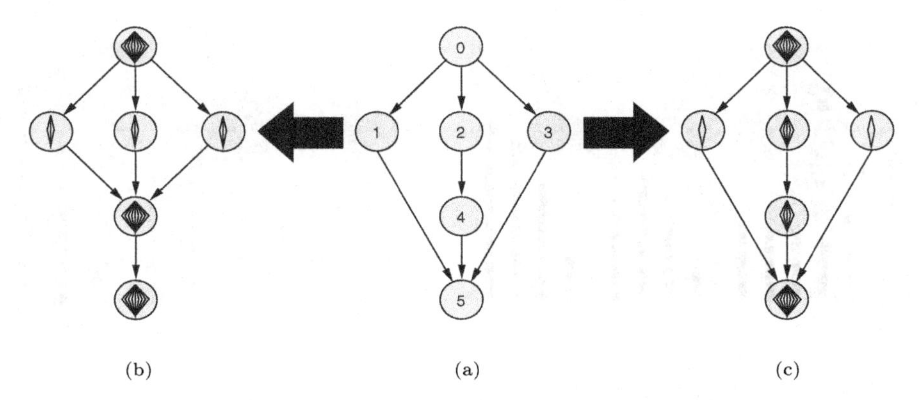

Fig. 3. (a) The 3-branch test program with (b) threads remapping and (c) restructured network

In each experiment the number of threads allocated for parallelism (for both levels) is fixed and threads are bound to processors so the number of generated threads represents the actual number of cpus allocated for the application. We chose different resource assignment policies (*thread assignment policy*), and for each choice the problem size was varied by squaring the automata grid starting from 64x64 up to 2048x2048 elements for each property.

In both set of experiments the data parallel version is obtained executing all network automata in the sequential order specified by the user, and the available threads execute the transition function on portions of the automaton grid obtained dividing each property along rows into equal–sized chunks.

For the first set of experiments the results are reported in figure 4(a). When exploiting the two–levels of parallelism, the thread allocation policy is to assign one thread to each branch of the CA network and, within each branch, to allocate and uniformly distribute the remaining threads for data parallel processing.

First we notice that, except for small problem sizes (up to 128x128), the two–level and the one–level parallelism approaches show speedups that increase with the number of threads once fixed the problem dimension, and the two–level approach performs better than the one–level approach. In fact, in the one–level approach all threads are assigned to each automaton for exploiting data parallelism, so for small problem size this results in poor workloads for concurrent threads; therefore thread creation and management overheads have significant costs. Exploiting also the control parallelism level means to assign the computations of different automata to different threads. When the number of available threads is fixed, thread data chunks result larger than the ones used only for the data parallelism. Once fixed the available resources, the two approaches show similar performances when the problem size increases because the thread management and synchronization overheads are less influent.

For the second set of experiments, the results are reported in figure 4(b), where data parallelism speedups are compared to those obtained exploiting two–levels of parallelism with three different parallel executions. In the first case

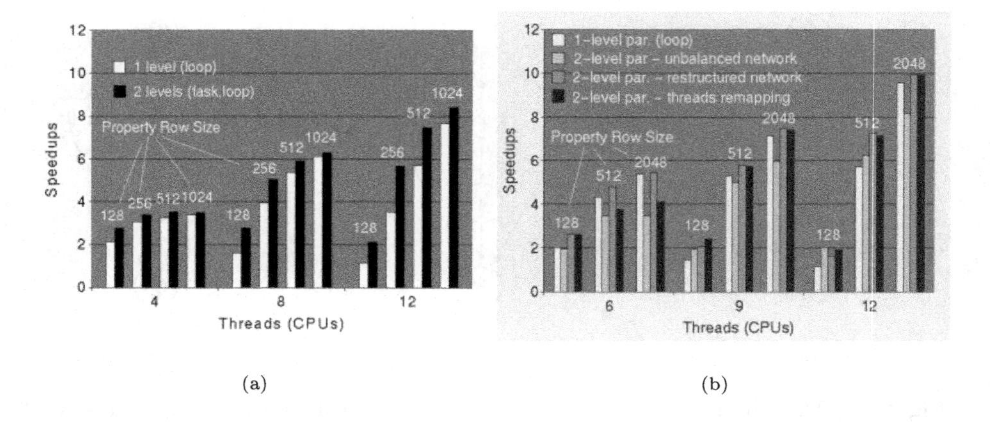

(a) (b)

Fig. 4. (a) 1-level versus 2-level parallelism in the 4-branch test program: (b) 2-level parallelism solutions in the 3-branch test program

(*unbalanced network*) each branch spawns, in turn, the same number of threads for data parallelism, although the branch in the middle has twice as much work to do (see figure 3(a)). In the second case (*restructured network*) automaton 4 has been shifted out from the middle branch into the main network branch (see figure 3(b)) obtaining a balanced network where threads can be equally distributed among the network branches. In the third case (*threads remapping*) a different number of threads is assigned to each network branch according to its cost (see figure 3(c)).

First consider the speedups of the three approaches for small problem sizes (e.g. 128x128). As we notice only with six threads the unbalanced network performs worse than the other cases. As the number of threads increases the three approaches give similar speedups that are low due to the grain size of computations carried out by each thread.

Moreover, for this problem size, the restructuring and the thread remapping policies do not win for a greater number of threads due to the way loop iterations are distributed over the threads: the property row size is divided by the number of used threads but the rest of the division is assigned to the last thread in the pool. This extra computation introduces unbalancing which is more costly when small properties are mapped to an increasing number of threads.

For mean and big problem sizes (e.g. 512x512 and 2048x2048) speedups increase with the number of threads because grain size is larger and parallelism gains overcome thread management overheads. Moreover, the unbalancing due to the rest of iterations coming from loop chunk definition has a relevant effect only when using 6 threads, since the rest is greater than in the other cases. This effect causes better performances of the restructuring policy towards the thread remapping one, while for a greater number of threads the two policies are equivalent.

5 Conclusions and Future Works

In this paper we presented our preliminary results obtained exploiting the two–level parallelism potentially offered by the CAN model. As we showed in the experimental results, when exploiting multi–level parallelism, CA network restructuring and thread assignment policies have to be adopted to better use the available parallel computational resources of the target machine. These policies can be driven by the user through the CANVIZ tool.

We plan to carry out more experiments on real CAN applications to study the feasibility to extract heuristics that could be adopted in choosing a network configuration and a thread scheduling policy for better parallelism exploitation. In this way it will be possible to reduce user interaction by making it automatic a possible restructuring of the CA network and the selection of the one that better matches the available parallel resources. At this purpose we plan to build a new module (based on CANVIZ tool) in the PECANS environment to provide a CA parallel programming environment that, with a limited user interaction, produces parallel code so that no particular skills in parallel computing area are required to users of the environment.

References

1. Carotenuto, L., Mele, F., Mango Furnari, M., Napolitano, R.: PECANS: A Parallel Environment for Cellular Automata Modeling. Complex Systems **10**, Complex Systems Publications, (1996), pp. 23–41
2. Di Napoli, C., Giordano, M., Mango Furnari, M., Napolitano, R.: CANL: a Language for Cellular Automata Network Modeling. Proc. of Parcella '96, Akademie Verlag, (1996), pp. 101–111
3. Di Napoli, C., Giordano, M., Mango Furnari, M., Napolitano, R.: A Portable Parallel Environment for complex systems simulation through cellular automata networks. Journal of Systems Architecture **42**, Elsevier Science, (1996), pp. 341–350
4. IEEE Computer Society: POSIX System Application Program Interface: Threads Extension [C Language] POSIX 1003.4A Draft8. Available from the IEEE Standard Dept.
5. Krishnamurthy, B.: Practical Reusable Unix Software. John Wiley Sons, (1997)
6. Laudon, J., Lenoski, D.: The SGI Origin: A ccNUMA Highly Scalable Server. Proc. of 24th Int'1 Symp. on Computer Architecture, (1997), pp. 253–251
7. Ousterhout, J.K: Tcl and the Tk Toolkit. Addison Wesley (1994)
8. Ramaswamy, S., Sapatnekar, S., Banerjee, P.: A Framework for exploiting Task and Data Parallelism on a Distributed Memory Multicomputers. IEEE Transactions on Parallel and Distributed Systems, Vol. **8** No **11**, (1997)
9. Toffoli, T., Margolus, N.: Cellular Automata Machines: A New Environment for Modeling. MIT Press (1987)
10. Welsh, B.B.: Practical Programming in TclTk. 2nd edition, Prentice Hall (1997)
11. Worsh, T.: Programming Environments for Cellular Automata. Proc. of ACRI '96. Springer, (1996) pp. 3–12

Orgel: An Parallel Programming Language with Declarative Communication Streams

Kazuhiko Ohno[1], Shigehiro Yamamoto[1],
Takanori Okano[2], and Hiroshi Nakashima[1]

[1] Toyohashi University of Technology
Toyohashi, 441-8580 Japan
[2] Current affiliation: Hitachi Systems & Services, Ltd.
{ohno,okanon,yamamoto,nakasima}@para.tutics.tut.ac.jp

Abstract. Because of the irregular and dynamic data structures, parallel programming in non-numerical field often requires asynchronous and unspecific number of messages. Such programs are hard to write using MPI/Pthreads, and many new parallel languages, designed to hide messages under the runtime system, suffer from the execution overhead.
Thus, we propose a parallel programming language Orgel that enables brief and efficient programming. An Orgel program is a set of agents connected with abstract channels called streams. The stream connections and messages are declaratively specified, which prevents bugs due to the parallelization, and also enables effective optimization. The computation in each agent is described in usual sequential language, thus efficient execution is possible.
The result of evaluation shows the overhead of concurrent switching and communication in Orgel is only 1.2 and 4.3 times larger than that of Pthreads, respectively. In the parallel execution, we obtained 6.5–10 times speedup with 11–13 processors.

1 Introduction

To obtain high performance using parallel machines, the means for brief and efficient programming is necessary. Especially in the non-numerical processing, existing programming methods require low-level specifications or suffer from large runtime overhead.

So, we propose a new parallel programming language called *Orgel*, which has both abstract description of parallelism and runtime efficiency. The programming paradigm of Orgel is *multi-agents* connected with abstract communication channels called *streams*. The agents run in parallel, passing messages via streams.

The computation of each agent is described with a usual sequential language. Thus the execution of each agent is efficient. The connection among agents and streams are declaratively described. This feature statically determines the parallel model of the program, thus prevents bugs in the communications and enables strong optimization using static analysis.

M. Valero et al. (Eds.): ISHPC 2000, LNCS 1940, pp. 344–354, 2000.

This paper is organized as follows: Section 2 describes the background. Section 3, 4 presents the language design of Orgel and the current implementation. Section 5 shows the result of evaluation, and in Section 6 we give the conclusion.

2 Background

For the non-numerical processing, automatic parallelization is extremely difficult. The dynamic and irregular structures like lists and trees cannot be statically divided, and the program structure is also irregular because of recursive calls. Therefore many parallel programming method have been proposed for this field.

One of the major method is to use message passing libraries (PVM [1], MPI [2]) or thread libraries (Pthreads [3]) on a sequential language like C, Fortran, etc. This way is easy to learn for the users accustomed to the sequential programming language, and efficient programming is possible by the low-level tuning. However, the order and number of communications are nondeterministic in many non-numerical programs. In such cases, the mismatch of corresponding sends/receives easily occurs if low-level communications are explicitly specified.

Another method is to design a new programming languages with parallel execution semantics (KL1 [4]). Parallelism can be naturally described in such languages. And owing to the abstraction of communications and synchronizations, the user can avoid timing bugs. However, such abstraction causes large overhead at runtime, and leads to inefficiency.

To reduce such overhead, the optimization schemes using static analysis have been proposed. For example, our optimization scheme for KL1 achieved remarkable speedup for typical cases [5, 6]. However, precise static analysis of dynamic behavior is difficult, thus the optimization is ineffectual in some cases.

So, the desirable parallel programming language should have the following features: 1) efficient and similar style with the usual sequential programming, and 2) abstract specification of parallelism and communications to reduce the burden of users. The specification also should help static optimization.

3 Language Design

3.1 Language Overview

Orgel is designed for non-numerical programming. Because automatic parallelization is difficult in this field, Orgel leaves the specification of parallelism and communications to the user, and supplies frameworks for such specification.

The execution unit of Orgel is called an *agent*. We also introduce an abstract message channel called a *stream*, for the inter-agent communication. Thus, the Orgel program is represented as a set of agents connected by streams. The agents run in parallel, passing messages via streams.

The syntax of Orgel is based on C. We added 1)declarations of stream/agent/ message/network connection; 2) statements for message creation/transmission/

```
stream StreamType [inherits StreamType1 [, ...]] {
   MessageType [(Type Arg :Mode [, ...])];
     ...
};
```

Fig. 1. Stream type declaration

dereference and agent termination; and 3) *agent member functions.* We also eliminated global variables and added *agent member variables.* Thus each function can be coded efficiently in usual sequential programming.

As we describe in Section 3.2, 3.3, the structure and behavior of streams and agents are defined as stream types and agent types. The instances of agents and streams are automatically created at runtime, according to the variables definition of stream types or agent types. They are automatically connected according to the connection declaration (see Section 3.4).

When an Orgel program is executed, a `main` agent is created and starts its execution. If the `main` agent type contains some variable definitions of agent/stream types, their instances are also automatically created [1], streams are connected to the agents, and the agents start execution in parallel. Thus, the network of agents and streams are built without operational creation nor connection.

Compared with other many multi-agent/object-oriented languages [7, 8], this declarative specification of network clears the parallel execution model of the program. It prevents the creation of unexpected network structure, and also enables precise static analysis which leads to effective optimization.

3.2 Stream

A *stream* is an abstract message channel, based on KL1's stream communication model [9]. Our stream has direction of message flow, and one or more agents can be connected to each end.

A stream type is declared in the form of Fig. 1. This declaration enumerates message types that a stream type *StreamType* accepts. An Orgel message type takes the form of a function: *MessageType* which is used as the message identifier, and a list of arguments with types and input/output mode (`in`/`out`).

3.3 Agent

An *agent* is an active execution unit, which sends messages each other while performing its computation.

To create an agent, an agent type declaration in the form of Fig. 2 is needed. The form of agent type declaration is similar to C function declaration. The arguments of an agent type are its input/output streams with types and modes.

Member functions are defined in the same way as usual C functions, except the function name is specified in the form: *AgentType*::*FunctionName*.

[1] For the efficient execution, the creation is delayed until a message is sent to them.

```
agent AgentType([ StreamType StreamName: Mode [,...] ] ){
  member function prototype declarations
  member variable declarations
  connection declarations
  initial Initializer;
  final Finalizer;
  task TaskHandler;
  dispatch (StreamName){
    MessageType: MessageHandler;
      ...
  };
};
```

Fig. 2. Agent type declaration

For *member variables*, independent memory areas are allocated to each agent instance. The scope of member variables is within the agent type declaration and all member functions of the agent type.

Agents and streams are logically created by the definition of member variables of the declared types. In this paper, we call these variables *agent variables* and *stream variables.* As explained in Section 3.1, physical creation is automatic without operational creation.

Connection declarations specify how to connect agents and streams defined as member variables. We will show details of this declaration in Section 3.4.

The last four elements: `initial`, `final`, `task`, `dispatch`; defines event handlers. The handlers are sequential C code, with extensions for message handling. *Initializer* and *Finalizer* are executed on the creation and destruction of the agent, respectively. *TaskHandler* defines the agent's own computation, and is executed when the agent is not handling messages. And `dispatch` declaration specifies the message handler to each message type of input stream *StreamName*, by enumerating acceptable message types *MessageType* and handler code *MessageHandler.* This declaration works as an framework for asynchronous message receiving.

3.4 Connection Declaration

The connection among agents and streams can be specified by a connection declaration in the following form:

```
connect [ Agent0.S0 Dir0 ] Stream Dir1 Agent1.S1 ;
```

The declaration takes a stream variable *Stream* and input/output streams *S0*, *S1* of agent variables *Agent0*, *Agent1*. A specifier `self` can be used in place of an agent variable, to connect the agent that contains the connection declaration. *Dir0*, *Dir1* are direction specifiers (`==>` or `<==`) that indicates the message flow.

If the agent/stream variable is an array, an array specifier in form of `[`*SubscriptExpression*`]` is needed. If the subscript expression is a constant expression,

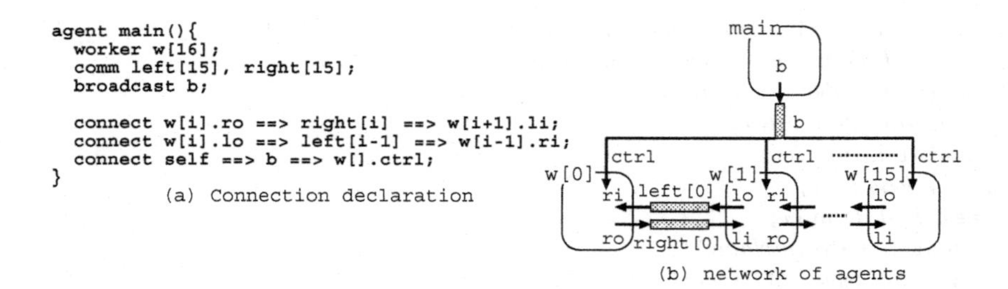
agent main(){
worker w[16];
comm left[15], right[15];
broadcast b;
connect w[i].ro ==> right[i] ==> w[i+1].li;
connect w[i].lo ==> left[i-1] ==> w[i-1].ri;
connect self ==> b ==> w[].ctrl;
}
main
b
b
ctrl
ctrl
ctrl
w[0]
w[1]
w[15]
ri
left[0]
lo
ri
lo
ro
right[0]
li
ro
li

(a) Connection declaration

(b) network of agents

Fig. 3. Example of connection declarations

the declaration argument means an element of the array. If the expression is omitted, the argument means all elements of the array.

The subscript expression may contain one identifier called *pseudo-variable.* It works as an variable whose scope is within the connection declaration, and represents every integer values with the restriction that each subscript expression does not exceed the array size.

By using array specifier, a set of one-to-one connections, or one-to-many/many-to-one connection, can be declared. A message to a stream with multiple receivers are multicasted to them, and messages to a stream from multiple senders are nondeterministically sequentialized.

An example is shown in Fig. 3(a). Here we regard that `worker` is an agent type, and `comm`, `broadcast` are stream types. In the first connection declaration, each subscript expression is restricted to the size of array `w` and `right`. Thus the value of pseudo-variable `i` is $0 \ldots 14$, and as shown in Fig. 3(b), the output stream `ro` of each agent is connected to the input stream `li` of the right neighbor agent, via each `right` stream. By the third declaration, the stream `b`'s sender side is connected to the agent `main`, and the receiver side is connected to every `worker` type agent. Thus this stream works as a broadcast network from `main` to all `worker` agents.

Because the `connect` declaration statically defines network model, the compiler can make static analysis precisely and can optimize scheduling and communications. Such optimization is much difficult with the operational stream connection in A'UM [10] or AYA [11]. Candidate Type Architecture [12] also offers declarative network configuration, but our stream model is more flexible.

3.5 Message

Message Variables Using message types declared in the stream type declaration, variables for messages can be defined. We call them *message variables.*

A message variable acts as a logical variable in logic programming languages. Its initial state is *unbound*, and changes to *bound* when assigned with a message object or other message variables. The bound state has two cases: if the variable is assigned to a message object, the state is *instantiated*; and if the variable is

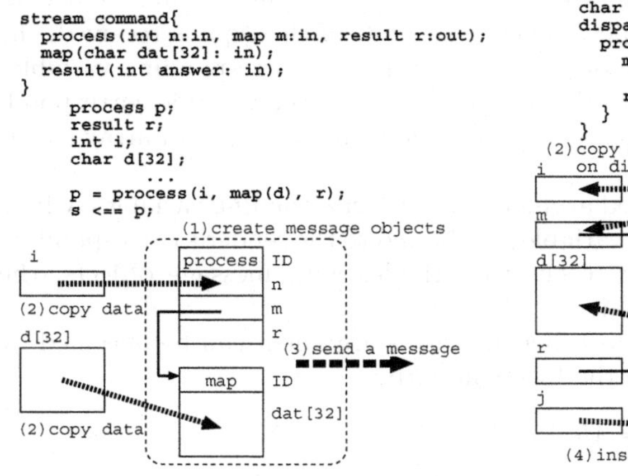

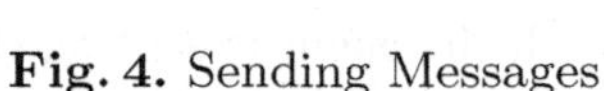
Fig. 4. Sending Messages

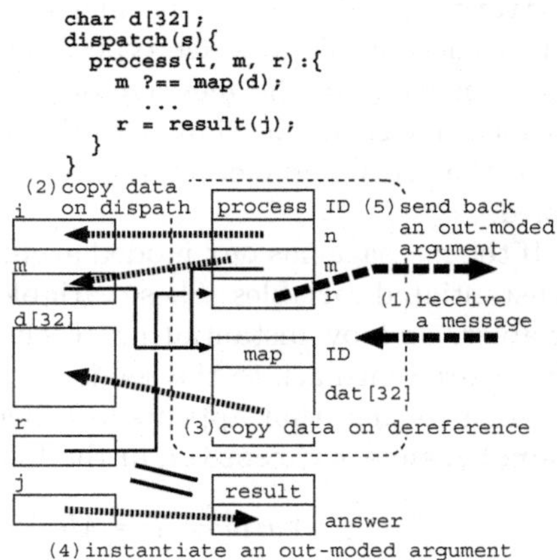

Fig. 5. Receiving Messages

assigned to other variable that is not instantiated, the state is *uninstantiated.* In the latter case, later assignment with a message object can change the state of every related variables to instantiated.

Creating and Sending Messages To create a message object, the message type and actual arguments is described in a functional form. For example, if a stream type is declared as shown in Fig. 4, a message variable can be defined and assigned to a message object as shown in Fig. 4(1).

The type of a message argument must be any C data type except pointer types, or any message type.

In the former case, the actual argument value is stored in the message object. If the argument type is an array, the actual argument should be a pointer indicating the head of array data of declared size. In the example of Fig. 4(2), arguments of `int` and an array of `char` are stored in the message.

In the latter case, the mode can be either `in` or `out`. If the mode is `in`, this argument is instantiated by the message sender. The actual argument must be a message object of the declared argument type. It can be an uninstantiated variable on the message creation, but must be instantiated by the message sender in time. If the mode is `out`, this argument is instantiated by the message receiver. The actual argument must be an uninstantiated message variable.

The created message object is sent to a stream by a *send statement* of the following form:

Stream `<==` *Message* ;

Receiving Messages An agent, connected to a stream's receiver side, receives and handles messages according to the `dispatch` declaration of the agent type.

`dispatch` declaration specifies one of the agent's input streams, and enumerates acceptable message types. When a message of the type is received, the argument value of the message object is stored in the corresponding variables specified in `dispatch`. Similar to the message creation, the variable corresponding to the array argument is regarded as a pointer, and the area of array size is copied.

If the message has `out` moded arguments, the corresponding variables will be uninstantiated variables. These variables are bound to the sender's corresponding variables, and by instantiating receiver's variables with message objects, the objects are sent back to the sender.

The messages that appears as other message's argument can be obtained by a `dereference expression` in the following form:

Variable `?==` *MessageType*`[(`*Arg1*`[, ...])];`

If a message variable *Variable* is uninstantiated, the agent executing this expression is suspended, and resumed when other agent instantiates the variable. If the message's type is *MessageType*, The expression returns non-zero and assigns *Arg1, ...* to the corresponding arguments. If the type differs, it returns zero.

Fig. 5 shows an example of receiving messages, sent in the example of Fig. 4. When a `process` message arrives (1), the receiver agent assigns the variable `i` to the value of argument `n` (2). Next, by a dereference expression, the message-type argument `m` is obtained and the value of argument `dat` is copied to `d` (3). And finally, by assigning a message object to the `out` moded argument `r` (4), the object is sent back to the sender of `process` message (5).

4 Implementation

Using Pthreads, the current implementation supports concurrent execution on a single-processor or parallel execution on shared-memory multi-processors.

The implementation consists of a Orgel compiler called Orc and Orgel runtime libraries which support agent management and stream communication. Orc is implemented as an Orgel-to-C translator. The automatically generated C program is compiled by a C compiler, linked with Orgel runtime libraries, and the executable file is generated.

4.1 Implementation of Agents

For each agent type declaration shown in Fig. 2, Orc generates *agent main function.* On the creation of an agent, a thread is created and starts execution of the corresponding agent main function.

In an agent main function, `initial` handler is called first. Then in the main loop, each message handlers are called according to the received message types, and the task handler is called if no message is received. A `terminate` statement is compiled into the code that breaks the main loop. And when the main loop ends, `final` hander is called before the thread terminates.

The agent member variables are translated into a C struct type that has each variables as members. The agent main function defines a variable of this struct type, and its pointer is added to the arguments of member functions. Each access to member variables are replaced with the access to the corresponding member of the struct. Thus, the instance of member variables are allocated for each agent instance, and can be accessed in any member functions.

To suppress the number of threads for efficiency, and to enable lazy creation of agents, an agent instance is represented by an agent record. Corresponding to the logical creation of agents, agent records are first created. And Orgel runtime schedules agents using these records, creating threads in case of need.

4.2 Implementation of Streams and Messages

A stream instance is represented as a stream record, which keeps connection information and a message queue.

Because the structure of an Orgel message is statically declared, it can be compiled into a C struct type. Every message struct type has a ID field to distinguish message types and a logical pointer of a message struct to form a message queue for streams. The C-data-type arguments of the message are compiled as members of the struct, and the message-type arguments are represented as a logical pointer to the corresponding struct type.

The messages are not always freed in the creation order, because of the message-type arguments. So they are allocated on a global heap and managed using garbage collection (GC). This heap has a 2-level structure; A message variable contains the index for heap entry table, and each entry has the address of corresponding message in the heap. Thus on GC, the runtime system can pack messages without changing the value of message variables.

4.3 Implementation of Sending/Receiving Messages

Message operations in agent type declarations and member functions are replaced by the calls of corresponding functions in the runtime library. The once-assignment rule for the message variables is assured by compile-time and runtime check. For the latter, Orc inserts some inline code to check the restriction.

The suspension/resumption of agents are implemented as follows: The dereference function checks if the message variable is instantiated. If it is uninstantiated, the function creates a hook record with a condition variable [3] and suspends the thread. When a message object is assigned to the variable, the inserted code finds the hook and sends a signal to resume the suspended thread.

5 Evaluation

We evaluated the prototype implementation using 2 programs: `nqueen` and `pia`. The latter is a multiple protein sequence alignment program by parallel iterative improvement method.

Table 1. Sequential Performance (On SS10+Solaris 2.5)

	C	Orgel	ratio
nqueen	66.54s	73.88s	1.11
pia	12.18s	12.78s	1.05

(a)

	C+Pthreads	Orgel	ratio
switching	18.50μs	22.58μs	1.22
communication	2.96μs	12.76μs	4.31

(b)

Table 2. Parallel Performance (On SPARCcenter + Solaris 2.6)

	sequential	parallel	speedup
nqueen	213.58s	21.24s	10.06
pia	50.24s	7.72s	6.50

5.1 Sequential Performance

We evaluated the efficiency on single processor, by the comparison with sequential C programs and concurrent programs using Pthreads library.

Table 1(a) shows the execution time of `nqueen` and `pia`. C version makes the computation, equivalent to Orgel agents, sequentially in loops. The result shows that the overhead using Orgel is only 5–11% compared to C.

Table 1(b) shows overhead of Orgel runtime, compared with directly using Pthreads in C. We used a benchmark program that repeats transferring an integer value between 2 threads. The C+Pthreads version uses shared variable for integer transfer, and uses semaphore for synchronization.

The thread switching overhead of Orgel is only 22% larger than C+Pthreads. To deal with many-to-many dependencies among threads, Orgel uses condition variables for synchronization. But its overhead is small enough.

Even for transferring just an integer, a message must be created and sent via stream in Orgel. But the overhead is only 4 times larger than using a shared variable. We regard it is small enough because the ratio of transmission overhead is smaller in practical programs, and the overhead using Pthreads grows larger when buffering data or transferring dynamic data structures. Still more, the current overhead of Orgel communication includes that of locking/unlocking stream records and the heap, which can be reduced using static analysis.

5.2 Parallel Performance

We executed the Orgel version of `nqueen` and `pia` on a shared-memory multiprocessor machine SPARCcenter (Solaris 2.6). The result is shown in Table 2.

`Nqueen` obtained 10.06 speedup using 11 threads, which is almost linear speedup. `Pia` obtained 6.50 speedup using 13 threads. The communications in `pia` are 1-to-many or many-to-1, and by optimizing the mutual exclusion on message transmission, the performance will be improved.

6 Conclusion

In this paper we proposed a new parallel programming language Orgel, and presented its design, implementation and evaluation.

Orgel is based on an execution model that multi-agents, connected with abstract message channels called streams, run in parallel. The distinctive feature of Orgel is declarative description of agent networks, which prevents communication/synchronization bugs and also enables precise static analysis for effective optimization. On the other hand, the computation of each agent is described sequentially, which enables to write efficient programs.

The evaluation on prototype implementation shows the overhead on single processor is small enough compared to C or Pthreads library, and promising speedup is obtained on a multi-processor machine.

We are currently working on the optimizer using static analysis. Supporting distributed-memory multi-processors is also our future work.

Acknowledgment

This research work is being pursued as a part of the research project entitled "Software for Parallel and Distributed Supercomputing" of Intelligence Information and Advanced Information Processing supported as a part of Research for the Future Program by Japan Society for the Promotion of Science (JSPS).

References

[1] V. S. Sunderam. PVM: A framework for parallel distributed computing. *Concurrency: Practice and Experience 2,* Vol. 2, No. 4, pp. 315-339, December 1990. 345
[2] Message Passing Interface Forum. *MPI: A Message-Passing Interface Standard,* June 1995. 345
[3] B. Nichols, D. Buttlar, and J. P. Farrell. *Pthreads Programming.* O'REILLY, 1998. 345, 351
[4] K. Ueda and T. Chikayama. Design of the kernel language for the parallel inference machine. *The Computer Journal,* Vol. 33, No. 6, pp. 494-500, 1990. 345
[5] K. Ohno, M. Ikawa, M. Goshima, S. Mori, H. Nakashima, and S. Tomita. Improvement of message communication in concurrent logic language. In *Proceeding of the Second International Symposium on Parallel Symbolic Computation PASCO'97,* pp. 156-164, 1997. 345
[6] K. Ohno, M. Ikawa, M. Goshima, S. Mori, H. Nakashima, and S. Tomita. Efficient goal scheduling in concurrent logic language using type-based dependency analysis. In *LNCS1345 Advances in Computing Science - ASIAN'97,* pp. 268-282. Springer-Verlag, 1997. 345
[7] Y. Shoham. Agent-oriented programming. *Artificial Intelligence,* Vol. 60, pp. 51-92, 1993. 346
[8] G. Agha. Concurrent object-oriented programming. *Communications of the A CM,* Vol. 33, No. 9, pp. 125-141, September 1990. 346

[9] T. Chikayama, T Fujise, and D. Sekita. *KLIC User's Manual.* ICOT, March 1995. 346

[10] K. Yoshida and T. Chikayama. aum - a stream-based concurrent object-oriented language -. *New Generation Computing,* Vol. 7, No. 2, pp. 127-157, 1990. 348

[11] ICOTTM-1206. *Introduction of AY A (version 1.0) (in Japanese).* 348

[12] G. A. Alverson, W. G. Griswold, C. Lin, D. Notkin, and L. Snyder. Abstractions for portable, scalable parallel programming. *IEEE Transactions on Parallel and Distributed Systems,* Vol. 9, No. 1, pp. 71-86, January 1998. 348

$\mathbf{BS\lambda}_p$: Functional BSP Programs on Enumerated Vectors

Frédéric Loulergue

LIFO
BP6759, 45067 Orléans Cedex 2, France.
loulergu@lifo.univ-orleans.fr

Abstract. The $\text{BS}\lambda_p$ calculus is a calculus of functional BSP programs on enumerated parallel vectors. This confluent calculus is defined and a parallel cost model is associated with a weak call-by-value strategy. These results constitute the core of a formal design for a BSP dialect of ML

1 Introduction

Some problems require performance that only massively parallel computers offer whose programming is still difficult. Works on functional programming and parallelism can be divided in two categories: explicit parallel extensions of functional languages – where languages are either non-deterministic or non functional – and parallel implementations with functional semantics – where resulting languages don't express parallel algorithms directly and don't allow the prediction of execution times. Algorithmic skeletons languages [5,9], in which only a finite set of operations (the skeletons) are parallel, constitute an intermediate approach. Their functional semantics is explicit but their parallel operational semantics is implicit. The set of algorithmic skeletons has to be as complete as possible but it is often dependent on the domain of application.

We explore this intermediate position thoroughly in order to obtain universal parallel languages where source code determines execution cost. This last requirement forces the use of explicit processes corresponding to the parallel machine's processors. A denotational approach led us to study the expressiveness of functional parallel languages with explicit processes [6] but is not easily applicable to BSP [12] algorithms. An operational approach has led to a BSP λ-calculus that is confluent and universal for BSP algorithms [8], and a library of Bulk Synchronous primitives for the Objective Caml language which is sufficiently expressive and allows the prediction of execution times [1].

Our goal is to provide a framework where programs can be proved correct, can be given an a priori execution cost and can be of course implemented. It is to notice that we want a model simpler enough to allow a programmer to predict the performance of a program from its source code and to know where to modify the source code to improve the execution time of the program (if possible)

However if the $\text{BS}\lambda$ calculus and the `BSMLlib` library share the same BSP primitives, there doesn't exist any formal connection between them. So it is

M. Valero et al. (Eds.): ISHPC 2000, LNCS 1940, pp. 355–363, 2000.

possible to write a BSλ program and prove its correctness and then write "the same" program with the `BSMLlib` library and use the BSP model to predict its performance. But as there is no formal connection between them we are not guaranteed that the BSλ term and the `BSMLlib` program will have the same behaviour as well as for its correctness as for its performance. To attain our goal we need a tower of formalisms where each level is correct with respect to the next one. The ground floor will be an abstract machine (formal description of the implementation) and the last floor will be the BSλ calculus. Chapter 5 of [7] presents two intermediate floors: the BSλ_p-calculus and a distributed semantics (called the distributed evaluation). Both are formalised as higher order rewrite systems. In this article we will present the BSλ_p-calculus in a more classical (and readable) manner and will provide an associated cost model.

The main gap between the BSλ-calculus and the `BSMLlib` library is that in the calculus, base parallel objects are expressed intensionally by a function f from processor names to values (the term πf represents the parallel vector where processor i holds the value $f\,i$), the network being potentially infinite, whereas the `BSMLlib` library runs on a network with p processors. The BSλ_p-calculus replaces the intensional vectors πf by enumerated ones ($\langle\, e_0 \,, \ldots \,, \, e_{p-1} \,\rangle$), the enumeration being finite and with fixed width. As a result, the parallel interpretation of reduction as well as the cost of such a reduction is more naturally expressed than for the BSλ-calculus. Moreover, it is still possible to express vectors in an intensional way, which is more convenient to write programs.

Section 2 presents the syntax and rules of the BSλ_p calculus. Section 3 is devoted to a weak call-by-value strategy. The proof of confluence of this strategy highlights the parallel interpretation of reduction which is given a parallel cost in section 4.

2 The BSλ_p-Calculus

In this section we introduce an extension of the λ-calculus called the BSλ_p-calculus. Its parallel data structures are flat and map directly to physical processors. This difference with certain languages, although apparently minor is crucial: BSλ_p programs require no flattening [3, ch. 10] and have thus complete control of the computation / communication ratio.

The calculus introduces operations for data-parallel programming but with explicit processes in the spirit of BSP. We now describe the BSλ_p syntax and its reduction with operational motivations. The reader is assumed to be familiar with the elements of BSP[11].

Syntax We consider a set $\dot{V}$ of *local* variables and a set $\bar{V}$ of *global* variables. Let $\dot{x}, \dot{y}, \ldots$ denote local variables and $\bar{x}, \bar{y}, \ldots$ denote global variables from now on. x will denote a variable which can be either local or global.

The syntax of BSλ_p begins with *local* terms t: λ-terms representing programs or values stored in a processor's local memory. The set $\dot{\mathcal{T}}$ of local terms is given by the following grammar:

$$t ::= \dot{x} \;\mid\; t\,t \;\mid\; \lambda\dot{x}.\,t \;\mid\; c$$

where $\dot{x}$ denotes an arbitrary local variable. We will abbreviate to $(t_1 \rightarrow t_2, t_3)$ the conditional term $t_1\ t_2\ t_3$. We assume for the sake of simplicity[1] a finite set $\mathcal{N} = \{0, \ldots, p-1\}$ which represent the set of processors names.

The principal BSλ_p terms E are called *global* and represent parallel vectors i.e. tuples of p local values where the ith value is located at the processor with rank i. The notation is: $\langle\ t_0\ , \ldots\ ,\ t_{p-1}\ \rangle$ where $t_0, \ldots, t_{p-1}$ are local terms.

The set $\bar{\mathcal{T}}$ of global terms is given by the following grammar[2]:

$$\begin{aligned} T ::= \bar{x}\ &|\ TT\ \ |\ \ Tt\ \ |\ \ \lambda\bar{x}.\,T\ \ |\ \ \lambda\dot{x}.\,T \\ &|\ \langle\, t\,, \ldots\,, t\,, \ldots\,, t\,\rangle\ \ |\ \ T\#T\ \ |\ \ T?T\ \ |\ \ (T \overset{t}{\rightarrow} T, T) \end{aligned}$$

and has the following denotational meaning.

Global terms denote parallel vectors (finite maps from $\mathcal{N}$ to local values) functions between them $(\lambda\bar{x}.\ T)$ or functions from local values to such vectors $(\lambda\dot{x}.\ T)$.

The forms $T_1 \# T_2$ and $T_1\,?\,T_2$ are called parallel application (*apply-par*) and *get* respectively. Apply-par represents point-wise application of a vector of functions to a vector of values. i.e. the pure computation phase of a BSP superstep. Get represents the communication phase of a BSP superstep: a collective data exchange with a barrier synchronization. In $T_1\,?\,T_2$, the resulting vector field contains values from T_1 taken at processor names defined in T_2 (Fig. 1, left).

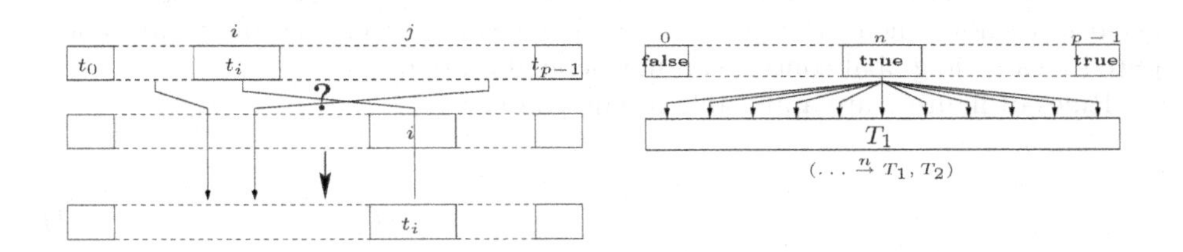

Fig. 1. Get and global conditional

The exact meanings of apply-par and get are defined by the BSλ_p rules (Fig. 2). The last form of global terms define synchronous conditional expressions. The meaning of $(T_1 \overset{n}{\rightarrow} T_2, T_3)$ (not to be confused with $(t_1 \rightarrow t_2, t_3)$) is that of T_2 (resp. T_3) if the vector denoted by T_1 has value **true** (resp. **false**) at the processor name denoted by n (Fig. 1, right).

In the following we will identify terms modulo renaming of bound variables and we will use Barendregt's variable convention[2]: if terms $t_1, \ldots t_n$ occur in a certain mathematical context then in these terms all bound variables are chosen to be different from free variables.

[1] This set $\mathcal{N}$ can be defined in a general way as a set of closed β-normal forms

[2] Terms of the form $(\lambda\bar{x}.T)t$ (resp. $(\lambda\dot{x}.T)T'$) are not β-contracted and constitute implicit errors because they present a local argument to a global→global function (resp. a global argument to a local→global function). In practice a two-level type system should eliminate them, but we will not discuss this here.

Rules We now define the reduction of BSλ terms.

The reduction of local terms is simply β-reduction, obtained from the local β-contraction rule (1) with the usual context rules.

The reduction of global terms is defined by syntax-directed rules and context rules which determine the applicability of the former. First, there are rules for global beta-equivalence (2) and (3).

There are also axioms for the interaction of the vector constructor with the other BSP operations (4) and (5) where for all $i \in \{0, \ldots, p-1\}$, n_i is a processor name belonging to $\mathcal{N}$. The value of $E_1 \mathbin{?} E_2$ at processor name n_i is the value of E_1 at processor name given by the value of E_2 at n_i. Notice that, in practical terms, this represents an operation whereby every processor receives one and only one value from one and only one other processor. This restriction can be lifted for the BSλ-calculus [7] and for the BSλ_p-calculus.

Next, the global conditional is defined by two rules (6) where n belongs to $\mathcal{N}$ and T is T_1 (resp. T_2) when S is **true** (resp. **false**). The two cases generate the following bulk-synchronous computation: first a pure computation phase where all processors evaluate the local term yielding to n; then processor n evaluates the parallel vector of booleans; if at processor n the value is **true** (resp. **false**) then processor n broadcasts the order for global evaluation of T_1 (resp. T_2); otherwise the computation fails. Those two rules are necessary to express algorithms of the form:

Repeat `Parallel Iteration` **Until** `Max of local errors < epsilon`
because without them, the global control can not take into account data computed locally, ie global control can not depend on data.

Rules of figure 2 are applicable in any context.

$$
\begin{aligned}
(\lambda\dot{x}.\, t)t' &\longrightarrow t[\dot{x} \leftarrow t'] && (1)\\
(\lambda\bar{x}.\, T)T' &\longrightarrow T[\bar{x} \leftarrow T'] && (2)\\
(\lambda\dot{x}.\, T)t' &\longrightarrow T[\dot{x} \leftarrow t'] && (3)\\
\langle\, t_0 \,, \ldots \,, t_{p-1} \,\rangle \# \langle\, u_0 \,, \ldots \,, u_{p-1} \,\rangle &\longrightarrow \langle\, t_0\, u_0 \,, \ldots \,, t_{p-1}\, u_{p-1} \,\rangle && (4)\\
\langle\, t_0 \,, \ldots \,, t_{p-1} \,\rangle ? \langle\, n_0 \,, \ldots \,, n_{p-1} \,\rangle &\longrightarrow \langle\, t_{n_0} \,, \ldots \,, t_{n_{p-1}} \,\rangle && (5)\\
(\langle\, t_0 \,, \ldots \,, \underbrace{S}_{n} \,, \ldots \,, t_{p-1} \,\rangle \xrightarrow{n} T_1, T_2) &\longrightarrow T && (6)
\end{aligned}
$$

Fig. 2. The BSλ_p calculus

Examples The first example shows that the intensionality is not lost. The π or parallel vector constructor of BSλ can be defined as :

$$\pi = \lambda f.\langle\, f\,0 \,, \ldots \,, f\,(p-1) \,\rangle$$

The second one is the direct broadcast algorithm which broadcasts the value held at processor 0: i.e `bcast0` $= \lambda\bar{x}\,.\,\bar{x} \mathbin{?} \pi(\lambda i.0)$. If applied to a vector it can be reduced as follows:

$$\begin{array}{l}\texttt{bcast0}\ \langle\ t_0\ ,\dots,\ t_{p-1}\ \rangle \\ \xrightarrow{(2)}\ \langle\ t_0\ ,\dots,\ t_{p-1}\ \rangle\ ?\ \pi(\lambda i.0) \quad \xrightarrow{(3)} \langle t_0,\dots,t_{p-1}\rangle\ ?\ \langle(\lambda i.0)\,0,\dots,(\lambda i.0)\,(p-1)\rangle \\ \xrightarrow{(1)}{}^{p}\ \langle\ t_0\ ,\dots,\ t_{p-1}\ \rangle\ ?\ \langle 0,\dots,0\rangle \ \xrightarrow{(5)}\ \langle\ t_0\ ,\dots,\ t_0\ \rangle\end{array}$$

Confluence The BSλ_p-calculus is confluent. [7] presents BSλ_p as a higher-order rewrite system, and its confluence comes from a general theorem on the confluence of such systems. The calculus presented here and the calculus expressed as an higher-order rewrite system are equivalent. The confluence is then obvious. By comparison the confluence of BSλ required a much longer proof.

3 Weak Call-by-Value Strategy

For the BSλ_p-calculus it is possible to define two different reduction strategies for the two levels of the calculus. We choose here the same strategy for local and global reduction: weak call-by-value strategy. With such a strategy, codings as Church numerals are no longer usable. The calculus has to be extended with new constants and rules to deal at least with numbers and booleans. We will omit such constants for the sake of clarity[3].

The strategy can be roughly described as follows: (1) reduction is impossible in the scope of a λ, the last two arguments of a global conditional cannot be reduced (2) for application, apply-par and get, the right argument is first evaluated, then the left one ; for global conditionals the local term is first evaluated then the condition (3) for application, apply-par and get, the non-context rules can be applied only if the arguments are in normal form; for global conditionals the rule can be applied when term representing the processor name and the condition are in normal-form.

More precisely we define the local values : constants, $\lambda\dot{x}.t$, the global values : $\lambda\bar{x}.T$, $\lambda\dot{x}.T$, $\langle\ v_0\ ,\dots,\ v_{p-1}\ \rangle$ where v_i is a local value for all $i\in\{0,\dots,p-1\}$. In the following set of rules $v, v_0,\dots, V$ are values and $n, n_0,\dots$ are processor names :

$$(S1)(\lambda\dot{x}.\ t)\ v \Rightarrow t[\dot{x}\leftarrow v] \qquad (S2)\frac{t\Rightarrow t'}{u\,t\Rightarrow u\,t'} \qquad (S3)\frac{t\Rightarrow t'}{t\,v\Rightarrow t'\,v}$$

$$(S4)\frac{t\Rightarrow t'}{\langle\ t_0\ ,\dots,\ t\ ,\dots,\ t_{p-1}\ \rangle \Rightarrow \langle\ t_0\ ,\dots,\ t'\ ,\dots,\ t_{p-1}\ \rangle}$$

$$(S5)(\lambda\bar{x}.\ T)T' \ \Rightarrow\ T[\bar{x}\leftarrow T'] \qquad (S6)(\lambda\dot{x}.\ T)t' \ \Rightarrow\ T[\dot{x}\leftarrow t']$$

$$(S7)\frac{T\Rightarrow T'}{U\,T\Rightarrow U\,T'} \qquad (S8)\frac{T\Rightarrow T'}{T\,V\Rightarrow T'\,V} \qquad (S9)\frac{t\Rightarrow t'}{T\,t\Rightarrow T\,t} \qquad (S10)\frac{T\Rightarrow T'}{T\,v\Rightarrow T'\,v}$$

$$(S11)\langle\ v_0\ ,\dots,\ v_{p-1}\ \rangle\#\langle\ v'_0\ ,\dots,\ v'_{p-1}\ \rangle \ \Rightarrow\ \langle\ v_0\,v'_0\ ,\dots,\ v_{p-1}\,v'_{p-1}\ \rangle$$

$$(S12)\frac{T\Rightarrow T'}{U\,\#T\Rightarrow U\,\#T'} \qquad (S13)\frac{T\Rightarrow T'}{T\#V\Rightarrow T'\#V}$$

[3] Using the formulation of BSλ_p as a higher order rewrite system, it is very simple to verify that the calculus is still confluent.

$$(S14)\langle\ v_0\ ,\dots,\ v_{p-1}\ \rangle ?\langle\ n_0\ ,\dots,\ n_{p-1}\ \rangle\ \Rightarrow\ \langle\ v_{n_0}\ ,\dots,\ v_{n_{p-1}}\ \rangle$$

$$(S15)\frac{T \Rightarrow T'}{U\,?\,T \Rightarrow U\,?\,T'} \qquad (S16)\frac{T \Rightarrow T'}{T\,?\,V \Rightarrow T'\,?\,V}$$

$$(S17)(\langle\ v_0\ ,\dots,\ \underbrace{\mathbf{true}}_{n},\dots,\ v_{p-1}\ \rangle \xrightarrow{n} T_1, T_2)\ \Rightarrow\ T_1$$

$$(S18)(\langle\ v_0\ ,\dots,\ \underbrace{\mathbf{false}}_{n},\dots,\ v_{p-1}\ \rangle \xrightarrow{n} T_1, T_2)\ \Rightarrow\ T_2$$

$$(S19)\frac{u \Rightarrow u'}{(T \xrightarrow{u} T_1, T_2)\ \Rightarrow\ (T \xrightarrow{u'} T_1, T_2)} \qquad (S20)\frac{T \Rightarrow T'}{(T \xrightarrow{v} T_1, T_2)\ \Rightarrow\ (T' \xrightarrow{v} T_1, T_2)}$$

Confluence On local terms, $\Rightarrow$ is a function. This property is easily proved by induction on the rules $(S1)$, $(S3)$ and $(S2)$.

Lemma 1. *On global terms $\Rightarrow$ is strongly confluent. If $T \Rightarrow T_1$ and $T \Rightarrow T_2$ then there exists T_3 such that $T_1 \Rightarrow T_3$ and $T_2 \Rightarrow T_3$*

Sketch of proof: By induction on $T \Rightarrow T_1$. Each operation excludes the others. So for example if $T \Rightarrow T_1$ is $(S11)$ then $T \Rightarrow T_2$ can only be $(S11)$, $(S13)$ or $(S12)$. But the arguments of # in T are values otherwise the rule $(S11)$ could not be applied: this excludes rule $(S13)$ and $(S12)$. Moreover the hypothesis must be the same, so $T_2 = T_3$.

This is similar for all operations: $\Rightarrow$ is also function on global terms but on parallel vectors. If $T \Rightarrow T_1$ is $(S4)$, for example at processor i, then $T \Rightarrow T_2$ can be $(S4)$ but at a different processor j. In this case, applying $(S4)$ at processor j on T_2 and $(S4)$ at processor i on T_1 will lead to the same term T_3. □

This proof highlights the fact that $\Rightarrow$ is non-deterministic only for parallel vectors. The non-determinism of $\Rightarrow$ corresponds to the asynchronous computation phase of the BSP model. Some operations like # are also implemented to such phases but this feature is not well captured by the $\Rightarrow$ strategy. That is why a *distributed* evaluation has been designed [7]. It is non-deterministic for all operations but those which correspond to the communication and synchronization phases, namely the get and global conditional. Nevertheless, the $\Rightarrow$ strategy is necessary to prove the properties of the distributed evaluation and each one offers a different point of view. BSλ_p corresponds to the macroscopic view of data-parallel programs while the distributed semantics corresponds to the microscopic view or implementation.

4 Parallel Reduction and Parallel Cost Model

Any reduction within the above system corresponds to a BSP computation as follows. A term $\langle\ t_0\ ,\dots,\ t_{p-1}\ \rangle$ is implemented by storing terms t_i on processor i for $i = 0, \dots, p-1$. The part of global terms that lie outside $\langle \dots \rangle$ are replicated on every processor and, by design of the BSλ_p rules, vector constructors $\langle \dots \rangle$ are never nested. As a result, we can associate a parallel cost to reductions. The cost is a vector of p numbers, the overall cost being the maximum of local costs.

1. A global application of $(S5)$ or $(S6)$, or an application of $(S1)$ outside the vector constructor is applied by every processor to its local terms. This counts for one local operation on every processor:
$\langle\, c_1 \,, \ldots \,, \, c_i \,, \ldots \,, \, c_{p-1} \,\rangle \hookrightarrow \langle\, c_1 + 1 \,, \ldots \,, \, c_i + 1 \,, \ldots \,, \, c_{p-1} + 1 \,\rangle$
2. An application of $(S1)$ to one of the p terms in a vector construction is local to one processor:
$\langle\, c_1 \,, \ldots \,, \, c_i \,, \ldots \,, \, c_{p-1} \,\rangle \hookrightarrow \langle\, c_1 \,, \ldots \,, \, c_i + 1 \,, \ldots \,, \, c_{p-1} \,\rangle$
3. An application of $(S11)$ involves one local operation on each processor:
$\langle\, c_1 \,, \ldots \,, \, c_i \,, \ldots \,, \, c_{p-1} \,\rangle \hookrightarrow \langle\, c_1 + 1 \,, \ldots \,, \, c_i + 1 \,, \ldots \,, \, c_{p-1} + 1 \,\rangle$
4. An application of $(S14)$ to a term $\langle\ t_0 \ ,\ldots,\ t_{p-1}\ \rangle\,?\,\langle\ n_0 \ ,\ldots,\ n_{p-1}\ \rangle$ generates: (1) the request for data: the communication cost is $g \cdot h$ where g is the parallel architecture's BSP parameter, and h the highest frequency of any integer within $\{n_0, \ldots, n_{p-1}\}$ (a message contain here an integer, so its length is 1) (2) a synchronization barrier: the cost is L, the parallel architecture's BSP parameter (3) the reception of data: the cost is $g \cdot h^s$, where $h^s = \max_{0 \leq i < p}\{s_i \cdot \#\{n | n = n_i\}\}$, s_i is the size of the local term t_i and $\#S$ is the cardinality of S (4) another synchronization barrier. This barrier could be suppressed because each processor requested a known number of data from other processors. So it is sufficient to count the number of incoming messages to known whether the superstep is completed or not. The Oxford BSPlib do not offer such zero cost barrier, so the current implementation uses 2 synchronization barriers. The cost is so either 0 or L. The BSP time required for an application of $(S14)$ is: $g \cdot h + g \cdot h^s + L_?$ where $L_? = L$ or $2L$ and g, L are the parallel architecture's BSP parameters:
$\langle\, c_1 \,, \ldots \,, \, c_i \,, \ldots \,, \, c_{p-1} \,\rangle \hookrightarrow \langle \ldots, \max_{0 \leq i < p} c_i \ + \ g \cdot h \ + \ g \cdot h^s \ + \ L_?, \ldots \rangle$
5. An application of $(S17)$ or $(S18)$ involves the broadcast of the value (**true** or **false**) to every other processor, followed by a local operation to realize the branching of control towards T_1 or T_2. The BSP time associated to this application is therefore: $g \cdot (p-1) + L + 1$ where g, L are the parallel architecture's BSP parameters:
$\langle\, c_1 \,, \ldots \,, \, c_i \,, \ldots \,, \, c_{p-1} \,\rangle \hookrightarrow \langle \ldots, \max_{0 \leq i < p} c_i \ + \ g_{\rightarrow} \cdot (p-1) \ + \ L_{\rightarrow} \ + \ 1, \ldots \rangle$

Following the above remarks it is possible to associate a BSP cost estimate to any reduction. We will illustrate the strategy and the cost model on a small example:

$$
\begin{array}{lll}
 & \texttt{bcast0}\ \langle t_0, \ldots, t_{p-1} \rangle & \langle\, 0 \,, \ldots, \, 0 \,\rangle \\
\overset{(S5)}{\Rightarrow} & \langle t_0, \ldots, t_{p-1} \rangle \,?\, \pi(\lambda i.0) & \langle\, 1 \,, \ldots, \, 1 \,\rangle \\
\overset{(S6)}{\Rightarrow} & \langle t_0, \ldots, t_{p-1} \rangle \,?\, \langle (\lambda i.0)\, 0, \ldots, (\lambda i.0)\, (p-1) \rangle & \langle\, 2 \,, \ldots, \, 2 \,\rangle \\
\overset{(S1)\ \text{at}\ i}{\Rightarrow} & \langle t_0, \ldots, t_{p-1} \rangle \,?\, \langle (\lambda i.0)\, 0, \ldots, 0, \ldots, (\lambda i.0)\, (p-1) \rangle & \langle\, 2 \,, \ldots, \underbrace{3}_{i}, \ldots, \, 2 \,\rangle \\
\overset{(S1)^{p-1}}{\Rightarrow} & \langle t_0, \ldots, t_{p-1} \rangle \,?\, \langle\, 0 \,, \ldots, \, 0 \,\rangle & \langle\, 3 \,, \ldots, \, 3 \,\rangle \\
\overset{(5)}{\Rightarrow} & \langle\ t_0 \ , \ldots, \ t_0 \ \rangle & \langle \ldots, 3 + g \cdot (p-1) \cdot s_0 + L, \ldots \rangle
\end{array}
$$

where s_0 is the size of the local term t_0.

It is very important to notice that such costing of reductions would be impossible without the data structure $\langle \ldots \rangle$ of BSλ_p and its parallel interpretation. The reason for this is that standard recursively-defined types like lists, do not refer to an explicit notion of process. When applying parallel evaluation strategies to lists there is, from the point of view of BSλ_p, excessive freedom in selecting redexes and parallelizing reductions. As a result, the exact parallel meaning (local on processor 0, local on processor 1, global, etc.) of a reduction is not fixed by the theory and therefore depends on the syntactic context of its application (at the beginning of a list, in the middle of it, etc.).

5 Conclusions and Future Work

We have design a calculus of BSP functional programs, a weak call-by-value strategy for this calculus and an associated parallel cost model. The formalism has the same advantages as our previous work BSλ [8] but the parallel interpretation and the cost model are far more easy to express. Moreover its distributed evaluation [7] realizes the correspondence between the programming model (macroscopic) and the execution model (microscopic) of data-parallel programs [4]. The distributed evaluation has been proved correct w.r.t the weak call-by-value strategy [7] but it remains to give a cost model to the distributed semantics and prove its equivalence to the cost model of the weak call-by-value strategy.

The last formal step will be a distributed abstract machine that will be correct w.r.t the distributed semantics. We will then obtain a complete formal basis for the design of a complete programming environment containing: a polymorphic strongly typed parallel functional language (Bulk Synchronous ML), tools for performance prediction, tools to help to prove programs correction and tools to derivate programs, the derivation being driven by the costs. Such an environment will be particularly well suited to implement skeletons libraries based on BSP [10] or complex BSP algorithms.

References

1. O. Ballereau, F. Loulergue, and G. Hains. High-level BSP Programming: BSML and BSλ. In P Trinder and G. Michaelson, eds., 1st *Scottish Functional Programming Workshop*, Heriot-Watt University, 1999 355
2. H. P. Barendregt. *The Lambda Calculus, Its Syntax and Semantics.* North-Holland, 1986. 357
3. G. E. Blelloch. *Vector Models for Data-Parallel Computing.* MIT Press, 1990. 356
4. L. Bougé. Le modèle de programmation à parallélisme de données: une perspective sémantique. *Technique et Science Informatiques*, 12(5), 1993. 362
5. M. Cole. *Algorithmic Skeletons: Structured Management of Parallel Computation.* MIT Press, 1989. 355
6. G. Hains, F. Loulergue, and J. Mullins. Concrete data structures and functional parallel programming. *Theoretical Computer Science*, Accepted for publication. 355

7. F. Loulergue. *Conception de langages fonctionnels pour la programmation massivement parallèle.* thèse de doctorat, Université d'Orléans, January 2000. **http://www.univ-orleans.fr/SCIENCES/LIFO/Members/loulergu**. 356, 358, 359, 360, 362
8. F Loulergue, G. Hains, and C. Foisy. A Calculus of Functional BSP Programs. *Science of Computer Programming*, 37(1-3):253–277, 2000. 355, 362
9. S. Pelagatti. *Structured Development of Parallel Programs.* Taylor & Francis, 1997. 355
10. D. B. Skillicorn, M. Danelutto, S. Pelagatti, and A. Zavanella. Optimising data-parallel programs using the BSP cost model. In *Europar'98*, volume 1470 of *Lecture Notes in Computer Science*, pages 698–715. Springer Verlag, 1998. 362
11. David Skillicorn, Jonathan M. D. Hill, and W. F. McColl. Questions and answers about BSP. *Scientific Programming*, 6(3):249–274, Fall 1997. 356
12. Leslie G Valiant. A bridging model for parallel computation. *Communications of the ACM*, 33(8):103, August 1990. 355

Ability of Classes of Dataflow Schemata with Timing Dependency

Yasuo Matsubara and Hiroyuki Miyagawa

Bunkyo University, Faculty of Information and Communications,
Namegaya 1100, Chigasaki-shi, 253-8550 Japan
matubara@shonan.bunkyo.ac.jp

Abstract. There are two classes of dataflow schemata DF and ADF. ADF is known to be equivalent to EF, but DF is not. ADF is given by strengthening with two devices compared with DF. One is recursion and the other is an arbiter which allows timing dependent processing. We are interested in whether both devices are necessary for ADF to have such a powerful expression ability. In this paper, we present relations between some dataflow schemata classes, and find that classes without recursion but with timing dependent processing are not powerful enough as ADF.

1 Introduction

A programming language of high expression ability is necessary to extract maximum power from high performance computers. Currently, to examine expression ability of a programming language, a comparison is made using the functional classes which can be realized by a program schemata class [1].

The class EF of the effective functionals[2] is known to have the greatest expression ability. Jaffe[3] investigated the ability of the dataflow schemata class for the first time, and showed the class DF to be equivalent to the class EF in the total interpretations.

However, Matsubara and Noguchi[4] have shown that DF is equivalent to the class EF^d of the deterministic effective functionals in the partial interpretations.

After that, Matsubara and Noguchi[5] proposed the class ADF of the dataflow schemata, and showed that the class is equal to EF in the partial interpretations. ADF is strengthened with 2 devices which are compared with DF. One is an arbiter and another is a recursion. The arbiter is used to introduce the dependency of action on timing. To allow ADF to acquire such expression ability, the question arises whether both of these are required or just one factor is sufficient.

To answer such a question, it is necessary to examine the expression ability of a class having just one kind of device.

Until now, the class RDF where the recursion is solely usable was investigated and it was found $RDF(1)$ and $RDF(\infty)$ are also equivalent to EF^d[6]. Further, it was also found that $ADF(\infty) \equiv EF$[7]. Here,'(1)' means that each arc holds one token at most, and '(∞)' means that arcs works as FIFO queues.

M. Valero et al. (Eds.): ISHPC 2000, LNCS 1940, pp. 364-373, 2000.

In this paper, we are concerned with the classes for which timing dependency is given and no recursion is included.

2 Program Schemata

In this paper, the function symbols shall be taken from the set $\mathcal{F} = \{F1,F2,...\}$ and the predicate symbols shall be taken from the set $\mathcal{P} = \{P1,P2,...\}$. For element e of $\mathcal{F}$ or $\mathcal{P}$, let the number of the arguments of e be expressed as Re. For the sake of simplicity, let RFi≥ 1 and RPi ≥ 1 with respect to arbitrary $i \geq 1$. On the other hand, variables shall be taken from the set $\mathcal{X} = \{X1,X2,...\}$.

The schema becomes a concrete program when interpretation is given. The program gives at most one result value by executing concrete calculation when input values are provided. Interpretation for a schema S gives the data domain D and gives maps for function symbols and predicate symbols.

The partial interpretation is an interpretation which give maps not necessarily totally defined. Hereunder, interpretation means a partial one.

Let the 'computation' composing an effective functional[2] be defined.

Definition 2.1. *(a) Let $\mathbf{X} \subset \mathcal{X}$ be a finite set of the input variables, let $\mathbf{F} \subset \mathcal{F}$ be a finite set of the function symbols, and let $\mathbf{P} \subset \mathcal{P}$ be a finite set of the predicate symbols. In this occasion, let Λ and Π be the minimum set sufficing the following conditions.*

I. $\mathbf{X} \subset \Lambda$

II. For each $f \in \mathbf{F}$ *and* $e_1, ..., e_{Rf} \in \Lambda$,
$f(e_1, ..., e_{Rf}) \in \Lambda$.

III. For each $p \in \mathbf{P}$ *and* $e_1, ..., e_{Rp} \in \Lambda$,
$p(e_1, ..., e_{Rp}) \in \Pi$ *and* $\neg p(e_1, ..., e_{Rp}) \in \Pi$

where '$\neg$' is the symbol expressing the negation.

The elements of Λ are called an expression concerning $\mathbf{X}$ and $\mathbf{F}$, whereas the elements of Π are called a proposition concerning $\mathbf{X}, \mathbf{F}$ and $\mathbf{P}$.

(b) The computation is a sequence comprised of an expression and proposition, and is finally terminated with the expression.

Assume an interpretation I is given and the elements of D are given to the individual input variables. If the expressions and propositions composing the computation are all defined and the propositions have all true values, the computation has a value. The said value is the one of the final expression.

Definition 2.2. *The effective functional S, an element of the class EF, is defined by the 4-tuple $S = <\mathbf{X}, \mathbf{F}, \mathbf{P}, \mathcal{T}>$. This is a recursively enumerable set of computations. Here,*

(1) Let $\mathbf{X} \subset \mathcal{X}, \mathbf{F} \subset \mathcal{F}$, and $\mathbf{P} \subset \mathcal{P}$ be finite sets of the input variables, function symbols, and predicate symbols.

(2) $\mathcal{T}$ indicate the Turing machine, and outputs the i-th computation $\mathcal{T}(i)$ with the positive integer i as an input under proper coding.

(3) Provided that the computation for more than 2 inputs, e.g. $\mathcal{T}(i)$ and $\mathcal{T}(j)$ has values when interpretation and input values are given, let it be assumed that they are the same values. The said value is the value of S.

As a sub-class of EF, let the class of the deterministic effective functionals be defined.

Definition 2.3. *The class of the deterministic effective functionals EF^d is comprised of the element $S = <\mathbf{X}, \mathbf{F}, \mathbf{P}, \mathcal{T}>$ of EF, having the properties shown below.*

When interpretation and input are given and when i is the smallest number such that $\mathcal{T}(i)$ has a value, for arbitrary k such that $k < i$ all the propositions or expressions in $\mathcal{T}(k)$ are defined until a proposition of the falsity value appears for the first time, assuming the propositions or expressions are screened successively from the left to right.

Here, we introduce the class P of ordinary program schemata[1], where simple variables can be used. Executable statements used in P are assignment statement, halt statement, goto statement and conditional statement. For detailed description see [8].

When an undefined value of a predicate or a function is to be evaluated during the execution, let it be considered that the execution entered an infinite loop and no result will be outputted.

Other than the class P, we must introduce the class P_C. In this class, other than normal variables, a control variable, which holds at most one truth-and-falsity value, can be used.

3 Dataflow Schemata

The dataflow schema is a kind of program schema, but different from ordinary program schema is that the way of execution is made in parallel in accordance with the flow of the data.

In the class of dataflow schemata, it is necessary to distinguish the case where an arc can hold at most one token and the case where an arc can hold arbitrary numbers of tokens. Either class is equal judged from a viewpoint of the syntax. When the arc holds arbitrary pieces of the tokens, the order of the tokens is maintained. That is to say, the arc plays the role of FIFO.

Definition 3.1. *The only kind of nodes included in the graph belonging to the class df_{arb} are depicted in Fig.1.*

A white arc is a control arc, a black one is a data arc, and a mixed one is either of these. The input arcs and the output arc of the schema are clearly depicted as Fig.1(o),(p).

In the description below, let the figure and statement be used according to requirement on the assumption that they are equivalent to each other.

To define the semantics of the graph, it is necessary to designate $\mu = 1$ or ∞, where $\mu = 1$ means that each arc can hold at most one token, and $\mu = \infty$ means that each arc can hold arbitrary numbers of tokens. We attach (μ) after the class name.

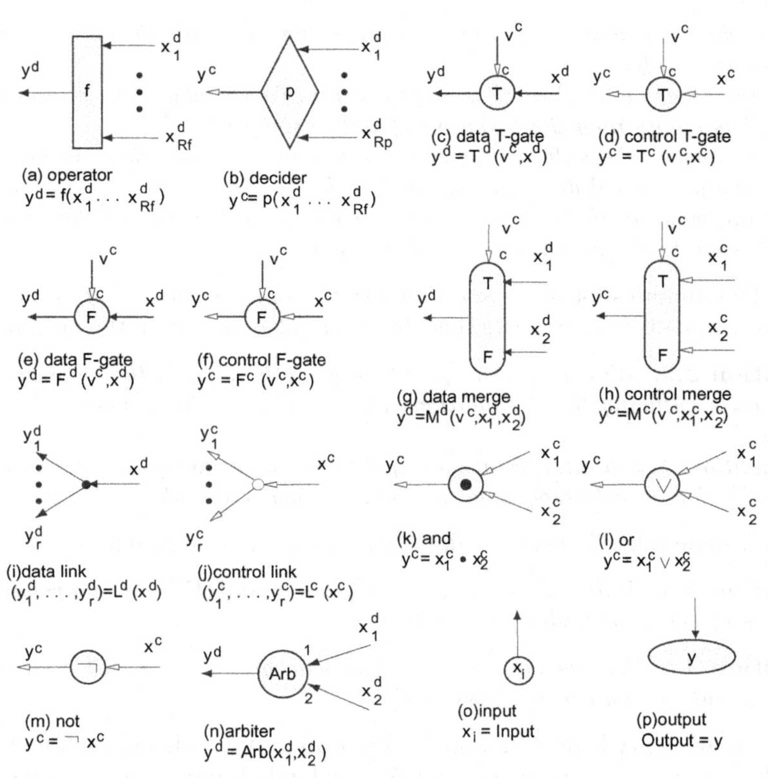

Fig. 1. Dataflow diagram and it's statement

Definition 3.2. *Semantics of $df_{arb}(1)$: the schema of $df_{arb}(1)$ is driven into an action as shown below when interpretation I is given and input data are provided to the individual input variables.*

Let it be assumed that the execution is made on a discrete time such as $t = 0$, 1, 2, ... On the assumption that u is a predicate symbol or function symbol and on the supposition that $d_1,\ldots,d_{Ru} \in D$, $\tau(u,d_1,\ldots,d_{Ru})$ gives the evaluation time of $u(d_1,\ldots,d_{Ru})$. A time function τ, that gives infinity evaluation time when the value of $u(d_1,\ldots,d_{Ru})$ becomes undefined and that gives the evaluation time of positive integer values when the value is defined in the interpretation I, is called a time function consistent with the interpretation I.

In the description hereunder, the action of the individual statements is explained.

$y=u(x_1^d,\ldots,x_{Ru}^d)$, where let u be a function symbol or predicate symbol:

When tokens $d_1,\ldots,d_{Ru}$ are placed on the individual input arcs and concurrently no tokens are placed on y at the time $t = \tau_0$, then at $t=\tau_0+\tau(u,d_1,\ldots,d_{Ru})$, the tokens on the individual input arcs are removed and the tokens of the evaluation result are placed on the output arc y.

$y^d = Arb(x_1^d,x_2^d)$:

If a token is placed on x_1^d *and no token is placed on* y^d*, then the token on* x_1^d *is transferred onto* y^d*.*

On the other hand when x_1^d *is empty and a token is placed on* x_2^d *and concurrently* y^d *is empty then the token on* x_2^d *is transferred to* y^d*.*

Other nodes take only a unit time to execute. Because there is not enough space, for the detailed description, please refer to [8].

Let the value of the token to be placed on the output arc in the schema for the first time be the output result of the schema.

As to the semantics for $df_{arb}(\infty)$, they are the same as that of $df_{arb}(1)$ but the action is made disregarding whether there is a token or not on the output arc(s).

Definition 3.3. *With respect to* μ *=1 or* ∞*,* $DF_{arb}(\mu)$ *is the class comprised of the ones sufficing the 2 conditions shown below in the schema of the class* $df_{arb}(\mu)$*.*

Condition: When interpretation I and input values are given, the same result is provided for an arbitrary time function τ *consistent with I.*

Here, some related classes of dataflow schemata are defined.

Definition 3.4. *With respect to* $\mu = 1, \infty$*, the class* $DF(\mu)$ *is comprised of the elements of* DF_{arb} *including no arbiters.*

Definition 3.5. *A Boolean graph is an acyclic graph comprised exclusively of logical operators and control links of Fig.1.*

For an arbitrary logical function, a Boolean graph realizing the function can easily be composed. As is shown in [4, 5], any kind of finite state machine can be composed by connecting the output arcs to the input arcs of a Boolean graph realizing an appropriate logical function.

4 $P \equiv DF(1)$

In this section, let the ability of $DF(1)$ be observed. Firstly, we observe that the relation $DF(1) \leq P_C$ is satisfied.

Because the schema S of $DF(1)$ does not contain an element dependent on timing, a sequential routine can be written not only to check the necessary and sufficient condition for a token to be outputted on the output arc, but also to evaluate the value of the token[4, 6].

As an example we observe the cases of T-gate :y = T(v^c,x).

Case A(When a token on y is demanded):
(a)A token on v^c is obtained.
(b)A token on x is obtained.
(c)If the token on v^c is 'T' then go to (e).
(d)Tokens on both input arcs are removed and then go to (a).
(e)The token on v^c is removed and the token on x is transfered to y.
(f) Return.

Case B(When v^c is demanded to be empty):
(a)y is demanded to be empty.
(b)A token on x is obtained.
(c)If the token on v^c is 'T' then go to (e).
(d)Tokens on both input arcs are removed and then go to (f).
(e)The token on v^c is removed and the token on x is transfered to y.
(f)Return.

Case C(When x is demanded to be empty):
(a)y is demanded to be empty.
(b)A token on v^c is obtained.
(c)If the token on v^c is 'T' then go to (e).
(d)Tokens on both input arcs are removed and then go to (f).
(e)The token on v^c is removed and the token on x is transfered to y.
(f)Return.

Other cases can be analogized from the above examples. These routines mutually call each other on necessity. While executing a routine, it is possible to come again to a routine about the same statement. As an example, in a routine to output a token from a statement $y = T(v^c, x)$, the routine calls the other routine to obtain the token on v^c if no token is there. While calling the other routine, it is possible for the statement to be demanded to make x empty. To meet the demand it is necessary to obtain the token on v^c. This means a contradiction and then no token should be outputted on the original schema S. The routines get into a infinite loop and then no result is outputted. Thus the action of S can be simulated.

Theorem 4.1. *($DF(1) \leq P_C$): For arbitrary schema S of $DF(1)$, we can compose the schema S' of P_C which is equivalent to S.*

Proof. The simulation in the above can be implemented by the schema S'. Each arc can be simulated by a pair of variables, one of which indicates whether there is a token or not and the other holds the value. By the fact that S includes only a finite number of statements, we can prepare a routine for each case and each statement. The return from each routine can be realized by a simple GOTO statement, since the point to which to return is definite for each routine.

The first statement should be the one calling the routine which obtains the token on the output arc of S. Therefore, S' can simulate S □

In the next step, we observe the relation $P_C \leq P$ is satisfied.

Theorem 4.2. *($P_C \leq P$):For each schema S of P_C, we can compose an equivalent schema of P.*

Proof. For arbitrary control variable u in the schema S, we can construct an equivalent schema S' not including the variable u. We prepare a pair of statements for each statement in S according to the truth-falsity values of u. As an example, for the statement'w ← u · v', two statements are given. On one hand, for the value 'T' of u, 'w ← v' is given. On the other hand, for the value 'F' of u, 'w ← F' is given.

The statement 'IF u THEN statement_1 ELSE statement_2' is divided into two statements 'statement_1' and 'statement_2' in accordance with the value of u.

We can repeat this modification until the schema includes no control variable. □

Now, we should indicate that $P \leq DF(1)$ is satisfied.

Theorem 4.3. *($P \leq DF(1)$):For an arbitrary schema S of P, we can construct an equivalent schema S' of $DF(1)$.*

Proof. S includes only a finite number of variables and also includes a finite number of statements. Therefore, it is not that difficult to simulate the sequential motion of S by S' which can act in parallel. As a matter of fact, S' can be constructed in a similar but easier manner than the procedure COMPUTE_1 in [5]. S' includes a finite state machine which controls the evaluating part. The evaluating part includes some latches which hold values of each variable, a function evaluator, a predicate evaluator and a transfer network which transfers data tokens between the components. The finite state machine simulates the action of each statement in a sequential manner by controlling the evaluating part. When a HALT statement is simulated, the value is outputted on the output arc of S'. □

Based on the three theorems above, we can present the relation that $P \equiv DF(1)$.

Theorem 4.4. *($P \equiv DF(1)$): $DF(1)$ and P are equivalent to each other.*

Proof. From the three theorems above, we can conclude that $P \leq DF(1) \leq P_C \leq P$. Therefore we can conclude the relation. □

Corollary 4.1. *($DF(\infty) > DF(1)$): $DF(\infty)$includes $DF(1)$ properly.*

Proof. It is known that $P < EF^d \equiv DF(\infty)$ [4]. Therefore, we can conclude the result from Theorem 4.4. □

5 $DF_{arb}(1) > P$

Since $DF_{arb}(1)$ can use an arbiter as a device which cannot be used by $DF(1)$, it is evident that $DF_{arb}(1) \geq DF(1)$. Therefore, it is easy to see that $DF_{arb}(1) \geq P$ by the relation $P \equiv DF(1)$. In this section, it is considered whether the inclusion is proper or not.

Here, we introduce an example which can be used for the test. Example A is indicated in Fig.2. When both of predicate $P_1(x)$ and $P_2(x)$ have the value T, the value of the output token is x, disregarding which predicate outputs the result faster than the other. Therefore, the example is an element of the class

$DF_{arb}(1)$.

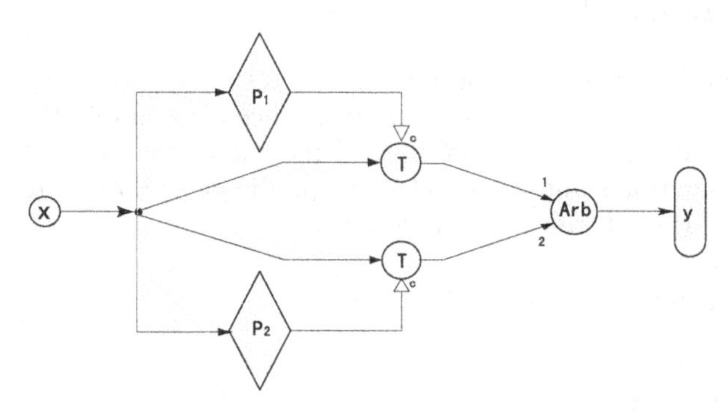

Fig. 2 Example A

Lemma 5.1. *Example A cannot be simulated by a schema of EF^d.*

Proof. Here, we assume that a schema S equivalent to example A belongs to the class EF^d. S should have the same sets of input variables, predicate symbols and function symbols, and the same output variable as example A.

We introduce three interpretations I_1, I_2 and I_3 which have the same data domain $D = \{a\}$.

On I_1, $P_1(a) = T$ and $P_2(a) =$ undefined.

On I_2, $P_1(a) =$ undefined and $P_2(a) = T$.

On I_3, $P_1(a) = P_2(a) =$ undefined.

On I_1, the first computation in S which has a value is assumed to be the i_1-th computation. On I_2, the i_2-th.

Firstly, we assume that i_1 equals i_2. This computation should not include $P_1(x)$ or $P_2(x)$. The reason is as follows: If $P_1(x)$ is included, then the computation is undefined on I_2. If $P_2(x)$ is included, then the computation is undefined on I_1.

If these predicates are not included, the computation should have a value on I_3. This contradicts the assumption that S is equivalent to example A.

Therefore i_1 should not equal i_2.

Next, we assume that $i_1 < i_2$. On I_2, the i_1-th computation should not have a value. Therefore, it should include $\neg P_2(x)$. This means that i_1-th computation becomes undefined on I_1. When $i_2 < i_1$ is assumed, a similar contradiction is deduced. □

Therefore, we can conclude the next theorem.

Theorem 5.1. *($DF_{arb}(1) > P$): The class $DF_{arb}(1)$ properly includes the class P.*

Proof. It is evident that example A cannot be simulated by a schema of P by Lemma 5.1 and the relation $EF^d \geq P$. Therefore we can conclude the result by $DF_{arb}(1) \geq P$. □

6 $EF > DF_{arb}(1)$

It is easy to show that the relation $EF \geq DF_{arb}(1)$ is satisfied. To reveal whether the inclusion is a proper one or not, we introduce an example B.

Example B:This is a schema of EF^d including x as the input variable, L and R as the function symbols and P as the predicate symbol. The computations enumerated by the schema are as follows:

1st:$< P(x), x >$
2nd:$< PL(x), x >$
3rd:$< PR(x), x >$
4th:$< PLL(x), x >$
5th:$< PLR(x), x >$
6th:$< PRL(x), x >$
$\vdots$

Here, we use an abbreviation for the propositions. As an example $PRL(x)$ is the abbreviation of $P(R(L(x)))$.

This schema evaluates the computations in a fixed order, and if it finds the computation having a value, then the value is outputted as the result of the schema. While evaluating a computation, if the value of a proposition or an expression becomes undefined, then the schema's value also becomes undefined.

Lemma 6.1. *There is no schema of $DF_{arb}(1)$ which is equivalent to example B.*

Proof. We assume S is a schema of the class $DF_{arb}(1)$ which is equivalent to example B. We can imagine an interpretation I which gives 'T' for a proposition of a length and gives 'F' for other propositions. If the length is greater than the number of arcs included, S cannot hold all the intermediate values. This is a contradiction. □

Theorem 6.1. $EF > DF_{arb}(1)$

Proof. If we assume that $EF \equiv DF_{arb}(1)$, then $DF_{arb}(1) > EF^d$ is concluded by $EF > EF^d$. This contradicts Lemma 6.1. □

7 Conclusions

We revealed the structure of inclusion relations between some dataflow schemata classes. Adding some equivalence relations [4–6], the whole structure is illustrated in Fig. 3. Here, $\alpha \rightarrow \beta$ means that α is properly included in β.

We list up important points which can be concluded from the figure as follows:

(1) Both timing dependency and recursion are necessary for the class of dataflow schemata to become equivalent to the class EF of effective functionals.

(2) Timing dependency and recursion take different roles from each other in dataflow schemata. Recursion is strong enough to make $RDF(1)$ stronger

than $DF(1)$, while it cannot make $RDF(\infty)$ stronger than $DF(\infty)$. Timing dependency makes $DF_{arb}(1)$ stronger than $DF(1)$.

(3) The value of μ takes rather a minor role. It has an effect only when recursion is not included.

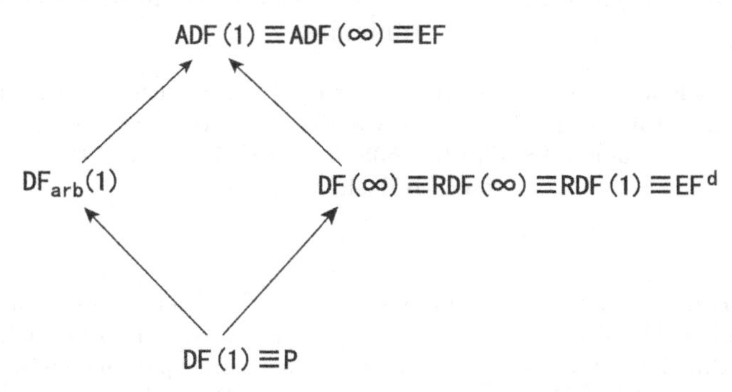

Fig. 3 Inclusion relations between classes.

We have investigated the expression abilities of dataflow languages by the classes of functionals which can be realized. It seems that the conclusions are very suggestive for the language design of any kind of parallel processing.

References

1. Constable , R. L. and Gries , D. , "On Classes of Program Schemata", SIAM Journal on computing , 1 , 1 , pp.66-118(1972).
2. Strong , H. R. , "High Level Languages of Maximum Power" , PROC. IEEE Conf. on Switching and Automata Theory , pp.1-4(1971).
3. Jaffe , J. M. , "The Equivalence of r.e. Program Schemes and Data flow Schemes" , JCSS , 21 , pp.92-109(1980).
4. Matsubara , Y. and Noguchi , S. , "Expressive Power of Dataflow Schemes on Partial Interpretations" , Trans. of IECE Japan , J67-D , 4 , pp.496-503 (1984).
5. Matsubara , Y. and Noguchi , S. , "Dataflow Schemata of Maximum Expressive power" , Trans. of IECE Japan , J67-D , 12 , pp.1411-1418(1984).
6. Matsubara, Y. "Necessity of Timing Dependency in Parallel Programming Language", HPC-ASIA 2000,The Fourth International Conference on High Performance Computing in Asia-Pacific Region. May 14-17,2000 Bejing,China.
7. Matsubara,Y. ,"$ADF(\infty)$ is Also Equivalent to EF", The Bulletin of The Faculty of Information and Communication , Bunkyo University. (1995).
8. Matsubara, Y. and Miyagawa, H. "Ability of Classes of Dataflow Schemata with Timing Dependency", http://www.bunkyo.ac.jp/ matubara/timing.ps

A New Model of Parallel Distributed Genetic Algorithms for Cluster Systems: Dual Individual DGAs

Tomoyuki Hiroyasu, Mitsunori Miki, Masahiro Hamasaki, and Yusuke Tanimura

Department of Knowledge Engineering and Computer Sciences, Doshisha University
1-3 Tatara Miyakodani Kyotanabe Kyoto 610-0321, Japan
Phone: +81-774-65-6638, Fax: +81-774-65-6780
tomo@is.doshisha.ac.jp

Abstract. A new model of parallel distributed genetic algorithm, Dual Individual Distributed Genetic Algorithm (DuDGA), is proposed. This algorithm frees the user from having to set some parameters because each island of Distributed Genetic Algorithm (DGA) has only two individuals. DuDGA can automatically determine crossover rate, migration rate, and island number. Moreover, compared to simple GA and DGA methods, DuDGA can find better solutions with fewer analyses. Capability and effectiveness of the DuDGA method are discussed using four typical numerical test functions.

1 Introduction

The genetic algorithm (GA) (Goldberg, 1987) is an optimization method based on some of the mechanisms of natural evolution. The Distributed Genetic Algorithm (DGA) is one model of parallel Genetic Algorithm (Tanese, 1989; Belding, 1995; Miki, et. al., 1999). In the DGA, the total population is divided into sub-populations and genetic operations are performed on several iterations for each sub-population. After these iterations, some individuals are chosen for migration to another island. This model is useful in parallel processing as well as in sequential processing systems. The reduced number of migrations reduces data transfer, so this model lends itself to use on cluster parallel computer systems. Moreover, DGA can derive good solutions with lower calculation cost as compared to the single population model (Gordon and Whitely, 1993; Whitley, et. al., 1997).

Genetic Algorithms (GA) require user-specified parameters such as crossover and mutation rates. DGA users must determine additional parameters including island number, migration rate, and migration interval. Although Cantu-Paz (1999) investigated DGA topologies, migration rates, and populations, the problems relating to parameter setting remain.

The optimal number of islands was investigated for several problems and it was found that a model with a larger number derives better solutions when the

M. Valero et al. (Eds.): ISHPC 2000, LNCS 1940, pp. 374-383, 2000.

total population size is fixed. Using this result, a new algorithm called Dual Individual Distributed Genetic Algorithm (DuDGA) is proposed. In DuDGA, there are only two individuals on each island. Since DuDGA has only two individuals per island, crossover rate, migration rate, and island number are determined automatically, and the optimum solution can be found rapidly. The capability and effectiveness of DuDGA and its automatic parameter setting and lower calculation cost are discussed using four types of typical numerical test functions. The results are derived using a sequential processing system.

2 Dual Individual Distributed Genetic Algorithms

Distributed Genetic Algorithms (DGAs) are powerful algorithms that can derive better solutions with lower computation costs than Canonical GAs (CGAs). Therefore, many researchers were studied on DGAs (Nang, et. al., 1994, Whitley, et. al., 1997; Munemoto, et. al., 1993, Gordon and Whitley, 1993). However, DGAs have the disadvantage that they require careful selection of several parameters, such as the migration rate and migration intervals, that affect the quality of the solutions.

In this paper, we propose a new model of Distributed Genetic Algorithm. This proposed new model of DGAs is called "Dual Individual Distributed Genetic Algorithms" (DuDGAs). DuDGAs have only two individuals on each island. The concept is shown in Figure 4.

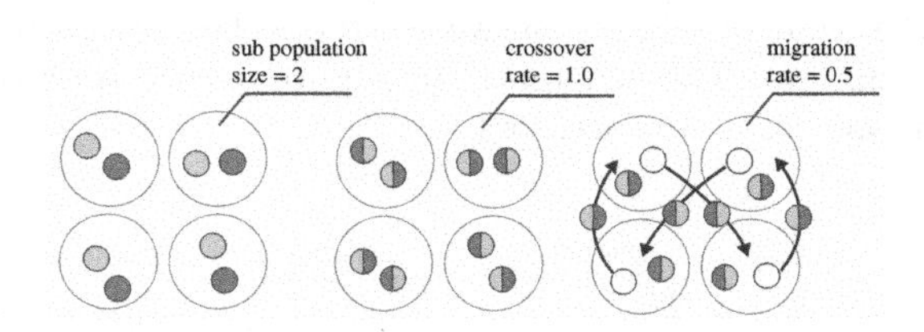

Fig. 1. Dual Individual Distributed Genetic Algorithms

In the proposed DuDGA model , the following operations are performed.

- The population of each island is determined (two individuals).
- Selection: only individuals with the best fit in the present and in one previous generation are kept.
- Migration method: the individual who will migrate is chosen at random.
- Migration topology: the stepping stone method where the migration destination is determined randomly at every migration event.

One of the advantages of the DuDGA is that users are free from setting some of the parameters. By limiting the population to two individuals on each island, the DuDGA model enables the following parameters to be determined automatically:

- crossover rate: 1.0
- number of islands: total population size/2
- migration rate: 0.5

However, because each island has only two individuals, several questions arise. Does the DuDGA model experience a premature convergence problem? Even when the DuDGA can find a solution, does the solution depend on the operation of mutation? The numerical examples clarify answers to these questions. The examples also demonstrate that the DuDGA model can provide higher reliability and achieve improved parallel efficiencies at a lower computation expense than the DGA model.

3 Parallel Implementation of DuDGA

The schematic of the parallel implementation of the DuDGA model is presented in Figure 5. This process is performed as follows:

1. The islands are divided into sub groups. Each group is assigned to one processor.
2. DuDGA is performed for each group. During this step, migration occurs within the group.
3. After some iterations, one of the islands in each group is chosen and is moved to the other group.
4. Step 2 is repeated for the newly formed groups.

Limiting the migration of islands between groups keeps network traffic (data transfer) at a minimum. The schematic in Figure 5 corresponds to a DuDGA implemented on two parallel processors.

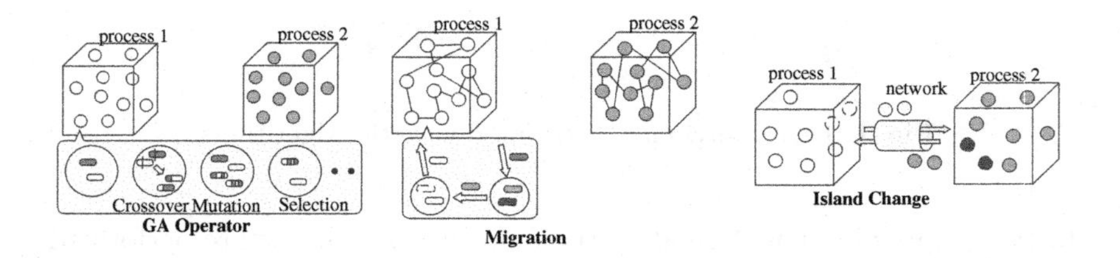

Fig. 2. Parallel Implementation of DuDGA

4 Numerical Examples

This section discusses numerical examples used to demonstrate the DuDGA model. The effects of the number of islands and population size on the performance of DuDGA are presented. The reliability, convergence, and parallel efficiency of the algorithm are discussed.

4.1 Test Functions and Used Parameters

Four types of numerical test functions (Equations 1-4) are considered.

$$F_1 = 10N + \sum_{i=1}^{N}\{x_i^2 - 10cos(2\pi x_i)\}$$
$$(-5.12 \leq x_i < 5.12) \qquad (1)$$

$$F_2 = \sum_{i=1}^{N}\{100(x_{2i-1} - x_{2i}^2)^2 + (x_{2i} - 1)^2\}$$
$$(-2.048 \leq x_i < 2.048) \qquad (2)$$

$$F_3 = 1 + \sum_{n=i}^{10} \frac{x_i^2}{4000} - \prod_{n=1}^{N} cos(\frac{x_i}{\sqrt{i}})$$
$$(-512 \leq x_i < 512) \qquad (3)$$

$$F_4 = \sum_{i=1}^{N}(\sum_{j=1}^{i} x_j)^2$$
$$(-64 \leq x_i < 64) \qquad (4)$$

The number of design variables (ND), the number of bits (NB) and the characteristics of the test functions are summarized in Table 1.

Table 1. Test Functions

	Function Name	ND	NB
F_1	Rastrigin	20	200
F_2	Rosenbrock	5	50
F_3	Griewank	10	100
F_4	Ridge	10	100

It is easy for GAs to derive solutions using the Rastrigin function (F1) because it is a linear function of the design variables. Conversely, it is difficult for GAs to find solutions using non-linear functions such as the Rosenbrock (F2) and Ridge (F4) functions. The degree of difficulty in finding solutions using the Griewank function (F3) is in the range between that for F1 and F2. Table 2 summarizes the parameters specified for the DGA and DuDGA operators.

The algorithm is terminated when the number of generations is more than 5,000. Results shown are the average of 20 trials. The DGA needs several parameters which users must set. However, since the DuDGA has only two individuals in its islands, with the exception of population size and the migration interval, the parameters are automatically determined.

Table 2. Used Parameters

	DGA	DuDGA
Crossover rate	1.0	1.0
Population size	240	240
Mutation rate	$1/L$	$1/L$
Number of islands	4, 8, 12, 24	120
Migration rate	0.3	0.5
Migration interval	5	5

L : Chromosome length

4.2 Cluster System

In this paper, the simulations are performed on a parallel cluster that is constructed with 16 Pentium II (400 Mhz) personal computers (PCs) and fast ethernets. This cluster is similar to a Beowulf cluster and has normal networks. Therefore, increase in network traffic decreases the parallel efficiency.

4.3 Effects of the Number of Islands

The effect of the number of islands on reliability and convergence of the DGA are discussed in this section. Reliability is the fraction of times during 20 trails that an optimum was found. The reliability of DGA for the four test functions and varying number of islands is shown in Figure 6.

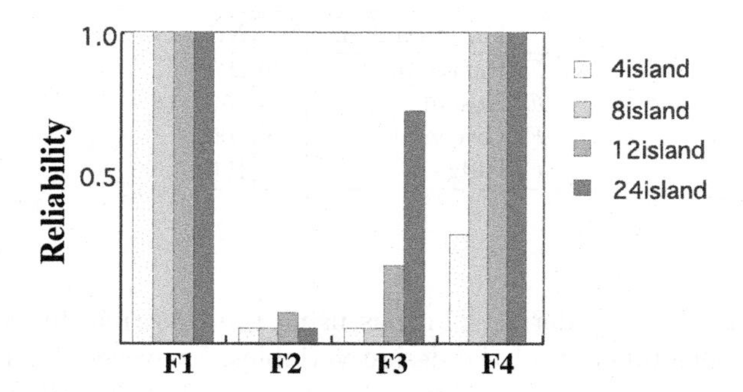

Fig. 3. Reliability of DGA

Figure 6 shows that the reliability of the DGA increases with the number of islands for test functions F1, F3, and F4. F2 is a problem that GAs are not good at finding solutions. Therefore, DGAs can find good results in F2.

Figure 7 shows the number of evaluations needed to located the optimum solution. A substantial portion of the computation effort is spent in evaluating fitness functions an hence a smaller number of calls for function evaluations is desirable.

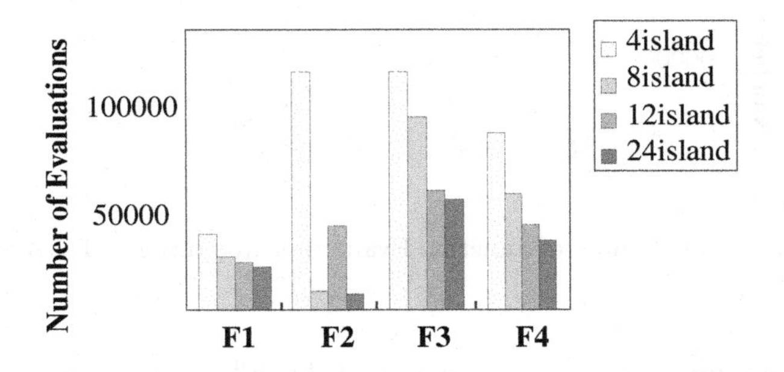

Fig. 4. Number of Evaluations

The results presented in Figure 7 indicate that the DGA requires the least number of function evaluations with highest number of islands. Hence, it can be concluded that the DGA should have as many islands as possible. DuDGA exploits this characteristic by maximizing the number of islands and minimizing the number of individuals.

4.4 Evaluation of DuDGA Performance

Reliability and Convergence Figures 8 and 9 show the reliability and the number of function evaluations for convergence of DuDGA.

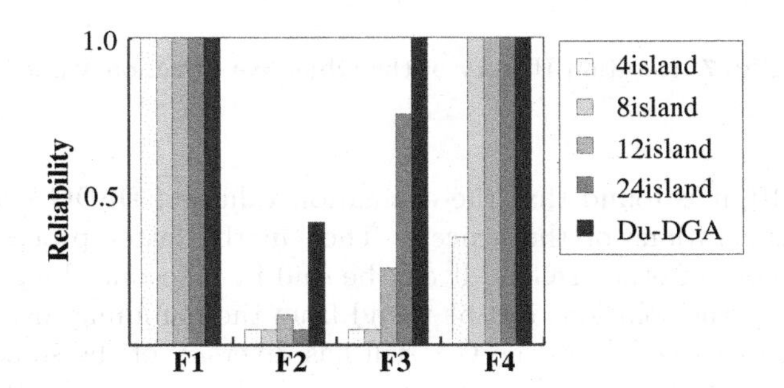

Fig. 5. Comparison of Reliability of DGA and DuDGA

Figures 8 and 9 show that DuDGA exhibited higher reliability and faster convergence for all four test functions when compared to DGA. Figures 10 and 11 show the iteration histories of the objective function and hamming distance

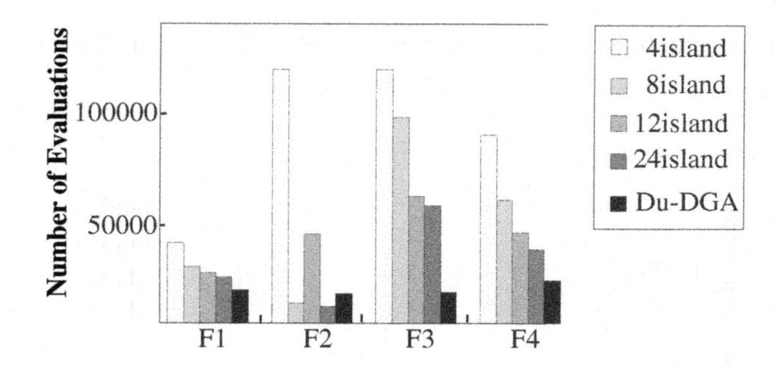

Fig. 6. Comparison of Number of Function Evaluations Required by DGA and DuDGA

values. The Hamming distance is a measure of the difference between two strings, in this case the binary coded chromosome.

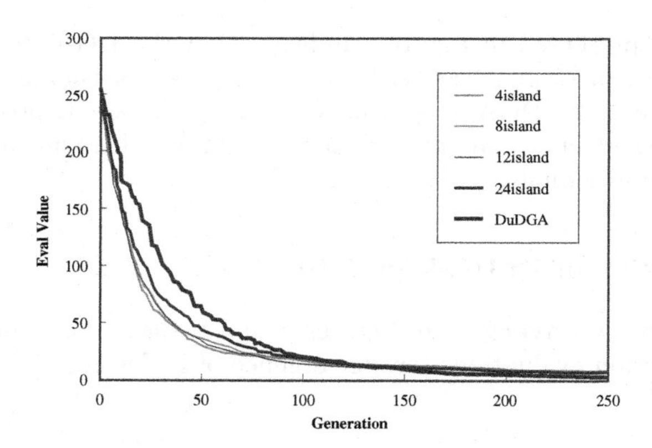

Fig. 7. Iteration History of the Objective Function Value

In figure 10, it is found that the evaluation values of DuDGA are not good at the first generations of the process. Then in the latter process, thoes are better than thoes of other DGAs. It can be said in the same thing in figure 11. The diversity of the solutions can be found from the hamming distances. When the hamming distance is big, the GA still has diversity of the solutions. In the first generations of the process, there is a high diversity in DuDGA. On the other hand, in the latter generations, solutions are converged to the point and DuDGA lost the diversity quickly. Compared to other DGAs, the convergence of DuDGA is slower during the first generations of the process. This is because the DuDGA is searching for global rather than local solutions. Later, when DuDGA is searching for local solutions, values converge quickly, and the model finds better solutions than did the other DGAs.

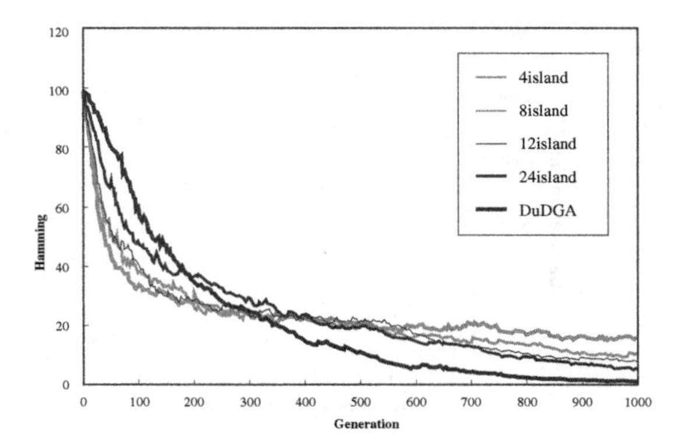

Fig. 8. Iteration History of the Hamming Distance to the Optimum Solution

Parallel Efficiency Calculation speed up of DuDGA for the Rastrigin function (F1) using multiple parallel processors are shown in Figure 13. Results are for fixed population size and number of islands but with varying number of groups.

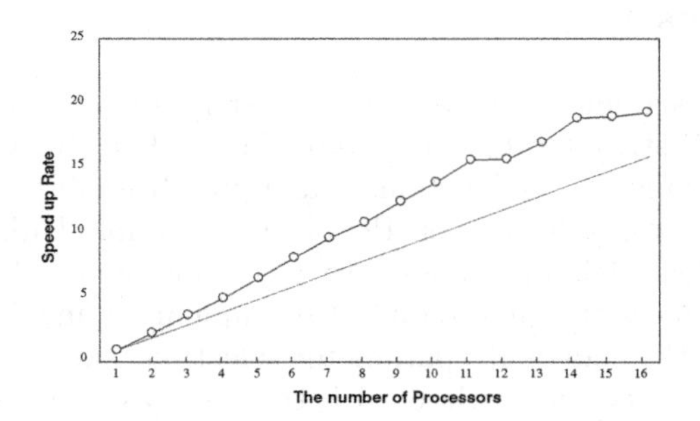

Fig. 9. Speed Up (one process per one processor)

The speed up of the DuDGA implemented on multiple processors is more than linear. The speed up is much higher initially (1-10 processors) and levels off later (14-16 processors). There are two significant reasons for this high parallel efficiency. First, DuDGA limits information flow between processes (groups) thereby reducing data transfer and traffic jams. The second reason is that the total number of calculations that must be performed is reduced by distributing the computation processes.

In Figure 14, computation times of the PC cluster system are shown. These results are obtained for the 16-group DuDGA problem. When there are two processors, each processor has 8 groups. When there are 8 processors, each processor has 2 groups. As Figure 14 shows, when the number of processors increases, calculation time decreases. When the PC cluster with fewer processors than groups is used, each processor must perform several processes. This is not an efficient

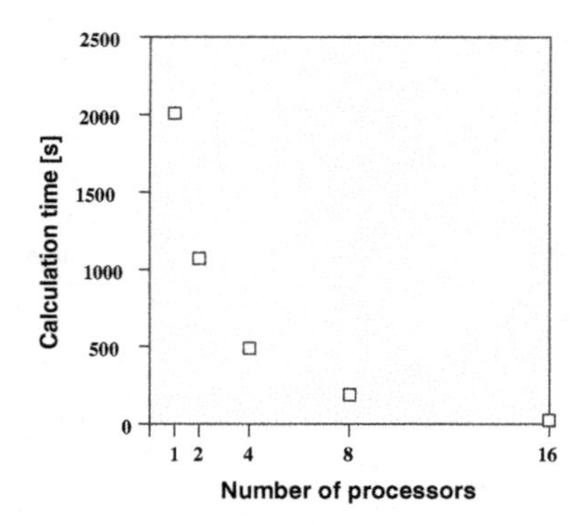

Fig. 10. Computation time (16 processes)

option when a Linux based PC cluster system is used. In order to maximize efficiency, the process threads should be parallelized or the user should use the same number of processors as groups.

5 Conclusions

This paper presents a new model of parallel distributed genetic algorithm called "Dual Individual Distributed Genetic Algorithms." The DuDGA model was applied using four typical test functions–Rastrigin, Rosenbrock, Griewank, and Ridge–to find optimum solutions on PC cluster systems. DuDGA's use of only two individuals in each island enables it to determine some GA parameters automatically. This reduces the time required to implement analyses, reduces problems associated with inadequate parameter selection, and decreases processing time. The evaluation of the method with the test cases examples leads to the following conclusions:

- When the total population size is fixed, the more islands there are, the faster the convergence. The DuDGA exploits this characteristic.
- Compared to the DGA where the number of islands is relatively small, DuDGA can derive better solutions with a smaller number of function evaluations.
- The DuDGA searches using a crossover operation; it cannot search effectively while using only the mutation operation.
- DuDGA performs global searches during the first generations. In the latter part of the analysis, convergence proceeds rapidly, and a local search is performed.
- When the population size is small, a standard GA cannot find an optimum solution due to premature convergence. When the population size is large, GA can derive an optimum solution. However, computation time is wasted. DuDGA does not waste much computational effort, even when the population size is large.

– Because of its high efficiency resulting from the reduced data transfer between groups, DuDGA is an effective method for performing Genetic Algorithms on distributed parallel cluster systems.

References

T. C. Belding (1995). The Distributed Genetic Algorithms Revised, Proc. of 6th Int. Conf. Genetic Algorithms,114-121.

E. Cantu-Paz (1999). Topologies, Migration Rates, and Multi-Population Parallel Genetic Algorithms, *Proceedings of GECCO 1999*,91-98.

D. E. Goldberg (1989) Genetic Algorithms in Search, *Optimization and Machine Learning*, Addison-Wesley, Reading, MA.

V. S. Gordon and D. Whitley (1993). Serial and Parallel Genetic Algorithms as Function Optimizers, *Proc. of 5th Int. Conf. Genetic Algorithms* .

M. MIKI, T. HIROYASU and K. HATANAKA (1999). Parallel Genetic Algorithms with Distributed-Environment Multiple Population Scheme, *Proc. 3rd World Congress of Structural and Multidisciplinary Optimization (WCSMO)*, Vol. 1,186-191.

M. Munemoto, Y. Takai and Y. Sato (1993). An efficient migration scheme for subpopulation based asynchronously parallel genetic algorithms, *Proceedings of the Fifth International Conference on Genetic Algorithms*, 649-159.

J. Nang and K. Matsuo (1994). A Survey on the Parallel Genetic Algorithms, *J. SICE*, Vol. 33, No.6, 500-509.

R. Tanese (1989). Distributed Genetic Algorithms, *Proc. of 3rd Int. Conf. Genetic Algorithms*, 432-439.

D. Whitley, et. al. (1997). Island Model Genetic Algorithms and Linearly Separable Problems, *Proc. of AISB Workshop on Evolutionary Computation*

An Introduction to OpenMP 2.0

Timothy G. Mattson

Intel Corp.
Microcomputer Software Laboratory

Abstract. OpenMP is a programming API for shared memory computers. It is supported by most of the major vendors of shared memory computes and has in the few years since it came out, become one of the major industry standards for parallel programming. In this paper, we introduce the latest version of OpenMP: version 2.0.

1 Introduction

OpenMP is an Application Programming Interface (API) for writing multi-threaded applications for shared-memory computers. OpenMP lets programmers write code that is portable across most shared memory computers.

The group that created OpenMP – the OpenMP architecture review board (or ARB) -- included the major players in the world of shared memory computing: DEC (later Compaq), HP, IBM, Intel, SGI, SUN, KAI, and to provide a user point of view, scientists from the U.S. Department of Energy's ASCI program. We started our work in 1996 and completed our first specification in late 1997 (Fortran OpenMP version 1.0). This was followed by OpenMP 1.0 for C and C++ in late 1998 and a minor upgrade to the Fortran specification in late 1999 (version 1.1).

OpenMP 1.1. is well suited to the needs of the Fortran77 programmer. Programmers wanting to use modern Fortran language features, however, were frustrated with OpenMP 1.1. For example, threadprivate data could only be defined for common blocks, not module variables. OpenMP 2.0 was created to address the needs of these programmers. We started work on OpenMP 2.0 in the spring of 1999. It should be complete by late 2000.

In this paper, we provide a brief introduction to OpenMP 2.0. We begin by describing the OpenMP ARB and the goals we use for our work. Understanding these goals, will make it easier for readers to understand the compromises we made as we created this new specification. Next, we provide a quick introduction to the major new features in OpenMP 2.0.

Throughout this paper, we assume the reader is familiar with earlier OpenMP specifications. If this is not the case, the reader should go to the OpenMP web site (www.openmp.org) and download the specifications.

M. Valero et al. (Eds.): ISHPC 2000, LNCS 1940, pp. 384-390, 2000.

2 The OpenMP ARB

OpenMP is an evolving standard. Rather than create the specifications and then go our separate ways, the creators of OpenMP formed an ongoing organization to maintain OpenMP, provide interpretations to resolve inevitable ambiguities, and produce new specifications as needed. This organization is called the OpenMP Architecture Review Board – or the ARB for short. The official source of information from the ARB to the OpenMP-public is our web site: www.openmp.org.

Companies or organizations join the ARB – not people. Members fall into two categories. Permanent members have a long term business interest in OpenMP. Typically, permanent members market OpenMP compilers or shared memory computers. Auxiliary members serve for one year terms and have a short term or non-business interest in OpenMP. For example, a computing center servicing several teams of users working with OpenMP might join the ARB so their users could have a voice in OpenMP deliberations. Regardless of the type of membership, each member has an equal voice in our OpenMP discussions.

To understand the design of OpenMP and how it is evolving, consider the ARB's goals:

- To produce API specifications that let programmers write portable, efficient, and well understood parallel programs for shared memory systems.
- To produce specifications that can be readily implemented in robust commercial products. i.e. we want to standardize common or well understood practices, not chart new research agendas.
- To whatever extent makes sense, deliver consistency between programming languages. The specification should map cleanly and predictably between C, Fortran, and C++.
- We want OpenMP to be just large enough to express important, control-parallel, shared memory programs -- but no larger. OpenMP needs to stay "lean and mean".
- Legal programs under an older version of an OpenMP specification should continue to be legal under newer specifications.
- To whatever extent possible, we will produce specifications that are sequentially consistent. If sequential consistency is violated, there should be documented reasons for doing so.

We use this set of goals to keep us focused on that delicate balance between innovation and pragmatic technologies that can be readily implemented.

3 The Contents of OpenMP 2.0

OpenMP 2.0 is a major upgrade of OpenMP for the Fortran language. In addition to the ARB goals discussed earlier, the OpenMP 2.0 committee had three additional goals for the project:

- To support the Fortran95 language and the programming practices Fortran95 programmers typically use.
- To extend the range of applications that can be parallelized with OpenMP.
- OpenMP 2.0 will not include features that prevent the implementation of OpenMP conformant Fortran77 compilers.

At the time this paper is being written, the specification is in draft form and open to public review. Any interested reader can download the document from the OpenMP web site (www.openmp.org). Here are some key statistics about the OpenMP 2.0 specification:

- The document is 115 pages long.
- 56 pages for the specification itself.
- 53 pages of explanatory appendices -- including 28 pages of examples.

Note that the specification itself is not very long. It can easily be read and understood in a single sitting. Another nice feature of the specification is the extensive set of examples we include in the appendices. Even with a simple API such as OpenMP, some of the more subtle language features can be difficult to master. We have tried to include examples to expose and clarify these subtleties.

The changes made to the specification in moving from OpenMP 1.1 to OpenMP 2.0 fall into three categories:

- Cleanup: Fix errors and oversights in the 1.1 spec and address consistency issues with the C/C++ spec.
- Fortran90/95 support.
- New functionality: New functionality to make OpenMP easier to use and applicable to a wider range of problems.

In the next three sections, we will briefly outline the OpenMP 2.0 features in each category. We will not define each change in detail, and refer the reader interested in such details to the draft specification itself (available at www.openmp.org).

4 OpenMP 2.0 Features: Cleanup

We made some mistakes when we created the Fortran OpenMP specifications. In some cases, we didn't anticipate some of the ways people would use OpenMP. In other cases, we forgot to define how certain constructs would interact with other parts of the Fortran language. In a few cases, we just plain "got it wrong".

In OpenMP 2.0, we fixed many of these problems. The main changes we made under this category are:

- Relax reprivatization rules
- Allow arrays in reduction clause
- Require that a thread cannot access another thread's private variables.
- Add nested locks to the Fortran runtime library
- Better define the interaction of private clauses with common blocks and equivalenced variables

- Allow comments on same line as directive.
- Define how a STOP statement works inside a parallel region.
- Save implied for initialized data in f77.

In figure 1, we show an example of some of these "cleanup-changes". Notice in line 6 of the program, we have an IF statement that has a STOP statement in its body. Most OpenMP implementations support the use of STOP statements, but the specification never stated how this statement worked in the context of OpenMP. In OpenMP 2.0, we have defined that a STOP statement halts execution of all threads in the program. We require that all memory updates occurring at the barrier (either explicit of implicit) prior to the STOP have completed, and that no memory updates occurring after the subsequent barrier have occurred.

A more important change is seen in the reduction clause. In OpenMP 1.1, we only allow scalar reduction variables. This was a serious oversight on our part. It doesn't matter to the compiler implementer if the reduction variable is a member of an array or a scalar variable. Given that many applications that use reductions need to reduce into the elements of an array, we changed the specification to allow arrays as reduction variables.

```
    real force(NMAX,NMAX)
    logical FLAG
    force(1:NMAX, 1:NMAX) = 0.0
C$OMP parallel private(fij, FLAG)
    call big_setup_calc (FLAG)
    if (FLAG) STOP
C$OMP do private (fij, i, j) reduction(+:force)
     do i=0,N
       do j=low(i),hi(i)
           fij = potential(i,j)
           force(i,j) += fij
           force(j,i) += -fij
       end do
    end do
C$OMP end parallel
```

Fig. 1. This program fragment shows several different features that we cleaned up in OpenMP 2.0. First, we defined the semantics of a STOP statement inside an OpenMP parallel region. Second, it is now possible to privitize a private variable (fij in this program fragment). Finally, it is now possible to use an array in a reduction clause

In OpenMP 1.1, we did not allow programmers to privitize a variable that had already been privatized. This was done since we thought a programmer would never do this intentionally and therefore, we were helping assure more correct programs with the re-privitization restriction. Well, in practice, this restriction has only irritated OpenMP programmers. Since there is no underlying reason compelling us to add this restriction to the language, we decided to drop it in OpenMP 2.0. You can see this in figure 1 where we privatized the variable fij on the PARALLEL construct and again on the enclosed DO construct.

5 OpenMP 2.0 Features: Addressing the Needs of Fortran95

The primary motivation in creating OpenMP 2.0 was to better meet the needs of Fortran95 programmers. We surveyed Fortran95 programmers familiar with OpenMP and found that most of them didn't need much beyond what was already in OpenMP 1.1. We met their needs by adding the following features to OpenMP:

- Share work of executing array expressions among a team of threads.
- Extend Threadprivate so you can make variables threadprivate, not just common blocks.
- Interface declaration module with integer kind parameters for lock variables.
- Generalize definition of reduction and atomic so renamed intrinsic functions are supported.

We show an example of work sharing array expressions in figure 2. This program fragment shows how a simple WORKSHARE construct can be used to split up the loops implied by array expressions between a team of threads. Each array expression is WORKSHARE'ed separately with a barrier implied at the end of each one. Notice that the WORKSHARE statement doesn't indicate how this sharing should take place. We felt that if such detailed control was needed, a programmer had the option to expand the loops by hand and then use standard "OMP DO" construct.

```
       Real, dimension(n,m,p) :: a, b, c, d
           ...
!$omp parallel
!$omp workshare
        a = b * c
        d = b - c
!$omp end workshare
!$omp end parallel
           ...
```

Fig. 2. A program fragment showing how to share the work from array statements among the threads in a team

In Figure 3, we show how the threadprivate construct can be used with variables (as opposed to named common blocks). This is an important addition to OpenMP 2.0 since Fortran95 programmers are encouraged to use module data as opposed to common blocks. With the current form of the threadprivate construct, this is now possible within OpenMP.

```
PROGRAM P
       REAL                    A(100),                  B(200)
       INTEGER IJK
!$THREADPRIVATE(A,B,IJK)
         ...
!$omp parallel copyin(A)  ! Here's an inline comment
         ...
!$omp end parallel
```

Fig. 3. This program fragment shows the use of the threadprivate construct with variables as opposed to common blocks. Notice the program also shows yet another new OpenMP feature: an in-line comment

6 OpenMP 2.0 Features: New Functionality

A continuing interest of the ARB is to make OpenMP applicable to a wider range of parallel algorithms. At the same time, we want to make OpenMP more convenient for the programmer. To this end, we added the following new functionality to OpenMP:

- OpenMP routines for MP code timing
- The NUM_THREADS() clause for nested parallelism.
- COPYPRIVATE on single constructs.
- A mechanism for a program to query OpenMP Spec version number.

In figure 4, we show several of these new features in action. First, note the use of the runtime library routines OMP_GET_WTIME(). These were closely modeled after those from MPI and they return the time in seconds from some fixed point in the past. This fixed point is not defined, but it is guaranteed not to change as the program executes. Hence, these routines can be used to portably find the elapsed wall-clock time used in a program.

Another important new feature in figure 4 is the NUM_THREADS clause. This clause lets the programmer define how many threads to use in a new team. Without this clause, the number of threads could only be set in a sequential reason preventing programs with nested parallel regions from changing the number of threads used at each level in the nesting.

```
      Double precision start, end
      start = OMP_GET_WTIME()
!$OMP PARALLEL NUM_THREADS(8)
     .... Do a bunch of stuff
!$OMP PARALLEL DO NUM_THREADS(2)
         do I=1,1000
            call big_calc(I,results)
         end do
!$OMP END PARALLEL
       end = OMP_GET_WTIME()
       print *,' seconds = ', end-start
```

Fig.4. This program fragment shows how a threads clause can be used to suggest a different number of threads to be used on each parallel region. It also provides an example of the OpenMP wallclock timer routines

Another new feature in OpenMP 2.0 is the COPYPRIVATE clause. This clause can be applied to a END SINGLE construct to copy the values of private variables from one thread to the other threads in a team. For example, in figure 5, we show how the COPYPRIVATE clause can be used to broadcast values input from a file to the private variables within a team. Without COPYPRIVATE, this could have been done, but only be using a shared buffer. The COPYPRIVATE is much more convenient and doesn't require the wasted space implied by the shared buffer.

```
      REAL A,B, X, Y
       COMMON /XY/ X,Y
!$OMP THREADPRIVATE (/XY/)
!$OMP PARALLEL PRIVATE (A, B)
!$OMP SINGLE
           READ (11) A, B, X, Y
!$OMP END SINGLE COPYPRIVATE( A,B, /XY/)
!$OMP END PARALLEL
       END
```

Fig.5. This program fragment shows how to use the copyprivate clause to broadcast the values of private variables to the corresponding private variables on other threads

7 Conclusion

Computers and the way people use them will continue to change over time. It is important, therefore, that the programming standards used for computers should evolve as well.

OpenMP is unique among parallel programming API's in that it has a dedicated industry group working to assure it is well tuned to the needs of the parallel programming committee. OpenMP 2.0 is the latest project from this group. The contents of this specification are being finalized as this paper is being written. We expect implementations of OpenMP 2.0 will be available late in the year 2000 or early in 2001.

To learn more about OpenMP 2.0 and the activities of the broader OpenMP community, the interested reader should consult our web site at www.openmp.org.

Acknowledgements

The Fortran language committee within the OpenMP Architecture Review Board created OpenMP 2.0. While I can't list the names of the committee members here (there are far too many of them), I want to gratefully acknowledge the hard work and dedication of this group.

Implementation and Evaluation of OpenMP for Hitachi SR8000

Yasunori Nishitani[1], Kiyoshi Negishi[1], Hiroshi Ohta[2], and Eiji Nunohiro[1]

[1] Software Development Division, Hitachi, Ltd.
Yokohama, Kanagawa 244-0801, Japan
{nisitani,negish_k,nunohiro}@soft.hitachi.co.jp
[2] Systems Development Laboratry, Hitachi, Ltd.
Kawasaki, Kanagawa 215-0013, Japan
ohta@sdl.hitachi.co.jp

Abstract. This paper describes the implementation and evaluation of the OpenMP compiler designed for the Hitachi SR8000 Super Technical Server. The compiler performs parallelization for the shared memory multiprocessors within a node of SR8000 using the synchronization mechanism of the hardware to perform high-speed parallel execution. To create an optimized code, the compiler can perform optimizations across inside and outside of a PARALLEL region or can produce a code optimized for a fixed number of processors according to the compile option. For user's convenience, it supports combination of OpenMP and automatic parallelization or Hitachi proprietary directive and also supports reporting diagnostic messages which help user's parallelization.
We evaluate our compiler by parallelizing NPB2.3-serial benchmark with OpenMP. The result shows 5.3 to 8.0 times speedup on 8 processors.

1 Introduction

Parallel programming is necessary to exploit high performance of recent supercomputers. Among the parallel programming models, the shared memory parallel programming model is widely accepted because of its easiness or incremental parallelization from serial programs. Until recently, however, to write a parallel program for shared memory systems, the user must use vendor-specific parallelization directives or libraries, which make it difficult to develop portable parallel programs.

To solve this problem, OpenMP[1][2] is proposed as a common interface for the shared memory parallel programming. The OpenMP Application Programming Interface(API) is a collection of compiler directives, library routines, and environment variables that can be used to specify shared memory parallelism in Fortran and C,C++ programs.

Many computer hardware and software vendors are supporting OpenMP. Commercial and non-commercial compilers[3][4][5] are available on many platforms. OpenMP is also being used by Independent Software Vendors for its portability.

M. Valero et al. (Eds.): ISHPC 2000, LNCS 1940, pp. 391-402, 2000.

We implemented an OpenMP compiler for Hitachi SR8000 Super Technical Server. The SR8000 is a parallel computer system consisting of many nodes that incorporate high-performance RISC microprocessors and connected via a high-speed inter-node network. Each node consists of several Instruction Processor(IP)s which share a single address space. Parallel processing within the node is performed at high speed by the hardware mechanism called Co-Operative Microprocessors in a single Address Space(COMPAS).

Most implementations such as NanosCompiler[3] or Omni OpenMP compiler[4] use some thread libraries to implement the fundamental parallel execution model of OpenMP. As the objective of OpenMP for the SR8000 is to control the parallelism within the node and exploit maximum performance of IPs in the node, we implemented the fork-join parallel execution model of OpenMP over COMPAS, in which thread invocation is started by the hardware instruction, so that the overhead of thread starting or synchronization between threads can be reduced and high efficiency of parallel execution can be achieved.

The other characteristic of our OpenMP compiler is as follows.

- Can be combined with automatic parallelization or Hitachi proprietary directive
  Our OpenMP compiler supports full OpenMP1.0 specifications. In addition, our compiler can perform automatic parallelization or parallelization by Hitachi proprietary directives. Procedures parallelized by OpenMP can be combined with procedures parallelized by the automatic parallelization or by the Hitachi proprietary directives. This can be used to supplement OpenMP by Hitachi proprietary directives or to parallelize whole program by automatic parallelization and then to use OpenMP to tune an important part of the program.
- Parallelization support diagnostic messages
  Our compiler can detect loop carried dependence or recognize variables that should be given a PRIVATE or REDUCTION attribute and report the results of these analysis as diagnostic messages. The user can parallelize serial programs according to these diagnostic messages. It also is used to prevent incorrect parallelization by reporting warning messages if there is a possibility that the user's directive is wrong.
- Optimization across inside and outside of the PARALLEL region
  In OpenMP, the code block that should be executed in parallel is explicitly specified by the PARALLEL directive. Like many implementations, our compiler extracts PARALLEL region as a procedure to make implementation simple. Each thread performs parallel processing by executing that procedure. We designed our compiler that the extraction of PARALLEL region is done after global optimizations are executed. This enables optimizations across inside and outside of the PARALLEL region.
- Further optimized code generation by -procnum=8 option
  -procnum=8 option is Hitachi proprietary option with the purpose of bringing out the maximum performance from the node of SR8000. By default, our compiler generates an object that can run with any number of threads, but

if -procnum=8 option is specified, the compiler generates codes especially optimized for the number of threads fixed to 8. This can exploit maximum performance of 8 IPs in the node.

In this paper, we describe the implementation and evaluation of our OpenMP compiler for the Hitachi SR8000. The rest of this paper is organized as follows. Section 2 describes an overview of the architecture of the Hitachi SR8000. Section 3 describes the structure and features of the OpenMP compiler. Section 4 describes the implementation of the OpenMP main directives. Section 5 describes the results of performance evaluation of the compiler. Section 5 also describes some problems about OpenMP specification found when evaluating the compiler. Section 6 concludes this paper.

2 Architecture of Hitachi SR8000

The Hitachi SR8000 system consists of computing units, called "nodes", each of which is equipped with multiple processors. Figure 1 shows an overview of the SR8000 system architecture.

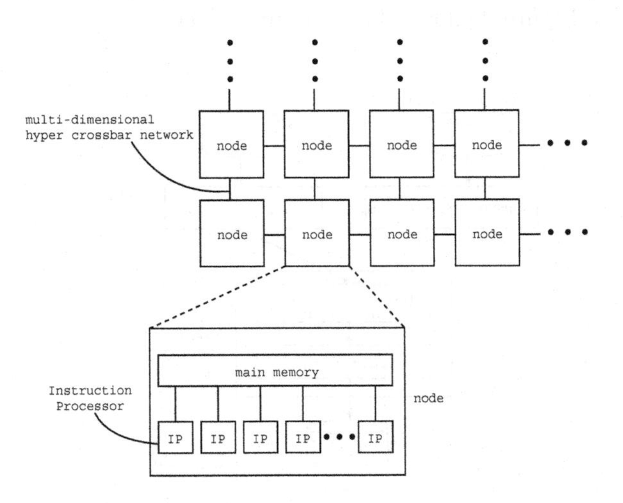

Fig. 1. Architecture of the SR8000 system

The whole system is a loosely coupled, distributed memory parallel processing system connected via high-speed multi-dimensional crossbar network. Each node has local memory and data are transferred via the network. The remote-DMA mechanism enables fast data transfer by directly transferring user memory space to another node without copying to the system buffer.

Each node consists of several Instruction Processor(IP)s. The IPs organize shared memory parallel processing system. The mechanism of Co-Operative Microprocessors in a single Address Space(COMPAS) enables fast parallel processing within the node.

The mechanism of COMPAS enables simultaneous and high-speed activation of multiple IPs in the node. Under COMPAS, one processor issues the "start instruction" and all the other processors start computation simultaneously. The "start instruction" is executed by hardware, resulting in high-speed processing. The operating system of SR8000 also has a scheduling mechanism to exploit maximum performance under COMPAS by binding each thread to a fixed IP.

As described above, SR8000 employs distributed memory parallel system among nodes and shared memory multiprocessors within a node that can achieve high scalability. The user can use message passing library such as MPI to control parallelism among nodes, and automatic parallelization or directive based parallelization of the compiler to exploit parallelism within the node. OpenMP is used to control parallelism within the node.

3 Overview of OpenMP Compiler

OpenMP is implemented as a part of a compiler which generates native codes for the SR8000. Figure 2 shows the structure of the compiler.

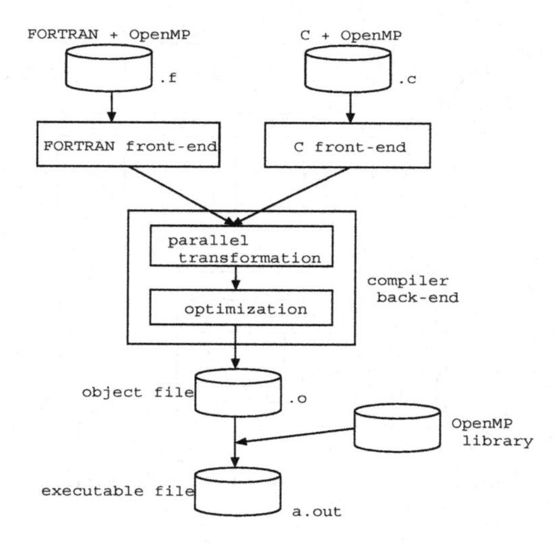

Fig. 2. Structure of the OpenMP compiler

The OpenMP API is specified for Fortran and C,C++. The front-end compiler for each language reads the OpenMP directives, converts to an intermediate language, and passes to the common back-end compiler. The front-end compiler for C is now under development.

The back-end compiler analyzes the OpenMP directives embedded in the intermediate language and performs parallel transformation. The principal transformation is categorized as follows.

- Encapsulate the PARALLEL region as a procedure to be executed in parallel, and generate codes that starts the threads to execute the parallel procedure.
- Convert the loop or block of statements so that the execution of the code is shared among the threads according to the work-sharing directives such as DO or SECTIONS.
- According to the PRIVATE or REDUCTION directives, allocate each variable to the thread private area or generate the parallel reduction codes.
- Perform necessary code transformation for the other synchronization directives.

The transformed code is generated as an object file. The object file is linked with the OpenMP runtime library and the executable file is created.

Our compiler supports full set of OpenMP Fortran API 1.0. All of the directives, libraries, and environment variables specified in the OpenMP Fortran API 1.0 can be used. However, nested parallelism and the dynamic thread adjustment are not implemented.

In addition, our compiler has the following features.

- Can be combined with automatic parallelization or Hitachi proprietary directives
  Procedures parallelized by OpenMP can be mixed together with procedures parallelized by automatic parallelization or Hitachi proprietary directives. By this feature, functions which do not exist in OpenMP can be supplemented by automatic parallelization or Hitachi proprietary directives. For example, array reduction is not supported in OpenMP 1.0. Then automatic parallelization or Hitachi proprietary directives can be used to parallelize the procedure which needs the array reduction and OpenMP can be used to parallelize the other procedures.
- Parallelization support diagnostic message
  Basically in OpenMP, the compiler performs parallelization exactly obeying the user's directive. However, when creating a parallel program from a serial program, it is not necessarily easy for the user to determine if a loop can be parallelized or if a variable needs privatization.
  For this reason, our compiler can perform the same analysis as the automatic parallelization even when OpenMP is used and report the result of the analysis as the diagnostic messages.

4 Implementation of OpenMP Directives

In this section, we describe the implementation of the main OpenMP directives.

4.1 PARALLEL Region Construct

In OpenMP, the code section to be executed in parallel is explicitly specified as a PARALLEL region. OpenMP uses the fork-join model of parallel execution. A program begins execution as a single process, called the master thread. When the master thread reaches the PARALLEL construct, it creates the team of threads. The codes in the PARALLEL region is executed in parallel and in a duplicated manner when explicit work-sharing is not specified. At the end of the PARALLEL region, the threads synchronize and only the master thread continues execution.

We implemented the PARALLEL region of OpenMP for SR8000 using COMPAS, so that the thread fork-join can be performed at high speed. Figure 3 represents the execution of PARALLEL region on COMPAS.

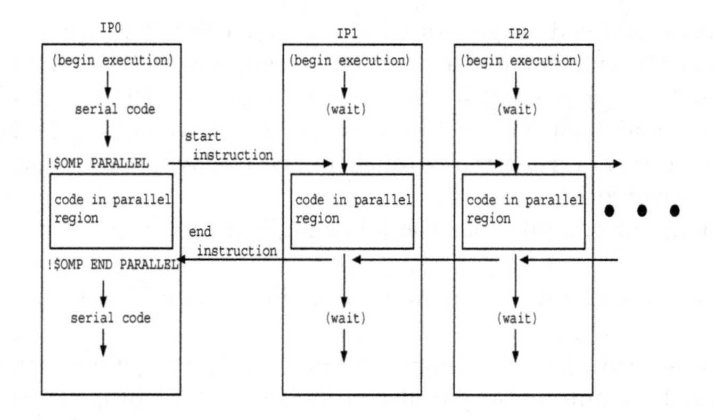

Fig. 3. Execution of PARALLEL region on COMPAS

Under the COMPAS mechanism, each thread is bound to a fixed IP. The execution begins at the IP corresponding to the master thread, and the others wait for starting.

When the master thread reaches the PARALLEL construct, the master IP issues the "start instruction" to the other IPs. This "start instruction" is executed by hardware resulting in high-speed processing. The IPs receiving the "start instruction" also receive the starting address of the PARALLEL region and immediately begin the execution simultaneously.

When PARALLEL region ends, each IP issues the "end instruction" and synchronizes. Then only the master IP continues the execution and the other IPs again enter the wait status.

Using COMPAS mechanism, the thread creation of OpenMP is performed by the hardware instruction to the already executing IPs, so that the overhead of thread creation can be reduced. Also as each thread is bound to a fixed IP, there is no overhead of thread scheduling.

The statements enclosed in PARALLEL region is extracted as a procedure inside the compilation process. The compiler generates a code that each thread runs in parallel by executing this parallel procedure. The PARALLEL region is extracted to a procedure to simplify the implementation. As the code semantics inside the PARALLEL region may differ from those outside the PARALLEL region, normal optimizations for serial programs cannot always be legal across the PARALLEL region boundary. Dividing the parallel part and non-parallel part by extracting the PARALLEL region allows normal optimizations without being concerned about this problem. The implementation of the storage class is also simplified by allocating the auto storage of the parallel procedure to the thread private area. The privatization of the variable in the PARALLEL region is done by internally declaring it as the local variable of the parallel procedure.

However, if the PARALLEL region is extracted as a procedure at the early stage of compilation, the problem arises that necessary optimization across inside and outside of the PARALLEL region is made unavailable. For example, if the PARALLEL region is extracted to a procedure before constant propagation, propagation from outside to inside of the PARALLEL region cannot be done, unless interprocedural optimization is performed.

To solve this problem, our compiler first executes basic, global optimization then extracts the PARALLEL region to a parallel procedure. This enables optimizations across inside and outside of the PARALLEL region. As described above, however, the same optimization as serial program may not be done as the same code in the parallel part and non-parallel part may have different semantics of execution. For example, the optimization which moves the definition of the PRIVATE variable from inside to outside the PARALLEL region should be prohibited. We avoid this problem by inserting the dummy references of the variables for which such optimization should be prohibited at the entry and exit point of the PARALLEL region.

4.2 DO Directive

The DO directive of OpenMP parallelizes the loop. The compiler translates the codes that set the range of loop index so that the loop iterations are partitioned among threads. In addition to parallelizing loops according to the user's directive, our compiler has the following features:

1. Optimization of loop index range calculation codes
   The code to set index range of parallel loop contains an expression which calculates the index range to be executed by each thread using the thread number as a parameter. When parallelizing the inner loop of a loop nest, the calculation of the loop index range will become an overhead if it occurs just before the inner loop. Our compiler performs the optimization that moves the calculation codes out of the loop nest if the original range of the inner loop index is invariant in the loop nest.
2. Performance improvement by -procnum=8 option

In OpenMP, the number of threads to execute a PARALLEL region is determined by the environment variable or runtime library. This means the number of threads to execute a loop cannot be determined until runtime. So the calculation of the index range of a parallel loop contains the division expression of the loop length by the number of threads. While this feature increases the usability because the user can change the number of threads at runtime without recompiling, the calculation of the loop index range will become an overhead if the number of threads used is always the same constant value.

Our compiler supports -procnum=8 option which aims at exploiting the maximum performance of the 8 IPs in the node. If -procnum=8 option is specified, the number of threads used in the calculation of the loop index range is assumed as the constant number 8. As the result, the performance is improved especially if the loop length is a constant as the loop length after parallelization is evaluated to a constant at the compile time and the division at runtime is removed.

3. Parallelizing support diagnostic message

   In OpenMP, the compiler performs parallelization according to the user's directive. It is the user's responsibility to ensure that the parallelized loop has no dependence across loop iterations or whether privatization is needed for each variable. However, it is not always easy to determine if a loop can be parallelized or to examine the needs of privatization for all the variables in the loop. Especially, once the user gives an incorrect directive, it may take a long time to discover the mistake because of the difficulties of debugging peculiar to the parallel program.

   For this reason, while parallelizing exactly obeying the user's directive, our compiler provides the function of reporting the diagnostic messages which is the result of the parallelization analysis of the compiler. The compiler inspects the statements in a loop and analyze whether there is any statement which prevents parallelization or variable that has dependence across loop iterations. It also recognizes the variables which need privatization or parallel reduction operation. Then if there is any possibility that the user's directive is wrong, it generates the dianostic messages. Figure 4 shows the example of the diagnostic message.

   Also the compiler can report information for the loop with no OpenMP directives specified whether the loop can be parallelized or each variable needs privatization. This is useful when converting serial program to parallel OpenMP program.

4.3 SINGLE and MASTER Directives

SINGLE and MASTER directives both specify the statements enclosed in the directive to be executed once only by one thread. The difference is that SINGLE directive specifies the block of statements to be executed by any one of the threads in the team while MASTER specifies the block to be executed by the master thread. Barrier synchronization is inserted at the end of the SINGLE

```
1:        subroutine sub(a,c,n,m)
2:        real a(n,m)
3: !$OMP PARALLEL DO PRIVATE(tmp1)
4:        do j=2,m
5:          do i=1,n
6:            jm1=j-1
7:            tmp1=a(i,jm1)
8:            a(i,j)=tmp1/c
9:          enddo
10:       enddo
11:       end

(diagnosis for loop structure)
   KCHF2015K
             the do loop is parallelized by
             "omp do" directive. line=4
   KCHF2200K
             the parallelized do loop contains
             data dependencies across loop
             iterations. name=A line=4
   KCHF2201K
             the variable or array in do loop is
             not privatized. name=JM1 line=4
```

Fig. 4. Example of diagnostic message

directive unless NOWAIT clause is specified, but no synchronization is done at the end of the MASTER directive and at the entry of the both directives.

The SINGLE directive is implemented that the first thread which reaches the construct executes the SINGLE block. This is accomplished by arranging SHARED attribute flag which means the block is already executed and accessed by each thread to determine whether the thread should execute the block or not. This enables timing adjustment if there is difference of execution timing among threads.

In contrast, the implementation of the MASTER directive is to add conditional branch that the block is executed only by the master thread.

The execution of the SINGLE directive is controled by the SHARED attribute flag, so the flag should be accessed exclusively by the lock operation. As the lock operation generally requires high cost of time, it becomes an overhead. As the result, in case that the execution timings of the threads are almost the same, using MASTER(+BARRIER) directive may show better performance than using SINGLE directive.

5 Performance

5.1 Measurement Methodology

We evaluated our OpenMP compiler by the NAS Parallel Benchmarks(NPB)[6]. The version used is NPB2.3-serial. We parallelized the benchmark by only inserting the OMP directives and without modifying the execution statement, though some rewriting is done where the program cannot be parallelized due to the limitation of current OpenMP specification(mentioned later). The problem size is class A.

The benchmark was run on one node of the Hitachi SR8000, and the serial execution time and parallel execution time on 8 processors are measured.

5.2 Performance Results

Figure 5 shows the result of performance.

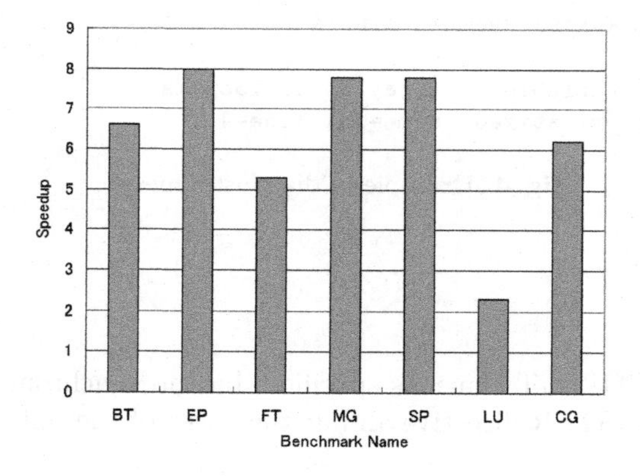

Fig. 5. Speedup of NPB2.3-serial

The figure shows that the speedup on 8 processors is about 5.3 to 8.0 for 6 benchmarks except LU. The reason why the speedup of LU is low is that it contains wavefront style loop which every loop in the loop nest has dependence across loop iterations as follows, and the loop is not parallelized.

```
1:      do j=jst,jend
2:        do i=ist,iend
3:          do m=1,5
4:            v(m,i,j,k) = v(m,i,j,k)
5:            - omega*(ldy(m,1,i,j)*v(1,i,j-1,k)
6:      >                +ldx(m,1,i,j)*v(1,i-1,j,k)
7:      >                + ...
```

```
8:              ...
9:            enddo
10:         enddo
11:      enddo
```

As the DO directive of OpenMP indicates that the loop has no dependence across loop iterations, this kind of loops cannot be parallelized easily.

The automatic parallelization of our compiler can parallelize such a wavefront style loop. We parallelized the subroutine with such a loop by automatic parallelization, parallelized the other subroutines by OpenMP, and measured the performance. The result on 8 processors shows the speedup of 6.3.

5.3 Some Problems of OpenMP Specification

When we parallelized the NPB or other programs by OpenMP, we met the case that the program cannot be parallelized or the program must be rewritten to enable parallelization, as some function do not exist in OpenMP. We describe some of the cases below.

1. Parallelization of a wavefront style loop
   As mentioned in section 5.2, we met the problem that a wavefront style loop cannot be parallelized in the LU benchmark of NPB. This was because OpenMP has only directives which mean that the loop has no dependence across loop iterations.
2. Array reduction
   In the EP benchmark of NPB, parallel reduction for an array is needed. However, in the OpenMP 1.0 specification, only the scalar variable can be specified as REDUCTION variable and arrays cannot be specified. This causes rewriting of the source code.
3. Parallelization of loop containing induction variable
   It is often needed to parallelize the loop that involves an induction variable as follows.

   ```
   K=...
   do I=1,N
     A(2*I-1)=K
     A(2*I)=K+3
     K=K+6
   enddo
   ```

   However, also in this case it cannot be parallelized only with the OpenMP directive and needs to modify the source code.

As these kind of parallelizations described above often appear when parallelizing real programs, it is desirable to extend OpenMP so that these loops can be parallelized only by OpenMP directives.

6 Conclusions

In this paper, we described the implementation and evaluation of the OpenMP compiler for parallelization within the node of Hitachi SR8000. This compiler implements the fork-join execution model of OpenMP using hardware mechanism of SR8000 and achieves high efficiency of parallel execution. We also made our compiler possible to perform optimization across inside and outside of the PARALLEL region or to generate codes optimized for 8 processor by -procnum=8 option. Furthermore, for user's convenience, we implemented parallelizing support diagnostic messages that help the user's parallelization, or enabled combination with automatic parallelization or Hitachi proprietary directives. We evaluated this compiler by parallelizing NAS Parallel Benchmarks with OpenMP and achieved about 5.3 to 8.0 speedup on 8 processors.

Through this evaluation of OpenMP, we found several loops cannot be parallelized because there are features which OpenMP lacks. These loops can be parallelized by automatic parallelization or Hitachi proprietary directives provided by our compiler. We intend to develop the extension to the OpenMP specification so that these loops can be parallelized only by directives.

References

[1] OpenMP Architecture Review Board. OpenMP: A Proposed Industry Standard API for Shared Memory Programming. *White paper*, Oct 1997.

[2] OpenMP Architecture Review Board. OpenMP Fortran Application Program Interface, Oct 1997 1.0.

[3] Eduard Ayguadé, Marc Gonzàlez, Jesús Labarta, Xavier Martorell, Nacho Navarro, and José Oliver. NanosCompiler. A Research Platform for OpenMP extensions. In *EWOMP'99*.

[4] Mitsuhisa Sato, Shigehisa Satoh, Kazuhiro Kusano, and Yoshio Tanaka. Design of OpenMP Compiler for an SMP Cluster. In *EWOMP'99*.

[5] Christian Brunschen and Mats Brorsson. OdinMP/CCp - A portable implementation of OpenMP for C. In *EWOMP'99*.

[6] NASA. The NAS Parallel Benchmarks. http://www.nas.nasa.gov/Software/NPB/.

Copyright Information

Performance Evaluation of the Omni OpenMP Compiler

Kazuhiro Kusano, Shigehisa Satoh, and Mitsuhisa Sato

RWCP Tsukuba Research Center, Real World Computing Partnership
1-6-1, Takezono, Tsukuba-shi, Ibaraki, 305-0032, Japan
TEL: +81-298-53-1683, FAX: +81-298-53-1652
{kusano,sh-sato,msato}@trc.rwcp.or.jp
http://pdplab.trc.rwcp.or.jp/Omni/

Abstract. We developed an OpenMP compiler, called Omni. This paper describes a performance evaluation of the Omni OpenMP compiler. We take two commercial OpenMP C compilers, the KAI GuideC and the PGI C compiler, for comparison. Microbenchmarks and a program in Parkbench are used for the evaluation. The results using a SUN Enterprise 450 with four processors show the performance of Omni is comparable to a commercial OpenMP compiler, KAI GuideC. The parallelization using OpenMP directives is effective and scales well if the loop contains enough operations, according to the results.

Keywords: OpenMP, compiler, Microbenchmarks, parkbench, performance evaluation

1 Introduction

Multi-processor workstations and PCs are getting popular, and are being used as parallel computing platforms in various types of applications. Since porting applications to parallel computing platforms is still a challenging and time consuming task, it would be ideal if it could be automated by using some parallelizing compilers and tools. However, automatic parallelization is still a challenging research topic and is not yet at the stage where it can be put to practical use.

OpenMP[1], which is a collection of compiler directives, library routines, and environment variables, is proposed as a standard interface to parallelize sequential programs. The OpenMP language specification came out in 1997 for Fortran, and in 1998 for C/C++. Recently, compiler vendors for PCs and workstations have endorsed the OpenMP API and have released commercial compilers that are able to compile an OpenMP parallel program.

There have been several efforts to make a standard for compiler directives, such as OpenMP and HPF[12]. OpenMP aims to provide portable compiler directives for shared memory programming. On the other hand, HPF was designed to provide data parallel programming for distributed or non-uniform memory access systems. These specifications were originally supported only in Fortran, but OpenMP announced specifications for C and C++. In OpenMP and HPF, the

M. Valero et al. (Eds.): ISHPC 2000, LNCS 1940, pp. 403–414, 2000.

directives specify parallel actions explicitly rather than as hints for parallelization.

While high performance computing programs, especially for scientific computing, are often written in Fortran as the programming language, many programs are written in C in workstation environments. We focus on OpenMP C compilers in this paper. We also report our evaluation of the Omni OpenMP compiler[4] and make a comparison between Omni and commercial OpenMP C compilers. The objectives of our experiment are to evaluate available OpenMP compilers including our Omni OpenMP compiler, and examine the performance improvement gained by using the OpenMP programming model.

The remainder of this paper is organized as follows: Section 2 presents the overview of the Omni OpenMP compiler and its components. The platforms and the compilers we tested for our experiment are described in section 3. Section 4 introduces Microbenchmarks, an OpenMP benchmark program developed at the University of Edinburgh, and shows the results of an evaluation using it. Section 5 presents a further evaluation using another benchmark program, Parkbench. Section 6 describes related work and we conclude in section 7.

2 The Omni OpenMP Compiler

We are developing an experimental OpenMP compiler, Omni[4] , for an SMP machine. An overview of the Omni OpenMP compiler is presented in this section.

The Omni OpenMP compiler is a translator which takes OpenMP programs as input and generates multi-thread C programs with run-time library calls. The resulting programs are compiled by a native C compiler, and then linked with the Omni run-time library to execute in parallel. The Omni is supported the POSIX thread library for parallel execution, and this makes it easy to port the Omni to other platforms. The platforms the Omni has already been ported to are the Solaris on Sparc and on intel, Linux on intel, IRIX and AIX.

The Omni OpenMP compiler consists of three parts, a front-end, the Exc Java tool and a run-time library. Figure 1 illustrates the structure of Omni.

The Omni front-end accepts programs parallelized using OpenMP directives that are specified in the OpenMP application program interface[2][3]. The front-end for C and FORTRAN77 are available now, and a C++ version is under development. The input program is parsed into an Omni intermediate code, called Xobject code, for both C and FORTRAN77.

The next part, the Exc Java tool, is a Java class library that provides classes and methods to analyze and transform the Xobject intermediate code. It also generates a parallelized C program from the Xobject. The representation of Xobject code which is manipulated by the Exc Java tool is a kind of Abstract Syntax Tree(AST) with data type information. Each node of the AST is a Java object that represents a syntactical element of the source code that can be easily transformed. The Exc Java tool encapsulates the parallel execution part into a separate function to translate a sequential program with OpenMP directives into a fork-join parallel program.

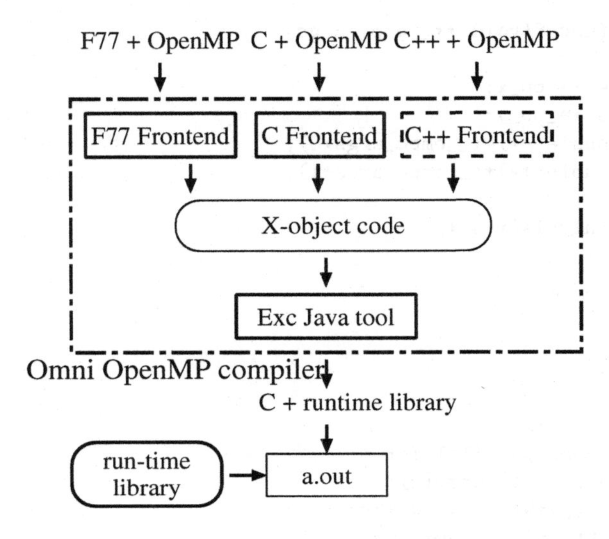

Fig. 1. Omni OpenMP compiler

Figures 2 and 3 show the input OpenMP code fragment and the parallelized code which is translated by Omni, respectively. A master thread calls the Omni

```
func(){
  ...
#pragma omp parallel for
  for(...){
    x=y...
  }
}
```

Fig. 2. OpenMP program fragment

run-time library, _ompc_do_parallel, to invoke slave threads which execute the function in parallel. Pointers to shared variables with auto storage classes are copied into a shared memory heap and passed to slaves at the fork. Private variables are redeclared in the functions generated by the compiler. The work sharing and synchronization constructs are translated into codes which contain the corresponding run-time library calls.

The Omni run-time library contains library functions used in the translated program, for example, _ompc_do_parallel in Figure 3, and libraries that are specified in the OpenMP API. For parallel execution, the POSIX thread library and the Solaris thread library on Solaris OS can be used according to the Omni com-

```
void __ompc_func_6(void **__ompc_args)
{
  auto double **_pp_x;
  auto double **_pp_y;
  _pp_x = (double **)*(__ompc_args+0);
  _pp_y = (double **)*(__ompc_args+1);
  {
   /* index calculation */
    for(...){
      __p_x=__p_y...
    }
  }
}
func(){
  ...
  {/* #pragma omp parallel for */
    auto void *__ompc_argv[2];
    *(__ompc_argv+0) = (void *)&x;
    *(__ompc_argv+1) = (void *)&y;
    _ompc_do_parallel(__ompc_func_6,__ompc_argv);
  }
```

Fig. 3. Program parallelized using Omni

pilation option. The Omni compilation option also allows use of the mutex_lock function instead of the spin-wait lock we developed, the default lock function in Omni. The 1-read/n-write busy-wait algorithm[13] is used as a default Omni barrier function.

Threads are allocated at the beginning of an application program in Omni, not at every parallel execution part contained in the program. All threads but the master are waiting in a conditional wait state until the start of parallel execution, triggered by the library call described before. The allocation and deallocation of these threads are managed by using a free list in the run-time library. The list operations are executed exclusively using the system lock function.

3 Platforms and OpenMP Compilers

The following machines were used as platforms for our experiment.

- SUN Enterprise 450(Ultra sparc 300MHz x4), Solaris 2.6, SUNWspro 4.2 C compiler, JDK1.2
- COMPaS-II(COMPAQ ProLiant6500, Pentium-II Xeon 450MHz x4), RedHat Linux 6.0+kernel-2.2.12, gcc-2.91.66, JDK1.1.7

We evaluated commercial OpenMP C compilers as well as the Omni OpenMP compiler. The commercial OpenMP C compilers we tested are:

- KAI GuideC Ver.3.8[10] on the SUN, and
- PGI C compiler pgcc 3.1-2[11] on the COMPaS-II.

KAI GuideC is a preprocessor that translates OpenMP programs into parallelized C programs with library calls. On the other hand, the PGI C compiler translates an input program directly to the executable code. The compile options used in the following tests are '-fast' for the SUN C compiler, '-O3 -malign-double' for the GNU gcc, and '-mp -fast' for the PGI C compiler.

4 Performance Overhead of OpenMP

This section presents the evaluation of the performance overhead of OpenMP compilers using Microbenchmarks.

4.1 *Microbenchmarks*

Microbenchmarks[6], developed at the University of Edinburgh, is intended to measure the overheads of synchronization and loop scheduling in the OpenMP runtime library. The benchmark measures the performance overhead incurred by the OpenMP directives, for example 'parallel', 'for' and 'barrier', and the overheads of the parallel loop using different scheduling options and chunk sizes.

4.2 Results on the SUN System

Figure 4 shows the results of using the Omni OpenMP compiler and KAI GuideC. The native C compiler used for both OpenMP compilers is the SUNWspro 4.2 C compiler with the '-fast' optimization option.

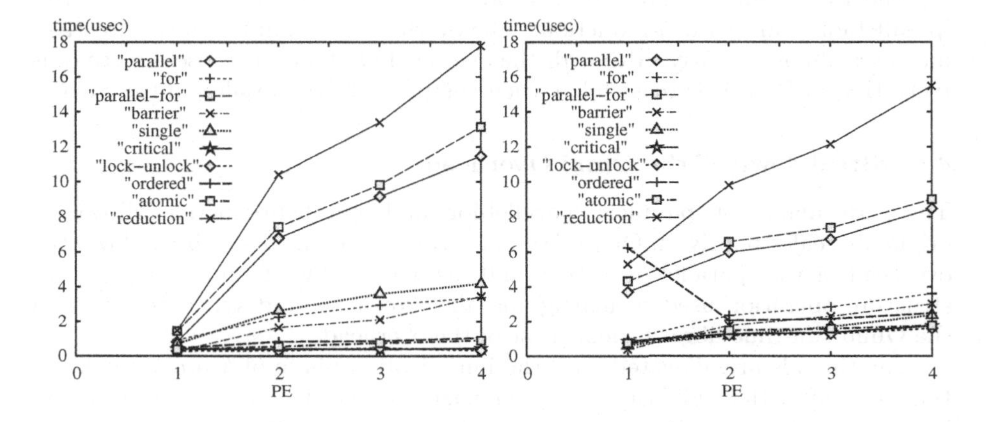

Fig. 4. Overhead of Omni(left) and KAI(right)

These results show the Omni OpenMP compiler achieves competitive performance when compared to the commercial KAI GuideC OpenMP compiler. The overhead of 'parallel', 'parallel-for' and 'parallel-reduction' is bigger than that of other directives. This indicates that it is important to reduce the number of parallel regions to achieve good parallel performance.

4.3 Results on the COMPaS-II System

The results of using the Omni OpenMP compiler and the PGI C compiler on the COMPaS-II are shown in Figure 5.

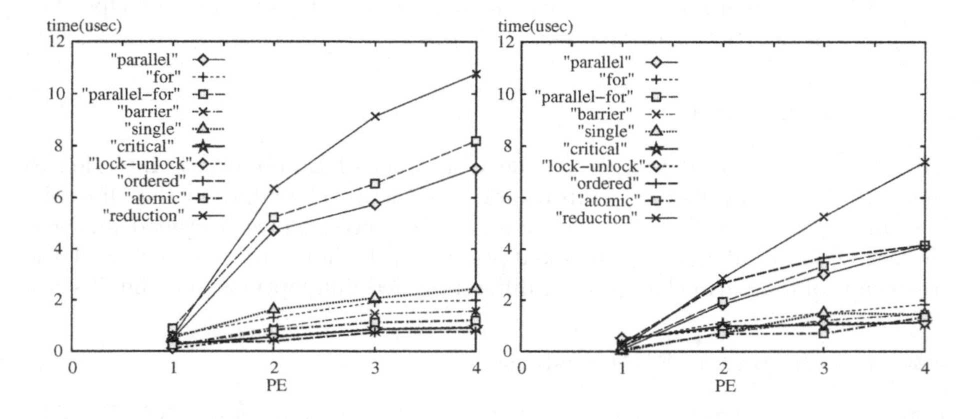

Fig. 5. Overhead of Omni(left) and PGI(right)

The PGI compiler shows very good performance, especially for 'parallel', 'parallel-for' and 'parallel-reduction.' The overhead of Omni for those directives increases almost linearly. Although the overhead of Omni for those directives is twice that of PGI, it is reasonable when compared to the results on the SUN.

4.4 Breakdown of the Omni Overhead

The performance of 'parallel', 'parallel-for' and 'parallel-reduction' directives originally scales poorly on Omni. We made some experiments to breakdown the overhead of the 'parallel' directive, and, as a result, we found that the data structure operation used to manage parallel execution and synchronization in the Omni run-time library spent most of the overhead.

The threads are allocated once the initialization phase of a program execution, and, after that, idle threads are managed by the run-time library using an idle queue. This queue has to be operated exclusively and this serialized queue operations. In addition to the queue operation, there is a redundant barrier synchronization at the end of the parallel region in the library. We modified the run-time library to reduce the number of library calls which require exclusive operation and eliminate redundant synchronization. As a result, the performance shown in Figures 4 and 5 are achieved. Though the overhead of 'parallel for' on the COMPaS-II is unreasonably big, the cause of this is not yet fixed.

Table 1 is the time spent for an allocation of threads and a release of threads and barrier synchronization on the COMPaS-II system.

Table 1. Time to allocate/release data(usec(%))

PE	1	2	3	4
allocation	0.40(43)	2.7(67)	3.5(65)	4.0(63)
release + barrier	0.29(31)	0.50(12)	0.56(10)	0.60(9)

This shows thread allocation still spent the most of the overhead.

5 Performance Improvement from Using OpenMP Directives

This section describes the performance improvements using the OpenMP directives.

We take a benchmark program from Parkbench to use in our evaluation. The performance improvements of a few simple loops with the iterations ranging from one to 100,000 show the efficiency of the OpenMP programming model.

5.1 *Parkbench*

Parkbench[8] is a set of benchmark programs designed to measure the performance of parallel machines. Its parallel execution model is message passing using PVM or MPI. It consists of low-level benchmarks, kernel benchmarks, compact applications and HPF benchmarks.

We use one of the programs, rinf1, in the low-level benchmarks to carry out our experiment. The low-level benchmark programs are intend to measure the performance of a single processor. We rewrote the rinf1 program in C, because the original was written in Fortran. The rinf1 program takes a set of common Fortran operation loops in different loop lengths. For the following test, we chose kernel loops 3, 6 and 16. Figure 6 shows code fragments from a rinf1 program.

5.2 Results on the SUN System

Figures 7, 8 and 9 show the results of kernel loops 3, 6 and 16, respectively, in the rinf1 benchmark program which was parallelized using OpenMP directives

```
 for( jt = 0 ; jt < ntim ; jt++ ){
   dummy(jt);
#pragma omp parallel for
   for( i = 0 ; i < n ; i++ )/* kernel 3 */
     a[i] = b[i] * c[i] + d[i];
 }
 ....
#pragma omp parallel for
   for( i = 0 ; i < n ; i++ )/* kernel 6 */
     a[i] = b[i] * c[i] + d[i] * e[i] + f[i];
...
```

Fig. 6. rinfl kernel loop

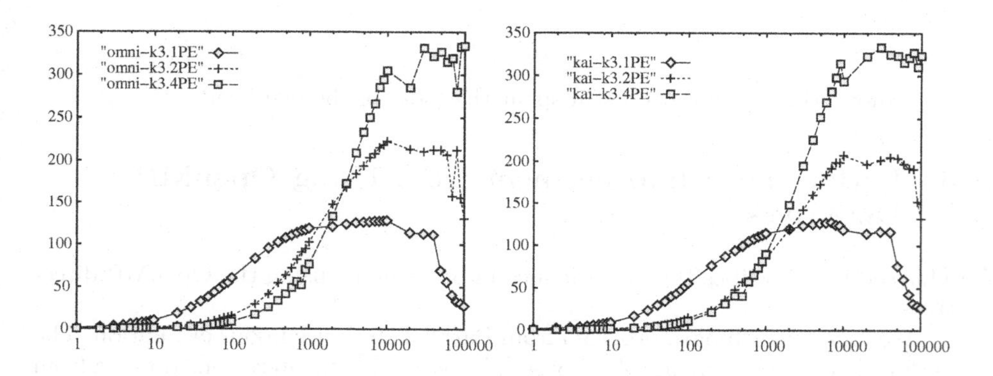

Fig. 7. kernel 3[a(i)=b(i)*c(i)+d(i)] on the SUN: Omni(L) and KAI(R)

executed on the SUN machine. In these graphs, the x-axis is loop length, and the y-axis represents performance in Mflops.

Both OpenMP compilers, Omni and KAI GuideC, achieve almost the same performance improvement, though there are some differences. The differences resulted mainly from the run-time library, because both OpenMP compilers translate to the C program with run-time library calls. KAI GuideC shows better performance for short loop lengths of kernel 6 on one processor, and the peak performance for kernel 16 on two and four processors is better than that of Omni.

5.3 Results on the COMPaS-II System

Figures 10, 11 and 12 are the results of kernel loops in the rinfl benchmark program which were parallelized using the OpenMP directive executed on the COMPaS-II. The x-axis represents loop length, and the y-axis represents performance in Mflops, the same as in the previous case.

The results show the PGI compiler achieves better performance than the Omni OpenMP compiler on the COMPaS-II. The PGI compiler achieves very

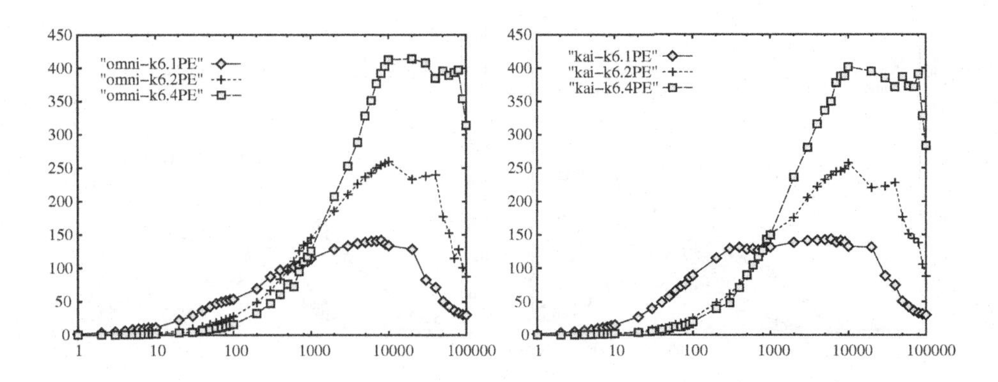

Fig. 8. kernel 6[a(i)=b(i)*c(i)+d(i)*e(i)+f(i)] on the SUN: Omni(L) and KAI(R)

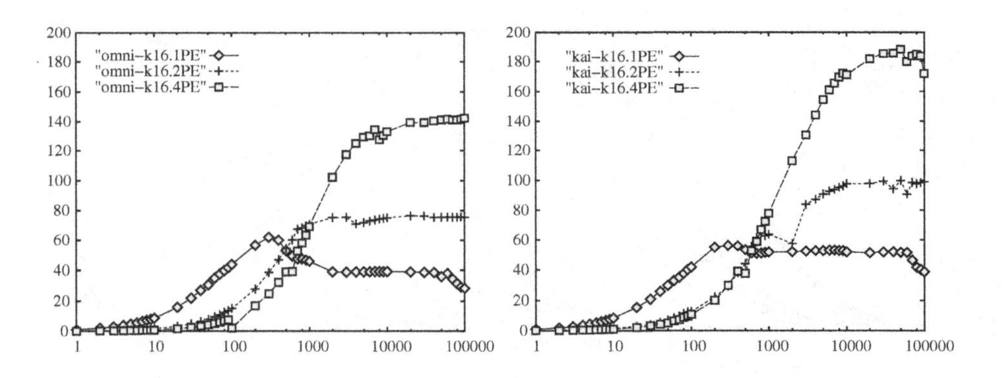

Fig. 9. kernel 16[a(i)=s*b(i)+c(i)] on the SUN: Omni(L) and KAI(R)

good performance for short loop lengths on one processor. The peak performance of PGI reaches about 400 Mflops or more on four processors, and it is nearly double that of Omni in kernels 3 and 16.

5.4 Discussion

Omni and KAI GuideC achieve almost the same performance improvement on the SUN, but the points described above must be kept in mind. The performance improvement of the PGI compiler on the COMPaS-II has different characteristics when compared to the others. Especially, the PGI achieves higher performance for short loop lengths than the Omni on one processor, and the peak performance nearly doubles for kernel 3 and 16. This indicates the performance of Omni could be improved on the COMPaS-II by the optimization of the Omni runtime library, though one must consider the fact that the backend of Omni is different.

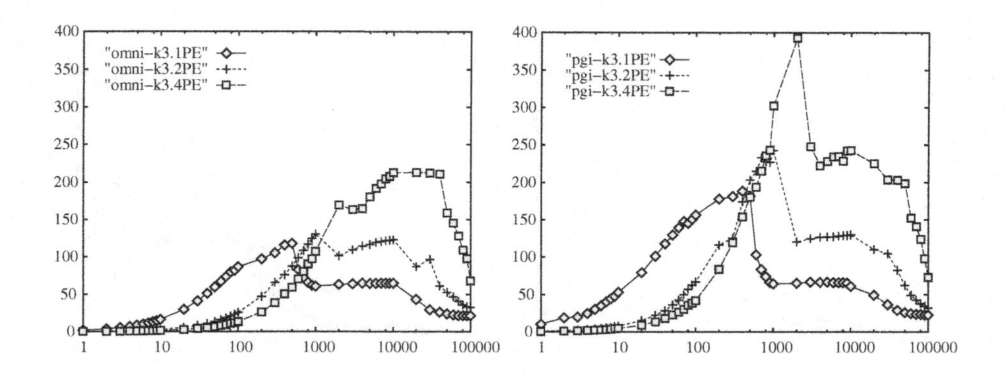

Fig. 10. kernel 3[a(i)=b(i)*c(i)+d(i)] on the COMPaS-II: Omni(L) and PGI(R)

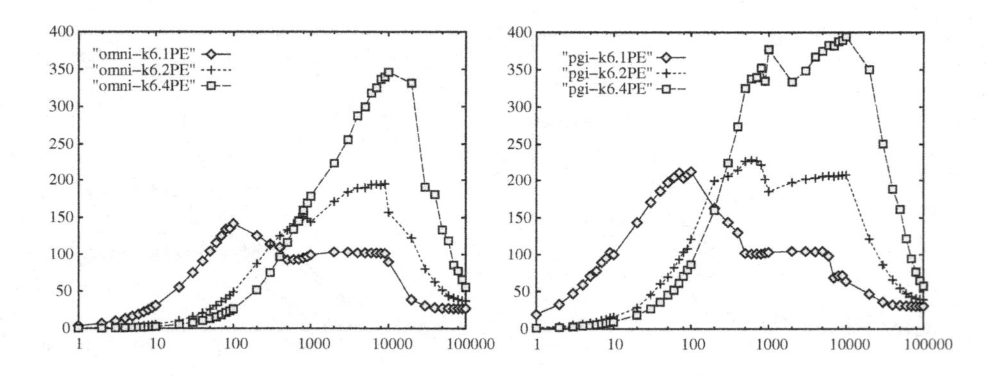

Fig. 11. kernel 6[a(i)=b(i)*c(i)+d(i)*e(i)+f(i)] on the COMPaS-II: Omni(L) and PGI(R)

Those results show that parallelization using the OpenMP directives is effective and the performance scales up for tiny loops if the loop length is long enough.

6 Related Work

Lund University in Sweden developed a free OpenMP C compiler, called OdinMP/CCp[5]. It is also a translator to a multi-thread C program and uses Java as its development language, the same as our Omni. The difference is found in the input language. OdinMP/CCp only supports C as input, while Omni supports C and FORTRAN77. The development language of each frontend is also different, C in Omni and Java in OdinMP/CCp.

There are many projects related to OpenMP, for example, research to execute an OpenMP program on top of the Distributed Shared Memory(DSM)

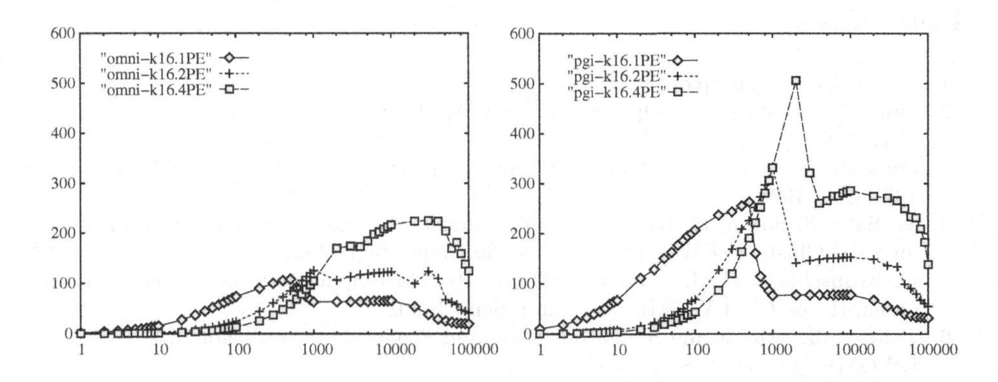

Fig. 12. kernel 16[a(i)=s*b(i)+c(i)] on the COMPaS-II: Omni(L) and PGI(R)

environment on a network of workstations[7], and the investigation of a parallel programming model based on the MPI and the OpenMP to utilize the memory hierarchy of an SMP cluster[9]. Several projects, including OpenMP ARB, have stated the intention to develop an OpenMP benchmark program, though Microbenchmarks[6] is the only one available now.

7 Conclusions

This paper presented an overview of the Omni OpenMP compiler and an evaluation of its performance. The Omni consists of a front-end, an Exc Java tool, and a run-time library, and translates an input OpenMP program to a parallelized C program with run-time library calls. We chose Microbenchmarks and a program in Parkbench to use for our evaluation. While Microbenchmarks measures the performance overhead of each OpenMP construct, the Parkbench program evaluates the performance of array calculation loop parallelized by using the OpenMP programming model. The latter gives some criteria to use to parallelize a program using OpenMP directives.

Our evaluation, using benchmark programs, shows Omni achieves comparable performance to a commercial OpenMP compiler, KAI GuideC, on a SUN system with four processors. It also reveals a problem with the Omni run-time library which indicates that the overhead of thread management data is increased according to the number of processors.

On the other hand, the PGI compiler is faster than the Omni on a COMPaS-II system, and it indicates the optimization of the Omni run-time library could improve its performance, though one must consider the fact that the backend of Omni is different

The evaluation also shows that parallelization using the OpenMP directives is effective and the performance scales up for tiny loops if the loop length is long enough, while the COMPaS-II requires very careful optimization to get peak performance.

References

1. http://www.openmp.org/ 403
2. OpenMP Consortium, "OpenMP Fortran Application Program Interface Ver 1.0", Oct, 1997. 404
3. OpenMP Consortium, "OpenMP C and C++ Application Program Interface Ver 1.0", Oct, 1998. 404
4. M. Sato, S. Satoh, K. Kusano and Y. Tanaka, "Design of OpenMP Compiler for an SMP Cluster", EWOMP '99, pp.32-39, Lund, Sep., 1999. 404
5. C. Brunschen and M. Brorsson, "OdinMP/CCp - A portable implementation of OpenMP for C", EWOMP '99, Lund, Sep., 1999. 412
6. J. M. Bull, "Measuring Synchronisation and Scheduling Overheads in OpenMP", EWOMP '99, Lund, Sep., 1999. 407, 413
7. H. Lu, Y. C. Hu and W. Zwaenepoel, "OpenMP on Networks of Workstations", SC'98, Orlando, FL, 1998. 413
8. http://www.netlib.org/parkbench/html/ 409
9. F. Cappello and O. Richard, "Performance characteristics of a network of commodity multiprocessors for the NAS benchmarks using a hybrid memory model", PACT '99, pp.108-116, Oct., 1999. 413
10. http://www.kai.com/ 407
11. http://www.pgroup.com/ 407
12. C. Koelbel, D. Loveman, R. Schreiber, G. Steele Jr. and M. Zosel,"The High Performance Fortran handbook", The MIT Press, Cambridge, MA, USA, 1994. 403
13. John M. Mellor-Crummey and Michael L. Scott, "Algorithms for Scalable Synchronization on Shared-Memory Multiprocessors", ACM Trans. on Comp. Sys., Vol.9, No.1, pp.21-65, 1991. 406

Leveraging Transparent Data Distribution in OpenMP via User-Level Dynamic Page Migration*

Dimitrios S. Nikolopoulos[1], Theodore S. Papatheodorou[1], Constantine D. Polychronopoulos[2], Jesús Labarta[3], and Eduard Ayguadé[3]

[1] Department of Computer Engineering and Informatics
University of Patras, Greece
{dsn,tsp}@hpclab.ceid.upatras.gr

[2] Department of Electrical and Computer Engineering
University of Illinois at Urbana-Champaign
cdp@csrd.uiuc.edu

[3] Department of Computer Architecture
Technical University of Catalonia, Spain
{jesus,eduard}@ac.upc.es

Abstract. This paper describes transparent mechanisms for emulating some of the data distribution facilities offered by traditional data-parallel programming models, such as High Performance Fortran, in OpenMP. The vehicle for implementing these facilities in OpenMP without modifying the programming model or exporting data distribution details to the programmer is user-level dynamic page migration [9,10]. We have implemented a runtime system called *UPMlib*, which allows the compiler to inject into the application a smart user-level page migration engine. The page migration engine improves transparently the locality of memory references at the page level on behalf of the application. This engine can accurately and timely establish effective initial page placement schemes for OpenMP programs. Furthermore, it incorporates mechanisms for tuning page placement across phase changes in the application communication pattern. The effectiveness of page migration in these cases depends heavily on the overhead of page movements, the duration of phases in the application code and architectural characteristics. In general, dynamic page migration between phases is effective if the duration of a phase is long enough to amortize the cost of page movements.

* This work was supported by the E.C. through the TMR Contract No. ERBFMGECT-950062 and in part through the IV Framework (ESPRIT Programme, Project No. 21907, NANOS), the Greek Secretariat of Research and Technology (Contract No. E.D.-99-566) and the Spanish Ministry of Education through projects No. TIC98-511 and TIC97-1445CE. The experiments were conducted with resources provided by the European Center for Parallelism of Barcelona (CEPBA).

M. Valero et al. (Eds.): ISHPC 2000, LNCS 1940, pp. 415–427, 2000.

1 Introduction

One of the most important problems that programming models based on the shared-memory communication abstraction are facing on distributed shared-memory multiprocessors is poor data locality [3,4]. The non-uniform memory access latency of scalable shared-memory multiprocessors necessitates the alignment of threads and data of a parallel program, so that the rate of remote memory accesses is minimized. Plain shared-memory programming models hide the details of data distribution from the programmer and rely on the operating system for laying out the data in a locality-aware manner. Although this approach contributes to the simplicity of the programming model, it also jeopardizes performance, if the page placement strategy employed by the operating system does not match the memory reference pattern of the application. Increasing the rate of remote memory accesses implies an increase of memory latency by a factor of three to five and may easily become the main bottleneck towards performance scaling.

OpenMP has become the de-facto standard for programming shared-memory multiprocessors and is already widely adopted in the industry and the academia as a simple and portable parallel programming interface [11]. Unfortunately, in several case studies with industrial codes OpenMP has exhibited performance inferior to that of message-passing and data parallel paradigms such as MPI and HPF, primarily due to the inability of the programming model to control data distribution [1,12]. OpenMP provides no means to the programmer for distributing data among processors. Although automatic page placement schemes at the operating system level, such as first-touch and round-robin, are often sufficient for achieving acceptable data locality, explicit placement of data is frequently needed to sustain efficiency on large-scale systems [4].

The natural means to surmount the problem of data placement on distributed shared-memory multiprocessors is data distribution directives [2]. Indeed, vendors of scalable shared-memory systems are already providing the programmers with platform-specific data distribution facilities and the introduction of such facilities in the OpenMP programming interface is proposed by several vendors. Offering data distribution directives similar to the ones offered by High-performance Fortran (HPF) [7] in shared-memory programming models has two fundamental shortcomings. First, data distribution directives are inherently platform-dependent and thus hard to standardize and incorporate seamlessly in shared-memory programming models like OpenMP. OpenMP seeks for portable parallel programming across a wide range of architectures. Second, data distribution is subtle for programmers and compromises the simplicity of OpenMP. The OpenMP programming model is designed to enable straightforward parallelization of sequential codes, without exporting architectural details to the programmer. Data distribution contradicts this design goal.

Dynamic page migration [14] is an operating system mechanism for tuning page placement on distributed shared memory multiprocessors, based on the observed memory reference traces of each program at runtime. The operating system uses per-node, per-page hardware counters, to identify the node of the system that references more frequently each page in memory. In case this node is other than the node that hosts the page, the operating system applies a competitive criterion and migrates the page to the most-frequently referencing node, if the page migration does not violate a set of resource management constraints. Although dynamic page migration was proposed merely as an optimization for parallel programs with dynamically changing memory reference patterns, it was recently shown that a smart page migration engine can also be used as a means for achieving good data placement in OpenMP without exporting architectural details to the programmer [9,10].

In this paper we present an integrated compiler/runtime/OS page migration framework, which emulates data distribution and redistribution in OpenMP without modifying the OpenMP application programming interface. The key for leveraging page migration as a data placement engine is the integration of the compiler in the page migration mechanism. The compiler can provide useful information on three critical factors that determine data locality: the areas of the address space of the program which are likely to concentrate remote memory accesses, the structure of the program, and the phase changes in the memory reference pattern. This information can be exploited to trigger a page migration mechanism at the points of execution at which data distribution or redistribution would be theoretically needed to reach good levels of data locality.

We show that simple mechanisms for page migration can be more than sufficient for achieving the same level of performance that an optimal initial data distribution scheme achieves. Furthermore, we show that dynamic page migration can be used for phase-driven optimization of data placement under the constraint that the computational granularity of phases is coarse enough to enable the migration engine to balance the high cost of coherent page movements with the earnings from reducing the number of remote memory accesses. The presented mechanisms are implemented entirely at user-level in *UPMlib* (User-level Page Migration library), a runtime system designed to tune transparently the memory performance of OpenMP programs on the SGI Origin2000. For details on the implementation and the page migration algorithms of *UPMlib* the reader is referred to [9,10]. This paper emphasizes the mechanisms implemented in *UPMlib* to emulate data distribution.

The rest of this paper is organized as follows. Section 2 shows how user-level page migration can be used to emulate data distribution and redistribution in OpenMP. Section 3 provides a set of experimental results that substantiate the argument that dynamic page migration can serve as an effective substitute for page distribution and redistribution in OpenMP. Section 4 concludes the paper.

2 Implementing Transparent Data Distribution at User-Level

This section presents mechanisms for emulating data distribution and redistribution in OpenMP programs without programmer intervention, by leveraging dynamic page migration at user-level.

2.1 Initial Data Distribution

In order to approximate an effective initial data distribution scheme, a page migration engine must be able to identify early in the execution of the program the node in which each page should be placed according to the expected memory reference trace of the program. Our user-level page migration engine uses two mechanisms for this purpose. The first mechanism is designed for iterative programs, i.e. programs that enclose the complete parallel computation in an outer sequential loop and repeat exactly the same computation for a number of iterations, typically corresponding to time steps. This class of programs represents the vast majority of parallel codes. The second mechanism is designed for non-iterative programs and programs which are iterative but do not repeat the same reference trace in every iteration. Both mechanisms operate on ranges of the virtual address space of the program which are identified as *hot* memory areas by the OpenMP compiler. In the current setting, hot areas are the shared arrays which are both read and written in possibly disjoint OpenMP `PARALLEL DO` and `PARALLEL SECTIONS` constructs.

The iterative mechanism is activated by having the OpenMP compiler instrument the programs to invoke the page migration engine at the end of every outer iteration of the computation. At these points, the page migration engine obtains an accurate snapshot of the complete page reference trace of the program after the execution of the first iteration. Since the recorded reference pattern will repeat itself throughout the lifetime of the program, the page migration engine can use it to place any given page in an optimal manner, so that the maximum latency due to remote accesses by any node to this page is minimized. Snapshots from more than one iterations are needed in cases in which some pages are ping-pong'ing between more than one nodes due to page-level false sharing. This problem can be solved easily in the first few iterations of the program by freezing the pages that tend to bounce between nodes [9].

The iterative mechanism makes very accurate page migration decisions and amortizes well the cost of page migrations, since all the page movement activity is concentrated in the first iteration of the parallel program. This is also the reason that this mechanism is an effective alternative to an initial data distribution scheme. *UPMlib* actually deactivates the mechanism after detecting that page placement is stabilized and no further page migrations are needed to reduce the rate of remote memory accesses. Figure 1 gives an example of the usage of the iterative page migration mechanism in the NAS BT benchmark. In this example, `u,rhs` and `forcing` are identified as hot memory areas by the compiler and monitoring of page references is activated on these areas via the `upmlib_memrefcnt()` call to the runtime system. The function `upmlib_migrate_memory()` applies a competitive page migration criterion on all pages in the hot memory areas and moves the pages that satisfy this criterion [9].

In cases in which the complete memory reference pattern of a program cannot be accurately identified, *UPMlib* uses a mechanism which samples periodically the memory reference counters of a number of pages and migrates the pages that appear to concentrate excessive remote memory accesses. The sampling-based

```
...
call upmlib_init()
call upmlib_memrefcnt(u, size)
call upmlib_memrefcnt(rhs,size)
call upmlib_memrefcnt(forcing,size)
...
do step=1,niter
    call compute_rhs
    call x_solve
    call y_solve
    call z_solve
    call add
    call upmlib_migrate_memory()
enddo
```

Fig. 1. Using the iterative page migration mechanism of *UPMlib* in NAS BT

page migration mechanism is implemented with a memory management thread that wakes up periodically and scans a fraction of the pages in the hot memory areas to detect pages candidate for migration. The length of the sampling interval and the amount of pages scanned upon each invocation are tunable parameters. Due to the cost of page migrations, the duration of the sampling interval must be at least a few hundred milliseconds, in order to provide the page migration engine with a reasonable time frame for migrating pages and moving the cost of some remote accesses off the critical path.

The effectiveness of the sampling mechanism depends heavily on the characteristics of the temporal locality of the program at the page level. The coarser the temporal locality, the better the effectiveness of the sampling mechanism. Assume that the cost of a page migration is 1 ms (typical value for state of the art systems) and a program has a resident set of 3000 pages (typical value for popular benchmarks like NAS). In a worst-case scenario in which all the pages are misplaced, the page migration engine needs 3 seconds to fix the page placement, if the memory access pattern of the program remains uniform while pages are being moved by the runtime system. Clearly, if the program has execution time or phases of duration less than 3 seconds, there is not enough time for the page migration engine to move the misplaced pages. The sampling mechanism is therefore expected to have robust behaviour for programs with reasonably long execution times or reasonably short resident sets with respect to the cost of coherent page migration by the operating system.

2.2 Data Redistribution

Data redistribution in data parallel programming models such as HPF requires identification of phase changes in the reference pattern of the programs. In analogy to HPF, a phase in OpenMP can be defined as a sequence of basic blocks in which the program has a uniform communication pattern among processors.

Each phase may encapsulate more than one OpenMP `PARALLEL` constructs. Under this simple definition the OpenMP compiler can use the page migration engine to establish implicitly an appropriate page placement scheme before the beginning of each phase.The hard problem that has to be addressed in this case is how can the page migration engine identify a good page placement for each phase in the program, using only implicit memory reference information available from the hardware counters.

UPMlib uses a mechanism called *record/replay* to address the aforementioned problem. This mechanism is conceptually similar to the record/replay barriers described in [6]. The record/replay mechanism handles effectively strictly iterative parallel codes in which the same memory reference trace is repeated for a number of iterations. For non-iterative programs, or iterative programs with non-repetitive access patterns, *UPMlib* employs the sampling mechanism outlined in Section 2.1. The record/replay mechanism is activated as follows. The compiler instruments the OpenMP program to record the page reference counters at all phase transition points. The recording procedure stores two sets of reference traces per phase, one at the beginning of the phase and one before the transition to the next phase. The recording mechanism is activated only during the first iteration. *UPMlib* estimates the memory reference trace of each phase by comparing the two sets of counters that were recorded at the phase boundaries. The runtime system identifies the pages that should move in order to tune page placement before the transition to a phase, by applying the competitive criterion to all pages accessed during the phase, based on the corresponding reference trace. After the last phase of the first iteration, the program can simply undo the page migrations executed at all phase transition points by sending the pages back to their original homes. This action recovers the initial page placement scheme. In subsequent iterations, the runtime system replays the recorded page migrations at the respective phase transition points.

Figure 2 gives an example of how the record/replay mechanism is used in the NAS BT benchmark. BT has a phase change in the routine `z_solve`, due to the alignment of data in memory, which is done along the `x` and `y` dimensions. The page reference counters are recorded before and after the first execution of `z_solve` and the recorded values are used to identify page migrations which are replayed before every execution of `z_solve` in subsequent iterations. The routine `upmlib_undo()` is used to undo the page migrations performed by `upmlib_replay()`, in order to recover the initial page placement scheme that is tuned for `x_solve,y_solve` and `add`.

With the record/replay mechanism, page migrations necessarily reside on the critical path of the program. The mechanism is sensitive to the granularity of phases and is expected to work in cases in which the duration of each phase is long enough to amortize the cost of page migrations. In order to limit this cost, the record/replay mechanism can optionally move the n *most critical* pages in each iteration, where n is a tunable parameter set experimentally to balance the overhead of page migrations with the earnings from reducing the rate of remote memory accesses. The n most critical pages are determined as follows: the pages

```
...
call upmlib_init()
call upmlib_memrefcnt(u, size)
call upmlib_memrefcnt(rhs,size)
call upmlib_memrefcnt(forcing,size)
...
do step=1,niter
    call compute_rhs
    call x_solve
    call y_solve
    if (step .eq. 1) then
      call upmlib_record()
    else
      call upmlib_replay()
    endif
    call z_solve
    if (step .eq. 1) then
      call upmlib_record()
    else
      call upmlib_undo()
    endif
    call add
enddo
```

Fig. 2. Using the UPMlib record/replay mechanism in NAS BT

are sorted in descending order according to the ratio $\frac{racc_{max}}{lacc}$, where $lacc$ is the number of local accesses from the home node of the page and $racc_{max}$ is the maximum number of remote accesses from any of the other nodes. The pages that satisfy the inequality $\frac{racc_{max}}{lacc} > thr$, where thr is a predefined threshold are considered as eligible for migration. Let m be the number of these pages. If $m > n$, the mechanism migrates the n pages with the highest ratios $\frac{racc_{max}}{lacc}$. Otherwise, the mechanism migrates the m eligible pages.

Similarly to data distribution and redistribution, the mechanisms described in Sections 2.1 and 2.2 can be combined effectively to obtain the best of the two functionalities in OpenMP programs. For example, the iterative page migration mechanism can be used in the first few iterations of a program to establish quickly a good initial page placement. The record/replay mechanism can be activated afterwards to optimize page placement across phase changes.

3 Experimental Results

We provide a set of experimental results that substantiate our argument that dynamic page migration is an effective substitute for page distribution and redistribution in OpenMP. Our results are constrained by the fact that we were able to experiment only with iterative parallel codes —the OpenMP implementations of the NAS benchmarks as provided by their vendors developers [5].

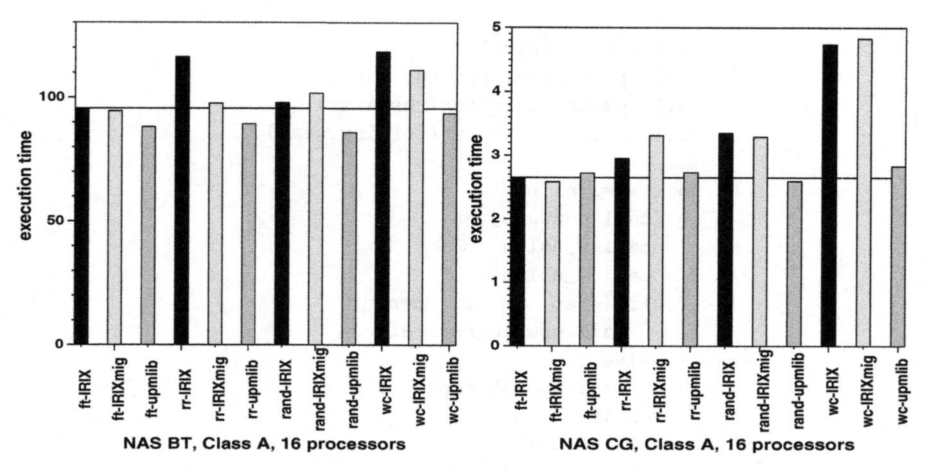

Fig. 3. Performance of UPMlib with different page placement schemes

Therefore, we follow a synthetic experimental approach for the cases in which the characteristics of the benchmarks do not meet the analysis requirements. All the experiments were conducted on 16 idle processors of a 64-processor SGI Origin2000 with MIPS R10000 processors running at 250 MHz and 8 Gbytes of memory. The system ran version 6.5.5 of the SGI IRIX OS.

3.1 Data Distribution

We conducted the following experiment to assess the effectiveness of the iterative mechanism of *UPMlib*. We used the optimized OpenMP implementations of five NAS benchmarks(BT,SP,CG,MG,FT), which were customized to exploit the first-touch page placement scheme of the SGI Origin2000 [8]. Considering first-touch as the page placement scheme that achieves the best data distribution for these codes, we ran the codes using three alternative page placement schemes, namely round-robin page placement, random page placement and worst-case page placement. Round-robin page placement could be optionally requested via an environment variable. Random page placement was hand-coded in the benchmarks, using the standard UNIX page protection mechanism to capture page faults and relocate pages, thus bypassing the default operating system strategy. Worst-case page placement was forced by a sequential execution of the cold-start iteration of each program, during which all data pages were placed on a single node of the system.

Figure 3 shows the results from executing the OpenMP implementations of the NAS BT and CG benchmarks with four page placement schemes. The observed trends are similar for all the NAS benchmarks used in the experiments. We omit the charts for the rest of the benchmarks due to space considerations. Each bar is an average of three independent experiments. The variance in all cases was negligible. The black bars illustrate the execution time with the dif-

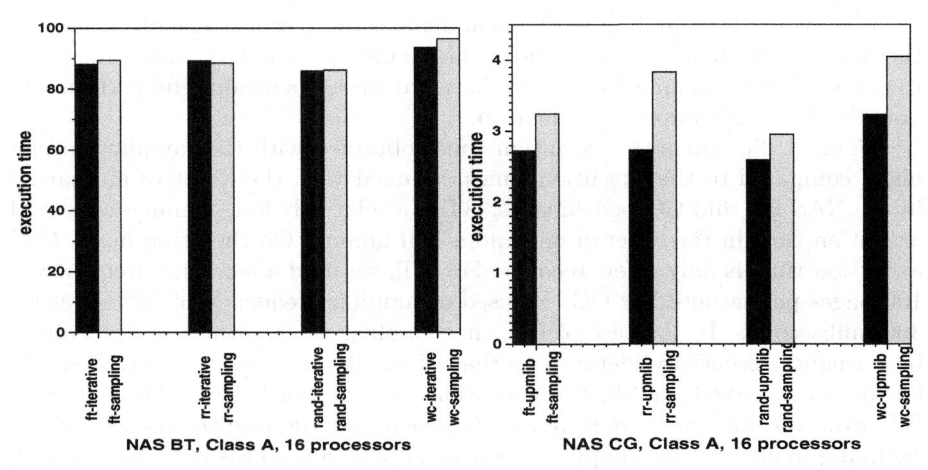

Fig. 4. Performance of the sampling page migration mechanism of UPMlib

ferent page placement schemes, labeled as `ft-IRIX`, `rr-IRIX`, `rand-IRIX` and `wc-IRIX`, for first-touch, round-robin, random, and worst-case page placement respectively. The light gray bars illustrate the execution time with the same page placement scheme and the IRIX page migration engine enabled during the execution of the benchmarks (same labels with suffix `-IRIXmig`). The dark gray bars illustrate the execution time with the *UPMlib* iterative page migration mechanism enabled in the benchmarks (same labels with suffix `-upmlib`). The horizontal lines show the baseline performance with the native first-touch page placement scheme of IRIX.

The results show that page placement schemes other than first-touch incur significant slowdowns compared to first-touch, ranging from 24% to 210%. The same phenomenon is observed even when page migration is enabled in the IRIX kernel, although page migration generally improves performance. On the other hand, when the suboptimal page placement schemes are combined with the iterative page migration mechanism of *UPMlib*, they approximate closely the performance of first-touch. When the page migration engine of *UPMlib* is injected in the benchmarks, the average performance difference between first-touch and the other page placement schemes is as low as 5%. In the last half iterations of the programs the performance difference was measured less than 1%. This practically means that the iterative page migration mechanism approaches rapidly the best initial page placement in each program. It also means that the performance of OpenMP programs can be immune to the page placement strategy of the operating system as soon as a page migration engine can relocate early poorly placed pages. No programmer intervention is required to achieve this level of optimization.

In order to assess the effectiveness of the sampling-based page migration engine of *UPMlib*, we conducted the following experiment. We activated the sampling mechanism in the NAS benchmarks and compared the performance obtained with the sampling mechanism against the performance obtained with the iterative mechanism. The iterative mechanism is tuned to exploit the structure of the NAS benchmarks and can therefore serve as a meaningful performance boundary for the sampling mechanism.

Figure 4 illustrates the execution times obtained with the sampling mechanism, compared to the execution times obtained with the iterative mechanism in the NAS BT and CG benchmarks. BT is a relatively long running code with execution time in the order of one and a half minute. On the other hand, CG's execution time is only a few seconds. For BT, we used a sampling frequency of 100 pages per second. For CG, we used a sampling frequency of 100 pages per 300 milliseconds. In the case of BT, the sampling mechanism is able to obtain performance essentially identical to that of the iterative mechanism. A similar trend was observed for SP, the execution time of which is similar to that of BT. However, for the short-running CG benchmark, despite the use of a higher sampling frequency the sampling mechanism performs consistently significantly worse than the iterative mechanism. The same happens with MG and FT. The results mainly demonstrate the sensitivity of the sampling mechanism to the execution characteristics of the applications. The sampling mechanism is unlikely to benefit short running codes.

3.2 Data Redistribution

We evaluated the ability of our user-level page migration engine to emulate data redistribution, by activating the record/replay mechanism of *UMPlib* in the NAS BT and SP benchmarks. Both benchmarks have a phase change in the execution of the `z_solve` function, as shown in Fig. 2. In these experiments, we restrict the page migration engine to move the n most critical pages across phases. The parameter n was set equal to 20.

Figure 5 illustrates the performance of the record/replay mechanism with first-touch page placement (labeled `ft-recrep` in the charts), as well as the performance of a hybrid scheme that uses the iterative page migration mechanism in the first few iterations of the programs and the record/replay mechanism in the rest of the iterations, as described in Section 2.2 (labeled `ft-hybrid` in the charts). The striped part of the bars labeled `ft-recrep` and `ft-hybrid` shows the non-overlapped overhead of the record/replay mechanism. For illustrative purposes the figure shows also the execution time of BT and SP with first-touch and the IRIX page migration engine, as well as the execution time with the iterative page migration mechanism of *UPMlib*.

The results indicate that applying page migration for fine-grain tuning across phase changes may be non-profitable due to the excessive overhead of page movements in the operating system. In the cases of BT and SP the overhead of the record/replay mechanism appears to outweigh the gains from reducing the rate of remote memory accesses. A more detailed analysis of the codes reveals that the parallel execution of `z_solve` in BT and SP on 16 processors takes approximately 130 to 180 ms. The recording mechanism of *UPMlib* identifies between 160 and 250 pages to migrate before the execution of `z_solve`. The cost of a page

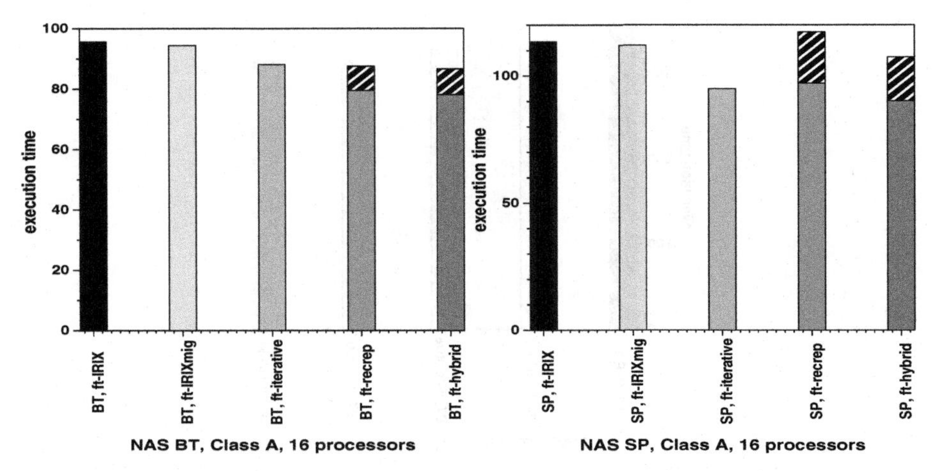

Fig. 5. Performance of the record/replay mechanism for NAS BT and SP

migration in the system we were experimenting with was measured to range between 1 and 1.3 ms, depending on the distance between the nodes that competed for the page. The total cost for moving all the pages identified by the recording mechanism as candidates for migration would exceed significantly the duration of the phase, making the record/replay mechanism useless.

Migrating the 20 most critical pages was able to reduce the execution time of useful computation by about 10% in the case of BT. However, the overhead of page migrations outweighed the earnings. The performance of the record/replay mechanism in the SP benchmark was disappointing. The limited improvements are partially attributed to the architectural characteristics of the Origin2000 and most notably the very low remote to local memory access latency ratio of the system, which is about 2:1 on the scale on which we experimented. The reduction of remote memory accesses would have a more significant performance impact on systems with higher remote to local memory access latency ratios. We have experimented with larger values for n and observed significant performance degradation, attributed again to the page migration overhead. The hybrid scheme appears to outperform the record/replay scheme (marginally for BT and significantly for SP), but it is still biased by the overhead of page migrations.

In order to quantify the extent to which the record/replay mechanism is applicable we executed the following synthetic experiment. We modified the code in NAS BT to quadruple the amount of work performed in each iteration of the parallel computation. We did not change the problem size of the program to preserve its locality characteristics. We rather enclosed each of the functions that comprise the main body of the computation (`x_solve,y_solve,z_solve,add`) in a loop. With this modification, we lengthened the duration of the parallel execution of `z_solve` to approximately 500 ms. Figure 6 shows the results from these experiments. It is evident that a better amortization of the overhead of page migrations helps the record/replay mechanism. In this experiment, the cost

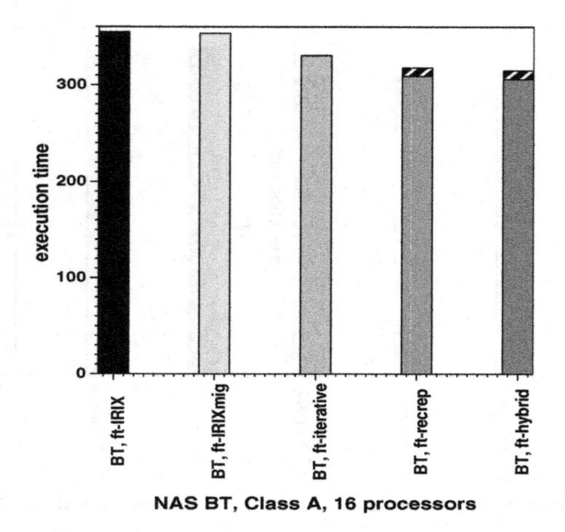

Fig. 6. Performance of the record/replay mechanism in the synthetic experiment with NAS BT

of the record/replay mechanism remains the same as in the previous experiments, however, the reduction of remote memory accesses achieved by the mechanism is exploited over a longer period of time. This yields a performance improvement of 5% over the iterative page migration mechanism.

4 Conclusion

This paper presented and evaluated mechanisms for transparent data distribution in OpenMP programs. The mechanisms leverage dynamic page migration as an oblivious to the programmer data distribution technique. We have shown that effective initial page placement can be established with a smart user-level page migration engine that exploits the iterative structure of parallel codes. Our results demonstrate clearly that the need for introducing data distribution directives in OpenMP is obscure and may not warrant the implementation and standardization costs. On the other hand, we have shown that although page migration may be effective for coarse-grain optimization of data locality, it suffers from excessive overhead when applied for tuning page placement at fine-grain time scales. It is therefore critical to estimate the cost/performance tradeoffs of page migration in order to investigate up to which extent can aggressive page migration strategies work effectively in place of data distribution and redistribution on distributed shared-memory multiprocessors. Since the same investigation would be necessary in data-parallel environments also, we do not consider it as a major restriction of our environment.

References

1. L. Brieger. *HPF to OpenMP on the Origin2000. A Case Study.* Proc. of the First European Workshop on OpenMP, pp. 19–20. Lund, Sweden, October 1999. 416
2. R. Chandra et.al. *Data Distribution Support for Distributed Shared Memory Multiprocessors.* Proc. of the 1997 ACM Conference on Programming Languages Design and Implementation, pp. 334–345. Las Vegas, NV, June 1997. 416
3. C. Holt, J. Pal Singh and J. Henessy. *Application and Architectural Bottlenecks in Large Scale Shared Memory Machines.* Proc. of the 23rd Int. Symposium on Computer Architecture, pp. 134–145. Philadelphia, PA, June 1996. 416
4. D. Jiang and J. Pal Singh. *Scaling Application Performance on a Cache-Coherent Multiprocessor.* Proc. of the 26th Int. Symposium on Computer Architecture, pp. 305–316. Atlanta, GA, May 1999. 416
5. H. Jin, M. Frumkin and J. Yan. *The OpenMP Implementation of NAS Parallel Benchmarks and its Performance.* Tech. Rep. NAS-99-011. NASA Ames Research Center, October 1999. 421
6. P. Keleher. *A High Level Abstraction of Shared Accesses.* ACM Transactions on Computer Systems, Vol. 18, No. 1, pp. 1–36. Feburary 2000. 420
7. C. Koelbel et.al. *The High Performance Fortran Handbook.* The MIT Press, Cambridge, MA, 1994. 416
8. J. Laudon and D. Lenoski. *The SGI Origin: A ccNUMA Highly Scalable Server.* Proc. of the 24th Int. Symposium on Computer Architecture, pp. 241–251. Denver, CO, June 1997. 422
9. D. Nikolopoulos et. al. *A Case for User-Level Dynamic Page Migration.* Proc. of the 14th ACM Int. Conference on Supercomputing, pp. 119–130. Santa Fe, NM, May 2000. 415, 417, 418
10. D. Nikolopoulos et. al. *UPMlib: A Runtime System for Tuning the Memory Performance of OpenMP Programs on Scalable Shared-Memory Multiprocessors.* Proc. of the 5th ACM Workshop on Languages, Compilers and Runtime Systems for Scalable Computers. Rochester, NY, May 2000. 415, 417
11. OpenMP Architecture Review Board. *OpenMP Specifications.* http://www.openmp.org, accessed April 2000. 416
12. M. Resch and B. Sander. *A Comparison of OpenMP and MPI for the Parallel CFD Case.* Proc. of the First European Workshop on OpenMP. Lund, Sweden, October 1999. 416
13. Silicon Graphics Inc. Technical Publications. *IRIX 6.5 man pages.* `proc(4)`, `mmci(5)`, `schedctl(2)`. http://techpubs.sgi.com, accessed January 2000.
14. B. Verghese, S. Devine, A. Gupta and M. Rosenblum. *Operating System Support for Improving Data Locality on CC-NUMA Compute Servers.* Proc. of the 7th Int. Conference on Architectural Support for Programming Languages and Operating Systems, pp. 279–289. Cambridge, MA, October 1996. 416

Formalizing OpenMP Performance Properties with ASL *

Thomas Fahringer[1], Michael Gerndt[2], Graham Riley[3], Jesper Larsson Träff[4]

[1] Institute for Software Technology and Parallel Systems
University of Vienna, tf@par.univie.ac.at
[2] Institute for Computer Science, LRR, Technical University of Munich
m.gerndt@in.tum.de
[3] Department of Computer Science, University of Manchester
griley@cs.man.ac.uk
[4] C&C Research Laboratories, NEC Europe Ltd.
traff@ccrl-nece.technopark.gmd.de

Abstract. Performance analysis is an important step in tuning performance critical applications. It is a cyclic process of measuring and analyzing performance data which is driven by the programmer's hypotheses on potential performance problems. Currently this process is controlled manually by the programmer. We believe that the implicit knowledge applied in this cyclic process should be formalized in order to provide automatic performance analysis for a wider class of programming paradigms and target architectures. This article describes the performance property specification language (ASL) developed in the APART Esprit IV working group which allows specifying performance-related data by an object-oriented model and performance properties by functions and constraints defined over performance-related data. Performance problems and bottlenecks can then be identified based on user- or tool-defined thresholds. In order to demonstrate the usefulness of ASL we apply it to OpenMP by successfully formalizing several OpenMP performance properties.

Keywords: performance analysis, knowledge representation, OpenMP, performance problems, language design

1 Introduction

Performance-oriented program development can be a daunting task. In order to achieve high or at least respectable performance on today's multiprocessor systems, careful attention to a plethora of system and programming paradigm details is required. Commonly programmers go through many cycles of experimentation involving gathering performance data, performance data analysis (a-priori and postmortem), detection of performance problems, and code refinements in slow progression. Clearly, the programmer must be intimately familiar

* The ESPRIT IV *Working Group on Automatic Performance Analysis: Resources and Tools* is funded under Contract No. 29488

M. Valero et al. (Eds.): ISHPC 2000, LNCS 1940, pp. 428-439, 2000.

with many aspects related to this experimentation process. Although there exists a large number of tools assisting the programmer in performance experimentation, it is still the programmer's responsibility to take most strategic decisions.

In this article we describe a novel approach to formalize performance bottlenecks and the data required in detecting those bottlenecks with the aim to support automatic performance analysis for a wider class of programming paradigms and architectures. This research is done as part of APART Esprit IV *Working Group on Automatic Performance Analysis: Resources and Tools* [Apart 99]. In the remainder of this article we use the following terminology:

Performance-related Data: Performance-related data defines information that can be used to describe performance properties of a program. There are two classes of performance related data. First, static data specifies information that can be determined without executing a program on a target machine. Second, dynamic performance-related data describes the dynamic behavior of a program during execution on a target machine.

Performance Property: A performance property (e.g. load imbalance, communication, cache misses, redundant computations, etc.) characterizes a specific performance behavior of a program and can be checked by a set of *conditions*. Conditions are associated with a *confidence value* (between 0 and 1) indicating the degree of confidence about the existence of a performance property. In addition, for every performance property a *severity figure* is provided that specifies the importance of the property.

Performance Problem: A performance property is a performance problem, iff its severity is greater than a user- or tool-defined threshold.

Performance Bottleneck: A program can have one or several performance bottlenecks which are characterized by having the highest severity figure. If these bottlenecks are not a performance problem, then the program's performance is acceptable and does not need any further tuning.

This paper introduces the APART Specification Language (ASL) which allows the description of performance-related data through the provision of an object-oriented specification model and which supports the definition of performance properties in a novel formal notation. Our object-oriented specification model is used to declare – without the need to compute – performance information. It is similar to Java, uses only single inheritance and does not require methods. A novel syntax has been introduced to specify performance properties.

The organization of this article is as follows. Section 2 presents in related work. Section 3 presents ASL constructs for specifying performance-related data and, as examples, classes specifying performance-related data for OpenMP programs. The syntax for the specification of performance properties is described in Section 4. Examples for OpenMP property specifications are presented in Section 5. Conclusions and Future work are discussed in Section 6.

2 Related work

The use of specification languages in the context of automatic performance analysis tools is a new approach. Paradyn [MCCHI 95] performs an automatic online analysis and is based on dynamic monitoring. While the underlying metrics can be defined via the *Metric Description Language* (MDL), the set of searched bottlenecks is fixed. It includes *CPUbound*, *ExcessiveSyncWaitingTime*, *ExcessiveIOBlockingTime*, and *TooManySmallIOOps*.

A rule-based specification of performance bottlenecks and of the analysis process was developed for the performance analysis tool OPAL [GKO 95] in the SVM-Fortran project. The rule base consists of a set of parameterized hypothesis with proof rules and refinement rules. The proof rules determine whether a hypothesis is valid based on the measured performance data. The refinement rules specify which new hypotheses are generated from a proven hypothesis [GeKr 97].

Another approach is to define a performance bottleneck as an event pattern in program traces. EDL [Bates 83] allows the definition of compound events based on extended regular expressions. EARL [WoMo 99] describes event patterns in a more procedural fashion as scripts in a high-level event trace analysis language which is implemented as an extension of common scripting languages like Tcl, Perl or Python.

The language presented here served as a starting point for the definition of the ASL but is too limited. MDL does not allow to access static program information and to integrate information for multiple performance tools. It is specially designed for Paradyn. The design of OPAL focuses more on the formalization of the analysis process and EDL and EARL are limited to pattern matching in performance traces.

Some other performance analysis and optimization tools apply automatic techniques without being based on a special bottleneck specification, such as KAPPA-PI [EsMaLu 98], FINESSE [MuRiGu 00], and the online tuning system Autopilot [RVSR 98].

3 Performance-related Data Specification

This section presents the performance-related data specification for OpenMP [DaMe 99]. Performance-related data are specified in ASL by a set of of classes following an object-oriented style with single-inheritance. In this article we present the classes as UML diagramms. More details on the specification including a number of base classe for all models can be found in [ApartWP2 99]

Several classes were defined that model static information for OpenMP programs. Class *SmRegion* is a subclass of the standard class *Region* and contains an attribute with data dependence information about the modeled region. *SmRegion* is then further refined by two subclasses *ParallelRegion* and *SequentialRegion* which, respectively, describe parallel and sequential regions. Parallel regions include a boolean variable *no_wait_exit* which denotes whether or not the region is terminated by an implicit exit barrier operation. A specific execution of a region corresponds to a region instance.

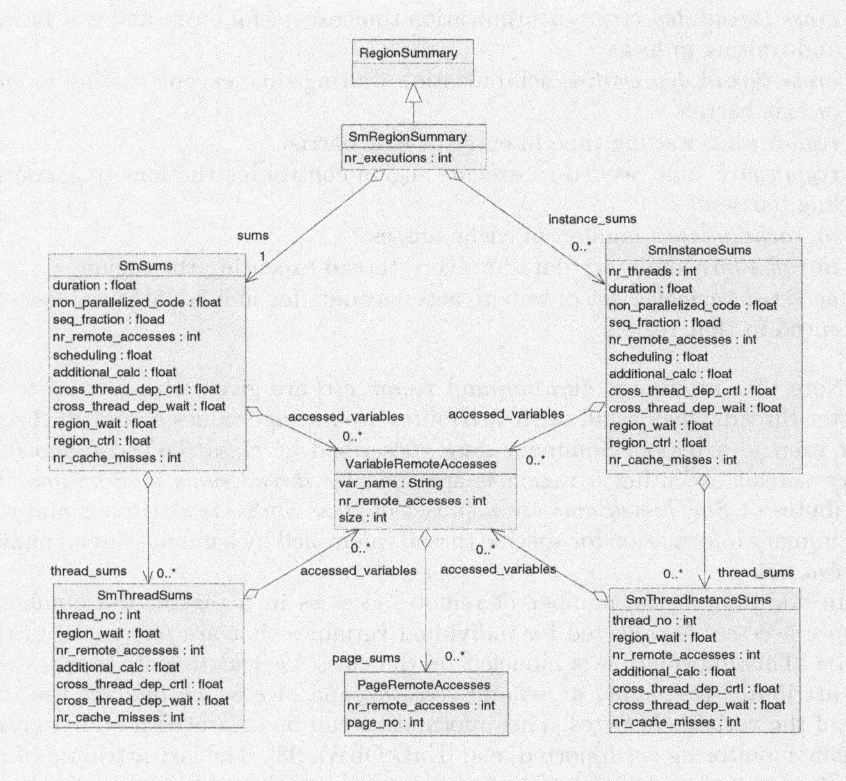

Fig. 1. OpenMP classes for dynamic information

Figure 1 shows the OpenMP class library for dynamic information. Class *SmRegionSummary* extends the standard class *RegionSummary* and comprises three attributes: *nr_executions* specifies the number of times a region has been executed by the master thread, *sums* describes summary information across all region instances, and *instance_sums* relates to summary information for a specific region instance. The attributes of class *SmSums* include:

- *duration*: time needed to execute region by master thread
- *non_parallelized_code*: time needed to execute non-parallelized code
- *seq_fraction*: $\frac{non\_parallelized\_code}{duration}$
- *nr_remote_accesses*: number of accesses to remote memory by load and store operations in ccNUMA machines
- *scheduling*: time needed for scheduling operations (e.g. scheduling of threads)
- *additional_calc*: time needed for additional computations in parallelized code (e.g. to enforce a specific distribution of loop iterations) or for additional computations (e.g. where it is cheaper for all threads to compute a value rather than communicate it, possibly with synchronization costs)

- *cross_thread_dep_ctrl*: synchronization time except for entry and exit barriers and waiting in locks
- *cross_thread_dep_wait*: synchronization waiting time except waiting in entry or exit barrier
- *region_wait*: waiting time in entry or exit barrier
- *region_ctrl*: time needed to execute region control instructions (e.g. controlling barriers)
- *nr_cache_misses*: number of cache misses
- *thread_sums*: summary data for every thread executing the region
- *accessed_variables*: set of remote access counts for individual variables referenced in that region

Note that attributes *duration* and *region_ctrl* are given with respect to the master thread, whereas all other attributes are average values across all threads that execute a region. Summary data (described by class *SmThreadSums*) for every thread executing a region is specified by *thread_sums* in *SmSums*. The attributes of *SmThreadSums* are a subset of class *SmSums* attributes and refer to summary information for specific threads identified by a unique thread number (*thread_no*).

In addition to the number of remote accesses in a region, the number of remote accesses is collected for individual variables that are referenced in that region. This information is modeled by the class *VariableRemoteAccesses* with the attributes *var_name*, *nr_remote_accesses*, and *size* which denotes the total size of the variable in bytes. This information can be measured if address range specific monitoring is supported, e.g. [KaLeObWa 98]. The last attribute of this class is *page_sums* which is a set of page-level remote access counters. For example, the remote access counters on SGI Origin 2000 provide such information. With the help of additional mapping information, i.e. mapping variables to addresses, this information can be related back to program variables. Each object of class *PageRemoteAccesses* determines the *page_no* and the number of remote accesses.

The second attribute of *SmRegionSummary* is given by *instance_sums* which is described by a class *SmInstanceSums*. This class specifies summary information for a specific region instance. *SmInstanceSums* contains all attributes of *SmSums* and the number of threads executing the region instance. Finally, class *SmThreadInstanceSums* describes summary information for a given region instance with respect to individual threads.

4 Performance Property Specification

A performance property (e.g. load imbalance, remote accesses, cache misses, redundant computations, etc.) characterizes a specific performance behavior of a program. The ASL property specification syntax defines the name of the property, its context via a list of parameters, and the condition, confidence, and severity expressions. The property specification is based on a set of parameters.

These parameters specify the property's context and parameterize the expressions. The context specifies the environment in which the property is evaluated, e.g. the program region and the test run. Details can be found in [ApartWP2 99].

The condition specification consists of a list of conditions. A condition is a predicate that can be prefixed by a condition identifier. The identifiers have to be unique with respect to the property since the confidence and severity specifications may refer to the conditions by using the condition identifiers.

The confidence specification is an expression that computes the maximum of a list of confidence values. Each confidence value is computed as an arithmetic expression in an interval between zero and one. The expression may be guarded by a condition identifier introduced in the condition specification. The condition identifier represents the value of the condition.

The severity specification has the same structure as the confidence specification. It computes the maximum of the individual severity expressions of the conditions. The severity specification will typically be based on a parameter specifying the ranking basis. If, for example, a representative test run of the application has been monitored, the time spent in remote accesses may be compared to the total execution time. If, instead, a short test run is the basis for performance evaluation since the application has a cyclic behavior, the remote access overhead may be compared to the execution time of the shortened loop.

5 OpenMP Performance Properties

This section demonstrates the ASL constructs for specifying performance properties in the context of the shared memory, OpenMP paradigm. Some global definitions are presented first. These are then used in the definitions of a number of OpenMP properties.

Global definitions

In most property specifications it is necessary to access the summary data of a given region for a given experiment. Therefore, we define the *summary* function that returns the appropriate *SmRegionSummary* object. It is based on the set operation *UNIQUE* that selects arbitrarily one element from the set argument which has cardinality one due to the design of the data model.

```
SmRegionSummary summary(Region r, Experiment e)=
                    UNIQUE({s IN e.profile WITH s.region==r});
```

The *sync* function determines the overhead for synchronization in a given region. It computes the sum of the relevant attributes (which are deemed by the property specifier to be components of synchronisation cost) in the summary class.

```
float sync(Region r, Experiment e)=summary(r,e).sums.region_wait +
                        summary(r,e).sums.region_ctrl +
                        summary(r,e).sums.cross_thread_dep_wait +
                        summary(r,e).sums.cross_thread_dep_ctrl ;
```

The *duration* function returns the execution time of a region. The execution time is determined by the execution time of the master thread in the OpenMP model.

```
float duration(Region r, Experiment e)=summary(r,e).sums.duration;
```

The *remote_access_time* function estimates the overhead for accessing remote memory based on the measured number of accesses and the mean access time of the parallel machine.

```
float remote_access_time(Region r, Experiment e)=
                            summary(r,e).sums.nr_remote_accesses
                            * e.system.remote_access_time);
```

Property specifications

The *costs* property determines whether the total parallel overhead in the execution is non-zero.

```
Property costs(Region r, Experiment seq,
               Experiment par, Region rank_basis){

LET
   float total_costs = duration(r,par) - (duration(r,seq)/
                          par.nr_processors);
IN
   CONDITION: total_costs>0;
   CONFIDENCE: 1;
   SEVERITY: total_costs / duration(rank_basis,par);
}
```

This property specifies that the speedup of the application is not simply the serial execution time divided by the number of processors, i.e. that the naive ideal linear speedup is not being achieved. It uses information from two experiments, a sequential run and a parallel run, to compute the costs of parallel execution. Those costs determine the severity of the property.

Performance analysis tools initially only help in analyzing costs for properties which are known to affect performance. After accounting for these properties, any remaining costs must be due new, as yet unencountered, effects. A region has the *identified_costs* property if the sum of the expected potential costs is greater than zero. The severity of this property is the fraction of those costs relative to the execution time of *rank_basis*.

```
Property identified_costs(Region r, Experiment e, Region rank_basis){
LET
   float costs = summary(r,e).sums.non_parallelized_code +
                 sync(r,e) +
                 remote_access_time(r,e) +
                 summary(r,e).sums.scheduling +
                 summary(r,e).sums.additional_calc;
IN
   CONDITION: costs>0;
   CONFIDENCE: 1;
   SEVERITY: costs(r,e)/duration(rank_basis,e);
}
```

The total cost of the parallel program is the sum of the identified and the unidentified overhead. The *unidentified_costs* property determines whether an unidentified overhead exists. Its severity is the fraction of this overhead in relation to the execution time of *rank_basis*. If this fraction is high, further tool-supported performance analysis might be required.

```
Property unidentified_costs(Region r, Experiment seq, Experiment e,
                            Region rank_basis){
LET
   float totcosts = duration(r,e) - (duration(r,seq)/e.nr_processors);
   float costs = summary(r,e).sums.non_parallelized_code+sync(r,e)+
                 remote_access_time(r,e)+summary(r,e).sums.scheduling
                 +summary(r,e).sums.additional_calc;
IN
   CONDITION: totcosts-costs>0;
   CONFIDENCE: 1;
   SEVERITY: totcosts(r,e)/duration(rank_basis,e);
}
```

Non-parallelized code is a very severe problem for application scaling. In the context of analyzing a given program run, its severity is determined in the usual way, relative to the duration of the rank basis. If the focus of the analysis is more on application scaling the severity could be redefined to stress the importance of this property.

```
Property non_parallelized_code(Region r, Experiment e,
                               Region rank_basis){
LET
   float non_par_code = summary(r,e).sums.non_parallelized_code>0;
IN
   CONDITION: non_par_code>0;
   CONFIDENCE: 1;
   SEVERITY: non_par_code/duration(rank_basis,e);
}
```

A region has the following *synchronization* property if any synchronization overhead occurs during its execution. One of the obvious reasons for high synchronization cost is load imbalance, which is an example of a more specific property: an application suffering the *load_imbalance* property is, by implication, suffering *synchronisation*.

```
Property synchronization(Region r, Experiment e, Region rank_basis){
   CONDITION: sync(r,e)>0;
   CONFIDENCE: 1;
   SEVERITY: sync(r,e)/duration(rank_basis,e);
}

Property load_imbalance( Region r, Experiment e, Region rank_basis) {
   CONDITION: summary( r, e ).sums.region_wait >0;
   CONFIDENCE: 1;
   SEVERITY: summary( r, e ).sums.region_wait/duration(r,e);
}
```

When work is unevenly distributed to threads in a region, this manifests itself in *region_wait* time (time spent by threads waiting on the region exit barrier). If the *region_wait* time cannot be measured, the property can also be derived as the execution time of the thread with the longest duration minus the average thread duration.

The *synchronization* property defined above is assigned to regions with an aggregate non-zero synchronization cost during the entire execution of the program. If the dynamic behaviour of an application changes over the execution time — load imbalance, for example, might occur only in specific phases of the simulation — the whole synchronization overhead might result from specific instances of the region. The *irregular_sync_across_instances* property identifies this case. The severity is equal to the severity of the synchronization property since the *irregular_sync_across_instances* property is only a more detailed explanation.

```
Property irregular_sync_across_instances
         (Region r, Experiment e, Region rank_basis){
LET
   float inst_sync(SmInstanceSums sum)=sum.region_wait +
         sum.region_ctrl + sum.cross_thread_dep_wait +
         sum.cross_thread_dep_ctrl ;

IN
   CONDITION: stdev(inst_sync(inst_sum)
                    WHERE inst_sum IN summary(r,e).instance_sums)
           > irreg_behaviour_threshold * sync(r,e)/r.nr_executions;
   CONFIDENCE: 1;
   SEVERITY: sync(r,e)/duration(rank_basis,e);
}
```

An important property for code executing on ccNUMA machines, *remote accesses*, arises from access to data in memory located on nodes other than that of the requesting processor. Remote memory access involves communication among parallel threads. Since, usually, only the number of accesses can be measured, the severity is estimated based on the mean access time.

```
Property remote_accesses(Region r, Experiment e, Region rank_basis){
   CONDITION: summary(r,e).nr_remote_accesses>0;
   CONFIDENCE: 1;
   SEVERITY: remote_access_time(r,e) / duration(rank_basis,e);
}
```

A previous property (*remote_accesses*) identifies regions with remote accesses. The next property, *remote_access_to_variable*, is more specific than this since its context also includes a specific variable. The property indicates whether accesses to a variable in this region result in remote accesses. It is based on address-range-specific remote access counters, such as those provided by the SGI Origin 2000 on a page basis. The severity of this property is based on the time spent in remote accesses to this variable. Since this property is very useful in explaining a severe remote access overhead for the region, it might be ranked with respect to this region, rather than with respect to the whole program, during a more detailed analysis.

```
Property remote_access_to_variable
   (Region r, Experiment e, String var, Region rank_basis)
{
LET
   VariableRemoteAccesses var_sum =
        UNIQUE({info IN summary(r,e).sums.accessed_variables
               WITH info.var_name==var});
IN
   CONDITION: var_sum.nr_remote_accesses > 0;
   CONFIDENCE: 1;
   SEVERITY:  var_sum.nr_remote_accesses * e.system.remote_access_time
              /duration(rank_basis,e);
}
```

6 Conclusions and Future Work

In this article we describe a novel approach to the formalization of performance problems and the data required to detect them with the future aim of supporting automatic performance analysis for a large variety of programming paradigms and architectures. We present the APART Specification Language (ASL) developed as part of the APART Esprit IV Working Group on *Automatic Performance Analysis: Resources and Tools.* This language allows the description of performance-related data through the provision of an object-oriented specification model and supports definition of performance properties in a novel formal notation.

We applied the ASL to OpenMP by successfully formalizing several OpenMP performance properties. ASL has also been used to formalize a large variety of MPI and HPF performance properties which is described in [ApartWP2 99].

Two extensions to the current language design will be investigated in the future:

1. The language will be extended by templates which facilitate specification of similar performance properties. In the example specification in this paper some of the properties result directly from the summary information, e.g. *synchronization* is directly related to the measured time spent in synchronization. The specifications of these properties are indeed very similar and need not be described individually.
2. Meta-properties may be useful as well. For example, synchronization can be proven based on summary information, i.e. synchronization exists if the sum of the synchronization time in a region over all processes is greater than zero. A more specific property is to check, whether individual instances of the region or classes of instances are responsible for the synchronization due to some dynamic changes in the load distribution. Similar, more specific properties can be deduced for other properties as well. As a consequence, meta-properties can be useful to evaluate other properties based on region instances instead of region summaries.

ASL should be the basis for a common interface for a variety of performance tools that provide performance-related data. Based on this interface we plan to develop a system that provides automatic performance analysis for a variety of programming paradigms and target architectures.

References

[Apart 99] *APART – Esprit Working Group on Automatic Performance Analysis: Resources and Tools*, Dr. Hans Michael Gerndt, Forschungszentrum Jülich, Zentralinstitut für Angewandte Mathematik (ZMG), D-52425 Jülich, http://www.kfa-juelich.de/apart, Feb. 1999.

[ApartWP2 99] T. Fahringer, M. Gerndt, G. Riley, J. Träff: *Knowledge Specification for Automatic Performance Analysis*, APART Technical Report, Workpackage 2, Identification and Formalization of Knowledge, Technical Report FZJ-ZAM-IB-9918, D-52425 Jülich, http://www.kfa-juelich.de/apart, Nov. 1999.

[Bates 83] P. Bates, J.C. Wileden: *High-Level Debugging of Distributed Systems: The Behavioral Abstraction Approach*, The Journal of Systems and Software, Vol. 3, pp. 255-264, 1983

[DaMe 99] L. Dagum, R.Menon: *OpenMP: An Industry-Standard API for Shared-Memory Programming*, IEEE Computational Science & Engineering, Vol.5, No.1, pp 46-55, 1998.

[EsMaLu 98] A. Espinosa, T. Margalef, E. Luque: *Automatic Performance Evaluation of Parallel Programs*, Sixth Euromicro Workshop on Parallel and Distribued Processing, 1998

[GeKr 97] M. Gerndt, A. Krumme: *A Rule-based Approach for Automatic Bottleneck Detection in Programs on Shared Virtual Memory Systems*, Second Workshop on High-Level Programming Models and Supportive Environments (HIPS '97), in combination with IPPS '97, IEEE, 1997

[GKO 95] M. Gerndt, A. Krumme, S. Özmen: *Performance Analysis for SVM-Fortran with OPAL*, Proceedings Int. Conf. on Parallel and Distributed Processing Techniques and Applications (PDPTA'95), Athens, Georgia, pp. 561-570, 1995

[KaLeObWa 98] R. Hockauf, W. Karl, M. Leberecht, M. Oberhuber, M. Wagner: *Exploiting Spatial and Temporal Locality of Accesses: A New Hardware-Based Monitoring Approach for DSM Systems*, In: D. Pritchard, Jeff Reeve (Eds.): Euro-Par'98 Parallel Processing / 4th International Euro-Par Conference Southampton, UK, September 1-4, 1998 Proceedings. Springer-Verlag, Heidelberg, Lecture Notes in Computer Science Vol.1470, 1998, pp. 206-215

[MCCHI 95] B.P. Miller, M.D. Callaghan, J.M. Cargille, J.K. Hollingsworth, R.B. Irvin, K.L. Karavanic, K. Kunchithapadam, T. Newhall: *The Paradyn Parallel Performance Measurement Tool*, IEEE Computer, Vol. 28, No. 11, pp. 37-46, 1995

[MuRiGu 00] N. Mukherjee, G.D. Riley, J.R. Gurd: *FINESSE: A Prototype Feedback-guided Performance Enhancement System*, Accepted for PDP2000, to be held in Rhodes in January 2000.

[RVSR 98] R. Ribler, J. Vetter, H. Simitci, D. Reed: *Autopilot: Adaptive Control of Distributed Applications*, Proceedings of the 7th IEEE Symposium on High-Performance Distributed Computing, 1998

[WoMo 99] F. Wolf, B. Mohr: *EARL - A Programmable and Extensible Toolkit for Analyzing Event Traces of Message Passing Programs*, 7th International Conference on High-Performance Computing and Networking (HPCN'99), A. Hoekstra, B. Hertzberger (Eds.), Lecture Notes in Computer Science, Vol. 1593, pp. 503-512, 1999

Automatic Generation of OpenMP Directives and Its Application to Computational Fluid Dynamics Codes

Haoqiang Jin [1], Michael Frumkin [1], and Jerry Yan [2]

[1] Computer Sciences Corporation, [2] NAS Systems Division
NASA Ames Research Center, Moffett Field, CA 94035-1000 USA
{hjin,frumkin,yan}@nas.nasa.gov

Abstract. The shared-memory programming model is a very effective way to achieve parallelism on shared memory parallel computers. As great progress was made in hardware and software technologies, performance of parallel programs with compiler directives has demonstrated large improvement. The introduction of OpenMP directives, the industrial standard for shared-memory programming, has minimized the issue of portability. In this study, we have extended CAPTools, a computer-aided parallelization toolkit, to automatically generate OpenMP-based parallel programs with nominal user assistance. We outline techniques used in the implementation of the tool and discuss the application of this tool on the NAS Parallel Benchmarks and several computational fluid dynamics codes. This work demonstrates the great potential of using the tool to quickly port parallel programs and also achieve good performance that exceeds some of the commercial tools.

1 Introduction

Porting applications to high performance parallel computers is always a challenging task. It is time consuming and costly. With rapid progressing in hardware architectures and increasing complexity of real applications in recent years, the problem becomes even more sever. Today, scalability and high performance are mostly involving handwritten parallel programs using message-passing libraries (e.g. MPI). However, this process is very difficult and often error-prone. The recent reemergence of shared-memory parallel (SMP) architectures, such as the cache coherent Non-Uniform Memory Access (ccNUMA) architecture used in the SGI Origin2000, show good prospects for scaling beyond hundreds of processors. Programming on an SMP is simplified by working in a globally accessible address space. The user can supply compiler directives to parallelize the code without explicit data partitioning. Computation is distributed inside a loop based on the index range regardless of data location and the scalability is achieved by taking advantage of hardware cache coherence. The recent emergence of OpenMP [13] as an industry standard offers a portable solution for implementing directive-based parallel programs for SMPs. OpenMP overcomes the portability issues encountered by machine-specific directives without sacrificing much of the performance and has gained popularity quickly.

M. Valero et al. (Eds.): ISHPC 2000, LNCS 1940, pp. 440-456, 2000.

Although programming with directives is relatively easy (when comparing to writing message passing codes), inserted directives may not necessarily enhance performance. In the worst cases, they can create erroneous results when used incorrectly. While vendors have provided tools to perform error-checking [10], automation in directive insertion is very limited and often failed on large programs, primarily due to the lack of a thorough enough data dependence analysis. To overcome the deficiency, we have developed a toolkit, CAPO, to automatically insert OpenMP directives in Fortran programs. CAPO is aimed at taking advantage of detailed interprocedural data dependence analysis provided by Computer-Aided Parallelization Tools (CAPTools) [4], developed at the University of Greenwich, to reduce potential errors made by users and, with nominal help from user, achieve performance close to that obtained when directives are inserted by hand. Our approach is differed from other tools and compilers in two respects: 1) emphasizing the quality of dependence analysis and relaxing much of the time constraint on the analysis; 2) performing directive insertion and preserving the original code structure for maintainability. Translation of OpenMP codes to executables is left to proper OpenMP compilers.

In the following we first outline the OpenMP programming model and give an overview of CAPTools and CAPO for generating OpenMP programs. Then, in Sect. 3 we discuss the implementation of CAPO. Case studies of using CAPO to parallelize the NAS Parallel Benchmarks and two computational fluid dynamics (CFD) applications are presented in Sect. 4 and conclusions are given in the last section.

2 Automatic Generation of OpenMP Directives

2.1 The OpenMP Programming Model

OpenMP [13] was designed to facilitate portable implementation of shared memory parallel programs. It includes a set of compiler directives and callable runtime library routines that extend Fortran, C and C++ to support shared memory parallelism. It promises an incremental path for parallelizing sequential software, as well as targeting at scalability and performance for any complete rewrites or new construction of applications.

OpenMP follows the *fork-and-join* execution model. A fork-and-join program initializes as a single lightweight process, called the *master thread*. The master thread executes sequentially until the first parallel construct (`OMP PARALLEL`) is encountered. At that point, the master thread creates a team of threads, including itself as a member of the team, to concurrently execute the statements in the parallel construct. When a work-sharing construct such as a parallel do (`OMP DO`) is encountered, the workload is distributed among the members of the team. An implied synchronization occurs at the end of the `DO` loop unless a „`NOWAIT`" is specified. Data sharing of variables is specified at the start of parallel or work-sharing constructs using the `SHARED` and `PRIVATE` clauses. In addition, reduction operations (such as summation) can be specified by the `REDUCTION` clause. Upon completion of the parallel construct, the threads in the team synchronize and only the master thread continues execution. The fork-and-join process can be repeated many times in the course of program execution.

Beyond the inclusion of parallel constructs to distribute work to multiple threads, OpenMP introduces a powerful concept of *orphan directives* that greatly simplifies the task of implementing coarse grain parallel algorithms. Orphan directives are directives outside the lexical extent of a parallel region. This allows the user to specify control or synchronization from anywhere inside the parallel region, not just from the lexically contained region.

2.2 CAPTools

The Computer-Aided Parallelization Tools (CAPTools) [4] is a software toolkit that was designed to automate the generation of message-passing parallel code. CAPTools accepts FORTRAN-77 serial code as input, performs extensive dependence analysis, and uses domain decomposition to exploit parallelism. The tool employs sophisticated algorithms to calculate execution control masks and minimize communication. The generated parallel codes contain portable interface to message passing standards, such as MPI and PVM, through a low-overhead library.

There are two important strengths that make CAPTools stands out. Firstly, an extensive set of extensions [5] to the conventional dependence analysis techniques has allowed CAPTools to obtain much more accurate dependence information and, thus, produce more efficient parallel code. Secondly, the tool contains a set of browsers that allow user to inspect and assist parallelization at different stages.

2.3 Generating OpenMP Directives

The goal of developing computer-aided tools to help parallelize applications is to let the tools do as much as possible and minimize the amount of tedious and error-prone work performed by the user. The key to automatic detection of parallelism in a program and, thus parallelization is to obtain accurate data dependences in the program. Generating OpenMP directives is simplified somehow because we are now working in a globally addressed space without explicitly concerning data distribution. However, we still have to realize that there are always cases in which certain conditions could prevent tools from detecting possible parallelization, thus, an interactive user environment is also important.

The design of the CAPTools-based automatic parallelizer with OpenMP, CAPO, had kept the above tactics in mind. The schematic structure of CAPO is illustrated in Fig. 1. The detailed implementation of the tool is given in Sect. 3. CAPO takes a serial code as input and uses the data dependence analysis engine in CAPTools. User knowledge on certain input parameters in the source code may be entered to assist this analysis for more accurate results. The process of exploiting loop level parallelism in a program and generating OpenMP directives automatically is summarized in the following three stages.

1) *Identify parallel loops and parallel regions.* The loop-level analysis classifies loops as parallel (including reduction), serial or potential pipeline based on the data dependence information. Parallel loops to be distributed with work-sharing directives for parallel execution are identified by traversing the call graph of the program from top to down. Only outer-most parallel loops are considered, partly due to the very limited support of multi-level parallelization in available OpenMP compilers. Parallel regions are then formed around the distributed parallel loops. Attempt is also made to identify and create parallel pipelines. Details are given in Sects. 3.1-3.3.

2) *Optimize loops and regions.* This stage is mainly for reducing overhead caused by fork-and-join and synchronization. A parallel region is first expanded as far as possible and may include calls to subroutines that contain additional (*orphaned*) parallel loops. Regions are then merged together if there is no violation of data usage in doing so. Region expansion is currently limited to within a subroutine. Synchronization optimization between loops in a parallel region is performed by checking if the loops can be executed asynchronously. Details are given in Sects. 3.2 and 3.4.

3) *Transform codes and insert directives.* Variables in common blocks are analyzed for their usage in all parallel regions in order to identify threadprivate common blocks. If a private variable is used in a non-threadprivate common block, the variable is treated specially. A routine needs to be duplicated if its usage conflicts at different calling points. Details are given in Sects. 3.5-3.7.

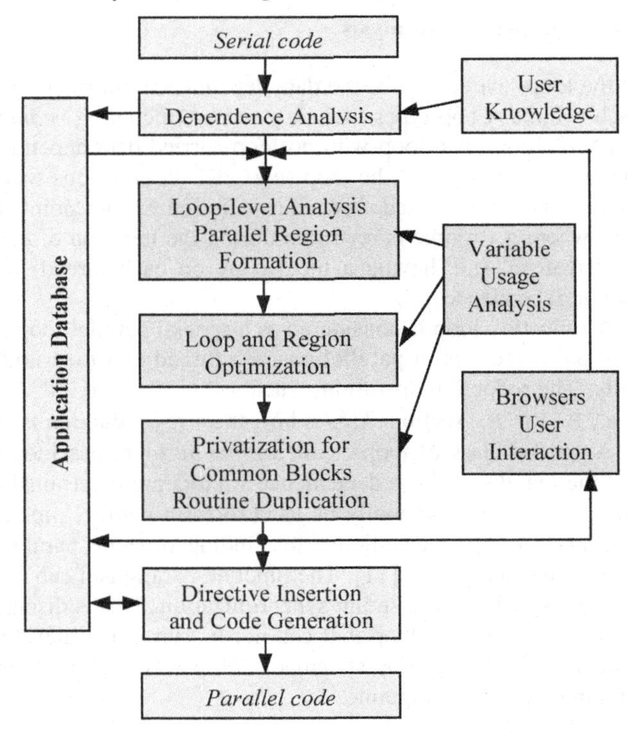

Fig. 1. Schematic flow chart of the CAPO architecture

By traversing the call graph one more time OpenMP directives are lastly inserted for parallel regions and parallel loops with variables properly listed. The variable usage analysis is performed at several points to identify how variables are used (e.g. private, shared, reduction, etc.) in a loop or region. Such analysis is required for the identification of loop types, the construction of parallel regions, the treatment of private variables in common blocks, and the insertion of directives.

Intermediate results can be stored into or retrieved from a database. User assistance to the parallelization process is possible through browsers implemented in CAPO (Directives Browser) and in CAPTools. The Directives Browser is designed to provide more interactive information from the parallelization process, such as reasons why loops are parallel or serial, distributed or not distributed. User can concentrate on areas where potential improvements could be made, for example, by removing false data dependences. It is part of the iterative process of parallelization.

3 Implementation

In the following subsections, we will give some implementation details of CAPO organized according to the components outlined in Sect. 2.3.

3.1 Loop-Level Analysis

In the loop-level analysis, the data dependence information is used to classify loops in each routine. Loop types include *parallel* (including *reduction*), *serial*, and *pipeline*. A parallel loop is a loop with no loop-carried data dependences and no exiting statements that jump out of the loop (e.g. `RETURN`). Loops with I/O (e.g. `READ`, `WRITE`) statements are excluded from consideration at this point. Parallel loop includes the case where a variable is rewritten during the iteration of the loop but the variable can be privatized (i.e. having a local copy on each thread) to remove the loop-carried output dependence.

A reduction loop is considered as a special parallel loop since the loop can first update partial results in parallel on each thread and then update the final result atomically. The reduction operations, such as "+", "-", "`min`", "`max`", etc. can be executed in a `CRITICAL` section. This is how the array reduction is implemented later on.

A special class of loops, called *pipeline loop*, has loop-carried true dependencies and the lengths of these dependence vectors are determinable and with the same sign. Such a loop can potentially be used to form parallel pipelining with an outside loop nesting. Compiler techniques for finding pipeline parallelism through affine transforms are discussed in [11]. The pipeline parallelism can be implemented in OpenMP directives with point-to-point synchronization. This is discussed in Sect. 3.3.

A serial loop is a loop that can not be run in parallel due to loop-carried data dependences, I/O or exiting statements. However, a serial loop may be used for the formation of a parallel pipeline.

3.2 Setup of Parallel Region

In order to achieve good performance, it is not enough to simply stay with parallel loops at a finer grained level. In the context of OpenMP, it is possible to express coarser-grained parallelism with parallel regions. Our next task is to use the loop-level information to define these parallel regions.

There are several steps in constructing parallel regions:

a) Traverse the call graph in a top-down approach and identify parallel loops to be distributed. Only outer-most parallel loops with enough granularity are considered. Parallel regions are then formed around the distributed parallel loops, including pipeline loops if no parallel loops can be found at the same level and parallel pipelines can be formed.
b) Expand each parallel region as much as possible in a routine to the top-most loop nest that contains no I/O and exiting statements and is not part of another parallel region. If a variable will be rewritten by multiple threads in the potential parallel region and cannot be privatized (the memory access conflict test), back down one loop nest level. A reduction loop is in a parallel region by itself.
c) Include in the region any preceded code blocks that satisfy the memory access conflict test and are not yet included in other parallel regions. Orphaned directives will be used in routines that are called inside a parallel region but outside a distributed parallel loop.
d) Join two neighboring regions to form a larger parallel region if possible.
e) Treat parallel pipelines across subroutine boundaries if needed (see next subsection).

3.3 Pipeline Setup

A potential pipeline loop (as introduced in Sect. 3.1) can be identified by analyzing the dependence vectors in symbolic form. In order to set up a parallel pipeline, an outer loop nest is required. If the top loop in a potentially parallel region is a pipeline loop and the loop is also in the top-level loop nesting of the routine, then the loop is further checked for loop nest in the immediate parent routine. The loop in the parent routine can be used to form an „upper" level parallel pipeline only if all the following tests are true: a) such a parent loop nest exists, b) each routine called inside the parent loop contains only a single parallel region, and c) except for pipeline loops all distributed parallel loops can run asynchronously (see Sect. 3.4). If any of the tests failed, the pipeline loop will be treated as a serial loop.

OpenMP provides directives (e.g. „`OMP FLUSH`") and library functions to perform the point-to-point synchronization, which makes the implementation of pipeline parallelism possible with directives. These directives and functions are used to ensure the pipelined code section will not be executed in a thread before the work in the neighboring thread is done. Such an execution requires the scheduling scheme for the pipeline loop to be `STATIC` and `ORDERED`. Our implementation of parallel pipelines with directives is started from an example given in the OpenMP Program Application Interface [13]. The pipeline algorithm is used for parallelizing the NAS benchmark LU in Sect. 4.1 and also described in [12].

3.4 End-of-Loop Synchronization

Synchronization is used to ensure the correctness of program execution after a parallel construct (such as `END PARALLEL` or `END DO`). By default, synchronization is added at the end of a parallel loop. Sometime the synchronization at the end of a loop can be eliminated to reduce the overhead. We used a technique similar to the one in [15] to remove synchronization between two loops.

To be able to execute two loops asynchronously and to avoid a thread synchronization directive between them we have to perform a number of tests aside from the dependence information provided by CAPTools. The tests verify whether a thread executing a portion of the instructions of one loop will not read/write data read/written by a different thread executing a portion of another loop. Hence, for each non-private array we check that the set of written locations of the array by the first thread and the set of read/written locations of the array by the second thread do not intersect. The condition is known as the Bernstein condition (see [9]). If Bernstein condition (BC) is true the loops can be executed asynchronously. The BC test is performed in two steps. The final decision is made conservatively: if there is no proof that BC is true it set to be false. We assume that the same number of threads execute both loops, the number of threads is larger than one and there is an array read/written in both loops.

Check the number of loop iterations. Since the number of the threads can be arbitrary, the number of iterations performed in each loop must be the same. If it cannot be proved that the number of iterations is the same for both loops the Bernstein condition set to be false.

Compare array indices. For each reference to a non-privatizable array in the left hand side (LHS) of one loop and for each reference to the same array in another loop we compare the array indices. If we can not prove for at least one dimension that indices in both references are different then we set BC to be false. The condition can be relaxed if we assume the same thread schedule is used for both loops.

3.5 Variable Usage Analysis

Properly identifying variable usage is very important for the parallel performance and the correctness of program execution. Variables that would cause memory access conflict among threads need to be privatized so that each thread will work on a local copy. For cases where the privatization is not possible, for instance, an array variable would partially be updated by each thread, the variable should be treated as shared and the work in the loop or region can only be executed in sequential (except for the reduction operation). Private variables are identified by examining the data dependence information, in particular, output dependence for memory access conflict and true dependence for value assignment. Partial updating of variables is checked by examining array index expressions.

With OpenMP, if a private variable needs its initial value from outside a parallel region, the `FIRSTPRIVATE` clause can be used to obtain an initial copy of the original value; if a private variable is used after a parallel region, the `LASTPRIVATE` clause can be used to update the shared variable.

The reduction operation is commonly encountered in calculation. A typical implementation of parallel reduction has a private copy of each reduction that is first created for each thread, the local value is calculated on each thread, and the global copy is updated according to the reduction operator. OpenMP 1.0 only supports reductions for scalar values. For array, we first transform the code section to create a local array and, then, update the global copy in a CRITICAL section.

3.6 Private Variables in Common Blocks

For a private variable, each thread keeps a local copy and the original shared variable is untouched during the course of updating the local copy. If a variable declared in a common block is private in a loop, changes made to the variable through a subroutine call may not be updated properly for the local copy of this variable. If all the variables in the common block are privatizable in the whole program, the common block can be declared as *threadprivate*. However, if the common block can not be thread-privatized, additional care is needed to treat the private variable.

The following algorithm is used to treat private variables in a common block. The algorithm identifies and performs the necessary code transformation to ensure the correctness of variable privatization. The following convention is used: R_INSIDE for routine called inside a parallel loop, R_OUTSIDE for routine called outside a parallel loop, R_CALL for routine in a call statement, R_CALLBY for routine that calls the current routine, and V (or VC, VD, VN) for a variable named in a routine.

```
TreatPrivate(V, R_ORIG, callstatement) {
    check V usage in callstatement
    if V is not used in the call (via dependences) || is on the command parse tree
      || is not defined in a regular common block in a subroutine along the call path
        return
    TreatVinCall(VC, R_CALL) (V is referred as VC in R_CALL) {
        if VC is in the argument list of R_CALL
            return VC
        if R_CALL is R_OUTSIDE {
            if VC is not declared in R_CALL {
                replicate the common block in which V is named as VN
                set VC to VN from the common block
                set V to VN in the private variable list if R_CALL==R_ORIG
            }
        } else {
            add VC to the argument list of R_CALL
            if VC is defined in a common block of R_CALL
                add R_CALL & VC to RenList (for variable renaming later on)
            else declare VC in R_CALL
            for each calledby statement of R_CALL {
                VD is the name of VC used in R_CALLBY
                TreatVinCall(VD, R_CALLBY) and set VD to the returned value
                add VD to the argument of call statement to R_CALL in R_CALLBY
            }
            for each call statement in R_CALL
```

```
                *TreatPrivate*(VC, R_CALL, callstatement_in_R_CALL)
            }
            return VC
        }
    }
    Rename common block variables listed in RenList.
```

The algorithm starts with private variables listed for a parallel region in routine R_ORIG, one variable at a time. It is used recursively for each call statement in the parallel region along the call graph. A list of routine-variable pairs (R_CALL,VC) is stored in RenList during the process to track where private variables appear in common blocks. These variables in common blocks are renamed at the end.

As an example in Fig. 2 the private array B is assigned inside subroutine SUB via the common block /CSUB/ in loop S2. Applying the above algorithm, the private variable B is added as C to the argument list of SUB and the original variable C in the common block in SUB is renamed to C_CAP to avoid usage conflict. In this way the local copy of B inside loop S2 will be updated properly in subroutine SUB.

```
S1   common /csub/ b(100),
   &              a(100,100)
S2   do 10 j=1, ny
S3     call sub(j, nx)
       do 10 i=1, nx
         a(i,j) = b(i)
 10    continue

S4   subroutine sub(j, nx)
S5   common /csub/ c(100),
   &              a(100,100)
     c(1) = a(1,j)
     c(nx) = a(nx,j)
     do 20 i=2, nx-1
       c(i) = (a(i+1,j) +
   &           a(i-1,j))*0.5
 20  continue
     end
```

```
    S1   COMMON/CSUB/B(100),A(100,100)
!$OMP PARALLEL DO PRIVATE(I,J,B)
    S2   DO 10 J=1, NY
    S3     CALL SUB(J, NX, B)
           DO 10 I=1, NX
             A(I,J) = B(I)
     10    CONTINUE
!$OMP END PARALLEL DO

    S4   SUBROUTINE SUB(J, NX, C)
    S5   COMMON/CSUB/C_CAP(100),A(100,100)
    S6   DIMENSION C(100)
         C(1) = A(1,J)
         C(NX) = A(NX,J)
         DO 20 I=2, NX-1
           C(I) = (A(I+1,J)+A(I-1,J))*0.5
     20  CONTINUE
         END
```

Fig. 2. An example of treating a private variable in a common block

3.7 Routine Duplication

Routine duplication is performed after all the analyses are done but before directives are inserted. A routine needs to be duplicated if it causes usage conflicts at different calling points. For example, if a routine contains parallel regions and is called both inside and outside other parallel regions, the routine is duplicated so that the original routine is used outside parallel regions and the second copy contains only orphaned directives without „OMP PARALLEL" and is used inside parallel regions. Routine duplication is often used in a message-passing program to handle different data distributions in the same routine.

4 Case Studies

We have applied CAPO to parallelize the NAS parallel benchmarks and two computational fluid dynamics (CFD) codes well known in the aerospace field: ARC3D and OVERFLOW. The parallelization started with the interprocedural data dependence analysis on sequential codes. This step was the most computationally intensive part. The result was saved to an application database for later use. The loop and region level analysis was then carried out. At this point, the user inspects the result and decides if any changes are needed. The user assists the analysis by providing additional information on input parameters and removing any false dependences that could not be resolved by the tool. This is an iterative process, with user interaction involved. As we will see in the examples, the user interaction is nominal. OpenMP directives were lastly inserted automatically.

In the case studies, we used an SGI workstation (R5K, 150MHz) and a Sun E10K node to run CAPO. The resulting OpenMP codes were tested on an SGI Origin2000 system, which consisted of 64 CPUs and 16 GB globally addressable memory. Each CPU in the system is a R10K 195 MHz processor with 32KB primary data cache and 4MB secondary data cache. The SGI's MIPSpro Fortran 77 compiler (7.2.1) was used for compilation with the „`-O3 -mp`“ flag.

4.1 The NAS Parallel Benchmarks

The NAS Parallel Benchmarks (NPB) were designed to compare the performance of parallel computers and are widely recognized as a standard indicator of computer performance. The NPB suite consists of five kernels and three simulated CFD applications derived from important classes of aerophysics applications. The five kernels mimic the computational core of five numerical methods used by CFD applications. The simulated CFD applications reproduce much of the data movement and computation found in full CFD codes. Details of the benchmark specifications can be found in [2] and the MPI implementations of NPB are described in [3].

In this study we used six benchmarks (LU, SP, BT, FT, MG and CG) from the sequential version of NPB2.3 [3] with additional optimization described in [7]. Parallelization of the benchmarks with CAPO is straightforward except for FT where additional user interaction was needed. User knowledge on the grid size (≥ 6) was entered for the data dependence analysis of BT, SP and LU. In all cases, the parallelization process for each benchmark took from tens of minutes up to one hour, most of the time being spent in the data dependence analysis. The performance of CAPO generated codes is summarized in Fig. 3 together with comparison to other parallel versions of NPB: MPI from NPB2.3, hand-coded OpenMP [7], and versions generated with the commercial tool SGI-PFA [17].

CAPO was able to locate effective parallelization at the outer-most loop level for the three application benchmarks and automatically pipelined the SSOR algorithm in LU. As shown in Fig. 3, the performance of CAPO-BT, SP and LU is within 10% to the hand-coded OpenMP version and much better than the results from SGI-PFA. The SGI-PFA curves represent results from the parallel version generated by SGI-PFA without any change for SP and with user optimization for BT (see [17] for details). The worse performance of SGI-PFA simply indicates the importance of accurate interprocedural dependence analysis that usually cannot be emphasized in a compiler. It should be pointed out that the sequential version used in the SGI-PFA study was not optimized, thus, the sequential performance needs to be counted for the comparison. The hand-coded MPI versions scaled better, especially for LU. We attribute the performance degradation in the directive implementation of LU to less data locality and larger synchronization overhead in the 1-D pipeline used in the OpenMP version as compared to the 2-D pipeline used in the MPI version. This is consistent with the result of a study from [12].

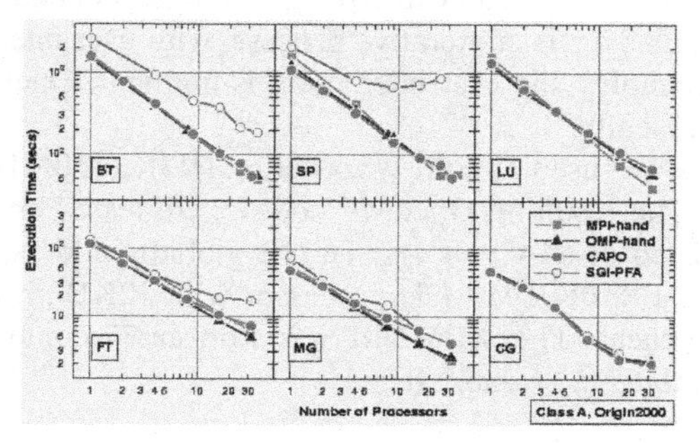

Fig. 3. Comparison of the OpenMP NPB generated by CAPO with other parallel versions: MPI from NPB2.3, OpenMP by hand, and SGI-PFA

The basic loop structure for the Fast Fourier Transform (FFT) in one dimension in FT is as follows.

```
      DO 10 K=1,D3
      DO 10 J=1,D2
        DO 20 I=1,D1
20      Y(I) = X(I,J,K)
        CALL CFFTZ(...,Y)
        DO 30 I=1,D1
30      X(I,J,K) = Y(I)
10    CONTINUE
```

A slice of the 3-D data (X) is first copied to a 1-D work array (Y). The 1-D FFT routine CFFTZ is called to work on Y. The returned result in Y is then copied back to the 3-D array (X). Due to the complicated pattern of loop limits inside CFFTZ, CAPTools could not disprove the loop-carried true dependences by the working array Y for loop K. These dependences were deleted by hand in CAPO to identify the K loop as a parallel loop.

The resulted parallel FT code gave a reasonable performance as indicated by the curve with filled circles in Fig. 3. It does not scale as well as the hand-coded versions (both in MPI and OpenMP), mainly due to the unparallelized code section for the matrix creation which was artificially done with random number generators. Restructuring the code section was done in the hand-coded version to parallelize the matrix creation. Again, the SGI-PFA generated code performed worse.

The directive code generated by CAPO for MG performs 36% worse on 32 processors than the hand-coded version, primarily due to an unparallelized loop in routine `norm2u3`. The loop contains two reduction operations of different types. One of the reductions was expressed in an IF statement, which was not detected by CAPO, thus, the routine was ran in serial. Although this routine takes only about 2% of the total execution time on a single node, it translates into a large portion of the parallel execution on large number of processors, for example, 40% on 32 processors. All the parallel versions achieved similar results for CG.

4.2 ARC3D

ARC3D is a moderate-size CFD application. It solves Euler and Navier-Stokes equations in three dimensions using a single rectilinear grid. ARC3D has a structure similar to NPB-SP but contains curve linear coordinates, turbulent models and more realistic boundary conditions. The Beam-Warming algorithm is used to approximately factorize an implicit scheme of finite difference equations, which is then solved in three directions successively.

For generating the OpenMP parallel version of ARC3D, we used a serial code that was already optimized for cache performance by hand [16]. The parallelization process with CAPO was straightforward and OpenMP directives were inserted without further user interaction. The parallel version was tested on the Origin2000.and the result for a 194x194x194-size problem is shown in the left panel of Fig. 4. The results from a hand-parallelized version with SGI multi-tasking directives (*MT by hand*) [16] and a message-passing version generated by CAPTools (*CAP MPI*) [8] from the same serial version are also included in the figure for comparison.

As one can see from the figure, the OpenMP version generated by CAPO is essentially the same as the hand-coded version in performance. This is indicative of the accurate data dependence analysis and sufficient parallelism that was exploited in the outer-most loop level. The MPI version is about 10% worse than the directive-based versions. The MPI version uses extra buffers for communication and this could contribute to the increase of execution time.

4.3 Overflow

OVERFLOW is widely used for airflow simulation in the aerospace community. It solves compressible Navier-Stokes equations with first-order implicit time scheme, complicated turbulence model and Chimera boundary condition in multiple zones. The code has been parallelized by hand [6] with several approaches: PVM for zonal-level parallelization only, MPI for both inter- and intra-zone parallelization, multi-tasking directives, and multi-level parallelization. This code offers a good test case for our tool not only because of its complexity but also its size (about 100K lines of FORTRAN 77).

In this study, we used the sequential version (1.8f) of OVERFLOW. CAPO took 25 hours on a Sun E10K node to complete the data dependence analysis. A fair amount of effort was spent on pruning data dependences that were placed due to lack of necessary knowledge during the dependence analysis. An example of false dependence is illustrated in the following code segment:

```
      NTMP2 = JD*KD*31
      DO 100 L=LS,LE
        CALL GETARX(NTMP2,TMP2,ITMP2)
        CALL WORK(L,TMP2(ITMP2,1),TMP2(ITMP2,7),...)
        CALL FREARX(NTMP2,TMP2,ITMP2)
  100 CONTINUE
```

Inside the loop nest, the memory space for an array TMP2 is first allocated by GETARX. The working array is then used in WORK and freed afterwards. However, the dependence analysis has reviewed that the loop contains loop-carried true dependences caused by variable TMP2, thus, the loop can only be executed in serial. The memory allocation and de-allocation are performed dynamically and cannot be handled by CAPO. This kind of false dependence can safely be removed with the Directives Browser included in the tool. Even so, CAPO provides an easy way for user to interact with the parallelization process. The OpenMP version was generated within a day after the analysis was completed and an additional few days were used to test the code.

The right panel of Fig. 4 shows the execution time per time-iteration of the CAPO-OMP version compared with the hand-coded MPI version and hand-coded directive (MT) version. All three versions were running with a test case of size 69×61×50, 210K grid points in a single zone. Although the scaling is not quite linear (when comparing to ARC3D), especially for more than 16 processors, the CAPO version out-performed both hand-coded versions. The MPI version contains sizable extra codes [6] to handle intra-zone data distributions and communications. It is not surprising that the overhead is unavoidably large. However, the MPI version is catching up with the CAPO-OMP version on large number of processors. On the other hand, further review has indicated that the multi-tasking version used a fairly similar parallelization strategy as CAPO did, but in quite number of small routines the MT version did not place any directives for the hope that the compiler (SGI-PFA in this case) would automatically parallelize loops inside these routines. The performance number seemed to have indicated otherwise.

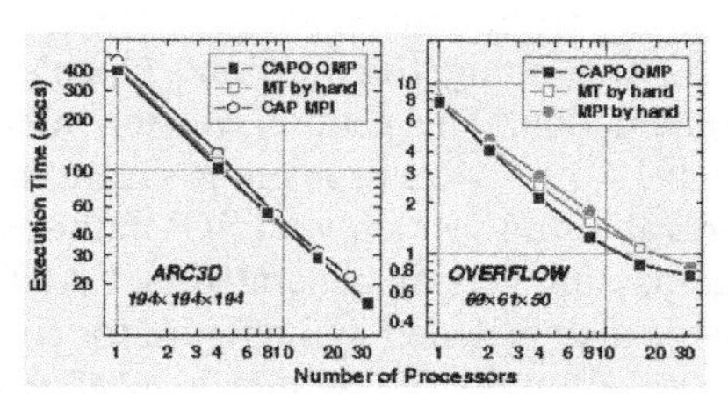

Fig. 4. Comparison of execution times of CAPO generated parallel codes with hand-coded parallel versions for two CFD applications: ARC3D on the left and OVERFLOW on the right

We also tested with a large problem of 1.5M grid points. The result was not included in the figure but CAPO's version has achieved 18-fold speedup on 32 processors of the Origin2000 (10 out of 32 for the small test case). It is not surprising that the problem with large grid size has achieved better parallel performance.

5 Related Work

There are a number of tools developed for code parallelization on both distributed and shared memory systems. The KAPro-toolkit [10] from Kuck and Associates, Inc. performs data dependence analysis and automatically inserts OpenMP directives in a certain degree. KAI has also developed several useful tools to ensure the correctness of directives insertion and help user to profile parallel codes. The SUIF compilation system [18] from Standard is a research product that is targeted at parallel code optimization for shared-memory system at the compiler level.

The SGI's MIPSpro compiler includes a tool, PFA, that tries to automatically detect loop-level parallelism, insert compiler directives and transform loops to enhance their performance. SGI-PFA is available on the Origin2000. Due to the constraints on compilation time, the tool usually cannot perform a comprehensive dependence analysis, thus, the performance of generated parallel programs is very limited. User intervention with directives is usually necessary for better performance. For this purpose, Parallel Analyzer View (PAV), which annotate the results of dependence analysis of PFA and present them graphically, can be used to help user insert directives manually. More details of a study with SGI-PFA can be found in [17].

VAST/Parallel [14] from Pacific-Sierra Research is an automatic parallelizing preprocessor. The tool performs data dependence analysis for loop nests and supports the generation of OpenMP directives.

Parallelization tools like FORGExplorer [1] and CAPTools [4] emphasize the generation of message passing parallel codes for distributed memory systems. These tools can easily be extended to handle parallel codes in the shared-memory arena. Our work is such an example. As discussed in previous sections, the key to the success of our tool is the ability to obtain accurate data dependences combined with user guidance. An ability to handle large applications is also important.

6 Conclusion and Future Work

In summary, we have developed the tool CAPO that automatically generates directive-based parallel programs for shared memory machines. The tool has been successfully used to parallelize the NAS parallel benchmarks and several CFD applications with CAPO, as summarized in Table 1 which included also information for another CFD code, INS3D, the tool was applied to.

By taking advantage of the intensive data dependence analysis from CAPTools, CAPO has been able to produce parallel programs with performance close to hand-coded versions in a relatively short period of time. It should be pointed out, however, that the results did not show the effort in cache optimization of the serial code, such as for ARC3D. Our approach is different from parallel compilers in that it spends much of its time on whole program analysis to discover accurate dependence information. The generated parallel code is produced using a source-to-source transformation with very little modification to the original code and, therefore, is easily maintainable.

Table 1. Summary of CAPO applied on six NPBs and three CFD applications

Application	***BT,SP,LU***	***FT,CG,MG***	***ARC3D***	***OVERFLOW***	***INS3D***
Code Size	~3000 lines benchmark	~2000 lines benchmark	~4000 lines	851 routines 100K lines	256 routines 41K lines
Code Analysis a)	30 mins to 1 hour	10 mins to 30 mins	40 mins	25 hours	42 hours
Code Generation b)	< 5 mins	< 5 mins	< 5 mins	1 day	2 days
Testing c)	1 day	1 day	1 day	3 days	3 days
Performance Compared to Hand-coded Version	within 5-10%	within 10% for CG 30-36% for FT,MG	within 6%	slightly better (see text in Sect. 4.3)	no hand-coded parallel version

a) *„Code Analysis" refers to wall-clock time spent on the data dependence analysis, for NPB and ARC3D on an SGI Indy workstation and for OVERFLOW and INS3D on a Sun E10K node.*

b) *„Code Generation" includes time user spent on interacting with the tool and code restructuring by hand (only for INS3D in four routines). The restructure involves mostly loop interchange and loop fuse that cannot be done by the tool.*

c) *„Testing" includes debugging and running a code and collecting results.*

For larger and more complex applications such as OVERFLOW, it is our experience that the tool will not be able to generate efficient parallel codes without any user interactions. The importance of a tool, however, is its ability to quickly pinpoint the problematic codes in this case. CAPO (via Directives Browser) was able to point out a small percentage of code sections where user interactions were required for the test cases.

Future work will be focused in the following areas:

- Include a performance model for optimal placement of directives.
- Apply data distribution directives (such as those defined by SGI) rather than relying on the automatic data placement policy, *First-Touch*, by the operating system to improve data layout and minimize number of costly remote memory reference.
- Develop a methodology to work in a hybrid approach to handle parallel applications in a heterogeneous environment or a cluster of SMP's. Exploiting multi-level parallelism is important.
- Develop an integrated working environment for sequential optimization, code transformation, code parallelization, and performance analysis.

Acknowledgement

The authors wish to thank members of the CAPTools team at the University of Greenwich for their support on CAPTools and Dr. James Taft and Dr. Dennis Jespersen at NASA Ames for their support on the CFD applications used in the study. This work is supported by NASA Contract No. NAS2-14303 with MRJ Technology Solutions and No. NASA2-37056 with Computer Sciences Corporation.

References

1. Applied Parallel Research Inc., „FORGE Explorer," http://www.apri.com/
2. D. Bailey, J. Barton, T. Lasinski, and H. Simon (Eds.), „The NAS Parallel Benchmarks," *NAS Technical Report RNR-91-002*, NASA Ames Research Center, Moffett Field, CA, 1991
3. D. Bailey, T. Harris, W. Saphir, R. Van der Wijngaart, A. Woo, and M. Yarrow, „The NAS Parallel Benchmarks 2.0," *RNR-95-020*, NASA Ames Research Center, 1995. NPB2.3, http://www.nas.nasa.gov/Software/NPB/
4. C. S. Ierotheou, S. P. Johnson, M. Cross, and P. Legget, „Computer Aided Parallelisation Tools (CAPTools) – Conceptual Overview and Performance on the Parallelisation of Structured Mesh Codes," *Parallel Computing*, 22 (1996) 163-195. http://captools.gre.ac.uk/
5. S. P. Johnson, M. Cross, and M.G. Everett, „Exploitation of symbolic information in interprocedural dependence analysis," *Parallel Computing*, 22 (1996) 197-226
6. D. C. Jespersen, „Parallelism and OVERFLOW, " *NAS Technical Report NAS-98-013*, NASA Ames Research Center, Moffett Field, CA, 1998
7. H. Jin, M. Frumkin and J. Yan., „The OpenMP Implementation of NAS Parallel Benchmarks and Its Performance," *NAS Technical Report*, NAS-99-011, NASA Ames Research Center, 1999
8. H. Jin, M. Hribar and J. Yan, „Parallelization of ARC3D with Computer-Aided Tools," *NAS Technical Report*, NAS-98-005, NASA Ames Research Center, 1998
9. C.H. Koelbel, D.B. Loverman, R. Shreiber, GL. Steele Jr., M.E. Zosel. „The High Performance Fortran Handbook," MIT Press, 1994, page 193
10. Kuck and Associates, Inc., „Parallel Performance of Standard Codes on the Compaq Professional Workstation 8000: Experiences with Visual KAP and the KAP/Pro Toolset under Windows NT," Champaign, IL; "Assure/Guide Reference Manual," 1997
11. Amy W. Lim and Monica S. Lam. "Maximizing Parallelism and Minimizing Synchronization with Affine Transforms," The 24th Annual ACM SIGPLAN-SIGACT Symposium on Principles of Programming Languages, Paris, France, Jan,. 1997
12. X. Martorell, E. Ayguade, N. Navarro, J. Corbalan, M. Gonzalez and J. Labarta, „Thread Fork/Join Techniques for Multi-level Parallelism Exploitation in NUMA Multiprocessors," 1999 ACM International Conference on Supercomputing, Rhodes, Greece, 1999
13. OpenMP Fortran/C Application Program Interface, http://www.openmp.org/
14. Pacific-Sierra Research, „VAST/Parallel Automatic Parallelizer," http://www.psrv.com/

15. E. A. Stohr and M. F. P. O'Boyle, „A Graph Based Approach to Barrier Synchronisation Minimisation," in the Proceeding of 1997 ACM International Conference on Supercomputing, page 156
16. J. Taft, „Initial SGI Origin2000 Tests Show Promise for CFD Codes," *NAS News*, July-August, page 1, 1997. (http://www.nas.nasa.gov/Pubs/NASnews/97/07/article01.html)
17. A. Waheed and J. Yan, „Parallelization of NAS Benchmarks for Shared Memory Multiprocessors," in Proceedings of High Performance Computing and Networking (HPCN Europe '98), Amsterdam, The Netherlands, April 21-23, 1998
18. Robert P. Wilson, Robert S. French, Christopher S. Wilson, Saman P. Amarasinghe, Jennifer M. Anderson, Steve W.K. Tjiang, Shih-Wei Liao, Chau-Wen Tseng, Mary W. Hall, Monica Lam, and John Hennessy, „SUIF: An Infrastructure for Research on Parallelizing and Optimizing Compilers," Computer Systems Laboratory, Stanford University, Stanford, CA

Coarse-grain Task Parallel Processing Using the OpenMP Backend of the OSCAR Multigrain Parallelizing Compiler

Kazuhisa Ishizaka, Motoki Obata, Hironori Kasahara
{ishizaka,obata,kasahara}@oscar.elec.waseda.ac.jp

Waseda University
3-4-1 Ohkubo, Shinjuku-ku, Tokyo, 169-8555, Japan

Abstract. This paper describes automatic coarse grain parallel processing on a shared memory multiprocessor system using a newly developed OpenMP backend of OSCAR multigrain parallelizing compiler for from single chip multiprocessor to a high performance multiprocessor and a heterogeneous supercomputer cluster. OSCAR multigrain parallelizing compiler exploits coarse grain task parallelism and near fine grain parallelism in addition to traditional loop parallelism. The OpenMP backend generates parallelized Fortran code with OpenMP directives based on analyzed multigrain parallelism by middle path of OSCAR compiler from an ordinary Fortran source program. The performance of multigrain parallel processing function by OpenMP backend is evaluated on an off the shelf eight processor SMP machine, IBM RS6000. The evaluation shows that the multigrain parallel processing gives us more than 2 times speed up compared with a commercial loop parallelizing compiler, IBM XL Fortran compiler, on the SMP machine.

1 Introduction

Automatic parallelizing compilers have been getting more important with the increase of parallel processing in a high performance multiprocessor system and use of multiprocessor architecture inside a single chip and for an upcoming home server for improving effective performance, cost-performance and ease of use. Current parallelizing compilers exploit loop parallelism, such as Do-all and Do-across[33, 3]. In these compilers, Do-loops are parallelized using various data dependency analysis techniques [4, 25] such as GCD, Banerjee's inexact and exact tests [33, 3], OMEGA test[28], symbolic analysis[9], semantic analysis and dynamic dependence test and program restructuring techniques such as array privatization[31], loop distribution, loop fusion, strip mining and loop interchange[32, 23].

For example, Polaris compiler[26, 6, 29] exploits loop parallelism by using inline expansion of subroutine, symbolic propagation, array privatization [31, 6] and run-time data dependence analysis[29]. PROMIS compiler[27, 5] combines Parafrace2 compiler[24] using HTG[8] and symbolic analysis techniques[9], and

M. Valero et al. (Eds.): ISHPC 2000, LNCS 1940, pp. 457-470, 2000.

EVE compiler for fine grain parallel processing. SUIF compiler parallelizes loop by using inter-procedure analysis [10, 11, 1], unimodular transformation and data locality optimization[20, 2].

Effective optimization of data localization is more and more important because of the increasing a speed gap between memories and processors. Also, many researches for data locality optimization using program restructuring techniques such as blocking, tiling, padding and data localization, are proceeding for high performance computing and single chip multiprocessor systems [20, 12, 30, 35].

OSCAR compiler has realized a multigrain parallel processing [7, 22, 19] that effectively combines the coarse grain task parallel processing [7, 22, 19, 16, 13, 15], which can be applied from a single chip multiprocessor to HPC multiprocessor systems, the loop parallelization and near fine grain parallel processing[17]. In the conventional OSCAR compiler with the backend for OSCAR architecture, coarse grain tasks are dynamically scheduled onto processors or processor clusters to cope with the runtime uncertainties by the compiler. As the task scheduler, the dynamic scheduler in OSCAR Fortran compiler, and distributed dynamic scheduler[21] have been proposed.

This paper describes the implementation scheme of a thread level coarse grain parallel processing on a commercially available SMP machine and its performance. Ordinary sequential Fortran programs are parallelized using by OSCAR compiler with newly developed OpenMP backend automatically and a parallelized program with OpenMP directive is generated. In other words, OSCAR Fortran compiler is used as a preprocessor which transforms a Fortran program into a parallelized OpenMP Fortran. Parallel threads are forked only once at the beginning of the program and joined only once at the end in this scheme to minimize fork/join overhead. Also, this OSCAR OpenMP backend realizes hierarchical coarse grain parallel processing only using ordinary OpenMP directives though NANOS Compiler uses customly made n-thread library[34].

The rest of this paper is composed as follows. Section 2 introduces the execution model of the thread level coarse grain task parallel processing. Section 3 shows the coarse grain parallelization in OSCAR compiler. Section 4 shows the implementation method of the multigrain parallel processing in OpenMP backend. Section 5 evaluates the performance of this method on IBM RS6000 SP 604e High Node for several programs like Perfect Benchmarks and SPEC 95fp Benchmarks.

2 Execution Model of Coarse Grain Task Parallel Processing in OSCAR OpenMP Backend

This section describes the coarse grain task parallel processing using OpenMP directives. Coarse grain task parallel processing uses parallelism among three kinds of macro-tasks(MTs), namely, Basic Block(BB), and Repetition Block(RB), Subroutine Block(SB) described in Section 3. Macro-tasks are generated by decomposition of a source program and assigned to threads or thread groups and executed in parallel.

In the coarse grain task parallel processing using OSCAR OpenMP backend, threads are generated only once at the beginning of the program, and joined only once at the end. In other words, OSCAR OpenMP backend realizes hierarchical coarse grain parallel processing without hierarchical child thread generation. For example, in Fig.1, four threads are generated at the beginning of the program, and all generated threads are grouped to one thread group(group0). Thread group0 executes MT1, MT2 and MT3.

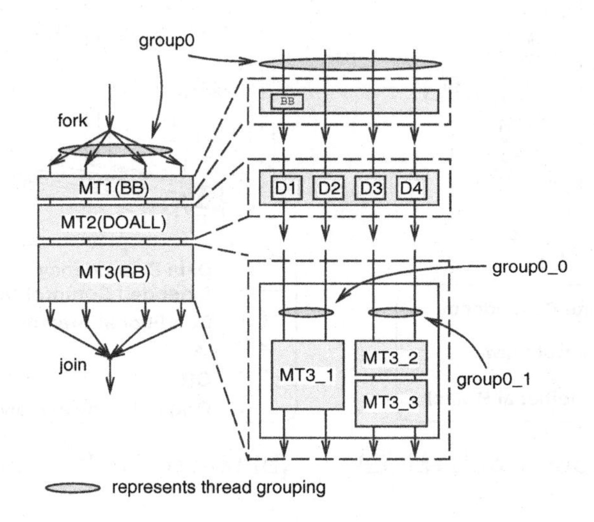

Fig. 1. execution image

When thread group executes a MT, threads in the group use parallelism inside a MT. For example, if MT is a parallelizable loop, threads in group use parallelism among loop iteration. In Fig.1, a parallelizable loop MT2 is distributed to four threads in the group. Also, nested parallelism among sub-MTs, which are generated by decomposition of body of a MT, is used. Sub MTs are assigned to nested(lower level) thread groups, that are hierarchically defined inside a upper level thread group. For example, MT3 in the Fig.1 is decomposed into sub-MTs(MT3_1,MT3_2 and MT3_3), and sub-MTs are executed by two nested thread groups, namely group0_0 and group0_1, each of which have two threads respectively. These groups are defined inside thread group0 which execute MT3.

3 Coarse Grain Parallelization in OSCAR Compiler

This section describes the analysis of OSCAR compiler for coarse grain task parallel processing. First, OSCAR compiler defines coarse grain macro-tasks from source program, and analyzes parallelism among macro-tasks. Next, the

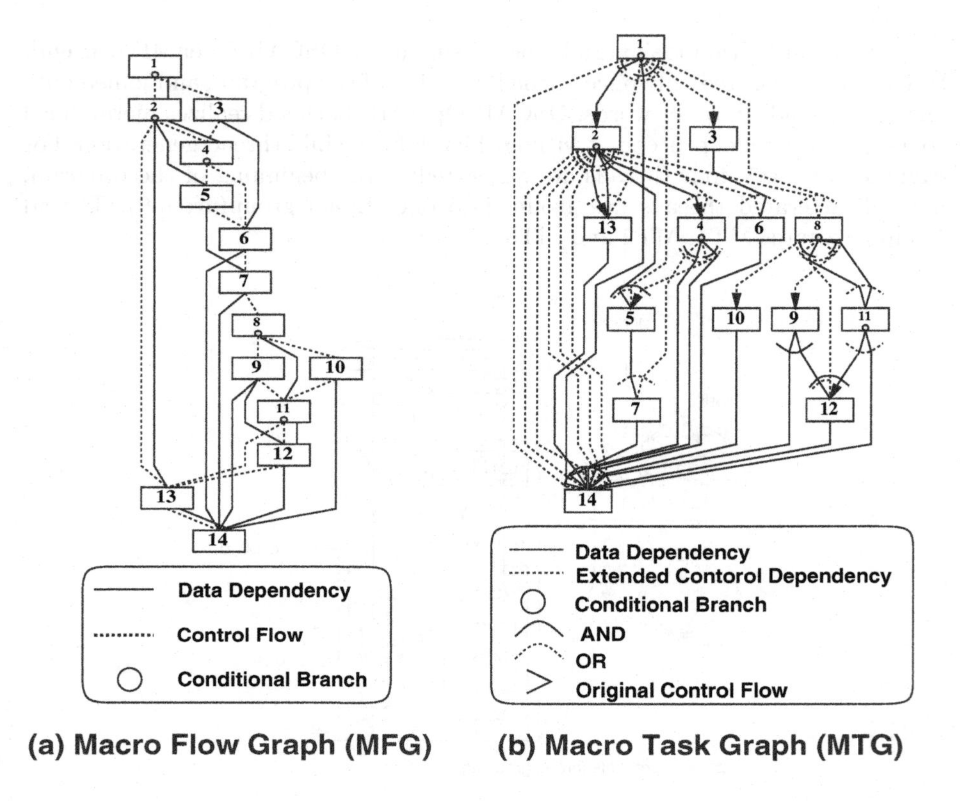

Fig. 2. Macro Flow Graph and Macro Task Graph

generated MTs are scheduled to thread groups statically at compile time or dynamically by embedded scheduling code generated by compiler.

3.1 Definition of Coarse Grain Task

In the coarse grain task parallel processing, a source program is decomposed into three kinds of MTs, namely, BB, RB and SB as mentioned above. Generated MTs are assigned to thread groups, and executed in parallel by threads in the thread group.

If a generated RB is a parallelizable loop, parallel iterations are distributed onto threads inside thread group considering cache size.

If a RB is a sequential loop having large processing cost or SB, it is decomposed into sub-macro-tasks and hierarchically processed by coarse grain task parallel processing scheme like MT3 in Fig.1.

3.2 Generation of Macro-Flow Graph

After generation of macro-tasks, the data dependency and control flow among MTs for each layer are analyzed hierarchically, and represented by Macro-Flow Graph(MFG) as shown in Fig.2(a).

In the Fig.2, nodes represent MTs, solid edges represent data dependencies among MTs and dotted edges represent control flow. A small circle inside a node represents a conditional branch inside the MT. Though arrows of edges are omitted in the MFG, it is assumed that the directions are downward.

3.3 Generation of Macro-Task Graph

To extract parallelism among MTs from MFG, Earliest Executable Condition analysis considering data dependencies and control dependencies is applied. Earliest Executable Condition represents the conditions on which MT may begin its execution earliest. It is obtained assuming the following conditions.

1. If MTi data-depends on MTj, MTi can not begin execution before MTj finishes execution.
2. If the branch direction of MTj is determined, MTi that control-depends on MTj can begin execution even though MTj has not completed its execution.

Then, the original form of Earliest Execution Condition is represented as follows;

(MTj, on which MTi is control dependent, branches to MTi) AND (MTk($0 \leq k \leq |N|$), on which MTi is data dependent, completes execution OR it is determined that MTk is not be executed), where N is the number of predecessors of MTi

For example, the original form of Earliest Execution Condition of MT6 on Fig.2(b) is

(MT1 branches to MT3 OR MT2 branches to MT4) AND (MT3 completes execution OR MT1 branches to MT4).

However, the completion of MT3 means that MT1 already branched to MT3. Also, "MT2 branches to MT4" means that MT1 already branched to MT2. Therefore, this condition is redundant and its simplest form is

(MT3 completes execution OR MT2 branches to MT4).

Earliest Execution Condition of MT is represented in Macro-Task Graph(MTG) as shown in Fig.2(b).

In MTG, nodes represent MTs. A small circle inside nodes represents conditional branches. Solid edges represent data dependencies. Dotted edges represent extended control dependencies. Extended control dependency means ordinary normal control dependency and the condition on which a data dependence predecessor of MTi is not executed.

Solid and dotted arcs connecting solid and dotted edges have two different meanings. A solid arc represents that edges connected by the arc are in AND relationship. A dotted arc represents that edges connected by the arc are in OR relation ship.

In MTG, though arrows of edges are omitted assuming downward, an edge having arrow represents original control flow edges, or branch direction in MFG.

3.4 Scheduling of MTs to Thread Groups

In the coarse grain task parallel processing, the static scheduling and the dynamic scheduling are used for assignment of MTs to thread groups.

In the dynamic scheduling, MTs are assigned to thread groups at runtime to cope with runtime uncertainties like conditional branches. The dynamic scheduling routine is generated and embedded into user program by compiler to eliminate the overhead of OS call for thread scheduling. Though generally dynamic scheduling overhead is large, in OSCAR compiler the dynamic scheduling overhead is relatively small since it is used for the coarse grain tasks with relatively large processing time.

In static scheduling, assignment of MTs to thread groups is determined at compile-time if MTG has only data dependency edges. Static scheduling is useful since it allows us to minimize data transfer and synchronization overheard without run-time scheduling overhead.

In the proposed coarse grain task parallel processing, both scheduling schemes are selectable for each hierarchy.

4 Code Generation in OpenMP Backend

This section describes a code generation scheme for the coarse grain task parallel processing using threads in OpenMP backend of OSCAR multigrain automatic parallelization compiler.

The code generation scheme is different for each scheduling scheme. Therefore, after a thread generation method is explained, the code generation scheme for each scheduling scheme is described.

4.1 Generation of Threads

In the proposed coarse grain task parallel processing using OpenMP, the same number of threads as the number of processors are generated by PARALLEL SECTIONS directive only once at the beginning of the execution of program.

Generally, to realize nested or hierarchical parallel processing, nested threads are forked by an upper level thread. However, in the proposed scheme, it is assumed that the number of generated thread, thread grouping and the scheduling scheme applied to each hierarchy are determined at compile-time. In other words, the proposed scheme realizes this hierarchical parallel processing with single level thread generation by writing all MT code or embedding hierarchical scheduling routines in each OpenMP SECTION between PARALLEL SECTIONS and END PARALLEL SECTIONS.

This scheme allows us to minimize thread fork and join overhead and to implement hierarchical coarse grain parallel processing without special extension of OpenMP.

4.2 Static Scheduling

If a Macro Task Graph in a target layer has only data dependencies, the static scheduling is applied to reduce data transfer, synchronization and scheduling overheads.

In the static scheduling, the assignment of MTs to thread groups is determined at compile-time. Therefore, each OpenMP SECTION needs only the MTs that should be executed in the predetermined order.

At runtime, each thread group should synchronize and transfer shared data to other thread groups in the same hierarchy to satisfy the data dependency among MTs. Therefore, the compiler generates synchronization codes using shared memory.

A code image for eight threads generated by OpenMP backend of OSCAR compiler is shown in Fig.3. In this example, static scheduling is applied to the first layer. In Fig.3, eight threads are generated by OpenMP PARALLEL SECTIONS directives. The eight threads are grouped into two thread groups, each of which has four threads. MT1 and 3 are statically assigned to thread group0 and MT2 is assigned to thread group1.

When static scheduling is applied, compiler generates different codes for the thread groups which include only task codes assigned to the thread group.

The assigned MTs to thread groups are processed in parallel by threads inside the thread group by using static scheduling or dynamic scheduling hierarchically.

4.3 Dynamic Scheduling

Dynamic scheduling is applied for a Macro Task Graph with runtime uncertainty caused by a conditional branch. In the dynamic scheduling, since each thread group has possibility to execute any MTs, the all MT codes are copied to every OpenMP SECTION. Each thread group executes MTs selectively according to the scheduling result.

For the dynamic scheduling, OpenMP backend can generate centralized scheduler codes or distributed scheduler codes to be embedded into user code for any parallel processing layer, or nested level. In Fig3, MT2 assigned onto thread group1 is processed by four threads in parallel using the centralized scheduler. In the centralized scheduler method, a master thread, assigned to a thread, assigns macro-tasks to the other three slave threads.

The master thread repeats the following steps.

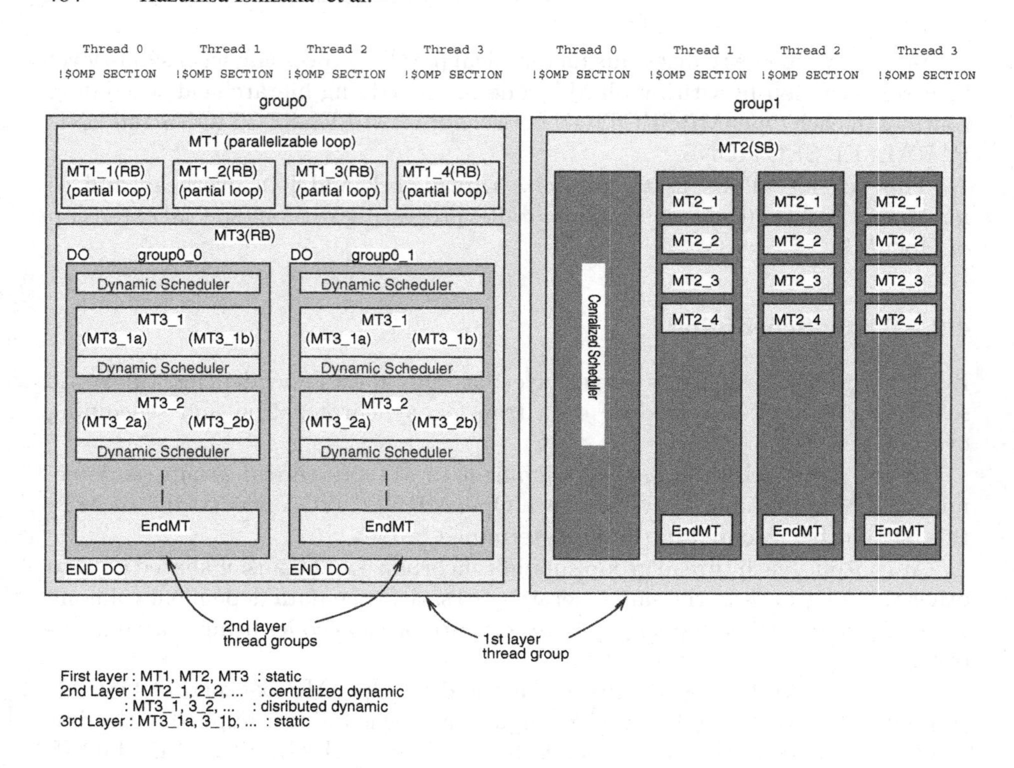

Fig. 3. Code image (four threads)

Behavior of Master thread

step1 Searc hexecutable, or ready, MTs of which Earliest Executable Condition(EEC) are satisfied by the completion or a branch of the preceding MT and enqueue the ready MTs to the ready queue.
step2 Choose a MT with highest priority and assigned it to a idle slave thread.
step3 Go back to step1

The behavior of slave threads is summarized in following.

Behavior of Slave thread

step1 Wait for the macro-task assignment by master thread.
step2 Execute assigned macro-task.
step3 Send signals to report to the master thread a branch direction and/or completion of the task execution.
step4 Go back to step1.

Also, the compiler generates a special MT called EndMT(EMT) in all OpenMP SECTIONS in each hierarchy. The assignment of EndMT shows the end of its hierarchy. In other words, if a EndMT is scheduled to thread groups, the groups finish execution of a hierarchy. As shown in the second layer in Fig.3, the EndMT is written at the end of layer.

In Fig.3, it is assumed that MT2 is executed by master thread(thread 4) and three slave threads.

Next, MT3 shows an example of distributed dynamic scheduling. In this case, MT3 is decomposed into sub-macro-tasks and assigned thread group0_0 and 0_1 defined inside thread group0. In this example, the thread group0_0 and 0_1 has two threads. Each thread group works as scheduler, which behave same as master thread described before, though distributed dynamic schedulers need mutual exclusion to access the shared scheduling data like EEC and ready queue.

The distributed dynamic scheduling routines are embedded into before each macro-task code as shown in Fig.3. Furthermore, Fig.3 shows MT3_1, 3_2 and so on are processed by two threads inside thread group0_0, or 0_1.

5 Performance Evaluation

This section describes the performance of coarse grain task parallelization by OSCAR Fortran Compiler for several programs in Perfect benchmarks and SPEC 95fp benchmarks on IBM RS6000 SP 604e High Node 8 processor SMP.

5.1 OSCAR Fortran Compiler

Fig.4 shows the overview of OSCAR Fortran Compiler. It consists of Front End(FE), Middle Path(MP) and Back Ends(BE). OSCAR Fortran Compiler has various Back Ends for different target multiprocessor systems like OSCAR distributed/shared memory multiprocessor system[18], Fujitsu's VPP supercomputer, UltraSparc, PowerPC, MPI-2 and OpenMP. The newly developed OpenMP Backend used in this paper, generates the parallelized Fortran source code with OpenMP directives. In other words, OSCAR Fortran Compiler is used as a preprocessor that transforms from an ordinary sequential Fortran program to OpenMP Fortran program for SMP machines.

5.2 Evaluated Programs

The programs used for performance evaluation are ARC2D in Perfect Benchmarks, SWIM, TOMCATV, HYDRO2D, MGRID in SPEC 95fp Benchmarks. ARC2D is an implicit finite difference code for analyzing fluid flow problems and solves Euler equations. SWIM solves the system of shallow water equations using finite difference approximations. TOMCATV is a vectorized mesh generation program. HYDRO2D is a vectorizable Fortran program with double precision floating-point arithmetics. MGRID is the Multi-grid solver in 3D potential field.

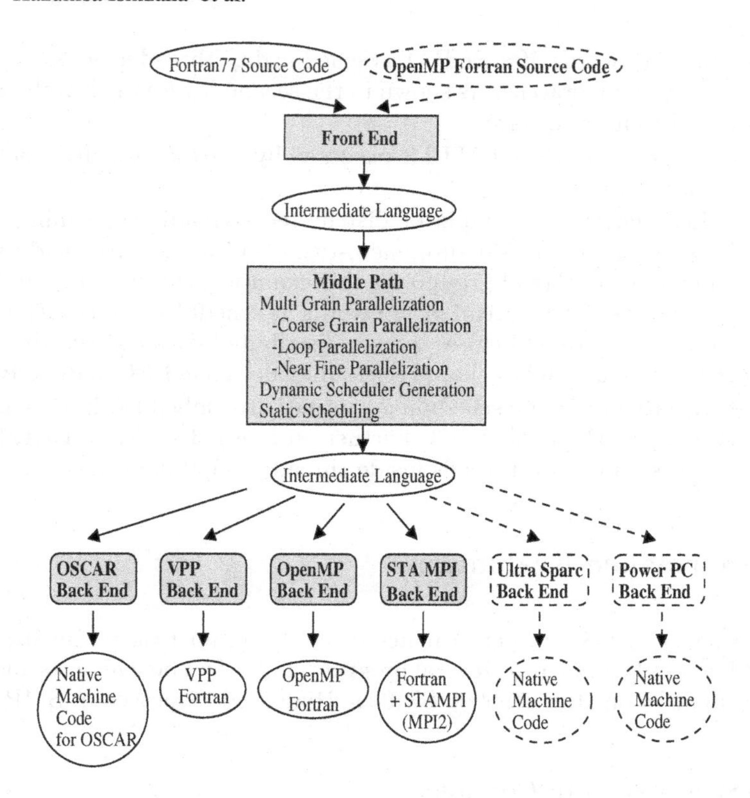

Fig. 4. Overview of OSCAR Fortran Compiler

5.3 Architecture of IBM RS6000 SP

RS6000 SP 604e High Node used for the evaluation is a SMP server having eight PowerPC 604e (200 MHz). Each processor has 32 KB L1 instruction and data caches and 1 MB L2 unified cache. The shared main memory is 1 GB.

5.4 Performance on RS6000 SP 604e High Node

In this evaluation, a coarse grain parallelized program automatically generated by OSCAR compiler is compiled by IBM XL Fortran compiler version 5.1[14] and executed on 1 through 8 processors of RS6000 SP 604e High Node. The performance of OSCAR compiler is compared with IBM XL automatic parallelizing Fortran compiler. In the compilation by a XL Fortran, maximum optimization option "-qsmp=auto -O3 -qmaxmem=-1 -qhot" is used.

Fig.5(a) shows speed-up ratio for ARC2D by the proposed coarse grain task parallelization scheme by OSCAR compiler and the automatic loop parallelization by XL Fortran compiler. The sequential processing time for ARC2D was 77.5s and parallel processing time by XL Fortran version 5.1 compiler using

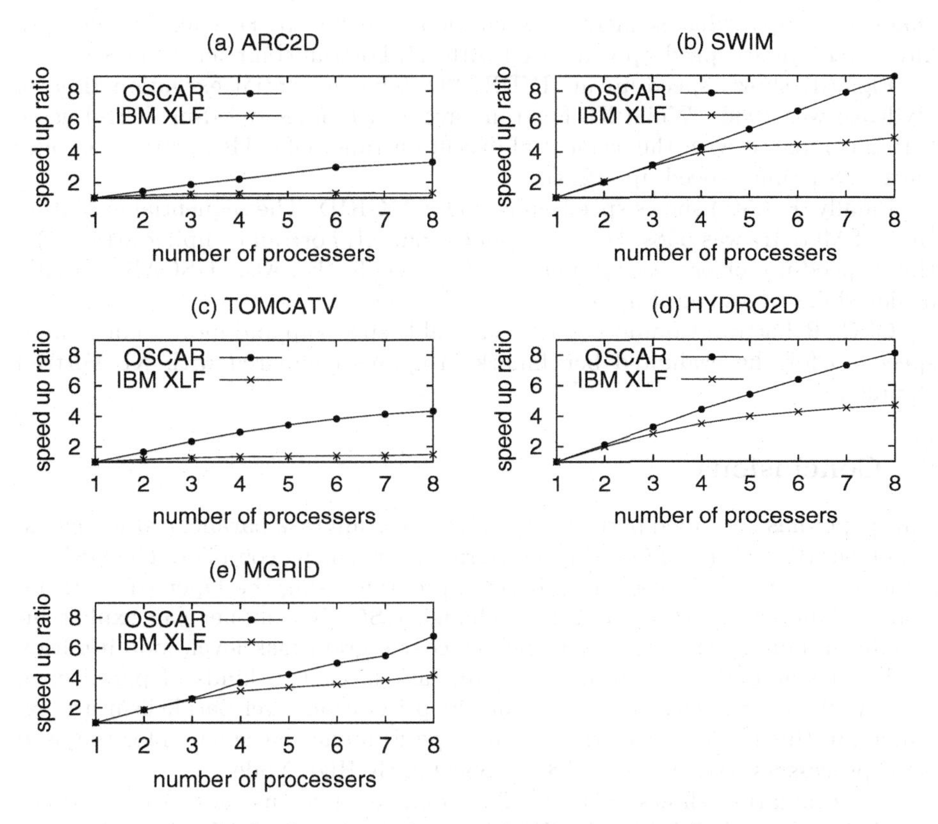

Fig. 5. Speed-up of several benchmarks on RS6000

8PEs was 60.1s. On the other hand, the execution time of coarse grain parallel processing using 8 PEs by OSCAR Fortran compiler with XL Fortran compiler was 23.3s. In other words, OSCAR compiler gave us 3.3 times speed up against sequential processing time and 2.6 times speed up against XL Fortran compiler for 8 processors.

Next, Fig.5(b) shows speed-up ratio for SWIM. The sequential execution time of SWIM was 551s. While the automatic loop parallel processing time using 8 PEs by XL Fortran needed 112.7s ,coarse grain task parallel processing by OSCAR Fortran compiler required only 61.1s and gave us 9.0 times speed-up by the effective use of distributed caches.

Fig.5(c) shows speed-up ratio for TOMCATV. The sequential execution time of TOMCATV was 691s. The parallel processing time using 8 PEs by XL Fortran was 484s and 1.4 times speed-up. On the other hand, the coarse grain parallel processing using 8 PEs by OSCAR Fortran compiler was 154s and gave us 4.5

times speed-up against sequential execution time. OSCAR Fortran compiler also gave us 3.1 times speed up compared with XL Fortran compiler using 8 PEs.

Fig.5(d) shows speed-up in HYDRO2D. The sequential execution time of Hydro2d was 1036s. While XL Fortran gave us 4.7 times speed-up (221s) using 8 PEs compared with the sequential execution time, OSCAR Fortran compiler gave us 8.1 times speed-up (128s).

Finally, Fig.5(e) shows speed-up ratio for MGRID. The sequential execution time of MGRID was 658s. For this application, XL Fortran compiler attains 4.2 times speed-up, or processing time of 157s, using 8 PEs. Also, OSCAR compiler achieved 6.8 times speed up, or 97.4s.

OSCAR Fortran Compiler gives us scalable speed-up and more than 2 times speed up for the evaluated benchmark programs compared with XL Fortran compiler

6 Conclusions

This paper has described performance of coarse grain task parallel processing using OpenMP backend of OSCAR multigrain parallelizing compiler. The OSCAR compiler generates a parallelized Fortran program using the OpenMP backend from a sequential Fortarn program. Though OSCAR compiler can exploit hierarchical multigrain parallelism, such as coarse grain task level, loop iteration level and statement level near fine grain task level, two kinds of parallelism, namely, the coarse grain task level and loop iteration level parallelism are examined in this paper considering machine performance parameters for the used eight processors SMP machine IBM RS6000 604e High Node.

The evaluation shows that OSCAR compiler gives us more then 2 times speedup compared with IBM XL Fortran compiler version 5.1 for several benchmark programs, such as Perfect Benchmarks ARC2D, spec95fp TOMCATV, SWIM, HYDRO2D, and MGRID.

The authors are planning to evaluate the performance of coarse grain parallel processing on various shared memory multiprocessor systems including SGI 2100, Sun Enterprise 3000 and so on using the developed OpenMP backend.

References

1. S. Amarasinghe, J. Anderson, M. Lam, and C. Tseng. The suif compiler for scalable parallel machines. *Proc. of the 7th SIAM conference on parallel processing for scientific computing*, 1995.
2. J. M. Anderson, S. P. Amarasinghe, and M. S. Lam. Data and computation transformations for multiprocessors. *Proceedings of the Fifth ACM SIGPLAN Symposium on Principles and Practice of Parallel Processing*, Jul. 1995.
3. U. Banerjee. Loop parallelization. *Kluwer Academic Pub.*, 1994.
4. U. Barnerjee. Dependence analysis for supercomputing. *Kluwer Pub.*, 1989.
5. Carrie J. Brownhill, Alexandru Nicolau, Steve Novack, and Constantine D. Polychronopoulos. Achieving multi-level parallelization. *Proc. of ISHPC'97*, Nov. 1997.

6. Rudolf Eigenmann, Jay Hoeflinger, and David Padua. On the automatic parallelization of the perfect benchmarks. *IEEE Trans. on parallel and distributed systems*, 9(1), Jan. 1998.
7. H. Kasahara et al. A multi-grain parallelizing compilation scheme on oscar. *Proc. 4th Workshop on Languages and Compilers for Parallel Computing*, Aug. 1991.
8. M. Girkar and C.D. Polychronopoulos. Optimization of data/control conditions in task graphs. *Proc. 4th Workshop on Languages and Compilers for Parallel Computing*, Aug. 1991.
9. Mohammad R. Haghighat and Constantine D. Polychronopoulos. *Symbolic Analysis for Parallelizing Compliers.* Kluwer Academic Publishers, 1995.
10. M. W. Hall, B. R. Murphy, S. P. Amarasinghe, S. Liao, , and M. S. Lam. Interprocedural parallelization analysis: A case study. *Proceedings of the 8th International Workshop on Languages and Compilers for Parallel Computing (LCPC95)*, Aug. 1995.
11. Mary W. Hall, Jennifer M. Anderson, Saman P. Amarasinghe, Brian R. Murphy, Shih-Wei Liao, Edouard Bugnion, and Monica S. Lam. Maximizing multiprocessor performance with the suif compiler. *IEEE Computer*, 1996.
12. Hwansoo Han, Gabriel Rivera, and Chau-Wen Tseng. Software support for improving locality in scientific codes. *8th Workshop on Compilers for Parallel Computers (CPC'2000)*, Jan. 2000.
13. H. Honda, M. Iwata, and H. Kasahara. Coarse grain parallelism detection scheme of fortran programs. *Trans. IEICE (in Japanese)*, J73-D-I(12), Dec. 1990.
14. IBM. *XL Fortran for AIX Language Reference.*
15. H. Kasahara. *Parallel Processing Technology.* Corona Publishing, Tokyo (in Japanese), Jun. 1991.
16. H. Kasahara, H. Honda, M. Iwata, and M. Hirota. A macro-dataflow compilation scheme for hierarchical multiprocessor systems. *Proc. Int'l. Conf. on Parallel Processing*, Aug. 1990.
17. H. Kasahara, H. Honda, and S. Narita. Parallel processing of near fine grain tasks using static scheduling on oscar. *Proc. IEEE ACM Supercomputing'90*, Nov. 1990.
18. H. Kasahara, S. Narita, and S. Hashimoto. Oscar's architecture. *Trans. IEICE (in Japanese)*, J71-D-I(8), Aug. 1988.
19. H. Kasahara, M. Okamoto, A. Yoshida, W. Ogata, K. Kimura, G. Matsui, H. Matsuzaki, and H.Honda. Oscar multi-grain architecture and its evaluation. *Proc. International Workshop on Innovative Architecture for Future Generation High-Performance Processors and Systems*, Oct. 1997.
20. Monica S. Lam. Locallity optimizations for parallel machines. *Third Joint International Conference on Vector and Parallel Processing*, Nov. 1994.
21. Jose E. Moreira and Constantine D. Polychronopoulos. Autoscheduling in a shared memory multiprocessor. *CSRD Report No.1337*, 1994.
22. M. Okamoto, K. Aida, M. Miyazawa, H. Honda, and H. Kasahara. A hierarchical macro-dataflow computation scheme of oscar multi-grain compiler. *Trans. IPSJ*, 35(4):513–521, Apr. 1994.
23. D.A. Padua and M.J. Wolfe. Advanced compiler optimizations for supercomputers. *C.ACM*, 29(12):1184–1201, Dec. 1986.

24. Parafrase2. http://www.csrd.uiuc.edu/parafrase2/.

25. P.M. Petersen and D.A. Padua. Static and dynamic evaluation of data dependence analysis. *Proc. Int'l conf. on supemputing*, Jun. 1993.

26. Polaris. http://polaris.cs.uiuc.edu/polaris/.

27. PROMIS. http://www.csrd.uiuc.edu/promis/.

28. W. Pugh. The omega test: A fast and practical integer programming algorithm for dependence alysis. *Proc. Supercomputing'91*, 1991.

29. Lawrence Rauchwerger, Nancy M. Amato, and David A. Padua. Run-time methods for parallelizing partially parallel loops. *Proceedings of the 9th ACM International Conference on Supercomputing, Barcelona, Spain*, pages 137–146, Jul. 1995.

30. Gabriel Rivera and Chau-Wen Tseng. Locality optimizations for multi-level caches. *Super Computing '99*, Nov. 1999.

31. P. Tu and D. Padua. Automatic array privatization. *Proc. 6th Annual Workshop on Languages and Compilers for Parallel Computing*, 1993.

32. M. Wolfe. Optimizing supercompilers for supercomputers. *MIT Press*, 1989.

33. M. Wolfe. *High Performance Compilers for Parallel Computing.* Addison-Wesley, 1996.

34. Nacho Navaro Xavier Martorell, Jesus Labarta and Eduard Ayguade. A library implementation of the nano-threads programing model. *Proc. of the Second International Euro-Par Conference, vol. 2*, 1996.

35. A. Yoshida, K. Koshizuka, M. Okamoto, and H. Kasahara. A data-localization scheme among loops for each layer in hierarchical coarse grain parallel processing. *Trans. of IPSJ*, 40(5), May. 1999.

Impact of OpenMP Optimizations for the MGCG Method

Osamu Tatebe[1], Mitsuhisa Sato[2], and Satoshi Sekiguchi[1]

[1] Electrotechnical Laboratory, 1-1-4 Umezono, Tsukuba, Ibaraki 305-8568 JAPAN
{tatebe, sekiguchi}@etl.go.jp
[2] Real World Computing Partnership, 1-6-1 Takezono, Tsukuba, Ibaraki 305-0032 JAPAN
msato@trc.rwcp.or.jp

Abstract. This paper shows several optimization techniques in OpenMP and investigates their impact using the MGCG method. MGCG is important for not only an efficient solver but also benchmarking since it includes several essential operations for high-performance computing. We evaluate several optimizing techniques on an SGI Origin 2000 using the SGI MIPSpro compiler and the RWCP Omni OpenMP compiler. In the case of the RWCP Omni OpenMP compiler, the optimization greatly improves performance, whereas for the SGI MIPSpro compiler, it does not affect very much though the optimized version scales well up to 16 processors with a larger problem. This impact is examined by a light-weight profiling tool bundled with the Omni compiler. We propose several new directives for further performance and portability of OpenMP.

1 Introduction

OpenMP is a model for parallel programming that is portable across shared memory architectures from different vendors[4, 5]. Shared memory architectures have become a popular platform and are commonly used in nodes of clusters for high-performance computing. OpenMP is accepted for portable specification across shared memory architectures.

OpenMP can parallelize a sequential program incrementally. To parallelize a **do** loop in a serial program, it is enough to insert a parallel region construct as an OpenMP directive just before the loop. Sequential execution is always possible unless a compiler interprets OpenMP directives.

This paper investigates how to optimize OpenMP programs and its impact. Each directive of OpenMP may cause an overhead. The overhead of each OpenMP directive can be measured by, for example the EPCC OpenMP microbenchmarks[2], however, the important thing is how these overhead affect the performance of applications. In this paper, we focus the impacts of the OpenMP source-level optimization to reduce overhead by creating and joining a parallel region and eliminating unnecessary barrier for the MGCG method.

M. Valero et al. (Eds.): ISHPC 2000, LNCS 1940, pp. 471-481, 2000.

2 MGCG Method

The MGCG method is a conjugate gradient (CG) method with a multigrid (MG) preconditioner and is quite efficient for the Poisson equation with severe coefficient jumps[3, 7, 8]. This method is important for not only several applications but also benchmarking because MGCG includes several important operations for high-performance computing.

MGCG consists of MG and CG. MG exploits several sizes of meshes from a fine-grain mesh to a coarse-grain mesh. It is necessary to efficiently execute parallel loops with various loop lengths. CG needs an inner product besides a matrix-vector multiply and a daxpy. It is also necessary to efficiently execute a reduction in parallel.

The major difficulties for the parallelization of MGCG are the matrix-vector multiply, the smoothing method, restriction and prolongation. In this paper, we use only regular rectangular meshes and a standard interpolation. That is for not only avoiding complexity but also its wide applicability for applications such as computational fluid dynamics, plasma simulation and so on. Note that MG with standard interpolation results in a divergence when a coefficient has severe jumps, while MGCG converges efficiently.

3 Optimization Techniques in OpenMP

Parallelization using OpenMP is basically incremental and straightforward. All we should do is to determine the time-consuming and parallelizable **do** (or **for**) loops and to annotate them by an OpenMP directive for parallel execution. For example, the following code is parallelized using the PARALLEL DO directive and REDUCTION clause.

```
      IP = 0.0
!$OMP PARALLEL DO REDUCTION(+:IP)
      DO I = 1, N
          IP = IP + A(I) * B(I)
      END DO
```

This easy parallelization may cover a large number of programs including dusty deck codes and achieve moderate performance, while the performance strongly depends on the program, compiler and platform. Generally, it is necessary that a parallelized loop spends most of the execution time and it has enough loop length or plenty of computation to hide an overhead of OpenMP parallelization overhead, thread management, synchronization and so on.

To get much higher performance in OpenMP, reducing these parallelization overheads are necessary. In order to reduce these overheads, the following optimizations are necessary.

1. Join several parallel regions. Each parallel region construct starts parallel execution that may contain an overhead of thread creation and join. At the extreme, it is desirable for an entire program to be one big parallel region.

2. Eliminate unnecessary barriers. Each work-sharing construct implies a barrier (and a flush) at the end of the construct unless `NOWAIT` is specified. Note that the `BARRIER` directive is necessary to ensure that each thread accesses updated data.
3. Privatize variables if possible by specifying `PRIVATE`, `FIRSTPRIVATE` and `LASTPRIVATE`.
4. Determine a trade-off of parallel execution including parallelizing overhead and sequential execution. Parallel execution of a loop without enough loop length may be slower than serial execution, since parallel execution needs an overhead of thread management and synchronization. For several independent loops with short loop length, each loop may be parallelized using the `SECTIONS` construct.

Fig. 1 is an excerpt of the MGCG method written in OpenMP. In this case, a parallel region includes three work-sharing constructs and one subroutine call. The subroutine `MG` called in the parallel region, includes several work-sharing constructs as an orphaned directive. Orphaned directives are interpreted as a work-sharing construct when the subroutine is called within a parallel region.

The second work-sharing construct computes an inner product, and it includes a `REDUCTION` clause. The shared variable `RR2` is summed up at the end of this work-sharing construct. In most cases, this kind of computation is a subroutine call. In this case, the subroutine includes an orphaned directive as follows:

```
      SUBROUTINE INNER_PRODUCT(N, R1, R, RR2)
!$OMP DO REDUCTION(+:RR2)
      DO I = 1, N
          RR2 = RR2 + R1(I) * R(I)
      END DO
      RETURN
      END
```

Since it is necessary to perform a reduction within the subroutine, `RR2` should not be privatized in the parallel region.

The MGCG method is parallelized in two ways. One is a version that is parallelized naively. Basically, almost all `DO` loops are parallelized using a `PARALLEL DO` directive. We call this a *naive* version. The other is an *optimized* version that exploits the above optimization techniques. In the optimized version, the entire MGCG program is a parallel region. In Fig. 1, the first and the last statements are in a serial region. The optimized version uses a `MASTER` directive followed by a `BARRIER` directive to execute serially within the parallel region. This version also eliminates all unnecessary barriers by specifying `NOWAIT`.

```
      RR2 = 0

!$OMP PARALLEL PRIVATE(BETA)

*     /*****  MG Preconditioning  *****/

!$OMP DO
      DO I = 1, N
          R1(I) = 0.0
      END DO
      CALL MG(...)

*     /*****  beta = (new_r1, new_r) / (r1, r)  *****/

!$OMP DO REDUCTION(+:RR2)
      DO I = 1, N
          RR2 = RR2 + R1(I) * R(I)
      END DO
      BETA = RR2 / RR1

*     /*****  p = r1 + beta p  *****/

!$OMP DO
      DO I = 1, N
          P(I) = R1(I) + BETA * P(I)
      END DO

!$OMP END PARALLEL

      RR1 = RR2
```

Fig. 1. An excerpt of the MGCG method in OpenMP

4 Evaluation of Optimization Impact

4.1 Overhead of Creating a Parallel Region

To investigate an overhead of a parallel region construct and a work-sharing construct, we consider two kinds of programs, one has several parallel regions (Fig. 2) and the other has one large parallel region with several work-sharing constructs (Fig. 3). For small N and large M, the program of Fig. 2 shows an overhead to create and join a parallel region including an overhead of a work-sharing construct and that of Fig. 3 shows an overhead of a work-sharing construct. We evaluate these programs on an SGI Origin 2000 with 16 R10000 195MHz pro-

```
      DO J = 1, M
!$OMP PARALLEL DO
         DO I = 1, N
            A(I) = A(I) + 2 * B(I)
         END DO
      END DO
```

Fig. 2. A naive program with several parallel regions

```
!$OMP PARALLEL
      DO J = 1, M
!$OMP DO
         DO I = 1, N
            A(I) = A(I) + 2 * B(I)
         END DO
      END DO
!$OMP END PARALLEL
```

Fig. 3. An optimized program with one parallel region and several work-sharing constructs

Table 1. Overhead of parallel region construct and work-sharing construct when $N = 8$ and $M = 100000$ using eight threads

	naive	optimized
MIPSpro	2.90	2.07
Omni	46.76	2.86

[sec.]

cessors with a developing version of the RWCP Omni OpenMP compiler[1, 6] and the SGI MIPSpro compiler version 7.3. The current version 1.1 of the Omni compiler only supports pthreads for the IRIX operating system; however, this developing version exploits **sproc** for further efficient execution on IRIX. The sproc system calls create a new process whose virtual address space can be shared with the parent process.

Table 1 shows the overhead of parallel and work-sharing constructs when $N = 8$ and $M = 100000$ using eight threads. Overhead for a parallel region construct is quite small using the MIPSpro compiler as also reported by EPCC OpenMP microbenchmarks, while it is quite large using the Omni OpenMP compiler. The Omni compiler basically translates an OpenMP program to a C program using the Omni runtime libraries. To reduce the overhead for the parallel

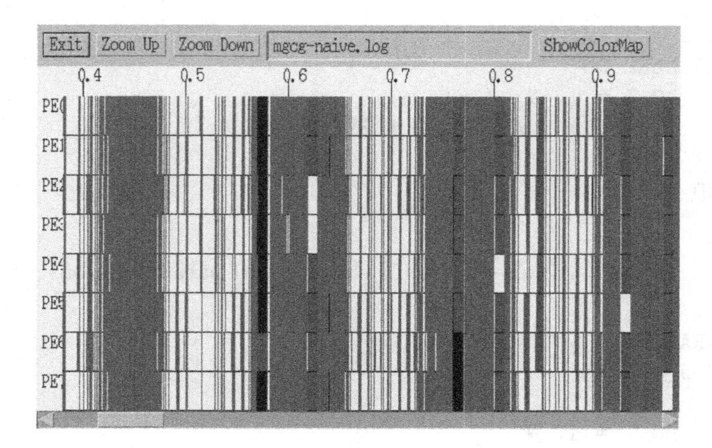

Fig. 4. Snapshot of `tlogview` for the naive version

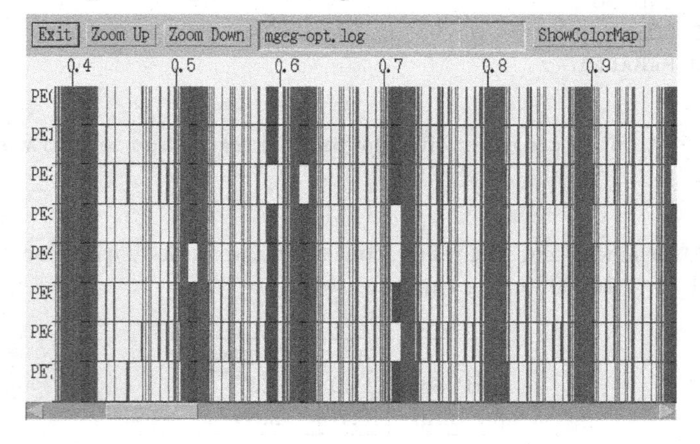

Fig. 5. Snapshot of `tlogview` for the optimized version

region construct, it is necessary to improve the performance of the runtime library for Origin 2000.

4.2 Performance Evaluation of MGCG Method

Profiling The RWCP Omni OpenMP compiler provides a very light-weight profiling tool and a viewer to investigate a behavior of parallel execution. All events related to OpenMP directives are time-stamped at runtime. Each event consists of only two double words, and the profiled log can be viewed by `tlogview`.

Naive and optimized versions of MGCG are profiled on an SGI Origin 2000 with the developing version of the Omni compiler. Figures 4 and 5 show profiled data of the MGCG program by **tlogview**. The white region denotes a parallel execution, the red (dark gray) region shows a barrier, the green (light gray) region shows a loop initialization, and the black region shows a serial part. The MGCG method roughly consists of two parts; CG and MG. The CG part includes a daxpy operation, an inner product and a matrix-vector multiply on the finest grid. Since each loop length is long, the CG part is white. The MG part includes a smoothing method (Red-Black Gauss-Seidel), restriction and prolongation in each grid level. Since MG needs operations on coarse grids that do not have long loop length, the MG part is almost green and red.

Figures 6 and 7 are snapshots of a finer scale. The upper picture of Fig. 6 shows the CG part and the lower picture shows the MG part. These two pictures show almost one iteration of MGCG. In this scale, a serial part stands out. Particularly, an overhead of MG part is conspicuous, since the MG part has loops with various loop length from long length to short length. On the other hand, the optimized version successfully reduces these overheads and serial part.

Performance Evaluation We evaluate naive and optimized MGCG programs using the Omni OpenMP compiler and the SGI MIPSpro compiler. We also compare these OpenMP programs with a program written in MPI. The MPI library is the SGI Message-passing toolkit version 1.3. Every program is compiled with the `-Ofast` optimization option.

Each program solves a two-dimensional Poisson equation with severe coefficient jumps. The two-dimensional unit domain is discretized by 512×512 meshes and 1024×1024 meshes. This kind of problem with severe coeffient jumps is difficult by MG, while it is quite efficiently solved by a combination of MG and CG.

Fig. 8 shows floating-point performance for the problem size 512×512. Using the Omni compiler, the floating-point performance of the optimized version is clearly better than that of the naive version, and it is almost double the performance for large numbers of processors. On the other hand, the difference of the two programs is quite small and both programs achieve good performance with the SGI MIPSpro compiler, since the overhead of a parallel region construct of the MIPSpro compiler is small. In both compiler cases, the performance does not scale with more than 8 processors because the problem size is not so large. The elapsed execution time of the optimized version is only 1.4 seconds with 8 processors.

The MPI program achieves quite good performance with 8 processors. In fact, each processor achieves 44 MFLOPS with 8 processors, while only 32 MFLOPS with 1 processor. The MGCG program parallelized by MPI needs approximately 32 MB of data when the problem size is 512×512. With 8 processors, each processor processes only 4 MB of data that fits into secondary cache. Unfortunately, OpenMP programs cannot take this advantage so much.

Fig. 9 is a result of 1024×1024. In this case, approximately 128 MB of data is necessary. Up to 8 processors, all programs but the naive version with the Omni compiler achieve almost the same good performance. With more processors, the

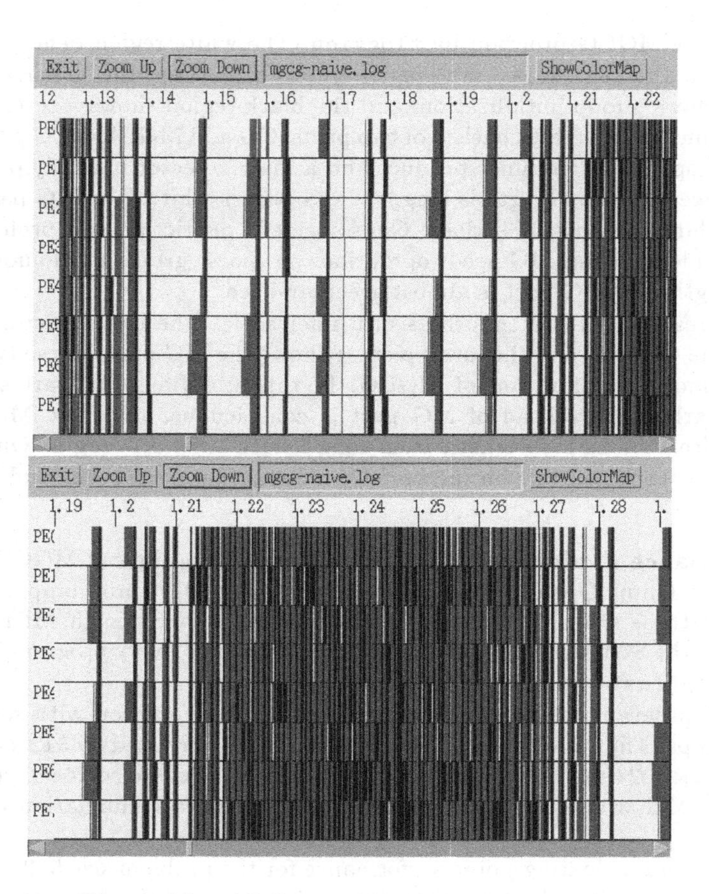

Fig. 6. Magnified snapshot of the naive version

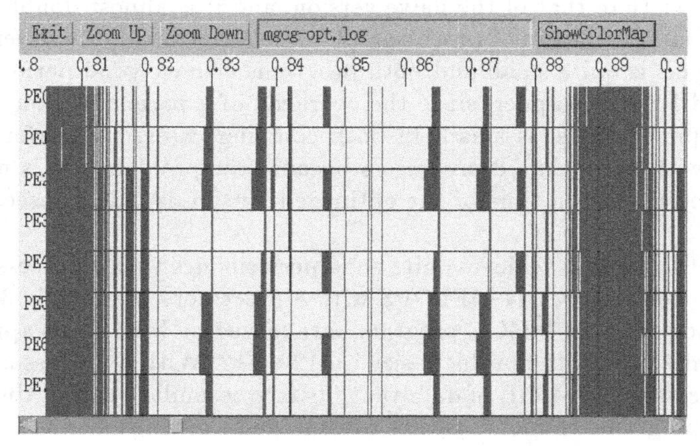

Fig. 7. Magnified snapshot of the optimized version

naive version with the MIPSpro compiler and the optimized version with the Omni compiler slightly degrade the performance, while the MPI program and the optimized version with MIPSpro scale well. That is because the ratio of computation on coarse grid, i.e., a loop with short loop length that tends to become the overhead of parallelization increases, when the number of processes increases. This difference mainly comes from the overhead of a parallel region construct or a work-sharing construct as evaluated by the previous subsection.

Comparing with the performance of the OpenMP programs of two problem sizes, the problem size 512×512 is better than 1024×1024 in the case of eight threads. This shows that a program parallelized by OpenMP can utilize memory locality slightly however it is no match for a program written in MPI.

5 Proposals for OpenMP Specification

When constant variables are privatized in a parallel region, both `FIRSTPRIVATE` and `LASTPRIVATE` need to be specified, since these constant variables become undefined at the end of the parallel region without `LASTPRIVATE`; however, this also may cause an overhead, since copying is necessary from a private variable to a shared variable at the end of the parallel region. This case is quite common. We propose a new data scope attribute clause; `READONLY`. This clause only copies from the shared variable to a private variable and ensures to maintain the shared variable after the parallel region.

We employed an iterative method for performance evaluation. The iterative method is quite sensitive to rounding error and the order of floating-point operations of reduction. Since the OpenMP specification allows any order of floating-point operations for reduction, an OpenMP program may compute a different inner product on successive execution even with the static loop scheduling, the same loop length and the same number of threads. Actually, even the number of iterations until convergence differs every time using the MIPSpro compiler. This may also happen with MPI since the MPI specification also allows any order of floating-point operations for reduction. The problem is that the difference with OpenMP is observed to be much larger than that with MPI. The situation is much worse using dynamic loop scheduling. Since reproducible results are quite important for debugging and testing, this should be controlled by a new directive or a new environmental variable.

6 Conclusions and Future Research

This paper showed several techniques to optimize OpenMP programs and their impact for the MGCG method. Impact is quite dependent on compilers. The Omni compiler showed attractive and almost double the performance compared with the naive version. The SGI MIPSpro compiler showed little difference of performance with small number of processors, however, the optimized version scales well with more processors.

Even though the SGI Origin 2000 supports distributed shared memory, the program written in MPI performs better than the OpenMP program especially with 8 processors and the problem size 512×512. That is because the MPI program exploits locality of memory explicitly. How to exploit the locality is also a big future research of the OpenMP compiler and the OpenMP specification.

Acknowledgment

We are grateful to every member of TEA group, joint effort of Electrotechnical Laboratory (ETL), Real World Computing Partnership, University of Tsukuba and Tokyo Institute of Technology for discussion about semantics and performance of OpenMP programs. We give great thanks for Dr. L. S. Blackford at University of Tennessee, Knoxville for precious comments and proofreading. We also give great thanks for Dr. K. Ohmaki, director of the computer science division at ETL for supporting this research.

References

1. *http://pdplab.trc.rwcp.or.jp/pdperf/Omni/home.html.*
2. BULL, J. M. Measuring synchronisation and scheduling overheads in OpenMP. In *Proceedings of First European Workshop on OpenMP (EWOMP'99)* (1999).
3. KETTLER, R. Analysis and comparison of relaxation schemes in robust multigrid and preconditioned conjugate gradient methods. In *Multigrid Methods*, W. Hackbusch and U. Trottenberg, Eds., vol. 960 of *Lecture Notes in Mathematics*. Springer-Verlag, 1982, pp. 502–534.
4. OPENMP ARCHITECTURE REVIEW BOARD. *OpenMP C and C++ Application Program Interface*, October 1998. version 1.0.
5. OPENMP ARCHITECTURE REVIEW BOARD. *OpenMP Fortran Application Program Interface*, November 1999. version 1.1.
6. SATO, M., SATOH, S., KUSANO, K., AND TANAKA, Y. Design of OpenMP compiler for an SMP cluster. In *Proceedings of 1st European Workshop on OpenMP (EWOMP'99)* (1999), pp. 32–39.
7. TATEBE, O. The multigrid preconditioned conjugate gradient method. In *Proceedings of Sixth Copper Mountain Conference on Multigrid Methods* (April 1993), NASA Conference Publication 3224, pp. 621–634.
8. TATEBE, O., AND OYANAGI, Y. Efficient implementation of the multigrid preconditioned conjugate gradient method on distributed memory machines. In *Proceedings of Supercomputing '94* (November 1994), IEEE Computer Society, pp. 194–203.

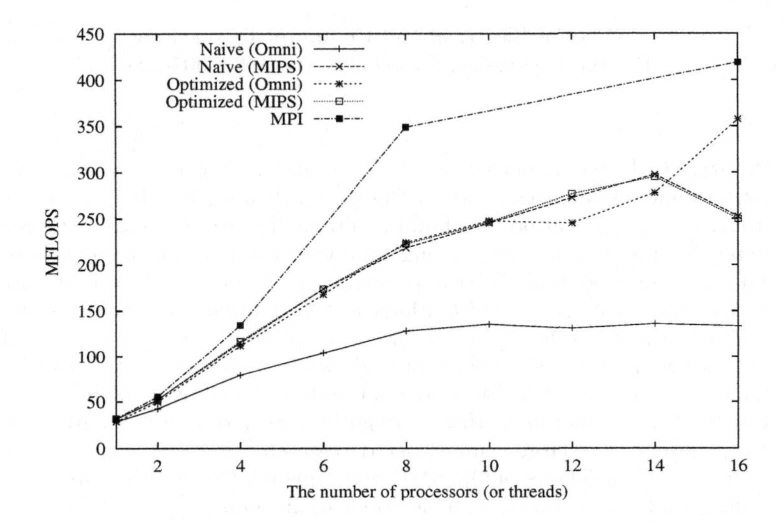

Fig. 8. Floating-point performance of MGCG (512 × 512)

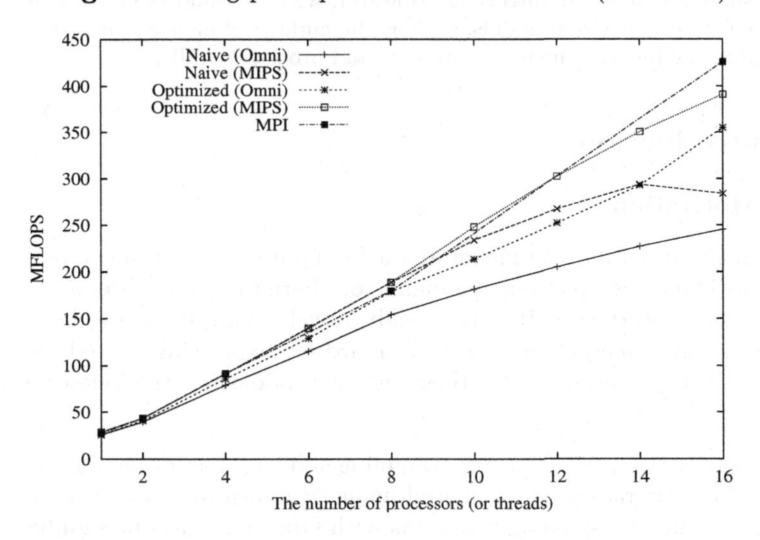

Fig. 9. Floating-point performance of MGCG (1024 × 1024)

Quantifying Differences between OpenMP and MPI Using a Large-Scale Application Suite*

Brian Armstrong, Seon Wook Kim, and Rudolf Eigenmann

School of Electrical and Computer Engineering
Purdue University, West Lafayette, IN 47907-1285

Abstract. In this paper we provide quantitative information about the performance differences between the OpenMP and the MPI version of a large-scale application benchmark suite, SPECseis. We have gathered extensive performance data using hardware counters on a 4-processor Sun Enterprise system. For the presentation of this information we use a *Speedup Component Model*, which is able to precisely show the impact of various overheads on the program speedup. We have found that overall, the performance figures of both program versions match closely. However, our analysis also shows interesting differences in individual program phases and in overhead categories incurred. Our work gives initial answers to a largely unanswered research question: what are the sources of inefficiencies of OpenMP programs relative to other programming paradigms on large, realistic applications. Our results indicate that the OpenMP and MPI models are basically performance-equivalent on shared-memory architectures. However, we also found interesting differences in behavioral details, such as the number of instructions executed, and the incurred memory latencies and processor stalls.

1 Introduction

1.1 Motivation

Programs that exhibit significant amounts of data parallelism can be written using explicit message-passing commands or shared-memory directives. The message passing interface (MPI) is already a well-established standard. OpenMP directives have emerged as a new standard for expressing shared-memory programs. When we choose one of these two methodologies, the following questions arise:

- Which is the preferable programming model, shared-memory or message-passing programming on shared-memory multiprocessor systems? Can we replace message-passing programs with OpenMP without significant loss of speedup?

* This work was supported in part by NSF grants #9703180-CCR and #9872516-EIA. This work is not necessarily representative of the positions or policies of the U. S. Government.

M. Valero et al. (Eds.): ISHPC 2000, LNCS 1940, pp. 482–493, 2000.

– Can both message-passing and shared-memory directives be used simultaneously? Can exploiting two levels of parallelism on a cluster of SMP's provide the best performance with large-scale applications?

To answer such questions we must be able to understand the sources of overheads incurred by real applications programmed using the two models.

In this paper, we deal with the first question using a large-scale application suite. We use a specific code suite representative of industrial seismic processing applications. The code is part of the SPEC High-Performance Group's (HPG) benchmark suite, SPEChpc96 [1]. The benchmark is referred to as SPECseis, or *Seis* for short. Parallelism in *Seis* is expressed at the outer-most level, i.e., at the level of the main program. This is the case in both the OpenMP and the MPI version. As a result, we can directly compare runtime performance statistics between the two versions of *Seis*.

We used a four-processor shared-memory computer for our experiments. We have used the machine's hardware counters to collect detailed statistics. To discuss this information we use the *Speedup Component Model*, recently introduced in [2] for shared memory programs. We have extended this model to account for communication overhead which occurs in message passing programs.

1.2 Related Work

Early experiments with a message passing and a shared-memory version of *Seis* were reported in [3]. Although the shared-memory version did not use OpenMP, this work described the equivalence of the two programming models for this application and machine class. The performance of two CFD applications was analyzed in [4]. Several efforts have converted benchmarks to OpenMP form. An example is the study of the NAS benchmarks [5,6], which also compared the MPI and OpenMP performances with that of SGI's automatic parallelizing compiler.

Our work complements these projects where it provides performance data from the viewpoint of a large-scale application. In addition, we present a new model for analyzing the sources of inefficiencies of parallel programs. Our model allows us to identify specific overhead factors and their impact on the program's speedup in a quantitative manner.

2 Characteristics of SPECseis96

Seis includes 20,000 lines of Fortran and C code, and includes about 230 Fortran subroutines and 120 C routines. The computational parts are written in Fortran. The C routines perform file I/O, data partitioning, and message passing operations. We use the 100 MB data set, corresponding to the *small* data set in SPEC's terminology.

The program processes a series of seismic signals that are emitted by a single source which moves along a 2-D array on the earth's surface. The signals are reflected off of the earth's interior structures and are received by an array of receptors. The signals take the form of a set of *seismic traces*, which are processed by applying a sequence of data transformations. Table 1 gives an overview of these data transformation steps. The seismic transformation steps are combined into four separate seismic applications, referred to as four phases. They include *Phase 1: Data Generation, Phase 2: Stacking of Data, Phase 3: Frequency Domain Migration*, and *Phase 4: Finite-Difference Depth Migration.* The seismic application is described in more detail in [7].

Table 1. Seismic Process. A brief description of each seismic process which makes up the four processing phases of *Seis*. Each phase performs all of its processing on every seismic data trace in its input file and stores the transformed traces in an output file. We removed the seismic process called `RATE`, which performs benchmark measurements in the official SPEC benchmark version of *Seis*

Process	Description
	Phase 1: Data Generation
`VSBF`	Read velocity function and provide access routines.
`GEOM`	Specify source/receiver coordinates.
`DGEN`	Generate seismic data.
`FANF`	Apply 2-D spatial filters to data via Fourier transforms.
`DCON`	Apply predictive deconvolution.
`NMOC`	Apply normal move-out corrections.
`PFWR`	Parallel write to output files.
`VRFY`	Compute average amplitude profile as a checksum.
	Phase 2: Stacking of Data
`PFRD`	Parallel read of input files.
`DMOC`	Apply residual move-out corrections.
`STAK`	Sum input traces into zero offset section.
`PFWR`	Parallel write to output files.
`VRFY`	Compute average amplitude profile as a checksum.
	Phase 3: Fourier Domain Migration
`PFRD`	Parallel read of input files.
`M3FK`	3-D Fourier domain migration.
`PFWR`	Parallel write to output files.
`VRFY`	Compute average amplitude profile as a checksum.
	Phase 4: Finite-Difference Depth Migration
`VSBF`	Data generation.
`PFRD`	Parallel read of input files.
`MG3D`	A 3-D, one-pass, finite-difference migration.
`PFWR`	Parallel write to output files.
`VRFY`	Compute average amplitude profile as a checksum.

The four phases transfer data through file I/O. In the current implementation, previous phases need to run to completion before the next phase can start, except for Phases 3 and 4, which both migrate the stacked data, and therefore only depend on data generated in Phase 2. The execution times of the four phases on one processor of the Sun Ultra Enterprise 4000 system are:

Data Generation Phase 1	Data Stacking Phase 2	Time Migration Phase 3	Depth migration Phase 4	Total
272s	62.2s	7.1s	1,201s	1,542s

More significant is the heterogeneous structure of the four phases. Phase 1 is highly parallel with synchronization required only at the start and finish. Phases 2 and 4 communicate frequently throughout their execution. Phase 3 executes only three communications, independent of the size of the input data set, and is relatively short.

Figure 1 shows the number of instructions executed in each application phase and the breakdown into several categories using the SPIX tool [8]. The data was gathered from a serial run of *Seis*. One fourth of the instructions executed in Phase 4 are loads, contributing the main part of the memory system overhead, which will be described in Figure 4. Note that a smaller percentage of the instructions executed in Phase 3 are floating-point operations, which perform the core computational tasks of the application. Phase 3 exhibits startup overhead simply because it executes so quickly with very few computation steps.

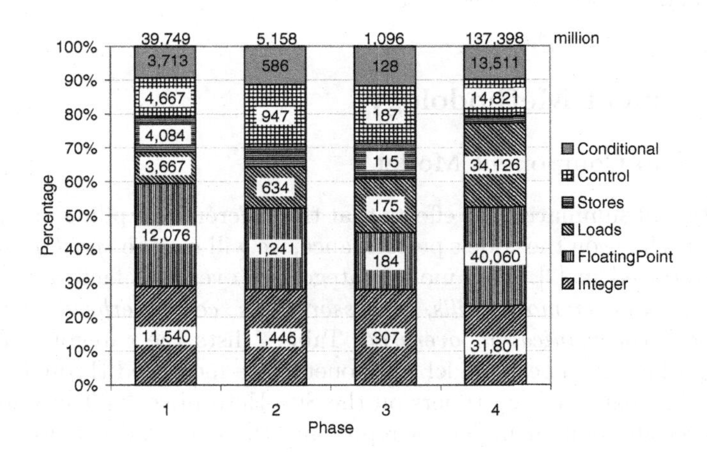

Fig. 1. The Ratio of Dynamic Instructions at Run-Time, Categorized by Type. Instructions executed for the four seismic phases from a serial run were recorded

Figure 2 shows our overall speedup measurements of MPI and OpenMP versions with respect to the serial execution time. The parallel code variants execute nearly the same on one processor as the original serial code, indicating that negligible overhead is induced by adding parallelism. On four processors, the MPI code variant exhibits better speedups than the OpenMP variant. We will describe reasons in Section 4.

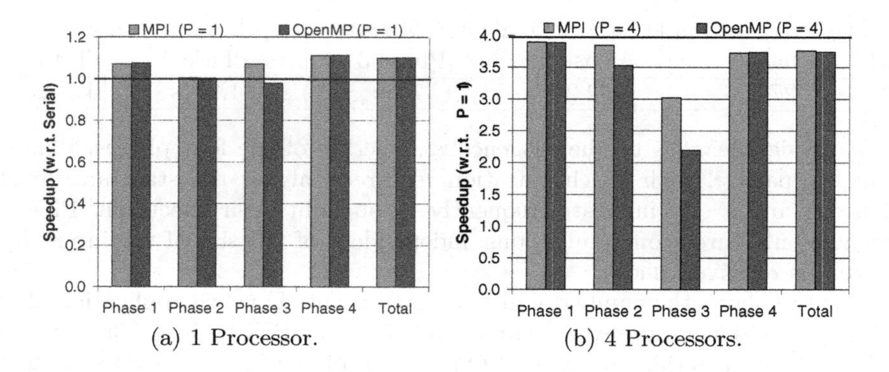

(a) 1 Processor. (b) 4 Processors.

Fig. 2. Speedups of the MPI and OpenMP Versions of *Seis*. Graph (a) shows the performance of each seismic phase as well as the total performance on one processor. Graph (b) shows the speedups on four processors. Speedups are with respect to the one-processor runs, measured on a Sun Enterprise 4000 system. Graph (a) shows that the parallel code variants run at high efficiency. In fact, parallelizing the code improves the one-processor execution of Phase-1 and Phase 4. Graph (b) shows that nearly ideal speedup is obtained, except with Phase 3

3 Experiment Methodology

3.1 Speedup Component Model

To quantify and summarize the effects that the different compiling and programming schemes have on the code's performance, we will use the *speedup component model*, introduced in [2]. This model categorizes overhead factors into several main components: *memory stalls, processor stalls, code overhead, thread management*, and *communication overhead.* Table 2 lists the categories and their contributing factors. These model components are measured through hardware counters (`TICK` register) and timers on the Sun Enterprise 4000 system [9].

The speedup component model represents the overhead categories so that they fully account for the performance gap between measured and ideal speedup. For the specific model formulas we refer the reader to [2]. We have introduced the communication overhead category specifically for the present work to consider the type of communication used in *Seis*. The parallel processes exchange data at regular intervals in the form of all-to-all broadcasts. We define the communication overhead as the time that elapses from before the entire data exchange (of all processors with all processors) until it completes. Both the MPI and the

Table 2. Overhead Categories of the Speedup Component Model

Overhead Category	Contributing Factors	Description	Measured with
Memory stalls	IC miss	Stall due to I-Cache miss.	HW Cntr
	Write stall	The store buffer cannot hold additional stores.	HW Cntr
	Read stall	An instruction in the execute stage depends on an earlier load that is not yet completed.	HW Cntr
	RAW load stall	A read needs to wait for a previously issued write to the same address.	HW Cntr
Processor stalls	Mispred. Stall	Stall caused by branch misprediction and recovery.	HW Cntr
	Float Dep. stall	An instruction needs to wait for the result of a floating point operation.	HW Cntr
Code overhead	Parallelization	Added code necessary for generating parallel code.	computed
	Code generation	More conservative compiler optimizations for parallel code.	computed
Thread management	Fork&join	Latencies due to creating and terminating parallel sections.	timers
	Load imbalance	Wait time at join points due to uneven workload distribution.	
Communication overhead	Load imbalance	Wait time at communication points.	timers
	Copy operations	Data movement between processors.	
	Synchronization	Overhead of synch. operations.	

OpenMP versions perform this data exchange in a similar manner. However the MPI version uses send/receive operations, whereas the OpenMP version uses explicit copy operations, as illustrated in Figure 3.
The MPI code uses blocking sends and receives, requiring processors to wait for the send to complete before the receive in order to swap data with another processor. The OpenMP code can take advantage of the shared-memory space and have all processors copy their processed data into the shared-space, perform a barrier, and then copy from the shared-space.

3.2 Measurement Environment

We used a Sun Ultra Enterprise 4000 system with six 248 MHz UltraSPARC Version 9 processors, each with a 16 KB L1 data cache and 1 MB unified L2 cache using a bus-based protocol. To compile the MPI and serial versions of the code

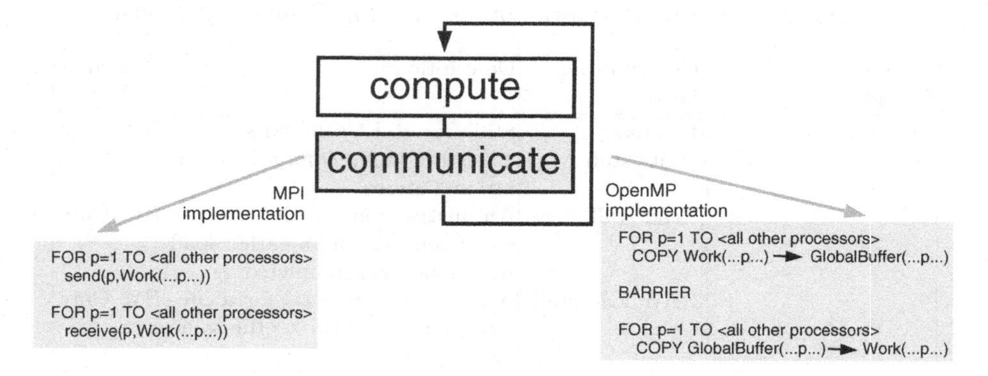

Fig. 3. Communication Scheme in *Seis* and its Implementation in MPI and OpenMP

we use the Sun Workshop 5.0 compilers. The message-passing library we used is the `MPICH 1.2` implementation of MPI, configured for a Sun shared-memory machine. The shared-memory version of *Seis* was compiled using the KAP/Pro compilers (`guidef77` and `guidec`) on top of the Sun Workshop 5.0 compilers. The flags used to compile the three different versions of *Seis* were `-fast -O5 -xtarget=ultra2 -xcache=16/32/1:1024/64/1`.

We used the Sun Performance Monitor library package and would make 14 runs of the application, gathering hardware counts of memory stalls, instruction counts, etc. Using these measurements we could describe the overheads seen in the performance of the serial code and difference between observed and ideal speedup for the parallel implementations of *Seis*. The standard deviation for all these runs was negligible, except in one case mentioned in our analysis.

4 Performance Comparison between OpenMP and MPI

In this section we first inspect the overheads of the 1-processor executions of the serial as well as the parallel program variants. Next, we present the performance of the parallel program executions and discuss the change in overhead factors.

4.1 Overheads of the Single-Processor Executions

Figure 4 shows the breakdown of the total execution time into the measured overheads. "OTHER" captures all processor cycles not spent in measured stalls. This category includes all productive compute cycles such as instruction and data cache hits, and instruction decoding and execution without stalls. It also includes stalls due to I/O operations. However we have found this to be negligible. For all four phases, the figure compares the execution overheads of the original serial code with those of the parallel code running on only one processor. The difference in overheads between the serial and single-processor parallel executions indicate performance degradations due to the conversion of the original code to parallel form. Indeed, in all but the Fourier Migration code (Phase 3) the parallel codes incur more floating-point dependence stalls than the serial code. This change is unexpected because the parallel versions use the same code generator that the serial version uses, except that they link with the MPI libraries or transform the OpenMP directives in the main program to subroutines with thread calls, respectively.

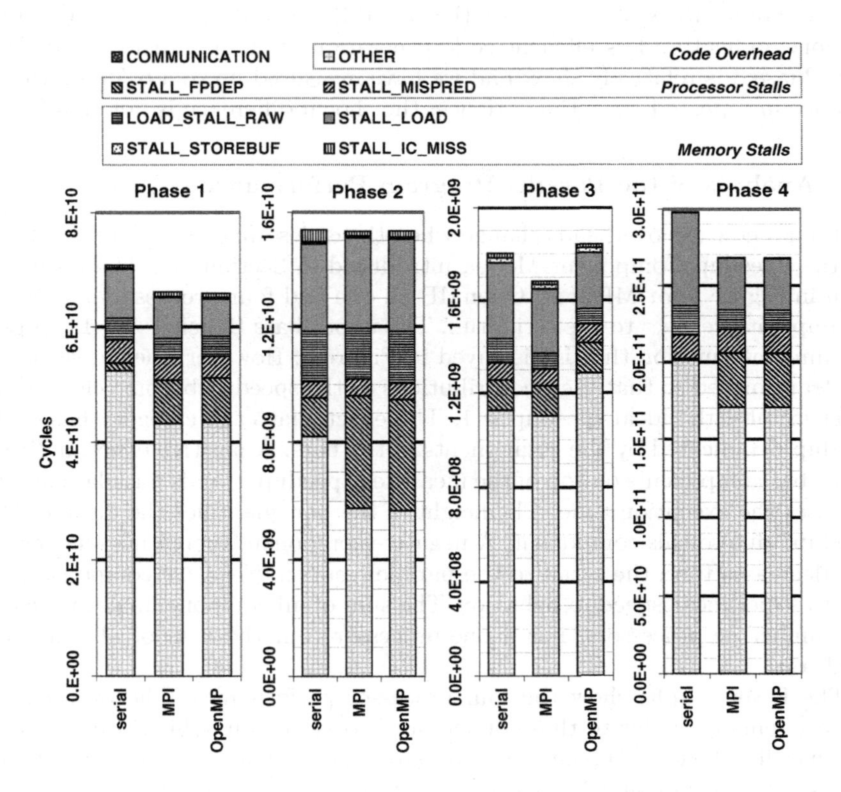

Fig. 4. Overheads for One-Processor Runs. The graphs show the overheads found in the four phases of *Seis* for the serial run and the parallel runs on one processor. The parallel versions of the code cause more of the latencies to be within the FP units than the serial code does. Also, notice that the loads in the Finite-Difference Migration (Phase 4) cause less stalls in the parallel versions than in the serial code. In general, the latencies accrues by the two parallel version exhibit very similar characteristics

Also from Figure 4, we can see that in Phases 1, 2, and 4 compiling with the parallel environment reduces the "OTHER" category. It means that the instructions excluding all stalls execute faster in the 1-processor run of the parallel code than in the serial code. This can be the result of higher quality code (more optimizations applied, resulting in less instructions) or in an increased degree of instruction-level parallelism. Again this is unexpected, because the same code generator is used.

In Phase 4, the parallel code versions reduce the amount of load stalls for both the one-processor and four-processor runs. The parallel codes change data access patterns because of the implemented communication scheme. We assume that this leads to slightly increased data locality.

The OpenMP and MPI programs executed on one processor perform similarly, except for Phase 3. In Phase 3, the OpenMP version has a higher "OTHER" category, indicating less efficient code generation of the parallel variant. However, Phase 3 is relatively short and we have measured up to a 5% performance variance in repeated executions. Hence, the shown difference is not significant.

4.2 Analysis of the Parallel Program Performance

To discuss how the overheads change when the codes are executed in parallel we use the Speedup Component Model, introduced in Section 3.1. The results are given in Figure 5 for MPI and OpenMP on one and four processors in terms of speedup with respect to the serial run. The upper bars (labeled "P=1") present the same information that is displayed in Figure 4. However, the categories are now transformed so that their contributions to the speedup become clear. In the upper graphs, the ideal speedup is 1. The effect that each category has on the speedup is indicated by the components of the bars. A positive effect, indicated by the bar components on top of the measured speedup, stands for a latency that increases the execution time. The height of the bar quantifies the "lack of ideal speedup" due to this component. A negative component represents an overhead that decreases from the serial to the parallel version. Negative components can lead to superlinear speedup behavior. The sum of all components always equals the number of processors. For a one-processor run, the sum of all categories equals one.

The lower graphs show the four-processor performance. The overheads in Phase 1 remain similar to those of the one-processor run, which translates into good parallel efficiency on our four-processor system. This is expected of Phase 1, because it performs highly parallel operations but only communicates to fork processes at the beginning and join them at the end of the phase. Phase 2 of the OpenMP version shows a smaller improvement due to the *code generation* overhead component, which explains why less speedup was measured than with the MPI version. Again, this difference is despite the use of the same code generating compiler and it shows up consistently in repeated measurements. Phase 3 behaves quite differently in the two program variants. However this difference is not significant, as mentioned earlier.

Figure 5 shows several differences between the OpenMP and the MPI implementation of *Seis*. In Phase 4 we can see the number of memory system stalls is less in the OpenMP version than in the MPI version. This shows up in the form

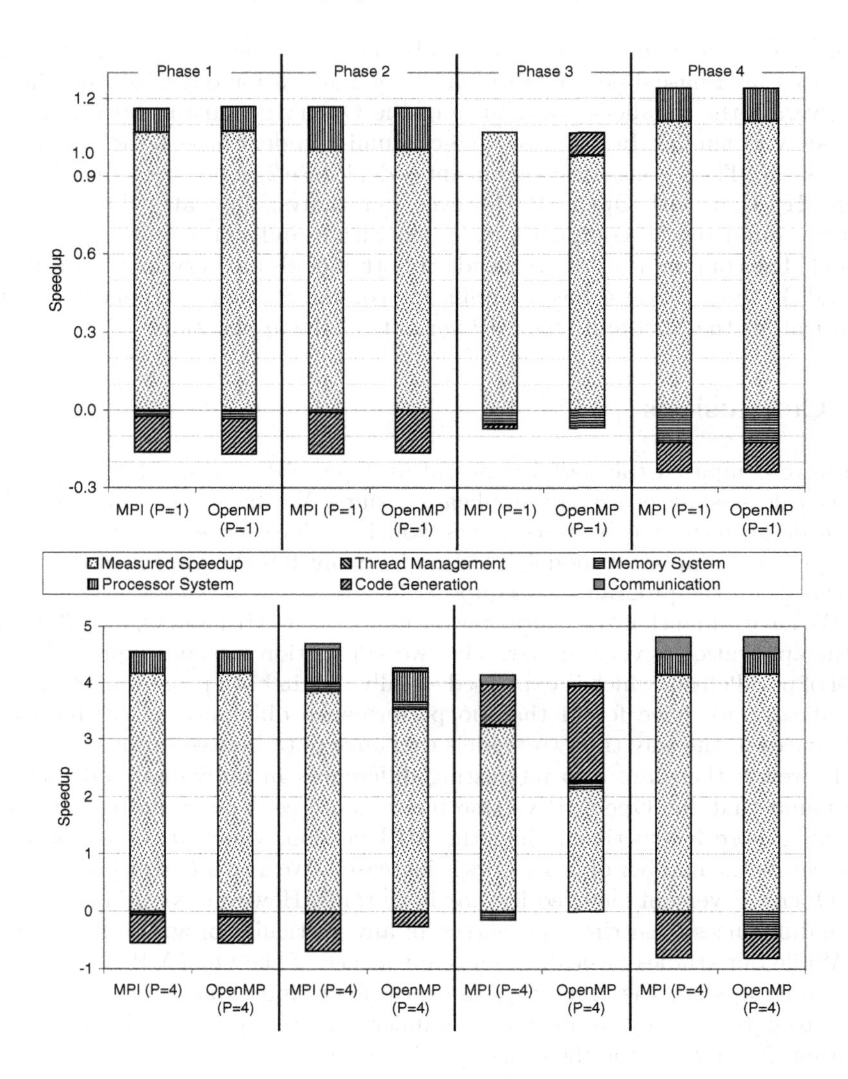

Fig. 5. Speedups Compared Model for the Versions of *Seis*. The upper graph displays the speedups with respect to the serial version of the code and executed on only one processor. The lower graph shows the speedups obtained when executing on four processors. An overhead component represents the amount that the measured speedup would increase (decrease for negative components) if this overhead were eliminated and all other components remained unchanged

of a negative memory system overhead component in the OpenMP versions. Interestingly, the MPI versions has the same measured speedup, as it has a larger negative code generation overhead component. Furthermore, the processor system stalls decrease in the 4-processor execution, however this gain is offset with an increase in communication overheads. These overheads are consistent with the fact that Phase 4 performs the most communication out of all the phases.

Overall, the parallel performance of the OpenMP and the MPI versions of *Seis* are very similar. In the most time-consuming code, Phase 4, the performance is the same. The second-most significant code, Phase 2, shows better performance with MPI than with OpenMP. However, our analysis indicates that the reason can be found in the compiler's code generation and not in the programming model. The communication overheads of both models are very small in Phases 1, 2, and 3. Only Phase 4 has a significant communication component and it is identical for the MPI and OpenMP variants of the application.

5 Conclusions

We have compared the performance of an OpenMP and an MPI version of a large-scale seismic processing application suite. We have analyzed the behavior in detail using hardware counters, which we have presented in the form of the speedup component model. This model quantifies the impact of the various overheads on the programs' speedups.

We have found that the overall performance of the MPI and OpenMP variants of the application is very similar. The two application variants exploit the same level of parallelism, which is expressed equally well in both programming models. Specifically, we have found that no performance difference is attributable to differences in the way the two models exchange data between processors.

However, there are also interesting differences in individual code sections. We found that the OpenMP version incurs more *code overhead* (e.g., the code executes more instructions) than the MPI version, which becomes more pronounced as the number of processors is increased. We also found situations where the OpenMP version incurred less *memory stalls*. However, we do not attribute these differences to intrinsic properties of any particular programming model.

While our studies basically show equivalence of the OpenMP and MPI programming models, the differences in overheads of individual code sections may point to potential improvements of compiler and architecture techniques. Investigating this potential is the objective of our ongoing work.

References

1. Rudolf Eigenmann and Siamak Hassanzadeh. Benchmarking with real industrial applications: The SPEC High-Performance Group. *IEEE Computational Science & Engineering*, III(1):18–23, Spring 1996. 483
2. Seon Wook Kim and Rudolf Eigenmann. Detailed, quantitative analysis of shared-memory parallel programs. Technical Report ECE-HPCLab-00204, HPCLAB, Purdue University, School of Electrical and Computer Engineering, 2000. 483, 486
3. Bill Pottenger and Rudolf Eigenmann. Targeting a Shared-Address-Space version of the seismic benchmark Seis1.1. Technical Report 1456, Univ. of Illinois at Urbana-Champaign, Cntr. for Supercomputing Res. & Dev., September 1995. 483

4. Jay Hoeflinger, Prasad Alavilli, Thomas Jackson, and Bob Kuhn. Producing scalable performance with OpenMP: Experiments with two CFD applications. Technical report, Univ. of Illinois at Urbana-Champaign, 2000. 483
5. Abdul Wahed and Jerry Yan. Code generator for openmp. http://www.nas.nasa.gov/Groups/Tools/Projects/LCM/demos/openmp_frame/target.html, October 1999. 483
6. Abdul Waheed and Jerry Yan. Parallelization of nas benchmarks for shared memory multiprocessors. In *Proceedings of High Performance Computing and Networking (HPCN Europe '98)*, Amsterdam, The Netherlands, apr 21–23 1998. 483
7. C. C. Mosher and S. Hassanzadeh. ARCO seismic processing performance evaluation suite, user's guide. Technical report, ARCO, Plano, TX, 1993. 484
8. Bob Cmelik and David Keppel. Shade: A fast instruction-set simulator for execution profiling. *Proceedings of the 1994 ACM SIGMETRICS Conference on the Measurement and Modeling of Computer Systems*, pages 128–137, May 1994. 485
9. David L. Weaver and Tom Germond. *The SPARC Architecture Manual, Version 9.* SPARC International, Inc., PTR Prentice Hall, Englewood Cliffs, NJ 07632, 1994. 486

Large Scale Parallel Direct Numerical Simulation of a Separating Turbulent Boundary Layer Flow over a Flat Plate Using NAL Numerical Wind Tunnel

Naoki Hirose[1], Yuichi Matsuo[1], Takashi Nakamura[1], Martin Skote[2], and Dan Henningson[2]

[1] Computational Science Division, National Aerospace Laboratory
Jindaiji-Higashi 7-44-1, Chofu-shi, Tokyo 182-8522, Japan
nahirose@nal.go.jp
[2] Department of Mechanics, KTH, S-100 44 Stockholm, Sweden

Abstract Direct numerical simulation(DNS) of fundamental fluid flow simulation using 3-dimensional Navier-Stokes equations is a typical large scale computing which requires high performance computer with vector and parallel processing. In the present paper a turbulent boundary layer flow simulation with strong adverse pressure gradient on a flat plate was made on NAL Numerical Wind Tunnel. Boundary layer subjected to a strong adverse pressure gradient creates a separation bubble followed by a region with small, but positive, skin friction. This flow case contains features that has proven to be difficult to predict with existing turbulence models. The data from present simulation are used for investigation of the scalings near the wall, a crucial concept with respect to turbulence models. The present analysis uses spectral methods and the parallelization was done using MPI(Message-Passing Interface). A good efficiency was obtained in NWT. To compare with other machine performances, previous computations on T3E and SP2 are also shown.

1 Introduction

The near wall scaling of the mean velocities are very important for the correct behavior of wall damping functions used when turbulence models are used in the Reynolds averaged Navier-Stokes equations (RANS). For a zero pressure gradient(ZPG) boundary layer, the damping functions and boundary conditions in the logarithmic layer are based on a theory where the friction velocity, u_τ, is used as a velocity scale. However, in the case of a boundary layer under an adverse pressure gradient (APG), u_τ is not the correct velocity scale, especially for a strong APG and moderate Reynolds number. In the case of separation this is clear since $u_\tau$becomes zero. The combination of a pressure gradient and moderate Reynolds number give a flow that deviates from the classical near wall laws. The near wall behavior of a

M. Valero et al. (Eds.): ISHPC 2000, LNCS 1940, pp. 494-500, 2000.

turbulent boundary layer close to separation is studied using direct numerical simulations (DNS) on massively parallel computers. The results are analyzed and can be used to improve the near wall behavior in turbulence models for flows with separation. This will be extremely important because the existing turbulence models which are used in CFD codes for aerodynamic design and analysis fail to predict separation or even a flow field under strong adverse pressure.

In the computational aspects, DNS of Navier-Stokes equations is a typical large scale computing best fitted for vector and parallel computations. In the past, DNS has been made for simple flows such as homogeneous isotropic turbulence, and it takes an enormous amount of computing time [1]. DNS for simple but realistic flow such as 2D flow on a flat plate was difficult because such flow is spacially developing and spectral methods can not be applied. In DNS, spectral methods have been used because of its high accuracy in time and space. In most cases, finite difference methods for physical space will not be applied because of their low accuracy. Merits of spectral methods are their periodicity characteristics. But it becomes demerits when spacially developing flow is the target. Such flow is not periodical at least in x or flow direction and spectral methods can not be applied in that direction. To overcome this difficulty, an idea called "fringe region technique" is incorporated in the present analysis and periodicity condition was realized in x-direction as well as in other directions. Thus, the paper treats flow filed with strong adverse pressure gradient enough to create reversed flow close to the flat plate wall, i.e. separation, and will get informations on turbulent characteristics of such flow field where the existing turbulence models have difficulty in prediction.

In the present analysis, spectral methods were applied and the parallelization was done using MPI(Message-Passing Interface). A good efficiency was obtained in NWT. The original code was developed at KTH and FFA using FORTRAN77 [2]. We did not try to convert to NWT-FORTRAN because of limited time available for joint work between NAL and KTH. The original 1D FFT subroutine was run on NWT. 3D-FFT subroutine is available for NWT-FORTRAN and it should give more faster performance. In the future work we may try it. To compare with other machine performances, previous computations on T3E and SP2 at KTH are also shown.

2 Numerical Method and Parallelization

The numerical approximation consists of spectral methods with Fourier discretization in the horizontal directions and Chebyshev discretization in the normal direction. Since the boundary layer is developing in the downstream direction, it is necessary to use non-periodic boundary conditions in the streamwise direction. This is possible while retaining the Fourier discretization if a fringe region is added downstream of the physical domain. In the fringe region the flow is forced from the outflow of the physical domain to the inflow. In this way the physical domain and the fringe region together satisfy periodic boundary conditions.

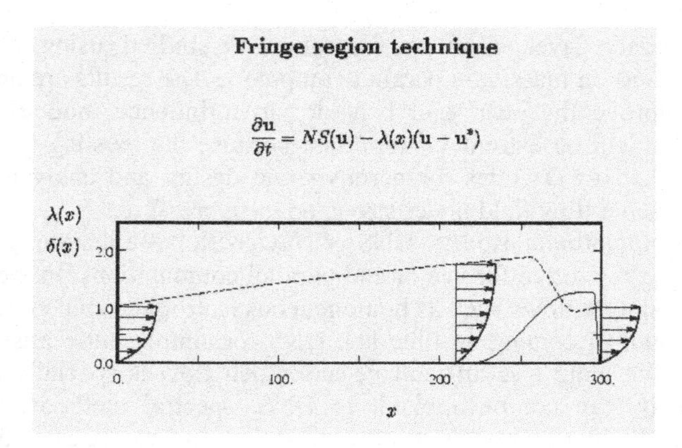

Figure 1 Fringe region technique

Time integration is performed using a third order Runge-Kutta method for the advective and forcing terms and Crank-Nicolson for the viscous terms. A 2/3-dealizing rule is used in the streamwise and spanwise direction. The numerical code is written in FORTRAN and consists of two major parts (figure 2), one linear part (**linear**) where the actual equations are solved in spectral space, and one non-linear part (**nonlin**) where the non-linear terms in the equations are computed in physical space. All spatial derivatives are calculated in the spectral formulation. The main computational effort in these two parts is in the FFT.

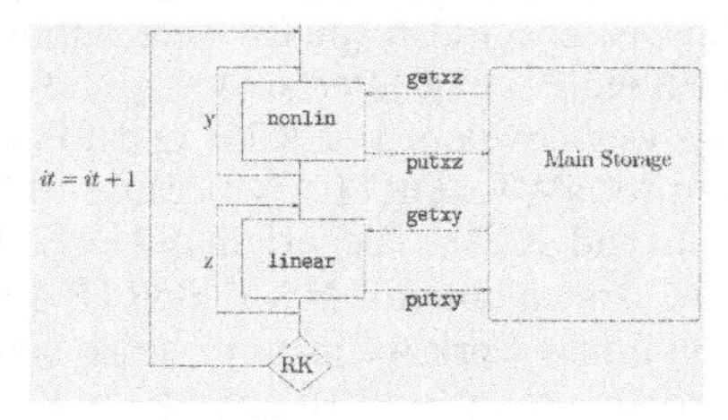

Figure 2 The main structure of the program

In the linear part one *xy*-plane is treated separately for each *z*-position. The field is transformed in the *y* direction to spectral space, a solution is obtained and then transformed to physical space in the *y* direction. This is performed with an loop over all *z* values where the subroutine **linear** is called for each *z*. The *xy*-planes are transferred from the main storage with the routine **getxy** to the memory where the actual computations are performed. The corresponding storing of data is performed with **putxy**.

In the non-linear part the treatment of the data is similar to that in the linear part. One *xz*-plane is treated separately for each *y*-position. The field is transformed in both the *x* and *z* directions to physical space where the non-linear terms are computed. Then the field is transformed in the *x* and *z* directions to spectral space. This is performed with a loop over all *y* values where the subroutine **nonlin** is called to at each *y*. The *xz*-planes are transferred from the main storage with the routine **getxz** to the memory where the actual computations are performed. The corresponding storing of data is performed with **putxz**.

Communication between processors is necessary in the two different parts of the code. The data set (velocity field) is divided between the different processors along the z direction, see figure 3a. Thus, in the linear part, no communication is needed since each processor has data sets for z-position(xy-planes). When the non-linear terms are calculated, each processor needs data for a horizontal plane (xz-planes). The main storage is kept at its original position on the different processors. In the non-linear part each processor collects the two dimensional data from the other processors, on which it performs the computations, and then redistributes it back to the main storage. Figure 3b shows an example of the data gathering for one processor.

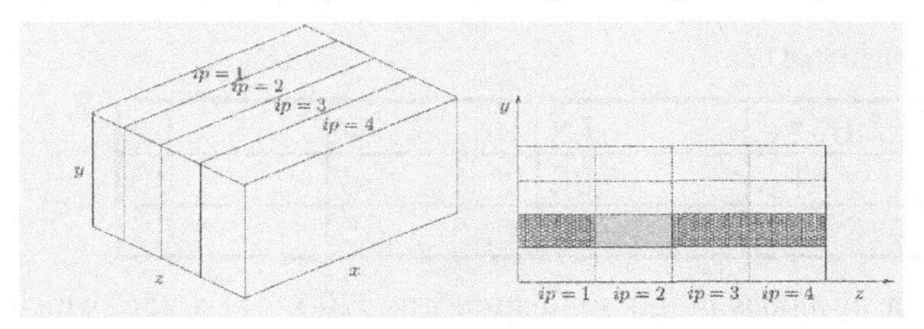

Figure 3 a)The distribution of the main storage on four processors *ip=1,...,4.* b)The gathering of data in the nonlinear part (**nonlin**) of the code for processor number two. The completely shaded area is local on the processor and need not to be received from the others, and the half-shaded area is sent to processor number two. The x-direction is omitted for clarity

3 Numerical Parameters

The first simulation was made on Cray T3E at NSC in Linköping, using 32 processors. The tuning of the pressure gradient for the desired flow situation was performed. After the design of the pressure gradient, a simulation with 20 million modes was performed on a IBM SP2 at PDC, KTH in Stockholm, using 32 processors. The same size of 20 million modes was made(maximum on SP2). Further large scale computation should be made to validate the 20 million result and possibly to obtain more refined resolution of the flowfield. Numerical Wind Tunnel(NWT) at NAL was selected for this purpose. A 40 million modes analysis was made on NWT using 64 nodes. The same code was used on all three computers, using MPI (Message-Passing Interface) for the communication between the processors. NWT is a parallel vector processor consisting of 166 nodes. Each node is 1.7GFLOPS. NWT uses vector processing with higher performances while the previous machines use super-scaler processors. For comparison between the three computers for the full simulation, see table 1.

The simulations start with a laminar boundary layer at the inflow which is triggered to transition by a random volume force near the wall. All the quantities are non-dimensionalized by the freestream velocity (U) and the displacement thickness (δ^*) at the starting position of the simulation ($x=0$) where the flow is laminar. At that position R_{δ^*}=400. The length (including the fringe), height and width of the computation box were 700 x 65 x 80 in these units(see table 2).

Table 1. The performance of the code given in Mflop/s for the 20 million mode simulation on T3E and SP2, and 40 million mode simulation on NWT

	T3E	SP2	NWT
peak processor performance	600	640	1700
code performance per processor	30	50	320
total performance on 64 processors	1900	3200	20500

Table 2. Computational box

Case	Lx	Ly	Lz
APG	700	65	80
SEP	700	65	80

The number of modes in this simulation was 720 x 217 x 256, which gives a total of 40 million modes or 90 million collocation points. The fringe region has a length of 100 and the trip is located at x=10. The simulations were run for a total of 7500 time units (δ^*/U) starting from scratch, and the sampling for the turbulent statistics was performed during the 1000 last time units. Actual NWT CPU Time is about 700 hours using 64 nodes(see table 3).

Table 3 Number of modes

Case	NX	NY	NZ	N
APG	512	193	192	19x10^6
SEP	720	217	256	40x10^6

4 Performance of the Code

In figure 4 the performance of the code on the two super scalar computers is shown as Mflop/s together with the optimal speed. This is a small test case and the performance is lower than for a real simulation on many processors. The scaling is better on the T3E, while the overall performance is better on the SP2, which is approximately twice as fast as the T3E. The NWT is over six times as fast as the SP2, which give a performance of 20 Gflop/s on 64 processors for the simulation with 40 million modes, table 1. Most fast record of FFT computation on NWT was made by a code BIGCUBE and it showed 90.3GFLOPS using 128 nodes. This is a record presented to SC94 Gordon Bell Prize Award[1]. The value corresponds to about 45.1 GFLOPS with 64 nodes. Note that on NWT, linear scalability is attained up to maximum nodes number. Present result of 20 GFLOPS is about a half less than the maximum record. The reason of this is that Gordon Bell record uses 3-D FFT and NWT FORTRAN.

Also pre-fetching and overlapping of data I/O was fully used. In the present computation, MPI was used for 1-D FFT. FORTRAN 77 with MPI was used. Since we do not have enough time for source code conversion to NWT FORTRAN, we did not change the code. Even with this not-optimized code condition, it is remarkable that NWT showed this performance.

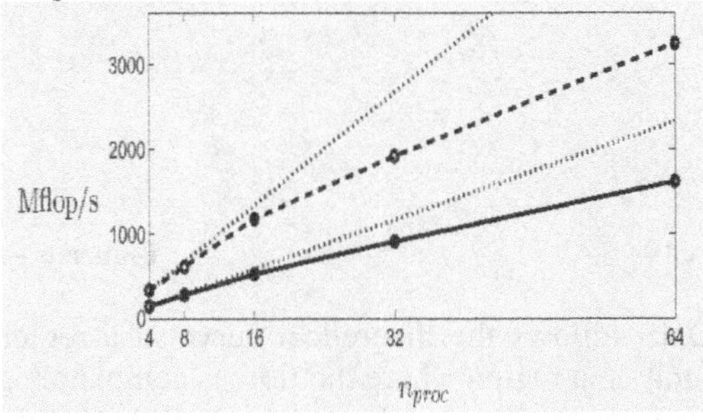

Figure 4 Mflop/s rates for different number of processors for a small test case. --- T3E - - SP2

5 Results

Results from smaller simulations with weaker pressure gradients have been fully analyzed and presented in [3]. These simulations were an important step towards the strong APG case presented here. The free stream velocity varies according to a power law in the down stream coordinate, $U \sim x^m$. In the present simulation the exponent m is equal to -0.25. The friction velocity, u_τ, is negative where separation, i.e. reversed flow, occurs. The boundary layer has a shear stress very close to zero at the wall for a large portion in the down stream direction as seen from figure 5. The separated region is located between $x=150$ to $x=300$.

For a zero pressure gradient (ZPG) boundary layer the velocity profile in the viscous sub-layer is described by $u^+=y^+$, where superscript + denotes the viscous scaling based on u_τ.Under a strong APG this law is not valid and from the equations the following expression can be derived,

$$u^p = 1/2\ y^{p2} - (u_\tau / u_p)^2\, y^p \ .$$

The viscous scaling based on $u_\tau$has to abandoned since u_τ becomes zero at separation. A different velocity scale, based on the pressure gradient, can be derived from the equations valid in the near wall region. This velocity, u_p , replaces u_τ as the scaling parameter and the scaled quantities are denoted by superscript p instead of +.

Comparing velocity profiles from DNS with the profile above is done in figure 6. This figure shows velocity profiles near the wall for x=250 and x=300 in pressure gradient scaling. The higher profile is located at x=250. Both profiles are from within the separated region. The solid lines are DNS data and the dashed are the profiles given by equation 1.

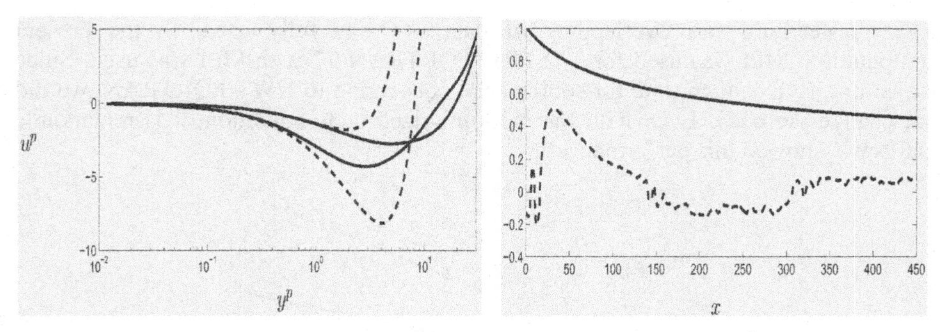

Figure 5 --- U - - u_τ x 10 **Figure 6** --- u^p - -

The data from DNS follows the theoretical curve in a region close to the wall. Equation 1 is valid only in a region where the flow is completely governed by viscous forces. This region is very small at low Reynolds numbers, hence the limited overlap in figure 6. The agreement of DNS data and the theoretical expression in the near wall region indicates that boundary conditions for turbulence models can be improved to give proper results even in a separated region. The near wall behavior is a crucial step in turbulence modeling, and the new result from this simulation has a potential to dramatically improve the wall damping functions.

References

1. H. Miyoshi, M. Fukuda, et al., Development and Achievement of NAL Numerical Wind Tunnel(NWT) for CFD Computations, Proc. IEEE SuperComputing 1994, 1994.
2. A. Lundbladh, S. Berlin, M. Skote, C. Hildings, J. Choi, J. Kim, and D. S. Henningson, An efficient spectral method for simulation of incompressible flow over a flat plate, Technical Report TRITA-MEK 1999:11, Royal Institute of Techmology, Stockholm, 1999.
3. M. Skote, R. A. W. M. Henkes, and D. S. Henningson, Direct numerical simulation of self-similar turbulent boundary layers in adverse pressure gradients, Flow, Turbulence and Combustion, 60:47-85, 1998.

Characterization of Disorderd Networks in Vitreous SiO2 and Its Rigidity by Molecular-Dynamics Simulations on Parallel Computers

Hajime Kimizuka[1], Hideo Kaburaki[1], Futoshi Shimizu[1], and Yoshiaki Kogure[2]

[1] Center for Promotion of Computational Science and Engineering,
Japan Atomic Energy Research Institute, Meguro, Tokyo, 153-0061, Japan
kaburaki@sugar.tokai.jaeri.go.jp
[2] Faculty of Science and Engineering,
Teikyo University of Science & Technology, Yamanashi, 409-0193, Japan

Abstract. Macroscopic elastic properties of materials depend on the underlying microscopic structures. We have investigated the topological structure of three-dimensional network glass, such as vitreous SiO2, and its effect on the rigidity, using a parallel molecular-dynamics (MD) approach. The topological analysis based on the graph theory is employed to characterize disordered networks in the computer generated model of vitreous SiO2. The nature of connectivity of the elementary units beyond the nearest-neighbor, which is related to the medium-range order structure of amorphous state, is described in terms of the ring distribution by the shortest-path analysis. In large-scale MD simulations, the task of detecting these rings from a large amount of data is computationally demanding. Elastic moduli of vitreous SiO2 are calculated with the fluctuation formula for internal stress. The quantitative relation between the statistics of rings for vitreous SiO2 and the elastic moduli are discussed.

Keyword: parallel molecular dynamics simulation, vitreous SiO2, network connectivity, elastic properties, fluctuation formula

M. Valero et al. (Eds.): ISHPC 2000, LNCS 1940, p. 501, 2000.

Direct Numerical Simulation of Coherent Structure in Turbulent Open-Channel Flows with Heat Transfer

Yoshinobu Yamamoto, Tomoaki Kunugi, and Akimi Serizawa

Department of nuclear engineering, Kyoto University,
Yoshida, Sakyo, Kyoto, Japan
{yyama,kunugi,serizawa}@nucleng.kyoto-u.ac.jp

Abstract. In the present study, turbulent heat transfer in open-channel flows has been numerically investigated by means of a Direct Numerical Simulations (DNSs) with a constant temperature at both free surface and bottom wall. The DNSs were conducted for two Prandtl number, 1.0 and 5.0 with a neutral (i.e., zero gravity) or stable stratification (Richardson number; 27.6), while a Reynolds number of 200, based on the friction velocity and flow depth. As the results, the coherent turbulent structures of fluid motion and thermal mixing, and the influence of Pr change for heat transfer, buoyancy effect for turbulent structures and hear transfer, and relationship among them, are revealed and discussed.

1 Introduction

Free surface turbulent flows are very often found in the industrial devices such as a nuclear fusion reactor and a chemical plant, not to speak of those in river and ocean. Therefore, to investigate the turbulent structures near free surface is very important to understand the heat and mass transport phenomena across the free surface. From like this viewpoint, some DNSs with scalar transport including a buoyancy effect were carried out [1], [2]. The interesting information about the relationship between turbulent motion, so-called "surface renewal vortex" [3], and the scalar transport, and also the interaction between buoyancy and turbulence were obtained. However, in these studies, molecular diffusivity of scalar was comparable to that of momentum, it is questionable whether these results can be used for higher Prandtl or Schmidt number fluid flows. The accuracy of experimental or DNS database near free-surface has not been enough to make any turbulence model at the free-surface boundary until now, especially considered the effects of the Prandtl number of fluids, the buoyancy and the surface deformation on the flow, heat and mass transfer.
The aims of this study are to clarify the buoyancy effect on the turbulent structure under various Prandtl number conditions and to reveal the turbulent heart transfer mechanism in open-channel flows, via DNS.

M. Valero et al. (Eds.): ISHPC 2000, LNCS 1940, pp. 502-513, 2000.

2 Numerical Procedure

2.1 Governing Equations

Governing equations are incompressible Navier-Stokes equations with the Boussinesq approximation, the continuity equation and the energy equation:

$$\frac{\partial u_i^*}{\partial t}+u_j^*\frac{\partial u_i^*}{\partial x_j}=\beta\theta^*\Delta T g\delta_{i2}-\frac{\partial}{\partial x_i}\left(\frac{P^*}{\rho}\right)+\nu\frac{\partial^2 u_i^*}{\partial x_j \partial x_j}, \tag{1}$$

$$\frac{\partial u_i^*}{\partial x_i}=0, \tag{2}$$

$$\frac{\partial \theta^*}{\partial t}+u_j^*\frac{\partial \theta^*}{\partial x_j}=\alpha\frac{\partial^2 \theta^*}{\partial x_j \partial x_j}. \tag{3}$$

where u_i^* is i th-component of velocity (i=1, 2, 3), $x_1(x)$ is a streamwise direction, $x_2(y)$ is a vertical direction, $x_3(z)$ is a spanwise direction, t is time, β is the thermal coefficient of volumetric expansion, a super script * denotes the instantaneous value, g is the gravitational force, p^* is the pressure, ρ is the fluid density, ν is the kinetic viscosity, and the dimensionless temperature is defined as $\theta^*=(T^*-T_{wall})/\Delta T$, temperature difference is defined as $\Delta T=T_{surface}-T_{wall}$, T_{surface} denotes free surface temperature, T_{wall} denotes wall temperature, α is the thermal diffusivity, respectively.

2.2 Numerical Method and Boundary Condition

Numerical integration of the governing equations is based on a fractional step method [4] and time integration is a second order Adams-Bashforth scheme. A second order central differencing scheme [5], [6] is adapted for the spatial discretization. The computational domain and coordinate system are shown in Fig. 1.

As the boundary conditions for fluid motion, free-slip condition at the free surface, no-slip condition at the bottom wall and the cyclic conditions in the stream- and the spanwise- directions are imposed, respectively. As for the equation of energy, temperatures at the free surface and the bottom wall are kept constant ($T_{\mathrm{surface}} > T_{\mathrm{wall}}$).

2.3 Numerical Method and Boundary Condition

Numerical conditions are tabled in Table 1, where $R_\tau=u_\tau h/\nu$ is a turbulent Reynolds number based on a friction velocity of the neutral stratification and the flow depth h, and $Ri=\beta g\Delta T h/u_\tau^2$ is a Richardson number. The computations were carried out for about 2000 non-dimensional time units (tu_τ^2/ν) and all statistical values were calculated by time and spatial averages over horizontal planes (homogeneous directions), after flows reached to a fully developed one. However, in case of the stable stratification (Ri=27.6) for Pr=1.0, a laminarization of the flow was appeared, so the computation was stopped at the 1200 non-dimensional time units from the initial turbulent condition of a fully-developed neutral stratification case for the passive scalar. All quantities normalized by the friction velocity of the neutral stratification, kinetic viscosity and the mean heat flux at the free surface, are denoted by the super script +.

3 Results and Discussion

3.1 Coherent Turbulent Structures in Open-Channel Flow

Figures 2-4 show the visualization of coherent turbulent structures in case of the neutral stratification. Figure 2 shows the iso-surface representation of a second invariant velocity gradient tensor $Q^{+}=1/2(\partial u_i^* / \partial x_j \cdot \partial u_j^* / \partial x_i)$. The iso-surface regions are corresponding to the strong vorticity containing regions. Near the bottom wall, the streamwise vortex stretched out the streamwise direction can be seen. This indicates that turbulence is generated near the wall. However, the free surface has no contribution to the turbulence generation in open-channel flows at low Reynolds number.

Figure 3-(b) shows a top view of fluid markers being generated along a line to the z-axis at $y^{+}=12.35$. Alternating high and low speed regions can be seen near the bottom wall. In these like, turbulence structure near the wall is as well as the wall turbulence of ordinary turbulent channel flows. Figure 3-(a) shows a side view of fluid markers being generated along a line to the y-axis at $z^{+}=270$. As well as the wall turbulence, the lift-up of low-speed streaks, so-called the "burst", are depicted. However, in open-channel flows, if this burst reaches to the free surface, a typical turbulence structure affected by the free surface could be appeared underneath the free surface. Near the free surface, the effect of the velocity gradient on the turbulent structure is reduced by a very large horizontal vortex as shown in Fig. 3-(c) as well as the effect of the flow depth scale. This horizontal vortex impinges onto the free surface and turns toward the wall. This motion is in good agreement with the flow visualization experiment [7], i.e., it may correspond to the "surface renewal vortex." It is also consistent with turbulent statistic results of neutral stratification [2], [3], [9]. The mean velocity, the turbulent statistics and the budget of Reynolds stresses are published in elsewhere [9].

Figure 4 shows an instantaneous temperature field. Since the lifted-up cold fluids near the bottom wall and the down-drafted warm fluids near the free surface caused by this vortex motion are observed, a thermal mixing between these motions has been conducted. These typical fluid motions could be the main reason of heat transfer enhancement in the turbulent open-channel flows despite the neutral or stable stratification.

3.2 Statistics of Turbulence

Mean velocity profiles are shown in Fig. 5. In the stable stratification and high *Pr* (=5.0) case, the flow laminarization was clearly observed near the free surface while it is no difference from the neutral case near wall region. On the other hand, in the stable stratification and low *Pr* (=1.0) case, the turbulence throughout the flow cannot be maintained.

The turbulent intensity profiles are shown in Fig. 6. Near the free surface, all components of turbulent intensity are constrained by the stable stratification. Reynolds-stress profiles are shown in Fig. 7. There is a slightly negative value near the free surface in case of the stable stratification.

Mean temperature profiles are shown in Fig. 8. The mean temperature gradient for the stable case (*Pr*=5.0, dotted line) is compared with the neutral case (*Pr*=5.0, solid line). This might indicate that a local heat transfer for the stable case may be promoted by the buoyancy effect. However, a bulk mean temperature of the stable case (*Pr*=5.0) is the lowest of all cases, total heart transfer itself seems to be constrained by the stable stratification.

Figure 9 shows the scalar flux and fluctuation profiles in case of neutral stratification. It can be seen that the scalar fluctuation is produced by the mean velocity gradient near the wall, and the mean scalar gradient near the free surface. However, these profiles are distinct from each other. Especially, in the neutral case (*Pr*=5.0), the turbulent scalar statistics amount to maximums at near free surface where typical turbulence structures are existent. These may suggest that if the Prandtl number is higher, the heat transfer is enhanced by turbulent structures near the free surface.

Wall-normal turbulent scalar flux profiles are shown in Fig. 10. In the neutral and lower *Pr* (=1.0) case, the profile of turbulent heat flux is almost symmetry, and in the neutral and higher *Pr* (=5.0) case, the profile leans toward the bottom wall. In the stable case (*Pr*=5.0), it leans toward the free surface caused by the buoyancy effect.

A scale difference between the neutral and the stable cases may be concerned with a normalization method based on the friction velocity of the neutral stratification, etc. Figures 11 and 12 show the budgets of the Reynolds shear stress and turbulent kinetic energy in the stable case (*Pr*=5.0). As for the Reynolds stress as shown in Fig. 11, a buoyancy production (solid line) is actively conducted near the wall. In the turbulent kinetic energy as shown in Fig. 12, a stable stratification does not affect the turbulent energy budget near the wall. It is shown the reason why the momentum boundary layer thickness is thinner than that of the thermal boundary layer caused by the above local heat transfer mechanism. These are consisting with the results of mean velocity and scalar profiles.

In the neutral case, instantaneous turbulent temperature fields near the free surface are shown in Fig. 13. A scalar field is transferred with the fluid motions, so-called a "surface renewal vortex." However, in case of *Pr*=5.0, the filamentous high temperature fragments are kept because the time scale of the fluid motion is so fast compared with the thermal diffusion time scale. This filamentous structure might be closely concerned with the local heat transfer and Counter-Gradient Flux (CGF) [10] as shown in Fig. 14. This indicates that we have to pay attention whether the Boussinesq approximation for high Prandtl or Schmidt number fluids can be assumed.

4 Conclusions

In this study, Direct Numerical Simulations of two-dimensional fully developed turbulent open-channel flows were performed. The main results can be summarized as follows:

(1) According to the flow visualization, Near the free surface, a large horizontal vortex as well as the flow depth scale affected by the presence of free surface is enhanced the heat transfer.

(2) If the Prandtl number is higher, turbulent structures near the free surface greatly impact on the scalar transport and the Reynolds analogy between a momentum and a scalar transports could not be applied in near free surface region. The reason is that the filamentous high temperature fragments are kept because the time scale of the fluid motion is so fast compared with the heat diffusion time scale.

(3) By the stable stratification effect, if the Prandtl number is lower, the flow could not maintain the turbulence and be impacted on the turbulence structures near the free surface.

(4) By the buoyancy effect, the wall-normal turbulent scalar flux in the stable case is locally enhanced near the wall and its statistical scalar profile is the opposite one of the neutral stratification case. Eventually, the total heat transfer itself was constrained.

Acknowledgments

This work was supported by Japan Science and Technology Corporation (JST) and Japan Atomic Energy Research Institute, Kansai-Establishment (JAERI).

References

1. Nagaosa and Saito, AIChE J., vol. 43, 1997.
2. Handler et al, Physics of Fluids, vol. 11, No.9, pp.2607-2625, 1999.
3. Komori et al., Int. J. Heat Mass Transfer, 25, pp.513-522, 1982.
4. A. J. Chorin, Math. Comp., vol. 22, pp.745-762, 1968.
5. Kawamura, The Recent Developments in Turbulence Research, pp.54-60, 1995.
6. Kajishima, Trans. JSME, Ser. B, vol. 60, 574, pp.2058-2063, 1994 (In Japanese).
7. Banerjee, Appl. Mech, Rev., vol.47, No. 6, Part 2, 1994.
8. Handler et al., AIAA J., vol.31, pp.1998-2007, 1993.
9. Yamamoto et al., Proc. of 8th European Turbulence conference, pp.231-234, 2000.
10. Komori et al., J. Fluid Mech., vol. 130, pp.13-26, 1983.

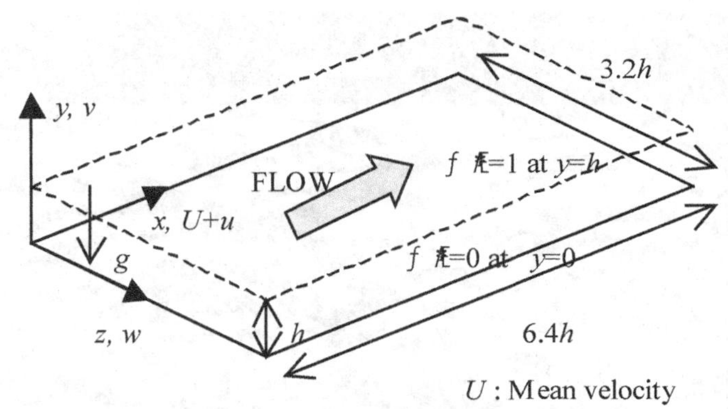

Fig.1 Computational domain and coordinate system

Table 1 Numerical condition

R_τ	Grid Number (x, y, z)	Resolution $(\Delta x^+, \Delta y^+, \Delta z^+)$	***Pr***	***Ri***
200	128,108,128	10, 0.5-4, 5	1.0	-
200	128,108,128	10,0.5-4, 5	1.0	27.6
200	256,131,256	5, 0.5-2, 2.5	5.0	-
200	256,131,256	5, 0.5-2, 2.5	5.0	27.6

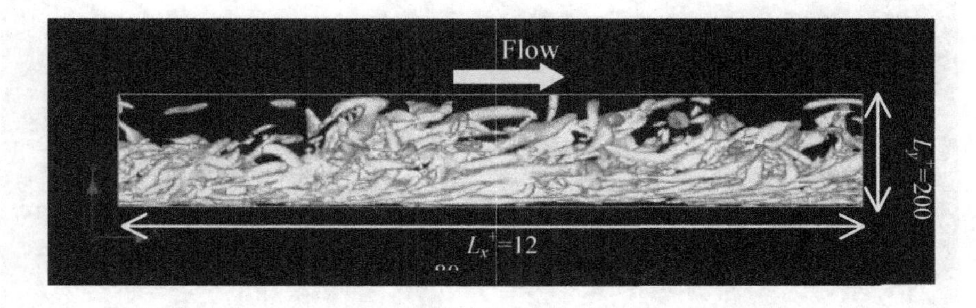

(a) Side view

Fig. 2 Surfaces of second invariant velocity gradient tensor Q^+=0.03

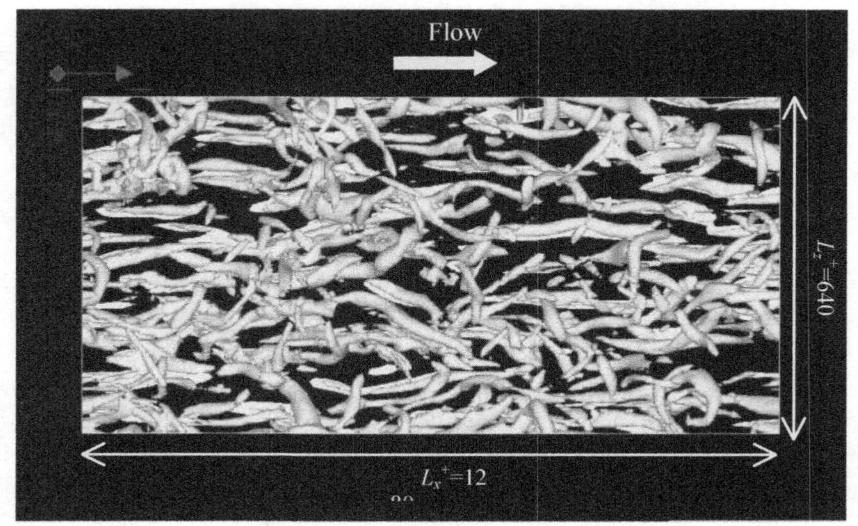

(b) Top view

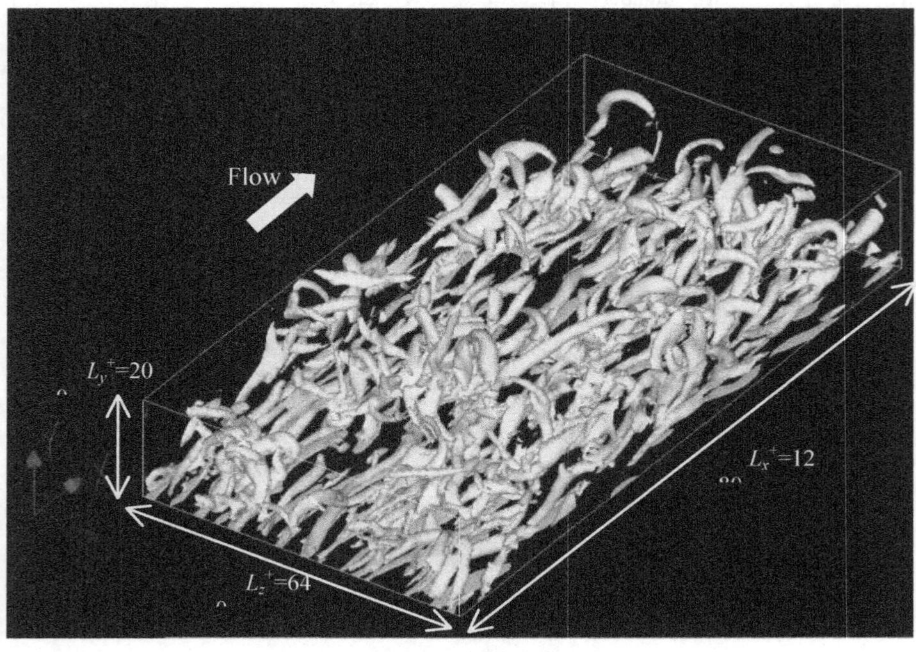

(c) Bird view

Fig. 2 (continue) Surfaces of second invariant velocity gradient tensor Q^+=0.03

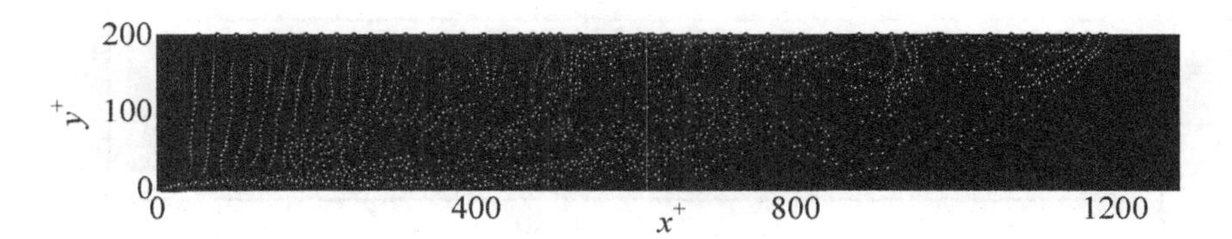

(a) Fluid markers are generated along a line to the *y*-axis

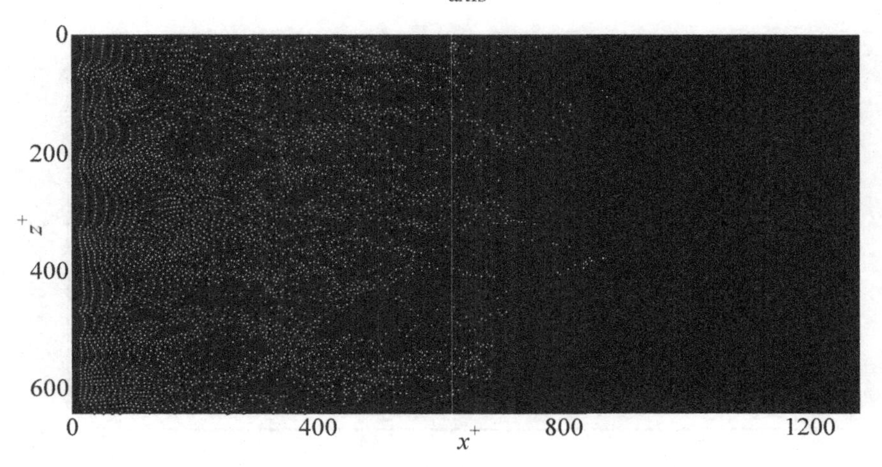

(b) Fluid markers are generated along a line to the *z*-axis
Near bottom wall, y^+=12.35 (Top view)

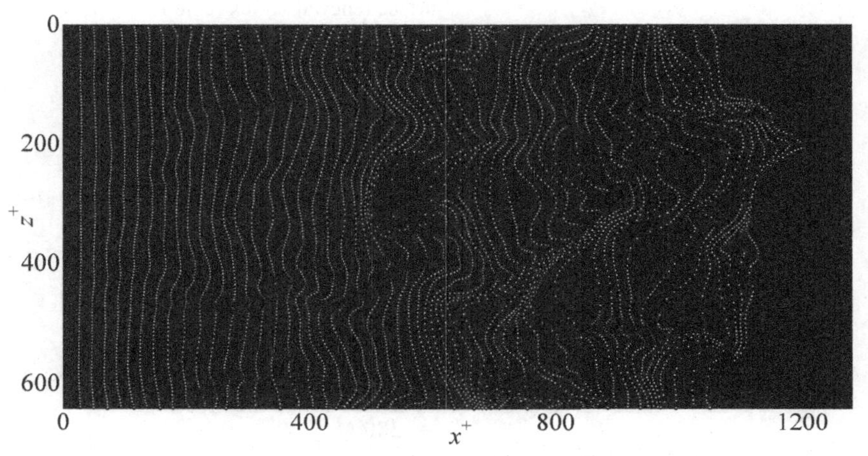

(c) Fluid markers are generated along a line to the *z*-axis
Near free-surface, y^+=194.1 (Top view)

Fig. 3 Visualization of coherent structures

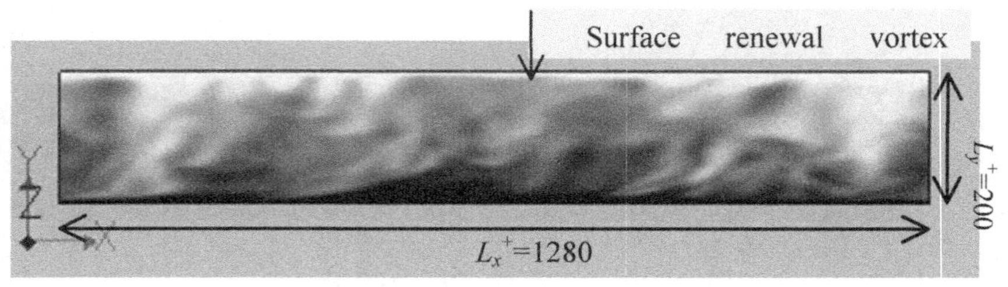

(a) Side view 0.0(Black)<θ^*<1.0(White) $z^+=270$

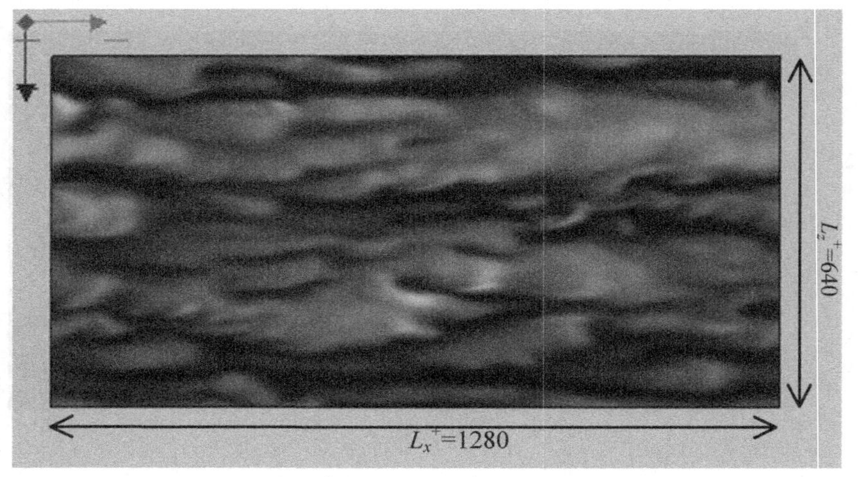

(b) Top view 0.1(Black)<θ^*<0.9(White) $y^+=12.35$ Near bottom wall

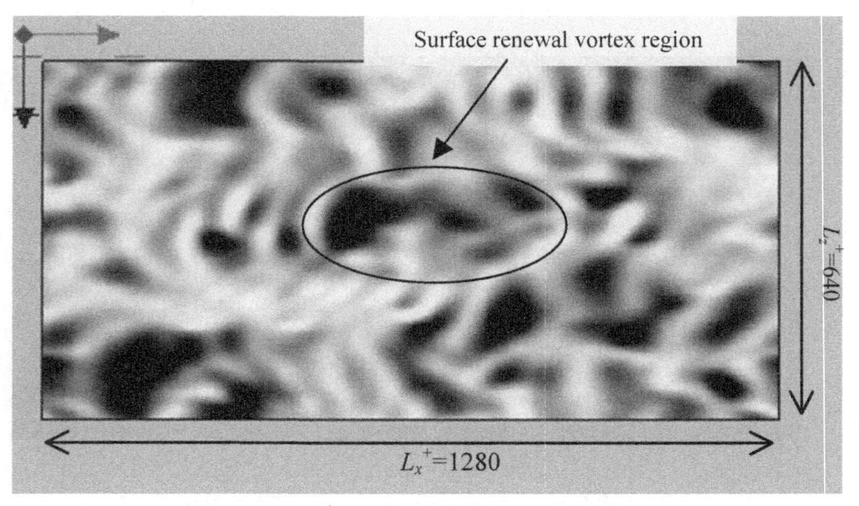

(c) Top view 0.7(Black)<θ^*<1.0(White) $y^+=196$□Near free-surface

Fig.4 Instantaneous scalar fields Pr=1.0

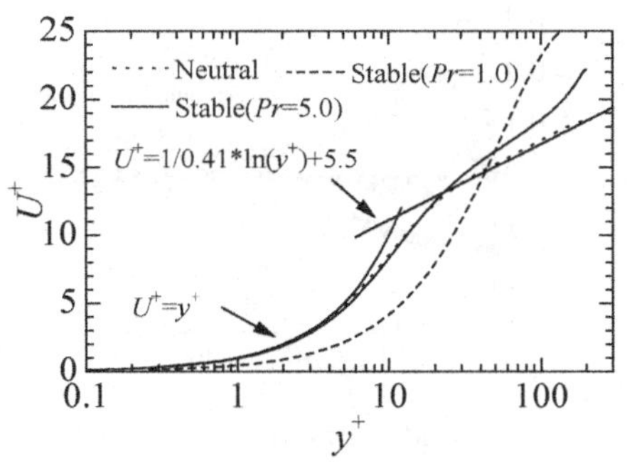

Fig.5 Mean velocity profiles

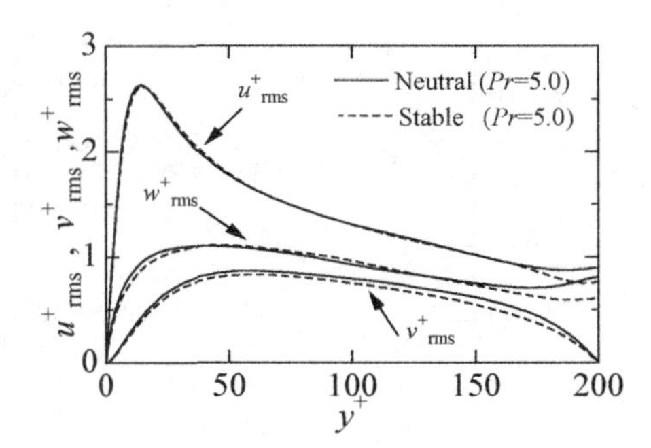

Fig.6 Turbulent intensity profiles

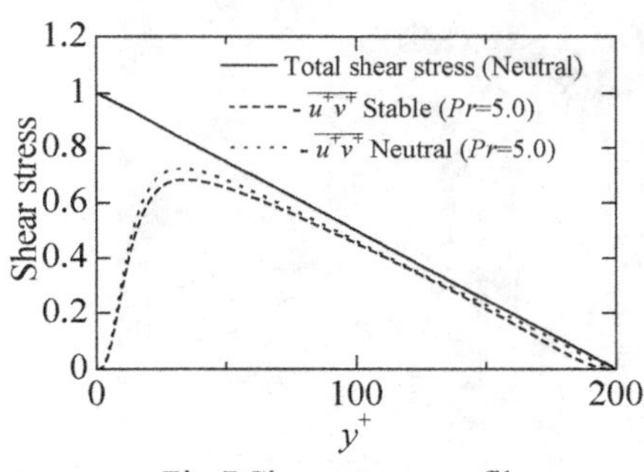

Fig.7 Shear stress profiles

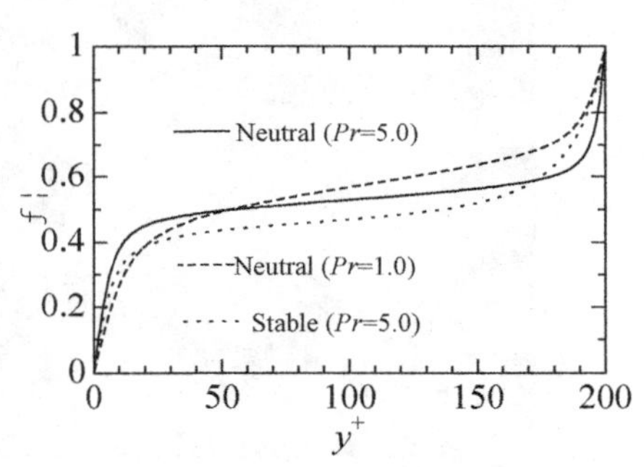

Fig. 8 Mean scalar profiles

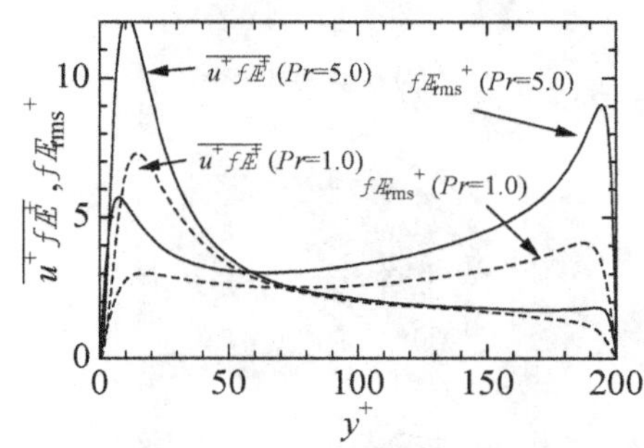

Fig.9 Scalar flux and fluctuation profiles (Neutral stratification)

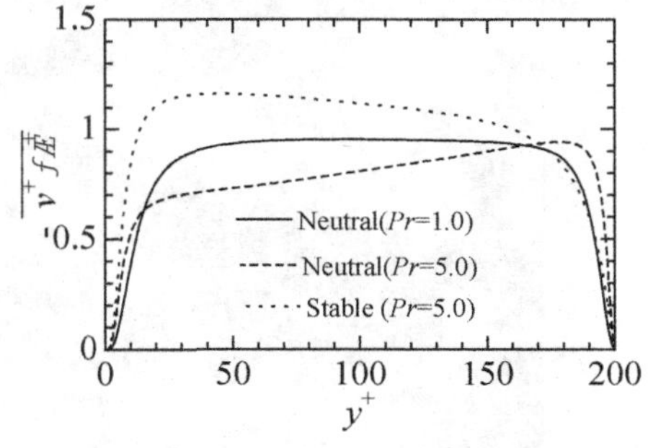

Fig.10 Wall-normal scalar profiles

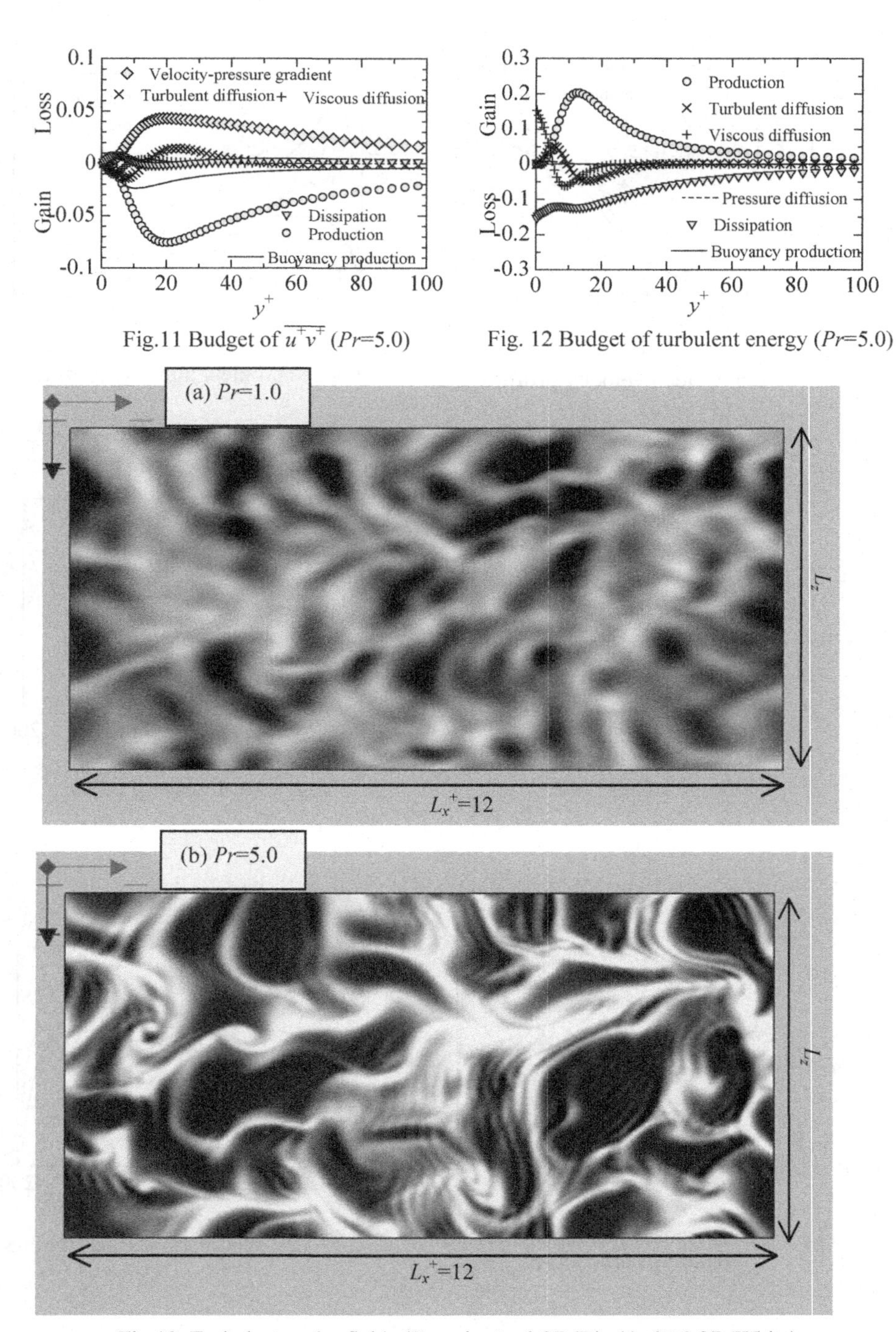

Fig.11 Budget of $\overline{u^+v^+}$ (Pr=5.0)

Fig. 12 Budget of turbulent energy (Pr=5.0)

Fig.13 Turbulent scalar fields (Top view), -0.27 (Black)$<\theta\leq$0.27 (White)

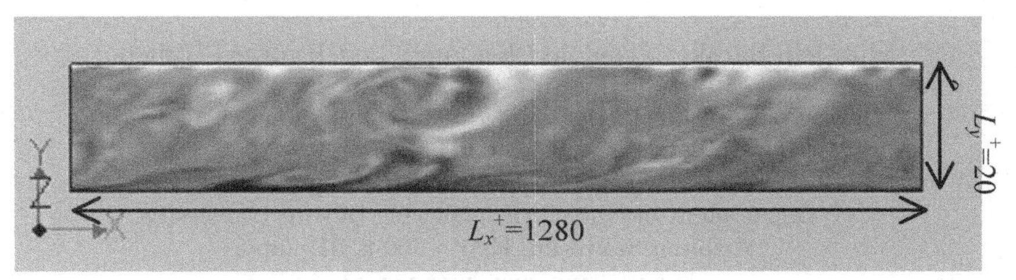

(a) Side view 0.1 (Black) < θ* < 0.9 (White)

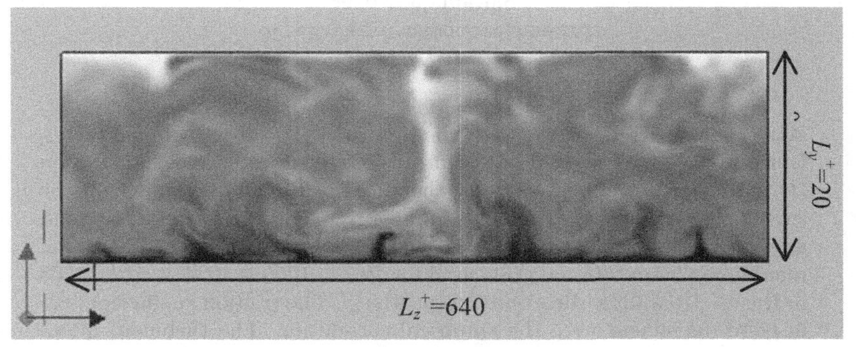

(b) End view 0.1 (Black) < θ* < 0.9 (White)

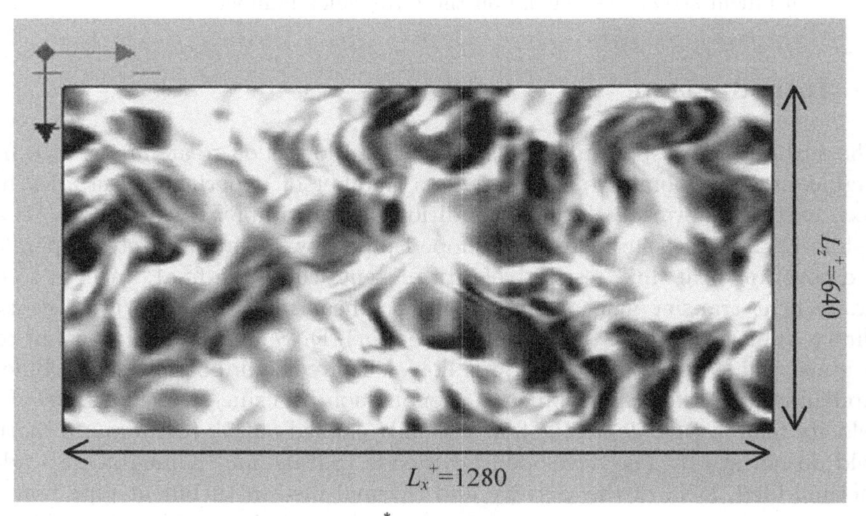

(C) Top view 0.4 (Black) < θ* < 0.9 (White)

Fig.14 Instantaneous scalar fields (Pr=5.0, Neutral stratification case)

High Reynolds Number Computation for Turbulent Heat Transfer in a Pipe Flow

Shin-ichi Satake[1], Tomoaki Kunugi[2], and Ryutaro Himeno[3]

[1] Department of Mechanical and Intellectual Systems Engineering, Toyama University, 3190 Gofuku, Toyama 930-8555, Japan
ssatake@eng.toyama-u.ac.jp
[2] Department of Nuclear Engineering, Kyoto University, Yoshida, Sakyo-ku, Kyoto 606-8501, Japan
kunugi@nucleng.kyoto-u.ac.jp
[3] RIKEN,Dept. of Computer and Information. 2-1 Hirosawa, Wako, Saitama,351-0198
himeno@postoman.riken.go.jp

Abstract. Turbulent transport computations for fully-developed turbulent pipe flow were carried out by means of a direct numerical simulation (DNS) procedure. To investigate the effect of Reynolds number on the turbulent sturcures, the Reynolds number based on a friction velocity and a pipe radius was set to be $Re_\tau = 150, 180, 360, 500, 1050$. The number of maximum computational grids used for $Re_\tau = 1050$ is $1024 \times 512 \times 768$ in the $z-$, $r-$and ϕ -directions, respectively. The friction coefficients are in good agreement with the empirical correlation. The turbulent quantities such as the mean flow, turbulent stresses, turbulent kinetic energy budget, and the turbulent statistics were obtained. It is found that the turbulent structures depend on these Reynolds numbers.

1 Introduction

The turbulent transport mechanism in a pipe flow is of great importance from engineering viewpoint. Up to present, a number of experimental studies have been carried out various heating conditions. For example, Hishida et al. [1] and Isshiki et al.[2] measured fully-developed pipe flow with constant temperature and constant heat flux, respectively. In present study, the final object is to elucidate heat transfer mechanism of constant heat flux in a pipe flow by using Direct Numerical Simulations (DNS, hereafter). DNS has played important roles in numerical study at turbulent pipe flow. Despite the importance of turbulence modelling and studying turbulence phenomenon by using DNS and DNS of the velocity filed in turbulent pipe flow was only carried out[5]; furthermore thermal field do not study. The developed DNS code (Satake and Kunugi,[6],[8]) adopt thermal filed, DNS of the velocity and thermal filed in turbulent pipe flow has been carried out. The heating boundary condition is imposed on the wall with constant heat flux on the circumferencially facing. The spatially distribution of the turbulent structure and scalar flux are presented and studied the transport

M. Valero et al. (Eds.): ISHPC 2000, LNCS 1940, pp. 514-523, 2000.

mechanism of thermal field from scalar flux budgets. In addition, Nusselt number as a macroscopic parameter is discussed as the problem of thermal-hydraulic design compared with experience equation.

2 Numerical Procedure

The DNS code can numerically solve the incompressible Navier-Stokes and continuity equations described in cylindrical coordinate using a second-order finite volume discretization scheme with the radial momentum flux formulation[9]. These equations are integrated in time by using the fractional-step method[10] with Crank-Nicholson and a modified third-order Runge-Kutta scheme[11]. Poisson equation for pressure is adopted for the direct FFT solver. The Poisson equation in Fourier space is written as

$$\frac{1}{r}\frac{\partial}{\partial r}\left(r\frac{\partial \delta\hat{p}}{\partial r}\right) - \left(k_z^2 + k_\phi^2\right)\delta\hat{p} = \frac{1}{2\beta_k \Delta t}\nabla\delta\hat{u},$$

$\mathrm{k}_z^2 = \frac{1}{\Delta z^2}\left[1-\cos\left(\frac{2\pi m}{N_z}\right)\right], m = 0, N_z - 1$

$\mathrm{k}_\phi^2 = \frac{1}{r^2\Delta\phi^2}\left[1-\cos\left(\frac{2\pi n}{N_\phi}\right)\right], n = 0, N_\phi - 1.(1)$

As shown in Fig. 1, the domain decomposition and the transpose algorithms were applied to Eq(1). Watanabe et al. [7]found that this code can be treated up to 6,300,000-grid system on 16 processors of the Fujitsu VPP machine. In case of 16 processors, the efficiency of parallelization for this code was around 46.6 %. In our previous study regarding the turbulent pipe flow (Satake and Kunugi,[6],[8]), this DNS code has been shown in good agreement with the existing DNS results.

3 Computational Conditions

The computational domain of the fully developed turbulent pipe flow is shown in Fig. 2. The number of grid points, the Reynolds number and grid resolutions are summarized in Table 1. To perform a DNS with the highest Reynolds number in this study, a highest grids size of 1024 $\times$ 512 $\times$ 768 is adopted for 84GB main memory for 64 PEs on a vector-parallel computer Fujitsu VPP 700E at RIKEN. A uniform heat-flux was applied to the wall as a thermal boundary condition. Prandtl number of the working fluid was set to be 0.71. Further details of the velocity boundary condition for pipe geometry DNS can be found in Satake and Kunugi [6].

4 Results and Discussions

The friction coefficients and Nusselt numbers are shown in Figs. 3 and 4. All results are excellent agreement with Blausius′s friction law and empirical correlation. The present Nusselt number for Re_τ = 180 obtained is 18.74. Isshiki et al. [2] obtained Nu=20.06. This value is in close agreement with the present result. The values for $Re_\tau = 150$ and 180 are also in good agreement with the empirical correlation by Gnielinski [12]. In comparison with the previous DNS and experimental data, the mean velocity profiles are shown in Fig. 5. At Re_τ =180 ($Re_b = u_b D/\nu$ =5300), the present results are in excellent agreement with the DNS data by Eggels et al. [5]and with the experimental data by Durst et al. [3]. Other present results except Re_τ =1050 also agree with the experimental data obtained by Durst et al. [3]. At $Re_\tau = 1050$ (Re_b =40000), the present DNS is in excellent agreement with those of the experimental data by Laufer [4].

The Reynolds shear stress profiles are shown in Fig. 6. All the total stress show the straight distribution. This is because the time averaging is sufficient. Figures 7 to 9 show the velocity fluctuation for each component. The fluctuations for radial and circumferential direction are energetic, and depends on the Reynolds number. The mean temperature profile is shown in Fig. 10. The results agree with the experimental data obtained by Isshiki [2] at Re_τ =180. At Re_τ =1050, the logarithmic region clearly observed. The distribution is coincident with $\Theta^+ = 2.18 \ln y^+ + 3.0$.

The total heat flux can be obtained as

$$-\overline{u_r'^+ \theta'^+} + \frac{1}{Pr}\frac{\partial \Theta^+}{\partial r^+} = \frac{2}{r^+}\frac{\int_0^{R^+} U_z r^+ dr^+}{U_b^+} \tag{2}$$

The above profile is computed and the result is shown in Fig.11. In the vicinity of the wall, the profile exists over unity owing to the effect of the circumferencial coordinate. Thus, the right hand side of Eq. (2) is multiplied by$\frac{1}{r^+}$. Figure 12 show the root-mean-square temperature fluctuation normalized by the friction temperature. The peak velocities of all slightly increase and move near wall region with increasing Reynolds number.The temperature fluctuation spreads to both the pipe center and the near wall region with increasing Reynolds number.

The correlation coefficients $R_{u_z u_r}$,$R_{u_z \theta}$ and $R_{u_r \theta}$ are defined as

$$R_{u_z u_r} = \frac{\overline{u_z u_r}}{\sqrt{\overline{u_z^2}}\sqrt{\overline{u_r^2}}} \tag{3}$$

$$R_{u_z \theta} = \frac{\overline{u_z \theta}}{\sqrt{\overline{u_z^2}}\sqrt{\overline{\theta^2}}} \tag{4}$$

$$R_{u_r \theta} = \frac{\overline{u_r \theta}}{\sqrt{\overline{u_r^2}}\sqrt{\overline{\theta^2}}}. \tag{5}$$

The cross correlation coefficients are shown in Fig. 13. The $R_{u_z \theta}$ is 0.96 near wall. For previous experiment in a pipe flow, Bremhost and Bullock [13] obtained $R_{u_r \theta} = -0.47$ and $R_{u_z u_r} = -0.43$. These value are in good agreement with the present data in logarithmic region. The close agreement between $R_{u_r \theta}$ and $R_{u_z u_r}$ shows that the mechanism of the wall-normal turbulent heat flux is similar to that of Reynolds shear stress one. These similarities exist owing to the molecular Prandtl number as 0.71.

The turbulent Prandtl number is defined as

$$Pr_t = \frac{\overline{u_z'^+ u_r'^+} \frac{\partial \Theta^+}{\partial r^+}}{\overline{u_r'^+ \theta^+} \frac{\partial U_z^+}{\partial r^+}}. \tag{6}$$

Note that the wall asymptotic value of Pr_t is independent molecular Prandtl number suggested by Antonia and Kim[14]. Kawamura[16] obtained the channel DNS data for various Prandtl numbers. The result indicated that Pr_t is close to 1 in the vicinity of the wall. The present results in Fig. 14 also shows to be unity and in good agreement with the results of other DNS[15][16]. Because the working fluid is air as 0.71.

The budget of the turbulent kinetic energy k^+ is written as

$$0 = \underbrace{-\frac{1}{r^+}\frac{\partial r^+ \overline{k^+ u_r'^+}}{\partial r^+}}_{Turbulent\ diffusion} \underbrace{-\overline{u_r'^+ u_z'^+}\frac{\partial U_z^+}{\partial r^+}}_{Production} \underbrace{-\frac{1}{r^+}\frac{\partial r^+ \overline{p'^+ u_r'^+}}{\partial r^+}}_{Pressure\ diffusion} + \underbrace{\frac{1}{r^+}\frac{\partial}{\partial r^+}\left(r^+ \frac{\partial k^+}{\partial r^+}\right)}_{Viscous\ diffusion}$$

$$\underbrace{-\left[\overline{\left(\frac{\partial u_z'^+}{\partial r^+}\right)^2} + \overline{\left(\frac{1}{r^+}\frac{\partial u_z'^+}{\partial \phi}\right)^2} + \overline{\left(\frac{\partial u_z'^+}{\partial z^+}\right)^2} + \overline{\left(\frac{\partial u_r'^+}{\partial r^+}\right)^2} + \overline{\left(\frac{1}{r^+}\frac{\partial u_r'^+}{\partial \phi} - \frac{u_\phi'^+}{r^+}\right)^2}\right.}_{Dissipation}$$

$$\underbrace{\left. + \overline{\left(\frac{\partial u_r'^+}{\partial z^+}\right)^2} + \overline{\left(\frac{\partial u_\phi'^+}{\partial r^+}\right)^2} + \overline{\left(\frac{1}{r^+}\frac{\partial u_\phi'^+}{\partial \phi} + \frac{u_r'^+}{r^+}\right)^2} + \overline{\left(\frac{\partial u_\phi'^+}{\partial z^+}\right)^2}\right]}_{Dissipation} . \tag{7}$$

The budgets of the above equation are shown in Fig. 15. The production and dissipation terms are dominant near wall region. The peak location of the production term slightly move to the wall with increasing of Reynolds number.

The budget of temperature variance $k_\theta = \theta'^{+2}/2$ is derived as

$$0 = \underbrace{-\frac{1}{r^+}\frac{\partial r^+ \overline{u_r'^+ \theta'^{+2}/2}}{\partial r^+}}_{Turbulent\ Diffusion} \underbrace{-\overline{u_r'^+ \theta'^+}\frac{\partial \Theta}{\partial r^+} + \overline{u_z'^+ \theta'^+}\frac{\partial \langle T \rangle^+}{\partial z^+}}_{Production}$$

$$\underbrace{+\frac{1}{Pr}\frac{1}{r^+}\frac{\partial}{\partial r^+}\left(r^+ \frac{\partial \overline{\theta'^{+2}/2}}{\partial r^+}\right)}_{Viscous\ Diffusion} \underbrace{-\frac{1}{Pr}\left\{\overline{\left(\frac{\partial \theta'^+}{\partial z^+}\right)^2} + \overline{\left(\frac{\partial \theta'^+}{\partial r^+}\right)^2} + \overline{\frac{1}{r^{+2}}\left(\frac{\partial \theta'^+}{\partial \phi}\right)^2}\right\}}_{Dissipation} . \tag{8}$$

The budget of these terms are shown in Fig. 16. In the viscous sublayer, the molecular diffusion and the dissipation are dominant and are increased with increasing Reynolds number.

The peak locations of the production term are almost the same for $Re_\tau =$ 360, 500, 1050. Note that the peak asymptotic value of turbulent production is 0.25 in the high Reynolds number flows. In Fig. 17, the result for $Re_\tau = 1050$ indicated that the peak asymptotic value of turbulent production is close to 0.25 at y^+ =15.

Figure 18 shows the contour of the low speed streaky structures (u^+ <-3.5)and high temperature region (θ^+ >3.5). They are normalized by ν and u_τ. The volume visualized is obtained as cutting volume (z^+ =985,$y^+ = R^+ - r^+$ =1050,$r^+\phi$=536) for Re_τ =1050 from full computational volume (L_z^+ =15750, $y^+ = R^+ - r^+$ =1050, $L_r^+\phi$=6597). The width of the large streaky structures is larger than $r^+\phi$=100 and located at away from the wall. A few streaks merged as "plate-like structures" at y^+=200 from the wall. The velocity and scalar fields shows strong analogy. The many tube like structures and large streaky structures exist in this volume. The second invariant of velocity gradient tensor (Q^+ <0.008), the ejection ($-U_z^+V^+$ <0, V^+ >0, $V^+ = -u_r^+$) and the sweep ($-U_z^+V^+$ <0, V^+ <0, $V^+ = -u_r^+$) for $Re_\tau = 1050, 500, 360, 180$ in the cross section perpendicular to the circumferential direction are shown in Fig. 19 (a)-(d), respectively. In low Reynolds number (Re_τ =180), the ejection and the sweep are observed very small region and located around vortical structures. However, in high Reynolds number ($Re_\tau = 360, 500, 1050$), the structures seem to be rather different from one in low Reynolds number. A characteristic sizes of the ejection and the sweep are even larger than the half of the pipe radius. Almost large structures located in y^+ >200 correspond to the wake region in the mean velocity profile. In the region, the scales of fluid motion are different from that of the near wall region.

5 Conclusion

DNS on a turbulent pipe flow with heated wall was carried out for five Reynolds numbers. The present results are in good agreement with the previous experimental data. The temperature bariance and the scalar-flux budget terms at Re_τ =1050 obtained are most enegetic close to the wall. From the numerical visualization results, it is found that the velocity streaks are merged at y^+=200. More details temperature vizualization result for Re_τ =1050 will be reported in the presentation.

References

1. Hishida, M., Nagano, Y. and Tagawa, M. Transport processes of heat and momentum in the wall region of turbulent pipe flow , *Proceedings of the 8th International Heat Transfer Conference, Tien, C.L. et al., Eds. Hemisphere Publishing Corp., Washington DC*, Vol.3, pp. 925-930, 1986.
2. Isshiki, S., Obata, T., Kasagi, N. and Hirata, M. *An experimental study on heat transfer in a pulsating pipe flow (1st report, time-averaged turbulent characteristics), Bulletin JSME Vol.59,* pp. 2245–2251 1993.

3. Durst, F., Kikura, H., Lekakis, I., Jovanovic and Ye, Q., Wall shear stress determination from near-wall mean velocity data in turbulent pipe and channel flows," *Experiments in Fluids* Vol.20, pp. 417–428,1996.
4. Laufer, J. ,The structure of turbulence in fully developed pipe flow, *NACA report 1174*,1954.
5. Eggels, J.G.M., Unger, F., Weiss, M.H., Westerweel, J. Adrian, R.J., Friedrich, R., and Nieuwstadt, F.T.M., Fully developed turbulent pipe flow: comparison between direct simulation and experiment, *J. Fluid Mech., Vol. 268*, pp. 175–209,1994.
6. Satake, S. and Kunugi, T. *Direct numerical simulation of turbulent pipe flow, Bulletin JSME Vol.64(in Japanense)*, pp. 65–70,1998.
7. Watanabe, H., Satake, S. and Kunugi, T. 1997, Proc. of 11 th Computational Fluid Dynamics Conf., in Tokyo, pp. 129-130.
8. Sakate, S. and Kunugi, T. *Direct numerical simulation of an impinging jet into parallel disks, Int. J. Numerical Methods for Heat and Fluid Flow*, 8, 768–780, 1998.
9. Verzicco, R. and Orlandi, P. A finite-difference scheme for three-dimensional incompressible flows in cylindrical coordinate ,*J. Comp.,Phys., Vol.123*, pp. 402–414, 1996.
10. Dukowicz, J. K. and Dvinsky, A. S. *Approximate factorisation as a high order splitting for the implicit incompressible flow equations, J. Comp.,Phys., Vol.102, No.2,*, pp. 336–347, 1992.
11. Spalart, P.R.., Moser, R. D. and Rogers, M.,M. Spectral methods for the Navier-Stokes equations with one infinite and two periodic directions, *J. Comp. Phys., 96*, pp. 297–324, 1991.
12. Glielinski, V., New equations for heat and mass transfer in turbulent pipe and channel flow, *Int. Chem. Eng., Vol. 16, No.2*, pp. 359–367, 1976.
13. Bremhorst, K. and Bullock, K.J., Spectral mesurement heat and momentum transfer in fully developed pipe flow, *Int. J. Heat and Mass Transfer, Vol.16*, pp. 2141-2154, 1973.
14. Antonia, R. and Kim, J., Turbulent Prandtl number in the near-wall region of a turbulent channel flow, *Int. J. Heat and Mass Transfer, Vol.34*, pp. 1905-1908, 1991.
15. Kasagi, N., Tomita, Y. and Kuroda, A. , Direct numerical simulation of passive scalar field in a turbulent channel flow, *Transaction ASME Journal Heat transfer*, Vol.114, pp. 598–606, 1992.
16. Kawamura, H., Ohsaka, K., Abe, H. and Yamamoto, K., DNS of turbulent heat transfer in channel flow with low to medium-high Prandtl number fluid,*Int. J. Heat and Fluid Flow, Vol.19*, 482-491, 1998.

Table 1. Computational condition

Case	Grid numbers	Δz^+	$R^+\Delta\phi$	Δr^+_{wall}	Δr^+_c
$Re_\tau = 150$	$256 \times 128 \times 128$	8.78	7.36	0.24	0.86
$Re_\tau = 180$	$256 \times 128 \times 128$	10.5	8.84	0.29	1.04
$Re_\tau = 360$	$384 \times 256 \times 256$	14.0	8.83	0.11	1.1
$Re_\tau = 500$	$512 \times 384 \times 384$	14.6	8.18	0.1	2.6
$Re_\tau = 1050$	$1024 \times 512 \times 768$	15.4	8.59	0.163	4.16

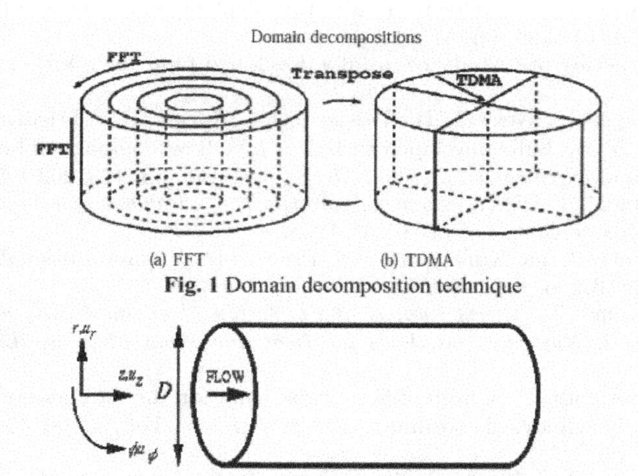

Fig. 1 Domain decomposition technique

Fig. 2 Computational domain

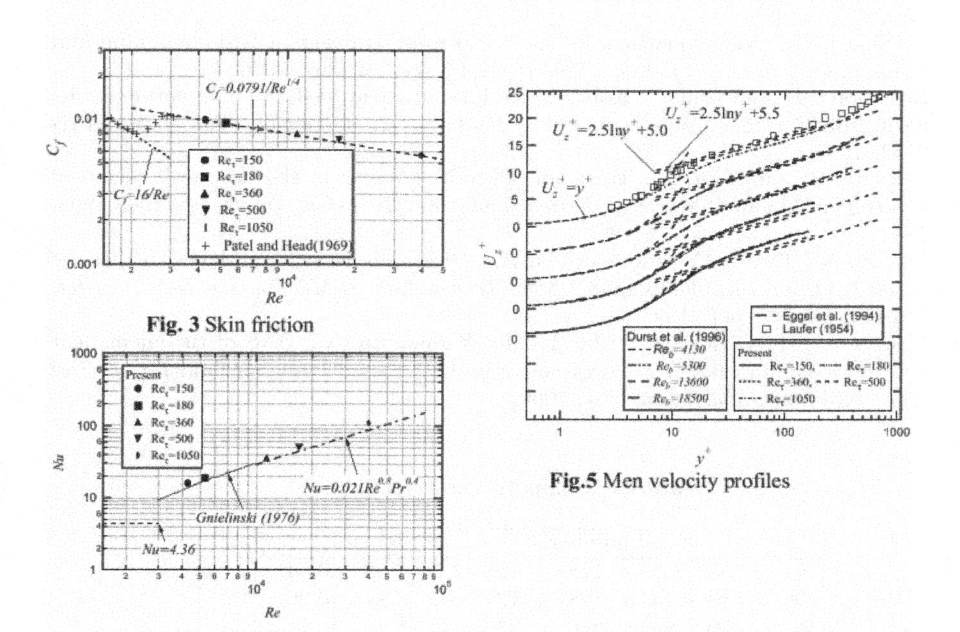

Fig. 3 Skin friction

Fig.5 Men velocity profiles

Fig. 4 Nusselt number

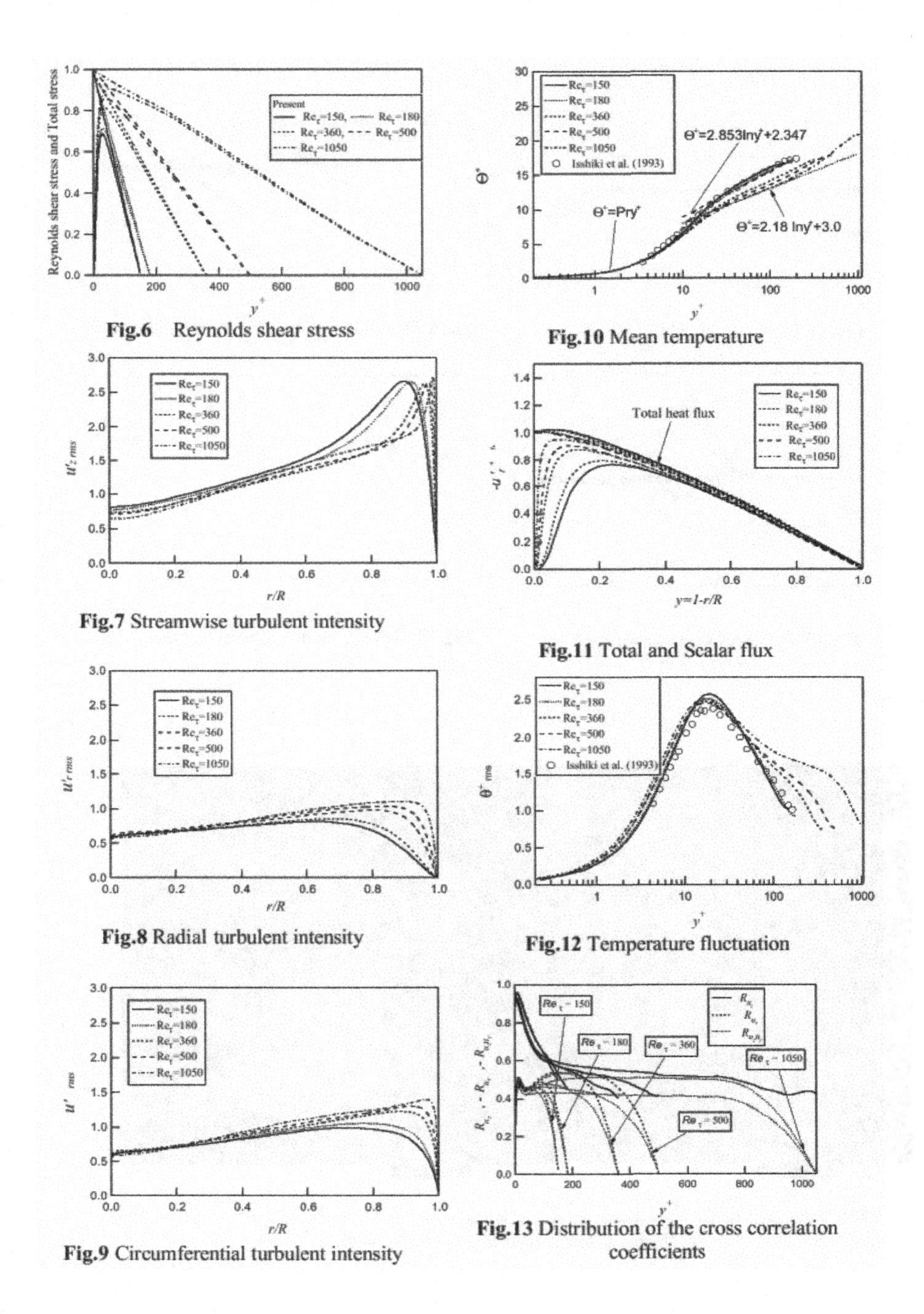

Fig.6 Reynolds shear stress

Fig.10 Mean temperature

Fig.7 Streamwise turbulent intensity

Fig.11 Total and Scalar flux

Fig.8 Radial turbulent intensity

Fig.12 Temperature fluctuation

Fig.9 Circumferential turbulent intensity

Fig.13 Distribution of the cross correlation coefficients

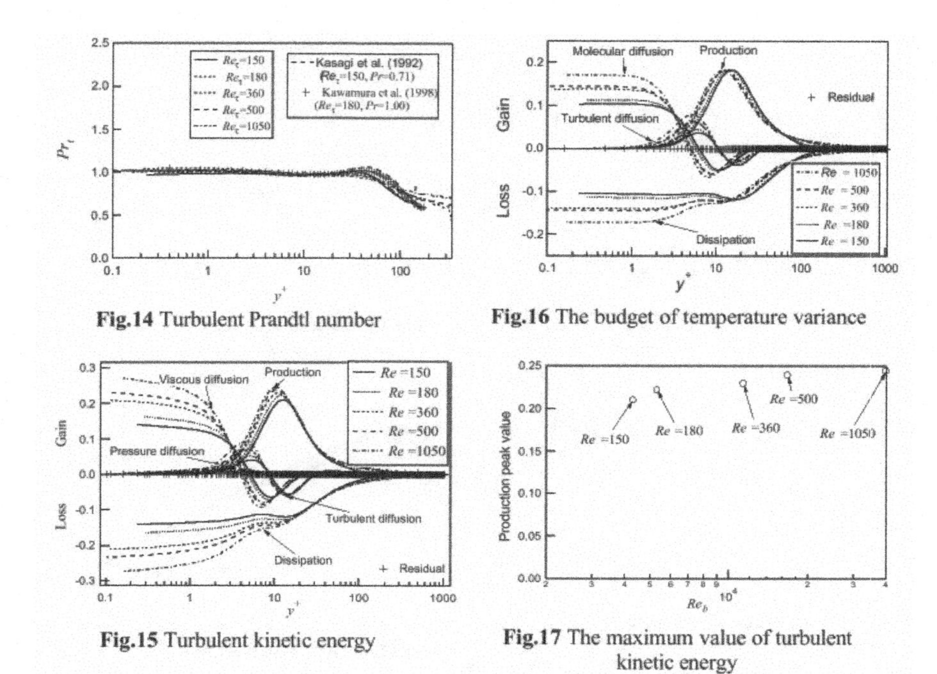

Fig.14 Turbulent Prandtl number

Fig.16 The budget of temperature variance

Fig.15 Turbulent kinetic energy

Fig.17 The maximum value of turbulent kinetic energy

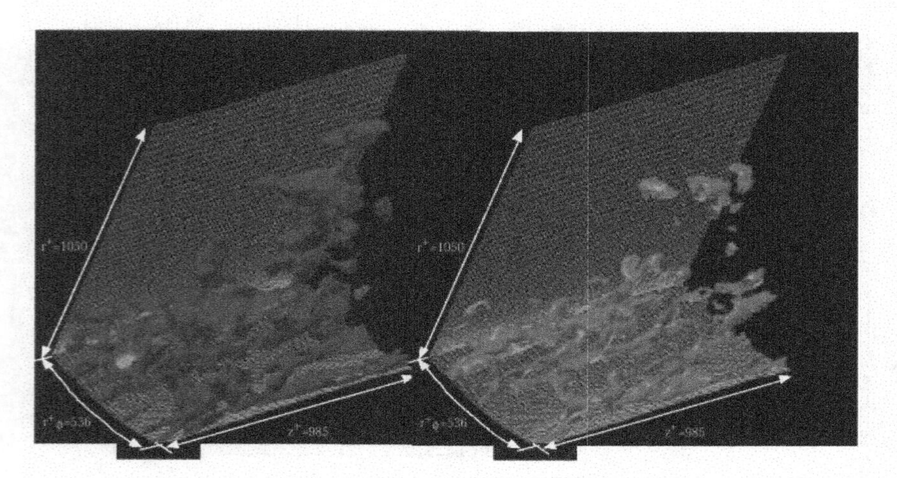

Fig. 18 the contour of the low speed streaky structures (u^+<-3.5); blue and the contour of the high temperature structures ($^+$>3.5); Red

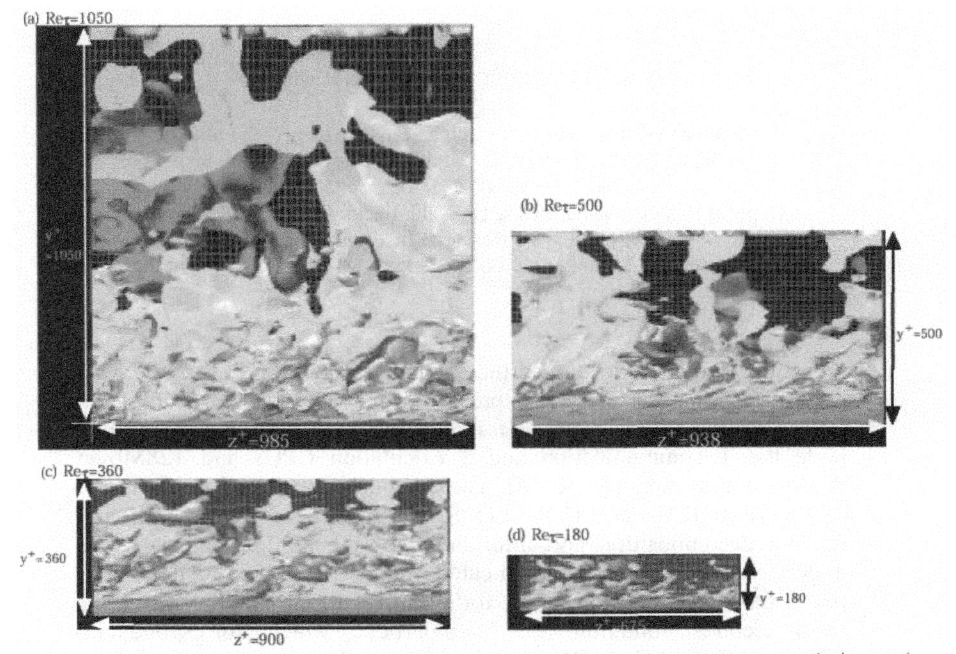

Fig.19 The second invariant of velocity gradient tensor (Q^+<0.008), the ejection ($U_z^+V^+<0$, $V^+>0$, $V^+=-u_r^+$) and the sweep ($U_z^+V^+<0$, $V^+<0$, $V^+=-u_r^+$) for Re =180,360,500,1050 in the cross section perpendicular to the circumferential direction

Large-Scale Simulation System and Advanced Photon Research

Yutaka Ueshima[1] and Yasuaki Kishimoto[2]

[1] Advanced Photon Research Center, Kansai Research Establishment,
Japan Atomic Energy Research Institute, Osaka, Japan
ueshima@apr.jaeri.go.jp
[2] Dept. of Fusion Plasma Research, Naka Fusion Research Establishment,
Japan Atomic Energy Research Institute, Ibaraki,Japan
kishimoy@fusion.naka.jaeri.go.jp

Abstract. We have developed the Progressive Parallel Plasma (P^3) 2D code tuned for the massively parallel scalar computer 'Intel Paragon XP/S 75MP834'. The computer is composed of 834 nodes and each node has 1 communication and 2 calculation CPUs and 128Mbyte random access memory (RAM). The computer has a total performance of 100GByte RAM and 120GFLOPS. The scheme for parallelization is domain decomposition and information of the particles crossing the node's boundary is communicated to 2 neighbor nodes. The performance of the calculation for plasma particle simulations is 53 nano seconds/simulation time step/particle, which corresponds to effectively 42GFLOPS. By using the P^3 2D code, a simulation including 1 Giga particles can be completed within a few days.

1 Introduction

The development of short-pulse ultra high intensity lasers[1] has opened a new regime in the study of laser-plasma interaction. Depending on the type of matter and laser parameters, various photon generation and particle acceleration mechanisms have been invoked in the different regimes of the laser plasma interaction. Recently, there has been a great deal of research devoted to generation of higher harmonics and X-ray by ultra high intensity lasers. Strong ion acceleration like as the Coulomb explosion [2] is associated with the break of the plasma quasineutrality when the electrons are expelled from a self-focusing radiation channel in the plasma after which the ions expand due to the repulsion of the noncompensated electrical charge [3]. The Coulomb explosion has also been invoked in order to describe the generation of fast ions during the interaction of high intensity laser pulses with clusters [4].

In the interpretation of the experimentally observed acceleration of the ions, it was assumed in Ref. [5] that the ions move radially with respect to the channel under the effect of the Coulomb explosion. However, the generation of fast ions not only radially but also in forward direction were observed in 2D Particle In Cell (PIC) simulations [6]. In the case of an overdense plasma the role of the channel is taken as

M. Valero et al. (Eds.): ISHPC 2000, LNCS 1940, pp. 524-534, 2000.

the hole bored by the laser pulse. In such plasmas, in addition to the plasma expansion in vacuum mentioned above and to the ion expulsion in the transverse direction due to the self-focusing channel, we also notice ion acceleration in the plasma resonance region [6] and forward ion acceleration if the laser pulse interacts with a thin foil [7]. The latter results were obtained via PIC simulations in the framework of a one dimensional planar model which is valid as long as the transverse size of the laser pulse is much larger than the acceleration length. Since in a one-dimension planar model the electrostatic potential diverges as the width of the ion cloud increases, we must perform at least two-dimensional simulations. In addition, an ultraintense laser pulse in a near-critical density plasma and in an overdense plasma, is subject to relativistic self focusing, the description of which also requires at least two dimensional PIC simulations.

The Numerical study of the interaction of an ultraintense laser with matter needs to be performed in at least two-dimensional simulations. However the computational cost of such a simulation is very high. For example, in the case of electron density of $6.2*10^{23}$ cm^{-3}, ie. Al $^{10+}$, the mesh size must be smaller than 2nm because the skin depth is 7nm. In a few 10's of femto seconds, ions and electrons expand to around 10μm from the irradiation point of the matter. Therefore, a 10000*10000=0.1G mesh simulation is required and 1 Giga particles must be contained in a simulation box to investigate properties of the accelerated ions and electrons, i.e. energy and spatial distribution etc.

2 Intel Paragon 75MP834

Parallel calculations with the use of over hundreds computers were performed in Japan since several years ago. The Intel Paragon XP/S 75MP834 <Kansai Research Establishment> and 15GP256 <Naka Fusion Research Establishment> were introduced as pioneers in Japan Atomic Energy Research Institute for the purpose of the massively parallel calculations for advanced photon and fusion research. Recently, a lot of parallel programs have been transplanted and newly produced to perform the parallel calculations with the computers. Therefore, there are some seeds of trouble in the massive parallel computing. When programs are developed under different computer and operating system, prudent directions and knowledge are needed. However, integration of knowledge and standardization of environment are quite difficult because of number of Paragon system. There are a few codes for massive parallel computing.

We have developed the P^3 2D Code tuned up for the massively parallel scalar computer 'Intel Paragon XP/S 75MP834'. The computer is composed of 834 nodes and every node has 1 communication and 2 calculation CPUs and 128Mbyte RAM. The Paragon has 26 I/O nodes independent of 834 calculation nodes. Each I/O node has 64Mbyte RAM and the effective I/O ability of 2.6Mbyte/s. Note that every I/O node cannot accept more than 32 I/O requests. The parallel file system (pfs) is constructed with 16 I/O nodes. The computer has total performance of 100GByte RAM and 120GFLOPS. The scheme for parallelization is domain decomposition and information of particles crossing the node's boundary is communicated to 2 neighboring nodes. The performance of the calculation for plasma particle simulations is 53 nano seconds/simulation time step/particle, which corresponds to effectively 42GFLOPS. By using the P^3 2D code, a simulation including 1 Giga particles can be completed within a few days.

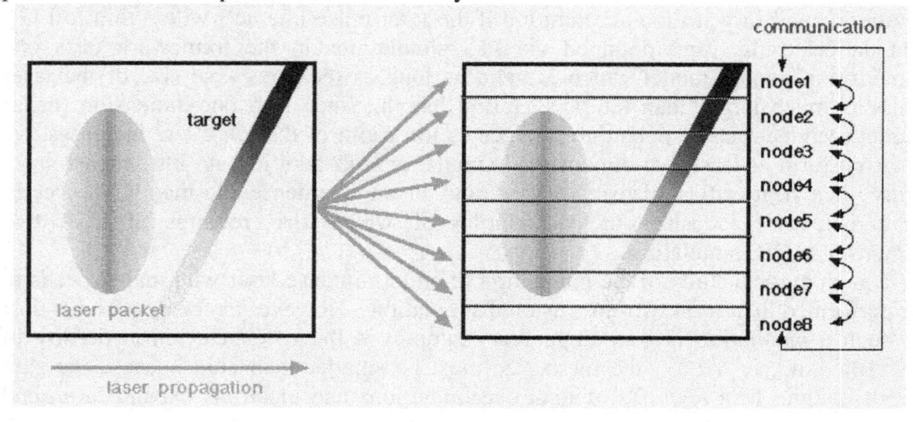

Fig. 1 The scheme for parallel, one dimensional domain decomposition and communication to 2 neighbor nodes

3 Progressive Parallel Plasma (P^3) 2D Code

To perform the simulation on a quite large scale, several techniques are used in the P^3 code. The first is a technique to use RAM effectively. The technique is called as "variable dimension", which is Fortran's programming in the old style. As shown in a Program.1 and Fig.2, a large matrix is defined in the main program and is divided into a lot of variable pieces in the each subroutine. The method has some advantages for RAM management, for example, saving of the needed RAM and systematically understanding the utilization of RAM. By the technique, scale of calculation is improved up to 300% and 1Giga-particle simulation has been realized.

```
dimension AT(1:4000)
ip1=1
ip2=1000
ip3=2000
ip4=3000
ip5=2000
ip6=2500
ip7=3200
ip8=3500
call sub1(AT(ip1),AT(ip2),AT(ip3),AT(ip4))
call sub2(AT(ip1),AT(ip2),AT(ip5),AT(ip6),AT(ip7))
stop
end
--------------------------------------------
```

```
subroutine sub1(FM1,FM2,TM3,TM4)
dimension  FM1(*),FM2(*),TM3(*),TM4(*)
      •
      •
return
end
-------------------------------------------
subroutine sub2(FM1,FM2,TM5,TM6,TM7)
dimension  FM1(*),FM2(*),TM5(*),TM6(*),TM7(*)
      •
      •
return
end
```

Program 1 Method of the variable dimension

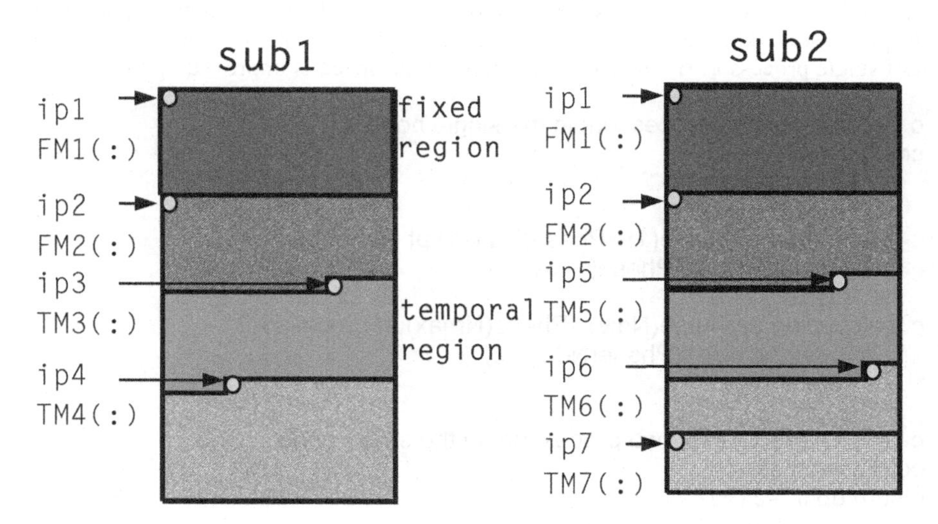

Fig. 2 Illustration of RAM mapping with the variable dimension

The second is a technique to prevent cash miss-hits. In the previous large-scale simulation, the programming for a vector processor was adopted. Because the Paragon has massively scalar processors, the programming is changed into that of a scalar processor as following,

```
Vecotr programming                 Scalar programming
do j = 1,Nmax                     do j = 1,Nmax
   kx = x(j)                      kx = phase(1,j)
   ky = y(j)                      ky = phase(2,j)
   ux = Px(j)                      ux = phase(3,j)
   uy = Py(j)                      uy = phase(4,j)
   uz = Pz(j)                      uz = phase(5,j)
      :                                 :
      :                                 :
enddo                            enddo
```

Program 2 Comparison of programs for vector and scalar computers

By the technique, calculation speed is improved up to 45%.

The third is a technique to parallelize two processors in the single node. As shown in Program.3, the calculations, especially access to RAM, in two processors should be independent each other. This is a unique technique for the Paragon MP series, although the technique may be used in the computers which have common RAM with multi-processors. The sample program is accelerated to about two times faster than that of single processor. By the technique, calculation speed is improved up to 45%.

```
c directive for two processors in the single node
cdir$l cncall
     do iprocessor =1,2
      if(i.eq.1) then
c  calculation of Phase(1)□Phase(Ndiv-1) at processor1
        call particle1(Phase(1),J1)
      else
c  calculation of Phase(Ndiv)□Phase(Nmax) at processor2
        call particle2(Phase(Ndiv),J2)
      end if
     enddo
c cancell directive for two processors in the single node
cdir$l nocncall
c sum up J1=J1+J2
     call sumup(J1,J2)
```

Program 3 Parallelization for two processors in the single node

The forth is a technique to parallelize I/O from a lot of calculation's nodes. The technique is based on Nx-library[8,9], which is made by INTEL for parallel computing. The Paragon XP834/MP has 834 calculation nodes and 26 I/O nodes. Each I/O node has 64Mbyte RAM and a speed of 2.6Mbyte/s. Every I/O node cannot accept more than 32 I/O requests. The parallel file system (pfs) is constructed with 16 I/O nodes. The I/O speed is achieved 40Mbyte/s with 16 splitting pfs and parallel writing at 800 node by the following program,

```
c irw=(number of nodes)/(number of I/O requests)
  irw=800/16
c gopen(logical unit,file path, option)
  call gopen(id,file_path,M_RECORD)
  do i= 1, irw
c control of 16 I/O requests in the I/O node.
      if ( mod(nodenumber,irw).eq.ir-1 )  then
c cwrite(logical unit,output variable, output data size)
        call cwrite(id, Phase, idatasize)
     end if
c barrier
     call gsync()
enddo
```

Program 4 Parallelization for I/O in the massively node

In the ordinary PIC simulation, a correction of electricfield by the fast Fourier transform (FFT) was needed to satisfy the Coulomb's law in the Maxwell's equations. However, the cost of needed RAM size and communication traffic is large with the use of the FFT. In the P^3 code, a rigid local solver [10] is adopted to overcome the difficulty.

As a result, the calculation speed of 53ns/step/particle is achieved in 800 nodes and the acceleration rate as a index of parallel computing is about 800. This is corresponding to 42GFLOPS, which is 35% of the maximum hardware specification,120GFLOPS. By the way, the calculation speeds in NEC-SX4-16CPU and Fujitsu-VPP300-1CPU are 340ns/step/particle and 4660ns/step/particle, respectively.

To simulate an interaction of a real solid with a relativistic laser, ionization, collision and radiation processes need to be calculated. In the code, ionization and collision processes were simulated by Monte-Carlo method. Bremsstrahlung and Larmor radiation were estimated by post process.

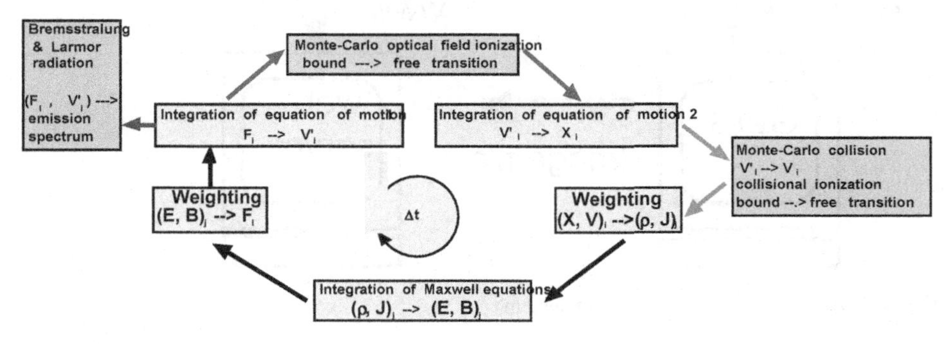

Fig. 3 The flow diagram in P^3 2D code

In the large-scale simulation, real-time visualization and steering system is thought as hopeful method of data analysis. This approach is valid in the fixed analysis at one time. In the simulation research for an unknown problem, it is necessary that the output data can be analyzed many times because profitable analysis is difficult at the first time. Consequently, output data should be filed to refer and analyze at any time. The pseudo-real-time visualization system is equipped in the P^3 code. The P^3 support system has the followed automatic functions,

1) make directory in the Paragon, file server and graphical work station
2) transport files from the Paragon to the file server
3) create CLI or V scripts of AVS5 or AVS Express
4) execute of AVS
5) convert image format and re-arrange them

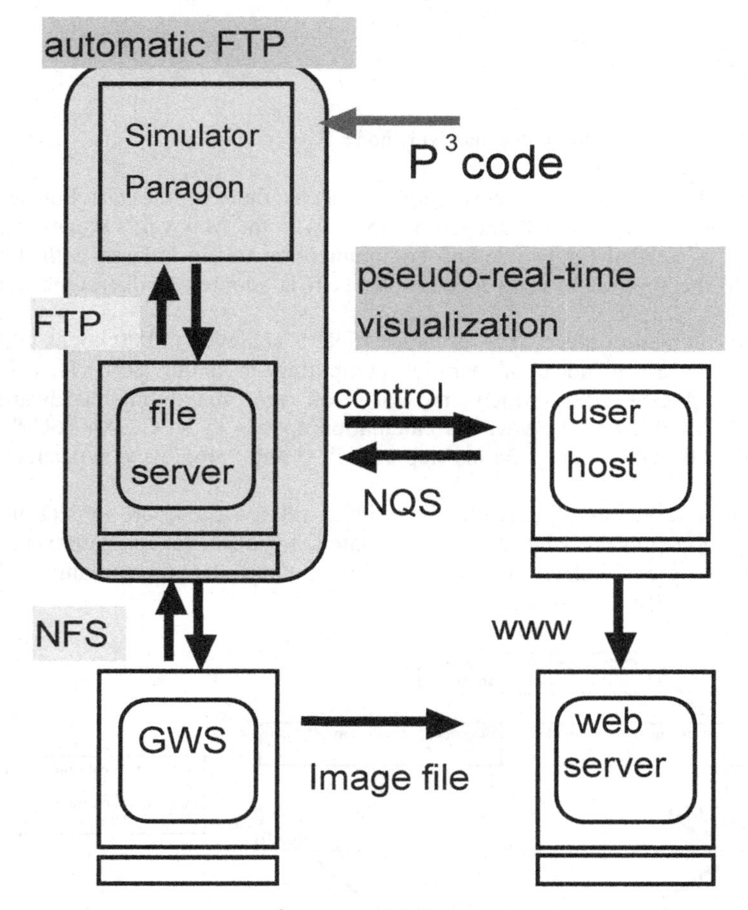

Fig. 4 Illustration of the P^3 support system

In the future work, we are going to add to the P^3 support system the following functions,

1) the files automatically file backup and restore
2) link database server and manage data and image files

4 Higher Harmonic Generation by Interaction of The Thin Solid Hydrogen with Use of The P^3 Code and Its Support System

Interaction of the thin solid hydrogen has been studied by the P^3 code and the support system. The simulation has be performed with the code in the condition as low temperature, (10 ~ 100eV) and higher harmonics (keV X-ray) observation. The laser pulse is gaussian along the x and y axes with full width half maximum 4μm. The ion density corresponds to the real solid hydrogen, $6.2*10^{23}$ cm^{-3}. The laser condition is set to be as shown in Fig.5. Figs.6 and 7 are the result of the 1G particle simulation of a interaction with relativistic laser a_0=10 with the P^3 2D code. Figs.6 and Fig.7 show the time variation of the electric field in the polarization direction and the spectrum (kx,ky) of the transmitted light back of the thin foil, respectively. In Fig.5, the directions of surface and depth of the foil are perpendicular and horizontal, respectively.

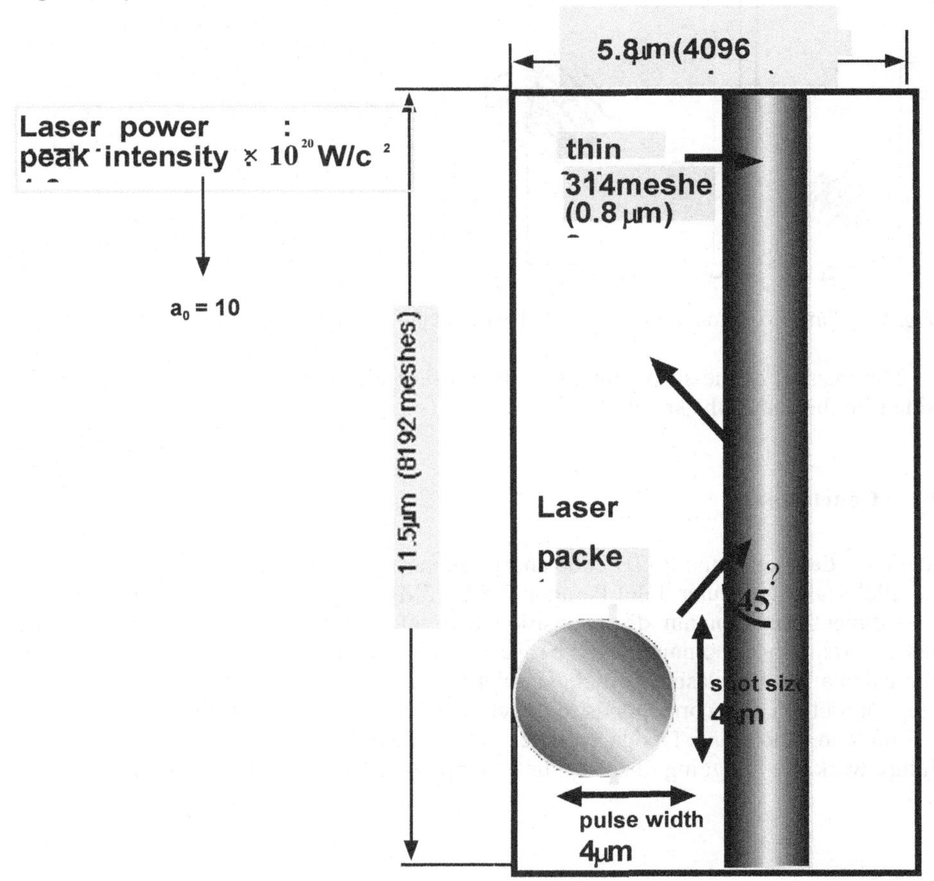

Fig. 5 Illustration of the geometry for the simulation with a hydrogen.

Fig.6 shows that the higher harmonics (~$10\omega_L$) in corresponding to the plasma frequency are generated strongly through the thin solid target. The strength of the higher harmonic field is a quite strong, which is tenth times as larger as that of the laser. As shown in Figs.7, higher harmonics below the plasma frequency are strongly suppressed in the transparent light. Only higher harmonics over twice as large as the plasma frequency are transmitted from the back of the foil. Separately from emissions from atomic processes induced by the intense laser, the radiation can become short-pulse and intense because of direct process in the interaction with the laser. The pulse length of the radiation is the same order as that of the laser. The result indicates that higher harmonics are generated in front of the foil. In fact, The spectrum of the reflected light in the front of the foil do not have the suppression.

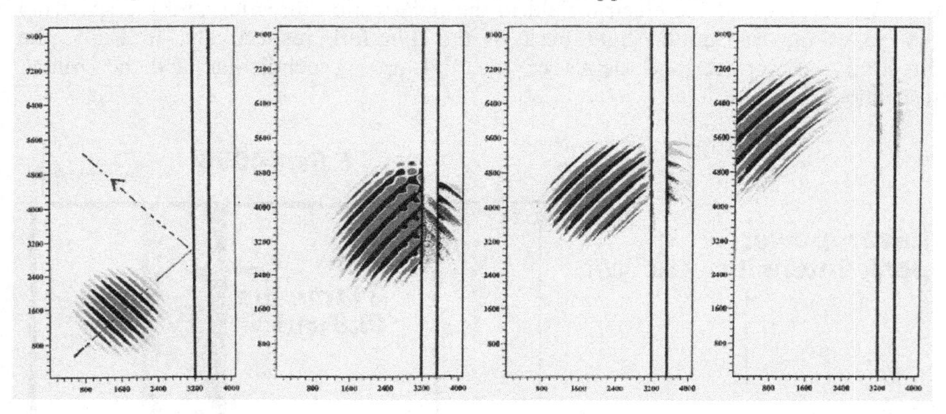

Fig. 6 Time evolution of the electric field (Ey) in the polarized direction.

The interval of the snap shots is 12 femto seconds, and the snap shot on the right side is at the end of the simulation.

5 Conclusions

We have developed the P^3 2D code and its support system tuned up for the massively parallel scalar computer 'Intel Paragon XP/S 75MP834'. The scheme for parallel was one dimensional domain decomposition and information of particles crossing the node's boundary is communicated between 2 neighboring nodes. The performance of the calculation for plasma particle simulations is 53 nano seconds / simulation time step /particle, which corresponds to effectively 42GFLOPS. By using the P^3 2D code, a simulation including 1 Giga particles can be completed within a few days. In the future work, we are going to add to the P^3 support system the following functions,

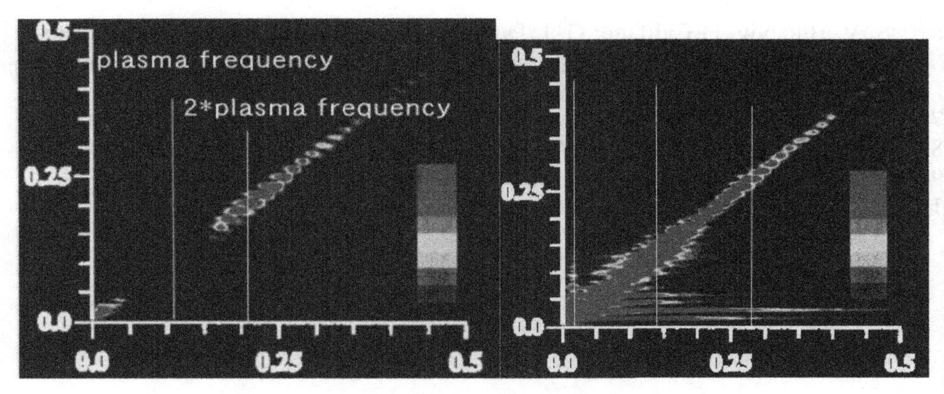

Figs. 7 Two dimensional wave number of the transmitted and reflected light.

1) files automatically backup and restore
2) link database server and managing data and image files

With the P^3 2D code and its support system, it is found that only higher harmonics over twice as large as the plasma frequency are transmitted from the back of the foil. The higher harmonics below the plasma frequency are strongly suppressed in the transparent light. The result indicates that higher harmonics are generated in front of the foil. Separately from emissions from atomic processes induced by the intense laser, the radiation can become short-pulse and intense. The pulse length of the radiation is the same order as that of the laser.

Acknowledgments

The authors would like to thank H. Ohno and Y. Kato for their interest and encouragement throughout this work. We would like to thank Arakawa and K.Nakagawa for their technical support and useful discussion.

References

1. G. A. Mourou, C. P. J. Barty, and D. Perry, Phys. Today 51, N1, 22 (1998).
2. A. V. Gurevich, L. V. Pariiskaya, and L. P. Pitaevskii, Sov. Phys. JETP 22, 449 (1966); A. V. Gurevich and A. P. Meshcherkin, Sov. Phys. JETP 53, 937 (1981).
3. G. S. Sarkisov, V. Yu. Bychenkov, V. T. Tikhonchuk, A.Maksimchuk, S.Y. Chen, R. Wagner, G. Mourou, and D. Umstadter, JETP Lett. 66, 828 (1997).
4. M. Lezius, S. Dobosz, D.Normand, and M. Schmidt, Phys. Rev. Lett. 80, 261 (1998).
5. G. S. Sarkisov, V. Yu. Bychenkov, V. N. Novikov, V. T. Tikhonchuk, A. Maksimchuk, S. Y. Chen, R. Wagner, G. Mourou and D. Umstadter, Phys.Rev.E 59, 7042 (1999).

6. S. V. Bulanov, F. Califano, G. I. Dudnikova, T. Zh. Esirkepov, F. F. Kamenets, T. V Liseikina, N. M. Naumova, F. Pegoraro, V. A. Vshivkov, Plasma Physics Reports 25, (1999).
7. S. V. Bulanov, N. M. Naumova, F. Pegoraro, Physics of Plasmas 1, 745 (1994).
8. PARAGON ™ Fortran Compiler User's Guide, INTEL.Co. (1996) .
9. PARAGON ™ Fortran System Calls Reference Manual, INTEL.Co. (1996)
10. J. Villasenor and O.Buneman. Comp.Physics.Comm. 69, 306 (1992).

Parallelization, Vectorization and Visualization of Large Scale Plasma Particle Simulations and Its Application to Studies of Intense Laser Interactions

Katsunobu Nishihara[1], Hirokazu Amitani[1], Yuko Fukuda[1], Tetsuya Honda[1], Y. Kawata[1], Yuko Ohashi[1], Hitoshi Sakagami[2] and Yoshitaka Sizuki[1]

[1]Institute of Laser Engineering, Osaka University,
Suita, Osaka 565-0871 Japan
koshu@ile.osaka-u.ac.jp
[2] Department of Computer Engineering, Himeji Institute of Technology,
Himeji, Hyogo 671-2201 Japan

Abstract. The use of a three-dimensional PIC (Particle-in-Cell) simulation is dispensable in the studies of nonlinear plasma physics, such as ultra-intense laser interactions with plasmas. The three-dimensional simulation requires a large number of particles more than 10^7 particles. It is therefore very important to develop a parallelization and vectorization scheme of the PIC code and a visualization method of huge simulation data. In this paper we present a new parallelization scheme suitable for a present day supercomputer and a construction method of scientific color animations to analyze simulation data. We also discuss the advantage of the Abe-Nishihara vectorization method for a large scale PIC simulations.

Most of supercomputers in present day consists of multi nodes and each node has multi processors with a sheared memory. We have developed a new parallelization scheme in which domain decomposition is applied among nodes and particle decomposition is used for processors within a nodes. The domain decomposition in PIC requires the exchange of two kinds of data between neighboring domains. One is particle data, such as particle position and velocity, when a particle crosses the boundary between the neighboring domains. The other is field data, such as electric field intensity and current density in the boundary region. In the three-dimensional Electro-magnetic PIC forty two-dimensional variables are transferred to the neighboring domain for each boundary surface. MPI (Message Passive Interface) has been used for the transmission of these data between the nodes. The particle and field data are respectively stored once in one-dimensional data and they are then sent to the other node. This reduces the number of communication. The particle decomposition is performed by using auto-parallelization of do-loop. We measured the scalability of the layered parallelization scheme for the particle number of 25,600,000 and the mesh number of 128x128x128 with the use of sixteen

M. Valero et al. (Eds.): ISHPC 2000, LNCS 1940, pp. 535-536, 2000.

processors of NEC SX-5. The layered parallelization is shown to provide a scalable acceleration of computation for the large system of PIC simulations.

In the PIC code it is necessary to calculate the current and charge densities from particle position and velocity. For a example, the calculation of the charge density of a mesh requires the assignment of particle charge on a mesh from its position. Since more than two particles may locate in the same mesh, the calculation of the assignment can not be vectorized. However for a large system of PIC simulation the possibility that more than two particles within a vector register locate in the same mesh becomes very small. Therefore even if the vector operation is enforced for the calculation of the charge assignment, the correction of the calculation is very seldom. The particle's number of which charge is overwritten by the forced vector operation can be easily found by checking the particle's number stored. This is the Abe-Nishihara method. We obtained 99% of vectoraization even for list vector calculation in PIC by using Abe-Nishihara method. As a result 0.73 sec per cycle is achieved for the number of particles, $2.56x10^7$ and meshes with $(128)^3$ by using sixteen processors of NEC SX5. Within our knowledge, this is the fastest speed achieved at present for this size of program.

Most of advanced visualization software requires interactive operations that is not suitable for drawing many pictures from simulation data to make scientific animations. It is desirable for drawing pictures to be performed by a batch job. We have developed a program that during the computer simulation geometry objects are constructed at the same time by using AVS library. Digital video files are constructed from the geometry objects with the use of the script function of AVS animation encoder on SGI by a batch job.

For modern high power lasers, an intensity of laser radiation can be of the order of $10^{19} - 10^{22}$ W/cm^2. In this ultra relativistic regime, when the quiver energy of an electron in the laser light becomes much greater than the electron rest mass, a novel nonlinear physics comes in our grasp. In this paper we present our resent studies on the ultra-intense laser interaction with overdense plasmas with the use of a three-dimensional PIC code, EMPAC-3D (Three-Dimensional Electro-Magnetic Particle Code). The three dimensional regime of the laser-plasma interaction reveals novel features of the laser light.

Fast LIC Image Generation Based on Significance Map

Li Chen[1,3], Issei Fujishiro[2,1], and Qunsheng Peng [3]

[1] Research Organization for Information Science & Technology
1-18-16 Hamamatsucho, Minato-ku, Tokyo 105-0013, Japan
chen@tokyo.rist.or.jp
[2] Department of Information Sciences, Faculty of Science, Ochanomizu University
2-1-1 Otsuka, Bunkyo-ku, Tokyo 112-8610, Japan
fuji@is.ocha.ac.jp
[3] State Key Lab. of CAD&CG, Zhejiang University, Hangzhou, 310027, P.R. China
{chen,peng}@cad.zju.edu.cn

Abstract. Although texture-based methods provide a very promising way to visualize 3D vector fields, they are very time-consuming. In this paper, we introduce the notion of "significance map", and describe how significance values are derived from the intrinsic properties of a vector field. Based on the significance map, we propose techniques to accelerate the generation of a line integral convolution (LIC) texture image, to highlight important structures in a vector field, and to generate an LIC texture image with different granularities. Also, we describe how to implement our method in a parallel environment. Experimental results illustrate the feasibility of our method.

1 Introduction

Visualizing 3D vector data fields in an intuitive and psychologically meaningful manner is a very challenging topic in scientific visualization. Traditionally, flow fields are usually displayed by inserting geometrical primitives such as arrows, particles, streamlines or stream surfaces. Due to inadequate sampling of vector fields, it often either results in cluttering images or fails to capture significant features of flows. Recently, Line Integral Convolution (LIC), which was proposed by Carbral and Leedom in 1993 [1], has been attracting much attention as a powerful texture-based vector field visualization method. Since a texture possesses shape information as well as color attributes, it has a great potential to visualize 3D vector fields. The texture deformation provides a straightforward cue for the direction of a vector field. Besides, since a texture is calculated at each pixel, the traditional sampling problem in flow visualization can be avoided.

LIC is a procedure that smears a given image along paths that are dictated by a vector field [1]. It is local, one-dimensional and independent of any predefined geometry or texture, and is capable of showing the vector directions even in the area where they change quickly. Much work has been done on extending its scope,

M. Valero et al. (Eds.): ISHPC 2000, LNCS 1940, pp. 537-546, 2000.

usefulness, quality and efficiency. Stalling and Hege [2] succeeded in a fast implementation of LIC. Their algorithm also allows the resolution of output images to be chosen independent of the size of vector fields. Forssell and Cohen [3] extended LIC for visualizing the flow on a 3D curvilinear grid. However, this method involves the defect of texture distortions caused by the non-isometric mappings between the computational space and the physical space. Mao, et al. [4] presented a method for performing the convolution operation directly in the physical space based on solid texturing and ray casting in order to generate LIC images without any artifacts due to misaligned local texture grids or the image interpolation during rendering. Shen, Johnson and Ma combined dye advection with 3D LIC to visualize global and local flow features at the same time [5]. Shen and Kao presented a new LIC method—UFLIC [6] for visualizing unsteady flows. By adopting a time-accurate value depositing scheme and a successive feed-forward method, UFLIC can produce highly coherent animation frames and trace the dynamic flow movement accurately. Okada and Kao [7] presented an enhanced LIC method, which uses post-filtering techniques to sharpen the LIC output and highlight flow features. Kiu and Banks [8] used multi-frequency noise inputs for LIC to enhance the contrasts among regions with different velocity magnitudes. Wegenkittl, Göller and Purgathofer [9] introduced a very expressive technique, which makes use of an asymmetric filter kernel with a low-frequency noise input texture rather than the typical high-frequency one to enhance the perception of the orientation of vector fields. Verma, Kao and Pang [10] presented the PLIC method, in which by adjusting a small set of key parameters, flow visualizations that span the spectrum of streamline-like to LIC-like images can be generated.

Most of the existing texture-based vector field visualization methods treat every pixel equally. This implies that details are uniformly distributed throughout the texture space, thus leading to a fixed detail over the entire texture space without any designated highlights. In fact, it is quite common for a flow field to have extremely heterogeneous distribution of details. Since each pixel has the same significance value in the previous methods, it is very time-consuming to generate a finer image if we want to visualize significant structures with enough precision. For the area containing less detail of a vector field, we need not take much time to calculate its texture very carefully. In this paper, we present a significance-driven texture-based vector field visualization method to ameliorate the problem. We introduce the "significance map", which is derived from both intrinsic properties of a vector field and user-guided highlights. We describe how the significance map is used to improve the generation of texture images, including accelerating the texture image generation, highlighting significant structures, and generating multi-granularity texture images. Finally, experimental results are given to illustrate the feasibility of our method.

2 Preliminaries

As the preliminaries to the description of the new algorithm in succeeding chapters, we give herein an overview of the original LIC algorithm presented by Cabral and Leedom [1].

Given a vector field, the LIC algorithm takes as input a white noise image of the same size with the vector field, and convolutes the image at each pixel with a 1D filter kernel defined along the local streamline in the vector field. As shown in Fig. 1, for a pixel $P(x,y)$, the local streamline is calculated by integrating forward and backward along the local vector direction.

The forward integration is performed in the following way:

$$\Delta s_{i-1} = \begin{cases} \min(D(P_{i-1}, P_{right}), D(P_{i-1}, P_{top})) & V^x_{\lfloor P_{i-1} \rfloor} \geq 0, V^y_{\lfloor P_{i-1} \rfloor} \geq 0 \\ \min(D(P_{i-1}, P_{left}), D(P_{i-1}, P_{top})) & V^x_{\lfloor P_{i-1} \rfloor} \leq 0, V^y_{\lfloor P_{i-1} \rfloor} \geq 0 \\ \min(D(P_{i-1}, P_{right}), D(P_{i-1}, P_{bottom})) & V^x_{\lfloor P_{i-1} \rfloor} \geq 0, V^y_{\lfloor P_{i-1} \rfloor} \leq 0 \\ \min(D(P_{i-1}, P_{left}), D(P_{i-1}, P_{bottom})) & V^x_{\lfloor P_{i-1} \rfloor} \leq 0, V^y_{\lfloor P_{i-1} \rfloor} \leq 0 \end{cases} \quad (1)$$

where $V_{\lfloor P_{i-1} \rfloor}$ is the vector at the grid point $(\lfloor P^x_{i-1} \rfloor, \lfloor P^y_{i-1} \rfloor)$, and $V^x_{\lfloor P_{i-1} \rfloor}, V^y_{\lfloor P_{i-1} \rfloor}$ are the x,y components of the vector $V_{\lfloor P_{i-1} \rfloor}$. $P_{right}, P_{left}, P_{top}, P_{bottom}$ are the intersections between the line passing through the point P_{i-1} along the direction of $V_{\lfloor P_{i-1} \rfloor}$ and the four edges of the cell $(\lfloor P^x_{i-1} \rfloor, \lfloor P^y_{i-1} \rfloor)$, respectively. $D(P_{i-1}, P_c)$ (*c*: *right, left, top, bottom*) are the distances from P_{i-1} to the intersections. If a zero velocity or a vector pointing back to the cell to be left is encountered, then the integration terminates at that point before reaching the specified streamline length.

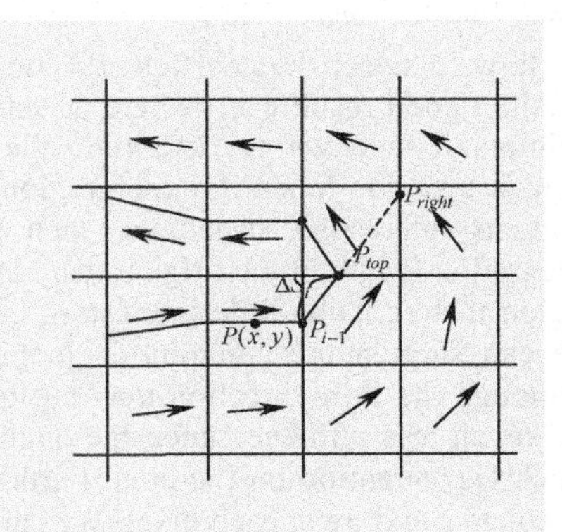

Fig. 1. Local streamline for a point P(*x,y*) in a 2D vector field

The backward integration P_i' is calculated simply by taking the opposite direction of the vector at each point. Now the output pixel value at the pixel $P(x, y)$ can be represented as follows:

$$F_{out}(P) = \frac{\sum_{i=0}^{l} F_{in}(\lfloor P_i \rfloor) h_i + \sum_{i=0}^{l'} F_{in}(\lfloor P_i' \rfloor) h_i'}{\sum_{i=0}^{l} h_i + \sum_{i=0}^{l'} h_i'} \quad (2)$$

$$h_i = \int_{s_i}^{s_i + \Delta s_i} k(w) dw$$

where $s_0 = 0, s_i = s_{i-1} + \Delta s_i$,

$k(w)$: The convolution kernel,

$F_{in}(\lfloor P_i \rfloor)$: The input pixel value at pixel $(\lfloor P_i^x \rfloor, \lfloor P_i^y \rfloor)$,

l, l' : The number of pixels which streamline passes through.

Note that the local streamline generated here depends only on the direction of vectors, but ignores the magnitude. More accurate streamline calculation is given in [2].

3 Significance Specification

3.1 Significance Values Derived from Topology Analysis of Flow Field

In the original LIC method, the selection of the convolution length for each pixel is crucial. If a long length is selected, much more calculation is needed. But if it is too short, the vector direction cannot be revealed. In most of the previous papers, the total convolution length is selected to be proportional to the vector magnitude at each pixel, that is, $length = k|v|$, where $|v|$ is the magnitude of vector v at a pixel. However, for how to select the coefficient *k,* no one gave the convincing answer. In order to obtain a good result everywhere, a long integration length is usually adopted. In our opinion, it is better to determine the integration length according to the relative significance coefficient for each region in the vector field. We regard the region that contains interesting structures, such as vortex, as significant area, and take its integration length in the original way to get a finer texture. On the other hand, for the region that contains little interesting detail, we regard it as insignificant area, where we can shorten the convolution length much to accelerate the image generation. Although the flow direction may not be very clear in the area on the final image, it has much less influence upon the quality of the entire image because the users have much less attention on the area. Furthermore, in the insignificant area, we need not calculate a texture at each pixel. We can calculate a smaller number of pixels to get an image with coarser texture granularities, which also can accelerate the LIC method.

In order to determine the significance value at each point in a given flow field, we employ the vector field topology analysis technique developed by Chong, et al. [11]. Topological concepts are very powerful in the analysis of flow fields. They are based on the critical point theory, which has been used widely to examine solution trajectories of ordinary differential equations. The topology of a vector field consists of critical points (where the velocity vector is zero), and integral curves and surfaces connecting these critical points. From this, we can infer the shape of other tangent curves, and hence to some extent the overall structure of a given vector field.

The positions of the critical points can be found by searching all cells in the flow field. They occur only in cells where all the three components of the vector pass through zero. The exact position of a critical point can be calculated by interpolation in case of a rectangular grid. The position of a critical point in case of a curvilinear grid can be calculated by recursively subdividing the cells or by a numerical method such as the Newton iteration.

Once the critical points have been found, they can be classified by approximating the velocity field in their neighborhood with the first order Taylor expansion. Thus, for a non-degenerate critical point (x_0, y_0, z_0), we can use the matrix of these derivatives, that is, the Jacobian matrix, to characterize the vector field and the behavior of nearby tangent curves [12].

$$\left[\frac{\partial(u,v,w)}{\partial(x,y,z)}\right]_{(x_0,y_0,z_0)} = \begin{bmatrix} \frac{\partial u}{\partial x} & \frac{\partial u}{\partial y} & \frac{\partial u}{\partial z} \\ \frac{\partial v}{\partial x} & \frac{\partial v}{\partial y} & \frac{\partial v}{\partial z} \\ \frac{\partial w}{\partial x} & \frac{\partial w}{\partial y} & \frac{\partial w}{\partial z} \end{bmatrix}_{(x_0,y_0,z_0)} \tag{3}$$

The eigenvalues and eigenvectors of this matrix determine the feature of the flow field near the critical point (x_0, y_0, z_0). Positive eigenvalues correspond to velocities away from the critical point (called as repelling nodes), and negative eigenvalues correspond to velocities towards the critical point (called as attracting nodes). Complex eigenvalues result in a focus. If the real part is non-zero, a spiral occurs (shown in Fig. 2(a), (d)), whereas if the real part is zero, concentric ellipses occur (shown in Fig. 2(f)). If we have both negative and positive real values at a critical point, the critical point is a saddle (shown in Fig. 2(c)).

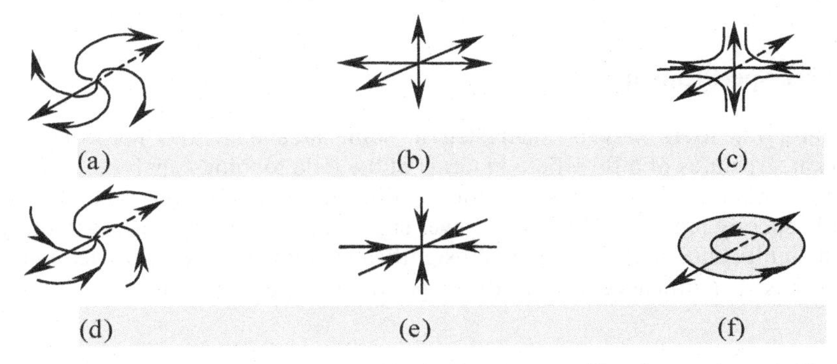

Fig. 2. Examples of three dimensional critical points (a) repelling focus, also repelling in the third dimension; (b) repelling node; (c) saddle, repelling in the third dimension; (d) attracting focus, repelling in the third dimension; (e) attracting node; (f) center, repelling in the third dimension

However, the flow topology analysis sometimes cannot locate all flow features in 3D data sets. Vortex core is known as a very important feature in flow fields, but up to date, it has no formal definition. Some researchers attempted to define it as the center of swirling flows. Obviously, some center points of swirling flows are not located at critical points because they have a velocity in the principle direction of swirling fields. Much work has been done on finding vortex lines [13][14][15]. Sujudi and Haimes [13] defined the vortex core line as the set of places where what they call reduced velocity is zero. The reduced velocity is defined as the velocity minus its component in the direction of the real eigenvector. In the method, only the regions having complex eigenvalues are considered, so there is always a single real eigenvalue, and the corresponding eigenvector direction is unique. But this method has problems with vortices that are curved. It succeeds only for very strong vortices or almost straight ones. Roth and Peikert [15] improved this method by using a higher-order derivatives to make it possible to handle the case of bent vortices.

Based on the above theory, we cannot find critical points in the 3D data set (that will miss some vortex cores). Therefore, we take an approach to find the critical points on 2D cross-sections, and then classify them by the flow topology analysis technique. This may exclude some points that have two zero components on the cross-sections without spiral property. Obviously, if some cross-sections of a flow field contain critical points that have two complex eigenvalues of its Jacobian matrix, the critical points must be very significant on the cross-sections of the flow field. Thereby, the areas near the critical points should contain much more details than others, and should be assigned relative higher significance values. The areas far away from all critical points are assigned relative lower significance values.

The steps for constructing the significance map according to the topology analysis are as follows. First, for each cross-section, find all the critical points. Then, classify these critical points. Find repelling focus, attracting focus and concentric critical points, and store them in an array *criticalpoint_on_section*. Finally for each cell, calculate the sum of distances from the cell to all points in the array *criticalpoint_on_section*, and map the sum to a significance value.

3.2 User Specification

Sometimes, the users may be interested in some area that does not contain any significant structures of a flow field in terms of the field topology analysis. Therefore, user intervention is also necessary for significance specification. The users can explicitly specify regions of higher or lower significance with a significance brush to define a significance map. User-defined significance may also be combined with the topology-based significance to take advantages of both specifications.

4 Texture Image Generation Based on Significance Map

4.1 Accelerating Texture Image Generation

From the significance map, we can get the significance value at each cell in the texture space. Then, we can shorten the convolution length in the areas with lower significance values. Our work could accelerate even for PLIC presented by Verma and Kao [10], which takes textures from the template directly instead of convolution. In the PLIC method, most of time is consumed on getting a streamline at each seed pixel. Therefore, we can accelerate it by tracing shorter streamlines in the areas with lower significance values. Also, we can control the seed points' positions and streamlines' thickness according to the significance map. Meanwhile, for the PLIC method, many streamlines intersect at the same pixel when the streamline is long and dense, so in insignificant areas, we can let each pixel contributed from a single streamline to get an inexact texture with a less computational cost. Since the number of critical points is usually very limited, our method can obviously accelerate the texture image generation with little loss of quality. If a flow field is very complex and has important details almost everywhere, this method is certainly of no effect.

4.2 Highlighting Significant Structures and Details

By the significance map, we can highlight significant structures and details. The input noise texture can be pre-multiplied by a function of the significance values to concentrate the highest input texture opacities in significant regions, and to reduce the input noise texture opacities to values approaching zero where the area is relatively insignificant.

4.3 Generating Multi-Granularity Texture Images

We can select different granularities of texture according to the significance map of a flow field. A coarser granularity is selected in the lower significance area, and a finer granularity is selected in the higher significance area. Since we can involve a smaller number of cells for the computation in the area with coarse granularity, this method also can accelerate the LIC image generation.

5 Implementation and Results

The present method is very easy to be implemented in a parallel environment because the feature analysis works on a cell by cell basis. The image is subdivided into several subdomains that are assigned to different PEs. First, each PE respectively finds the critical points, classifies them, and gets the coordinates of focus and concentric critical points. Then, all the slave PEs send their results to the master PE. The master PE collects all the focus and concentric critical points, and sends them back to each slave PE. Next, each PE calculates the significance map in its subdomain according to all focus and concentric critical points. Finally, each PE determines pixel intensities. For the pixels near the boundary of their subdomains, the PEs need to access vector field data from neighboring subdomains. To minimize the communication, we replicate some boundary vector data.

We have applied the method to a 3D irregular flow field, which simulates a flow passing through an ellipsoidal cylinder. The 3D data set has 25,790 grid cells. Fig. 3(a) is a vertical cross-section texture image with the resolution of 450×300, which was generated by the original LIC method. The cross-section has three critical points in terms of the field topology analysis. We marked them on Fig. 3(b), one of which is a saddle critical point located on the horizontal centerline of the image, and two of which are attracting focus critical points, symmetrically located on the two sides of the horizontal centerline. Because the saddle points have no focus, we have not considered them as significant critical points. In Fig. 3(c), we mapped the significance value at each pixel to a gray value. The darker point corresponds to lower significance value. Fig. 3(d) was generated by our significance-driven LIC method. We adopted very short convolution lengths in the regions with low significance values. The speed was 5 times faster than using the original LIC method. Although the image is a little blurred in the areas far away from the vortices, we keep almost the same quality of visualization. Meanwhile, the important structures of the flow field highlighted with the significance map can attract users' much attention. Fig. 3(e) is the result by taking different texture granularities according to the significance map. We can see that almost no detailed structures in the vector field were lost although we take coarser texture granularities in some insignificant regions. We generated this image almost 4 times faster than using the original method.

6 Conclusions and Future Work

In this paper, we have presented a fast LIC image generation method using the significance map. We employed the topology analysis technique for vector fields to generate the significance map. The significance map can also be combined with users' specification. By using the significance map as the basic reference structure, we developed the techniques to accelerate the texture image generation, to highlight the significant structures, and to generate a texture image with different texture granularities. Experimental results proved the feasibility and effectiveness of our method.

LIC method often fails for 3D volume due to its dense 3D texture. In future, we will use the significance map to control the transfer functions for comprehensible volume rendering, highlighting the significant regions and neglecting insignificant information.

References

[1] Cabral, B., Leedom, C.: Image Vector Field Using Line Integral Convolution. In: Computer Graphics Proceedings, Annual Conference Series. ACM SIGGRAPH, New York (1993) 263-272

[2] Stalling, D., Hege, H.: Fast and Resolution Independent Line Integral Convolution. In: Computer Graphics Proceedings, Annual Conference Series. ACM SIGGRAPH, New York (1995) 249-256

[3] Forssell, L. K., Cohen, S. D.: Using Line Integral Convolution for Flow Visualization: Curvilinear Grids, Variable-speed Animation and Unsteady Flows. IEEE Transactions on Visualization and Computer Graphics, Vol.1, No.2 (1995) 133-141

[4] Mao, X., et al.: Line Integral Convolution for Arbitrary 3D Surfaces Through Solid Texturing. In: Lefer, W., Grave, M. (eds.): Visualization in Scientific Computing '97. Springer-Verlag, Wien (1997) 57-69

[5] Shen, H.-W., Johnson, C., Ma, K.-L.: Visualizing Vector Fields Using Line Integral Convolution and Dye Advection. In: Proceedings of 1996 Symposium on Volume Visualization. ACM SIGGRAPH, New York (1996) 63-70

[6] Shen, H.-W., Kao, D.: UFLIC: Line Integral Convolution Algorithm for Visualizing Unsteady Flows. In: Proceedings of IEEE Visualization '97. ACM Press, New York (1997) 317-322

[7] Okada, A., Kao, D.: Enhanced Line Integral Convolution with Flow Feature Detection. In: SPIE, Vol. 3017, Visual Data Exploration and Analysis IV (1997) 206-217

[8] Liu, M.-H., Banks, D. C.: Multi-Frequency Noise for LIC. In Proceedings of IEEE Visualization '96. ACM Press, New York (1996) 121-126
[9] Wegenkittl, R., Groller, E., Purgathofer, W.: Animating Flow Fields: Rendering of Oriented Line Integral Convolution. In: Proceedings of Computer Animation'97. IEEE Computer Society Press, Los Alamitos (1997) 15-21
[10] Verma, V., Kao, D., Pang, A.: PLIC: Bridging the Gap Between Streamlines and LIC. In: Proceedings of IEEE Visualization'99. ACM Press, New York (1999) 341-348
[11] Chong, M. S., Perry, A. E., Cantwell, B. J.: A General Classification of 3D Flow Fields. Physics of Fluids Ann., Vol. 2, No.5 (1990) 765-777
[12] Helman, J., Hesselink, L.: Visualizing Vector Field Topology in Fluid Flows. IEEE Computer Graphics and Applications, Vol.11, No.3 (1993) 36-46
[13] Sujudi, D., Haimes, R.: Identification of Swirling Flow in 3-D Vector Fields. AIAA-95-1715 (1995)
[14] Kenwright, D.: Automatic Vortex Core Detection. IEEE Computer Graphics and Applications, Vol. 18, No. 4 (1998) 70-74
[15] Roth, M., Peikert, R.: A Higher-Order Method for Finding Vortex Core Lines. In: Proceedings of IEEE Visualization '98. ACM Press, New York (1998) 143-150

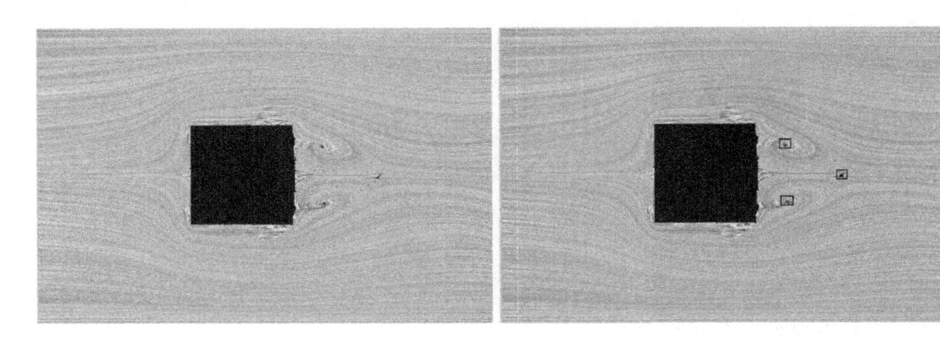

Fig. 3(a). A cross-section texture image generated by the original LIC method

Fig. 3(b). Three critical points on the cross-section: One is a saddle critical point, and the other two are attracting focus critical points

Fig. 3(c). A gray-value coding of the significance map

Fig. 3(d). A texture image generated by our significance-driven LIC method, which not only accelerates the image generation, but also highlights the vortices

Fig. 3(e). A texture image generated with different texture granularities

Fast Isosurface Generation Using the Cell-Edge Centered Propagation Algorithm

Takayuki Itoh, Yasushi Yamaguchi, and Koji Koyamada

IBM Research, Tokyo Research Laboratory,
1623-14 Shimotsuruma, Yamato-shi, Kanagawa, 242-8502, JAPAN
itot@trl.ibm.co.jp
yama@graco.c.u-tokyo.ac.jp
koyamada@soft.iwate-pu.ac.jp

Abstract. Isosurface generation algorithms usually need a vertex-identification process since most of polygon-vertices of an isosurface are shared by several polygons. In our observation the identification process is often costly when traditional search algorithms are used. In this paper we propose a new isosurface generation algorithm that does not use the traditional search algorithm for polygon-vertex identification. When our algorithm constructs a polygon of an isosurface, it visits all cells adjacent to the vertices of the polygon, and registers the vertices to polygons inside the visited adjacent cells. The method does not require a costly vertex identification process, since a vertex is registered in all polygons that share the vertex at the same time, and the vertex is not required after the moment. In experimental tests, this method was about 20 percent faster than the conventional isosurface propagation method.

1 Introduction

Isosurface generation is one of the most effective techniques for extracting features of a scalar field in a volume data, such as the results of numerical simulation or medical measurement. Discussion of efficient isosurfacing methods has therefore been very active. Many approaches have been reported for the acceleration of isosurface generation, such as parallelization [1], graphics acceleration by generating triangular strips [2], and geometric approximation [3]. The most popular approach is to skip non-isosurface cells. Many reported algorithms sort or classify cells according to their scalar values [4,5,6,7]. Other algorithms that use the spatial-subdivision algorithms have been also proposed [8,9]. The present authors have proposed extrema-based algorithms [10,11] that efficiently search for isosurface cells. Starting from the extracted isosurface cells, an isosurface is generated by recursively traversing adjacent cells [12].

The above-mentioned algorithms have drastically reduced the unnecessary cost of visiting non-isosurface cells. Table 1 shows the cost of generating 20 isosurfaces with

M. Valero et al. (Eds.): ISHPC 2000, LNCS 1940, pp. 547-556, 2000.

different iso-values in an unstructured volume consisting of tetrahedral cells. The experimental test compares a straightforward algorithm (ST) that visits all cells and the volume thinning algorithm (VT) [11]. Here,

-- N_c and N_n denote the numbers of cells and nodes in a volume.

-- N_t and N_v denote the total numbers of triangular polygons and vertices in the 20 isosurfaces.

-- T_1 denotes the computational time of visiting non-isosurface cells.

-- T_2 denotes the computational time of visiting isosurface cells and constructing the topology of polygons.

-- T_3 denotes the computational time of calculating polygon-vertex data, such as positions and normal vectors.

-- T_{total} denotes the total time of generating isosurfaces.

Table 1. Computational times of processes in generating isosurfaces

Dataset	1	1	2	2
Method	SF	VT	SF	VT
N_c	61680	61680	346644	346644
N_n	11624	11624	62107	62107
N_t	80995	80995	135358	135358
N_v	43158	43158	71358	71358
T_1 (sec.)	8.30	0.09	42.23	0.57
T_2 (sec.)	3.80	3.45	6.25	5.88
T_3 (sec.)	0.76	0.75	1.14	1.15
T_{total} (sec.)	12.86	4.29	49.62	7.60

The results indicate that above-mentioned acceleration algorithm archived the great reduction of T_1. In other words, other approaches that reduce T_2 or T_3 are needed to develop more efficient isosurfacing algorithms.

In our observations, a polygon-vertices identification process occupies the largest part of the computational time in constructing the topology of polygons. Though it would be possible to implement the isosurfacing algorithm without the polygon-vertex identification process, the process is desirable, because it reduces the amount of polygon-vertex data calculation and the memory-space. In Table 1, the number of vertices N_v would be 3 N_t --- about six times greater --- without the identification process. In this case, the computational time T_3 would be greater than the polygon construction time T_2, and the memory-space would be about three times greater. Moreover, the identification process is necessary if isosurfaces are used for applications that require the topology of polygons, such as parametric surface reconstruction, mesh compression, or mesh simplification.

An example of implementation of the vertex identification process is described in Doi and Koide [13]. Their implementation uses a hash-table to search shared vertices, however, such traditional search algorithm occupies the large computation time in our observation. Isosurfacing process would be accelerated if the vertex-identification process could be implemented without the costly search algorithm.

In this paper we propose an isosurface propagation algorithm that efficiently identifies shared polygon-vertices. When our algorithm constructs a polygon of an isosurface, it visits all cells adjacent to the vertices of the polygon, and registers the vertices to polygons inside the visited adjacent cells. The method does not require a costly vertex identification process, since a vertex is registered in all polygons that share the vertex at the same time, and the vertex is not required after the moment.

2 Related Work

2.1 Polygon-Vertex Identification

When polygons in an isosurface are generated by the conventional Marching Cubes method [14], all their vertices lie on cell-edges, and mostly shared by several polygons. If a volume data structure contains all cell-edges data, the shared vertices are immediately extracted. However, cell-edge data is not usually preserved in a volume data structure, owing to the limited memory-space.

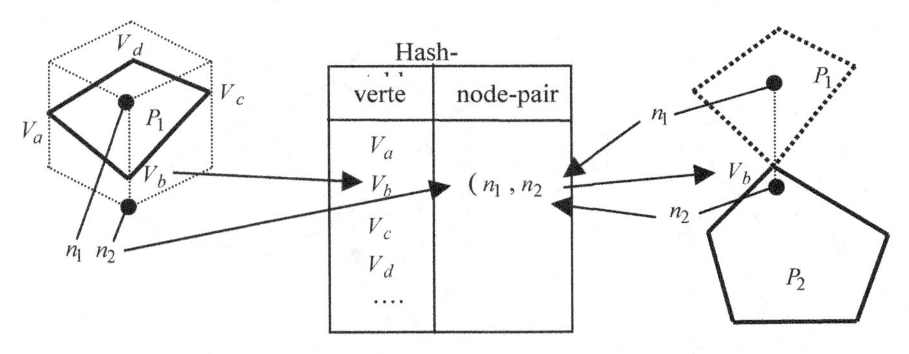

Fig. 1. Polygon-vertex identification using a hash-table

An example of a vertex identification process is described in Doi and Koide [13]. Given an iso-value of an isosurface C, the implementation first determines the sign of $S(x,y,z)-C$ at nodes of a cell, where $S(x,y,z)$ denotes the scalar value at a node. If all the signs are equal, the cell is not an isosurface cell and the calculation of the action value is skipped. Otherwise, the process extracts isosurface cell-edges of a cell. Here, a cell-edge is represented as a pair of nodes. When a polygon-vertex is generated on a cell-edge, it is registered to the hash-table with the pair of nodes that denotes the cell-edge. When another isosurface cell that shares the same cell-edge is visited, the polygon-vertex is extracted from the hash-table, by inputting the pair of nodes. In the implementation, all isosurface cell-edges are registered to the hash-table with the polygon-vertices of an isosurface.

Fig. 1 shows an example of this process. When polygon P_1 is first constructed, vertices V_a, V_b, V_c, and V_d are registered in a hash-table with pairs of nodes. For example, vertex V_b is registered with a pair of nodes, n_1 and n_2, that denotes a cell-edge that V_b lies on. When polygon P_2 is then constructed, vertex V_b is extracted from the hash-table, by inputting the pair of nodes, n_1 and n_2.

2.2 Isosurface Propagation

An isosurface is efficiently generated by recursively visiting adjacent isosurface cells. Such recursive polygonization algorithms were originally proposed for efficient polygonization of implicit functions [15,16], and have been then applied to a volume datasets [12].

In a typical isosurface propagation algorithm, isosurface cells are extracted by a breadth-first traverse. In the algorithm, several isosurface cells are first inserted into a FIFO queue. They are then extracted from the FIFO, and polygons are generated inside them. Isosurface cells adjacent to the extracted cells are then also inserted into the FIFO. This process is repeated until the FIFO queue becomes empty, and finally the isosurface is constructed.

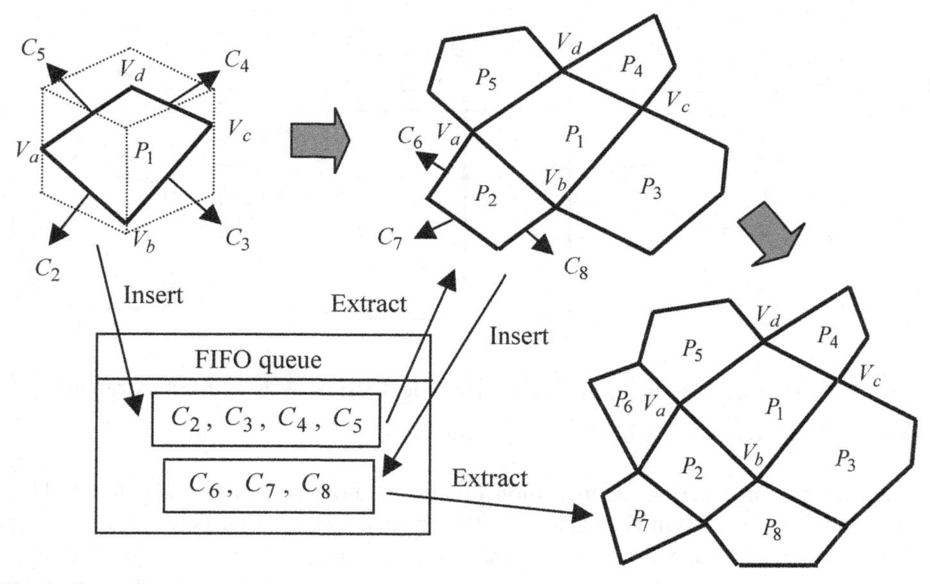

Fig. 2. Isosurface propagation

Fig. 2 shows an example of a typical isosurface propagation algorithm. When polygon P_1 is first constructed, four adjacent isosurface cells, C_2, C_3, C_4, and C_5 are inserted into the FIFO. When these cells are extracted from the FIFO, four polygons, P_2, P_3, P_4, and P_5, are constructed. When P_2 is constructed, adjacent isosurface cells, C_6, C_7, and C_8 are similarly inserted into the FIFO. Polygons P_6,

P_7, and P_8 are then similarly constructed when C_6, C_7, and C_8 are extracted from the FIFO.

The propagation algorithm has the great advantage of reducing the number of visiting non-intersecting cells. However, it also has a problem that the starting isosurface cells must first be specified. Efficient automatic extraction of the starting cells was previously difficult, especially when the isosurface was separated into many disconnected parts.

The authors have proposed a method for automatically extracting isosurface cells in all disconnected parts of an isosurface [10,11]. The method first extracts extremum points of a volume, and then generates a skeleton connecting all extremum points. The skeleton consists of cells, and every isosurface intersects at least one cell in the skeleton. The method efficiently generates isosurfaces by searching for isosurface cells in the skeleton and then applying the isosurface propagation algorithm [12].

Our method [10,11] requires less than $O(n)$ computational time for isosurfacing process, since the cost of searching for isosurface cells is regarded as $O(n^{1/3})$ on average, unless the number of extremum points is enormous. The computational time of pre-processing in the volume thinning method [11] is always regarded as $O(n)$.

Remark that the vertex-identification process in the isosurface propagation algorithm still needs a vertex search algorithm. For example, polygon-vertices of P_1, V_a, V_b, V_c, and V_d, are registered into a hash-table when P_1 is generated. The polygon-vertex V_b is then extracted from the hash-table when P_2, P_3, and P_8 are generated.

3 Cell-Edge Centered Isosurface Propagation

3.1 Algorithm Overview

In this paper we propose an isosurface generation algorithm that does not need a search algorithm in its vertex identification process. Fig. 3 shows the overview of the new method.

The method assumes that at least one isosurface cell is given. It first generates a polygon P_1 inside the given cell, and allocates its polygon-vertices, V_a, V_b, V_c, and V_d. It then visits all cells that are adjacent to polygon-vertices of P_1. In Fig. 3, cells that are adjacent to V_b are visited, and polygons P_2, P_3, and P_4 are generated. Remark that polygon-vertices of the new three polygons are not allocated at that time. It then assigns V_b to the three polygons. V_b is no more required in this algorithm, because all polygons that share V_b have been generated at that time. It means that the search algorithm is not necessary for the vertex-identification in the method. Similarly, in Fig. 3, cells that are adjacent to V_a are then visited. Polygons P_5 and P_6 are generated at that time, and V_a is assigned to them.

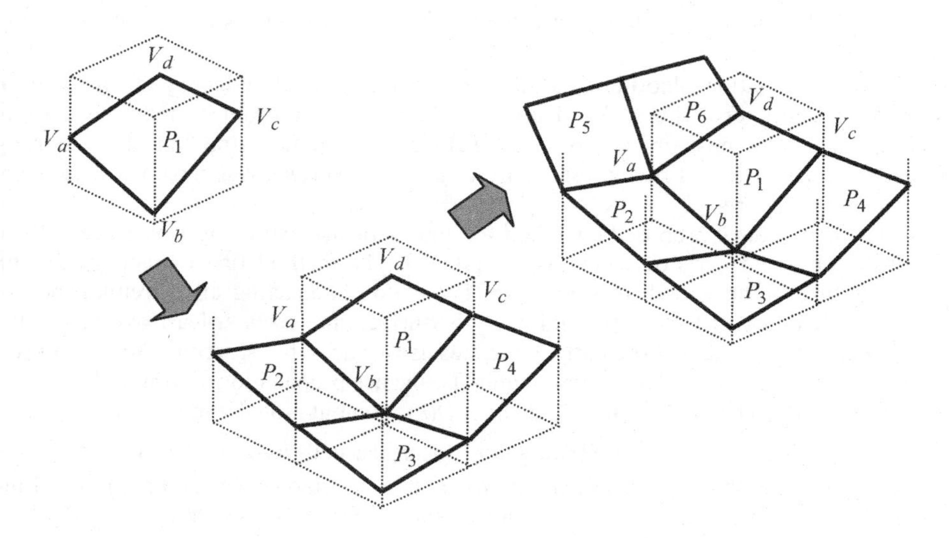

Fig. 3. Overview of the new method proposed in this paper

3.2 Combination with the Volume Thinning Method

The new method assumes that at lease one isosurface cell is given, so the method should be combined with an isosurface cell extraction method. We applied the volume thinning method [11] in order to extract isosurface cells in all disconnected parts of an isosurface.

The volume thinning method first generates an extrema skeleton, consists of cells, in a pre-processing. The extrema skeleton has a feature that every isosurface intersects it, so isosurface cells can be always extracted by traversing the extrema skeleton.

Fig. 4 shows the pseudo-code of our implementation. The implementation first extracts isosurface cells from the extrema skeleton, and inserts them into a FIFO queue. It then extracts an isosurface cell C_i from the FIFO, and constructs the polygon P_i inside C_i. At the moment, though the number of polygon-vertices of P_i is specified, each polygon-vertex is not allocated. The implementation then extracts the isosurface cell-edges of C_i. If a polygon-vertex V_n is not allocated on an isosurface cell-edge E_n at that time, the implementation allocates V_n and registers to the polygon P_i, and visits all cells that share the cell-edge E_n by using the connectivity of cells. If a polygon is not constructed in the visited cell C_j, the implementation constructs a polygon P_j in C_j. The polygon-vertex V_n on E_n is registered into the polygon P_j. If the visited cell C_j has not been inserted into the FIFO, the implementation also inserts C_j into the FIFO at that time. The above process repeats until the FIFO becomes empty.

In this algorithm, most of isosurface cells are several times visited by cell-edge-centered process (the for-loop (3) in Fig. 4), and a polygon is constructed at the first visit. The cells are also visited when they are extracted from the FIFO (the for-loop (1) in Fig. 4), and all polygon-vertices of the polygons inside the extracted cells are set at the moment. The method processes a cell several times; however, our experimental tests show that its computational time is less than the conventional methods.

```
void Isosurfacing() {

    for(each cell C_i in an extrema skeleton) {
       if( C_i is an isosurface cell) { insert C_i into FIFO; }
    }

    /* for-loop (1) */
    for (each cell C_i extracted from FIFO ) {
       if(polygon P_i in C_i is not constructed) { Construct P_i in C_i ; }

       /* for-loop (2) */
       for(each intersected edge E_n ) {
          if(a polygon-vertex V_n on E_n is not added into P_i ) {
             Allocate V_n on E_n ;
             Register V_n into P_i in C_i ;

             /* for-loop (3) */
             for(each cell C_j which share E_n ) {
                if( P_j in C_j is not constructed) { Construct P_j in C_j ; }
                Register V_n into P_j in C_j ;
                if ( C_j has never been inserted into FIFO) { insert C_j into FIFO; }
             } /* end for-loop(3) */
          } /* end if(there is not V_n ) */
       } /* end for-loop(2) */
    } /* end for-loop(1) */

    for(each polygon-vertex V_n ) { Calculate position and normal vector; }
```

Fig. 4. Algorithm of the cell-edge-centered isosurface propagation method

4 Experimental Results

This section compares the experimental results given by the cell-edge centered propagation method with those given by the conventional propagation method. The experiments were carried out on an IBM PowerStation RS/6000 (Model 560). Four datasets for unstructured volumes consisting of tetrahedral cells, which contain the results of numerical simulations, were used for the experiments.

Table 2 shows the results of experiments in which a series of 20 isosurfaces were generated for each volume, with various scalar values. Here,

-- N_c and N_n denote the numbers of cells and nodes in a volume.

-- N_t and N_v denote the total numbers of triangular polygons and vertices in the 20 isosurfaces.

-- T_1 denotes the computational time of generating 20 isosurfaces by the conventional propagation method.

-- T_2 denotes the computational time of generating 20 isosurfaces by the cell-edge centered propagation method.

-- T_{p1} and T_{p2} denote the computational times of the polygon construction processes of the two propagation methods.

In these experiments, the volume thinning method [11] extracts the starting cells of the propagation.

Table 2. Computational times of processes in generating isosurfaces

Dataset	1	2	3	4
N_c	61680	346644	458664	557868
N_n	11624	62107	80468	97943
N_t	80995	135398	494480	1164616
N_v	43158	71358	251506	588796
T_{p1} (sec.)	3.45	5.88	21.35	49.80
T_{p2} (sec.)	2.53	4.01	15.72	36.20
T_1 (sec.)	4.29	7.60	26.65	60.81
T_2 (sec.)	3.35	5.52	20.61	46.80

The results show that the polygon construction process in the cell-edge centered propagation method is about 25 percent faster than the conventional propagation method, and the total isosurfacing process is about 20 percent faster.

5 Conclusion

In this paper we proposed an isosurface generation algorithm that does not use a vertex search algorithm for the vertex-identification process. The algorithm visits all cells sharing an isosurface cell-edge at the same time, and the vertex that lies on the cell-edge is registered to all the polygons inside the visited cells. The vertex is no more required in the process, and the vertex search algorithm is not therefore necessary in our method. Our experimental tests showed that the method is about 20 percent faster than the conventional implementation.

In future, we would like to implement this method for hexahedral cells, and to measure the computational time of isosurfacing processes.

References

[1] Hansen C. D., and Hinker P., Massively Parallel Isosurface Extraction, Proceedings of IEEE Visualization '92, pp. 77-83, 1992.

[2] Howie C. T., and Blake E. H., The Mesh Propagation Algorithm for Isosurface Construction, Computer Graphics Forum (Eurographics), Vol. 13, No. 3, pp. C-65-74, 1994.

[3] Durkin J. W., and Hughes J. F., Nonpolygonal Isosurface Rendering for Large Volume, Proceedings of IEEE Visualization '94, pp. 293-300, 1994.

[4] Giles M., and Haimes R., Advanced Interactive Visualization for CFD, Computer Systems in Engineering, Vol. 1, No. 1, pp. 51-62, 1990.

[5] Gallagher R. S., Span Filtering: An Optimization Scheme for Volume Visualization of Large Finite Element Models, Proceedings of IEEE Visualization '91, pp. 68-74, 1991.

[6] Livnat Y., Shen H., and Johnson C. R., A Near Optimal Isosurface Extraction Algorithm Using the Span Space, IEEE Transactions on Visualization and Computer Graphics, Vol. 2, No. 1, pp. 73-84, 1996.

[7] Shen H., Hansen C. D., Livnat Y., and Johnson C. R., Isosurfacing in Span Space with Utmost Efficiency (ISSUE), Proceedings of IEEE Visualization '96, pp. 287-294, 1996.

[8] Welhelms J., and Gelder A. Van, Octrees for Fast Isosurface Generation, ACM Transactions on Graphics, Vol. 11, No. 3, pp. 201-227, 1992.

[9] Silver D., and Zabusky N. J., Quantifying Visualization for Reduced Modeling in Nonlinear Science: Extracting Structures from Data Sets, Journal of Visual Communication and Image Representation, Vol. 4, No. 1, pp. 46-61, 1993.

[10] Itoh T., and Koyamada K., Automatic Isosurface Propagation by Using an Extrema Graph and Sorted Boundary Cell Lists, IEEE Transactions on Visualization and Computer Graphics, Vol. 1, No. 4, pp. 319-327, 1995.

[11] Itoh T., Yamaguchi Y., and Koyamada K., Volume Thinning for Automatic Isosurface Propagation, Proceedings of IEEE Visualization '96, pp. 313-320, 1996.

[12] Speray D., and Kennon S., Volume Probe: Interactive Data Exploration on Arbitrary Grids, Computer Graphics, Vol. 24, No. 5, pp. 5-12, 1990.

[13] Doi A., and Koide A., An Efficient Method of Triangulating Equi-valued Surfaces by Using Tetrahedral Cells, IEICE Transactions, Vol. E74, No. 1, pp. 214-224, 1991.

[14] Lorensen W. E., and Cline H. E., Marching Cubes: A High Resolution 3D Surface Construction Algorithm, Computer Graphics, Vol. 21, No. 4, pp. 163-169, 1987.
[15] Wyvill G., McPheeters C., and Wyvill B., Data Structure for Soft Objects, The Visual Computer, Vol. 2, No. 4, pp. 227-234, 1986.
[16] Bloomenthal J., Polygonization of Implicit Surfaces, Computer Aided Geometric Design, Vol. 5, No. 4, pp. 341-355, 1988.

Fast Ray-Casting for Irregular Volumes

Koji Koyamada

Department of Software and Information Science, Iwate Prefectural University
152-52, Sugo, Takizawa-mura, Iwate-ken, 020-0193, Japan
koyamada@soft.iwate-pu.ac.jp
http://www.iwate-pu.ac.jp/

Abstract. This paper proposes a fast cell traverse method for volume rendering of irregular volume datasets. All cells of an irregular volume are subdivided into a set of tetrahedral cells for our algorithm. The number of calculations required to find the intersections of a ray and irregular volume is reduced by using the exterior faces of cells rather than the ray as a basis for processing. An efficient new method of computing the integration of the brightness equation along a ray takes advantage of the linear distribution of data within a tetrahedral cell. Benchmark tests proved that the proposed method significantly improves the performance of volume rendering.

1 Introduction

The volume rendering describes a given volume dataset as semi-transparent density clouds, whose appearance can be easily modified by specifying a transfer function for mapping scalar data to color (brightness) and opacity (light attenuation). The specification is performed by the user so that data values are related to meaningful colors, the part of the volume data most interesting to him/her is exposed, and the part that is not interesting to him/her is transparent. Images are formed from the resulting colored semi-transparent volume by blending together volume cells projected onto the same pixel on the picture plane. This projection can be performed in either image order or object order.

The object order approach has been in many ways preferable to the image order approach. It can take advantage of coherence when a voxel projects into many neighboring pixels. Methods have been developed for rendering the projection of volume cells by using Gouraud-shaded and partially transparent polygons [Wilh91,Shir90,Will92a,Laur91,Hanr90,Schr91,Koya92b]. But, when we deal with a very large volume dataset, the projected volume cell may be smaller than the pixel size. In that case, since the avarage number of viewing rays that intersect a single cell is small, we can expect that the image order approach gives a better performance than the object order approach.

The image order approach is generally called "ray-casting," because it scans the display screen and, by casting a viewing ray, determines for each pixel what volume cells affect it. The opacities and shaded colors encountered along the ray are summed to find the opacity and color of the pixel. Whereas in ray-tracing,

M. Valero et al. (Eds.): ISHPC 2000, LNCS 1940, pp. 557–572, 2000.

viewing rays bounce off when they hit reflective objects, in ray-casting, they continue in a straight line until the opacity encountered by the ray sums to unity or the ray reaches the exterior of the volume data. No shadows or reflections are generated. Descriptions of ray-casting approaches appear in various studies [Levo88,Upso88,Sabe88,Dreb88].

Initially, attention in volume rendering was focused on medical imaging. This promoted the development of techniques for regular volume datasets. Currently, we have a PC-based volume rendering hardware which can render a 128^3 voxel dataset at a rate of thirty frames per second [Pfis99]. Recently, efforts have been made to support the rendition of data stored in curvilinear volume datasets [Wilh90a,Hong99] and in irregular volume datasets [Koya90,Garr90]. These techniques can be used to convert a non-regular volume dataset into a regular volume dataset. However, a ray-casting approach generally takes a lot of computational time. Its most computationally intensive portions are testing for intersections of ray and a non-regular volume, and integration of brightness along rays.

In this paper, we propose a new cell traverse method to eliminate these two bottlenecks in computation. It reduces the number of intersection tests by using image coherence, and interpolates data efficiently along a ray by taking advantage of the characteristics of tetrahedral cells.

2 Overview of the Image Order Approach

To render a given volume dataset, it is necessary to calculate Eq. 1 for each viewing ray that passes through an eye position and a pixel position on an image screen.

$$B = \sum_{i=1}^{n} c_i \times (\alpha_i \prod_{j=1}^{i-1} (1 - \alpha_j)). \tag{1}$$

Here, B represents a total intensity along the ray, c_i means a luminosity in the i-th, and α_i denotes an opacity in the i-th subdomain along the ray.

In the image order approach, for each viewing ray, a scalar value s_i and a scalar gradient ∇s_i are computed at the center of basically equal-sized volume sample segments along the ray in front-to-back order, by interpolating from the scalar values and scalar gradients at the node points of the tetrahedral cell that includes the sampling location. The pseudo-color and opacity α_i at the sampling location are determined by reference to pre-defined color and opacity lookup tables from the sampled scalar value s_i. A shaded color c_i is then computed by shading the pseudo-color, using some shading model such as Phong's model [Phon75], where the surface normal is the normalized scalar gradient at the location. An implementation example is as follows:

- Cast a ray, which is equivalent to a pixel position, into a volume dataset.
- Interpolate a scalar value and a scalar gradient at a sampling point along the ray.

- Calculate the luminosity and opacity by using transfer functions and some shading model.
- Composite the color values by using the opacity for the final pixel values.

The corresponding algorithm is as follows:

For a pixel position (p)

For a sampling point (m)

1. $s_m^p \leftarrow$ interpolate-in-cell $(s^{node1}, s^{node2}, s^{node3}, s^{node4})$
2. $\nabla s_m^p \leftarrow$ interpolate-in-cell $(\nabla s^{node1}, \nabla s^{node2}, \nabla s^{node3}, \nabla s^{node4})$
3. $\alpha_m^p \leftarrow opacity(s_m^p)$
4. $c_m^p \leftarrow luminosity(color(s_m^p), \nabla s_m^p)$
5. $B_m^p \leftarrow B_{m+1}^p \times (1 - \alpha_m^p) + c_m^p \times \alpha_m^p$

where

- B_m^p is the brightness resulting from integrating Eq. 1 from the nth sampling point to the mth sampling point ($n \geq m$):

$$B_m^p = \sum_{i=m}^{n} c_i^p \times (\alpha_i^p \prod_{j=m}^{i-1} (1 - \alpha_j^p)). \tag{2}$$

  The final brightness is $B^p = B_1^p$.
- $s_m^p, \nabla s_m^p, \alpha_m^p$, and c_m^p denote a scalar value, a scalar gradient, an opacity, and a luminosity at the mth sampling point along the pth viewing ray (pixel position).
- interpolate-in-cell(,,,) is an operator that interpolates a value from the values at four node points (node1, node2, node3, and node4) that define a cell (in this case, a tetrahedral cell).
- opacity() is a function for transferring a scalar value to an opacity value.
- color() is a function for transferring a scalar value to its pseudo-color, which generally has three components of red, green, and blue.
- luminosity(,) means a shading model. For simplicity, we assume that the model is a function of pseudo-colors and a scalar gradient.

One major advantage of the image order approach is that geometric data, such as polygonally defined objects, can be easily integrated into the above procedure. An integrated rendering algorithm will be given in the next section. On the other hand, a well-known disadvantage of this approach is that the computation cost is huge because it does not take advantage of coherence within the volume data sets. In the following sections, we will describe two techniques for

- reducing the cost of ray-face intersection tests and
- reducing the cost of interpolation of the sampled values

3 Reducing the Cost of Ray-Face Intersection Tests

3.1 Related Work

In order to calculate the brightness of Eq. 1 for a given viewing ray, an exterior face that intersects the viewing ray is first searched for. Then, the cells along the ray are traversed until it exits from another exterior face. An additional calculation to check for reentry of the ray into the volume is required in handling a nonconvex mesh.

The cost of ray-face intersection tests strongly affects the total computational expense of volume rendering. Rubin and Whitted reported that most of the total computation time of a ray-tracing program is spent on intersection tests [Rubi80]. Various algorithms for reducing the number of intersection tests have been developed. Some of them use simple bounding volumes: Rubin and Whitted replaced exhaustive searching for intersections by checking with simple bounding volumes [Rubi80], while Weghorst et al. examined how bounding volumes can be selected in such a way as to reduce the computational cost of the intersection tests [Wegh84]. Another type of improvement employs techniques for subdividing three-dimensional space. Glassner investigated the effects of partitioning the space with an octree data structure [Glas84]. Fujimoto et al. compared octrees with a rectangular linear grid. A shortcoming of this type of technique is that some rays must be cast that do not pass through any face [Fuji85].

The above algorithms take no account of the knowledge that neighboring rays are very likely to intersect the same exterior face. Obviously, a new technique that exploits image coherence must be found in order to decrease the computational expense. Weghorst et al. created an item buffer to improve the first ray-intersection test [Wegh84]. This concept can be applied to the test for the intersection of a ray and the nearest (or furthest) exterior face of a tetrahedral model. Moreover, it can be extended to an efficient ray-face intersection algorithm, if we assume that refraction, reflection, and shadow rays are not considered in the ray tracing of a tetrahedral model.

Scan-Conversion of Exterior Faces One promising idea for such an intersection algorithm is to cast viewing rays into a volume dataset from each exterior face, which is processed either from back to front (BTF) or from front to back (FTB). In BTF, only back-facing exterior faces are processed. In FTB, only front-facing exterior faces are processed. Note that we process about half of the exterior faces in either case. Before casting rays, we need to calculate the priority of the back- or front-facing exterior faces by depth sorting. The intersection of a viewing ray and an exterior face can be easily determined by scan-converting it on the screen. If the visualized volume is convex, the priority need not be calculated, because there is no overlapping of back- or front-facing exterior faces when they are scan-converted on the screen. In this case, exterior faces can be scan-converted at random. The intersections can be incrementally calculated by using the digital differential analyzer (DDA) approach, because they exist within the face projection on the screen. If a vertex that is shared by triangles on the screen coincides with a pixel position, duplicate viewing rays may be cast. This leads to a visual artifact in the generated image. The coordinates of the vertex can be perturbed infinitesimally to eliminate such degenerate cases.

We start the cell traverse from a position that is incrementally decided. Reference to the cell adjacency component of the tetrahedral model enables us to search for cells that intersect the viewing ray successively until the ray reaches an exterior face. When it exits, the calculated brightness is stored at the corresponding pixel position of the frame buffer. If no other exterior face is scan-converted to this pixel, the value becomes a final one; otherwise, the result is used as an initial value for calculating the brightness. The whole volume, either convex or nonconvex, is finally traversed when all the exterior faces have been processed in order of priority, as shown in Figure 1.

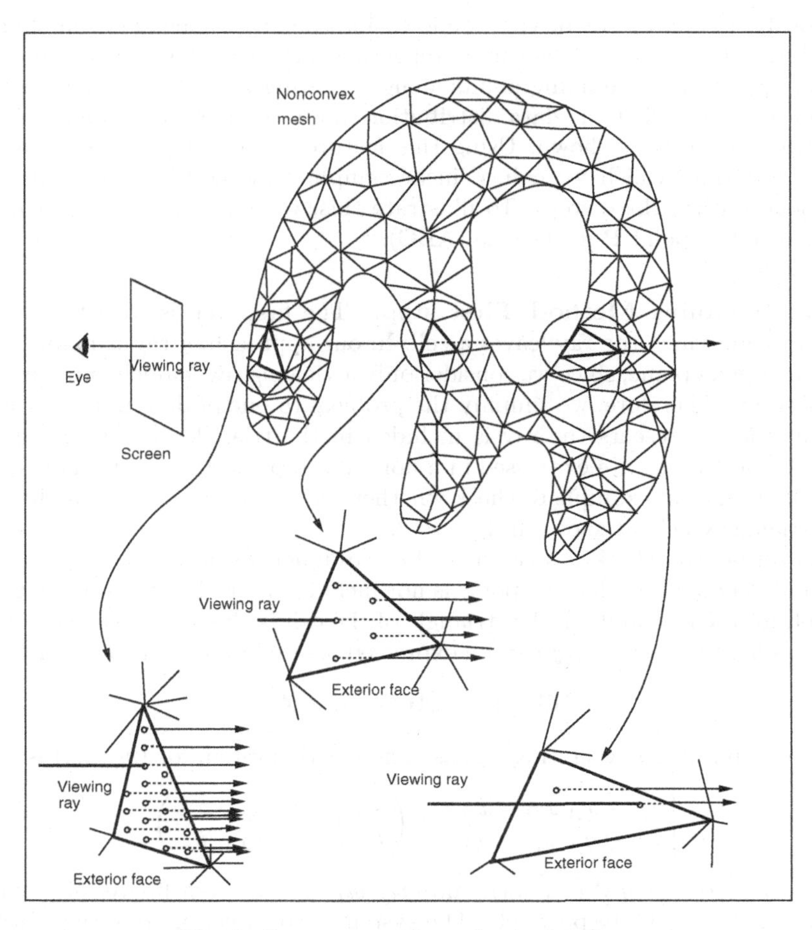

Fig. 1. Scan-conversion of exterior faces

3.2 Reducing the Cost of Interpolation of the Sampled Values

Interpolation in a Tetrahedral Cell In order to calculate the luminosity and opacity at a sampling point rapidly, it is very important that data, such as scalar data and scalar gradients, should be efficiently interpolated along a viewing ray. In general, a scalar value S_x at a point X in a tetrahedral cell (ABCD) is interpolated as

$$S_x = N_A \times S_A + N_B \times S_B + N_C \times S_C + N_D \times S_D \tag{3}$$

where

- N_A, N_B, N_C, and N_D are the interpolation functions of the tetrahedral cell at node points A, B, C, and D, respectively, and the equation $N_A + N_B + N_C + N_D = 1$ holds.
- S_A, S_B, S_C, and S_D are the scalar values at A, B, C, and D, respectively.

Obviously, it takes a lot of CPU time to interpolate scalar values at multiple sampling points in the cell by simply repeating this method. Our concern is with the interpolation at sampling points along a viewing ray. In this connection, we should remember that the data distribution in a tetrahedral cell is linear in any direction, as we have shown. Using this feature, we developed a new method of interpolation, which we call a linear sampling method. The procedure for interpolation has three steps. To illustrate this, we assume that a viewing ray enters a cell at point P in triangle (ABC).

Linear Sampling Method First step: The first step is to determine the face through which the ray leaves a cell. In our approach, a ray is mapped to a point in a pixel plane, since we consider only a viewing ray, namely a ray from an eye position. Therefore, we simplify the problem to one of determining whether a point that represents the ray is included in the triangle on the screen. The triangle and the ray are expressed in a normalized projection coordinate system (NPC). We use an example to check whether the ray leaves the cell at the point Q in triangle CDA, as shown in Figure 2.

Note that the check is performed in two-dimensional space. If the vectors $\boldsymbol{AC}$ and $\boldsymbol{AD}$ are parallel, the point is not included in it. The ray does not leave the cell from CDA, and another triangle of the cell is checked. If the vectors are not parallel, the vector $\boldsymbol{AQ}$ can be expressed as a linear combination of them:

$$\boldsymbol{AQ} = s_Q \times \boldsymbol{AC} + t_Q \times \boldsymbol{AD}, \tag{4}$$

where s_Q and t_Q are weighting values. The matrix system for (s_Q, t_Q) is

$$\begin{bmatrix} x_C - x_A & x_D - x_A \\ y_C - y_A & y_D - y_A \end{bmatrix} \times \begin{pmatrix} s_Q \\ t_Q \end{pmatrix} = \begin{pmatrix} x_Q - x_A \\ y_Q - y_A \end{pmatrix}, \tag{5}$$

where (x_C, y_C),(x_D, y_D),(x_A, y_A), and (x_Q, y_Q) are the coordinates of points A, B, C, and P in NPC, respectively. The system is solvable, because the matrix is regular. If s_Q and t_Q satisfy the following conditions:

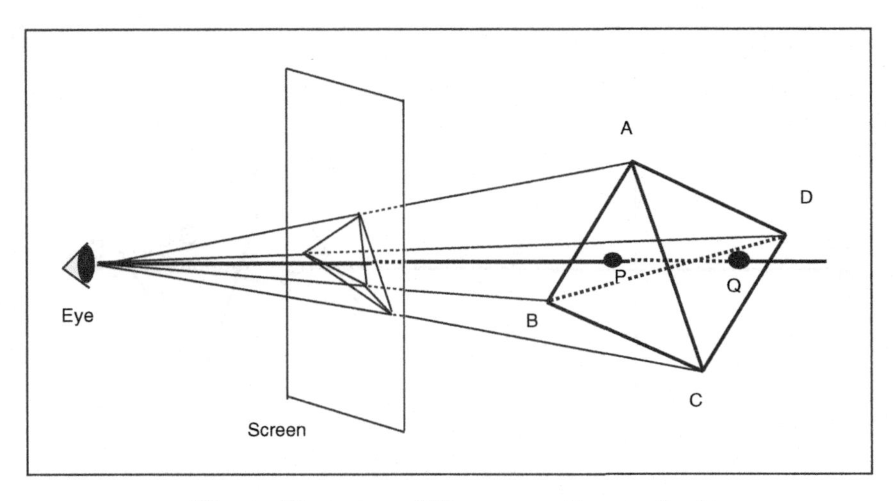

Fig. 2. First step of linear sampling method

- $s_Q \geq 0$
- $t_Q \geq 0$
- $s_Q + t_Q \leq 1$,

the point Q is inside the triangle CDA and checking need not be done at further candidate faces. Otherwise, the point is outside the triangle and another triangle is checked. Our approach checks two triangles on average.

In contrast, the conventional approach first calculates the distances along the ray at which it intersects the planes of three candidate exit triangles and then selects as the exit face the triangle that has the minimum distance value. The triangle and the ray are expressed in a view reference coordinate system (VRC). This approach has the merit of being able to handle various other rays in addition to viewing rays. However, it always needs to check three triangles in three-dimensional space, which is computationally inefficient in comparison with our approach.

Second step: The second step is to interpolate data (S_Q) at the intersection (point Q) of a cell and a viewing ray on the face determined in the first step, as shown in Figure 3. By replacing X with Q in Eq. 3, the data (S_Q) are interpolated as

$$S_Q = s \times S_C + t \times S_D + (1 - s - t) \times S_A, \tag{6}$$

where $s = N_C$, $t = N_D$, and $N_B = 0$. For this step, we propose two interpolation techniques that differ according to the way in which the weighting values (s,t) are calculated: an accurate technique and an approximate technique.

In the accurate technique, the weighting values can be obtained by solving matrix systems similar to Eq. 5, not in an NPC but in a VRC. All coordinates should be inversely transformed from an NPC into a VRC. Before the transformation, the z-coordinate of the point Q is calculated by using Eq. 6. Without

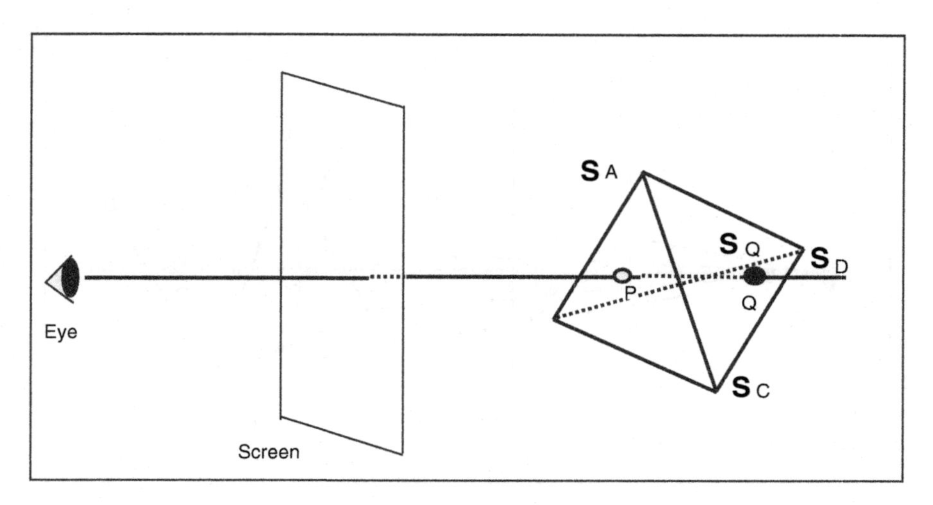

Fig. 3. Second step of linear sampling method

losing generality, we can assume that the triangle CDA is not parallel to the z-axis, because a regular triangle cannot be parallel to all three orthogonal axes x, y, and z. This assumption means that the parallel projection of the triangle CDA onto the xy-plane is not a straight line. As the point Q is in the triangle CDA,

$$\boldsymbol{AQ} = s_{QE} \times \boldsymbol{AC} + t_{QE} \times \boldsymbol{AD}, \tag{7}$$

where s_{QE} and t_{QE} are weighting values. Note that the values s_{QE} and t_{QE} are generally different from s_Q and t_Q of Eq. 5, respectively. Taking account of the x-components and y-components, the matrix system for (s_{QE}, t_{QE}) is

$$\begin{bmatrix} x_{CE} - x_{AE} & x_{DE} - x_{AE} \\ y_{CE} - y_{AE} & y_{DE} - y_{AE} \end{bmatrix} \times \begin{pmatrix} s_{QE} \\ t_{QE} \end{pmatrix} = \begin{pmatrix} x_{QE} - x_{AE} \\ y_{QE} - y_{AE} \end{pmatrix}, \tag{8}$$

where $(x_{CE}, y_{CE}), (x_{DE}, y_{DE}), (x_{AE}, y_{AE})$, and (x_{QE}, y_{QE}) are the coordinates of points C, D, A, and Q in the VRC, respectively. For efficient calculation, the coordinates of the nodal data component in the tetrahedral model should be stored in both the NPC and the VRC. If the triangle is parallel to the z-axis, the y-components and z-components, or the z-components and x-components, could be considered instead, according to whether the triangle is parallel to the x-axis or y-axis, respectively. The calculated weighting values (s_{QE}, t_{QE}) are used for (s,t) in Eq. 6.

In the approximate technique, the weighting values (s,t) are approximated by using the values of (s_Q, t_Q) calculated in the first step. No calculation for the inverse transformation is required in this technique. The technique ensures C_0 continuity across edges shared with neighboring triangles. However, it is not accurate, because it depends on the viewing parameters, such as the location of the viewing point. For example, when a relatively large triangle is rendered by this technique, a noticeable artifact appears in the generated image. Since we can assume that the tetrahedral cells handled are relatively small, we do not expect such an artifact to cause any problems. This expectation is confirmed by two images generated by using the two interpolation techniques given in the section on the performance evaluation.

Third step: The third step is to interpolate data at a sampling point, labeled X, along the ray by using the data (S_P, S_Q), as shown in Figure 4. We may assume that S_P has been previously interpolated. Using an interior division ratio r, defined as

$$r = \frac{\bar{PX}}{\bar{PQ}}, \tag{9}$$

we can calculate the data at a sampling point labeled X (S_X) as

$$S_X = r \times S_Q + (1 - r) \times S_P, \tag{10}$$

because they are linearly distributed along the segment $\bar{PQ}$. When interpolating the data at another sampling point along the ray in the same cell, we have only to repeat the third step.

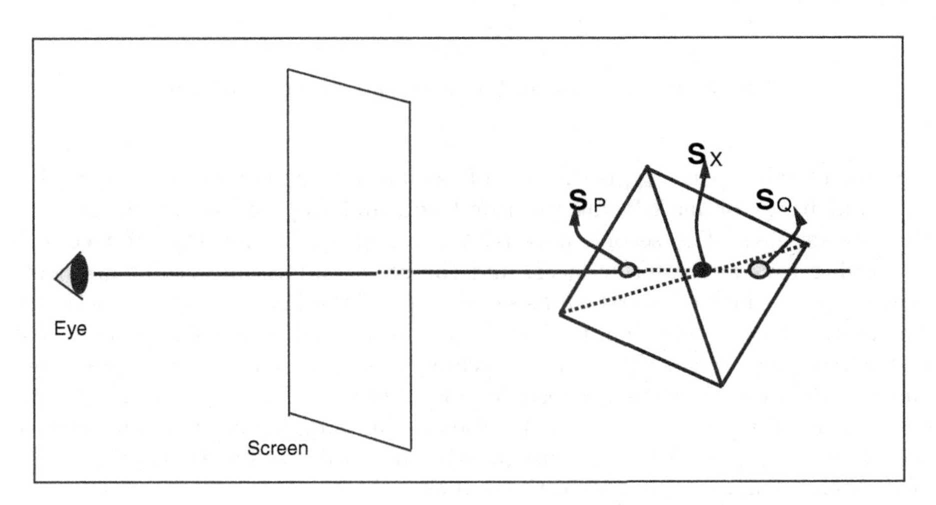

Fig. 4. Third step of linear sampling method

4 Performance Evaluation

4.1 Cost Functions

The algorithm for calculating the final image is given in Figure 5.

We divide the total cost, which excludes the cost of depth sorting of exterior faces, into three parts. The cost of depth sorting will be discussed in the next

Calculate the priority of exterior-faces
For an exterior face
1. Scan-convert the face
2. Calculate a pixel position by DDA
    For a viewing ray that is cast from the position
    a. Calculate the color and opacity
    b. Composite them into a final brightness
    c. Find the first cell
        For a cell
        1) Search for the exit triangle
        2) Calculate the weighting values (s, t)
        3) Interpolate the scalar and gradient
            For a sampling point
            a) Calculate the ratio r
            b) Interpolate the scalar and gradient
            c) Calculate the color and opacity
            d) Composite them into a final brightness
        4) Search for the next cell
    d. Search for the exterior face
    e. Calculate the color and opacity
    f. Composite them into a final brightness

Fig. 5. Fast algorithm for the image order approach

section. The first part, t_{re}, is the cost of calculating the intersections of viewing rays and front- or back-facing exterior faces, and the colors and opacities at the intersections. The second part is the cost of calculating the intersections of viewing rays and interior faces, and the scalar values and scalar gradients at the intersections (the first and second steps of the linear sampling method). An interior face is one that is shared by another cell in a volume. This cost is denoted as t_c^{ac} when the accurate technique is employed as the second step, and t_c^{ap} when the approximate technique is employed. The third part, t_s, is the cost of calculating the colors and opacities at sampling points (the third step of the linear sampling method). Consequently, the total cost functions (T_{AC}, T_{AP}) of this rendering algorithm are estimated as

$$T_{AC} = N_{re} \times t_{re} + N_C \times t_C^{AC} + N_S \times t_S, \tag{11}$$

$$T_{AP} = N_{re} \times t_{re} + N_C \times t_C^{AP} + N_S \times t_S, \tag{12}$$

where

- N_{re} is the number of times that viewing rays intersect front- or back-facing exterior faces.
- N_C is the number of times that viewing rays visit tetrahedral cells.
- N_S is the number of sampling points along viewing rays.

To demonstrate the efficiency of the algorithm, let us compare it with another algorithm, which has the following features:

- The linear sampling part is the same. (The accurate technique is employed as the second step.)
- The calculation of ray-face intersection uses bounding volumes whose shape is a frustum of a quadrangular pyramid, so that the intersection tests can be reduced to simple comparisons against the limits of the volumes on a pixel plane. (Exterior faces are projected onto the plane, as in the proposed algorithm.)

In this algorithm, given in Figure 6, a fourth cost, t_r, is added. This is for testing the bounding volume for intersections. The first cost, t_{re}, in this algorithm includes the cost of searching among all the candidate faces whose bounding volumes intersect a viewing ray for the face that the ray intersects. If the cost is denoted as $t_{re'}$, the cost function (T_{bv}) is

$$T_{bv} = N_{re} \times t_{re'} + N_C \times t_C^{AC} + N_S \times t_S + N_r \times N_e \times t_r, \tag{13}$$

where

- N_e is the number of front- or back-facing exterior faces.
- N_r is the number of viewing rays that hit the volume. $N_r \leq N_{re}$, because a viewing ray may intersect more than one front- or back-facing exterior face.

The main difference between the two algorithms is the object that controls the outermost loop: an exterior face for the former and a viewing ray for the latter. A ray-by-ray approach for intersection tests such as that in the latter algorithm requires a kind of "exhaustive search" for intersections. Indeed, the cost t_r of one test is trivial, because the test is a simple comparison, but the number of tests, $N_r \times N_e$, is very large. Therefore, the product $N_r \times N_e \times t_r$ is not negligible. A face-by-face approach such as the former algorithm is more efficient, because it makes use of coherence to avoid unnecessary intersection tests.

5 Result and Discussion

To demonstrate the efficiency of our algorithm, four irregular data sets of CFD results were rendered. These data sets, which involve cavities or holes, are categorized as nonconvex meshes. All the images were computed on an IBM 3090 at a resolution of 512 × 512 pixels.

Table 1 contains statistics on the test images.

From the values of T_{ac} in four datasets, the cost components of the total cost function in Eq. 11 are calculated as

$$t_{re} = 3.98 \times 10^{-4} \tag{14}$$
$$t_C^{AC} = 6.77 \times 10^{-5} \tag{15}$$
$$t_S = 1.29 \times 10^{-5}. \tag{16}$$

Calculate the priority of exterior-faces
Calculate the bounding volumes of the faces
For a viewing ray
  For an exterior face
  1. Find the bounding volumes that the ray hits
    For an exterior face surrounded by the volume
    1. Find the exterior faces that the ray hits
      For an exterior face that the ray hits
      a. Calculate the color and opacity
      b. Composite them into a final brightness
      c. Find the first cell
        For a cell
        1) Search for the exit triangle
        2) Calculate the weighting values (s, t)
        3) Interpolate the scalar and gradient
          For a sampling point
          a) Calculate the ratio r
          b) Interpolate the scalar and gradient
          c) Calculate the color and opacity
          d) Composite them into a final brightness
        4) Search for the next cell
      d. Search for the exterior face
      e. Calculate the color and opacity
      f. Composite them into a final brightness

Fig. 6. Conventional algorithm for the image order approach

Table 1. Statistics on benchmark tests for the image order approach

Dataset name	*no. 1*	*no. 2*	*no. 3*	*no. 4*
N	10167	61680	372500	557868
N_r	89694	165677	101107	127616
N_{re}	91740	194408	105018	132190
N_e	888	2716	6023	8839
N_c	2001800	7364650	6842093	11782659
N_s	2794063	7475567	2828874	4012530
Total time				
T_{bv}	520.1	2125.4	2374.5	4188.1
T_{ac}	211.5	661.3	539.0	906.7
T_{ap}	151.0	483.6	359.7	606.9
(seconds)				
T_{sort}	0.3	2.6	14.2	29.2
(seconds)				

The validity of these costs can be confirmed by using other datasets. By using the values of T_{bv} in four datasets and the calculated costs (t_{re}, t_C^{AC}, t_S), the costs, t_r and $t_{re'}$, of Eq. 13 are calculated as

$$t_r = 2.80 \times 10^{-6} \quad (17)$$

$$t_{re'} = 1.49 \times 10^{-3}. \quad (18)$$

The value of $t_{re'}$ is more than that of t_{re}. This is attributed to the cost of checking the exterior faces that are not intersected by a ray but that are surrounded by the bounding volumes hit by the ray. The value of t_S is less than that of t_C. This means that the cost at sampling points has less effect on the total cost than that at cell intersections. In other words, the total cost is not doubled even if twice as many sampling points are placed in such a way as to increase the image quality. By using the values of T_{ap} in four datasets, and the calculated costs, the cost, t_C^{AP}, in Eq. 12 is calculated as

$$t_C^{AP} = 4.12 \times 10^{-5}. \quad (19)$$

This cost is about forty percent lower than that of the accurate technique. Moreover, we cannot observe any noticeable difference between the two images, which were generated by using the two interpolation techniques, because the size of each tetrahedral cell is very small in relation to the extent of the image plane. Therefore, an approximate technique is preferable as the second step of the linear sampling method.

In the first step of the linear sampling method, we tested the ray individually against each candidate triangle to see if the ray intersected with the triangle. On the other hand, Hong treats the candidate triangles together as a group and uses a ray-crossing technique to find which triangles intersect with the ray [Hong99]. To compare our technique with Hong's technique, we measure a cost to find an intersected triangle and calculate the weighting values, s_Q and t_Q, using a single tetrahedral cell mapped onto an image plane. To consider finite precision of the measurement function, we repeat the first step thirty thousand times. Given an entry-point to the tetrahedral cell, Hong's technique calculates the ray-crossing number for each of the six triangle edges and the weighting values for a triangle the ray-crossing number of which is exactly one.

Table 2. Time cost for the first step (seconds)

No. of checked triangles	*1*	*2*	*3*
Ours	0.46	0.86	1.27
Hong's	1.67	1.68	1.70

The cost of our technique is dependent on the location of the entry point in the tetrahedral cell. On the other hand, that of Hong's technique is independ on the location. In the worst case, our technique calculates the weighting values three times. Therefore, we consider three cases in which our technique checks one, two, and three triangles. From Table 2, we confirm the effectiveness of our technique over Hong's technique even in the worst case.

The total time costs of dataset no. 4 were calculated by substituting these calculated costs and the statistics in Table 1 into Eqs. 11-13, as shown in Figure 7. The results of the calculation show the following:

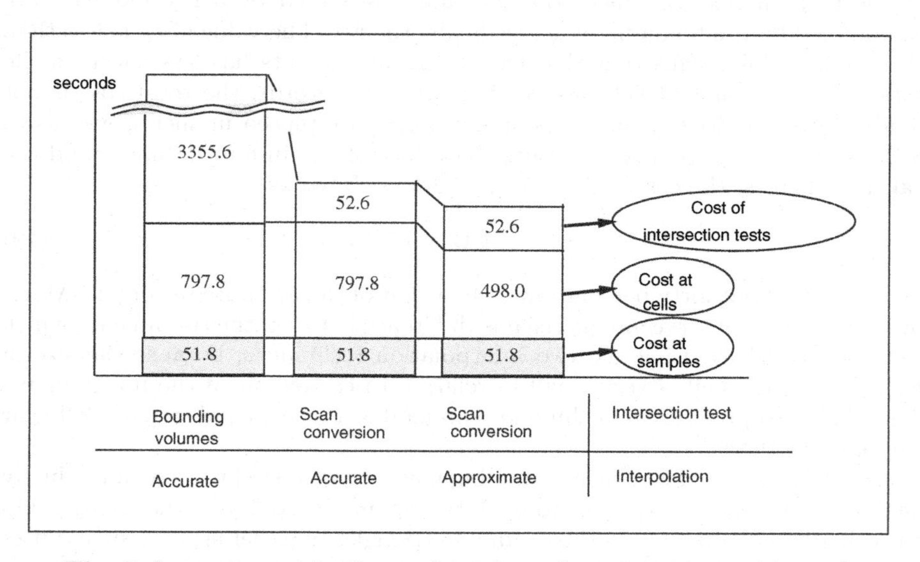

Fig. 7. Improvement in the performance when dataset no. 4 is used

1. If we adopt the intersection test based on bounding volumes, most of the total time cost is for the calculation of ray-face intersection tests.
2. If we adopt the intersection test based on scan conversion, the time required for the tests is very greatly reduced.
3. Interpolation at intersections of rays and cells still occupies a large part of the total computation time, although employment of an approximate interpolation technique reduces it significantly.

The total time cost is $O(N_C)$ in a rough estimate from Eqs. 11 and 12. The number N_C can be described as

$$N_C = N_C^{ray} \times N_{re}, \tag{20}$$

where N_C^{ray} is the average number of volume cells that a viewing ray visits from the entry point until the exit point. This number is estimated as $O(N^{1/3})$, where N is the number of volume cells. Therefore, the performance of our method except for depth sorting of front- or back-facing exterior faces is approximately O($N^{1/3}$).

Our algorithm requires depth sorting of front- or back-facing exterior faces as a pre-process. In the above test, an approximate priority was calculated by simply sorting the z-coordinates of the face-centroids. The priority resulting from the approximation may be incorrect and may lead to noticeable artifacts in the image. However, even this approximate sorting can generate a correct priority in many cases, because, unlike interior faces, the front- or back-facing exterior faces that overlap on a screen are likely to be sufficiently distant from each other for the sorting. The time cost, T_{sort}, of the sorting is small for the test datasets, as shown in Table 1. For larger datasets, this cost might not be negligible, because it is $O(N_e^2)$. In order to improve the performance, we can use a more efficient sorting algorithm such as radix sorting, whose performance is estimated as $O(N_e)$ [Suth74]. For correct sorting, we could employ the list-priority algorithm proposed by Newell [Newe72]. Although an exterior face may be split in this case, our algorithm can be made to work simply by replacing a component of an original exterior face with components of the split exterior faces in the tetrahedral model.

6 Summary

Although they can be expensive, volume ray tracing techniques are very useful for rendering irregular volumes with cavities or voids. The bottleneck of the overall process, which was confirmed by using some irregular data sets, is the ray-face intersection testing. To solve the problem, we incorporate a projection approach into the ray-casting process. Moreover, our technique can be used to speed up a conversion process to reconstruct a voxel dataset from a given irregular volume dataset.

References

Dreb88. R Drebin, L Carpenter, and P Hanrahan, "Volume Rendering," Computer Graphics, Vol. 22, No. 4, pp. 65-74, 1988. 558

Fuji85. A Fujimoto and I Kansei, "Accelerated Ray Tracing," Proceedings of Computer Graphics Tokyo '85, 1985. 560

Garr90. M P Garrity, "Raytracing Irregular Volume Data," Computer Graphics, Vol. 24, No. 5, pp. 35-40, 1990. 558

Glas84. A S Glassner, "Space Subdivision for Fast Ray Tracing," IEEE Computer Graphics and Applications, Vol. 4, No. 1, pp. 15-22, 1984. 560

Hanr90. P Hanrahan, "Three-Pass Affine Transformation for Volume Rendering," Computer Graphics, Vol. 24, No. 5, pp. 71-78, 1990. 557

Hong99. L Hong and A. E. Kaufman, "Fast Projection-Based Ray-Casting Algorithm for Rendering Curvilinear Volumes," IEEE TVCG, Vol. 5, No. 4, pp.322-332, 1999 558, 569

Koya90. K Koyamada, "Volume Visualization for the Unstructured Grid Data," SPIE/SPSE Symposium on Electronic Imaging Conference Proceedings, Vol. 1259, pp. 14-25, 1990. 558

Koya92a. K Koyamada, "Fast Traverse of Irregular Volumes," CG International '92 Proceedings, pp. 295-312.

Koya92b. K Koyamada, S Uno, A Doi, and T Miyazawa, "Fast Volume Rendering by Polygonal Approximation," Journal of Information Processing, Vol. 15, No. 4, pp. 535-544. 557

Laur91. D Laur and P Hanrahan, "Hierarchical Splatting: A Progressive Refinement Algorithm for Volume Rendering," Computer Graphics, Vol. 25, No. 4, pp. 285-288, 1991. 557

Levo88. M Levoy, "Display of Surfaces from Volume Data," IEEE Computer Graphic and Applications, Vol. 8, No. 3, pp. 29-37, 1988. 558

Newe72. M E Newell, R G Newell, and T L Sancha, "A New Approach to the Shaded Picture Problem," Proc. ACM National Conf., 1972. 571

Pfis99. H. Pfister, J. Hardenbergh, J. Knittel, H. Lauer, and L. Seiler, "The VolumePro Real-Time Ray-casting System," Computer Graphics, Vol. 24, No. 5, pp.63-70, 1999. 558

Phon75. B T Phong, "Illumination for Computer-Generated Pictures," Communications of ACM, Vol. 18, No. 6, pp. 311-317, 1975. 558

Rubi80. S M Rubin and T Whitted, A Three-Dimensional Representation for Fast Rendering of Complex Schemes," Proceedings of SIGGRAPH '80, pp. 110-116, 1980. 560

Sabe88. P Sabella, "A Rendering Algorithm for Visualizing 3D Scalar Fields," Computer Graphics, Vol. 22, No. 4, pp. 51-58, 1988. 558

Schr91. P Schroder and J B Salem, "Fast Rotation of Volume Data on Data Parallel Architecture," IEEE Visualization '91, pp. 50-57, 1991. 557

Shir90. P Shiley and A Tuchman, "A Polygonal Approximation to Direct Scalar Volume Rendering," Computer Graphics, Vol. 24, No. 5, pp. 63-70, 1990. 557

Suth74. I E Sutherland, R F Sproull, and R A Schumacker, "A Characterization of Ten Hidden-Surface Algorithms," Computer Surveys, Vol. 6, No. 1, 1974. 571

Upso88. C Upson and M Keeler, "The V-Buffer: Visible Volume Rendering," Computer Graphics, Vol. 22, No. 4, pp. 59-64, 1988. 558

Wegh84. H Weghorst, G Hooper, and D P Greenberg, "Improved Computational Methods for Ray Tracing," ACM Transactions on Graphics, Vol. 3, No. 1, pp. 52-69, 1984. 560

Wilh90a. J Wilhelms, J Challinger, and A Vaziri, "Direct Volume Rendering of Curvilinear Volumes," Computer Graphics, Vol. 24, No. 5, pp. 41-47, 1990. 558

Wilh91. J Wilhelms and A V Gelder, "A Coherent Projection Approach for Direct Volume Rendering," Computer Graphics, Vol. 25, No. 4, pp. 275-284, 1991. 557

Will92a. P L Williams, "Interactive Splatting of Nonrectilinear Volumes," IEEE Visualization '92 Proceedings, pp. 37-44, 1992. 557

A Study on the Effect of Air on the Dynamic Motion of a MEMS Device and Its Shape Optimization

Hidetoshi Kotera, Taku Hirasawa, Sasatoshi Senga, and Susumu Shima

Department of Mechanical Engineering, Kyoto University
Sakyou-ku, Kyoto, Japan
kotera@mech.kyoto-u.ac.jp

Abstract. We propose a new design concept for controlling the deflection of a micro-membrane with the aid of its thickness distribution for realizing a prescribed design in the MEMS. As an example, we treat a micro air pump that comprises a micro-membrane. The membrane is actuated by an electrostatic force. The membrane deflects and thus the deflection is influenced by the air pressure and the electrostatic field. This is a highly complicated system. To find out a proper thickness distribution, we use the genetic algorithm that is appropriate to reduce the searching space of solution.

1 Introduction

Recently, a few types of micro-electro-mechanical system (MEMS), e.g., micro optical mirrors, micro valves, micro pump and micro actuators, have been developed and used in practice (1-3). These devices are actuated by electrostatic force and/or fluid pressure. In developing the MEMS, the shape and material for the parts should be designed properly so that their motion would provide a desired function. We must consider that the dynamic response of the micro-membrane actuator is strongly affected by its shape and its mechanical properties. When a membrane is subjected to air pressure, electrostatic force or other external force, it would deflect depending on its material properties and boundary conditions. The deflection of the membrane in turn influences the airflow or pressure and electrostatic force. If the membrane is of a constant and uniform thickness, a particular external force gives a particular deflection to the membrane. If, however, the thickness is not uniform, the deflection may be varied. In other words, the deflection would be controlled by its thickness distribution. In fabricating a micro-membrane by physical vapor deposition or chemical vapor deposition, it is easy to distribute the thickness of the thin films. It would thus be possible to develop a new mechanical structure for the desired performance.

However, this is a highly complicated system and we should develop a new scheme for obtaining an optimal shape or thickness distribution of the membrane. In our previous study, we proposed a method of analyzing the dynamic response of the micro-membrane actuated by an electric field (4).

M. Valero et al. (Eds.): ISHPC 2000, LNCS 1940, pp. 573-584, 2000.

The purpose of this study is to develop an optimization method for calculating a thickness distribution of the micro-membrane actuated by an electrostatic force. Since the genetic algorithm is appropriate to reduce the searching space of solution, it is employed to find out an optimum shape (5). As an example, the thickness distribution of the micro-membrane of a micro-air-pump actuated by an electrostatic force is optimized to realize the prescribed response. We will show the numerical method for optimization based on the genetic algorithm and we discuss the convergence of the solution and efficacy of the developed method.

2 Dynamic Response of Membrane

As an example of MEMS actuated by an electrostatic force, we treat a model of a micro-air-pump as shown in Fig. 1. This composes a micro-membrane, or an upper electrode, and a base, or a lower electrode. The size of the micro-pump is 400μm long, 200μm wide and 20μm high. In this pump, the outlet is located not at the center, but 150μm apart from it. The air is pumped out according to the deflection of the micro-membrane actuated by an electrostatic force.

Pulled by the electrostatic force, the micro membrane undergoes deformation in both in-plane and out-of-plane directions. Therefore, stress equilibrium equations of in-plane stress and bending are used. As the micro membrane deflects, its distance from the lower electrode changes. Thus, the change of actuation force due to the micro-membrane deflection should be considered. We used the following equations for simulation.

1) Electric field

The Laplace equation for the electrostatic field where a distance charge between the upper and lower electrode does not exist is written as

$$\nabla^2 \phi = 0 \qquad [1]$$

where ϕ is an electrostatic potential. The strength of electrostatic field E is given by

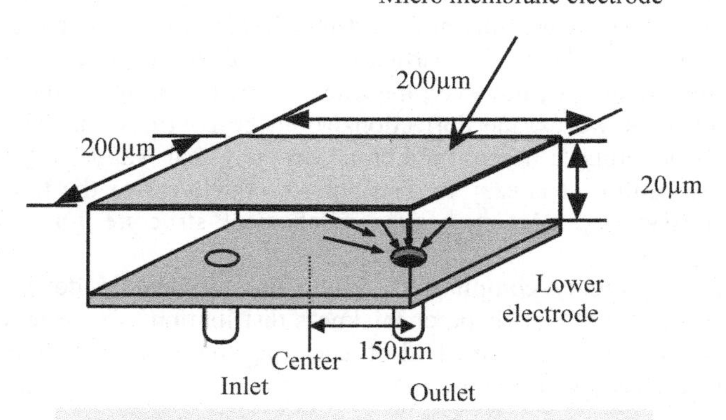

Fig.1 Schematic view of micro pump

$$E = -\nabla\phi = -\frac{\partial\phi}{\partial n} \qquad [2]$$

where n is the outward normal to the surface of the area concerned. The electrostatic stress is written as

$$f_e = -\varepsilon\frac{E^2}{2} \qquad [3]$$

where ε is dielectric constant of the fluid.

2) Stress equilibrium

The stress equilibrium equation for in-plane deformation of the membrane is written as

$$\frac{\partial\sigma_x}{\partial x} + \frac{\partial\tau_{xy}}{\partial y} + f_x = 0, \qquad \frac{\partial\sigma_y}{\partial y} + \frac{\partial\tau_{yx}}{\partial x} + f_y = 0 \qquad [4]$$

where f_x and f_y are body forces in x and y directions, respectively.

The membrane is bent by the dynamic pressure of the fluid and electrostatic force. The stress equilibrium equation for bending is expressed by

$$\begin{aligned} &D_{xx}\frac{\partial^4 w}{\partial x^4} + 2(D_{xy} + D_{ss})\frac{\partial^4 w}{\partial x^2\partial y^2} + D_{yy}\frac{\partial^4 w}{\partial y^4} \\ &-t_m\sigma_x\frac{\partial^2 w}{\partial x^2} - t_m\sigma_y\frac{\partial^2 w}{\partial y^2} = p - p_a + f_e + t_m\rho\frac{\partial^2 w}{\partial t^2} \end{aligned} \qquad \square\,[5]$$

with

$$D_{yy} = \frac{E_y t_m^{\ 3}}{12(1-\nu_x\nu_y)},\ D_{xx} = \frac{E_x t_m^{\ 3}}{12(1-\nu_x\nu_y)},\ D_{xy} = \frac{(\nu_y E_x + \nu_x E_y)t_m^{\ 3}}{24(1-\nu_x\nu_y)},$$

$$D_{ss} = \frac{G_{xy}t_m^{\ 3}}{6}, G_{xy} = \frac{E_x E_y}{(1+2\nu_y)E_x + (1+2\nu_x)E_y}$$

where w is membrane deflection; D_{xx}, D_{xy}, D_{yy} and D_{ss} are flexural rigidity; t_m is the thickness of micro-membrane; p pressure of fluid; p_a atmospheric pressure; ρ is density of micro-membrane; f_e is electrostatic stress; E_x and E_y are Young's moduli of micro-membrane in x and y direction respectively; ν_x and ν_y are Poisson's ratio. Since, the thin film is deposited by physical and/or chemical vapor deposition, the mechanical properties would be anisotropic. However, we assumed the material to be isotropic in the present simulation.

3) Fluid equation

The fluid, that is, the air in the pump is squeezed out by the micro-membrane deflection. To simulate the fluid flow and the pressure distribution in the micro-pump accurately, it is necessary to consider the fluid flow, a pressure loss and blowout of the fluid at the outlet. To do this, the Navier-Stokes equation should be solved. However, considering that the aspect ratio of the micro-pump cavity is more than ten and the height is in the order of 20 μm, we may estimate the performance of the micro-pump from the pressure distribution on the deflected micro-membrane and from the volume change of the pump cavity without taking account of these factors. Therefore, we use the modified Reynolds equation as the fluid equation. The outlet is a hypothetical one without a through hole.

The compressible fluid-pressure is expressed by the modified Reynolds equation considering the slip on the material surface as

$$\left\{\frac{\partial}{\partial x}\left(ph^2\frac{\partial p}{\partial x}\right)+\frac{\partial}{\partial y}\left(ph^2\frac{\partial p}{\partial y}\right)\right\}+6\lambda_a p_a\left\{\frac{\partial}{\partial x}\left(ph^2\frac{\partial p}{\partial x}\right)+\frac{\partial}{\partial y}\left(ph^2\frac{\partial p}{\partial y}\right)\right\}=6\mu_a\left\{V_x\frac{\partial(ph)}{\partial x}+V_y\frac{\partial(ph)}{\partial y}\right\}+12\mu_a\frac{\partial(ph)}{\partial t} \qquad [6]$$

where h is the distance between the micro-membrane and the lower electrode, λ_a is the molecular mean free path of the air, t refers to time, V_x and V_y are velocity of the surface of the micro-membrane in x and y directions, respectively, and μ_α is viscosity of the fuild. h is given by $h = h_0 + w$, where h_0 is initial gap between the micro-membrane and the lower electrode. V_x and V_y are calculated by the velocity of the micro-membrane deflection.

Equations [4], [5] and [6] should be solved simultaneously to analyze the deflection of the micro-membrane (4). The distance between the micro-membrane and the lower electrode in the present case is large enough so that the effect of molecular mean free path of the air may be negligibly small.

In the coupled analysis, first, the electric field equation is solved by the boundary element method. Second, derivatives with respect to time involved in equations [5] and [6] are calculated by the implicit method. Finally, the stress equilibrium equations [4] and [5] and the modified Reynolds equation [6] are solved until the deflection w becomes unchanged. This calculation was carried out iteratively. According to the deflection of the micro-membrane, the boundary elements of the electrostatic field are modified.

3 Coding and De-coding of Membrane Thickness for Optimization

We use the genetic algorithm to find out an optimum thickness distribution of the micro-membrane so that the pressure of the fluid at the vicinity of the outlet becomes a maximum. In the present method, the membrane consists of elements with two different thicknesses.

In the genetic algorithm, the thickness of the micro-membrane were coded to 0 or 1. As an example, the coded genes are as follows,

1110000011000011110000011000011110000011000011110000011000011110000011000011110000011000011111

Each component of the gene refers to the thickness of the membrane of the finite element; '1' means t_1 and '0' means t_2, as shown in Fig.2. The thickness of the neighboring four meshes is of the same value.

For optimization by the genetic algorithm, a fitness for each population should be calculated and verified for all population in each generation. The fitness function should be defined that is proper to express the characteristics of the phenomena for optimization. As a demonstration of the proposed method, we optimize the thickness distribution of the micro-membrane used for the micro-air-pump as shown in Fig.1. If the thickness of the micro-membrane actuated by the electrostatic force is uniform, the deflection of the micro-membrane is the largest at the center. The fluid pressure at the center also becomes a maximum, while the pressure at the outlet is low. Therefore, the performance of the micro-pump composed of the micro-membrane of a uniform thickness may be low. The maximum deflection point may better be changed according to the thickness distribution of the micro-membrane. Since the pressure distribution depends on the distance between the micro-membrane and the lower electrode, we utilized the distance at the outlet as a fitness function. Further, to increase the flow at the outlet, we considered the volume change in the pump cavity due to the micro-membrane deflection. We define the fitness function f as,

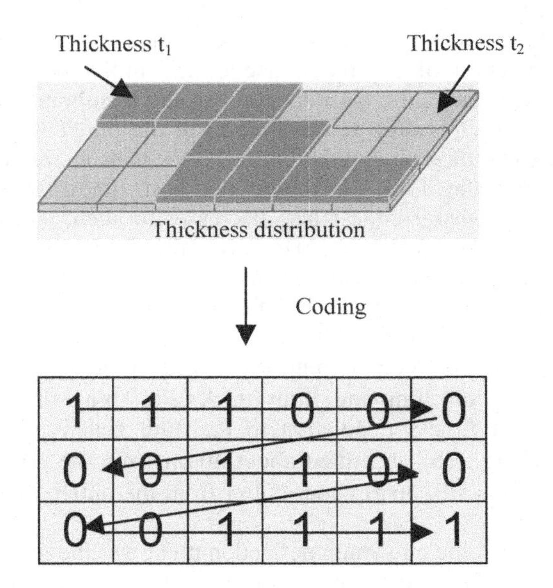

Fig.2-Coding method of thickness distribution

$$f = 0.7\sqrt{\frac{(x_p - x_h)^2 + (y_p - y_h)^2}{(x_c - x_h)^2 + (y_c - y_h)^2}} + 0.3\left|\frac{V_{def}}{V_{init}}\right| \qquad [7]$$

where (x_h,y_h) is the outlet position, (x_p,y_p) a maximum deflection point, (x_c,y_c) the center of the membrane, V_{def} the internal volume of the micro-pump cavity after the micro-membrane deflection, and V_{init} is the initial volume of the pump cavity. In this case the lowest possible value of "fitness" is the goal.

We used the one point mutation for cross over and elite strategy. The cross over rate was 0.9 and the mutation rate was 0.005. We used the tournament method for selection with population of 20 up to the 40th generation (5). In each generation, the micro-membrane deflection was calculated by the coupled analysis for 20 populations. The thickness distribution of each population for the following generation was calculated from the results of the previous one.

4 Results and Discussion

The material of the micro-membrane is silicon. The material constants for calculation are summarized in Table 1. In the calculation, the outlet is assumed to be located 150μm apart from the center of the micro-pump or 50μm from the side-wall. As a first demonstration of optimization, 2μm (t_1) and 3μm (t_2) thick elements were distributed to move the point at which the deflection gives a maximum at the outlet and to increase flow volume at the same time. Figure 3 shows the thickness distribution and deflection of the membrane of the initial population. Since the thickness is distributed randomly, the deflection of the membrane is a maximum at the center. Therefore, the pressure at the vicinity of the outlet is lower than at the center. The fitness value of each population decreases with increasing generation as shown in Fig.4. After the 15th generation, the best fitness for 20 populations decreases to 0.55. The average fitness also decreases to about 0.6. The best fitness does not change after 20th generation. However, the flow volume increases with increasing generation gradually. This means that the thickness distribution is optimized to move the point at which the deflection gives maximum before the 15th generation and to increase the flow volume after the 15th.

To move the maximum deflection point further near to the hypothetical outlet, we distribute the elements of 2μm and 4μm thick. Fig. 6(a) shows the thickness distribution of the best fitness population in the 40th generation. The maximum deflection point, see Fig. 6(b), is shifted about 40μm from the previous result with 2μm and 3μm thick. It is still apart about 20μm from the outlet, that is, 70μm apart from the side wall.

In the second example, the maximum deflection point was moved toward the outlet comparing with that in the first; the pump out volume rate was decreased by about 30%. This would be due to the fitness function that treats the deflection point and the volume as given by equation [7].

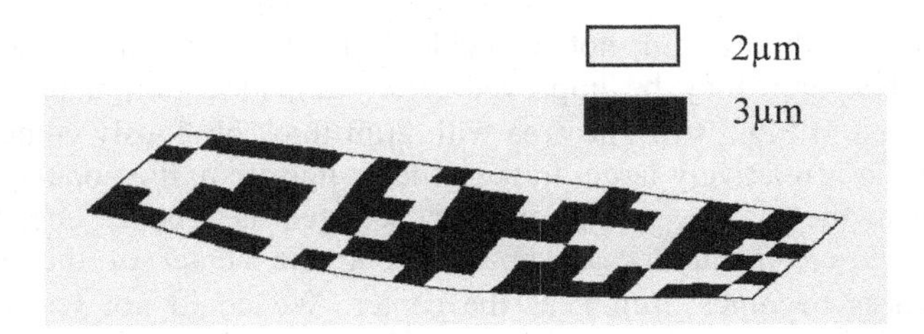

(a) Thickness distribution and deflection of membrane

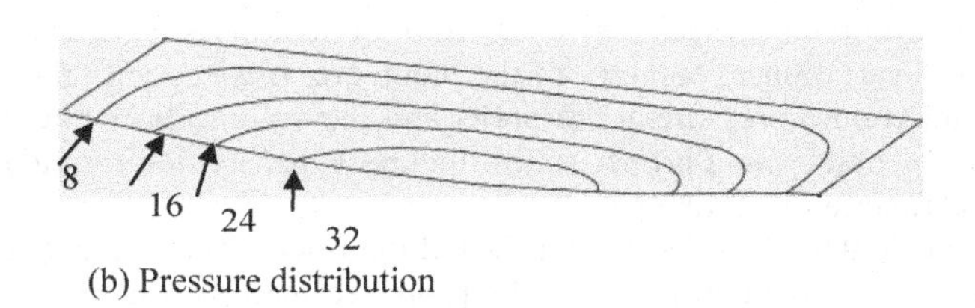

(b) Pressure distribution

Fig.3 Thickness and pressure distribution of initial generation

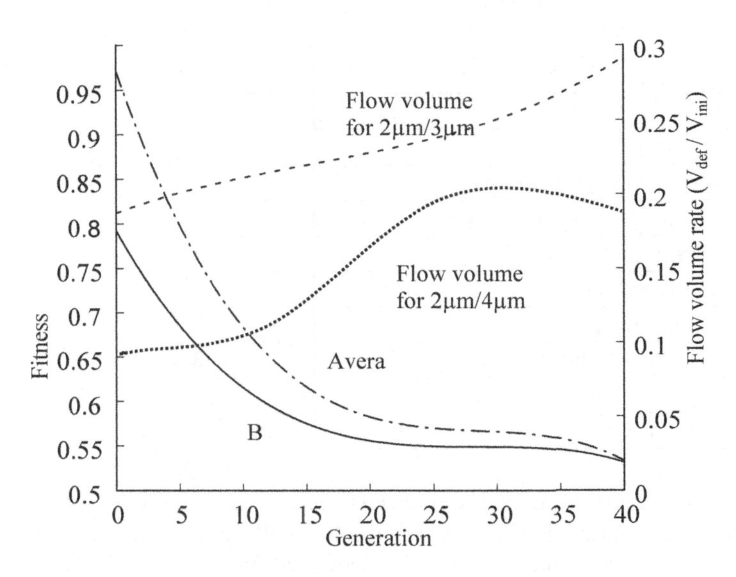

Fig.4 Fitness and flow volume in each generation

Figure 5(a) shows the thickness distribution of the best fitness individual in the 40th generation. As seen in Fig. 5(b), the maximum deflection point of the micro-membrane is still 60μm apart from the outlet. As shown in Fig. 5(c), the point at which the pressure gives a maximum is shifted toward the outlet. Although, the response of the membrane did not coincide with the prescribed design, the performance of the pump may be improved. The strain distribution in the micro-membrane is plotted in Fig.5 (d). The area with 2μm thick obviously coincides with that where the strain is relatively large. In the genetic algorithm, the population that is easier to deform at the vicinity of the outlet is selected as a better one. Since the deflection of the micro-membrane is suppressed at the center of the pump, the membrane thickness becomes 3μm near the center. We could not reach the best thickness distribution that provided the desired deflection. This may be due to that the calculation was undertaken for the defined thickness variation, 2μm or 3μm and the geometrical configuration of the pump.

It is obvious that the optimized results depend on the fitness function and the assumption in the simulation, such as, the ratio of thickness of the membrane, number of thickness variations, geometry of the pump etc. If we specify the geometry of the pump, the maximum pressure at the outlet and the volume flow rate, we could be able to succeed by obtaining a proper set of thickness distribution provided that an appropriate fitness function is defined.

To evaluate the calculated results, we fabricated the micro-membrane actuator that the thickness distribution was optimized by the proposed method. Figure 7 shows the fabrication process. The thickness of the silicon wafer was reduced by the wet etching with the SiO2 as a etching mask. The membrane is bonded on the glass basement on

Table 1 Material of micro-membrane and geometrical properties for analysis

Length (μm)	400
Width (μm)	200
height (μm)	20
Thisckness of the membrane (μm)	2 or 3
Mass density of the membrnae (kg/m^3)	2330
Young's modulus (GPa) Ex , Ey	150
Poisson's ratio	0.3
Viscosity (μPa s)	17.6
Molecular mean free path (μm)	0.064
Atmospheric pressure (MPa)	0.101
Permittivity (F/m)	8.854E-12

which the ITO is deposited as a lower electrode. The gap size between the lower electrode and the membrane is 10μm. When the 160 voltage was charged between the lower electrode and upper electrode that the deposited on the backside of the silicon micro-membrane, The membrane deflected as plotted in Fig.8. The small gray square blocks mean the thick area of the micro-membrane and the white area means thin area. The upper side of the figure is the measured result and the lower the calculated result by the proposed method that the hypothetical maximum deflection point is sifted 2.5mm from the center of membrane. The experimental result agrees very well with the calculated one.

5 Concluding Remarks

This study treats a new concept for designing the MEMS to control the micro-membrane deflection by the distribution of element thickness. We developed a method in an attempt to optimize by the genetic algorithm the thickness distribution in the micro-membrane actuated by an electric force to perform a desired deflection. As an application of our developed method, we optimized the thickness distribution of the micro-membrane actuated by the electrostatic force. It was shown that the deflection pattern depends very much on the thickness distribution.

By the present method, the dynamic motion of a micro membrane can be varied and therefore, the performance of MEMS would be optimized.

Acknowledgments

This work was supported by Grant-in-Aid for Scientific Research (C)(No. 10650258) by the Ministry of Education, Science, Sports and Culture.

References

1. Hiroshi Toshiyoshi, Journal of Microelectromechanical systems, **5**, 4, p231, (1996)
2. Xing Yang, Micro Electro Mechanical Systems January, p114, (1997)
3. Michael Koch et.al., Sensors and Actuators, **70A**, p98, (1998)
4. H. Kotera, Y. Sakamoto, T. Hirasawa, S. Shima and Robert W. Dutton, MICRO SYSTEM technologies, p91, (1998)
5. David E. Goldberg, Genetic Algorithms in Search Optimization and Machine Learning. Addison Wesley. (1989)

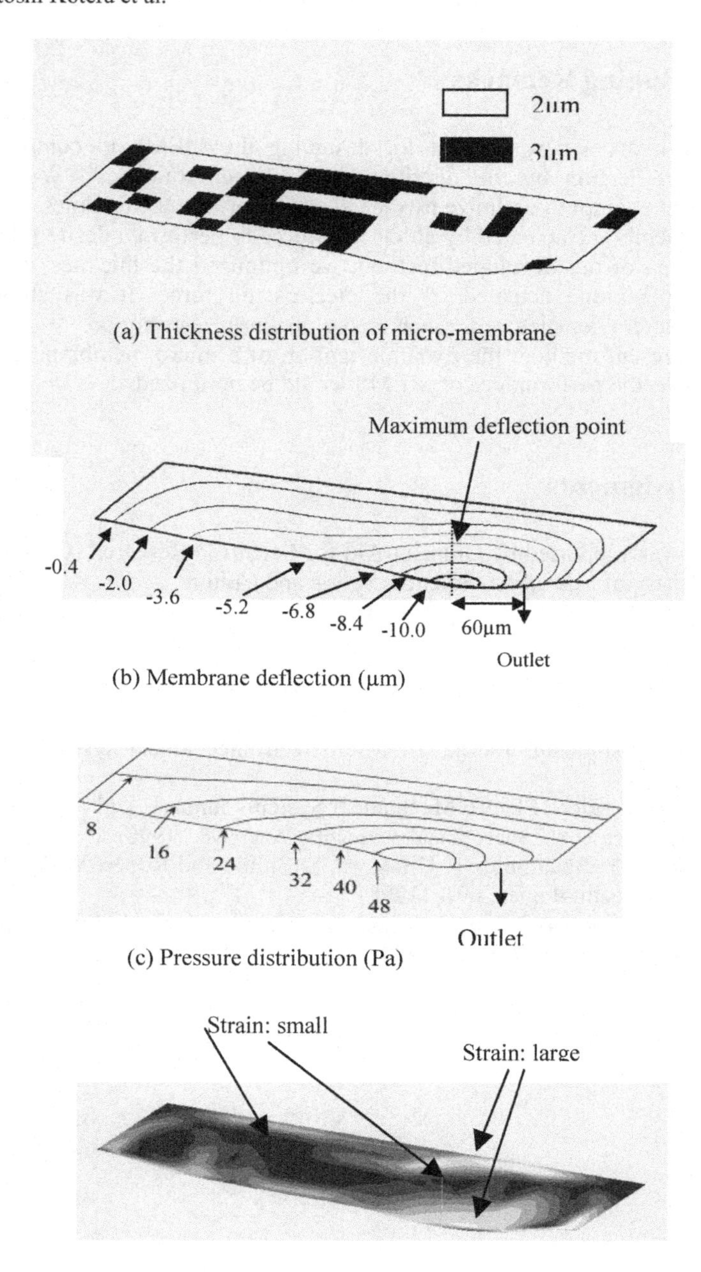

Fig.5 Thickness distribution and deflection of the membrane with 2μm and 3μm thick elements in 40th generation

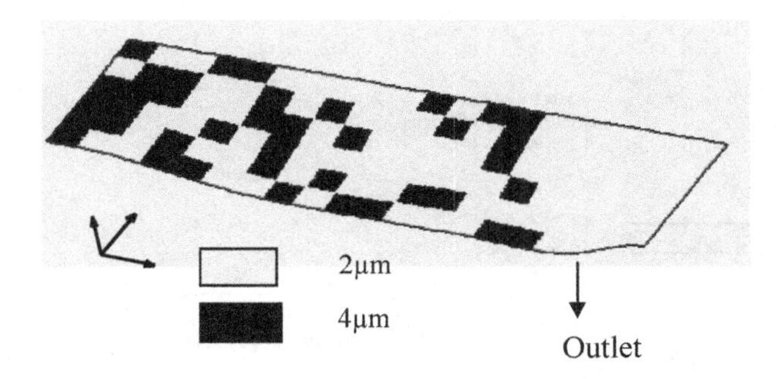

(a) Thickness distribution of micro-membrane

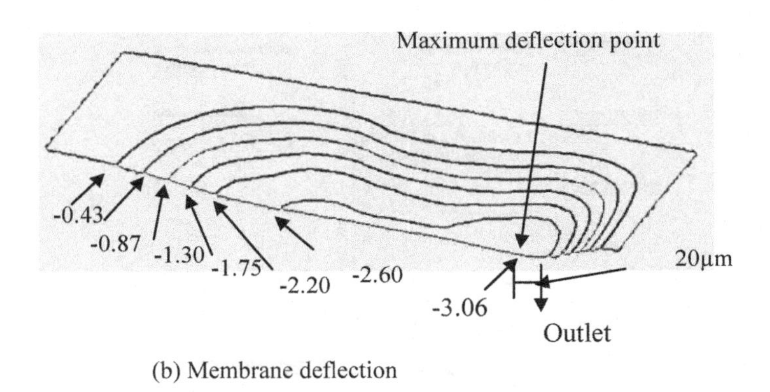

(b) Membrane deflection

Fig.6 Deflection of 2μm and 4μm thick element in 40th generation

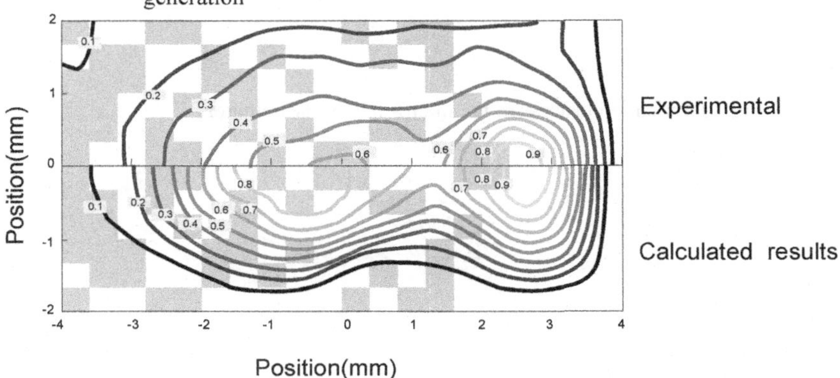

Fig.8 Deflection of the micro-membrane that its thickness distribution is optimized

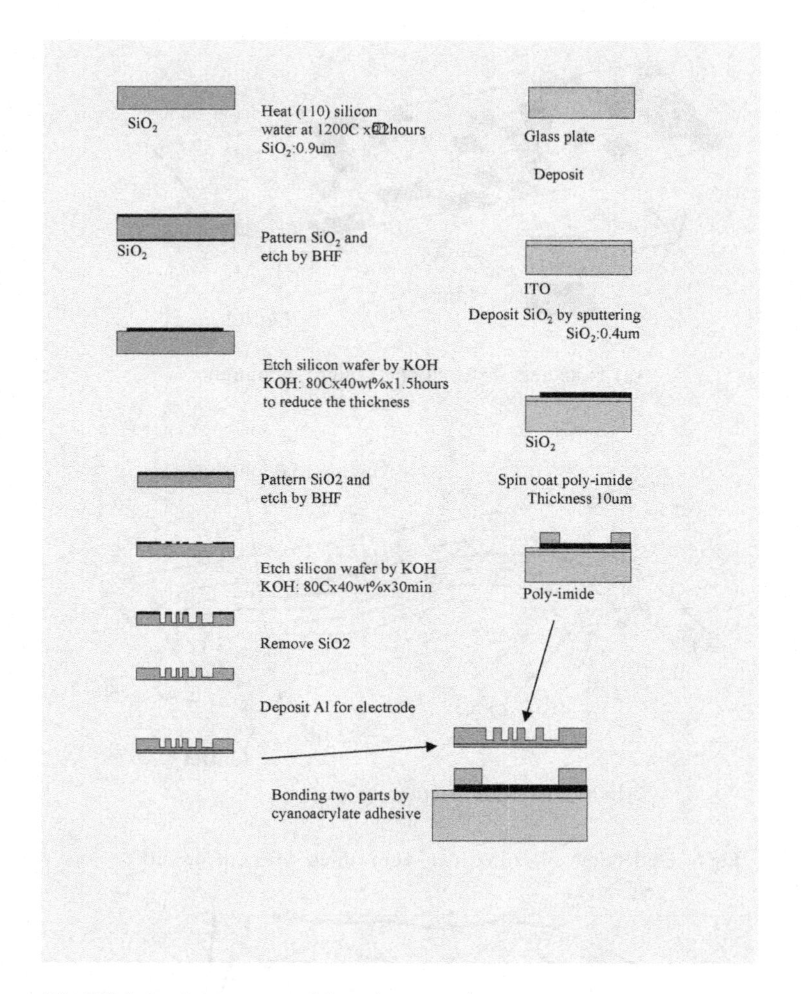

Fig.7 Fabrication process of the micro-membrane actuator

A Distributed Rendering System "On Demand Rendering System"

Hideo Miyachi[1], Toshihiko Kobayashi[1], Yasuhiro Takeda[1], Hiroshi Hoshino[2], Xiuyi Jin [2]

[1]Visualization Development Division, KGT Inc.,2-8-8 Shinjuku,Shinjuku-ku,Tokyo 160-0022,Japan)
{ miyachi, tosihiko, ytakeda}@Kubota.co.jp
http://www.kgt.co.jp
[2]Advanced Software Technology & Mechatronics Research institute of KYOTO, 17 Minamimachi Chudoji, Simogyou-ku, KYOTO 600-8813, Japan
{hoshino,jin}@astem.or.jp
http://www.astem.or.jp

Abstruct. In parallel computing, providing the way for visualizing a large scale dataset becomes demanding. To provide a necessary levels of performance, there have been several software-based rendering system developed for general purpose parallel architectures. We have been developing the distributed rendering system focused on reducing network traffic to visualize a large scale dataset especially 3D geometric data. This paper describes the design of our distributed parallel rendering server (called On Demand Rendering System), and the possibility for applying this system to the visualization of high performance computing.

1 Introduction

A large scale parallel numerical simulation has become very popular in any scientific field. As the power and availability of general-purpose parallel computer systems have grown, the computer graphics community has become increasingly interested in exploiting them to support sophisticated rendering methods and complex scenes. In order to meet the demands of parallel simulation and visualization, several software-based renderers have been developed these days[1] .

Software-based renderers are very effective especially massive datasets which is produced by large-scale scientific applications. They can easily be hundreds of megabytes in size, and time-dependent simulations might easily increases the output.

M. Valero et al. (Eds.): ISHPC 2000, LNCS 1940, pp. 585-591, 2000.

Sending such an large output data to other workstation is not practical for visualization. Almost all software-based renderers have API's to be embedded in parallel applications to produce renderable output for clients[2]. Fig1 shows the basic architecture of ordinary software-based renderers for parallel processing.

On the other hand, we have been developing distributed rendering system called "On Demand Rendering System"[3]. This system is also software-based renderers for 3D geometrical models. However, our rendering system is focusing on gathering 3D geometrical models (ex. VRML) stored in distributed databases on the network and serve them to the client as a rendered images (shown in Fig2) . In other words, former software-based parallel renderer represented Fig1 is very tightly connected with architecture in one parallel or cluster computer, but our system is network based parallel renderer.

In this paper, we describe the design and implementation of the On Demand Rendering System, and discuss the possibility of network parallel rendering for visualizing massive datasets.

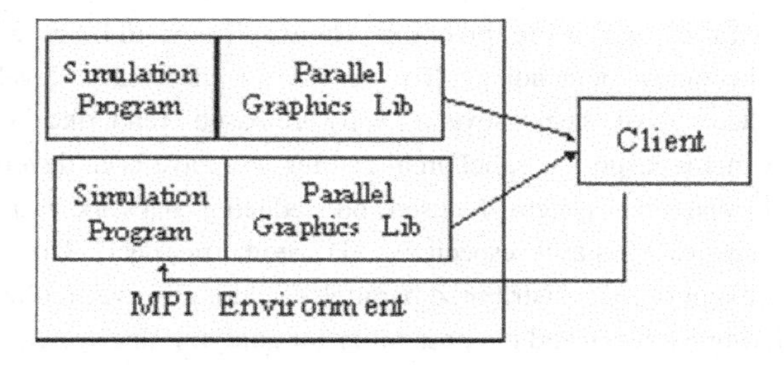

Fig. 1. MPI based Parallel Visualization System

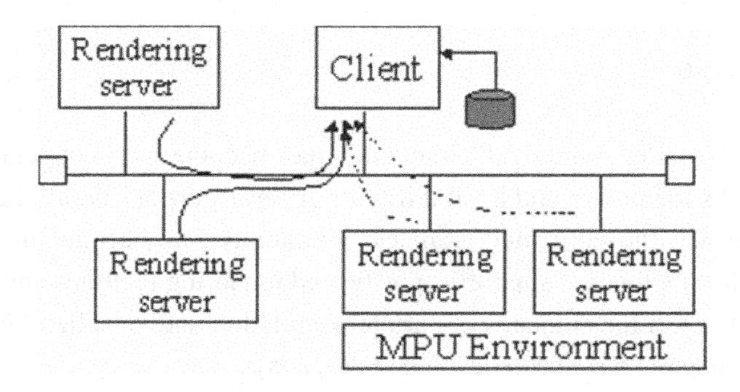

Fig. 2. On Demand Rendering System architecture

2 Design of On Demand Rendering System

The On Demand Rendering System is designed as a Server-Client system. It is basically the image based rendering system. However, considering that the system runs under various network and computer environments, there are several features to communicate between the server and client. Fig3 illustrates the architecture of the On Demand Rendering Server framework, which consists of the following parts:

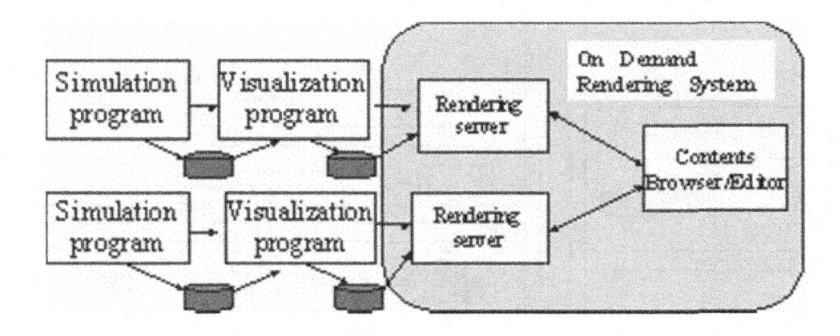

Fig. 3. A Sample Data flow of On Demand Rendering System

2.1 Rendering Server

The Rendering Server is the application that receives the geometrical data from computer disks or database and renders it as the client's requests. There are three steps defined for rendering a 3D geometrical data. They are "transformation" (defined as "rendering step 0"), " rasterising" (defined as "rendering step 1"), and "assembling" (defined as "rendering step 2"). The transformation process translates 3D geometrical information into 2D vector information with depth information (We call it 2.5D vector information), the rasterising process translates 2.5D vector information into 2.5D image information, and the assembling process gathers them in a scene. This three steps are taken charge of between server and client. The Rendering Server can be installed in distributed computers that are connected via network.

2.2 Contents Editor

Contents Editor is the client application that receives the rendering information from plural Rendering Server, and arranges them in 3D space. The result of the arrangement is described in the tag-based ASCII text file named Layout Description File. A client can request the distribution policy of the three steps to the server. Fig.4 illustrates the case of assembling two models. In this case, server rendering step are set to "transformation and rasterization". In short, two rendering steps out of three is performed at the server side, and the assembling step which compare the depth information of the images and arrange them, is processed at the client.

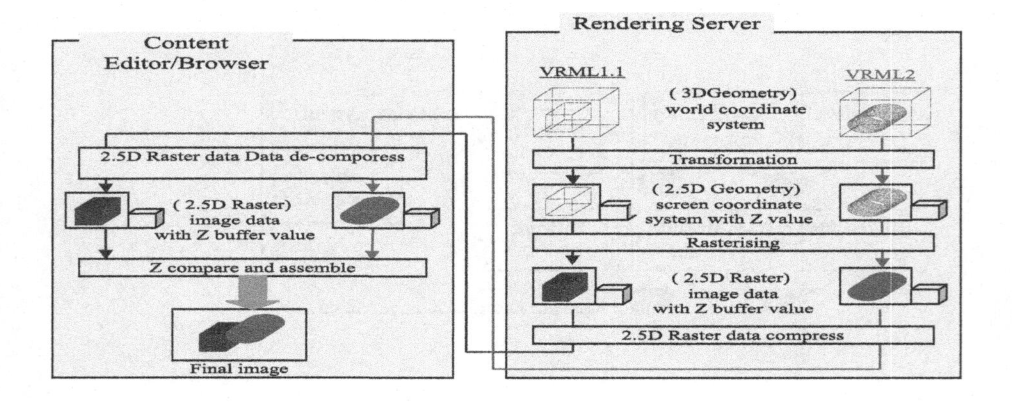

Fig. 4. An Example of Rendering Steps between Server and Client.

2.3 Contents Browser

Contents Browser interprets the Layout Description File and receives the rendering results from appropriate Rendering Server, and displays them on the screen.

3 Sample Analyses

Fig. 5. An Example of Final Rendering Image at the Client

Table 1. Benchmark Result

	Polygon [triangle]	VRML size [Mbyte]	Server Rendering Step [-]	Update Time [sec/frame]	Server [sec]	Transmission [sec]	Client [sec]	Size in transmission [Kbyte]
Building	5K	0.4	0	4.5	0.7	2.4	1.3	294
			2	7.4	1.5	4.6	1.3	104
Terrain	27K	1.2	0	8.0	0.9	5.4	1.8	814
			2	9.3	3.1	4.9	1.2	169
Robot	207K	4.2	0	32.0	1.7	28.1	2.2	2899
			2	9.8	3.9	4.5	1.4	80

Table.1 shows the performance examined by three kinds of geometry model. Fig.5 shows the model, Building, Terrain and Robot. Building is constructed by about 5000 triangles, the data size in VRML is about 400 K Bytes. Terrain is middle size and Robot is the largest. The case of no rendering on the server machine is described in server rendering step 0 column. In this case, the compressed 3D geometry data is passed to client. In the case of server rendering step 2, the most of rendering process are processed on the server, then, the image with Z value is passed to client. The image data is compressed in JPEG format. Other types, Z value, 3D geometry data are compress by Run-length algorithm. The size of compressed data is shown in the "size in transmission". The final rendered image size is 500x500 pixel in all case. "Update Time" means the update time per a frame when rotating the object. "Transmission" includes the data compress and de-compress process. The server side program is worked on a SGI indy (R5000), the client side programs are worked on a PC PentiumII 450MHz.

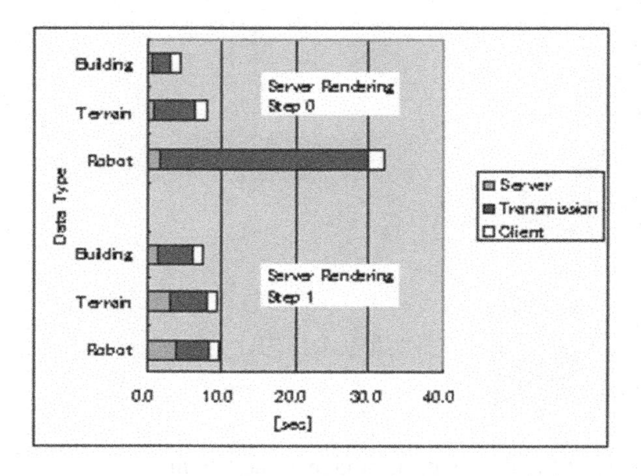

Fig. 6. Benchmark Result in Graph

Fig.6 shows the same data in bar chart. As the 3D geometry model can not be compressed effectively, the large data requires much cost for the transferring in the case of server step 0. However, using server step 2, the cost for transferring dose not depend on the model size. This approach, assembling the images with Z value on a client machine, is popular in other parallel rendering system. But, this system allows user to select the transferring policy. Using isosurface technique for visualization, the number of generated polygon is not predicted. If it is small, 3D isosurface can be downloaded. If it is large, 3D model is kept on the server, only the image can be assembled to the client images.

Once assembling the models, the layout can be stored in a file called "Layout Description File". The assembled images are recovered by the accessed server when the client loading this file. The time for recovering is 28 sec for three models selecting server step 2. This is almost the same as the sum total, because all server process is running on a machine in the case of this test.

4 Conclusion

We have developed the distributed rendering system of client initiative. This system enables to choose a 3D geometry model, a 2.5D geometry model, or 2.5D image data dynamically, when transmitting data to a client from a server. Rendering processing benchmark was performed using the model of three kinds of data sizes, and it was shown that it is effective to change the transmission method according to network load and the processing capability of a server and a client.

Although the system by which rendering processing became independent of a simulation program has a limit in the improvement in the whole processing performance, it is effective in the construction of a visualization system which employed efficiently the existing assets which do not correspond to distributed environment. When performing interactive processing, in order to reduce the amount of communications and the number of times of communication between client-servers, We are planning to implement more intellectual functions in the client side application. We would expect to apply this system to numerical computation area especially whose environment is not combined closely, for examples, coupling simulation.

Acknowledgement

A part of this work was financially supported by IPA(Information Technology Promotion Agency, Japan).

References

1. Thomas W. Crockett: An Introduction to Parallel Rendering, Parallel Computing, Vol.23 No.7 (1997) 819-843
2. Thomas W. Crockett: PGL: A Parallel Graphics Library For Distributed Memory Applications, NASA Contract No. NAS1-19480 (1997)
3. Hideo Miyachi: A Development of "On Demand Rendering Server", The JSME Annual Meeting, No.99-1 (1999)

Author Index

Lecture Notes in Computer Science

For information about Vols. 1–1865
please contact your bookseller or Springer-Verlag

Vol. 1902: P. Sojka, I. Kopeček, K. Pala (Eds.), Text, Speech and Dialogue. Proceedings, 2000. XIII, 463 pages. 2000. (Subseries LNAI).

Vol. 1903: S. Reich, K.M. Anderson (Eds.), Open Hypermedia Systems and Structural Computing. Proceedings, 2000. VIII, 187 pages. 2000.

Vol. 1904: S.A. Cerri, D. Dochev (Eds.), Artificial Intelligence: Methodology, Systems, and Applications. Proceedings, 2000. XII, 366 pages. 2000. (Subseries LNAI).

Vol. 1905: H. Scholten, M.J. van Sinderen (Eds.), Interactive Distributed Multimedia Systems and Telecommunication Services. Proceedings, 2000. XI, 306 pages. 2000.

Vol. 1906: A. Porto, G.-C. Roman (Eds.), Coordination Languages and Models. Proceedings, 2000. IX, 353 pages. 2000.

Vol. 1907: H. Debar, L. Mé, S.F. Wu (Eds.), Recent Advances in Intrusion Detection. Proceedings, 2000. X, 227 pages. 2000.

Vol. 1908: J. Dongarra, P. Kacsuk, N. Podhorszki (Eds.), Recent Advances in Parallel Virtual Machine and Message Passing Interface. Proceedings, 2000. XV, 364 pages. 2000.

Vol. 1910: D.A. Zighed, J. Komorowski, J. Żytkow (Eds.), Principles of Data Mining and Knowledge Discovery. Proceedings, 2000. XV, 701 pages. 2000. (Subseries LNAI).

Vol. 1912: Y. Gurevich, P.W. Kutter, M. Odersky, L. Thiele (Eds.), Abstract State Machines. Proceedings, 2000. X, 381 pages. 2000.

Vol. 1913: K. Jansen, S. Khuller (Eds.), Approximation Algorithms for Combinatorial Optimization. Proceedings, 2000. IX, 275 pages. 2000.

Vol. 1914: M. Herlihy (Ed.), Distributed Computing. Proceedings, 2000. VIII, 389 pages. 2000.

Vol. 1916: F. Dignum, M. Greaves (Eds.), Issues in Agent Communication. X, 351 pages. 2000. (Subseries LNAI).

Vol. 1917: M. Schoenauer, K. Deb, G. Rudolph, X. Yao, E. Lutton, J.J. Merelo, H.-P. Schwefel (Eds.), Parallel Problem Solving from Nature – PPSN VI. Proceedings, 2000. XXI, 914 pages. 2000.

Vol. 1918: D. Soudris, P. Pirsch, E. Barke (Eds.), Integrated Circuit Design. Proceedings, 2000. XII, 338 pages. 2000.

Vol. 1919: M. Ojeda-Aciego, I.P. de Guzman, G. Brewka, L. Moniz Pereira (Eds.), Logics in Artificial Intelligence. Proceedings, 2000. XI, 407 pages. 2000. (Subseries LNAI).

Vol. 1920: A.H.F. Laender, S.W. Liddle, V.C. Storey (Eds.), Conceptual Modeling – ER 2000. Proceedings, 2000. XV, 588 pages. 2000.

Vol. 1921: S.W. Liddle, H.C. Mayr, B. Thalheim (Eds.), Conceptual Modeling for E-Business and the Web. Proceedings, 2000. X, 179 pages. 2000.

Vol. 1922: J. Crowcroft, J. Roberts, M.I. Smirnov (Eds.), Quality of Future Internet Services. Proceedings, 2000. XI, 368 pages. 2000.

Vol. 1923: J. Borbinha, T. Baker (Eds.), Research and Advanced Technology for Digital Libraries. Proceedings, 2000. XVII, 513 pages. 2000.

Vol. 1924: W. Taha (Ed.), Semantics, Applications, and Implementation of Program Generation. Proceedings, 2000. VIII, 231 pages. 2000.

Vol. 1925: J. Cussens, S. Džeroski (Eds.), Learning Language in Logic. X, 301 pages 2000. (Subseries LNAI).

Vol. 1926: M. Joseph (Ed.), Formal Techniques in Real-Time and Fault-Tolerant Systems. Proceedings, 2000. X, 305 pages. 2000.

Vol. 1927: P. Thomas, H.W. Gellersen, (Eds.), Handheld and Ubiquitous Computing. Proceedings, 2000. X, 249 pages. 2000.

Vol. 1929: R. Laurini (Ed.), Advances in Visual Information Systems. Proceedings, 2000. XII, 542 pages. 2000.

Vol. 1931: E. Horlait (Ed.), Mobile Agents for Telecommunication Applications. Proceedings, 2000. IX, 271 pages. 2000.

Vol. 1658: J. Baumann, Mobile Agents: Control Algorithms. XIX, 161 pages. 2000.

Vol. 1766: M. Jazayeri, R.G.K. Loos, D.R. Musser (Eds.), Generic Programming. Proceedings, 1998. X, 269 pages. 2000.

Vol. 1791: D. Fensel, Problem-Solving Methods. XII, 153 pages. 2000. (Subseries LNAI).

Vol. 1799: K. Czarnecki, U.W. Eisenecker, Generative and Component-Based Software Engineering. Proceedings, 1999. VIII, 225 pages. 2000.

Vol. 1932: Z.W. Raś, S. Ohsuga (Eds.), Foundations of Intelligent Systems. Proceedings, 2000. XII, 646 pages. (Subseries LNAI).

Vol. 1933: R.W. Brause, E. Hanisch (Eds.), Medical Data Analysis. Proceedings, 2000. XI, 316 pages. 2000.

Vol. 1934: J.S. White (Ed.), Envisioning Machine Translation in the Information Future. Proceedings, 2000. XV, 254 pages. 2000. (Subseries LNAI).

Vol. 1935: S.L. Delp, A.M. DiGioia, B. Jaramaz (Eds.), Medical Image Computing and Computer-Assisted Intervention – MICCAI 2000. Proceedings, 2000. XXV, 1250 pages. 2000.

Vol. 1937: R. Dieng, O. Corby (Eds.), Knowledge Engineering and Knowledge Management. Proceedings, 2000. XIII, 457 pages. 2000. (Subseries LNAI).

Vol. 1938: S.Rao, K.I. Sletta (Eds.), Next Generation Networks. Proceedings, 2000. XI, 392 pages. 2000.

Vol. 1939: A. Evans, S. Kent, B. Selic (Eds.), «UML» – The Unified Modeling Language. Proceedings, 2000. XIV, 572 pages. 2000.

Vol. 1940: M. Valero, K. Joe, M. Kitsuregawa, H. Tanaka (Eds.), High Performance Computing. Proceedings, 2000. XV, 595 pages. 2000.

Vol. 1942: K. Masanori, R. Popescu-Zeletin (Eds.), Active Networks. Proceedings, 2000. XI, 424 pages. 2000.

Vol. 1945: W. Grieskamp, T. Santen, B. Stoddart (Eds.), Integrated Formal Methods. Proceedings, 2000. X, 441 pages. 2000.

Vol. 1948: T. Tan, Y. Shi, W. Gao (Eds.), Advances in Multimodal Interfaces – ICMI 2000. Proceedings, 2000. XVI, 678 pages. 2000.